Becoming-One Papers

– Official Edition –

Challenge to Mindsets

on

Religion and Science

Two-Column Format

(3 books in 1 volume)

by

Walter R. Dolen

President of the Becoming-One Church

Becoming-One Publications

Pennsylvania USA

BeComing-One Papers

New Two-Column Format

(3 Books in 1 volume)

A Cumulative Hardback Edition

ISBN: 9781619180659

All photos, graphics, charts and tables were created by the author, Walter R. Dolen, unless otherwise stated. No AI software was used to write this work.

Last Printed November 2025

(Books/Papers included)

BP: *Beginning Papers*

1. GP: *My God is the BeComing-One*

[aka: *God Papers*]

2. NM: *New Mind Papers*

3. PR: Prophecy Papers

All books also sold separately

BeComingOne Publications

b1publ.com | becoming-one.org/books.htm

books can be bought on the web

Barnes and Noble, Amazon, Walmart.com, etc

Visit our website: https://becoming-one.org

https://beone.ws

May Grace Abound to All

"**All silencing of discussion is an assumption of
infallibility** ... But the peculiar evil of silencing the
expression of opinion is, that it is robbing the human
race ... If the opinion is right, they are deprived of the
opportunity of exchanging error for truth; if wrong, they
lose, what is almost as great a benefit, the clearer
perception and livelier impression of truth, produced by
its collision with error."

(John Stuart Mill, On Liberty,
Chapter 2; see the quote here:
[https://www.beone.ws/resources/OnLiberty-40.pdf])

Profound Questions

Why is there anything?

Was there always something?

Or did something come from nothing?

Was there a beginning of the complex Universe?

Can something complex come from something simple?

Can evolution really evolve from a few substances to complexity?

Where did the first simple substances get the ability to turn into complexity?

If a Great Power created the universe, where did He come from?

If this Power is good and all powerful, why is there evil?

How do we learn what is **good** or **evil** for us?

Can we know both of these qualities?

Without learning about each?

And comparing both?

Mankind took from the tree of **knowledge** of Good and Evil

Instead of the tree of Life

Why?

Acknowledgment

I thank my wife Shirley Clare and others for their help with editing the grammar, spelling and for my wife's patience in the long hours I spent on my projects. I also thank all biblical scholars who wrote helps (concordances, interlinear Bibles, grammars, computer programs, creation v. evolution books, etc.) and critiques of doctrine, for they made my work easier. Lastly, I thank all scholars of serious works (philosophy, science, etc) for their work for no one person can think through all opinions pertaining to subjects: we need to compare our knowledge with others in order to ascertain the truth of the matter. Spiritually, I want to thank our creator for the Spirit and knowledge given to me, for without this I would not have recognized the obvious hints throughout the Bible.

Walter R. Dolen

2025

About the Author

Walter Dolen is an author/editor of several books, using the scientific method[1] including: *My God is the BeComingOne: God Papers*; *New Mind Papers*; *Chronology Papers*; *BeComing-One Bible*; *Harmony of the Gospels*; *Harmony of the Good News*; *Male & Female*; *Prophecy Papers*; *Einstein: Light, Time and Relativity, Male & Female*; *6000 Years of Mankind*; etc. These books were researched and written between 1969 and 2025. Walter has worked with his hands (carpenter/builder), with his mind (publisher/ writer/ building designer) and with his soul (President of the Becoming-One Church).

For more information about the author see his web site:

https://www.walterdolen.com or https://www.walterdolen.ws

---

[1] (1) Perceive a problem; (2) examine and analyze all the available evidence; (3) examine and imagine different hypotheses in attempt to solve the problem in a logical manner; (4) form a theory that answers the problem; (5) test the theory; (6) always have an open mind for better theories or answers to the problem; (7) change the theory if new evidence is inconsistent to your prior theory.

Books available by the Author:

My God is the Becoming-One
New Mind and Christianity
Prophecy Papers
6000 Years of Mankind: Chronology Papers
Harmony of the Good News
Becoming-One Papers
Becoming-One Bible: Old and New Testament

Male & Female: Complementary Partnership
Einstein: Light, Time and Relativity

Available at https://beone.ws/books.htm

Available on the World Wide Web:
Amazon, Barnes and Noble, Walmart, etc.

Also available in digital formats:
https://b1-church.org/products

Books/Papers included:

All books also sold separately

Becoming-One Papers
by Walter R. Dolen
General Contents

Documentation

When you see, "The God, all in all" (1 Cor 15:28), it means that this is a quote from the New Testament letter called First Corinthians, chapter 15, verse 28. If you see "2 Cor" it would mean the *second* letter of the Corinthians. If you see "2 Cor 11:4" it would mean we quoted from the second letter of the Corinthians, the 11th chapter, and the 4th verse. But sometimes you will see a documentation such as "(1 Pet 2:4)" after a sentence that has no quotes. This kind of documentation is used in order to *support* the previous sentence or sentences, or to *point out other similar or related views* of the previous sentence or sentences, or to *add new light* to the previous sentence or sentences.

When you see reference to "PR7" it means more information can be found in *Prophecy Papers*, Part 7. When you see "GP2" this means more information can be found in the *God Papers*, part 2.

BP	= *Beginning Papers*
NM	= *New Mind Papers*
GP	= *God Papers*
PR	= *Prophecy Papers*
CP	= *Chronology Papers*
cf or cf.	= confer or compare
p. or pp.	= page or pages
w/	= with

Beginning Papers

Introduction to the BeComingOne Papers

By

Walter R. Dolen

BP1: Beginning Papers – the Introduction to the BeComingOne Papers

Bad News of Old Age

Good News of New Age

How can there be a God, when there is Evil?

bp1» Is there a God – the highest Power, the greatest good, the all powerful, the creator of the universe? Isn't God's power everywhere? Isn't God good? If God is all good, why is there evil? If we look around us it is clear that there is evil, so how can God be all powerful and good, yet allow evil? How can evil be in the universe when God is all powerful and good?

bp2» Why, if God is good and all powerful, is there suffering? Why do innocent children suffer through no fault of their own. Why do we die? It is said that God made the angels immortal, why didn't he create us immortal? Why do we get diseases? Since he created the world, why did he place the possibility of diseases in the world? Sure, we bring on some of these things ourselves by eating poorly, using poor hygiene or by risky behavior. But what about those diseases caused by mosquitoes or microscopic germs or genetics? Didn't God create mosquitoes, germs and genetics? Didn't God create everything? Isn't God good? Or is the idea of God just some superstitious idea based on fear or naivety, an idea made obsolete by the scientific era? Are people more intelligent if they do not believe in God? Or are atheists 'fools' as the Bible relates (Psa 14:1)? An atheist or an agnostic could say, "you cannot prove there is a God" and, "we do not see anything being created or any miracles by God." While the believers could say, "you do not have any real proof of evolution – where are the billions of missing links?" Today each side would be right – no observable proof exists for either side. Today we do not see any Moses parting the seas or anyone being resurrected from the dead by God as reported in the Bible. Today we don't see any species being made nor do we see the billions of missing links being found. We see many things that are good today and in the past; we also see death, diseases and destruction all throughout history. In addition, we see the hypocrisy of those who claim to believe in God as well as those who do not believe.

Typical statement by agnostics/atheists and non-believers

bp3» Many see the negativity of religion. Mark Twain[1] was disillusioned with Christianity and religion because he only saw the paradoxes and the hell-damnation of religiosity. So he wrote the following in a book not published until after his death:

> "A God who could make good children as easily as bad, yet preferred to make bad ones; who could have made every one of them happy, yet never made a single happy one; who made them prize their bitter life, yet stingily cut it short; who gave his angels eternal happiness unearned, yet required his other children to earn it; who gave his angels painless lives, yet cursed his other children with biting miseries and maladies of mind and body; who mouths justice and invented hell – mouths mercy and invented hell – mouths Golden Rules, and forgiveness multiplied by seventy times seven, and invented hell." [Mark Twain, *The Mysterious Stranger*, Chap. 11]

This perception of the inexplicable paradoxes and negativity found in religion, or the emphasis upon such, is one-sided and unfair, for such negativity was superseded by Christ's teaching on the system of Love.

bp4» Jesus Christ, for whom Christianity is named, changed the way some perceived God. Unfortunately, Jesus' words were taken over by those who didn't understand and they changed Christ's teachings of forgiveness and love into those of hell and damnation. Because of this, we are forced to review in detail the doctrines of Christianity because the negativity of the world has been interjected into

---

[1] A pen name for Samuel Clemens, one of America's best known writers

Christianity. This negativity projects something about man's mind in this age, which we call the "old mind" or the "other mind." But Christ announced a new mind, a new spirit, and a new commandment – the commandment of love.

God or Evolution

bp5» But wait. Doesn't the belief in God, Jesus Christ, or the Jewish Messiah depend on the reliability of the book called the Bible? Yes and no. It also depends on the belief in creation or evolution. Either God or something like evolution "created" the universe and life on the earth. The Bible has been misrepresented or even distorted by many in the name of academia. One reason for this is that academia limits itself to *natural* scientific study, not the study of the invisible-*super*natural. God being spiritual is invisible to the human eyes. But this spiritual Being is not invisible to *indirect* methodological observation. For example, the wind is invisible, but we know it is there by indirect observation – we see the effects of the wind, such as leaves and branches moving. Academia ignores the fact that many of their most famous theories use indirect methods to demonstrate the possibility of their theories. After all, the study of evolution is pre-historical. If it ever occurred, it happened before human eyes, thus before any scientific observation. There is no direct confirmation of evolution through observation, because no intelligent being was alive to observe and record it. Any study of evolution uses indirect methods with a lot of conjectures and suppositions. Therefore indirect study pertaining to a supreme God-Being is as valid, if not more so, as the indirect study of evolution. Indirect study of the God-Being can detect God's existence by the great complexity of life (the human cell is as complex as a modern large city and the brain as complex as the universe), by spiritual revelation, by miracles, by prophecy coming true, by intelligence built into life (the DNA code is

more complex than any software) and so forth.

bp6» **Evolution and the big bang?** The big bang theory, is where the universe supposedly came into existence out of a huge explosion. It assumes there was gravity, energy and matter before the big bang. This theory relies on billions of years of matter, gravity and energy randomly working together without a design, without intelligence. Intelligence and complexity from random interaction? Isn't this something like saying: if monkeys can bang on a typewriter randomly for a long enough time they can write an intelligent book? But even this example is not realistic. After all, monkeys have brains and intelligence, but evolution does not have intelligence. Evolution is more like rain drops falling on typewriter keys, with millions or billions of keys with different markings on each key, and from that an intelligent book is created. All this before there were any laws of nature, any chemical bonding, any gravity, any energy, any physical matter or molecules.

bp7» **The God choice**. Evolution is not intelligent, but merely random interaction of pre-existing matter. The other option — the God Power — must innately have intelligence and substance, be it that the "substance" is invisible to our eyes. Therefore, the God Being must have created life through its own being and essence. Who or what is this God Being? (We examine this in our *God Papers* book) From the beginning of recorded history people have believed in a higher power(s) that they called God or gods. Why did they believe this? One historical document speaks about this God Being. It's what the Jews, Christians and others call the Bible. The Bible is a complex historical document with thousands of details about God, gods, religious cults, real people, real cities, diverse cultures, kingdoms, kings, genealogy, and chronology. In the last two hundred years important confirmation of these details have been unearthed by archeology, ancient written documents, words carved on stone, and libraries containing thousands of cuneiform clay tablets. They all point out that the Bible records are real history not myth. The Moses of the Bible could write and read; his people could write and read. Some because of this evidence and other evidence believe that the Bible is the very Word of God, as I do. (See the Bible Paper)

bp8» **It is either God or evolution.** Which idea is the winner? The real winner, and proof of, is yet to come. Neither side can claim victory in a provable way that both sides will accept. That real proof for both sides is in the future. But we should ask the real difficult question: why is there anything at all? Or this question: can intelligence come from random interaction of matter (which in fact has the evidence of intelligence within it)? Or this question: can something come from nothing? The Bible says that everything came out of God, not out of nothing as some mistakenly think.

Good and Evil

bp9» **Good and Evil.** To those who believe in the Bible and even those who don't – why did God put the tree of knowledge of good and *evil* in the garden of Eden? And why after Adam and Eve took from this tree of *knowledge* did God say: "The man has now become like one of us, knowing good *and* evil." So God has the knowledge of good and *evil* and now after the sin in the garden mankind has become like God? How can a God that is supposed to be all good, know about evil? Is there a reason and purpose for evil? Stop. Think. Both good and evil are comparative qualities. Can anyone know good without knowing evil? How can you know what is good without something to compare it to?:

bp10» "There can be nothing more inept than the people who suppose that good could have existed without the existence of evil. Good and evil being antithetical, both must needs subsist in opposition, each serving, as it were, by its contrary pressure as a prop to the other. No contrary, in fact can exist, without its correlative contrary. How could there be any meaning in 'justice,' unless there were such things as wrongs? What *is* justice but the prevention of injustice? What could anyone understand by 'courage,' but the antithesis of cowardice? Or by 'continence,' but for that of self-indulgence? What room for prudence, unless there was imprudence? Why do not such men in their folly go on to ask that there should be such a thing as truth, and not such a thing as falsehood? The same may be said of good and evil, felicity and inconvenience, pleasure and pain. There things are tied, Plato puts it, each to the other, by their heads: if you take away one, you take away the other." [*Chrysippus, Fragment* 1169. On the problem of evil. Barrett, p. 64]

We cover the reason why God allowed evil in this age and all the important ideas about God in our book: *My God is the BeComingOne*. But for now let's move on to other important subjects.

Creation or evolution: who or what created the universe?

bp11» Is the universe billions and billions of years old? Or could it be that our measurement of time is based on a false foundation? Is evolution a fact, or are there thousands of holes in this theory that we are not aware of?

Two Views: Earth is old; Earth is young
Earth is Old Theory

bp12» There are two general views of history that are polarized. One is that the cosmos is old, very old, billions of years old. The other view is that the cosmos is young, very young, only thousands of years old.

bp13» Those who believe that the earth is billions of years old have various theories to "prove" that the earth is billions of years old. They speak of the Uranium to Lead method of dating, or the Thorium to Lead method of dating. They speak of bones that they say are millions of years old. When you are educated in an environment that dogmatically indicates that the earth is billions of years old it is ludicrous to believe that the earth is only thousands of years old. To believe that the earth is thousands of years old is to be uneducated or ignorant, and you are ripe for belittling by the "educated." But every belief system has it foundations. The "earth is old" system of belief is related to the "evolutionary" system of belief. Those who believe in evolution *must* have an old earth. The magic of evolution needs billions of years of "natural selection" in order to work its miracles. But all methods of dating events and materials billions, or even millions of years old, are baseless, illusionary, and arbitrary.

Foundations for the "Earth Is Old System"

(1) Theory of Evolution

bp14» **(1)** The theory of evolution is the first foundation for the "earth is old" theory. Evolution needs an old earth for its development. There are numerous works that examine the theory of evolution (see list later in the *Beginning Papers*). Because the theory of evolution needs an old earth, it found an old earth through selective perception. Any method that indicates a great age is an acceptable method for evolutionists. Any method that indicates a young earth is a rejectable method for evolutionists. Evolutionists don't even feel a need to examine other points of view. Their minds are made up. They have a mindset. Their selective perception reaffirms to them each day that evolution is correct. Thus, any method that proves an old earth is correct; any method that proves the contrary is foolishness.

(2) Radioactive Dating Methods

bp15» **(2)** The radioactive dating method is the second foundation for the "earth is old" theory. All radioactive dating methods start with a parent element which through radioactive decay turns into a daughter element. The decay rate is measured in half lives. The half life of Uranium 238 is said to be about 4.5

billion years. A unit of Uranium 238 turns into ½ lead and ½ Uranium after about 4.5 billion years. The Uranium 238 is the parent element and Lead 206 is the end or final daughter element. There are other daughter elements between Uranium 238 and Lead 206. For example, Uranium 238 first decays into the daughter element Thorium 234 after about 4.5 billion years, and then after about 25 days turns into Protactinium 234, then after 1 minute turns into Uranium 234, then after 300,000 years turns into Thorium 230, then after 80,000 years turns into Radium 226, then after 1600 years turns into Radon 222, then after 4 days turns into Polonium 218, and continues its decay until it reaches Lead 206." (Krauskopf and Beiser, *Fundamentals of Physical Science*, 5th Ed., p. 252, see p. 562)

bp16» If the rate of decay is constant, then we have a clock in which to tell time, **if, and only if**, we know the ratio of Uranium 238 in the earth compared to Lead when the earth was formed/created, either by God or by the magical evolution. Because the decay rate of Uranium 238 is so slow compared to the decay rates of other elements in the series only the amount of the end daughter, Lead 206, is considered when ascertaining the age of the rock. The earth is believed to be about 5 billion years old according to evolutionists. But, of course, 5 billion years ago there was no man to observe the ratio of Uranium in the rocks compared to Lead. It is nothing but guesswork and nothing else when someone arbitrarily says that at the beginning there was such and such ratio of Uranium as compared to Lead. Guesswork is not scientific work.

Constant Decay Rates?

bp17» Furthermore it was believed at first that these decay rates were constant.

"Radioactivity was discovered by Becquerel in 1896. In 1906, Millikan stated, 'Radioactivity has been found to be independent of all physical as well as chemical conditions. The lowest cold or greatest heat does not appear to affect it in the least. Radioactivity seems to be as unalterable a property of the atoms of radioactive substances, as is weight itself.' This state of mind established the modern view, which is quite generally held today.... The electroscope and spinthariscope were used in early study of radioactive alpha-decay rates. The inherent limitations of these early instruments led to erroneous conclusions:

1. That radioactive decay rates are constant.

2. That these rates cannot be altered by change of the energy state of the electrons orbiting the nucleus.

3. That radioactivity results from processes which involve only the atomic nucleus.

Refinements in electronics resulted in the development of sophisticated counting apparatus. The equipment was used in the demonstration by several investigators (1949-73) of rather easily induced changes in the disintegration rates of 14 radionuclides, including ^{14}C, ^{60}Co, and ^{137}Cs. **The observed variations in the decay rates, (changes in the half life) were produced by changes in pressure, temperature, chemical state, electric potential stress of monomolecular layers, etc. ... The decay 'constant' is now considered to be a variable.**" [H.C. Dudley, "Is There Ether?,"*Industrial Research*, Nov 15, 1974, p. 42; my emphasis]

bp18» Even a small amount of variation in the decay rate can make a big difference in the assumed age of the rock:

> "Measurement of nuclear disintegration parameters has been done for about fifty years. To my knowledge no major research effort has been mounted to determine whether nuclear decay parameters vary at all with time [he is speaking of time not pressure, chemical state, etc]. Once values of the decay index for a particular nuclide are obtained and a particular value is agreed upon, this value is generally accepted. Usually no further measurements are taken....
>
> If a small amount of exponential variation occurs in the nuclear decay index, then the half lives of the radiometric nuclides are drastically reduced — orders of magnitude. In the case of U 238 the half life is reduced by a factor of 10^5" (Theodore W. Rybka, *ICR Impact Series* No. 106)

Decay Rates not Constant

bp19» As we see above temperature, pressure, chemical state, and other factors do change the decay rate of radioactive elements, and this drastically changes the so-called clock of radioactivity. Atomic clocks even seem to change their rates of decay by the direction in which they travel in an airplane. Those going westward gained time; those going eastward lost time (Hafele, Keating, 1972, "Around-the-world atomic clocks," *Science* 177 [4044]).

Radiohalos

bp20» Robert V. Gentry's work on radiohalos has cast a shadow on the premise that the decay rates are constant. "Radiohalos" are microscopic, ring-like discolorations caused by radioactivity in certain minerals. Early work seemed to indicate that the radiohalos exhibited dimensions predictable on the basis of modern decay rates. But Gentry who worked at the Chemistry Division of the Oak Ridge National Laboratory in the 1960's "set out to review previous work on the subject, then began his own painstaking study of thousands of halos in rocks from around the world. Almost immediately he found that all was not in order in this long neglected field. Gentry discovered that, although uranium halos, for example, are readily identifiable by the number and relative rough diameters of their rings, their actual dimensions often vary substantially, even within a single crystal" (Ralph E. Juergens, "Radiohalos and Earth History," *Kronos*, III:1, pp 7 ff; read article, and Gentry's articles noted in footnotes). Gentry has shown that the "halos furnish no proof that [the decay constant] is constant" (Gentry, *Science*, April 5, 1974, pp 62-66; Also see Don B. De Young, "The Precision of Nuclear Decay Rates," CRSQ, Vol. 13, No 1 [1976]; and John Lynde Anderson and George W. Spangler, "Radiometric Dating: Is the 'Decay Constant' Constant?," *Pensee*, Vol 4 No. 4 [1974]; *Scientific Creationism*, 2nd Ed., 1985, Chapter VI; and other works.).

Dubious Premises

bp21» Evolutionists use the elements with the slowest rates of decay to measure the age of the earth, and they use the highest ratio of the parent element to daughter element at the time of formation/creation in order to give a high age. Remember there were no human observations made at formation/creation to help establish the correct ratio. The ratio may have been low. Thus, even if the Uranium-Lead method is correct, the earth is still young since there was a low ratio at first.

bp22» Also there are other elements that decay at much higher rates. At the far extreme from Uranium 238 is Astatine 216 with a half-life less than a second as is Polonium 214. Why didn't they choose a faster decaying element to clock the earth's age? It is because they assumed a great age for the earth, therefore they arbitrarily chose an element with a slow decay rate along with a high parent to daughter radio instead of a low parent to daughter ratio, so as to self-fulfill their view.

Different Methods of Dating Don't Agree

bp23» There is also the problem of variation of the ages arrived at by using various elements and methods to date the earth. One system of dating gives one date, another gives a contradictory date. Or one set of rocks gives one age, while another set of rocks gives a different age for the earth. What does the believer in the "earth is old" theory do? With the Carbon 14 dating method (C14) they merely pick the result they wanted to begin with, "If a C14 date supports our theories, we put it in the main text. If it does not entirely contradict them, we put it in a footnote. And if it is completely 'out of date,' we just drop it" (T. Save-Soderbergh, "Carbon 14 and Egyptian Chronology," *Nobel Symposium 12 Radiocarbon Variations and Absolute Chronology*, Stockholm, Almquist and Wikwell, p. 35; quotes from R.D. Long, CRSQ, Vol 10, No 1, p. 19; *Science, Scripture, and the Yound Earth*, 1989 Edition, pp.42ff). This is the way some quote the Bible. If a verse agrees with a belief it is quoted, if not it is ignored. And this is like the "identification game" used in astronomical retro-calculations (see CP2).

(3) Great Distances in Space

bp24» **(3)** Great distances in space is the third foundation for the "earth is old" theory. The "earth is old" group believes in such things as the "big-bang" theory. (Although lately there have been articles critical of this theory.) All matter came from a big explosion and has been spreading out ever since. Since the earth to them *must* be billions of years old, then the matter in the universe has been traveling after the explosion for billions of years. Matter has spread out great distances since the beginning; the larger the universe, the older the universe. Thus they look for methods that "prove" great distances in space. Of course they have come up

with methods for great distance and have rejected any method or theory that may show a small universe or young universe. They use the red-shift method of dating, which is partly based on Einstein's relativity theories. But others have shown the shaky foundation of this red-shift method as well as the unstable foundation of the special theory of relativity and consequently the general theory of relativity (See *Science Papers*; Field, Arp, and Bahcall, *Red-Shift Controversy*,1973; Herbert Dingle, *Science at the Crossroads*, 1972; Walter R. Dolen, *Einstein: Light, Time and Relativity*; etc.).

Foundation for the "Earth is Young System"

(1) No Scientific Evidence For Evolution

bp25» The first foundation for the "earth is young" theory is the lack of real evidence that the earth is old. There is sound evidence against the red-shift method for ascertaining distances in space, against radioactive dating methods, and all other methods of dating the earth as old (see my *Science Papers*; Jeremy Rifkin, *Algeny*, 1983; Field, Arp, and Bahcall, *Red-Shift Controversy*, 1973; John C. Whitcomb and Henry M. Morris, *The Genesis Flood*, 1961; etc.).

(2) Proof that the Earth is Young

bp26» The second foundation for the "earth is young" theory are the *many* methods that prove the earth is young. They are at least 76 methods that prove the earth can not be older than 500 million years and of these 24 indicate that the earth is no older than 20,000 years. These methods include such things as the influx of titanium, or cobalt, or zinc, or mercury, or silver, or copper, or gold, or silicon, or nickel into ocean via rivers. If the earth was billions of years old, the ocean would be a soup of pollution without any life in it. And such methods as the influx of meteoritic dust from space, or development of total human population, or lack of vast amounts of ancient cultural debris, the decay of the earth's magnetic field, the decay of C-14 in pre-Cambrian wood, the growth of active coral reefs, the formation of river deltas, decay of short-period comets, and the instability of rings of Saturn show a young earth. These 76 methods are based on the assumption that there were constant rates, no initial daughter components, and all were in a closed system. These methods lead to an even younger age for the earth if, for example, there were some initial daughter components at the beginning (see *Scientific Creationism*, 2nd Ed., Chapter VI; Harold S. Slusher, *Age of the Cosmos*; Henry M. & John D. Morris, *Science, Scripture, and the Young Earth*, 1989 Edition, Chapter 8; Henry M. Morris, "The Young Earth," *ICR Impact Series*, No. 17).

(3) Biblical Chronology

bp27» The third foundation of the "earth is young" theory is the belief in the Biblical chronology, or in creation without mixing the false theory of evolution into the picture. And this in turn is based on the proof that the Bible is a sound document, more sound than any other ancient document ("Bible Paper" [BP3]). And this in turn is the belief in a powerful God, not a belief in a powerless and mystical God or the false belief in the magical evolution. Read our book, *6000 Years of Mankind: Chronology Papers*, to under the real time line of history

Your Mindset Limits You

bp28» What system you believe in depends on your belief, your research on *both* belief systems, your biases, your world view, and your mindset (perceptual set). The more you research different points of view, the more you see that the world sees through filters that color its perception of reality. To the Evolutionist the world is old. To the Creationists the world is young. It is difficult for either group to prove their case to the other group. Since the only witness to the Beginning (Creation) was either the powerful God or the magic of Evolution, it is only through inductive thinking that we can come to a conclusion. We must piece evidence upon evidence. But for most of us our "mindset" or "world view" interferes with our judgment. We see what we want to see and subconsciously disregard what we do not want to see.

Earth is Fragile; Life in Danger
Who can save it?

bp29» We live on the planet earth which teems with life. The Earth is like a jewel in a desert, for as far as we can look into the universe we see no other life forms. The Earth may in fact be unique in the universe. Yet we are in danger, we are in jeopardy. Life is fragile, amazing, unique, exciting, but sometimes painful. We love life; we sometimes hate it. What is life? Where did it come from? Did it appear through evolution that took billions and billions of years? Or did it suddenly appear?

How much proof is there for billions and billions of years of evolution?

bp30» Whether you believe in evolution or creation, we can agree that life is very important. In fact, it is everything. Without life we would know nothing, we would have nothing, we would be nothing.

bp31» But all is not well. There is much danger. It is now possible for mankind to destroy all life on earth, and with it all life in the universe, at least all known life. Atomic weapons, biological weapons, artificial bio-lab creations, and war can now destroy life, even all physical life on earth . This is very important. At no other time has it been possible for mankind to destroy all life on the Earth. We live at the crossroads of mankind. You and I must do all we can to preserve life. But further, you and I must decide if there is more to life than it being a mere chance occurrence. Is there meaning to life? Is there a purpose to life? We believe that there is good news coming. There is purpose and meaning to life. But why is there so much misery, suffering and pain? Before we look at the good news, we will examine some of the bad things.

Bad News
Magnetic Fields, Their Disintegration

bp32» It basically doesn't matter if life is billions of years old or thousands of years old. Life will end on this earth before the year 3900 AD. There is nothing you or I can do about this. I know most of you have not heard of this. Let me quickly explain. There is a magnetic field that protects the earth from harmful solar radiation. If there were not this magnetic field, life on earth would be destroyed by the harmful solar wind and cosmic rays.

> "Perhaps the most drastic impact would come from solar cosmic rays, ultra high energy protons spewed from the Sun during periodic, gigantic eruptions known as solar flares.... Today, Earth's magnetic field deflects the vast majority of these destructive particles.... During a magnetic decline, the onslaught of solar flare particles would be more global, with potentially fatal consequences for terrestrial organisms" (p. 15, *Frontiers of Time*, Time-Life Books, 1991).

But the magnetic field is decaying in strength. About each 1400 years the magnetic field loses half its strength. A 1967 US government publication concludes that the magnetic field "will vanish in A.D. 3991" (McDonald, Keith L. and Robert H. Gunst. July, 1967. "An analysis of the earth's magnetic field from 1835 to 1965," ESSA Techical Rept. IER 46-IES 1. U.S. Government Printing Office, Washington, D.C., p. 1.). This report was based on a linear decay rate not a more likely exponential decay rate. Even an exponential decay rate would leave the earth with a weak and worthless magnetic field by the year 4000 AD. But far before this, by the year 3000 AD, the earth will be suffering from the effects of increased solar radiation and its consequential mutations.

bp33» "The magnetic field is decaying at a rate of 32 gamma per year at the magnetic poles, 16 gamma per year at the magnetic equator, and at intermediate rates everywhere in between the equator and the poles" (p. 43, *Origin and Destiny of the Earth's Magnetic Field*, by Thomas G. Barnes, D. Sc., 1973).

Atomic Inferno

bp34» There are about 10 thousand atomic weapons on earth today. Most of these can destroy a city. Already two of these weapons have been used against the cities of Hiroshima and Nagasaki in Japan during World War II. They were small weapons compared to today's huge atomic weapons. Today there are atomic bombs 100 to 1000 times larger in power. They can be launched in missiles that can hit cities anywhere in the world in less than half an hour. At no time in history has such destructive power existed. One of today's air bombers has more destructive power than all the armies of the Roman Empire. The great nations of yesterday were weaklings compared to today's nations. Their power was pathetic compared to today's atomic-equipped countries. The USSR's (Russia's) Typhoon nuclear submarine (or newer Borei-class) is said to be able to obliterate any country within 5,000 miles of its position in the ocean. The United States of America also has its nuclear submarines that can do the same thing. The USSR (Russia) and the United States *each* have atomic bombs in their submarines' ballistic missiles equal to 1.4 billion tons of power. It only takes one small atomic weapon to completely destroy a city of 250,000. The city of Hiroshima in Japan was destroyed by a 20-kiloton bomb in 1945. This means that the total nuclear power in the submarines of Russia and the United States' can destroy 140,000 cities of 250,000 people.

bp35» Because of these weapons' frightening destructive power, we try to ignore these weapons, but we must be on guard. The whole earth and maybe life itself is in danger. Today you hear about the theory of the "nuclear winter," which, according to some, may be an effect of an atomic war. But a more threatening and immediate event could occur. Most have not read or even heard of this great danger since the 1950s, not because it can't happen, but because of the head-in-the-sand mentality of humans. Some of today's larger weapons exploded together could conceivably ignite the atmosphere and destroy mankind by burning all of us alive. Horace C. Dudley, a former Professor of Radiation Physics at the University of Illinois Medical Center, wrote in 1975 the book, *The Morality of Nuclear Planning?*, wherein he wrote:

> "During WWII the eminent scientists of that era offered two options to President Roosevelt. These were in effect: Accept the possible slavery of the Nazi Axis or develop and explode atomic bombs.

There was a third option that was kept under wraps, *TOP SECRET*, discussed only behind closed doors, although sometimes guardedly by the lower echelons of 'The Manhattan Project.' This was the possibility of triggering a vast nuclear accident when and if a fission device was detonated" (Dudley p.29).

bp36» Dudley quotes Pearl S. Buck from the *American Weekly* (March 8, 1959) who then wrote:

> "And if hydrogen, what about the hydrogen in sea water? Might not the explosion of the atomic bomb set off an explosion of the ocean itself? Nor was this all that Oppenheimer feared. The nitrogen in the air is also unstable, though less in degree. Might not it too, be set off by an atomic explosion in the atmosphere?

> 'The earth would be vaporized?,' I said [Pearl S. Buck questioned]. 'Exactly,' Compton said, and with what gravity! 'It would be the ultimate catastrophe.... If, after calculation, he said, it were proved that the chances were more than approximately three to one million that the earth would be vaporized by the atomic explosion, he would not proceed with the project.

Calculation proved the figures *slightly* less — and the project continued" (Dudley p. 29). [Compton here is the Nobel Prize winner Arthur H. Compton]

"What if Oppenheimer, Fermi, Compton *et al.* were right in 1945, and the odds were 3 to one million of a world-wide conflagration? But now the bombs are a thousand times as powerful. Does this lower the odds to 3,000 per one million or properly 3 in 1,000" (Dudley p. 29).

bp37» At the time the first Atomic Bomb was scheduled to go off in New Mexico, S. Groueff writes in *Manhattan Project*, p. 352:

"There were altogether too many excited people around giving him [Oppenheimer] advice on what he should do. Groves was annoyed, too, with Fermi, who was making bets with his colleagues on whether the bomb would ignite the atmosphere, and if so, whether it would destroy only New Mexico — or the entire world" (quoted by Dudley, p. 30).

bp38» At the end of the 1989 movie called *Fat Man And Little Boy*, just before the atomic device was exploded, the betting about whether the bomb would ignite the atmosphere and destroy New Mexico or the whole world is also mentioned.

bp39» A nuclear winter would be bad enough, but an immediate worldwide inferno would be even more dreadful, and it would be permanent. Life would end on earth. One mad leader could ignite such a war. Scientists do not know how many atomic bombs it would take to ignite the atmosphere, if exploded together. There is a new branch of science called the science of Chaos (James Gleick, *Chaos: Making a New Science*, 1987). The science of Chaos manifests that it may be impossible to predict how huge an atomic explosion would be needed to ignite the atmosphere. Past calculations are obsolete and hampered by naive linear thinking. Enough atomic bombs exploded together may be equivalent to the needle that broke the camel's back. There is a limit to the amount of intense heat that can be applied to the atmosphere before it will ignite and burn up. It may only take a few large explosions in a certain limited area before the atmosphere will ignite. We must not let this happen. But sometimes events take on a life of their own outside of common sense.

Biological Catastrophe

bp40» In the San Francisco Bay Area there are biogenetic laboratories creating new biological forms. It is no leap of imagination to think of new biogenetic materials being created that could exponentially spread throughout the earth and destroy it. There are also biological weapons being created. It may take only one terrorist or one careless laboratory mistake before a dangerous biological form escapes into the environment.

"Lately, concerns have grown about the potential ecological, social and economic effects of world commerce in engineered seeds, organisms and biotech products.... Some fear that engineered microbes or plants will disrupt local ecologies and undermine traditional farming practices. Others have focused on perceived, albeit unproven, health threats from eating genetically engineered grains or cereals" (*Washington Post*, Feb 14, 1999).

Old Age

bp41» We live in the old age. Not all in this old age is as bad as some of the worst news stories depict, but there is enough of it to make life a bitter pill in many ways. It's an age with a purpose that very few understand. Yes, we believe it does have a purpose. But the old age is about over. The Old Age will have a great and rightly deserved end. The *BeComingOne Papers* examines the old age's confusion, its source of confusion, and the coming New Age promised by a spiritual Being we call God.

New Age

bp42» There is good news, and lots of it. Irrespective of the world around you today, the universe is BeComingOne. You and I, our best friends and our worst enemies, are headed towards one goal. That one goal is BeComingOne. We are now in the age of confusion and hate, but a New Age is coming wherein we will all live together in harmony. There is a reason for the present age. The confusion isn't futile, even though it may seem so. You and I aren't perfect now, but we will be.

Time is headed towards a goal. The goal is harmony. The goal is BeComingOne.

bp43» This is the "Introduction Paper" [BP1] of The *BeComingOne Papers*. In these papers we seek the answers to the mysteries of life and the answers to the contradictions of religion and science. Yes, we know there are many others who have promised you answers, but how many of them really delivered the answers to you? I have been promised answers to life many times myself, but I have been repeatedly dissatisfied with the answers. Why evil? Is there a God? Who or what is God? Why death? Why pain? Does science have the answers? Does religion have the answers? Does a combination of both have the answer? How can God be good and all powerful, and yet allow evil? Who am I? Where am I going? What is it all about? Is life really meaningless as some say, or is there meaning? What about after death? The *BeComingOne Papers* does its best to answer these questions. The answers make sense even though they are different from traditional answers. Traditional answers have too much mystery and contradiction in them. We do away with mystery and contradictions against the law of contradiction.

bp44» This old age is full of opinion. Everyone seems to have their own opinion on what is and what is not evil. Many call actions evil when in fact the only evil is in the eyes of the beholder. Many call actions right when in fact they are destructive. There are many opinions on what is a good man, a good Christian, a good Islamic, a good agnostic, or a good.... There are so many opinions on who or what is God, or even if there is a God. In other words, in this old age there is much confusion. And with this confusion there is disillusion, cynicism, hate, jealousy, conceit, false-knowledge, and the rest.

bp45» Too many have cried "truth, truth." Too many have wrongly followed. But the truth will stand or fall on its own. There is a goal in the universe: **the universe will Become One**. You won't do it, nor can you prevent it; I can't do it, nor can I prevent it. All will Become One. But, how will all become ONE?

BP1: Review

bp46» Misguided leadership, depraved behavior, murder in its many forms, perceptual blindness, and all other forms of impersonal and dysfunctional behavior happen in one form or other throughout the world in each nation and in all cultures. Everyday throughout the world some husbands physically beat their wives, and wives verbally beat their husbands. Everyday parents, some drunk on wine or drugs, beat their children physically or mentally, and children defy their parents. Everyday some employers mistreat their workers, and workers defraud their employers. Everyday some form of natural catastrophe happens. And maybe even worst of all small children suffer and die each day. There is one word in the English language to describe the aforementioned behavior: *evil.* Contrary to what some say there is evil in the world. This is not to say that all or most of the world's behavior is evil or wrong, but that there is too much wrong behavior. Why? And what about natural catastrophes? And why do young children get sick, get mistreated, and even are killed? Did they do something wrong? Or, if there is a God, and if he is good, and if he is all powerful, then, yes, why do children have to suffer? Why does anyone have to suffer? (1) Is there something behind the world's wrong behavior and is there a reason and a purpose for this behavior? (2) Do natural catastrophes and even the sickness and death of young children also have their purpose?

BP2: Mindset Paper

Ptolemy's Theory

Brain Cell Problem

bp47» We are born into a world of traditions. The traditions that we are born into have sets of rules, written and non-written. We are taught or influenced by our parents, teachers, environment, mind(s), the language(s) we speak, and our biology to believe in certain things and act in certain ways. From this we form a belief system, or mindset. A "mindset" is a perceptual set and through this set we perceive the world. A mindset acts like a filter. It filters out any mental conceptions or realities that do not fit our mindset.

A person who strongly believes that victimizers are victims themselves, does not see the crime a victimizer commits in the same way as one who believes that everyone is totally at fault for their crimes, or the way the victim sees the crime, or in the way that I do.

bp48» A person who does not know anything about the game of baseball who overhears someone talking about Smith stealing second base, may think that Smith committed a crime.

bp49» As our knowledge and background filters our perception of the words, "Smith stole second base," so too with almost everything else. Words have different meaning to different people. The word "liberal" means something different to a liberal than to a conservative. The word "communist" means something different to a communist than to a capitalist. The word "Catholic" means something different to a Catholic than to a Protestant. The word "evolution" means something different to an evolutionist than to a creationist. A peaceful countryside, where a nuclear plant is planned, means something different to environmentalists than to the owner or builder of the nuclear plant.

Ptolemy's Mathematical-Geocentric Theory

bp50» One of the biggest examples of a mindset was the geocentric theory in which the earth was the center of the universe. The geocentric theory is the idea that the earth is the center of the universe while the sun, moon, planets, and stars made a complete revolution around the earth each day. This theory was represented well by Claudius Ptolemy. Claudius Ptolemy's work commonly known as the *Almagest* was actually called "Mathematical Systematic Treatise" in the Greek version because it was a mathematical system. Ptolemy believed that mathematics was the highest form of science:

> "that only mathematics can provide sure and unshakeable knowledge to its devotees, provided one approaches it rigorously. For its kind of proof proceeds by indisputable methods, namely arithmetic and geometry" (G.J. Toomer, *Ptolemy's Almagest*, [1984] p 36).

bp51» Today the public makes light of the *Almagest* by thinking of it as some naive theological or church backed doctrine. But instead it was the most scientific work of its day containing abundant mathematical proof with tables and charts, with premises from Greek philosophy, not church doctrine. "One of the most influential scientific works in history, and a masterpiece of technical exposition in its own right" (G.J. Toomer, p. vii). Yes, today the geocentric theory seems preposterous, since after all, we know that the earth is not the center of the universe, and in fact that the earth makes one revolution around the sun each year. We believe this even though it *appears* (empirical evidence) from our eyesight that the sun, planets, and stars revolve around the earth each day.

Ptolemy and his Treatise

bp52» "His name was Claudius Ptolemaeus ... he lived from approximately A.D. 100 to approximately A.D. 175, and that he worked in Alexandria, the principal city of Greco-Roman Egypt, which possessed, among other advantages, what was probably still the best library in the ancient world. ... As is implied by its Greek name, ... , 'mathematical systematic treatise,' the Almagest is a complete exposition of mathematical astronomy as the Greeks understood the term" (Toomer, p. 1). By the "fourth century (and probably much earlier), when Pappus wrote a commentary on it, the Almagest had become the standard textbook on astronomy which it was to remain for more than a thousand years.... It was dominant to an extent and for a length of time which is unsurpassed by any scientific work except Euclid's *Elements*.... " (Toomer p. 2-3)

bp53» "Then, in the second century A.D., came Claudius Ptolemaeus, an Egyptian -- the great Ptolemy who was to be the uncontested monarch of astronomy for a millennium and a half. He restored the harmonious cosmos Hipparchus had shattered. Ptolemy was a theoretician of such superior qualities that only Newton can be considered his peer. A universal mind, he perfected Greek mathematics and Greek natural science in general. His achievement appears all the more impressive when we compare it with the ordinary science of his time, which was hopelessly bogged down in speculation.

bp54» "Ptolemy called his principal work on astronomy the Great System (*Megale Syntaxis tes Astronomias*, later known as *Almagest* from the Arabic translation). This somewhat arrogant title was fully justified, for he had examined every problem in astronomy, and solved every one with Euclidean precision. Ptolemy created the first complete scientific system — a structure so vast and coherent that not even the comprehensive mind of an Aristotle could have conceived it, let alone worked it out.

bp55» "Toward the solution of the chief problem, the apparently irregular velocities of the planets, he made a crucial discovery. Ptolemy drew an overlapping circle near Apollonius' circle.... The second circles came to be known as Ptolemy's epicycles. From the center of the epicycle the motion around Apollonius' eccentric circle appeared to be uniform. The system was extremely complicated, but it worked; Ptolemy could use it to calculate any future position of Mars... Ptolemy could justly boast that he had laid the keystone of Greek astronomy.... Mathematically speaking, this was true; henceforth, everything was calculable.... The planets now traveled in loops, that is to say, around an imaginary point that for unknown reasons itself revolved around the Earth....

bp56» "A man named Kepler, fifteen hundred years after Ptolemy, at last was able to refute the last word of Greek astronomy, and overcome the mathematical sovereignty of

nonsense by even more exact mathematics, which once more made everything meaningful" (Rudolf Thiel, *And There was Light*, trans. by Richard and Clara Winston, pp. 49-51).

bp57» Ptolemy's system had the earth as the center with the stars, moon, planets, and even the sun circling the earth each day. Ptolemy used the wrong and illusionary concept of epicycles to explain the apparent movement of the planets in the night. He further used mathematics to predict the future movement of planets. His system worked to a remarkable degree. It had a mathematical system to back it up. His book was well written and seemed quite logical. After all even today the planet, sun, moon, and stars do *apparently* circle the earth.

compelling. It rules all. Since 1984 English readers have been able to read Ptolemy's work, as translated by G.J. Toomer, *Ptolemy's Almagest.* In this translation you can see the apparent logic to the whole work. You can see the massive amount of tables, observations, and mathematics to back Ptolemy's theory.

bp58» How can a work so logical, based on so many observations, and backed up by mathematics be wrong? It was wrong because it was based on some faulty thinking (the

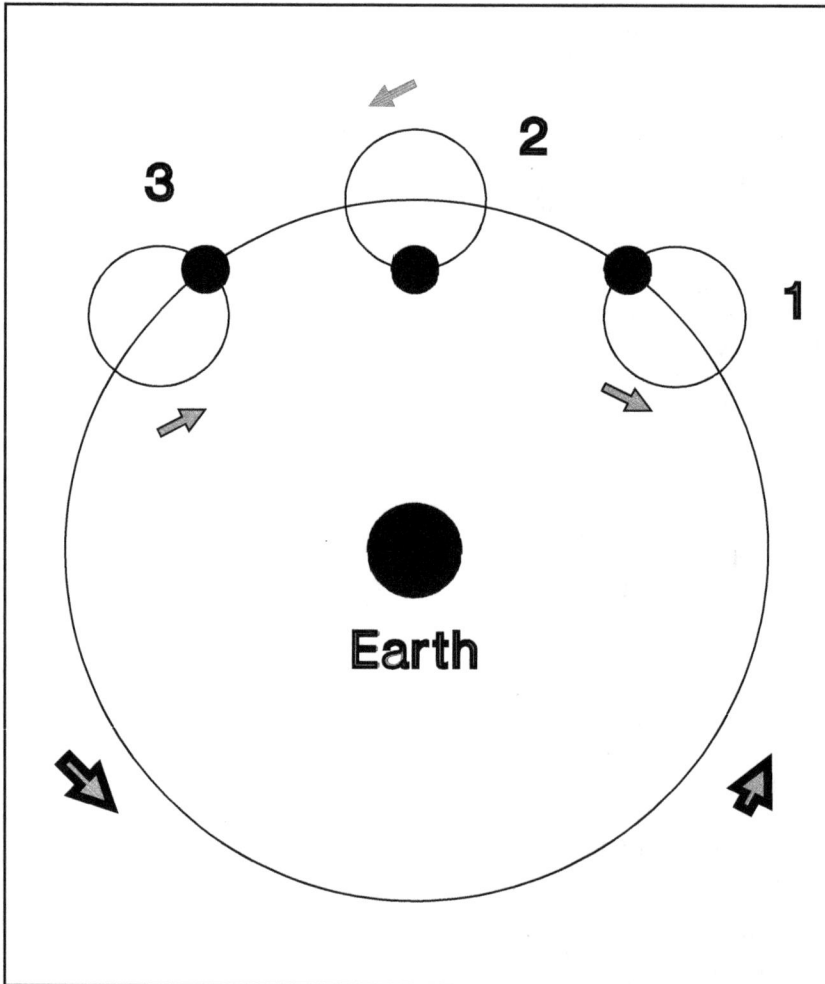

Ptolemy system made sense out of wandering stars (planets). It predicted future positions of planets. It was the great system. It lasted for almost 1500 years. Apparently it was the perfect system. It was backed by mathematics. It was apparently backed by observation. But it was wrong. How wrong can you be to think that the massive sun circles the earth *each* day? But because of the prevailing mindset Ptolemy remained king. A mindset can be very

enormous sun going around the smaller earth would have to move at an unbelievable rate), because Ptolemy was a charlatan that cheated on his mathematical figures and cheated on his observations (Newton, *The Crime of Claudius Ptolemy*), and because he had a mindset that told him that all heavenly objects were perfect and god-like, they moved in perfect circles, he thus placed epicycles into his system:

"The heaven is spherical in shape, and moves as a sphere; the earth too is sensibly spherical in shape ... in position it lies in the middle of the heavens very much like its center.... [Toomer, p.38] The following considerations also lead us to the concept of the sphericity of the heavens....[p. 39] We think that the mathematician's task and goal ought to be to show all the heavenly phenomena being reproduced by uniform circular motions.." (Toomer, p. 140).

Ptolemy got his mindset about the orbits having to be perfect circular orbits from the Greeks such as Aristotle:

"There must be some substance which is eternal and immutable.... But motion cannot be either generated or destroyed, for it always existed.... But there is no continuous motion except that which is spatial, and of spatial motion only that which is circular... There are other spatial motions – those of the planets – which are eternal (because a body which moves in a circle is eternal...).... for the nature of the heavenly bodies is eternal (Aristotle, *Metaphysics* Book XII [Loeb Classical Lib. No. 287], pp. 141 & 155).

Ptolemy was so overly influenced by the Grecian philosophy that he fabricated a mathematical system to help prove his preposterous belief: "We think that the mathematician's task and goal ought to be to show all the heavenly phenomena being reproduced by uniform circular motions.." (Toomer, p. 140).

bp59» Today math is used extensively to prove likewise absurd theories. (See my book, *Einstein: Light, Time & Relativity*) They do not appear preposterous to most today only because of today's mindsets which filter reality. Mathematics are wrongly used today in this so-called scientific age. Today mathematics are blinding otherwise intelligent people into believing in absolutely paradoxical and nonsensical theories on the cosmos, physics, and biology. Today much of what is called science exists inside of a mindset that shuts out the truth.

Mindset, A Brain Cell Problem

bp60» The main problem with a mindset occurs when you try to communicate with someone with a different mindset. Sometimes it is almost impossible. A Catholic trying to convert a Protestant has a terrible time trying to communicate his point of view, and vice versa. Many times even trying to communicate your different point of view will be met with a harsh reaction and sometimes even a violent reaction. Why?

bp61» One book tried to explain this. Daniel Cohen, in a 1982 book, called *Re:Thinking*, put it this way:

"Once a pattern — an idea or belief — becomes fixed in our neurological pathways, it is extremely hard to alter it. The more basic the belief, the more we refer to it in our thoughts, the more well worn is that particular neural pathway — and thus the harder it is to change the idea, even when it is wrong" (p. 70).

"Our memories and beliefs are stored in our brains in the form of nerve cell patterns. When you argue with someone you are pitting your nerve cell patterns against his. The beliefs and opinions you hold are not the result of some abstract intellectual process. They are the result of your total life experience. But your opponent's beliefs and opinions are the same. For both of you, changing these deeply held beliefs is hard and painful.

Since we all want to avoid pain, we all want to avoid changing these beliefs and opinions. The mind will work very hard to defend them.

If an argument isn't a means of persuading people, what is it? It's a fight, an attack, and you react to it as such" (p. 118).

bp62» With our mindset we see only what our mindset allows us to see. It acts like a filter and filters out any pattern not belonging to the sets of rules we have etched in our brain cells.

BP3: Bible Paper

Typical Criticism

Three Tests to Give

bp63» The *BeComingOne Papers* uses the Bible. Now to some the Bible is a book full of tales that isn't worth the paper it is written on. But this is a very biased, unfair and incorrect view. It is a view of a mindset. Like other mindsets it is imbedded in brain cells, thus, making it very difficult to change. Just as Ptolemy's geocentric theory of the universe seemed to be very sound to the generations of the past (see "Mindset Paper" [BP2]), today's myths and mistaken theories seem scientific to today's generation. Just because those of Ptolemy's generations believed in his theory, doesn't mean they were less intelligent. Many were quite smart, but nevertheless quite wrong. Irrespective of today's mindsets and biases concerning the Bible, it is a very worthwhile book.

bp64» The Bible is a historical document. Its history goes back to the beginning of mankind. It is an accurate document. Especially in the last hundred and eighty years, archeology has repeatedly confirmed facts recorded in the Bible that previously had no other confirmation. In comparison to other ancient writings, the Bible is by far, let me repeat, the Bible is by far the most accurate historical document in the world, especially considering the volume of words in the Bible.

bp65» The Bible is filled with specific place names, proper names, topographical descriptions, descriptions of ancient customs and nations, descriptions of ancient artifacts, temples, religions, and human behavior. Until the last 180 years the cynics used to call many of the nations, cultures, and customs described in the Bible myth, or just oral traditions that have lost their truth. But archaeological finds have made a mockery out of such outdated skepticism. Mythical books do not have the abundance of specific information as does the Bible. Details after details are abundant throughout the Bible. Today it is ludicrous to call the Bible anything but an accurate historical document. Yet, as a mindset filters a person's perception so does the mindset of anti-Biblical scholars. It is amazing to me how they can ignore the archaeological finds of the last 180 years and still cling to naive views about the Bible. I read in year 1988 a news story about a group of "scholars" voting on various drafts of a revision on what were Christ's real sayings. They refused to believe that the words written in the Bible were really spoken by him, and they had the naivete to *vote* on a new version. This is the height of arrogance, for no other ancient character has more proof for his person or words than Christ (*He Walked Among Us: Evidence for the Historical Jesus,* Josh McDowell & Bill Wilson, 1988). There is far more proof of Christ's existence than Alexander the Great, Julius Caesar, Plato, Aristotle, Homer, etc.

bp66» I have had different mindsets concerning the Bible. First I believed in the Bible because I was reared to believe in it, at least as the church interpreted it for me (the Catholic view). But I never studied it when I had that mindset. Next through reading too many biased liberal books, I thought that the Bible was of little significance. It was only after I had an epiphany in 1969 that I began to study the Bible itself and read other pro-Biblical views that I came to a different view: The Bible is a very important book, an accurate book, a historical book, a revealing book, etc.

bp67» Unlike what cynical people biased by their mindset attest, no other document comes close to the legitimacy of the Bible. The Bible has the oldest manuscripts of any other ancient document of its size. The intra-cohesiveness of these old manuscripts prove that today's Bible very closely, if not exactly, reflects the original documents. Remember there were no copy machines when the manuscripts of the Bible were handed down. The information age is a very recent phenomenon. The copying of old manuscripts was done by hand. Because it is almost impossible to copy a document the size of the Bible without some mistakes, there are some variations between the ancient manuscripts and today's, but most of these variations concern different spelling of words or omission of words or concern words or phrases that were added by scribes so as to clarify the meaning of the text. Very little to none of the variations affect the meaning or doctrine derived from the Bible, especially if you believe that the antitypical sense of the Bible is the real sense of it (see "Duality Paper" [BP4]).

bp68» Among other things, the Bible is a book of the history of man and man's relationship with God up to Israel (Jacob), from there a history of Israel up to Christ, and from there a history of the Church of Christ up to around 40-70 AD. The Bible is mainly concerned with the behavior of mankind and his relationship with God and the coming of the Messiah. In this the Bible is a very different book when compared to most other ancient books, or for that matter contemporary books. The Bible actually thinks there is something called evil in mankind's behavior, and that there is a God or Power that cares about mankind's behavior. The Bible indicates that wrong behavior actually causes grief and death. This is hard for some today to believe. Some who have a humanistic mindset, believe in one form or another that there is no or little real evil in mankind, but just some form of mis-education. To this mode of thinking it is society who is to blame, not the individual.

bp69» The Bible starts out describing the CREATION of the heavens and earth, or as we call it today — the universe. But what about evolution? Did an intelligent Power (God) create or did evolution create? Contrary to what most schools of today teach, the "creation by God" answer is more scientific than evolution. Evolution is nothing but a faith — a faith based on the ludicrous chance that the *complex* universe somehow evolved. I know how dogmatic the scientists seem, but if you read their journals you know that they too have dissimilar views on many aspects of life, and that the foundations of many of their views are very slim and in contradiction to each other. Some of today's most celebrated theories are based on a very thin film of evidence and on a very precarious set of conclusions based on this evidence (see my, *Science Papers*).

bp70» There are forms of mysticism in science today as there are in many if not most religions. But because we are *not* taught to analyze the foundations of branches of science, we have a mindset that cannot perceive the mysticism in science. *In fact there is a close parallel between mysticism, science, and religion.* See the list of Creation books and organizations in the back of this paper if you want greater details on the evolution versus creation controversy. There is more to the idea of Creation than to the mindlessness of evolution.

bp71» **Christianity, Judaism and Islam base their belief and knowledge of God on information found in the Bible.** The non-believers think the Bible is too legendary and therefore cannot be the word of God. To the disbeliever the Bible is full of exaggerated stories orally passed on through generations.

Bible's Rich Metaphorical Word Usage

bp72» The Bible is a historical document that includes poetry and a rich use of figures of speech. The Bible uses similes, "his eyes were as a flame of fire" (Rev 1:14). The Bible uses metaphors, "tell that fox" (Luke 13:32). The Bible uses metonyms, "if the house be worthy" (Mat 10:13). The Bible uses synecdoches, "all the world should be taxed" (Luke 2:1). The Bible uses personifications, "the earth mourns and fades away" (Isa 24:4). The Bible uses apostrophes, "O death, where is thy sting?" (1Cor 15:55) The Bible uses hyperboles, "the light of the sun shall be sevenfold" (Isa 30:26). The Bible uses allegories, "this Hagar is Mount Sinai in Arabia."(Gal 4:24) The Bible uses parables, "behold, a sower went forth to sow" (Mat 13:3). The Bible also uses irony, riddles, and fables (1Kings 18:27; Rev 13:18; and Judg 9:8 ff & 2Kgs 14:9 ff). So we see that the Bible is rich in its use of language. (The serpent did not literally speak to Eve, only figuratively did the serpent speak to Eve.) Yes, the Bible does have a few fables, riddles and metaphorical serpents talking within its pages. The Israelis were creative writers. Figures of speech are used to draw attention and interest to the meaning of the words, and to aid in the remembrance of the text. A text of poetry is easier to remember than a boring academic document. The fact that the Bible used colorful word usage to convey its message does not mean it does not convey a truthful picture of history and important philosophical and theological messages from God. It may just as well mean that God used man's colorful ways of expression to convey his word so as to better brand the message into the mind of man. Figures of speech can also breed misunderstanding if the hearer/reader takes literally a story that was only meant to teach a lesson. Trees clapping their hands and snakes talking are metaphorical, not literal. Poetry and metaphors grace the Bible throughout.

Bible, an Ancient Text with Abundance of Details

bp73» The Bible's history goes back thousands of years. Especially in the last hundred and eighty years, archeology has confirmed facts recorded in the Bible that previously had no other confirmation. In comparison to other ancient writings, the Bible is as accurate, if not more accurate than any other historical document in the world (See my *Chronology Papers*). Most ancient historians give a skewed view to make their ethnic group look better than they did in reality. Not so with the writers of the Bible. They wrote, not only of the glory, but of the foibles of their people.

bp74» Again, we restate, the Bible is filled with specific place names, proper names, topographical descriptions, descriptions of ancient customs and nations, descriptions of ancient artifacts, temples, religions, and human behavior. Until the last couple of centuries the skeptics used to call many of the nations, cultures, and customs described in the Bible – myth, or just oral traditions that had lost their truth. But archaeological finds have helped to alleviate some of this skepticism.

Typical Criticism

bp75» *Typical criticism: The Bible is a mythological book that contains orally transmitted myths that were passed down through generations until about the time of Ezra who compiled most of the Old Testament. Moses did not write five books of the Bible because for one thing, there were few in his day who could write: the Hebrews used oral tradition and/or he was illiterate and so could not write it.*

bp76» **First about Moses:** I don't see anywhere in the Bible where it specifically says that Moses wrote every single word of the first five books of the Bible. Of course he compiled sections from other writings and placed them within his books. He may have had scribes helping him; Jeremiah had a scribe to help him. I don't see in the Bible where it states specifically who actually penned each book. Some of the Torah (first five books of the Bible) was composed after Moses' death because it mentions his death and other facts that were impossible for him to write himself. The Torah also included information about the creation, the flood, Abraham, and genealogy that Moses or his scribe (at his direction) copied from older writings. I also don't see any proof that Moses did not know how to read or write, after all, he was brought up by the Pharaoh's daughter in the palace, so of course, he was taught to read and write. Evidence has come forward lately of an alphabetic script and inscriptions on stone that predate the time of Moses and that very much looks like early Hebrew (Douglas Petrovich, *The World's Oldest Alphabet* [2016]; The movie, *Patterns of Evidence: The Moses Controversy* [2019], Etc.). The general criticisms of Moses and the Bible are sometimes petty,

merely trying to find fault, and not giving the authors the benefit of the doubt. While others' criticism seems to be mere scholarly exercises, although they do point out apparent paradoxes in the text and in its depiction of the Hebrew God. Books like Richard Simon's, *A Critical History of the Old Testament* [1682, English Trans., (find at www.archive.org)], seem to be anti-Jewish in tone by attempting to prove that the caretakers of the Hebrew text made many mistakes in copying, while the Isaiah scroll from the Dead Sea Scrolls is proof of the immense care they took in preserving the Hebrew Bible. [Go *here* (digital version) for info about the Isaiah scroll.] To make his case Simon seems to point out every trivial criticism he could think of (the text repeats itself too many times, the text uses synonyms, it wasn't written in a style he appreciates or understands, laws are written with different words at different places within the text and so forth). Notice that Simon's book goes back two centuries before the 19th century criticism.

bp77» The general criticism is not that solid especially when we examine archaeological finds of the last few centuries. For example, the Ebla tablets, discovered in the 1970's prove that there was written text before Moses at least back to about 2250-2000 BC[1] (see *my Chronology Papers*). In the 1975 season over 15,000 tablets were found, about 18,000 complete clay tablets were eventually found. The language of the tablets was Sumerian script and the Eblaite language, the earliest known Semitic language. Personal names, geographic names, lists of animals, professions, names of officials, vocabularies, sacrificial systems, rituals, proverbs, hymns, and so forth were found. Most of the tablets dealt with economic matters such as bills of sale, receipts, tariffs, contracts of sale, etc. Among the tablets were copies of treaties, one was between Asshur and Ebla. Asshur is mentioned in the 10th chapter of Genesis. The language of Ebla was Semitic and the closeness to Hebrew is striking. The vocabularies were the oldest found so far in history, about 500 years earlier than any previously known. There are tablets with case law on them. This proves that hundreds of years before Moses there was written law. Moses didn't invent law, he merely put it in a Hebrew form. What is unique about Moses's

---

[1]The contemporary estimate at the time I wrote this.

law is the patterns in it and its God. These tablets named the five cities of the plain mentioned in the book of Genesis of the Bible, proving these cities were not mythological. The tablets reflect the culture of the patriarchal period and even mention people's names that appear in the book of Genesis. (see Beld, Hallo, and Michalowski, *The Tablets of Ebla: Concordance and Bibliography*, 1984; Giovanni Pettinato, *The Archives of Ebla*, *1981*; Clifford Wilson, *Ebla Tablets*, 1977; Benner, Jeff A., "The Archives of Ebla and the Bible," http://www.ancient-hebrew.org/bible_ebla.html etc.)

Because these tablets were found in Syria near the modern city of Aleppo, apparently the information that ties these tablets to the Hebrews is being censored by Syria because of the fear of giving any credence to the Jews' rights to the ancient land of Israel.

Three Tests

bp78» There are three tests we can use to determine the reliability of the Bible. **(1) Bibliographical Test**: Not having the original documents of the Bible, how reliable are the copies we have? **(2) Internal Evidence Test**: Is the written record credible? **(3) External Evidence Test**: Does other historical material confirm or deny the material in the Bible?

Bibliographical Test

bp79» How reliable are the copies we have in regard to the number of manuscripts and the interval of time between the original and the surviving copy? Concerning New Testament manuscripts there are about 22,000 copies of manuscripts with at least partial contents of the New Testament. The closest ancient work next to the Bible is the Homer's *Iliad* (700?? BC), but it only has about 643 manuscripts. Such works as Aristotle (*c.* 340 BC) have only about five manuscripts for any one of his works, the earliest copy is dated about 1100 AD, about 1400 years after he lived and wrote his work. The history of Thucydides (*c.* 460-400 BC) has just eight manuscripts and the earliest copy is from about 900 AD. Pliny the Younger's History has only 7 copies, the earliest copy from about 850 AD. Plato's work has only 7 copies, the earliest from about 900 AD. Livy's work has only 20 copies. Contrariwise the New Testament manuscripts are about 22,000 in number, with one of the earliest (John Ryland MSS) dating from about 130 AD, about a century after Christ. The *Chester Beatty Papyri* located in the Beatty Museum in Dublin has three

manuscripts containing major parts of the New Testament. Two of these papyri manuscripts are dated in the second half of the third century (250-300 AD). But manuscript ^{p46}, which was originally dated about 200 AD has since been dated to 100 AD on paleographical grounds (*Biblica* 69:2 [1988], pp. 248-257). "Paleography (literally, old writing) is the study of the manuscripts themselves rather than the text they contain. In attempting to date manuscripts, paleographers are especially concerned with the script, i.e., the style of the letters used. We have so many papyri from Egypt that a definite progression in the style of script from one period to the next can be seen" (Darrell Hannah, "New Testament Manuscripts," *Bible Review*, Feb. 1990, p. 7). [Some of this paragraph's info was taken from Josh McDowell, *New Evidence that Demands a Verdict*, 800 pages, 1999.]

bp80» Until the discovery of the Dead Sea Scrolls the oldest Old Testament manuscript was dated about 900 AD. This was about a 1300-1400 year gap from when the Bible was completed. Because of the reverence for the scriptures, the Jewish community went to great lengths in making new copies of the Old Testament as accurate and perfect as humanly possible. "Besides recording varieties of reading, tradition, or conjecture, the Massoretes undertook a number of calculations which do not enter into the ordinary sphere of textual criticism. They numbered the verses, words, and letters of every book. They calculated the middle word and the middle letter of each. They enumerated verses which contained all the letters of the alphabet ... These trivialities ... had yet the effect of securing minute attention to the precise transmission of the text; and they are but an excessive manifestation of a respect for the sacred Scriptures..." (Frederic Kenyon, *Our Bible and the Ancient Manuscripts*, 1941). Because of this meticulous care of the Jewish caretakers of the Bible, it has been believed the Bible copies were highly accurate. The Dead Sea Scrolls helped to confirm this belief.

bp81» The Dead Sea Scrolls are made up tens of thousands of inscribed fragments from over 900 texts. The texts can be divided into three groups: Biblical manuscripts (copies from the Hebrew Bible) make up about 40% of the total; Apocryphal texts, which make up about 30% of the total; and Sectarian manuscripts. They are dated from about 150 BC to 70AD. One complete scroll of the Old Testament book of Isaiah was found among the Dead Sea Scrolls.

According to Gleason Archer, the Isaiah scroll "proved to be word for word identical with our standard Hebrew Bible in more than 95% of the text, but in 1QIs[b] [a partial text about 1/3 of Isaiah], (ca. 75 B.C.) the preserved text is almost letter for letter identical with the Leningrad Manuscript. The 5% of variation consisted chiefly of obvious slips of the pen and variations in spelling" (Gleason Archer, *A Survey of the Old Testament*, 1994, p. 29).

Internal Evidence Test

bp82» When you analyze the Bible itself you must be fair. To use what some call Aristotle's dictum:[1] "the benefit of the doubt is to be given to the document itself, and not arrogated by the critic to himself." You should not assume fraud or error unless you find contradictions of known fact.

> "Giving "benefit of the doubt" until further evidence is uncovered and investigation undertaken is hardly incompatible with a healthy skepticism. Extreme incredulity is no more inherently virtuous or useful than extreme credulity. Indeed both represent a mindset not conducive to honest and fair examination of a particular claim....
>
> It is no coincidence that atheists, and skeptics come down on the side of the burden of proof falling upon the document while Conservative Christian scholars come down on the side of the burden of proof falling to the critic.... the burden of proof issue often says more about the person examining a particular text than about the text itself. It often reveals the presuppositions and philosophical assumptions of the contemporary historian.
>
> "Those who accept the empirical claims of a historical text bear the burden of proof just as much as those who assert their falsehood; in the absence of such proof we should suspend judgment. Empirical uncertainty thus forms the middle ground between the claim that empirical claims are certainly true and the claim that empirical claims are certainly false." [Jeff Lowder][2]

bp83» The biggest problem that the secular intellectuals find with scriptures is God and his supernaturalness. According to their system of thinking any supernaturalness is automatically thrown out. But at the same time the magic of evolution, the cosmic non-intelligent soup that by some miracle created the universe, is not thrown out. This is the result of a mindset. The writers of the New Testament were eyewitnesses (Luke 1:1-3; John 19:35; 1 John 1:3; 2 Peter 1:16; etc). They spoke to others who were eyewitnesses (Acts 2:22; 26:24-28; etc.). At first they did not believe in Christ's resurrection, and admitted this very thing in their writings (Mark 16:11; Luke 24:11, 25; John 20:24-29). But later they saw the resurrected Christ and believed (Luke 24:48; John 20:19-20; Acts 1:8; 2:24,32; 3:15; 4:33; 5:32; 10:39, 41; 13:31; 22:15; 26:16; 1 Cor 15:4-9, 15; 1 John 1:2). Later many of them died because of this belief (Acts 7:58-60; 9:1; Rev 6:11; Heb 11:35-12:1). Tradition has it that 11 of the apostles were martyred for their belief. If it was all a lie, if they made it up, why did they allow themselves to die for it? Even when they lived they gained nothing materially from their belief. They must therefore have believed it because they *saw* the things they wrote about.

bp84» Sir William Ramsay, one of the great archaeologists, is another witness to the Bible's accuracy:

> "He was a student of the German historical school that taught that the Book of Acts was a product of the mid-second century A.D. and not the first century as it purports to be. After reading modern criticism about the Book of Acts, he became convinced that it was not a trustworthy account of the facts of that time (A.D. 50) and therefore was unworthy of consideration by a historian. So in his research on the history of Asia Minor, Ramsay paid little attention to the New Testament. His investigation, however, eventually compelled him to consider the writing of Luke. He observed the meticulous accuracy of the historical details, and his attitude toward the Book of Acts began to change. He was

---

[1] I could find no evidence that Aristotle actually said this, but the idea is still worthy to note.

[2] www.theologyweb.com, June 22, 2003, by markg

forced to conclude that 'Luke is a historian of the first rank ... this author should be placed along with the very greatest of historians.'" (J. McDowell, *He Walked Among Us*, p. 110)

More could be said on the internal evidence, but we will let other books speak on this matter (see book lists below).

External Evidence Test

bp85» Does other historical material confirm or deny the testimony in the Bible? For one thing the names and descriptions of kings, cities, geography, customs, events, wars, and so forth are well attested and confirmed by secular findings such as archeology. In our *Chronology Papers* we give some evidence of this. The books in the book list below as well as the evidence and books referenced within these books also attest to this. Joseph P. Free, in his *Archaeology and Bible History*, said "Archaeology has confirmed countless passages which have been rejected by critics as unhistorical or contradictory to known facts" (p.1). Read the many books available on this subject.

bp86» The following short list of books will help you in your search:

- Josh McDowell, *New Evidence that Demands a Verdict*, 800 pages, 1999

- F.F. Bruce, *The New Testament Documents: Are They Reliable?*, 2009

- Josh McDowell & Bill Wilson, *He Walked Among Us: Evidence for the Historical Jesus*, 1988, 2011

- Merrill F. Unger, *Archaeology and the Old Testament*, 1954, 2009

- J. Pritchard, *Ancient Near Eastern Texts Relating to the Old Testament*, 1969, 2010

- Jack Finegan, *Archaeological History of the Ancient Middle East*, 1979, 1996

See our website for the latest info on proof that the Bible is a historical document given to man from God:

https://beone.ws

BP4: Duality Paper

Type and Antitype

Visible Projects Invisible

Examples

Look to Higher Meaning

For God speaks once, yet twice, though people do not perceive it (Job 33:14).

bp87» All of the papers in the *BeComingOne Papers* project the duality of the Bible. **If there is a secret to understanding the Bible, it is the duality of the Bible or the type and antitype of the Bible.** There are events and words in the Bible that have dual meanings. One meaning is the physical meaning; the other meaning is the spiritual meaning. The physical meaning is the typical rendition. The spiritual meaning is the antitypical rendition. By the time you read all the *BeComingOne Papers* you will understand, or should understand, what we mean when we say the Bible is dual. But in this short paper we will give you a beginning towards an understanding of the duality of the Bible.

Type and Antitype

bp88» The duality of the Bible consists of "types" and "antitypes." A "type" is an event, person, thing, or symbol in the Bible that represents some Spiritual Truth. The Spiritual Truth is the antitype of the type. For example, in the Old Testament it describes the Passover lamb. In the New Testament it tells us the True or Real Passover lamb is Jesus Christ. (1 Cor 5:7) The Old Testament's Passover lamb is a type of the New Testament's lamb Passover, which is Jesus Christ. (see "God's Appointed Times" paper [NM16]) The Old Testament's Passover foreshadowed the New Testament's Passover.

bp89» Paul of the New Testament, in his letter called Hebrews, tried to explain the duality of the Bible. He didn't use the word "duality" when he tried to explain it, but nevertheless he was explaining the duality of the Bible. Paul in Hebrews speaks of a "sanctuary that is a copy and shadow of what is in heaven. This is why Moses was warned when he was about to build the tabernacle: 'See to it that you make everything [in the tabernacle] according to the pattern shown you on the mountain.'" (Hebrews 8:5; Exo 25:9, 40) Paul is saying that the tabernacle that Moses built was a *pattern* of the tabernacle in heaven. What does this mean?

bp90» When you see the word "heaven" used in the Bible, you can think of it as *spiritual*, for both "heaven" and "spiritual" are used interchangeably in the Bible. (compare "heaven" and "spiritual" in 1 Cor 15:44-49) Thus Paul is saying that Moses made his tabernacle (the physical one) according to the pattern of the heavenly or spiritual tabernacle.

bp91» Paul explains that Christ didn't go into the physical tabernacle, but the "true tabernacle" or the "more perfect tabernacle that is not man-made," "for Christ did not enter a man-made sanctuary that was only a copy of the true one; He entered heaven itself [the spiritual dimension itself], now to appear for us in God's presence" (Hebrews 8:2; 9:11, 24). The physical tabernacle built by Moses was merely a copy of the Real or True tabernacle. Paul tells us that "the law [much of the Old Testament is called the law] is only a shadow of the good things that are coming — not the realities themselves" (Heb 10:1). The law and things of the Old Testament were merely shadows of the good things, the real things, to come. The Old Testament and the things in it are only the *types* of the *antitypes*. The antitype being the Real and True — the Spiritual fulfillment of the type. Paul tells us that the things written in the Old Testament were *types* or *examples* for us, that is, types or examples for us Christians. (1 Cor 10:11)

Visible projects the Invisible

bp92» Paul tells us that the invisible qualities of God can be understood by the things that God has made. (Rom 1:19-20) And in our papers you will see how many aspects of this world, like males and females, which God made, are types of the antitype. Marriage, being born, women, water, stars, and so forth all have higher meaning: they all point to the Spiritual plan of God; they are all types of the Real or True, which is the antitype. For example, "stars" are representative of angels. (see Rev 1:20) And even "water" foreshadows the Spirit (John 7:38-39).

Examples

bp93» Even New Testament rituals like water baptism are types of the antitypes. Water baptism represents spiritual baptism. (see "Baptism Paper" [NM4]) All of the Bible is dual: type and antitype. This includes the Old as well as the New Testament. Even the physical creation is representative of a higher or spiritual meanings. (Rom 1:20) The physical creation (the type) is representative of the spiritual creation (the antitype). For example the days of the week are seven. The week was instituted right after the creation. (see Genesis, chapter 1) But this week is a type. It represents the antitypical week. To God each day is 1000 years. (2 Pet 3:8) God, Who is Spiritual, has a week that lasts seven 1000 year periods, for each of His days lasts 1000 years. Therefore the week (the type) is representative of the 7,000 year week (the antitype). The first seven days of creation, which are described in the book of Genesis, are the physical or typical rendition of the creation. But since the Bible is dual, we know that these seven days of creation have corresponding antitypical or Spiritual days. According to Peter a day of God is 1000 years. (2 Peter 3:8) God being Spiritual has different days than us. God's days are each 1000 years. Therefore, with the awareness of duality, we know that the seven days in Genesis was speaking *first* of the typical seven physical days of creation, and *second* was speaking of the seven Spiritual days of creation, which are each 1000 years long.

bp94» Thus, we see that from Peter's mentioning that a day to God was like 1000 years, we can ascertain the antitypical meaning or higher meaning of the week. This is duality. The typical week is our week of seven literal days. But the antitypical week has 1000 years to each day.

Female and Male Language; Type and Antitype Language

The two sexes use the same language and understand the same language in slightly different ways. The same words or sentences have different meanings to each sex (*Male/Female Language*, by Mary Ritchie Key, 1996) because of their biosocial differences (see my book, *Male & Female*). Just as women and men can get two different meanings from the same words (a sex/gender difference), people also understand the Bible in two different ways: its physical and its Spiritual meaning. In all my books pertaining to the

Bible I manifest this and attempt to explain this phenomenon.

Look To The Higher Meaning

bp95» Even such things as "salt" and "light" have higher or antitypical meaning. (Matt 5:13-16) "Clean" and "unclean" have a higher meaning (Matt 15:2,11,15-20). "Yeast" has a higher meaning. (Matt 16:5-12) In fact, Christians are to look for the higher meanings. The enemies of Christ have their minds or thoughts on earthly things. (Phil 3:19) But Real Christians are to have their thoughts on the things above — the heavenly or Spiritual thoughts and meanings. As explained in the "Other Mind Paper" [NM21] all of us in this old age have the other-mind in us misleading us towards confusion. This other-mind is actually another living being — a spiritual being of confusion and evil. Peter had "Satan" inside of him. (Mark 8:33) That is, Peter had the spiritual influence of Satan inside of him. (see the "Other Mind Paper" [NM21]) When Christ spoke to this "Satan" inside of Peter, He said: "You do not have in mind the things of God, but the things of men" (Mark 8:33). That is, Satan has in his mind the things of men — the physical — instead of the things of God, which are the Spiritual. Satan and his spiritual power look on the lower things instead of the higher things. But Christians are to look upon the higher, or Spiritual things, which are the antitypical things.

bp96» Christians have in their mind the Real or True tabernacle when they read about the tabernacle in the Old Testament. Christians have in mind the Real or True water (the Spirit) when they read about the water baptism or living waters in the Bible. Christians have in mind the Real or True bread (Christ, His Spirit) when they read about the breaking of bread or the bread of life in the Bible. Christians think about the Real or True circumcision (being Spiritually circumcised) when they read about circumcision in the Bible. Christians think about the Real or True faith (Faith of the Spirit) when they read about faith in the Bible. Christians think about the Real or True light (God's Spirit) when they read about light in the Bible. Christians are those with the New Mind, which is called by some the Spirit of God, that is, they have the mind of God, they think like the God thinks. With this mind of God Christians look upon the *higher* meaning of people, events, and things they read about in the Bible. They just don't look to the physical things, they look upon the higher or Spiritual meaning of the physical things.

bp97» Christians have the New Mind — the Spirit of God leading them. As Paul said: "God has revealed it to us by his Spirit. The Spirit searches all things, even the deep things of God . . .. This is what we speak, not in words taught us by human wisdom but in words taught by the Spirit, expressing spiritual truths in spiritual words. The man without the Spirit does not accept the things that come from the Spirit of God, for they are foolishness to him, and he cannot understand them, because they are Spiritually discerned" (1 Cor 2:10,13-14). You have to have the Spirit of God to truly understand the higher meanings (the Spiritual meanings) that the Bible has hidden in its words. When you have the Spirit of God, you can see the type and antitype of the Bible. You can see the physical as well as the spiritual. You can see the duality of the Bible.

Even prophecy is dual.

bp98» As you can see by reading our *Prophecy Papers*, there is type and antitype in prophecy. All the ceremonial laws of the Old Testament have higher or Spiritual meaning. (see book of Hebrews) Even the things Christ did had dual meaning. Christ did most, if not all, of his healing on the Sabbath. The higher meaning of this is that Christ will do his True healing on the antitypical Sabbath, which is the 1000 years. Do read the rest of the papers we have to offer you. Look up the scriptures and begin to understand the duality of the Bible. If there is a secret to the Bible, it is the duality of the Bible.

==

BP5: Premises For Belief

bp99» The *BeComingOne Papers* were written to inform. All "Papers" were written in one form or another over the last 47 years (1970-2017). Together they are called — *BeComingOne Papers*. To every set of beliefs there are premises. In this paper we amplify on our premises.

How To Read These Papers

bp100» Because tradition is so strong and because each of us has a "mindset," we must be deliberate when reading the *BeComingOne Papers*. The *BeComingOne Papers* concern serious matters and must be read with this in mind. First, read the papers through once. Next, read them looking up our documentation. After this, read all the papers again. When you follow these three steps you should have a good idea what we are trying to convey to you.

bp101» Be careful. Because you have learned your views on God, the Bible, Evolution, Christianity, Science, and the meaning to life from tradition, you will be shocked by some of the papers in the *BeComingOne Papers*. Think. Question. There is something happening on earth. There is a reason for evil. God, the Power, is much more powerful than most think: "Jesus said to them, You err, not knowing the Scriptures, nor the POWER of God" (Matt 22:29). The might of the Power, the God of the universe, will make all *become one*.

bp102» "*Faith* is the substance of things hoped for, the evidence of things not seen." (Heb 11:1) Real *faith* is not naive, but based on substance. We seek real *faith* based on evidence.

How can we believe in the Bible? How can we know there is a God who created the universe? What can we base this faith on?

Premises

1. There is a God

bp103» A. The evolutionary Theory always starts with laws like those of heat, energy, motion, gravity, etc. Law projects order and intelligence. The genetic code of life found in our DNA also projects this high intelligence. *How can law evolve*? The laws of the universe must have come from somewhere. God is the creator of law. (Isa 33:22)

bp104» B. The law of biogenesis says that life comes from life. You can't get life from dead matter. God is the life giver. (Gen 1:20-31)

bp105» C. The fact of radioactivity indicates there was no eternity of matter. If matter always existed, without a starting point, then the "life" period of the radioactive elements would have long ago run its course, that is, the radioactive elements would have run down and there would not be any radioactive elements left. Thus, there was a *beginning* of matter. God created matter. (Gen 1:1)

bp106» D. The ecosystems of the earth, sea, and heavens indicate everything was planned, for the theory of evolution maintains that life was a hit and miss adventure ("natural selection" or "mutation," etc.). If this was so, then the universe would be filled with the misses (the inferior products of the evolutionary process) that were not superior enough to continue. There should be fossils of the inferior products of the evolutionary process. In other words, the rejections of the evolutionary process should be polluting the universe. But there are NO fossils of inferior or half-made species. Yet when one studies the ecosystems, one sees that all the substances in the ecosystem are needed in order to maintain these ecosystems even though some may seem at first observation to be not needed.

2. God Created The Heavens and The Earth.

Design

bp107» One proof of creation is that the theory of evolution can't be proven — it is full of contradictions. But what the Bible reveals about creation answers all the problems about the origin of life. How can life come from a mindless soup? The intelligence or mind or order of the universe presupposes or manifests a great Mind Power, the Great God. See the

Creationist Book List for more details on creation.

3. The Bible Is The Word Of God That Reveals The Purpose Of The Creation.

bp108» **A. *The Bible in its original languages*** is a doctrine given to mankind through spiritual influence in the minds of many individuals over approximately a 1500 year period. (Heb 1:1-2; 2 Peter 1:20-21; 2 Tim 3:16; see "Bible Paper" [BP3]; see below)

bp109» **B. *How was the Bible written and how was it inspired?*** It was written by humans in situations that were planned and predestinated [see NM8 & NM9]. The writers' languages (with its vagueness) their societies (with their complexities), their environments (micro and macro), and their biology (with its direct and indirect influences) were all planned, predestinated, and carried out by The Power. Each complete thought written had its type and antitype. They may have only intended the physical or typical meaning, but The Power predestinated its antitypical meaning (note 1 Cor 9:9-10; see "Duality Paper" [BP4]; etc.). The Power made it possible for the writings to be copied and passed down to our time in an accurate enough form (not too accurate so as to make it obvious that we were being manipulated, yet accurate enough). The Power is in *full* control because he is *all powerful*. He thinks of and controls many trillions of things and their interconnections. We think of a few interconnecting things. No cosmic soup can produce intelligence. We must be intelligent about this. There is something going on. Life and death have their purposes.

bp110» **C. *Any apparent contradiction is just that***, it only appears at first examination to be a contradiction. A total understanding of the Bible answers all apparent contradictions in a logical way. God is a God of logic: "Come now, and let us reason together says the LORD" (Isa 1:18).

bp111» **D. *The Bible interprets itself***; it interprets its own symbols.

(1) For example, the Bible interprets the symbolic meaning of stars by telling us that stars are symbolic of angels. (Rev 1:20)

(2) A second example is how the Bible interprets the symbolic heads on the beast of Revelation 17: "the seven heads are seven mountains" (Rev 17:9). Then it describes these seven mountains as seven kings. (Rev 17:10) Thus, these seven heads or mountains are seven kings.

(3) Notice that the beast of the book of Revelation has ten horns. (Rev 17:7) What are "horns" symbolic of? "And the ten horns ... are the ten kings" (Dan 7:24).

bp112» **E. *The Bible shows its reader HOW to read it*** in order to understand it. The Bible says it is "profitable for doctrine, for reproof, for correction, for instruction" (2 Tim 3:16). But one must know how to read it to get this information. There are *four* principles to know about when reading the Bible before one can understand it.

bp113» **(1) *Here a little, there a little***. "Whom shall he teach knowledge? and whom shall he make to understand doctrine? them that are weaned from the milk, and drawn from the breasts. For precept must be upon precept, precept upon precept; line upon line, line upon line; here a little, and there a little: For with stammering lips and another tongue will he speak to this people" (Isa 28:9-11).

bp114» A most important principle of the Bible is "here a little, and there a little." God has spoken to man in a stammering tongue. God through his Word (the Bible) speaks in a strange tongue to the world. Readers have noticed that the Bible in places puts information together in no apparent order. It seems that some of the writers of the Bible were in some kind of dream when they wrote. The Bible speaks on almost every subject, but it never totally gives all the information about that subject in one place in the Bible. A person must search throughout the Bible if he wants to find the complete information about any one subject.

bp115» An example of the principle shown in Isaiah 28:9-11 is if someone wants to ascertain exactly what happened shortly after Christ was resurrected from the dead. To find this information one must compare *all* four of the books of Matthew, Mark, Luke, and John with each other, *and* any other book of the Bible that may have any information in it concerning the resurrection of Christ. Only when you put all these scriptures together ("here a little, and there a little") will one have the correct information about Christ's resurrection (*Chronology Paper* [CP4]).

bp116» There is a reason why God is not going out of his way to "save" people or to inform people of his plan. Jesus Christ said he spoke in parables so people would *not* understand, but only those who had Spiritual ears would understand (Matt 13:10-17). God is not trying to save everyone Spiritually now, that comes later (see "All Saved Paper" [NM13] and the *New Mind Papers*).

bp117» **(2)** *Duality*. Next one must know that the Bible is DUAL — type and antitype. There are types of the real; or shadows of the real; or foreshadows of the real events; or symbols of the real. The REAL of these types *is* the antitypical thing or event.

bp118» An example of the type and antitype is that the physical is a *type* of the Spiritual. But the Spiritual is the *antitype* of the physical. The holy days mentioned in the Old Testament like the Passover are fore-shadows of the real or antitypical event. The physical Passover pictured the sacrificed lamb, which prefigured Christ, the Lamb of God — the real Passover. Another example is stars which are symbols of the antitype — the angels. (Rev 1:20)

bp119» God does things in twos: "the dream was doubled unto Pharaoh *twice*; it is because the thing is established by God" (Gen 41:32). God created male and female, he created the spiritual and the physical dimensions. He used a new and old Covenant when dealing with mankind (Hebrew, chaps. 8, 9, & 10). He even had Moses design the tabernacle as a shadow of the heavenly one.

bp120» God made "the law having a shadow of the good things to come, and not the very image of the thing" (Heb 10:1). He even created the physical world so that we could understand the spiritual dimension. (study, Rom 1:19-20) The physical is a *type* or shadow of the Spiritual dimension. God works in twos — type and antitype. This is duality. See "Duality Paper" [BP4].

bp121» **(3)** *Spirituality*. Knowing then that the Bible is a creation of God through his influence, we must learn how to read the Bible besides taking "here a little, and there a little," and besides understanding that the Bible is dual — type and antitype.

bp122» God says through his Word that : "God is Spirit: and they that worship him must worship him in Spirit and in truth." (John 4:24) And, "the words that I speak unto you, they are Spirit, and they are life" (John 6:63). Thus, God's words are Spiritual. They have a Spiritual meaning. They are not just physical words, they have a higher or antitypical meaning. There is actually a Spiritual language in the Bible.

bp123» And through God's Word it tells us to look away from the earthly (the physical) meaning *to* the higher or heavenly or Spiritual meaning. (Phil 3:19; Col 3:1-2; see "Duality Paper" [BP4]) We compare "Spiritual things with Spiritual" (1 Cor 2:13).

bp124» **Important:** The Spiritual or antitypical meaning of the Bible is the aspect of the Bible that is 100 % accurate all the time in every respect. The physical or typical meaning of the Bible is as close to accurate as human languages gets. But the typical language or meaning uses hyperboles, metaphors, and other language phenomena that may not be "literally" 100% true.

bp125» **(4)** *Effect of every vision*. The last principle we need to know is the one shown in Ezekiel 12:22-23: "Son of man, what is the proverb that you have in the land of Israel, saying, the days are prolonged, and every vision fails? Tell them therefore, Thus says the Lord GOD; I will make this proverb to cease, and they shall no more use it as a proverb in Israel; but say unto them, *the days are at hand, and the effect of every vision*."

> **bp126»** This means that the prophecy up to "the days are at hand" has not come true to its fullest degree. The visions have failed. Israel has not been completely restored, Christ (the Messiah) has not returned, etc. But when the "days are at hand," then will come "the effect of every vision." If we can locate when "the days are at hand," then we will know when the effects of every vision will happen, and then we will know when prophecy will be fulfilled to 100% of the words uttered in the Bible.
>
> **bp127»** Now in Revelation 1:3 it indicates that when someone finally comes to rightly read the mysterious book of Revelation, then the "time is at hand."

bp128» ***The proof of premise 3*** ("the Bible is the word of God that reveals the purpose of creation") is that logical answers concerning the creation and its purpose have been ascertained. And these solutions remove all apparent Biblical contradiction. For example, it explains how God, who is love (1 John 4:8), can also be a killer. (Deut 32:39) Love has nothing to do with killing, and since God is love, how can he also be a killer? That is surely a Biblical contradiction. How can God be love and a killer? The fact is that people do not know something about God that explains this "contradiction." The True God is love and won't and can't kill, but God is a killer! Now what kind of nonsense is that? Read the *God Papers* to find out who or what is God, and how this Biblical "contradiction" is logically removed. There is something most people don't know about God and the creation that easily explains this and other apparent contradictions concerning God's nature and person.

Creation Book List Information

- *Scientific Creationism*, Edited by Henry M. Morris (Master Books, San Diego, CA.)
- *Modern Creation Trilogy* (3 vol.) by Henry M. Morris & John D. Morris (Master Books, 1996)
- *The Creation - Evolution Controversy*, by R.L. Wysong (Inquiry Press, Box 1766, East Lansing, Mich., 1976)
- *Algeny*, by Jeremy Rifkin, 1983, The Viking Press, NY.

bp129» These books will give you a good idea why Creationism is a better explanation than evolution. Book two is more technical than book one, but goes much deeper into the question. Book three gives a refutation on Evolution from someone who is not a creationist.

Also see Web Sites such as: Institute for Creation Research; Creation Research Society; Creation Science.

Go to https://beone.ws for Internet links to creation sites..

bp130» I am not saying that everything in these publications is correct. I am saying these publications give good evidence on the side of creation in the creation-evolution debate.

My God is the BeComingOne

God Papers

aka: **God: God is the BeComingOne**

by

Walter R. Dolen

"the is, the was, the Coming One, the Almighty"
(Rev 1:8; Mat 11:3)

or in Hebrew: אֵלִיָּהוּ

Becoming-One Publications
B1Publ.com

This book (aka the *God Papers*) is a corrected, rewritten, enlarged version of a written work which was included in part in a 1970-71 non-published work and included in part in the 1977, 1989, 1996, 1999 and 2000 published and copyrighted books. Included in this 3rd edition is a new Introduction section added after the 2000 and 2012 edition.

A list of the author's other books can be found at:

www.walterdolen.com

This book:

ISBN-13: 978-1-61918-0543

April 2023 Printing

(Last part of GP1 moved to Appendix)

Other formats --

ISBN-13: 978-1-877981-16-6

(7 x 10 Trade Paper 12p font)

ISBN-13: 978-1-61918-003-1

(Matte Lamination Hardback)

ISBN-13: 978-1-61918-006-2

(Kindle or e-Book on Apple's iBooks)

ISBN-13: 978-1-61918-007-9

(iBooks Author edition)

Becoming-One Publications

b1publ.com | becoming-one.org/books.htm

books can be bought on the web

Barnes and Noble, Amazon, etc.

Important Principles to Understand

Logic: "The most certain principle of all is that regarding which it is impossible to be mistaken; for such a principle must be both the best known ... and non-hypothetical. For a principle which every one must have who understands anything that is, is not a hypothesis; and that which every one must know who knows anything ... Evidently then such a principle is the most certain of all ... **It is, that the same attribute cannot at the same time belong and not belong to the same subject and in the same respect.**" [Aristotle in *Metaphysics*]

Good and Evil: "There can be nothing more inept than the people who suppose that good could have existed without the existence of evil. Good and evil being antithetical, both must needs subsist in opposition, each serving, as it were, by its contrary pressure as a prop to the other. No contrary, in fact can exist, without its correlative contrary. How could there be any meaning in 'justice,' unless there were such things as wrongs? What *is* justice but the prevention of injustice? What could anyone understand by 'courage,' but the antithesis of cowardice? Or by 'continence,' but for that of self-indulgence? What room for prudence, unless there was imprudence? Why do not such men in their folly go on to ask that there should be such a thing as truth, and not such a thing as falsehood? The same may be said of good and evil, felicity and inconvenience, pleasure and pain. There things are tied, Plato puts it, each to the other, by their heads: if you take away one, you take away the other." [*Chrysippus, Fragment* 1169. On the problem of evil. Barrett, p. 64]

Introduction

This book attempts to demonstrate in a logical way who or what God is, assuming a few premises. Although most people believe in God or gods, some do not. Why, if God is good and all powerful, is there suffering? Why do innocent children suffer through no fault of their own. Why do we die? God made the angels immortal, why didn't he create us immortal? Why do we get diseases? Since he created the world, why did he place the possibility of diseases in the world? Sure, we bring on some of these things ourselves by eating poorly, using poor hygiene or by risky behavior. But what about those diseases caused by mosquitoes or microscopic germs or genetics? Didn't God create mosquitoes, germs and genetics? Didn't God create everything? Isn't God good? Or is the idea of God just some superstitious idea based on fear or naivety, an idea made obsolete by the scientific era? Are people more intelligent if they do not believe in God? Or are atheists 'fools' as the Bible relates (Psa 14:1)? An atheist or an agnostic could say, "you cannot prove there is a God" and, "we do not see anything being created or any miracles by God." While the believers could say, "you do not have any real proof of evolution – where are the billions of missing links?" Today each side would be right – no observable proof exists for either side. Today we do not see any Moses parting the seas or anyone being resurrected from the dead by God as reported in the Bible. Today we don't see any species being made nor do we see the billions of missing links being found. We see many things that are good today and in the past; we also see death, diseases and destruction all throughout history. In addition we see the hypocrisy of those who claim to believe in God as well as those who do not believe.

I believe in a powerful God. With all the suffering and pain in this world, how can I or anyone believe in the good almighty God? How can such a being be good, if he allows so much misery? My answer is, I do not believe in the stereotypical gods of most religions because they fail to answer the hard questions and paradoxes pertaining to their gods. After years of study I wrote this book to explain the results of my study. Instead of describing God in a few words, I had to use tens of thousands of words because of the confusing and self-contradictory ideas people have about God. I do not *have* to believe in God merely because I was raised to believe in God. I was also reared as a member of a certain religious group which I no longer believe in or follow. I have a relatively independent mind. If there is no God, why would I have to believe in him?

I believe in the scientific method. In my research for this book I used the scientific method as far as possible. I studied contrary ideas on God; I formed new hypotheses; I tested and analyzed each hypothesis with all available facts: I was looking for the answer that would solve the paradoxes pertaining to God that would make logical sense and that would be in harmony with available evidence. Since God is invisible, I could only use indirect methods. But, whether you know it or not, modern Science also uses indirect methods to study the universe and its origin (See my Science Papers). Fundamental and extensive 'scientific' theories important to science (the big bang theory, the age of the universe, the quantum theory, the string theory, and the special and general theory of relativity) were formed through indirect methods using mathematical formulas and thought experiments – not through direct observation. Since God is invisible (spiritual) we cannot see God, except indirectly, and only as much as he wishes us to "see" him. People do not need to know if God exists; humans also do not need to know how the universe came into existence. But something inside of many of us needs to know, as we need to know who our parents were, or where our ancestors came from, or who invented the first watch, or who first discovered America, etc. An orphan has an inner need to know who his parents were; many of us have an inner need to know where the world/universe came from. Why, or how, is there life? Is death permanent? How can God, if there is one, be all powerful, all good, yet allow evil, misery and suffering?

Although we cannot yet prove (as some define proof) in a scientific manner that there is an all powerful God who created the universe, we also cannot prove or show in a scientific method any other alternative for the genesis of the universe. No one can prove in a scientific manner that evolution created the universe because, apparently, we have no witnesses (observers) for the beginning of the cosmos. (The Bible mentions that the angels witnessed the creation [Job 38:7[1]].) Evolution is really a historical theory based on insufficient evidence, not a scientific theory based on observation.

I believe God did create the universe and here are a few reasons why I do

Law. The evolutionary theory always starts with, and assumes, the eternal existence of laws like those of mass, energy, motion, gravity, conservation, chemical bonding and so forth. Laws, in and of themselves, *are* systematic order and project intelligence and power outside of the law itself. The genetic code of life found in DNA also projects high intelligence and power. *How can* the *code of DNA evolve* or any law such as gravity or chemical bonding evolve? How can any code or law itself have any power? What gives a code power? I am speaking about the code itself, the order of the elements within the code. How can the *arrangement* of the code itself have power? The apparent connection between the code and its effect on a body or plant projects, or strongly suggests some kind of force or power *behind* the law. The code itself doesn't do anything, just as the letters in this book don't do anything by themselves. If you change the arrangement of the letters of the code or a word, it has a different result or may not have any. A seed grows into a certain kind of flower, not because of the code per se, but because of the power behind the code. The basic laws of the universe must have come from somewhere and the power behind these laws must have some connection to the law. Evolution has yet to explain the source of the power behind the universal laws. Science can only *describe* gravity (through mathematical formulas) and partially *describe* the code of life, but it has no idea how the power of gravity works or how or where the code of DNA gets its power. I believe that God, as described in this book, is the creator and power behind all universal laws. And I believe it is more naive to believe in a cosmic soup theory (evolution) than in a powerful God, although I agree that common descriptions of God are naive and do not explain the paradoxes pertaining to God.

Beginning. Radioactivity and laws of thermodynamics indicate there was no eternity of matter and it corollary: there was a beginning of matter. If matter always existed, without a starting point, then the "life" period of the

radioactive elements would have long ago run its course and the whole universe would be the same temperature (thermodynamic laws). The radioactive elements would have run down and there would not be any radioactive elements left; the whole universe should be the same temperature. Thus, there was a *beginning* of matter, and it wasn't that long ago, since there are still radioactive elements. The "science" of evolution cannot explain energy or matter or its source nor will it ever because it has no witnesses and has no real explanation for their beginning. A mathematical description of energy doesn't explain it, it only describes what it does in a quantitative manner in *our* solar system. God created matter and energy and in some way God is matter and God is energy as we attempt to explain in this book.

Life. The relative harmonic-symbiosis of the ecosystems, from the biochemical cell to the earth-sea-heavens, projects design. There is a co-operation, interaction and mutual dependence among life forms; one species cannot live well, or at all, without mutual-beneficial interaction of the whole: the flowers need the birds and insects for pollination in order to continue to exist and vis versa; the seed needs its DNA, the dirt with its nutriments, water and the power behind the DNA for it to grow. Our bodies need a heart, lungs, liver, intestines and so forth in order to exist: we need our whole factory of body parts and a compatible earth in order to live. The whole cannot live without the parts; the parts cannot exist without the whole. The theory of evolution maintains that life is arbitrary, for life came from a hit and miss adventure ("natural selection" or "mutation," etc.). If life is arbitrary, then the universe would be filled with the inferior products of this evolutionary process, and the inferior and half-made life-forms would greatly outnumber the surviving species. There should be fossils of the inferior products of the evolutionary process in all strata, in the rocks everywhere. In other words, the rejections of the evolutionary process should be polluting the universe. Where are the fossils of these inferior life-forms? For that matter, where are the masses of missing links in the evolutionary process? Where? Life came from God, not from the mindless soup of evolution.

The Proof. The big bang theory and other theories need to explain where the material and energy for the big bang theory came from. God, the all powerful Being, by definition, must have always been there, or else there is nothing and

---

[1] Compare Greek text with Hebrew cf. metaphorical usage of 'sons of God'

we are nothing and so this dialogue doesn't exist. Either the all powerful god of Evolution (mindless soup) was there at the beginning or the all powerful Being was there. Of course we cannot prove God by definition, but there is a way to settle this disagreement:

- The evolutionists can prove the universe came into existence through evolution by physically demonstrating evolution. For example, a new species being spontaneously 'created' before our eyes, or at very least finding the massive amount of missing links in the fossils record and logically explaining where laws get their power;

- The believers in the God can prove to others that there is an all powerful God by people seeing God create a new heaven and earth or by seeing God resurrect the dead back to life. Such is the prophecy recorded in the Bible: all will see the resurrection of the dead and the creation of the new heaven and earth, as apparently the angels witnessed the creation of the present universe at the beginning of the present heaven and earth.

"For out of Him [God], and through Him, and into Him, all things" (Rom 11:36)

"And Beginning at Moses and all the prophets, He expounded unto them in all the Scriptures the things concerning Himself" (Luke 24:27)

GP 1: God's Paradoxes and Name

Views on God

Paradoxes on God

Law of Contradiction

Attributes of God

Problem of Evil

Titles / Names of God

The Name of God

"I AM" Doctrine

Unchangeableness of God

God, Gods

One God (YHWH)

Views and Paradoxes on God

Gods of Science

gp1» God, gods, and idols come in all sizes, shapes, and powers. All cultures have their gods. Even science has its god. In Robert Wright's *Three Scientists and Their Gods* (1988), Wright writes:

> Some people find it hard to believe that a heartless, brainless, spineless bacterium floating around in the primordial ooze could have evolved into a multi-billion-celled animal... Given enough time ... unlikely things will come to pass—such as strands of DNA that make copies of themselves.

> But other scientists ... think that the first form of life owed its existence to some as-yet-undiscovered law of thermodynamics... This unformed law, says Bennett, has "taken over one of the jobs formerly assigned to God" (pp 205-206).

The god of science is the theory of evolution with its life-creating "black holes" and its invisible "anti-matter." Evolution does everything that the religious god does. Science thinks of itself as holy and worthy of praise, but it and its priests have created city-killing bombs, experimented on live humans, injected animal and human victims with drugs, diseases, plagues, and even theorized extermination of whole sets of people in the name of science.

Gods of the Aztecs

gp2» In the past most were "religious." To appease their gods, mankind built great stone altars. On these altars, sometimes located on high hills or pyramids, they built fires. In these fires some sacrificed their children and virgins. According to eyewitnesses with Cortez,[1] in the Aztecs' barbaric culture, on top of the pyramid the high priest dressed in black would cut open a live human victim pull out the live, bloody and beating heart, extend his bloody hand to the heavens while squeezing out all the heart's blood. Then the victim was pushed down the pyramid, the heartless body would tumble over pointed and jagged rocks that ripped it all the time it fell to the ground where others would cut off the victims arms and legs, which were later eaten by the populace, and then the priests discarded the remaining flesh of the victim to the waiting half-starved animals, who were kept near the bottom of the pyramid, to eat the bloody remains that the populace would not or could not eat.

Bizarre Gods of Yesterday

gp3» In contrast, some more "humane" societies only sacrificed animals: sheep, goats, and birds. Around their holy hills they sold animals for sacrifices. Temple prostitution was present in many cultures. Some walked on fire, wrapped poisonous snakes around their necks, and beat and disfigured themselves with whips and knives. Others prayed in various ritualistic ways to their gods with pious and disfigured

---

[1] (Bernal Diaz, *The True History of the Conquest of New Spain*, Pub. 1568; Francisco Lopez de Gomara, *Cortes: The Life of the Conqueror by His Secretary*, Pub. 1552)

faces, hoping that their gods would listen to them and grant their request. Kings assumed for themselves godhood and had their subjects worship them as gods. In their kingship they robbed and humiliated their subjects. These god-kings started wars, raped, killed, and destroyed cities and nations.

Today's Gods

gp4» Today there are many theories on who or what is God. Depending on your education and mindset, some of the explanations of God are serious while others are chaotic, if not ludicrous. Although there are few remnants of killing-sacrifices today, there are financial sacrifices, jihad, ritual prayers, asceticism, hedonism as well as plenty of rituals for the gods: free-form to rigid-formal as well as masochistic/sadistic rituals (ritual whipping).

Gods, Creation and Science

gp5» Did evolution, with its cosmic and non-intelligent soup, create the universe, did the god of modern religiosity create it, or did the all-powerful Being create it? The cosmic-soup theory (evolution) is omnipotent; it is like God: it creates matter; from it all life evolved; it's all-powerful. Although some theologians speak of God as all-powerful, for many God needs the magic of the cosmic-soup to create the universe and mankind. And for many God's power is tempered in someway because he is struggling for good against a surprisingly powerful anti-god, the Devil. For this "all-powerful" god of religiosity, there is the "problem of evil."

But there is evidence against both the magical cosmic-soup, and against the weak god of religions. The intelligence, design and complexity of the cosmos cannot come from a non-intelligent soup or a weak god. The genetic code of life that exists in each of our cells is one proof of the intelligence of life, complexity of life and design of life. This code of life and the complexity of life **must** have come from a highly intelligent Power not from a non-intelligent cosmic soup. I find the arguments against 'design' naive, since any man-made design has intelligence behind it. We assume intelligence behind all our design (inductive logic). Yet the design and complexity of the universe has no intelligence behind it? A non-intelligent soup created our universe? The vastness, complexity and design of the universe are evidence for a *powerful* and *intelligent* creative being. A great intelligent Power must have created the universe, not a non-intelligent soup. Science cannot and never will acknowledge a powerful

God, because the very definition of Science rules out the supernatural: "science" was instituted to negate the overbearing influence of religion on knowledge, but if the true answer to origins includes the acts of an invisible Power, "science" by its very definition[1] will be blind to this truth. See my *Science Papers* for my analysis and critique of Science.

Who or What is the Creation Power?

gp6» Considering the improbability of life coming from a non-intelligent soup, the question should be: who or what is the Power that created the universe? If this power is God, then where did God come from? Why should there be anything at all? Why not nothing? Of course there is something, there is life. We are the proof. We are the witness to life as well as to death. What is God? Is it even possible to know? Why is this power invisible? Or is he invisible? Why is there evil? Isn't God supposed to be good? If so, why is there evil? Doesn't the creator have responsibility for his creation? Is God a he, a she or an it? Do these terms even apply?

Premise for this Study

gp7» If an intelligent Power created the universe did he leave us a way to ascertain his essence? Is it even possible to prove his existence? Wouldn't you think in some manner he may have revealed his essence or presence to us? I have come to the conclusion that the Power *has* revealed his essence. In this book, *God Papers*, we (the reader and I) will examine God, the great Power, scripturally. This means, we will use the Bible to study God because I believe the Bible reveals the essence of God. I believe the Bible reveals the essence of God because of the Bible's uniqueness, its history, its inner cohesiveness, its fulfilled prophecy,[2] its continuing confirmation by archeology, and its honesty in pointing out the hypocrisy and fallibility of mankind and the paradoxes[3] pertaining to God. Remember science, in and of

---

[1] The activity encompassing the systematic study of the structure and behavior of the physical and natural world through observation and experiment.

[2] see *Prophecy Papers, Encyclopedia of Biblical Prophecy*, etc.

[3] We will see in this paper, there can be no 'good' God/Power without the apparent paradoxes.

itself, will always rule out the supernatural because science is only the study of the natural. If the true answer to origins includes the acts of a invisible God, 'science' by its very definition and practices will ignore a *super*natural God.

'Problems' with the Bible

The main problem with finding truth in the Bible is that it wasn't written as a scholarly text, but as a collection of writings that included history, poetry, ritual, fables, prophecy,[1] written in different styles by different people, often with metaphorical word usage, describing events and peoples over thousands of years, showing the foibles of humans as well as describing their unique view of their God and their hope for the coming messiah. There is an uniqueness and greatness to the Bible. After studying the Bible it was of great interest to me, not only what the Bible said about God, but what the religions that were supposedly based on the Bible chose not to teach. God in the Bible shows his other side, so to speak, through Biblical paradoxes. Religions do not admit these paradoxes. They ignore and even hide and deny them, sometimes even mistranslating words to hide them. For example, the word translated "forever" throughout the Bible does not mean forever, but merely a time of unknown length. This mistranslation, in of itself, changes the whole picture of doctrine taken from the Bible (see *Age Paper* NM7). The paradoxes pertaining to God were some of the evidence that helped to convince me that the Bible was written to manifest the real God, not the God of religiosity. There can be no *all powerful* God without these paradoxes. So what are these paradoxes?

Paradoxes on God

gp8» The Bible *seems* to be highly contradictory. How can God be love (1John 4:8), and also a killer? In scripture the LORD says, "I kill and I make alive; I wound, and I heal" (Deut 32:39; 1Sam 2:6). Yet the Bible says that God is good to all (Psa 145:9). How can God be good to all and also a killer? How can God predestinate some to wrath and destruction (Rom 9:21-23; Jude 1:4; Prov 16:4; 1Peter 2:8), and some to mercy and glory (Rom 9:21-23; Eph 1:4-5; etc.)? Not only is God love, but He is all-powerful (Gen 17:1; Rev 1:8). In his all-powerfulness He even *created* evil: "I make peace, and create evil: I the LORD do all

these things" (Isa 45:7). These are some of the Biblical paradoxes of God. Just how can God be love and also a killer, or how or why has He created evil? According to the Biblical definition of love (1Cor 13:4-8), killing or evil isn't one of the qualities of love. Yet, according to the Bible, God is love and in someway has killed and in someway has created evil.

Many attempts to negate these paradoxes of God have failed. Some call the problem of these paradoxes, the "problem of evil." But the only true description of the true God must explain these paradoxes.

Our Goal

gp9» **The goal of this book** is to <u>define</u> God through scripture without real contradictions using the paradoxes of God to help illuminate and explain. But this will not be easy. Christ even said: "no one knows who the Son is, but the Father; and who the Father is, but the Son, and to whom the Son reveals" (Luke 10:22). Theologians have for almost 2,000 years been studying the essence of Jesus and his Father and have come up with differing views, even more paradoxical and self-contradictory views (Trinity). **Are contradictions the proof that people's views about God are mistaken? Or are the contradictions a key in ascertaining the truth?**

Two Basic Laws and One Fact: God Cannot Lie

Law of Contradiction and Law of Knowledge

gp10» There are two basic laws of reasoning and knowledge. These laws are so elementary that most people know them only intuitively. Only a few such as Aristotle and the stoic writer Chrysippus have attempted to put these laws into words. By amplifying these two laws we project a logical reason why the all-powerful Being, the Real God, has "allowed" evil to exist in his creation, or in His own words why He, "created evil" (Isa 45:7).

- One law, the **Law of Contradiction**, shows us the only sure way of ascertaining the truth from known facts.

- The other law, the **Law of Knowledge**, shows us *why* God has allowed evil to exist.

---

[1] some say one-third of the Bible is prophecy

We will explain the Law of Contradiction now; in GP 7 of this book we will explain the Law of Knowledge.

God Does Not Lie

gp11» Along with these two laws of reasoning and knowledge must go the important fact that the true God does not lie. God cannot go back on his word (Isa 46:11). In fact, it is *impossible* for God to lie (Heb 6:17-18; 1John 5:18; etc.). With these three things we will be able to understand who or what God was/is/will-be.

Law Of Contradiction

What is the Law of Contradiction?

gp12» There is no greater principle in thinking than the Law of Contradiction. You cannot know anything, I repeat, you cannot know anything if the Law is not true. What is the Law?:

- "Now the best established of all principles may be stated as follows: The same attribute cannot at the same time belong and not belong to the same subject in the same respect ... This I repeat, is the most certain of all principles...." [Aristotle in *Metaphysics*]

- "There is a principle in existing things about which we cannot make a mistake; of which, on the contrary, we must always realize the truth — that the same thing cannot at one and the same time be and not be, nor admit of any other similar pair of opposites...." [Aristotle in *Metaphysics*]

- "The most certain principle of all is that regarding which it is impossible to be mistaken; for such a principle must be both the best known ... and non-hypothetical. For a principle which every one must have who understands anything that is, is not a hypothesis; and that which every one must know who knows anything ... Evidently then such a principle is the most certain of all ... **It is, that the same attribute cannot at the same time belong and not belong to the same subject and in the same respect.**" [Aristotle in *Metaphysics*]

Aristotle is reported to have written this in his *Metaphysics*. Aristotle further said that "everyone in argument relies upon this ultimate law, on which all others rest." He said this principle or law of logic "must be known if one is to know anything at all." He also said, "if everything is and at the same time is not, all opinions must be true."

If everything is and at the same time is not ...

gp13» Aristotle was right. There is no greater principle in thinking than the Law of Contradiction. Something cannot be all black and at the same time be all white. But a wall can be all white at noon time, and be all black at one hour past noon, because it was painted black shortly after noon time. Or for that matter, something cannot appear to be *all* white to a certain individual, and *at the same time* appear to the same certain individual as any other color. Either the object at that time was *all* white or it was not. But for those who ignore the Law, they say without blinking their eyes:

- the wall is all black at the same time it is all white, or the wall is simultaneously all black and all white.

You protest. You say, no one would say that a wall can be simultaneously all black and all white? Do read on.

At the same time ...

gp14» A man cannot be legally married and *not* be legally married at the same time. But a man named Joseph can be married at noon time on Tuesday, and not be married at two minutes past noon time because his wife died at one minute past noon. But this Joseph was not: married and not married *at the same time*. Although you can say that on Tuesday Joseph was single, he was married, and he was widowered; Joseph was not single, married, or widowered *at the same time* even though on the same day he was all three.

Good and Evil at the same time or ...

gp15» A man cannot be good (in the truest sense of the word) and yet *at the same time* commit murder. But John could *have* killed Joseph last year, yet today be good because he has changed from his former behavior. He is a reformed murderer. In the English language, you can still call this John a killer because in the *past* he killed Joseph, and you at the same time could call John, "good," because he has reformed. But you cannot say that John was good *when* he murdered Joseph. **Time** has an important part to play in the Law of Contradiction. Your general behavior cannot be good and evil at the same time, but your general behavior could have been bad in the past, and yet you have now changed your general behavior to that which may be called good.

An Example of Paradoxes and Time

gp16» In testimony at a trial, three witnesses testified that they saw illegal drugs being sold from a certain house on a certain day. (All houses on the block looked the same, had no street numbers, but did have different colored garage doors.) Each witness described the house, but each witness described the color of the garage door at the house as being a different color. One said it was brown, one said it was red, and one said it was green. This contradiction almost led to the home owners (husband and wife) being freed, except for the last witness. The last witness, who lived across the street from the house in question, explained that the normal color of the garage door was brown, but at 11 am on the day in question the owner came out and sprayed it red. His wife came home from shopping that same day at 12 pm and the witness could hear the man and woman arguing. She apparently didn't like the color. So the husband at 1 pm that same day came out of the house and sprayed the garage door green. On the same day the color of the garage door was brown, red, and green, but never was the garage door all three colors at the same time.

gp17» What at first appeared to be a real contradiction, later just turned out to be explainable. Time played an important part in this story. At one time the garage door was brown. Later it became red. Still later it became green. The garage door was **not** brown, red, and green at the same time even though on the same day the door was all three colors. On this same day, in time, the door **became** different colors. **Time** played a significant role in this story, as does time play an important role in the understanding of the apparent paradoxes pertaining to God.

Same time in the same respect

gp18» Because of the Law of Contradiction, you cannot be physically present on First Street in San Jose, California at 1:30 PM on April 20 and *at the same time* be physically present on First Street in New York, New York. Of course those who play word games could say that at

the same time you were *mentally* in San Jose, you were *physically* in New York. Notice the change in the sense of *being* in a place. For those who play word games, Aristotle qualified his statement: "the same attribute cannot at the same time belong and not belong to the same subject *in the same respect*." His qualification, "in the same respect," means that you cannot be, in the *same* sense, in San Jose and New York at the same time.

"If everything is and at the same time is not, all opinions must be true"

gp19» *If* the Law of Contradiction is not correct, you could say that John murdered Joseph at 1:30 PM, or just as truthfully say that the same John did not murder the same Joseph at 1:30 PM on the same day. Both of these contrary statements can be truthful at the same time, *if* the Law of Contradiction is not true. Again, *if* the Law of Contradiction is not valid, you could say and be 'correct': "I am alive physically, yet in the same sense and at the same time that I am alive — I am also dead." But you protest again. No one you say in their right mind would say he is alive and dead at the same time in the same respect. But –

Word Games or Lies

gp20» The Law of Contradiction is so obviously valid that few say it isn't true, yet there are many who act as if the Law of Contradiction is not true by their belief in contrary theories. In fact, impossible contradictions are taught as truth each day in the fields of religion, politics, law, and "science." If contradictions are taught by "respected" people, they are accepted by some, even though at some level of thought they see the contradiction. Authority and tradition are strong — so strong that real contradictions are taught as the absolute truth. Many dogmas use obviously false statements such as claiming:

- "The simultaneity of Jesus's death and immortality" (Hugh Ross, *Beyond the Cosmos*, p. 108).

gp21» How can Jesus be immortal and simultaneously experience death? There is a way to move beyond the paradox of Jesus being God, yet Jesus dying, without tossing out the Law of Contradiction. In order to know anything we must hold on to the Law of Contradiction. The theologians are making a mistake in their beliefs that force them to ignore and degrade the Law of Contradiction. You cannot find the Truth without using the Law of Contradiction.

Do words have meaning?

gp22» Look again at the statement from the astronomer Hugh Ross, a person with a Ph.D in a astronomy:

- "The simultaneity of Jesus's death and immortality" (Hugh Ross, *Beyond the Cosmos*, p. 108).

Ross is not simple. But because Ross and others believe that Jesus is God, and that God is not mutable or changeable,[1] then in order for Jesus to die on the cross, he must have been dead and alive at the same time. Instead of examining their immutable theory they insist on saying that God was alive and dead at the same time.

gp23» Do words have meaning? Apparently not for some theologians. Berkhof wrote:

- "In view of all this [scripture] it may be said that, according to Scripture, physical death is a termination of physical life by the separation of body and soul. It is never an annihilation... Death is not a cessation of existence, but a severance of the natural relations of life. Life and death are not opposed to each other as existence and non-existence, but are opposites only as different modes of existence. It is quite impossible to say exactly what death is. We speak of it as the cessation of physical life, but then the question immediately arises, Just what is life? And we have no answer." [Berkhof, *Systematic Theology*, p. 668].

I do not believe that Berkhof does not understand what death is. He merely doesn't want to believe it because of some view he holds. In order for some to believe in certain theories they must either change the normal meaning of words (death is not death) or diffuse its meaning. How can death be a different mode of *existence* as Berkhof maintains? He completely negates the meaning of death by asserting this. This is a ploy used by those who do not wish to look the truth in the eye. When their theory on the nature of God cannot hold up, they merely change the meaning of words, or make preposterous statements that claim and maintain:

- "The simultaneity of Jesus's death and immortality" (Hugh Ross, *Beyond the Cosmos*, p. 108).

Knowledge cannot exist outside the Law of Contradiction

gp24» The Law of Contradiction is true. Once explained and understood it is the most obvious law. It is the basis on which we judge what is true and what is not true. It is the basis on which courts judge whether a person committed a crime or not. Either the murderer was at the crime scene at the same time as the crime or he was not. He could not, be there *and* not be there, at the same time in the same respect.

Summarize the Law of Contradiction

gp25» The Law of Contradiction is the basis from which we reason:

- something or some specific action cannot *at the same time* be and not be.

But there are some, as Aristotle noted, that foolishly argue against this law. But I ask, how can anyone not believe in this law? If someone does not believe in this law, he cannot prove or disprove anything (at any one time something could be or could not be true); he cannot believe in anything (for what he believes in could just as well not be true).

---

[1] See "Unchangeableness of God" in this part for more information

Attributes Of God

Now that we know the importance of the Law of Contradiction, we now can continue with our search for the real essence of God by studying the main attributes attributed to God. How is God described in the Bible? Are there contradiction? If so, how can they be explained?

God Is Life

gp26»

- 'For as *the Father has life in Himself*, so He has granted the Son to have life in Himself' [John 5:26 NKJV]

  - " 'For *in Him we live and move and have our being*,' as also some of your own poets have said, 'For we are also His offspring.' " [Acts 17:28 NKJV]

God Has All Knowledge

gp27»

- Great is our LORD, and mighty in power; *His understanding is infinite*. [Psa 147:5 NKJV]:

  - For if our heart condemns us, God is greater than our heart, and *knows all things*. [1 Jo 3:20, NKJV].

God Is Everywhere

gp28» But will God indeed dwell on the earth? Behold, heaven and the heaven of heavens cannot contain You. How much less this temple which I have built! [1Ki 8:27, NKJV]

- "Can anyone hide himself in secret places, so I shall not see him?" says the LORD; "do I not fill heaven and earth?" says the LORD [Jer 23:24, NKJV].

- Where can I go from Your Spirit? Or where can I flee from Your presence? 8 If I ascend into heaven, You are there; If I make my bed in hell, behold, You are there. [Psa 139:7, NKJV]

- So that they should seek the LORD, in the hope that they might grope for Him and find Him, though He is not far from each one of us; 28 for in Him we live and move and have our being, as also some of your own poets have said, 'For we are also His offspring.' [Acts 17:27, NKJV]

There Is Nothing Else Besides God

gp29» "That they may know from the rising of the sun to its setting that there is none besides Me. I am the LORD, and there is no other." [Isa 45:6, NKJV]

gp30» This scripture does not say there is not any *like* God, but it does say there is none besides God, "I am YHWH, and there is no other." Of course, if there is none besides God, then it follows there is also none *like* God. In a sense, the true God is everything; there is nothing beside Him. This may make little sense now, but after you read *all* this book, you may come to understand.

God Is Invisible

gp31» As we have just seen, God's presence and/or spirit and/or power is everywhere. But up to the present, most, if not all, have not seen God in a physical way (Although some can "see" God in a Spiritual sense. See Chap2). This is because God in this age is invisible to human eyes:

- When he [God] passes me, *I cannot see him*; when he goes by, I cannot perceive him. [Job 9:11, NIV]

- He is the image of the *invisible God*... [Col 1:15, NIV]

- No one has ever seen God... [John 1:18]

See GP 2 and the rest of this book to further understand this.

God Is Almighty

gp32»

- When Abram was ninety-nine years old, the LORD appeared to Abram and said to him, 'I am *Almighty God*; walk before Me and be blameless.' [Gen 17:1, NKJV]

- Both riches and honor come from You, and You reign over all. In Your hand is power and might; in Your hand it is to make great and to give strength to all. [1Ch 29:12, NKJV]

- and said: "O LORD God of our fathers, are You not God in heaven, and do You not rule over all the kingdoms of the nations, and in Your hand is there not power and might, so that no one is able to withstand You? [2Ch 20:6, NKJV]

- Thus says the LORD, your Redeemer, and He who formed you from the womb: I am the LORD, who makes all things, Who stretches out the heavens all alone, Who spreads abroad the earth by Myself; [Isa 44:24, NKJV]

- You will say to me then, "Why does He still find fault? For who has resisted His will?" [Rom 9:19, NKJV]

All Things Possible for God

gp33»

- And He [Christ] said, "Abba, Father, all things (are) possible for You. Take this cup away from Me; nevertheless, not what I will, but what You will" [Mar 14:36, NKJV]

This "all things (are) possible" is qualified by Matt 26:39, Luke 22:42, and Mark 14:35. It is qualified by, "if it were possible" and "not as I will, but as you will." Everything *was* possible *before* God sent forth his will, or his word. But once God wills something, God does not go back on his word (See below under "God Keeps His Word."). Also notice that in Mark 14:36 there is no verb ("are") in the Greek text; therefore, all things *were* possible (to take away the death of Jesus) to the true God *before* he gave his word or before God predestinated Jesus Christ's death as the true Lamb of God (Acts 4:27-28; 2:23: 3:18).

See chapter 5, Jesus Christ the God, under "With God Nothing Shall Be Impossible" for more detailed information on this subject.

Creator Makes All Things

gp34» God has all the power in the whole universe. In fact God is the creator of the whole universe.

- In the beginning God created the heavens and the earth [Gen 1:1, NKJV].

- As you do not know what is the way of the wind, or how the bones grow in the womb of her who is with child, so you do not know the works of *God who makes all things* [Ecc 11:5, NKJV].

Problem Of Evil

gp35» The scripture we just studied tells us that God is almighty. With His great power God created all. God made all things. But do you understand what *all* includes? "All" not only includes the good, but "all" also includes the wicked, their evil, and even the waster or spoiler and his destruction (Isa 54:16). It is Impossible for the God to have created good without in some way also having created evil, for good and evil are comparative qualities which need each other in order for anyone to know either quality (See GP 7; NM19; NM9). *All* power not only includes all the power of good, but also, somehow or in someway, all the power of evil. Therefore God cannot be almighty without having power over evil. Yet at the same time God cannot be good and still execute evil. This is "the problem of evil" that the theologians write about. The power over evil is somehow included in God's power as scripture indicates, for God (YHWH) in someway or somehow even kills and wounds (Deu 32:39, see below), and even created evil (Isa 45:7).

God's Connection with Good **and** Evil

gp36» Job said to his wife: "shall we receive good at the hand of God, and shall we not receive evil? In all this Job did not sin with his lips." (Job 2:10)

- The LORD has made everything for its purpose, even the wicked for the day of trouble. [Prov 16:4]

- I form the light and create darkness, I make peace and create evil [Hebrew - *ra* Strong's # 7451]; I, the LORD, do all these things.' [Isa 45:7]

- Behold, I have created the blacksmith who blows the coals in the fire, who brings forth an instrument for his work; and I have created the spoiler to destroy. [Isa 54:16, NKJV]

- Now see that I, even I, am He, and there is no God besides Me; I kill and I make alive; I wound and I heal; nor is there any who can deliver from My hand. [Deu 32:39, NKJV; also note 1Sam 2:6]

Paradoxical Sides of God

Right and Left Sides

gp37» Notice that not only did the God create light, but he also created darkness (Isa 45:7; Gen 1:1-4). Notice that not only did God create peace, good, and life (Isa 45:7; Gen 1:31; 1Sam 2:6; Gen 1:24), but he also created evil and killed (Isa 45:7; Gen 1:1-2; Deut 32:39; 1Sam 2:6). There are two opposite aspects of God. You can call these two facets of God, **God's right and left hand or sides**. The Hebrew word for right hand (*yamin*) also means right side; the Hebrew word for left hand (*semovl*) also means left side.

Right Side or Positive Aspects of God

God Is Good

gp38» First let us look at the positive aspects of God – God is good.

- So He said to him, "Why do you call Me good? *No one is good but One, that is, God....*" [Mat 19:17; Mark 10:19; Luke 18:19]

- God's Name, Word, Spirit is good (Psa 54:6 [8]; Isa 39:8; Jer 29:10; Heb 6:5; Psa 143:10)

Not only is the one true God good, but God is or will be good to all:

- The LORD (is) good to all, And His tender mercies are over all His works. [Psa 145:9, NKJV]

When is God good to all:

- God (YHWH) for good and mercy in olam [see Hebrew text:1Ch 16:34; 2Ch 5:13; 7:3; Ezra 3:11; Psa 100:5; 106:1; 107:1; 118:1,29; (135:3)136:1; Jer 33:11]

God Is Love

gp39» He who does not love does not know God, for *God is love*. [1John 4:8, NKJV]

Love Is

gp40» Love suffers long and is kind; love does not envy; love does not parade itself, is not puffed up; 5 does not behave rudely, does not seek its own, is not provoked, thinks no evil; 6 does not rejoice in iniquity, but rejoices in the truth; 7 bears all things, believes all things, hopes all things, endures all things. 8 Love never fails. But whether there are prophecies, they will fail ['become ineffective' — because they will have been completed]; whether there are tongues, they will cease; whether there is knowledge, it will vanish away. [1 Co 13:4-8]

- The entire law is summed up in a single command: "Love your neighbor as yourself." [Gal 5:14]

Love is Not

gp41»

- Among other things Love is not: fornication, impurity, licentiousness, 20 idolatry, sorcery, enmities, strife, jealousy, anger, quarrels, dissensions, factions, 21 envy, drunkenness, carousing, and things like these. I am warning you, as I warned you before: those who do such things will not inherit the kingdom of God. [Galatians 5:19-21]

God Keeps His Word; He Does Not Lie

gp42» It is impossible for God to lie (Heb 6:18, NIV; see, Titus 1:2):

- So is my WORD that goes out from my mouth: it will not return to me empty, but will accomplish what I desire and achieve the purpose for which I sent it (Isa 55:11, NIV).

- What I have said, that will I bring about; what I have planned, that will I do (Isa 46:11, NIV).

- The WORD is gone out of my mouth in righteousness, and shall not return (Isa 45:23).

- My covenant I will not break, Nor alter the word that has gone out of My lips (Psa 89:34, NKJV).

gp43» God does not lie, therefore all that comes out of his mouth, or all his words, are the truth. God's words are found in the Bible. Thus,

- For assuredly, I say to you, till heaven and earth pass away, one jot or one tittle will by no means pass from the law [Old Testament books] till all is fulfilled (Matt 5:18, NKJV).

[This scripture does not mean that in our copies of the Hebrew text that there would not be any variant (even the smallest) when compared to the originals, but it means that it would be easier for heaven and earth to pass away than for the smallest word of God to fail. Note *Figures of Speech Used in the Bible*, by Bullinger, page 678, 1984 Baker printing.]

If what God has said has yet to happen, it *will* happen. The scripture cannot be broken (John 10:35).

Left Side or Negative Aspects of God:

gp44» An honest reading of the Bible manifests to us negative aspects of God. Here follows some of them:

- killing kings [Psa 135:10; 136:18; 145:20]

- of bringing evil on Job [Job 42:11; 1:6-12; 2:1-8]

- somehow causing drought, or floods [Job 12:15]

- destroying nations, and making the leaders of the world go mad [Job 12:23-25; Dan 4:28-35; Deut 28:28]

- sending curses and confusion; He plagues some with diseases, and so on [Deut 28:15-68]

- killing Er, Onan, the firstborn of Egypt, the Pharaoh and his army, Korah his family and men, Israelites, Amorites, Uzzah, and so forth for various reasons [Gen 38:7; 38:9-10; Exo 12:29; 14:16-19, 24-27; Num 16:1-35; Num 16:41-50; 2Sam 24:1-15; Josh 10:6-12; 2Sam 6:6-7]

- And God said to Noah, 'The end of all flesh has come before Me, for the earth is filled with violence through them; and behold, *I will destroy them* with the earth.' [Gen 6:13]

- The "**anger** of the LORD" or the "**wrath** of the LORD," or the "**jealousy**" God, or some "**angel**" of the LORD" destroyed the people and are pictured in the Old Testament scripture as bringing "all the curses that are written in this book [the Bible]," and destroying such cities as Sodom and Gomorrah and even destroying 70,000 Israelites [Deut 29:20; Gen 19:24-29 with Deut 29:23,20; 2Sam 24:1, 15-16; Nah 1:2; see "God's Wrath" paper (PR 4)].

gp45» Outside of the question of natural disasters, some of the evil God somehow brings upon mankind is because of mankind's behavior (Deut chap 28; Josh 24:20; "God's Wrath" PR4 to PR6; etc.). We are not saying here that the evil brought on each man is directly proportional to each man's sin (Luke 13:1-5).

Anger of God or Wrath of God?

gp46» We just saw a list of negative facets of God, and in it we saw the "anger of God" ("his anger"), or the "wrath of God" ("his wrath"), or "jealousy of God" ("his jealousy") that destroyed Sodom and Gomorrah and others (Gen 19:24-29; Deut 29:23,20). What does the Bible mean when it speaks about the "anger of God" or the "wrath of God"? First look at 2Samuel 24:1,15-16:

- "Now again the **anger of the LORD burned against Israel**, and it incited David against them to say, "Go, number Israel and Judah.... So the LORD sent a pestilence upon Israel from the morning until the appointed time, and seventy thousand men of the people from Dan to Beersheba died. 16 When the **angel stretched out his hand toward Jerusalem to destroy it**, the LORD relented from the calamity and said to the angel who destroyed the people, "It is enough! Now relax your hand!" And the **angel of the LORD** was by the threshing floor of Araunah the Jebusite."

gp47» Notice that it was an **angel of the LORD** (YHWH) that did the destroying. By doing a computer search for the words "anger of the LORD" we see the following verses also speak of the anger or wrath of LORD destroying and killing (Ex 4:14; 32:11,22; Num 11:1,10,33; 12:9; 25:4; 32:13,14; Deut 6:15; 7:4; 9:19; 11:17; 29:20; 29:23,27; 31:29; Joshua 1; 23:16; Jud 2:14,20; 3:8; 10:7; 14:19; 2Sam 6:7; 24:1; 1Kings 16:7; 22:53; 2Kings 13:3; 24:20; 1Chron 13:10; 12:12; 25:15; 28:11; Psa 6:1; 21:9; 106:40; Isa 5:25; 30:27; 66:15; Jer 4:8; 7:20; 12:13; 23:20; 25:37; 30:24; 42:18; 51:45; 52:3; Lam 1:12; 2:1,6; Ezek 25:14; 38:18; Zeph 2:2-3; 3:8; Zech 10:3; etc.).

Anger of God, Destroying Angel, and Satan

gp48» From the Bible we know there are two kinds of angels: one good; one evil (GP3). What kind of angel of God, destroys? Who is the destroyer? There is a parallel verse to 2Samuel 24 found in 1 Chron 21:1,12:

- "Then **Satan stood up** against Israel and moved David to number Israel.... pestilence in the land, and **the angel of the LORD destroying** throughout all the territory of Israel." (1Chron 21:1,12)

gp49» It is Satan that moved David to Number the Israelites against God's will (cf. 2Sam 24:1-2 with 1Chron 21:1-2). By comparing both versions and other scripture in the Bible, we see that the "anger" of the LORD is an angel called Satan, who goes about destroying, "the devil, prowls around like a roaring **lion**, seeking someone to devour" (1Pet 5:8).

gp50» Look at another verse that says the same thing:

- Because he [Balaam] was going, began burning the anger of God, and **an angel of the LORD** took his stand in the way as **an adversary [Satan] against him**. Now he was riding on his donkey and his two servants were with him." (Num 22:22; see Hebrew text)

In some way Satan is an "angel of the LORD" who destroys (1Chron 21:1,12). How can Satan be an "angel of the LORD"?

Evil Angel's Fate

gp51» It is this evil angel and his angels, who are on the left hand or side of God, that will be put in the fire at the end of the age for their evil deeds:

- "But when the Son of Man comes in His glory, and all the angels with Him, then He will sit on His glorious throne. 32 "All the nations will be gathered before Him; and He will separate them from one another, as the shepherd separates the sheep from the goats; 33 and **He will put the sheep on His right, and the goats on the left**. 34 Then the King will say to those on His right, 'Come, you who are blessed of My Father, inherit the kingdom prepared for you from the foundation of the world.'... Then **He will also say to those on His left**, 'Depart from Me, accursed ones, **into the aeonian fire which has been prepared for the devil and his angels**." (Matthew 25:31-34,41)

- "**And angels** who did not keep their own domain, but abandoned their proper abode, He has kept in eternal bonds under darkness **for the judgment of the great day**," (Jude 1:6)

- "Then I saw an angel coming down from heaven, holding the key of the abyss and a great chain in his hand. 2 And **he laid hold of the dragon, the serpent of old, who is the devil and Satan**, and bound him for a thousand years; 3 and **he threw him into the abyss**, and shut *it* and sealed *it* over him, so that he would not deceive the nations any longer, until the thousand years were completed; after these things he must be released for a short time." (Revelation 20:1-3 cf 20:10)

Right and Left Side Metaphor

gp52» By comparing various verses we see that the abyss is the great lake of fire, and it is this fire that will burn up the evil of the world (Mat 3:10-12; 13:40; NM24). As the above scriptures indicate this evil is so to speak on the "left hand" or "left side" of God. In other words, the Bible uses a metaphor that compares the right side or hand of God with goodness, and conversely compares the left side or hand of God with evil. It is the left hand that is cut off and sent to the fire. Notice the principle of the following pertinent verse:

- "If your **hand** causes you to stumble, cut it off; it is better for you to enter life crippled, than, having your two hands, to go into hell, into the unquenchable fire." (Mark 9:43)

gp53» The all powerful God has the power of all good and all evil, or else he is not all powerful. What Mark 9:43 is telling us along with Matthew 25:41 and other verses, is that the God will cut off the power of his left hand or side at the end of the age and put it in the hell-fire for punishment of sins.

God has Power over Satan

gp54» Notice that the LORD does indeed have power over Satan:

- Job 1:6 – Now there was a day when the sons of God came to present themselves before the LORD, and Satan also came among them. 7 The LORD said to Satan, "From where do you come?" Then Satan answered the LORD and said, "From roaming about on the earth and walking around on it." 8 The LORD said to Satan, "Have you considered My servant Job? For there is no one like him on the earth, a blameless and upright man, fearing God and turning away from evil." 9 Then Satan answered the LORD, "Does Job fear God for nothing? 10 "Have You not made a hedge about him and his house and all that he has, on every side? You have blessed the work of his hands, and his possessions have increased in the land. 11 "But put forth Your hand now and touch all that he has; he will surely curse You to Your face." 12 **Then the LORD said to Satan, "Behold, all that he has is in your power**, only do not put forth your hand on him." So Satan departed from the presence of the LORD. (Job 1:6-12)

- "And the Lord said to Satan, Behold, he [Job] is in your hand; but save his life" (Job 2:6).

So the LORD does have power over Satan, as He must, if He indeed is all powerful. The scriptures we are studying are hints, from which we will be able to understand and answer the "problem of evil."

Two Sides of God

gp55» As we are seeing there are two sides of God, or two facets of God that work together to create good and evil: one side creates good; one side evil. Both sides work together to create as the right and left side of our brain work together to form our knowledge, our speech, and our personality.

Evil *Never* a part of the True God

gp56» Does this mean that the real God now is in some way evil? No! God cannot be good and evil at the same time. Since the one true God is good, the real God can never be evil. Since God is all powerful, God in someway does have control over evil. But the real God now, is not doing evil. It is what we call the left side of God that is now doing evil. This evil "side" is not now the one true God. Evil will never be a part of true God. But evil is being "allowed" in this age through predestination as we will see. As we will see in this book, predestation, time, and God's real Name answer the paradoxes pertaining to God. Do read on.

God Predestinates Wrath and Mercy before Creation

Scripture shows God predestinating some to evil and wrath:

gp57» (Remembering that predestination occurred before creation [1 Pet 1:19-20; nm170ff (Compare 1 Pet 1:19-20; John 1:29; Rev 13:8; Isa 53:7-8; Matt 12:18; 1Pet 2:4; Isa 49:7; John 14:10; Rom 1:4]):

■ Does not the potter have power over the clay, from the same lump to make one vessel for honor and another for dishonor? 22 What if God, wanting to show His wrath and to make His power known, endured with much longsuffering the *vessels of wrath prepared for destruction*, 23 and that He might make known the riches of His glory on the vessels of mercy, which He had prepared beforehand for glory, [Rom 9:21-23, NKJV]

■ For *certain men* have crept in unnoticed, who long ago *were marked out for this condemnation*, ungodly men, who turn the grace of our God into licentiousness and deny the only Lord God and our Lord Jesus Christ. [Jud 1:4, NKJV]

■ The LORD has made all things for Himself, Yes, even the wicked for the day of doom. [Pro 16:4, NKJV]

■ And a stone of stumbling and a rock of offense. They stumble, being disobedient to the word, to which *they also were appointed.* [1 Pe 2:8, NKJV]

Some chosen to be good:

gp58»

■ Eph 1:4 -- Just as He chose us in Him before the foundation of the world [cosmos], that we should be holy and without blame before Him in love, 5 *having predestined us to adoption as sons* by Jesus Christ to Himself, according to the good pleasure of His will [Eph 1:4-5, NKJV; see Rom 9:21-23 above and "Predestination" paper (NM8)].

All generations chosen:

gp59»

■ (from Hebrew text): [LORD] who has appointed and done, calling forth the generations from the beginning. [Isa 41:4]

Predestination is very difficult to understand

gp60»

■ Paul said: "It does not, therefore, depend on man's desire or effort, but on God's mercy. For the Scripture says to Pharaoh: 'I raised you up for this very purpose, that I might display my power in you and that my name might be proclaimed in all the earth.' Therefore God has mercy on whom he wants to have mercy, and he hardens whom he wants to harden. One of you will say to me: 'Then why does God still blame us? For who resists his will?'" [Rom 9:16-19, NIV]

No one resists God's will. As we said this is very difficult to understand. But after you have read all of this book, it will be easier for you to understand.

The Great Paradox

gp61» God has ALL the power. This all-powerfulness must somehow include all the powers of evil. If God does not have in someway the power of both good and evil, then of course he does not have *all* the power.

gp62» But the true God does have <u>all</u> the power. Thus, he has in someway both the power of good and the power of evil. Yet somehow God is good and God is love, and God will give good to all. This is a great paradox. How can one be <u>good</u> and *at the same time* predestinate some to evil? How can God be <u>good</u> and yet *at the same time* kill and destroy? How can God be <u>love</u> and *at the same time* kill and destroy? It would be impossible for God to be <u>love</u> and *at the same time* kill and destroy. Or it would be impossible at the same time God is love to also predestinate some to destruction. It would be impossible because it would be against the most fundamental law of reasoning: the Law of Contradiction (see Law of Contradiction above). But it is within the Law of Contradiction for God to predestinate some for mercy and some for destruction, if they were predestinated <u>before</u> creation (as we know it), <u>before</u> time (as we know it), <u>before</u> good (as we know it), <u>before</u> evil (as we know it), <u>before</u> law (as we know it), and consequently <u>before</u> sin (as we know it).

Time Answers The Paradoxes

gp63» The key to these paradoxes and most, if not all, paradoxes concerning the true God has to do with predestination, time, and God's Name. There is a secret to understanding God. When you know this secret the paradoxes concerning God are answerable in a logical way. The answer to these paradoxes has to do with the phenomenon of **time**, as well as **when** God planned and gave power for evil in his creation, and lastly the fact that the one true God cannot be good and evil **at the same time**. All this plus the meaning of God's Name, which carries time within it (the was, is, will be one), is the answer to the paradox about God being love and God creating evil. The secret of "time" is hidden in God's NAME. There is a time element in God's Name. This will not make sense now, until you understand the meaning and significance of God's NAME. But before we learn about his NAME of names, we should learn about some of his other names and titles.

Titles or Names Of God

gp64» Names or titles of God:

- Holy One [Isa 43:15; 48:17; 49:7]
- Creator [Isa 45:18; 48:13; 51:13]
- Savior [Isa 45:15, 21; 49:26; 60:16]
- Father [Isa 63:16]

- Husband of Israel [Isa 54:5; Jer 3:14; Hos 2:19]
- Shepherd [Psa 23:1]
- Redeemer [Isa 48:17; 49:7, 26; 60:16]
- Rock [Isa 26:4; Deut 32:4]
- First and Last [Isa 44:6; 48:12]
- Mighty One [Isa 49:26; 60:16]
- God Almighty [Gen 17:1]
- King [Psa 10:16; 89:18; 5:2]
- King of Israel [Isa 43:15; 44:6; 1Sam 12:12]
- King of Kings (that is, King of the whole earth) [Psa 47:2, 7; Zech 14:9]
- King of Glory [Psa 24:10]
- King of *olam* [Psa 29:10; Jer 10:10]
- King above all gods [Psa 95:3]
- Lord of kings [Dan 2:47]
- God of gods [Josh 22:22;"Gods of gods" in Hebrew; see Psa 136:2 & Deu 10:17]
- The Great God [Deu 10:17]
- Lord(s) of lords [Deut 10:17; Psa 136:3]
- Lord(s) above all gods [Psa 135:5]
- Most High [(Heb, *'elion* or *'lyown*) is used as a title of God (Gen 14:18-22; Num 24:16; Deut 32:8; etc.). But this Hebrew word (*'elion*) is also used when not speaking about God. It is translated as "uppermost" in Gen 40:17; "upper" in 2Kings 18:17; "high" in 2Chron 23:20; etc.]

These could be called titles or names of God. These are not all of God's titles or names. But none of these are the real God's NAME. God has one NAME he has chosen to best represent himself.

gp65» There is something very important that we must know about God. By knowing the true NAME of God we will be able to understand God much better, and we will better understand the paradoxes concerning God. The true NAME of the God allows TIME to negate the paradoxes concerning God, and helps to answer the problem of evil.

Importance of a Name

Personal Names had Meaning

gp66» Names of people in the Bible had more meaning to them than personal names have for us. To Israel personal names generally expressed some personal characteristic, some incident connected with birth, some hope, desire, or wish of the parents. The Biblical Hebrews had a tendency to play on names and find analogies or contrasts in them (see Ruth 1:20; 1Sam 25:3, 25; Rom. 9:6; etc.). For example the following play on the name "Dan."

- "Dan ['judge'] shall judge his people" (Gen 49:16).

gp67» Personal names given at birth were sometimes changed later in life for various reasons. Sometimes the names given at birth expressed the time of birth, Hodesh (new moon). Sometimes the names indicated the place of birth, Zerubbabel (born in Babylon). Sometimes the condition of the mother called for a certain name for the child, Benoni (son of my pain). Sometimes the name of the child indicated the appearance of the child, Esau (hairy). Religious names were frequently given, the most simple being expressive of thanks to God for the gift of a child, Mahalaleel (praise to God).

gp68» Some names of people were changed by God to indicate what God was going to do with or through that person:

- Abram's name ("exalted father") was changed to Abraham ("father of many") because God was going to make him a father of many nations (Gen 17:5);

- Sarai's name ("Jah is Prince") was changed to Sarah ("princess") because God was going to make her a mother of nations and kings of peoples would come from her (Gen 17:15-16);

- and Jacob's name ("supplanter" or *heel* catcher) was changed to Israel ("ruling with God" or "contender or soldier or prince of God") after he wrestled with the angel (Gen 32:28).

The word "Israel" comes from two words: Sarah ("prince" or ruler or commander) and el ("god"). Princes had their names changed on their accession to the throne (2Kings 23:34; 24:17; note information under "name" in Unger's Bible Dictionary, The International Standard Bible Encyclopaedia, etc.).

gp69» In the New Testament names also were of a more distinctive nature than they are today. Names in the New Testament times, at least among the Biblical Jews, represented certain aspects of the person. For example, "Jesus" is the English translation of the Greek word "Iesous" which is the equivalent of the Hebrew "Joshua" (Jehoshua) meaning: "Jehovah (is) salvation." Thus, "she shall bring forth a Son, and thou shalt call his name Jesus, for he shall *save* his people from their sins" (Matt 1:21).

gp70» In the New Testament names were also changed during one's life time for various reasons. For example, Simon's name was changed to Peter and Saul's name was changed to Paul.

Dual Meaning Of Names

gp71» A name of a single person or quality can also refer to a whole nation or all those with that single quality:

- *Israel*, the individual, or Israel, the nation (see "Seed Paper" [PR 1]).

- *Christ*, the individual, or Christ, the whole Body of Spiritual people in Christ's Spirit (see *New Mind Papers*).

- *Seed*, the individual (Christ), or Seed in the sense of all those in the true Seed (see "Seed Paper" [PR 1]).

- *God's Spirit*, as individually distinctive versus other kinds of spirit, or any to all Spirits of the same nature as God's.

- *Satan*, as the individual, or any to all the spirits or angels of the same nature as Satan's.

- *Beast*, the individual, or the system of the Beast (see Beast Papers [PR 2, PR 3]).

A name of a person can also have a physical and Spiritual meaning: There is a physical Israel and a Spiritual Israel (see "Seed Paper" [PR 1]).

Great Significance of the NAME

The NAME in Scripture

gp72» In the Bible there was a great significance placed on the NAME of the true God. God revealed His NAME to Moses when Moses asked Him for His NAME (Ex 3:13-16). His NAME was a memento or memorial to all generations (Exo 3:15). Moses spoke in God's NAME (Exo 5:23). God spoke to Moses and told him that Abraham, Isaac, and Jacob knew God as "God Almighty" for God had not revealed His NAME to them (Exo 6:2-3). God declared His NAME to the people of the earth (land) by showing His great power against Egypt during the Hebrews' exodus from Egypt (Exo 9:13-16). God warned the Hebrews about taking His NAME in vain (Exo 20:7). God said He would bless the Hebrews in every place in which He caused His NAME to be remembered

(Exo 20:24). God proclaimed His NAME to Moses (Exo 33:19; 34:6).

gp73» Before the Hebrews went into the promised land God instructed them to seek the place where God shall choose to put His NAME (Deut 12:1-5). The Levites were chosen by God to stand and to minister in the NAME of God (Deut 18:l, 5). Aaron and his sons were to put God's NAME on the Israelites (Num 6:27). God's NAME is called on Israel (Deu 28:10; 2Chron 7:14; Isa 56:5; Dan 9:19). False prophets caused Israel to forget God's NAME and use the name of Baal ("Lord") instead (Jer 23:27). Israel would profane the NAME of God among the other nations (Ezek 36:21-22). Jews in Egypt would also forget God's NAME (Jer 44:26). But the God delivers for his NAME's sake (Psa 23:3; 25:11; 143:11; Isa 48:9). Since God's NAME was called on Israel, if Israel was totally destroyed, God's NAME would not have remained (Josh 7:9; Isa 48:9). Therefore, God for his holy NAME's sake, promises to give Israel a new heart and a new spirit so they can keep God's law and thus not profane God's NAME (Ezek 36:21-27). God told Moses that He was going to raise up a prophet to the Israelites from among their brothers, and that God would put His words in the month of the prophet (note, John 12:49), and that this prophet would speak in God's NAME (Deut 18:15-19).

gp74» God told David through a messenger that David's seed would build a house for God's NAME (2Sam 7:1-13). Solomon gave directions for the construction of the house for God's NAME (1 Kings 5:5-6). After Solomon finished building the house, God appeared to him and said to Solomon that His NAME would be put there (1 Kings 9:3). The temple was the house for God's NAME (1Kings 8:15-20). God's NAME was on Jerusalem and its temple (Jer 3:17; 2Kings 21:4, 7). The NAME was on mount Zion (Isa 18:7).

gp75» Jesus Christ came in his Father's NAME (John 5:43; John 10:25; Mat 21:9; etc.). Jesus Christ in a Spiritual sense was the true temple of God (note John 2:19, 21; compare with 1Cor 6:19; 3:16-17; etc.). Jesus Christ's Father is God (John 8:54; see GP 2). God the Father gave His NAME to Jesus (John 17:11-12, NIV, see Greek text; see Jer 23:5-6; 33:14-16). This is Jesus Christ's *new* NAME (Rev 3:12). Jesus Christ's *new* NAME is better than the angels (Heb 1:3-4). Jesus did his work in his Father's NAME (John 10:25). Jesus said that whatsoever a follower of him should ask in his NAME He would do it (remember Jesus was in his Father's NAME) (John 15:16).

gp76» After Jesus died, and then rose up to life again, it was said that those believing that Jesus was the Christ (the Messiah) would have life in Jesus' NAME (John 20:31). After this, people were baptized in the NAME of Jesus Christ (Acts 2:38; 8:16). Those who were baptized in the NAME of Jesus are in effect in the NAME of Jesus and are said to be in the NAME of Jesus (1 Cor. 5:4). Those in God's NAME are saved, have life, are justified, preach boldly, their sins are forgiven and they receive God's Spirit, and signs and wonders are done by them (Acts 4:12; John 20:31; 1Cor 6:11; Acts 9:27, 29; Acts 2:38; 10:43; 1John 2:12; Acts 4:30). These are called in a Spiritual sense the "temple of God" (1Cor 6:19; 3:16-17; 2Cor 6:16).

gp77» The Father, the Son, and the Holy Spirit have the same NAME (Matt 28:19). The 144,000 have the NAME written on their foreheads (Rev 14:1). *Remember* those in the NAME of Jesus Christ are in the NAME of God because God gave His NAME to Jesus Christ (John 17:11-12, NIV; Phil 2:9; see Jer 23:5-6; 33:14-16).

gp78» God is taking out of the nations a people for his NAME (Acts 15:2, 12-14; Amos 9:11-12). In fact all nations shall be gathered to the NAME (Jer 3:17; 4:2). God has sons and daughters from the ends of the earth who will be called by His NAME, "whom I [LORD, *YHWH*] created for my glory, whom I formed and made" (Isa 43:6-7, 21, NIV). After God's judgment he will change the people's speech and call all of them by the NAME of God: "For then will I turn to the people a pure language, to call them all by the NAME of the LORD [YHWH], to serve Him with one consent (Zeph 3:9, see Hebrew text; see YLT; see Eph 3:15). All people will be in His NAME, and call or pray in His NAME. If you can call in someone's name, you can be called by that name.

gp79» All through the Bible one can find where people call upon the NAME of God and trust in His NAME. By looking "name" up in Young's concordance or in Strong's concordance you can see how important God's NAME was to His people.

But what is God's NAME?

THE NAME OF GOD

gp80» As we've just seen there is great significance placed on God's NAME in the Bible. The importance placed on God's NAME has little to do with the pronunciation of the NAME. Unlike today in many nations, the Hebrews placed more significance on the *meaning* of names.

This is very important. We must not only take care to understand what is God's NAME, more importantly we must understand the real meaning of God's NAME. The paradoxes of God and the problem of evil can only be understood by knowing the true meaning and significance of God's NAME.

gp81» For some persons what follows is too detailed and repetitive, for others it is not detailed enough. We will repeat some things many times in order to make our point as clear as possible because we must break through a prevalent mindset imposed by tradition. See "More Details" at the end of GP in the GP: Appendix for more specific information on some topics.

What Is God's NAME?

gp82» We must go back to the book of Exodus to find God magnifying and revealing His NAME to Moses:

- Then Moses said to God, "Indeed, when I come to the children of Israel and say to them, 'The God of your fathers has sent me to you,' and they say to me, '**What is His name**?' what shall I say to them?" (Exo 3:13)

And God answered the question:

- "**I will be** that **I will be**" (Exo 3:14).

[Hebrew = אֶהְיֶה אֲשֶׁר אֶהְיֶה]

gp83» This is the literal English translation from the Hebrew text. But in the *King James Version* it reads: "**I am** that **I am**." The majority of English Bibles translates it this way. But this traditional translation is incorrect (See "I am" below). I repeat, the "I am" translation is incorrect. Look at the following examples:

- In the note for Exodus 3:14 in the *American Standard Version* it correctly says the verse is: **I will be that I will be.**

- In a footnote for *The NIV Study Bible*, it has **I will be what I will be.**

- In most Hebrew lexicons it shows that this phrase in Exodus 3:14 should be translated, **I will be that I will be,** or **I will be who I will be.**

- In the *Englishman's Hebrew-English Old Testament*, by Joseph Magil (printed by Zondervan in 1974), Exodus 3:14 reads: **I will be that I will be.**

- According to *The Pentateuch And Haftorahs: Hebrew Text, English Translation And Commentary*, edited by Dr. J. H. Hertz, C. H (former Chief Rabbi), published by Soncino Press, London (1956), in its commentary it states: "Most moderns follow Rashi in rendering [Hebrew - *ehyeh asher ehyeh*] '**I will be what I will be.**'"

    [But even though this is close to how Exodus 3:14 should be translated J. D. Hertz still allowed the traditional rendering of Exodus 3:14 to be used in the book's English translation of the verse.]

- According to The *International Standard Bible Encyclopedia* (1915 Edition) under "God, names of," page 1266, we see that it should be translated: **I will be that I will be.**

- By looking up the Hebrew words in The *Analytical Hebrew and Chaldee Lexicon*, by Benjamin Davidson we see that the correct translation is: **I will be that I will be.**

- Even the Bible in *Today's English Version*, published by the American Bible Society in 1976, has in a note for Ex 3:14, **I will be who I will be.**

- And in the *New International Version* (1978) it has a note for Exodus 3:14, "**I will be what I will be.**"

- And from the *Brown, Driver, Briggs, Gesenius Hebrew and English Lexicon*, "**I shall be the one who will be.**"

gp84» The "that," or "who," or "what," in "I will be ... I will be" is a relative pronoun, *'asher* (# 834), which can be translated in several ways such as: "that" or "who" or "what" or "when," etc (see Lexicon).

Exodus 3:12 v. Exodus 3:14

gp85» To transliterate **I will be that I will be** from Exodus 3:14 into English without the vowels we get:

- 'hyh 'shr 'hyh.

[Hebrew = אֶהְיֶה אֲשֶׁר אֶהְיֶה]

gp86» The root form of the Hebrew verb translated into **I will be** in Exodus 3:14 is *hyh*, a *to be* verb (Strong's # 1961). With the addition of ' [א] to *hyh* [היה] the word becomes, *'hyh* [אֶהְיֶה], and is now in the imperfect, first person, and singular form (*Analytical Hebrew and Chaldee Lexicon*, note Table N; *Gesenius' Grammar*, §40a-c; *The Essentials of Biblical Hebrew*, by Yates, p.41).[1]

gp87» This is the same verb as in Exodus 3:12: "**I will be** with you." Most English versions of the Bible translate Exodus 3:12 as, **I will be**, even the versions that translate Exodus 3:14 as, **I am**. This is important, so I'll repeat:

- *'hyh* [אֶהְיֶה] appears in both Exodus 3:12 and 3:14. In 3:12 it is translated, "**I will be** with you." But for some reason it is translated as, "**I am** " in Exodus 3:14 when pertaining to God's NAME. In most other places in the Bible in most translations it is translated, "I will be." In fact, in 41 other places in the Bible in most English translations it is mostly translated as, "I will be." (See below, "I will be in Context," gp180)

gp88» Notice the *Kings James Version* of Exodus 3:12 as compared to Exodus 3:14:

- And he said, Certainly **I will be** [אֶהְיֶה] with thee; and this shall be a token unto thee, that I have sent thee: When thou hast brought forth the people out of Egypt, ye shall serve God upon this mountain. [Exodus 3:12]

- And God said unto Moses, **I am that I am** [אֶהְיֶה אֲשֶׁר אֶהְיֶה]: and he said, Thus shalt thou say unto the children of Israel, **I am** [אֶהְיֶה] hath sent me unto you. [Exodus 3:14]

gp89» Do you see it? The same Hebrew word translated into *I am* in Exodus 3:14 is translated *I will be* in Exodus 3:12. Furthermore, this same word is translated into *I will be* dozens of other times in the Bible (See "I will be in Context" below). But why is it traditionally translated **I am**? Yes, something very strange is going on here with this common mistranslation of **I am**, and that something has to do with the influence of

Grecian philosophy on Biblical study, as well as the real reason — the "other-mind." We'll examine more on Grecian philosophy later.

Yehowah: God Revealed His NAME To Moses

God Restates His NAME

gp90» Right after God told Moses that his NAME was **I will be that I will be**, and for Moses to tell Israel that **I will be** had sent him (Exo 3:14), God rephrased his NAME and said unto Moses:

- "You shall say to the children of Israel that **Yehowah** [יְהֹוָה] ... has sent me [Moses] to you [Israel]" (Exo 3:15).

- "and say to them, '**Yehowah** [יְהֹוָה]the God of your fathers, the God of Abraham, of Isaac, and of Jacob, appeared to me [Moses]'" (Exo 3:16).

gp91» After Moses asked God his NAME, He answered with **I will be** repeating it twice, then He told Moses to tell Israel that his NAME was *I will be*, and right after this He told Moses to tell Israel that his NAME was **Yehowah** [יְהֹוָה]. Going back 1000s of years, in an ancient Hebrew script, the spelling of God's NAME without the vowels looked something like this:

God's NAME is Emphasized – He will be!

gp92» It is known that when words are repeated in Hebrew it has the effect of *emphasizing* the word (see Introduction in the *Emphasized Bible*, and *Gesenius' Hebrew Grammar*, § 133 k,l). For example in Genesis 2:17, the Hebrew word for "death" is repeated twice, and can be literally translated, "dying, you shall die." But when translated into English it becomes "you shall *surely* die." Or in Exodus 26:33 in Hebrew it has, "holy of the holies," and is translated as "the most holy" or "the most holy place." Therefore when God repeated his NAME twice (**I will be** that **I will be**), He was giving *emphasis* to his NAME.

gp93» God repeated his NAME twice, He again says that his NAME is **I will be**. He then changes

---

[1] The Hebrew *hyh* is a *to be* verb (Strong's # 1961). The Hebrew *'hyh* *perfectly* conforms to the rules of an imperfect verb when a verb is united with its personal pronoun fragment. When the first-person-pronoun fragment (א) is attached to the verb היה (*hyh*) together (אהיה) they mean, *I will-be.*

it to **Yehowah** only because this is the only grammatically correct way for Moses or anyone else to address God. Moses couldn't grammatically say, "**I will be** has sent me," but he could correctly say, "**Yehowah** has sent me." Because **Yehowah is** an imperfect **to be verb** in the masculine gender, except that it is in the 3rd person (see *BDBG Hebrew and English Lexicon pp. 217-218; Gesenius' Gram.* § 40 & § 75s; see below), **literally God was telling Moses to say to the nation of Israel:**

> "*He (who) will be* has sent me,"

God's NAME is an imperfect verb used as a noun.

gp94» In Hebrew verbs were used as nouns. Without its vowels, Yehowah is spelled YHWH. **Yehowah** as with "I will be" of Exodus 3:14 is an imperfect *to be* verb in the masculine gender, except that it is in the 3rd person (see *BDBG Hebrew and English Lexicon pp. 217-218; Gesenius' Gram.* § 40 & § 75s; see below). It is not a noun *per se*, but because it is used in the Bible as a proper noun because it is God's NAME as manifested in Exodus 3:14-16 (*Gesenius' Gram.* §125d; §§ 79, 83a, 116f).

What is an imperfect verb?

gp95» Hebrew has two different verbs: perfect and imperfect. God's NAME is in the imperfect. To understand what an imperfect verb is in Hebrew, we will contrast it with the perfect. Some call the Hebrew imperfect verb a future tense word, but this is not correct. From *Gesenius' Hebrew Grammar* (Oxford, 1980 reprint) we see that:

> ■ "The Hebrew (Semitic) **Perfect** **denotes** in general that which is *concluded*, *completed*, and *past*, that which is *represented* as accomplished, even though it is continued into present time or even be actually still future. The **Imperfect** denotes, on the other hand, the **beginning**, the **unfinished**, and the **continuing**, that which is just happening, which is conceived **as in process of coming to pass**, and hence, also, that which is yet future; likewise also that which occurs repeatedly or in a continuous sequence in the past (Latin Imperfect)" (*Gesenius* § 47.1, note 1).

gp96» More on the Hebrew Imperfect verb from S.R. Driver's *Hebrew Tenses*:

> ■ "It emphasizes the process introducing and leading to completion, it expresses what may be termed *progressive continuance*" (Driver, p. 27).

Meaning Contrary to "I AM" Doctrine

gp97» The meaning of God's NAME (beginning, unfinished, continuing, or coming to pass; see also Rev 1:8) is contrary to the "I AM" doctrine and the immutability doctrine. We will examine these traditional doctrines later. But for now remember that God's NAME is a verb, used as a noun, in the imperfect tense. For more information on this see read further in this chapter and see GP: Appendix in the back of this book.

Hebrew Words Written Without Vowels

gp98» At first the Hebrew language was written only with consonants and was written from right to left. When the Hebrews read, they added the vowels in their mind to the words. In Moses' time there was apparently no method of writing vowels in Hebrew. Two thousand years after Moses a system of vowel points was developed that was added below, between, and sometimes on top of the letters:

> ■ "The present pronunciation of this consonantal text, its vocalization and accentuation, rest on the tradition of the Jewish schools, as it was finally fixed by the system of punctuation (§ 7 h) introduced by Jewish scholars about the seventh century A. D." [*Gesenius' Hebrew Grammar*, p. 12]

Therefore when Moses wrote down God's NAME he did not write any vowels.

Is the Correct Pronunciation of the NAME Possible?

gp99» As we have just manifested, Moses did not write down the vowels for God's NAME, since in his time there was no method to write vowels. But it is said that the correct vowels for God's NAME were passed down orally through the years and are preserved in today's vowel point system. But it is unlikely that the exact sound of the Biblical Hebrew has been preserved for us today because there were different schools with different methods and interpretations, and there were Jews with different ways of pronouncing the Hebrew words (*Gesenius' Grammar*, p. 38, footnote 2; see § 7 *i*; § 8 "Preliminary Remark"; p. 42 footnote 3; etc.).

gp100» Because the Jews themselves pronounced words differently, depending on where they lived, it is debatable how one should pronounce God's NAME. It is only a guessing game. In order to write something with vowels we shall pick the spelling of **Yehowah**, which is the spelling found in some Jewish-Hebrew texts of the Old Testament (See "More Details" below). But Nehemia Gordon makes good arguments for Yehovah as maybe the original spelling.

(www.nehemiaswall.com/nehemia-gordon-name-god)

Different Spelling of the NAME

gp101» Now the Hebrew word "Yehowah" is sometimes translated into English as Jehovah or as the LORD (small caps). Some even translate the Hebrew word into Yahweh, Jehovah, LORD, and Yahweh, etc. The spelling of the Hebrew word YHWH as recorded in some Hebrew texts with vowel points is **Yehowah** (#3068) except when it is found with *'adhonay* (#136), then it is spelled, **Yehowih** (#3069). One text from about 1000 A.D. has it, **Yehwah**. As of the end of 2020, no Hebrew text that Nehemia Gordon has examined has it **Yahweh**.

gp102» The spelling of *Yehowah* for God's NAME is found in *The Pentateuch and Haftorahs*, edited by J.H. Hertz, Chief Rabbi, and published by the Soncino Press, 1956; the spelling is found in the *Interlinear Hebrew-English Old Testament* (Genesis-Exodus), by George R. Berry; the spelling is found in the C.D. Ginsburg's Hebrew Bible; the spelling is also found in some verses of the *Biblia Hebraica Stuttgartensia* (BHS), such as Gen 3:14; 9:26; Ex 3:2; 13:3,9,15; 14:1,8; etc. For the reason Yehowah is translated into LORD in some English translations, and for sufficient and qualifying details on the vowels used in God's NAME, you must read, "Yehowah or Yahweh or Jehovah or LORD." This is included in GP: Appendix of this book.

Gesenius admits the spelling "Yehowah" fits the evidence

gp103» Gesenius, the famous 19th century expert in Oriental literature, apparently popularized the theory that Yahweh was the true spelling of God's NAME instead of Yehovah or Yehowah. But at the same time Gesenius made this argument for the spelling, being Yahweh, he also wrote, "**Also those who consider that Yehowah was the actual pronunciation, are not altogether without ground on which to defend their opinion. In this way can the abbreviated syllables Yeho and Yo, with which many proper names begin, be more satisfactorily explained**." As the editor of *Gesenius' Hebrew and Chaldee Lexicon* [1949, Eerdmans Pub, 1857 Eng Ed.] said, "This last argument goes a long way to prove the vowels Yehowah to be the true ones" (p. 337). See GP Appendix for more info on this subject

NAME Pronounced

gp104» Keeping the above qualifications in mind, the NAME is pronounced with the vowels, **y^e hō wäh** [the "o" is a long o]. Or **Y^e ho väh**. The "**w**" in Yehowah came from *Gesenius's Grammar*, German language, but since the Germans pronounce their "**w**" like the English pronounce their "**v**," then Yehovah may be correct.

[More information - *https://www.youtube.com/watch?v=wRsbSLU9oFA* and *https://nehemiaswall.com/nehemia-gordon-name-god*]

God's NAME: BeComingOne

To Review and Conclude

gp105» As shown above, God said that his NAME was, "I will be." He repeated it twice in a row for emphasis. But to others God's NAME is "He-will-be" or "He (who) will be" or thus "Yehowah" or "Yehovah." We do not address God as, "I will be." To be grammatically correct we must call Him, "Yehowah" or "He (who) will be." As shown above, the Hebrew word "Yehowah" is from a verbal stem. "Yehowah" if used as a verb means, He-will-be, or He-will-become, or He-will-come-to-be. But when used as a noun "Yehowah" means, He-(who)-will-be, or He-(who)-will-become, or the **Becoming-One**. In The *Emphasized Bible*, page 26, it says the "Becoming-One" is a proper translation for YHWH. Many translations insist on using "LORD" in translating YHWH even though it is based on a mistaken Greek translation that used *Kurios* ("Lord") when the Hebrew YHWH was translated into Greek.

gp106» **BeComingOne** is a better translation than "He-(who)-will-be" since it indicates that

"Yehowah" exists now, but somehow is not yet perfected or completed or fully finished: He is *Becoming*. Since "Yehowah" is an imperfect verb (used as a noun), it signifies an incomplete state, it indicates something that is becoming, it indicates something that is in the process of coming-to-be, it indicates something that will be, yet is somehow now in existence. Thus, the translation, "BeComingOne," fits the Hebrew word "Yehowah" best for the English language. The meaning of God's NAME indicates that at some point in time the BeComingOne will come to be, or at that time will have become, or at that time will exist in his truest form or meaning.

NAME in the New Testament

gp107» In the New Testament please note the Lord God Almighty is the one "who is, and who was, and *who is to come*" (Rev 1:4, 8; 4:8; 11:17; 16:5). **The BeComingOne (YHWH) is the almighty God, the one "who is, and who was, and who is to come."** This is a good translation of the meaning of the Hebrew imperfect verb Yehowah, which is God's NAME. Or we can translate Revelation 1:8: "Lord, the God, the is, the was, and the coming-one, the almighty." God Almighty is to come, or He is the COMING-ONE, who is now, and who was; He is the BeComingOne.

gp108» With our knowledge that God's NAME was an imperfect verb, and that it was in the corhortative form, we can conclude that:

- ■ YHWH means one existing in someway in an incomplete state who yet will, without any doubt, come to be, or come to exist, in the fullest sense.

Hereafter in this book we will use the correct translation of YHWH — BeComingOne — instead of "LORD."

No Problem with the NAME, But with Immutability Theory

Yes I know that God's NAME is against the immutability theory, but the problem is not with His NAME, but with the false immutability theory.

"I Am" Doctrine

Grecian Mindset

gp109» The Hebrew word translated "**I Am**" in many of today's translations of Exodus 3:14 is an incorrect translation because the Hebrew word is a verb in the *imperfect* tense. The translation of "I am" doesn't give the full meaning of God's NAME. The translation, "I am," does not take into consideration that it was translated from a Hebrew *imperfect* verb. The "I am" translation is not only a wrong translation from the Hebrew text, but also was influenced by a mistaken Greek translation (*Septuagint*) made in Egypt.

Greek Translation of God's NAME: "The Being"

gp110» The much used Greek translation of the Old Testament, called the *Septuagint* (LXX or seventy), because it was translated by about 70 translators, was translated in Egypt in the third century BC for Ptolemy II, a king of Egypt. In this Greek translation, instead of "**I will be** that **I will be**," the Greek (*Septuagint*) has "**I am the Being**" and "**The Being** has sent me to you" for Exodus 3:14.

- ■ LXE Exodus 3:14: And God spoke to Moses, saying, **I am The Being**; and he said, Thus shall ye say to the children of Israel, **The Being** has sent me to you. [English of Greek text]

"The Being" was Egypt's God

gp111» It is important to point out the Greek version, the *Septuagint*, was made in Egypt and a notable Egyptian's god, Osiris, was addressed in their prayers as "the Being":

- ■ "At a later period, however, the Egyptians put their trust in Osiris himself, and addressed their prayers directly to him as **the Being**." (p. 151, *The Gods of the Egyptians*, Vol 1, by W.A. Wallis Budge, emphasis mine)

From this corruption of the Hebrew Bible, later translations intermingled the Hebrew and Greek translation in order to get: "I am that I am."

Bible Written in Hebrew Not Greek

gp112» But the Old Testament was written in Hebrew, not Greek. Besides the mistranslation of Exodus 3:14, the *Septuagint* mistranslates the Hebrew word, YHWH. For YHWH it substitutes the Greek word for "Lord," which is *Kurios* (# 2962). From this early Greek translation we see many translations that use "LORD" instead of "Yehowah" or as commonly misspelled, "Jehovah" or "Yahweh."

Catholic Church's Bias Toward the Greek Text

gp113» It was the "fathers" of the Catholic Church such as Augustine that were insistent on using translations from the Greek text instead of the Hebrew text:

- "There have, of course, been other translations of the Old Testament from Hebrew into Greek. We have versions by Aquila, Symmachus, Theodotion, and an anonymous translation which is known simply as the 'fifth edition.' Nevertheless, the Church [Catholic] has adopted the Septuagint as if it were the only translation.... From the Septuagint a Latin translation has been made, and this is the one which the Latin churches use. This is still the case despite the fact that in our own day the priest Jerome, a great scholar and master of all three tongues, has made a translation into Latin, not from Greek but directly from the original-Hebrew. The Jews admit that his [Jerome's] highly learned labor is a faithful and accurate version, and claim, moreover, that the seventy translators [Septuagint] made a great many mistakes in their version. Christ's Church [Catholic], however, thinks it inadvisable to choose the authority of any one man [Jerome] as against the authority of so many men — men hand-picked, too, by the high priest Eleazar for this specific task. [Augustine here speaks of the myth of the 70 or so translators of the Greek text (*The Canon of Scripture*, F.F. Bruce, pp 43ff).] For, even supposing that they [the 70] were not inspired by one divine Spirit, but that, after the manner of scholars, the Seventy merely collated their versions in a purely human way and agreed on a commonly approved text, still, I [Augustine] say, no single translator should be ranked ahead of so many. The truth is that there shone out from the Seventy so tremendous a miracle of divine intervention that anyone translating the Scriptures from the Hebrew into any other language will, if he is a faithful, translator, agree with the Septuagint; if not, we must still believe that there is some deep revealed meaning in the Septuagint." [*City of God*, by Augustine, book 18, chapter 43]

NAME Forgotten by Judah

gp114» It is very significant that Judah was prophesied to not pronounce God's NAME:

- "Behold, I have sworn by My great NAME, says Jehovah, that My NAME shall no more be named in the mouth of any man of Judah in all the land of Egypt." [Jer 44:26, King James II Version]

gp115» The *Septuagint* translation was done in Egypt, and it was in Egypt that the Jews were to forget God's NAME: they began to use the Greek equivalent for "Lord" instead of the Hebrew YHWH or Yehowah ("Jehovah"). *The International Standard Bible Encyclopaedia* (1915 A.D.) speaks about the translation:

- "It is one of the outstanding results of the breaking-down of international barriers by the conquests of Alexander the Great and the dissemination of the Greek language .... The Jewish commercial settlers at Alexandria forced by circumstances to abandon their language, clung tenaciously to their faith; and the translation of the Scriptures into their adopted language, produced to meet their own needs, had the further result of introducing the outside world to a knowledge of their history and religion.... The LXX [Septuagint] was also the Bible of the early Greek Fathers, and helped to mold dogma; it furnished proof-texts to both parties in the Arian controversy." [under "Septuagint"]

Greek Mindset

gp116» If God's Being is what or like what others say it is, then God's very NAME should have been written or spoken with a *perfect* verb.

- "A Hebrew perfect verb is "concluded, completed [they say that nothing can be added to God, he is eternal, not changeable, etc] ... even though it is continued into the present time or even be actually still future." [*Gesenius' Gram*, § 47.1, note 1]

gp117» But God's NAME was written and spoken with an *imperfect* verb, "I will be."

- "The imperfect does not imply *mere* continuance as such ... it emphasizes the process introducing and leading to completion, it expresses what may be termed *progressive* continuance." [Driver, *Hebrew Tenses*, p. 27]

gp118» If God's Being is what others say it is, then God's NAME should have been written with the Hebrew *participle active*, which indicates *mere* continuance and not *progressive* continuance (Driver, p. 27, 35ff; *Ges. Gram.*, §116a,c). The Hebrew imperfect indicates progressive continuance. (See "More Details" in the GP: Appendix)

Greek Mindset: God <u>had</u> to be Changeless

gp119» According to the Grecian mindset, which was influenced by Plato and Aristotle, God's NAME and its meaning could never, no *never* be from an imperfect verb, because an imperfect verb is one that is beginning, unfinished, and continuing. Plato in *Timaeus* makes the distinction between that which has existed always and that which is becoming:

- "We must in my opinion begin by distinguishing between that which always is and never becomes from that which is always becoming but never is....In addition, everything that becomes or changes must do so owing to some cause; for nothing can come to be without a cause." [Plato: Timaeus and Critias, trans. Desmond Lee, Penguin Classics, p. 40; see also *Plato*, volume IX in the Loeb Classical Library (No. 234), which gives a slightly different translation, p. 49 & p. 113]

gp120» God to the Grecian mindset could not be **becoming** in any sense, since He must be the First Cause, the One that cannot be caused in anyway; He must have existed always; He must have been perfect and complete always.

- "Moreover, life belongs to God. For the actuality of thought is life, and God is that actuality; and the essential actuality of God is life most good and eternal. We hold, then, that God is a living being, eternal, most good; and therefore life and a continuous eternal existence belong to God; for that is what God is." [Aristotle, *Metaphysics*, Loeb Classical Lib. #287, p. 151]

gp121» To the Greek philosophers it was God who was "the Cause wherefor He that constructed it constructed Becoming and the All" (*Plato*, volume IX in the Loeb Classical Library, p. 55).

God in no way could have been in anyway "becoming" to the Greek mindset.

gp122» Their Grecian mindset was unable to translate the Hebrew imperfect word for God into a Greek imperfect. Instead they translated Exodus 3:14 into, "the Being," which is a present participle in the Greek translation. Plato's God was:

- "the ever-existing God"
- someone who "existed always"
- had "no beginning of generation"
- He must have "constructed Becoming and the All"
- "'Was' and 'will be' on the other hand, are terms properly applicable to the Becoming ... but it belongs not to that which is ever changeless." [pp. 65, 51, 55, 77, Plato's *Timaeus*, Loeb Classical Library, No. 234, Harvard Univ. Press]

gp123» According to Plato, God was eternal, always existed, and since he was good, then any change must be change for the worse (Plato, *Republic*, Book II, 381B). Because God to the great Grecian philosophers was changeless, his special NAME could not have been translated, "I will be" or "He will be," but had to have been translated, "I am" and "The Being." Yet in Revelation 1:8 it reads, "Lord the God, the is, the was, and the Coming-One" or "the one who is, who was, and who is coming." The real NAME for God and its meaning is absolutely contrary to the Grecian mindset.

gp124» Because this Grecian mindset of a changeless God was passed on to the "fathers" of the Catholic Church, and from them to our day through tradition, modern translations of God's NAME as revealed in Exodus 3:14-15 are faulty.

Hebrew verbs are different from English verbs

gp125» Not only did the Greek culture make it difficult for some to translate God's NAME correctly, but the differences between Hebrew and other languages also make it difficult to translate God's NAME correctly. It should be noted here that it is difficult, if not impossible, to translate verbs from Hebrew to English:

- "There is no tense [past, present, future] in the Hebrew verb. The student is only kidding himself when he continually translates the Hebrew perfect into the English past, and the Hebrew imperfect into the English future. After a while, he unconsciously begins to believe it. The perfect state is really talking only about an action which is completed. The imperfect state speaks of an incomplete action. Both of these actions (completed and incomplete) can occur in the past, present or future. The only way you can tell the tense [past, present, future] in the Hebrew language is by the context.... So, when you find the tenses in your English Old Testament, don't lean too hard on them. You might be counting on what might be a translator's precarious guess. Don't blame the translators for putting those tenses in, however; you cannot write English without them." [*Do It Yourself Hebrew And Greek*, by Edward W. Goodrick, Pub. 1976, pages 15.4 & 15.5; *Hebrew Tenses*, S. R. Driver, ch. 1]

Didn't Jesus say "I am"?

gp126» According to the Trinitarians, because Christ said "I am" [ἐγώ εἰμί] (John 8:58; 4:26; 6:35; 8:12; 10:7; 10:11; 11:25; 13:13; 14:6 15:1; 18:8), "He thus identified Himself with the covenant name of Jehovah in the Old Testament" (p. 39, *All the Messianic Prophecies of the Bible*, Herbert Lockyer). The problem here is that God's NAME is not, "I am." God's NAME is, "I will be," as we have seen in chapter 1 of this book. First the Trinitarians use a false name for God ("I am") obtained from a false Old Testament translation of Exodus 3:14, then to prove their falsehood they quote a few times from the New Testament of the Bible where Jesus said the words, "I am."

gp127» Since God's real NAME is not "I am" [ἐγώ εἰμί] it means nothing that Jesus said "I am" a few times in the New Testament. Others in the New Testament also said, "I am." [ἐγώ εἰμί]:

■ The apostles said the same "I am"[1] when asking a question, "I am Lord?" In English we would say, "am I he Lord? (Mat 26:22).

■ Judas said the same "I am"[1] when asking a question, "I am Master?" In English we would say, "am I he Master?" (Mat 26:25).

■ The healed blind man said the same "I am"[1] when identifying himself, while we would say, "I am *he*."(John 9:9)

■ Peter said the same "I am"[1] when identifying himself, but in English we would say, "I am *he*."(Acts 10:21).

■ Paul said the same "I am"[1] when identifying himself as a Jew, "I am [exist] as a male Jew (Acts 22:3) or when identifying the way he existed, "such as I am [exist]" (Acts 26:29).

■ Paul said, "by the grace of God I am what I am" (1Cor 15:10)

This last verse is almost exactly how most English translations translate Exodus 3:14. Does this mean Paul is God? Of course not, but it further proves the nonsense of those who believe in the "I am" theory.

Thus to the Trinitarians' mode of thinking, the apostles, including Judas and Paul, are "I am."[2] Of course, this is nonsense, since Christ was not saying he was the very Jehovah when he said "before Abraham was, I AM" (John 8:58).

gp128» By studying how "I am"[1] is used in the New Testament, we see that it may mean either:

■ (1) "I am *he*"

■ or (2) "I exist" or "I existed"

What Jesus Christ was saying in John 8:58 was that he existed before Abraham: "**before Abraham was, I existed.**" In some way he existed before Abraham. This was true because the *Spirit* (not the flesh) of Christ did exist before Abraham (see, GP 3-5). In this scripture Christ was **not** saying he was the Jehovah or YHWH, by saying, "I am," even though we know through other scripture that he indeed is Jehovah (YHWH) after he went to the Father.

gp129» When Christ said he came in his Father's NAME ("I come in my Father's name," John 5:43), he was saying he was coming in the real NAME of God; he was coming in the NAME of the One who said his NAME was, "I will be." But the places in the New Testament where Christ said "I am" (John 8:58, etc.) had nothing to do with identifying Christ with Jehovah, for one reason God's NAME is not "I am," and for another reason others in the New Testament also said "I am" or used the phrase similarly to the way that Christ used it.

---

[1] ἐγώ εἰμί

[2] ἐγώ εἰμί

Unchangeableness of God

gp130» God's NAME tells us that God is in someway moving and changing towards his completed "state," for God is the BECOMINGONE, for God said his NAME is, *I will be that I will be*, He is Yehowah — He (who) will be. But the book of Malachi said that Yehowah does not change (Mal 3:6). Others speak about the "immutability" of God.

> ■ "The immutability of God is a necessary concomitant of His aseity [self-existence]. It is that perfection of God by which He is devoid of all change, not only in His Being, but also in His perfections, and in His purposes and promises. In virtue of this attribute He is exalted above all becoming, and is free from all accession or diminution and from all growth or decay in His Being or perfections. His knowledge and plans, His moral principles and volitions remain forever the same. Even reason teaches us that no change is possible in God, since a change is either for better or for worse. But in God, as the absolute Perfection, improvement and deterioration are both equally impossible." [*Systematic Theology*, Berkhof, p. 58]

The fathers of the Church took the "immutability of God" theory from Greek philosophers like Plato and Aristotle. Plato believed that God was always perfect and any change was for the worse. Aristotle thought that God could not change because it would prove that God was not completely actualized in all His potentialities (Note *Logic and the Nature of God*, by Davis, pp. 41-42). But as noted by Davis, "now the 'God' Plato speaks of in his writings is different in several respects from the Christian God ... Again, Aristotle's God is not the same thing as the Christian God" (pp. 41 & 42). The immutability of God doctrine has more to do with Grecian philosophy than with the Bible.

gp131» The champions of the immutability of God theory say, "this immutability of God is clearly taught in such passages of scripture as Ex 3:14; Ps 102:26-28; Isa 41:4; 48:12; Mal. 3:6; Rom 1:23; Heb. 1:11,12; Jas. 1:17" (Berkhof, p. 58-59). Yet when you study these scriptures you do not see anything that compares with the descriptions of the immutability doctrine just quoted from Berkhof's book (p.58). Shockingly, we see the immutability doctrine is described in almost the same words used by Plato and Aristotle when they characterize their God(s).

Immutable God Taught by Greeks

gp132» Plato's God was:

- ■ the ever-existing God.

- ■ one who existed always,

- ■ one who had no beginning of generation.

- ■ one who must have constructed Becoming and the All.

- ■ 'Was' and 'will be' on the other hand, are terms properly applicable to the Becoming ... but it belongs not to that which is ever changeless (pp. 65, 51, 55, 77, Plato's *Timaeus*, Loeb Classical Library, No. 234, Harvard Univ. Press).

gp133» Aristotle wrote in his *Metaphysics*:

- ■ "Moreover, life belongs to God. For the actuality of thought is life, and God is that actuality; and the essential actuality of God is life most good and eternal. We hold, then, that God is a living being, eternal, most good; and therefore life and a continuous eternal existence belong to God; for that is what God is. Those who suppose, as do the Pythagoreans and Speusippus, that perfect beauty and goodness do not exist in the beginning ... are mistaken in their view." [Aristotle, *Metaphysics*, Loeb Classical Lib. #287, p. 151]

gp134» Plato wrote in his *The Republic*:

- ■ "But think, God and what is God's is everywhere in a perfect state... if he does alter. Does he change himself for the better and more beautiful, or for the worse and more ugly than himself? He must change for the worse...." [Book II, 381B]

Therefore, according to this way of thinking, God does not change because he is already perfect, and any change would have to be "for the worse." But the theory ignores the Law of Knowledge among other things and limits what God can do. For one thing, change in and of itself is not negative. With the immutability theory God cannot create something new or change at all. Anything that cannot change is actually dead. Those who propagate an immutable God are describing a dead god, not the live God of the Bible. The immutability theory, when you understand the Law of Knowledge, is nothing but a naive theory, not very well thought out. But we cannot explain this until you yourself understand the fundamental Law of Knowledge, which we cover in chapter 7 of this book.

Immutable God or BeComingOne God?

gp135» **This unchangeable or immutable "God" of the great Grecian thinkers is not the one found in the Bible.** The Grecian mindset could not and did not admit that God in any way at all could be **becoming.** Thus they refused to translate God's NAME correctly. But God said His very NAME was "He (who) Will-Be" or the "BeComingOne." The true God emphasized His NAME over and over in scripture. Names in the Bible were used to describe certain important aspects of people. The true God said He was **He will be,** that he was **Yehowah,** or the **BeComingOne.** Some important aspect of Him is becoming. As explained previously, the real God used an imperfect Hebrew verb for His NAME:

> ■ "The *Imperfect* denotes ... the *beginning*, the *unfinished*, and the *continuing*, that which is just happening, which is **conceived as in process of coming to pass**, and hence, also, that which is yet future" (*Gesenius' Hebrew Grammar*).

Serious Subject

gp136» If God is becoming, then He is not immutable in the sense that the Grecian mindset taught. What the Bible teaches about God is not what the Grecian mindset teaches about God. The essence of God is called a "mystery" because hundreds of scriptures are being overlooked that would teach us what God's essence really is. Do we wish to believe what the Bible teaches about the essence of God, or do we wish to continue being blinded by the Grecian mindset? This is serious. We must pay attention to scripture, not to the theological courses taught inside the Grecian mindset.

One sense of God's changeability

gp137» One sense of God's changeability is that throughout the Bible it shows God changing his actions toward people depending on the people's good or bad behavior (Psa 18:25-26; Prov 3:32-35; Lev 26:3ff, 14ff, 40ff; Exo 32:9-13; Jer 18:7-10; etc). If Israel follows God's commandments they receive a just reward. If Israel does not follow

God's commandments, they receive a judgment (note Deut chap 28; etc.). The same applies to others besides Israel, for the true God is the God of all (Rom 3:29; Eph 4:6). The true God judges according to the *ways* of people: "the soul that sins, it shall die ... the righteousness of the righteous shall be upon him, and the wickedness of the wicked shall be upon him" (Ezek 18:20). Another sense of God's changeability is manifested in this book. But this change in no way diminishes the Power of God. We cannot speak of this change yet. Do read on.

Real Unchangeableness of God

gp138» Scripture indicates that the unchangeableness of God is his unchangeable words, his **unchangeable truth** (Isa 31:2; Heb 6:17-18; Isa 46:11; Isa 55:11; etc.) and his **all mighty power** (Gen 17:1; 1Chron 29:12; Isa 44:24; etc). God gave his Word that he will not totally consume Israel (note Isa 65:8-9; Exo 32:13, 9-13; 33:1; Lev 26:44-45), because it is through Israel that the true Seed or Savior was to come, so for the sake of His word and His NAME Israel is not consumed (note Ezek 36:21-22ff; Isa 48:9). The statement of Malachi ("I change not; therefore ye sons of Jacob are not consumed") merely indicates that God's *word* does not change, for he has promised that the true SEED would come from this nation. The word translated "change" in Malachi 3:6 is Strong's #8138 which has more to do with duplicity or changing one's promises than changing one's nature or power. To keep his word, to not lie, God must not consume the nation before the SEED came. Read the "Seed Paper" [PR 1] to understand more about God's promises to Israel and how God kept these promises.

gp139» Jesus Christ is not the same "forever" as Hebrews 13:8 in some English translations say, for this is incorrectly translated since it should be "Jesus Christ the same [or the very one], and into the ages" (see Greek text; see "Age Paper" [NM7]). What is unchangeable about God (or Jesus Christ) is his words, his love, his promises, and his power. These things are unchangeable because God does not lie, and he has all the power and life in his hands. In fact God is life (John 5:26; Acts 17:28). The fact that God is life does not change. The fact that God is all-powerful does not change. The fact that God does not lie does not change. But since God is the BeComingOne, then something about God is now changing. What is changing about God was manifested in the Bible. This book will also manifest the becomingness of God. Do read on.

gp140» In Psalm 55:19 it speaks of those who do evil as not changing: "they do not change" (NKJV). Does this mean they are immutable? Of course not. Those who use the "I change not" in Malachi 3:6 to prove their immutability of God theory are taking scripture out of context and using it to infuse the Greek theory of immutability into Christianity. They are not using scripture to find out who or what the God is, but want to hold on to myth instead of finding the truth. The very NAME of God is "He (who) Will-Be." Thus, in some way God is changing. This book will expound on this.

Immutability: One Conclusion.

gp141» In Stephen T. Davis's *Logic and the Nature of God*, he admits,

> ■ I believe the route for the Christian philosopher to follow is happily to admit that there are senses in which God does indeed change, i.e. alter... . In fact, it is not easy to read the Bible without forming the strong impression that the God revealed there does indeed change in some senses. To pick an obvious case, very typically God is at one moment angry with someone (the person has sinned) and at a later moment forgives that person (the person has repented)....What was the classical doctrine of divine immutability designed to protect? I believe the answer is this: as I noted earlier, it was designed to preserve the view that God is faithful in keeping his promises... . [p.47]

This "classical doctrine of divine immutability" that Davis is writing about is the Grecian influenced ideas, which are not Biblical.

There are ways in which God changes over time, but one thing that does not change is His power and the fact that God cannot lie (Heb 6:17-18; 1John 5:18; Isa 46:11). The true God has all the power. But in someway God does change. This book will amplify on the nature of these changes.

God, Gods

gp142» *In English*, most use the word "God" to describe the supreme being. But the word "god" in English can mean either: the almighty, supreme being; or "any of various beings conceived of as supernatural, immortal, and having special powers over lives and affairs of people" (*Webster's New Word Dictionary*). There can be one god, or many gods. The word "god" is not a proper name for the Supreme Being. The word "god" is a *generic* name for God: it can represent a *class* of beings. In Hebrew and Greek the same applies.

gp143» *In Hebrew*, "elohim," "eloah," "elah," and "el" are the Old Testaments words for god or God. As with the English word "god" these Hebrew words are generic names for god or God.

gp144» *Elohim* was translated into the English word "god" about 2555 times in the KJV. In about 2310 instances "elohim" is translated into "God," thus indicating the supreme God. For example in Genesis 1:1, "In the beginning *elohim* created...." But in some 245 cases "elohim" is translated into lower senses of the word. "Elohim" has been translated in the KJV into such words as:

- *gods* (Gen 3:5);

- strange *gods* (Gen 35:2,4);

- "I have made you [Moses] a *god* to Pharaoh" (Exo 7:1);

- *gods* of Egypt (Exo 12:12);

- *gods* of silver, *gods* of gold (Exo 20:23);

- *judges* (Exo 22:8[7], 9[8]);

- their *gods* (Exo 34:15);

- molten *gods* (Exo 34:17);

- *goddess* (1Kings 11:5,3);

- "I have said, you, *gods* and all of you sons of the most high, but you shall die as man.." [Psa 82:6-7; see John 10:34-36]

gp145» We see that the Hebrew word, *elohim* was translated in many different ways beside being translated as "God." *Elohim* can indicate *gods*, *gods* of silver and gold, *judges*, a *goddess* (like the female god, Ashtoreth) even indicate *Moses* (Exo 7:1) or *mankind* (Psa 82:6-7; see John

10:34-36). Notice that *elohim* is translated in the singular AND plural (god and god*s*). WHY?

Elohim Is Plural

gp146» The Hebrew word *elohim* is a plural noun as the lexicons indicate and as some of the translations above indicate. "Eloh*im*" has the ending "*im*." This indicates that it is a simple plural word (sec. 87a, *Gesenius' Hebrew Grammar*, 1980 printing). The correct nominal suffix is used for the plural *elohim*.

[Compare in the Hebrew text *their gods* (elohim), *my God* (elohim), and *our God* (elohim) in Exo 34:15; Isa 25:1,9 with table A, section I in the Tables of Paradigms of the *Analytical Hebrew and Chaldee Lexicon*.]

Thus the Hebrew word "elohim" itself is an ordinary plural noun.

God's NAME is Yehowah Not Elohim

gp147» The other names or titles of God can refer to others, but the NAME Yehowah only refers to the true God (*Gesenius' Gram.* §125*d*). "And let them [God's enemies, v.2] know that you, your NAME, Yehowah, you alone the Most High over all the earth" (Psa 83:18; see Exo 6:3).

From *Girdlestone's Synonyms of the Old Testament* we read:

> ▪ "The Hebrew may say *the* **Elohim**, the true God, in opposition to all false Gods; but he never says *the* **Jehovah**, for Jehovah is the name of the true God only. He says again and again *my God*, but never *my* **Jehovah**, for when he says 'my God' he means Jehovah. He speaks of *the God of Israel*, but never *the* **Jehovah** *of Israel*, for there is no other Jehovah. He speaks of *the living God*, but never of *the living* **Jehovah**, for he cannot conceive of Jehovah as other than living." [pp. 36-37, Jehovah = Yehowah]

Yehowah is the God's proper NAME. In Hebrew "Yehowah" means the BECOMINGONE, or He who will be. Thus, God is the BeComingOne.

Israel's Gods is One YHWH

gp148» But why is *elohim*, an ordinary plural word, translated into the English singular "God" when representing the TRUE God? The main reason for this is that the plural *elohim*, when referring to the TRUE God, is used as if it where a singular noun. "Although plural in form, the name is generally used with a singular verb when it refers to the true God" (p. 19, *Synonyms of the Old Testament*). Gesenius called this phenomenon the *plural of majesty* or *plural of excellence* (*Ges. Heb. Gram.* § 145h, § 124g).

gp149» When the Old Testament was written, the nations around Israel worshiped *god*S, deitie*S*, and idol*S*. These nations did not worship just ONE God, but many god*S*; their religion was not monotheistic. When the nations around Israel spoke of their deity, they called them "our gods," and they meant more than one kind of god; they spoke of gods who had different attributes. There were gods of fire, of heaven, of the sea, of love, of fertility, of maternity, of the moon, of the sun, of planets, etc (see *Unger's Bible Dict.*, under "gods false"; *The Gods of the Egyptians*, by E.A. Wallis Budge; etc.).

One YHWH, Not One Elohim

gp150» *One Yehowah.* But to Israel there was only ONE deity, and his NAME was/is Yehowah (YHWH) or as popularly spelled today, Jehovah or Yahweh or LORD.

> ▪ "Here Israel, Yehowah our *elohim*, Yehowah (is) ONE." [Deut 6:4, literal trans.]

gp151» *One NAME.* As we see, it is Yehowah (YHWH) who is ONE, not *elohim* (gods) who are ONE. But as Deut 6:4 says, Yehowah was Israel's Gods (elohim): "our Gods." But it is Yehowah who is ONE; his NAME one:

> ▪ "In that day there shall be ONE Yehowah, and his NAME ONE." [Zech 14:9]

gp152» **Israel's Gods** (elohim) was Yehowah and He was ONE; He had ONE NAME. Thus, Moses calls Yehowah, *our Gods*:

> ▪ "Yehowah, Gods [*elohim*] of Israel." [1Kings 8:20]

> ▪ "Moses began to explain this law, saying: *Yehowah, our Gods* [elohim] spoke to us in Horeb...." [Deut 1:5,6]

gp153» Yehowah, himself, tells Israel:

- "and you shall be afraid of your Gods, for *I Yehowah, your Gods*." [Lev 25:17, see Hebrew text]

gp154» What kind of Gods are or is Yehowah?:

- "God [el] of gods [elohim] (is) Yehowah." [Josh 22:22]

The expression "god of gods" means: greatest god. Thus, Yehowah is the greatest God, or the great God:

- For the LORD [YHWH] your Gods [elohim] is Gods of gods and Lords of lords, the great God [el]... [Deut 10:17]

gp155» Not only is Yehowah the greatest God, the God of Gods, but He *alone* dwells as or sits as *the* cherubim and *the* Gods, and he *alone* created the universe:

- "And Hezekiah prayed before Yehowah, and said, Yehowah, Gods [elohim] of Israel, who dwells [or sits as] the cherubim [plural], you alone <u>the</u> Gods [elohim], by yourself alone, for all the kingdoms of the earth, you have made the heavens and the earth." [2Kings 19:15]

יְהוָה אֱלֹהֵי יִשְׂרָאֵל יֹשֵׁב הַכְּרֻבִים אַתָּה־הוּא
וַיִּתְפַּלֵּל חִזְקִיָּהוּ לִפְנֵי יְהוָה וַיֹּאמַר

הָאָרֶץ אַתָּה עָשִׂיתָ אֶת־הַשָּׁמַיִם וְאֶת־הָאָרֶץ:
הָאֱלֹהִים לְבַדְּךָ לְכֹל מַמְלְכוֹת

gp156» In Malachi 2:10 it speaks of the one Father the one God who created us. The "God" here is "el" the singular case of the Hebrew "elohim." Remember the One YHWH is the God of Gods, or the greatest God. It is YHWH who is the true God, the real God, the greatest God.

gp157» Therefore, the nations around Israel had their gods (*elohim*), but *each* of these gods had different qualities or attributes. But Israel's God(s) (*elohim*) was one — there was a oneness to Israel's God(s). And the ONE NAME of Israel's God(s) was "Yehowah."

One God: Old and New Testament

gp158» The New Testament also speaks of One God, but the New Testament does not use God's NAME as manifested in the Old Testament. There is some evidence that at least some of the New Testament was written in Hebrew or Aramaic (Jerome, see "God's NAME in Greek ..." gp718). There also have been Greek texts of the New

Testament found that had God's NAME written in Hebrew or Aramaic instead of the word "Lord" as we see in today's New Testament's translations. One place where "Lord" should be translated into Yehowah is in Mark 12:29. In this scripture it speaks about the One Lord, but since it is a quote from the Old Testament (Deut 6:4) it should read, One Yehowah. So even in the New Testament it is One Yehowah when speaking of the true One God. In Mark 12:32 it should not read "for there is one God," but "for there is one." Other places in the New Testament Bible where it speaks of "one Father, the God," or "one the God," or "no one, but God," or "God is one," or "one God and Father of all," or "one God," or "the God is one,"[1] all point to the Old Testament God, who was/is/will be, He is the BeComingOne (YHWH). It was in the Old Testament that God revealed his NAME and said it was the NAME that was one; it was Yehowah that was one (Deut 6:4; Zech 14:9). It is Yehowah who is God of gods, the great God, the true God.

One Yehowah

gp159» As we have just seen Israel's deity is the most powerful God, he is the Great God, He is Yehowah (YHWH), He is ONE. How is he *one*?

One in History

gp160» In the past "one" was not even considered a number, but "unity." Plato even put unity (one) and numbers into separate categories: "To what class do unity and number belong?" (Smith, *History of Mathematics*, Vol II, p. 27, quoting Plato's *Republic*). Smith in his *History of Mathematics* lists numerous other mathematicians that agree that one (unity) was not a number (pp. 26-29).

- "Not until modern times was unity considered a number. Euclid defined number as a quantity made up of units, and in this he is followed by Nicomachus. **Unity was defined by Euclid as that by which anything is called 'one'** " (Smith, *History of Mathematics*, Vol II, p. 26-27). Euclid who wrote the famous book on Geometry called *Elements* lived around 300 B.C.

- "Number is a multitude brought together or assembled from several units,

---

[1] John 8:41; Rom 3:30; 1Cor 8:4, 6; Gal 3:20; Eph 4:6; 1Tim 2:5; James 2:19

always from two at least, as in the case of 2, which is the first and the smallest number. **Unity is that by virtue of which anything is said to be one**" (*The First Printed Arithmetic*, Treviso, Italy, 1478).

■ "A Living Creature perfect and whole, with all its parts perfect; and next, that it might be One, inasmuch as there was nothing left over out of which another Creature might come into existence... He fashioned it to be **One single Whole, compounded of all wholes**, perfect and ageless... Now for that Living Creature which is designed to embrace within itself all living creatures...."

[From *Timaeus* found in, *Plato* volume IX in the Loeb Classical Library [No. 234], p. 61; see also, Plato: *Timaeus* and *Critias*, trans. Desmond Lee, Penguin Classics, p. 43, which gives a slightly different translation]

This last item shows that even Plato believed that One equaled wholeness or unity, especially when speaking of the "one universe." One question here is at the time the Trinity doctrine was formulated, what was the prevailing idea of one? Was it also unity? Yet as seen by studying Augustine's almost 1600-year-old book called, *On the Trinity*, the Trinitarian belief indeed had something to do with three in one, not three in unity, even though they spoke of the "unity of the Trinity." You can see Augustine struggling with this problem and that is why he (and all of the Trinitarians) calls it a mystery.

One In Hebrew

gp161» The *Hebrew* word translated One in Deut 6:4 and Zech 14:9 is *'echad*. It means *one* as well as *united* or *unified*.

[Strong's number 259, 258; also Gesenius (7) under, *'echad*; note use in Judges 20:8 & 1Sam 11:7, KJV; *"in one"* translated as "together" in Ezra 2:64; 3:9; 6:20; and "alike" in Ecc 11:6]

One In Greek

gp162» The Greek word one (*heis*) means according to *Thayer's Greek Lexicon*:

■ "a cardinal numeral, *one* ... in opposition to a division into parts ... to be united most closely (in will, spirit) ..."

■ According to the *Analytical Greek Lexicon* "heis" means: one, one virtually by union, etc.

■ The Greek text of the Old Testament used the Greek word *heis* for the Hebrew *'echad* in Deut 6:4.

One In English

gp163» In English the word "one" means according to *The Synonym finder*, by Rodale under "one": "single person or thing, unit...," and under "oneness," "has quality of being one, unity, singleness, sameness..."

■ In *Webster's Collegiate Thesaurus* under "unity," we find "the condition of being or consisting of one."

■ In *Roger's International Thesaurus*, 3rd ed. we find under, "89. Unity,": "state of Being One. — Nouns 1. unity, oneness, singleness..."

■ In a translation of Aristotle's *Metaphysics* by John Warrington (Everyman's Library No 1000) the words "unity" and "one" are used interchangeably (p. 117).

■ In *Webster's New Word Dictionary*, College Edition, under "unit": "1. the smallest whole number; one." And under "unity": "1. the state of being one; oneness; singleness; being united." The English word "unity" comes from the Latin word *unitas* which means: oneness.

One Versus Only

gp164» Thus we see in three different languages that "one" has very similar meanings. One means one, as in **singular** (one thing), and one means **unity**. "One" does not mean "only." Hebrew has a special word for only, *yachiyd* (Strong's #3173). This Hebrew word is mostly translated as "only" in the Old Testament (Gen 22:2, 12, 16; Jud 11:34; Zech 12:10; etc.). In Greek there is also a word for only, *monos* (Strong's #3441). And of course English has a word for only.

Many in One

gp165» The ONE Yehowah does *not* mean only or alone. Scriptures such as "let **US** create man in **OUR** image" (Gen 1:26) indicate, there are more than a single person or entity in Yehowah. Other scripture project to us the same thing that there are more than one (single in number) in Yehowah (YHWH). The following plurals are correctly translated from the Hebrew and project the many-in-oneness of the God:

- "Yehowah, God**S**, look! the man has become like one of **US**" — Gen 3:22

- "Come, let **US** go and mix up their language" — Gen 11:7

- "the voice of the Lord**S** saying, Whom shall I send and who will go for **US**" — Isa 6:8

- "Yehowah, our God**S**, one Yehowah" — Deut 6:4

- "Yehowah, he, the God**S**" — Deut 4:35, 39; 7:9; 1Kings 18:39

- "Yehowah, you the God**S**" — 2Sam 7:28

- "Yehowah, he is God**S** in heaven above and earth below, there is none else" — Deut 4:39

- "that great (is) Yehowah and our Lord**S** above (#4480) all gods" — Psalm 135:5

- "your Creator**S**" — Eccl 12:1

- "Let Israel rejoice in his Maker**S**" — Psalm 149:2

- "For your husband, your Maker**S**, Yehowah of hosts" — Isa 54:5

- "knowledge of the Holie**S**" — Prov 9:10; 30:3

- "Yehowah God**S**, Holie**S** is he" — Joshua 24:19

- "Almightie**S**" or "Power**S**" — Gen 17:1; etc.

- "most High**S**" — Dan 7:18, 22, 25, 27

- "my lord**S**, Yehowah" — Isa 10:23; 25:8; 40:10; Jer 2:22; see Amos 5:14; Gen 18:27; Exo 4:10; Isa 6:1;'*adonay*="my lords"

Nation as One Man

gp166» The fact that in the Bible nations and groups of people are looked upon "as one man" helps us to understand the God's many-in-oneness:

- Then all the people of Israel came out, from Dan to Beersheba, including the land of Gilead, and the congregation assembled **as one man** to the LORD at Mizpah (RSV Judges 20:1).

- So all the men of Israel gathered against the city, united **as one man** (RSV Judges 20:11).

- When the seventh month came, and the sons of Israel were in the towns, the people

gathered **as one man** to Jerusalem (RSV Ezra 3:1).

- And all the people gathered **as one man** into the square before the Water Gate; and they told Ezra the scribe to bring the book of the law of Moses which the LORD had given to Israel (RSV Nehemiah 8:1).

Birth of One Son, as Birth of New Nation

gp167» The fact that the Bible looks upon the birth of one male child as the birth of a whole nation helps us to understand the many-in-oneness of the God:

- "[7] Before she was in labor she gave birth; before her pain came upon her **she was delivered of a son.** [8] Who has heard such a thing? Who has seen such things? Shall a land be born in one day? **Shall a nation be brought forth in one moment?** For as soon as Zion was in labor she brought forth her sons (RSV Isaiah 66:7-8).

Many in the One Body of Christ

gp168» This above mentioned use of ONE in "one Yehowah" and its meaning of, "unity" — or of many being united in the same spirit or quality, is also manifested to us in scripture about the **ONE body of Christ**:

- For as the body is one and has many members, but all the members of that one body, being many, are one body, so also is Christ. 13 *For by one Spirit we were all baptized into one body* — whether Jews or Greeks, whether slaves or free — and have all been made to drink into one Spirit. [1Cor 12:12, NKJV]

- Now you are the body of Christ, and members individually [1 Cor 12:27, NKJV]

- There is neither Jew nor Greek, there is neither slave nor free, there is neither male nor female; *for you are all one in Christ Jesus.* [Gal 3:28, NKJV]

gp169» This use of the word ONE also explains how Jesus Christ and God the Father are ONE and how real Christians are ONE in God and ONE in Christ:

- *I and My Father are one.* [John 10:30, NKJV]

- At that day you will know that I am in My Father, and you in Me, and I in you. [John 14:20, NKJV]

- And the glory which You gave Me I have given them, that *they may be one just as We are one*: [John 17:22, NKJV]

- If we love one another, *God abides in us* … [1John 4:12, NKJV]

- God is love, and he who abides in love abides in God, and God in him.. [1John 4:16, NKJV]

- By this we know that we abide in Him, and He in us, *because He has given us of His Spirit.* [1John 4:13, NKJV]

- "But by ONE Spirit we were all baptized into one body … Now you are the body of Christ …." [1Cor 12:13, 27]

- "But to us ONE God the Father, out of whom the all and we into Him, and ONE Lord Jesus Christ, through whom the all and we through him." [1Cor 8:6, from Greek text]

- "ONE Lord, ONE Faith, ONE baptism, ONE God and Father of all, the one upon all and through all and in all." [Eph 4:5-6, from the Greek]

Therefore: God, Jesus Christ, and Christians are ONE because they have the ONE Spirit of God — they are *united* (one) with the same Spirit.

gp170» Today, as in those days, we use "one" to mean "one in unity" as well as one as in singular of number. Yet because of tradition the so-called theologians seem to be unable to perceive the "one" Yehowah in any other way than singular of number. Because of this there is confusion concerning the nature of the God. But as we have seen there is some form of plurality in the unity or oneness of the true God, YHWH, the BeComingOne, who is our God(s).

Only God

gp171» Notice that Jesus Christ the man called his Father [YHWH, see GP 2] the "*only* true God" (John 17:3). But how is it that Jesus Christ is now the "*only*" God? (1Tim 1:17, Jude 1:4, 25) Jesus Christ in his own times will be the "*only*" ruler (1Tim 6:15) and now he is he "who *alone* ['only' — *monos*] has immortality" (1Tim 6:16). But also Jesus Christ was/is the "*only begotten* son" of God (John 1:18; see John 3:16, 18; 1John 4:9), but this should be translated, *one-of-a-kind* Son, because its first meaning is: *1) single of its kind.* Jesus was "unique" or "one-of-a-kind" because he is represented in the Holy of Holies, which was set apart (most holy) from all other aspects of God's

temple. As we see in the *New Mind Papers*, and as most Christians believe, there will be others who have and will obtain immortality and be born or begotten of God. Christ may at this time (2021) be the only one with immortality, but in time all others will be given immortality. The "only" aspect of God has meaning only in time and one's definition of who or what God is. Christ may be "only" now in some sense, but in **time** the Only One will share his qualities, so the only God will be all in all (cf 1:Cor 15:28). Remember, Jesus Christ is the "firstborn of all creation" (Col 1:16), he is the "firstborn from the dead" (Col 1:18 see 1Cor 15:20), he is the "beginning of the creation of the God" (Rev 3:14), he is "the beginning" (Col 1:18), he is the "first fruits Christ" (1 Cor 15:20, 23), and he is the "firstborn among many brethren" (Rom 8:29). Thus, Jesus was the first of many to come (GP 6). **Yet Jesus Christ the man, who was separate in a sense from his Father when he was a man on earth before his going to the Father (GP 4), is NOW the "only God"** (uniquely born– GP 5). Jesus Christ NOW is the only God. But look:

- "Jesus answered them, Is it not written in your law, I said, You are Gods? If he called them Gods, unto whom the word of God came, and the scripture cannot be broken..." [John 10:34-35; cf. Psa 82:5; 97:7]

gp172» There are/will-be more than one individual in the only ONE true God:

- "Yehowah" is the God(s). He is ONE. That is, ONE in Spirit. This ONE is the only true God (John 17:3). But He is not singular in number or as one individual. He is many in ONE Spirit. As Jesus Christ the man went into his Father (GP 5), who was, and is, and will-be the "only true God" (John 17:3; 1Cor 8:4, 6), and Jesus became *one* with that only true God, and thus became the only God (1Tim 1:17), so too will Christians and all others go into the Father and thus into the Son, at their appointed times (GP 6). Thus, all will go into the Spiritual Body of Christ and into the ONE Yehowah so that God will be all in all (1 Cor 15:28; Eph 1:23, 10; Phil 3:21; Col 1:20; see GP 6). Yehowah (YHWH) is the only true God (John 17:3 w/ GP 2; 2Kings 19:15). He alone knows the hearts of mankind (2Ch 6:30). He alone created the universe and everything in it and gave them life (Neh 9:6). He alone has the NAME Yehowah (Psalm 83:18). He alone dwells the cherubs (Isa 37:16). But he is not just single or alone as scripture in the Old Testament clearly point out in its original language: He is many in ONE. He is many in Unity as the Body of Christ is many in One. He is Yehowah the Gods (See above).

This may make little sense to you now, but after you read the rest of this book you will understand, especially with the New Mind.

Yehowah, Elohim

gp173» Before we continue let me explain something about the use of *Elohim* and *Yehowah* in the Bible. Remember, the Hebrew word "elohim" is the simple plural word for "el." The word elohim means godS. Most of the places in the KJV English Bible where you see "God," should read "Gods" since it was translated from the Hebrew word "elohim" which means godS. The first scripture in the Bible is, "In the beginning Gods created the heavens and earth," *not* "God created the heavens and the earth." But since in other verses of the Bible it says that Yehowah (LORD or Jehovah) created the heavens and earth (Isa 40:28; Ex 20:10), then the Hebrew word *elohim* has something to do with Yehowah.

gp174» In Christian D. Ginsburg's *Introduction to the Massoretico-Critical Edition of the Bible*, pages 368 to 369, he shows that in parallel verses in 2Samuel 5 and 1 Chronicles 14 that the words Yehowah ["LORD"] and Elohim [Gods] are interchangeable. 2Samuel 5 uses "Yehowah" while 1 Chronicles 14 uses "Elohim." Also in the book of Psalms the same phenomenon is detectable. And we can see throughout the Old Testament Yehowah ["LORD"] and Elohim are used together as follows: "LORD God" (KJV), but in the Hebrew it reads *Yehowah Elohim*. The literal translation of this would be the "*BeComingOne* (of) *Gods*," or "*He-(who)-Will-Be, Gods*."

gp175» We thus see that the "BeComingOne" is somehow connected with Gods. Now Gesenius, the great Hebrew grammarian, insisted that these two words (Yehowah Elohim) should not be translated as "Yehowah *of* Elohim" (*Gesenius' Lexicon*, under "YHWH"). But we see little difference between this usage and "Yehowah of HostS," or as in some English translations, "LORD of HostS," and "Yehowah of Elohim" or "BeComingOne of Gods."

gp176» As we mentioned above, Israel's *elohim* (gods) were/was the ONE Yehowah (YHWH). This is another reason the Hebrew word *elohim* (gods) is closely associated with Yehowah (YHWH).

gp177» The reason we are discussing this whole subject of God's names may not be clear to you now, but as you read on you will come to understand it, and by the time you finish this book it should make more sense.

Predestination, Time, NAME, and the Paradoxes

gp178» As mentioned earlier in this Part [GP 1], our awareness of predestination, time, and God's NAME gives us the secrets to understanding the paradoxes of God. Because of the Law of Contradiction we know that God cannot *at the same time* be love and also a creator of evil or a killer. We have learned that God's NAME — the BeComingOne (YHWH) — is from an imperfect or incomplete Hebrew verb. God's NAME tells us that the God is Becoming, that He-will-be, that His full essence is not yet complete. Therefore, in <u>time</u> the true God will come to be; and in <u>time</u> all that is said about the YHWH (the BeComingOne) in the Bible **will-be**, or will happen. Thus, it is possible, because of the true meaning of God's NAME, that God has/will have created evil and was/is/will-be all good without being evil and without being all good *at the same time*. God's NAME allows God, through his predestinated power, to create evil before creation and separate it through time as different sides of God until the end when the BeComingOne has become, or until the BeComingOne has been made complete, or until the full essence of God comes to be, or until God is all in all (GP6). Remember it is the scriptures that have said that YHWH made evil, killed, etc. But it is also scripture that says God predestinated events <u>before</u> the cosmos (Eph 1:4; 1Pet 1:19-20; 2Tim 1:9; Titus 1:2) and therefore before time (as we know it), before good (as we know it), before evil (as we know it), before law (as we know it), and consequently before sin (as we know it). So <u>before</u> creation (as we know it) when God predestinated good things and evil things, there was no sin because there was no law and no creation. You therefore cannot put sin on God because of predestination. Do read on.

We Will Use "BeComingOne" in GP

gp179» Before we begin the next part of this book let me mention first something about the NAME of God. The NAME of God as we have shown was *Yehowah* from the Hebrew, which has the meaning of, the "BeComingOne." In many English translations of the Bible it has the "BeComingOne" translated as either "LORD" or "Jehovah." For example in the King James Version (KJV) of the Bible it translates God's NAME as "LORD" (usually small capital letters). Since this book uses the King James Version for some of its quotation of Biblical scriptures, when you see "LORD," instead of "Lord," in this paper you know it is the very NAME of God, that is, it was translated from the Hebrew word *Yehowah*, which means: the BeComingOne. Hereafter, in this set of papers we will translate God's NAME as the "BeComingOne." We thus translate the *meaning* of God's NAME, for the meaning of God's NAME is the secret in answering the paradoxes of God. Do read on!

I Will Be in Context

gp180» From the Hebrew text, the Hebrew word, אֶהְיֶה , should always have been translated into English as, "I will be." The following English quotes were taken from the King James Version (KJV) of the Bible which proves that most of the time this word was translated as "I will be" except in Exodus 3:14. Never was it translated in the KJV as "I am" except in Exodus 3:14.

Exod 3:12 "I will be" אֶהְיֶה
Exod 3:14 "I am" or "I am that I am" [should be, "I will be that I will be"]

אֶהְיֶה אֲשֶׁר אֶהְיֶה

Exod 4:12 "I will be" אֶהְיֶה
Exod 4:15 "I will be" אֶהְיֶה
Deut 31:23 "I will be" אֶהְיֶה
Jos 1:5 "I will be" אֶהְיֶה
Jos 3:7 "I will be" אֶהְיֶה
Jdg 6:16 "I will be" אֶהְיֶה
Jdg 11:9 "I will be" אֶהְיֶה
Ruth 2:13 "I am not" from the Hebrew "not I will be" לֹא־אֶהְיֶה
1Sam 18:18 "I should be" אֶהְיֶה
1Sam 23:17 "I shall be" אֶהְיֶה
2Sam 7:14 "I will be his Father" אֶהְיֶה
2Sam 15:34 "I will be" אֶהְיֶה
2Sam 16:18 "I will be" or "will I be" אֶהְיֶה
2Sam 16:19 "will I be" אֶהְיֶה
1Chr 17:13 "I will be his Father" אֶהְיֶה
1Chr 28:6 "I will be his Father" אֶהְיֶה
Job 3:16 "I had not been" from the Hebrew "not I will be" לֹא־אֶהְיֶה
Job 10:19 "I had not been" from the Hebrew "not I will be" לֹא־אֶהְיֶה
Job 12:4 "I am" from the Hebrew "I will be" אֶהְיֶה
Job 17:6 "I was" should be from the Hebrew "I will be" אֶהְיֶה
Ps. 50:21 "I will" אֶהְיֶה
Song 1:7 "should I be" אֶהְיֶה
Isa 3:7 "I will not" from the Hebrew "not I will be" לֹא־אֶהְיֶה
Isa 47:7 "I shall be" אֶהְיֶה
Jer 11:4 "I will be your God" אֶהְיֶה
Jer 24:7 "I will be their God" אֶהְיֶה
Jer 30:22 "I will be your God" אֶהְיֶה
Jer 31:1 "I will be God of all" אֶהְיֶה
Jer 32:38 "I will be their God" אֶהְיֶה
Ezek 11:20 "I will be their God" אֶהְיֶה
Ezek 14:11 "I may be their God" אֶהְיֶה
Ezek 34:24 "And I Yehowah will be their God" אֶהְיֶה
Ezek 36:28 "And I will be your God" אֶהְיֶה
Ezek 37:23 "And I will be their God" אֶהְיֶה
Hos 1:9 "I will not" from the Hebrew "not I will be" לֹא־אֶהְיֶה
Hos 14:5 "I will be" אֶהְיֶה
Zech 2:5 "I ... will be" אֶהְיֶה
Zech 8:8 "And I will be their God" אֶהְיֶה

Web Page Links to Biblical Language Aids/Helps (in the digital books):

Link to S.R. Driver's Hebrew Tenses (2rd Ed)

Note: The 3rd Ed. was used in this book; page numbers are different

Link to Gesenius' Hebrew and Chaldee Lexicon

Link to Gesenius' Hebrew Grammar

Link to International Standard Bible Encyclopedia

Link to Hebrew Bible Biblia Hebraica Stuttgartensia

Link to Hebrew Bible Westminster Leningrad Codex

Link to Introduction to the Masoretico-Critical Edition of the Hebrew Bible by Christian D. Ginsburg

Review of GP 1

gp181» In GP 1 we started our search: who or what is God? From the Bible we learned about the apparent paradoxes of God: "I make peace, and create evil: I the Lᴏʀᴅ do all these things" (Isa 45:7). God who is Love (1John 4:8) has somehow and for some reason created evil; He has even killed (Deut 32:39). But how can God be Love and also a killer?

We next learned that there are two basic laws and one basic fact we must understand in order to rightly perceive the true nature of God: the Law of Contradiction and the Law of Knowledge plus the fact that the God cannot lie.

We then went on and explained the Law of Contradiction.

We further showed the many attributes and titles of God and put forth that "time" is very important in our understanding of the paradoxes of God.

We also showed you the very Nᴀᴍᴇ of the true God: ʏʜᴡʜ, or Jehovah, or Yehowah, or He (who) will-be, or the BeComingOne, or the One who was, who is, and who is coming. God's Nᴀᴍᴇ and its meaning is the real secret in revealing the answer to the Paradoxes of God. God's Nᴀᴍᴇ is an *imperfect* (incomplete) verb and not as would be expected a *perfect* (complete) verb or a noun. Names are very important in the Bible and many times describe some facet of a person. The true Nᴀᴍᴇ of the true God is important for it is the secret in explaining the apparently unexplainable scriptures about God.

In GP 1 we also looked into the meaning of "with God all things are possible," the "*one Yehowah*," the so-called unchangeableness of God, and other matters concerning the God. What GP 1 does is set the stage in our search for who or what is God.

GP 2: God The Father

Jesus Christ's Father

gp182» Who is the BeComingOne (YHWH) of the Old Testament, and who is God the Father? We must note again that the translation of "LORD God" in the Kings James Version of the Bible and other translations of the Bible is incorrect. Transliteration from Hebrew should read *Yehowah Elohim* in most cases. A translation of the literal meaning would be the "*BeComingOne* (of the) *Gods,*" or "*BeComingOne, (the) Gods,*" or "He (Who) will-be, (the) Gods" (see GP 1).

First Proof

gp183» Jesus was speaking to some Jews who had accused him of being possessed with a demon and making himself greater than Abraham by his words. Christ's answer is significant, for he reveals something important in it:

- "Jesus answered, If I honor myself, my honor is nothing: it is my Father that honors me; of whom you say, that he is your God" (John 8:54).

gp184» Notice Christ says his Father is the God that they, the Jews, say is their God. Now the Jews believe that their God was the "BeComingOne God(s)" or "Yehowah Elohim" or as mistranslated by some "LORD God" of the Old Testament (Psalm 140:6; Lev 18:30; 1Chron 29:10). And Jesus said his Father is that God (John 8:54; cf. Rom 15:6; 1Cor 8:6; 2Cor 1:3; 11:31; Eph 1:17; Phil 2:11; 1Peter 1:3). Therefore Jesus Christ's Father was the God of the Jews, and the Old Testament called the God of the Jews, Yehowah (YHWH).

Six More Proofs

gp185» Let's continue to prove that the BeComingOne of the Old Testament was the Father and is the ONE BeComingOne (Deut 6:4). We will give six more proofs besides John 8:54 that show that the BeComingOne of the Old Testament is Christ the man's Father.

God Swore By Himself

gp186» **(1)** "For when God made promise to Abraham, because he could swear by no greater, he swore by himself" (Heb 6:13). Now the God Paul was speaking about here was the BeComingOne (Gen 22:16; Isa 45:23). Paul said there was no greater than the BeComingOne of the Old Testament. He, the BeComingOne, was the greatest. Of course the BeComingOne was the greatest, for he was Jesus Christ the man's Father (John 8:54). Jesus Christ the man said his Father was the greatest of ALL, even greater than Jesus the man: "my Father who has given them to Me is greater than all ... I am going to the Father, for my Father is greater than I" (John 10:29; 14:28).

Throne

gp187» **(2)** Christ the man by a statement in Matthew 5:34 said God's (implying his Father's) throne was heaven, and in Isaiah 66:1 we see the BeComingOne calling heaven his throne. This is another proof that Christ's Father and the BeComingOne of the Old Testament were one and the same.

Prayer

gp188» **(3)** Now Christ taught that we should pray to our Father in heaven (Matt 6:6, 9-15). And Christ said his Father was the God of the Old Testament (John 8:54). Thus, we see Daniel praying to the BeComingOne, "And I [Daniel] prayed unto the BeComingOne my God and made my confession..." (Dan 9:4). Daniel and the rest of the others of the Old Testament prayed to the BeComingOne (note Jer 32:16-18), for he was in a sense their Father (Isa 63:16). We (Spiritual Israel) pray to our Father, who is the BeComingOne, the true God mentioned in the Old Testament, as physical Israel prayed to the BeComingOne, who was their Father (see # 5 below).

God The Father Chose

gp189» **(4)** In the New Testament it speaks of God the Father choosing people to be his sons through Jesus Christ (Eph 1:3-5). And since it is the Father who chooses, so does the BeComingOne of the Old Testament, for both the Father and the BeComingOne are the same being (note Isa 44:1-2; 43:10; 49:7; Psa 89:3; 105:43; 106:4, 5, 23; see "Predestination Paper" [NM8]).

YHWH *Of Old Testament Is The Father*

gp190» **(5)** The BeComingOne is called the Father in the Old Testament, and calls himself the Father: "You, O BeComingOne, art our *Father*, our Redeemer"(Isa 63:16). "For I am a *Father* to Israel, and Ephraim is my first-born" (Jer 31:9). "But now, O BeComingOne, you art our *Father*; we are the clay, and you our potter; and we all are the work of your hand" (Isa 64:8). "Thus, says the BeComingOne, the Holy One of Israel, and his Maker, Ask me of things to come concerning my *sons*, and concerning the work of my hands command you me" (Isa 45:11). "And David said, Blessed be you, BeComingOne of Israel our *Father*, from the age and to the age"(1 Chron 29:10). "He shall cry unto me [the BeComingOne], You art my *Father* my God, and the rock of my salvation" (Psa 89:26). Compare this with such verses as John 20:17. "I will be his *Father*" (2Sam 7:14, 1-29). Compare in context Psalms 2:7 with Hebrews 1:1, 5 and Psalms 110:1 with Hebrews 1:1, 13.

"See" The Father?

gp191» **(6)** Now some will say that God the Father could not be the God of the Old Testament, for scripture says that no one has seen the Father (John 1:18; 5:37). Since some did "see" the God of the Old Testament (Moses "saw," Deut 5:4; 34:10), this is proof that Jesus Christ's Father is not the BeComingOne of the Old Testament. But this is wrong, for did Christ say *no* one had seen his Father?

gp192» "And the Father himself, which has sent me, has borne witness of me. *You* have neither heard his voice at any time, nor seen his shape. And you have not his word abiding in you: for whom He has sent, him you don't believe" (John 5:37-38).

gp193» Notice verse 38 that the ones ("You") Jesus was speaking to didn't have the word abiding *in* them. Now in 1 John 2:14 we see that real Christians do have the word of God in them. John is writing to Spiritual Christians and says, "I have written unto you, young men, because you are strong, and the word of God abides *in* you" (1John 2:14). Hence, we know that Jesus was speaking to non-Spiritual people when he spoke in John 5:37-38.

gp194» Notice carefully: "Not that any man has seen the Father, except he which is of God, he has seen the Father" (John 6:46). "If you had known me, you should have known my Father also: and from henceforth you know him, and have seen him" (John 14:7). "He that is of God hears God's words: you therefore hear them not, because you are not of God" (John 8:47). We see that those of God are able to "see" the Father, at least in a Spiritual sense. Because God is spirit, then those of God can/will "see" God at least Spiritually. And soon they will see God as he is, and in the truest possible sense (see GP 10). Those who "saw" God in the Old Testament saw him in a vision or transfiguration (Deut 34:10; Num 12:8).

Outside of visions, no one had seen the true God in a physical sense (except to see Jesus Christ, who is the image of God, see GP 5, GP 10), because the true God is Spiritual and because the true God is the BeComingOne [He (who) Will-Be], whose completeness is yet to be manifested. Also see "Can we see spirits?" in GP 3. Eventually the true God will incorporate the entire new creation into Himself (1Cor 15:28).

gp195» God the Father is the BeComingOne (YHWH) of the Old Testament as shown herein.

GP 3: Angels, Spirits, and the WORD of God

What are Angels?
Two Kinds of Angels
Angels Closely Associated with God
NAME given to the Angel
Cherubs and the Name
Word of God
Can we see Spirits?

gp196» In order to continue our study on God we need to know something about angels, spirits, and the WORD of God. What are angels? What is spirit? Can we see spirit? What was the WORD (John 1:1) in the age before the resurrection of Jesus Christ? We say what was he before Christ's resurrection, for after it a new dimension was added to the WORD's make-up (GP 5).

What are Angels?

Angels are Spiritual Messengers or Word Carriers

gp197» The word "angel" is translated from a Hebrew word (*malak* # 4397) that means *messenger* and from a Greek word (*aggelos* # 32) that means *messenger*. An angel is a messenger. An angel brings the words of someone else. An angel or messenger is a spiritual being (Heb 1:7). An angel is a spiritual being who brings messages or words from someone else. All angels or spirits were created by God the Father (Psa 148:2,5; Col 1:16; Heb 12:9). A few verses seems to indicate that the angels ("sons of God") existed at the beginning of creation (Job 38:4-7; Gen 1:14-19; Rev 1:20).

Two Kinds of Angels: Good & Evil

gp198» There are two kinds of angels:

■ **Good Angels:** The good angels are holy angels (Mat 25:31; Mark 8:38). They are called elect angels (1Tim 5:21). They cannot die (Luke 20:36). They have great power and serve God's will by listening to his voice and carrying out his commandments (Psa 103:20-21). The age before Christ was subjected to angels (Heb 2:2,5; Acts 7:53) and they spoke through the fathers and prophets of Israel (Heb 1:1). They are spirit, or that is, made from spirit (Heb 1:7) and thus are invisible (Job 4:15-16), except when seen in visions (see below). They will come with Christ at the end of the age to resurrect God's saints (Mat 25:31; 24:31; Mark 8:38; Luke 9:26; Rev 14:3-4). There was a special angel mentioned in the Bible that carried the very words of the BeComingOne (see below).

■ **Evil Angels:** There are evil angels who carry words and thoughts of evil and they are testing mankind in the old age (NM20, "Other Mind"). These angels were appointed to a hell-fire judgment before the cosmos began (Mat 25:41; Jude 1:6,13; Rom 9:22; Prov 16:4; 1Pet 2:8; 2Pet 2:4; NM8; NM9; NM24). Adam and Eve were tested by an evil angel (Gen 3:4-7; Rev 20:1-2; 12:9). An evil angel called Satan destroyed Israel and tested Job (1Chron 21:1,12; Job 1:6-12; 2:6). For some reason this evil angel is called "angel of the LORD" (1Chron 21:12). Satan even tries to pass himself off as "an angel of light" (2Cor 11:14). But this angel is associated with God's anger or God's wrath, and as shown in chapter 1 and the God's Wrath Papers (PR4 to PR5) This angel is in some way the "left side" of God who was predestinated <u>before</u> creation began, and in that sense he is an "angel of the LORD" carrying out the works of the "left side" of God. As the good angels are invisible, so are the evil angels (Num 22:22-31).

gp199» Before we continue our study on angels we need to remember that "angels" were associated with the WORD of God in the Old Testament including the giving of the Ten Commandments (Heb 2:2; Acts 7:38,53, Gal 3:19). Apparently the old world before Jesus Christ was subject to angels (Heb 2:1-5). In the new age we are to be subjected to Christ (Heb 2:5-8). As of now (before Christ's return) all are not yet subjected to Christ (Heb 2:8), but in the near future all will be put under Christ and his rule (Psa 110:1; Acts 2:32-36; Heb 2:8-10; 1Cor 15:23-28; see GP6). There is confusion about this. We need to understand how the WORD was given through angels in the Old Testament, and the connection of these angels to the LORD, or Jehovah, or the BeComingOne of the Old Testament.

gp200» There was a very close connection between the angels of the Bible and the BeComingOne (YHWH) or the "LORD" of the Bible. Sometimes it is very difficult to see the difference between the "angel of the BeComingOne and the BeComingOne himself.

Angels Closely Associated with God

Angel of the Lord

gp201» Who was the "angel of the LORD [YHWH]" or more correctly the "angel of the BeComingOne" in the Old Testament, and in a few places in the New Testament of the Bible, the "angel of the Lord"? In the New Testament, the word "Lord," the Greek *Kurios*, is the word used instead of the "BeComingOne" or YHWH. The "angel of *Kurios*" translated as the "angel of the Lord" in many versions of the New Testament equals the "angel of the BeComingOne" of the Old Testament. One of the reasons *Kurios* was used in the New Testament instead of YHWH, was because of the *Septuagint* (See GP 1 under "More Details").

Hagar and the Angel

gp202» Now the *angel* of the BeComingOne was talking to Hagar, and right after the angel had spoken to her the verse reads: "And she called the name of the BeComingOne [YHWH] that spoke to her, *You God sees me*: for she said, Have I also here looked at him that sees me?" (Gen 16:13) Check the English translation like the Moffatt, NEB, etc. and the Greek Septuagint versions, they all say the same thing. Literally from Hebrew, "she called the NAME, YHWH, the one speaking to her: you God [el] of vision."

gp203» After the angel spoke to her, Hagar called the NAME of the BeComingOne who spoke to her *El roi* or "God of seeing," and asked herself if she had looked on the God of seeing or vision. She also named the well she was standing by, "*Beerlahairoi*," or "the well of the living one who sees me." In these scriptures the angel of the BeComingOne, the God of sight, and the NAME of the BeComingOne are closely connected. Why?

gp204» "And God [elohim] heard the voice of the lad; and the *angel* of God [elohim] called to Hagar out of heaven, and said unto her, What ails you, Hagar? fear not; for God [elohim] has heard the voice of the lad where he is" (Gen 21:17). The angel of God? God heard, and the angel of God called. In these verses the angel of God and God are closely connected.

Abraham and the Angel

gp205» "And the *angel* of the BeComingOne called unto Abraham out of heaven the second time. And said, By myself I have sworn" (Gen 22:15, 16). What? The angel of the BeComingOne called to Abraham and said, "By myself I have sworn." Notice in the New Testament where Paul describes the same event: "For when God made promise to Abraham, because he could swear by no greater, he swore by himself" (Heb 6:13). Does Paul call this *angel* of the BeComingOne, God? The answer is no, as we will see. But by comparing both verses it appears that way. Here again the angel and God are closely connected.

Jacob and the Angel

gp206» Now Jacob was blessing Ephraim and Manasseh when he said: "the *angel* which redeemed me from all evil bless the lads" (Gen 48:16). Jacob is asking an angel to bless his lads; he says this angel is "the redeemer." Genesis 48:16 in the KJV is a mistranslation. It should read: "*the* angel *the* redeemer of me from all harm, may he bless *the* boys." "The redeemer" in Gen 48:16 is the same Hebrew word as Isa 49:7, "Thus says the BeComingOne, (the) Redeemer of Israel," except with the additional article, "the." But who is *the* redeemer? "I have mercy on you says the BeComingOne your Redeemer" (Isa 54:8). "I will help you [Jacob], says the BeComingOne, your Redeemer, the Holy One of Israel" (Isa 41:14). Jacob in Gen 48:16 apparently spoke of an angel which redeemed him from evil, but the BeComingOne said in Isaiah 41:14 that He was that redeemer. Since the BeComingOne is the redeemer (Isa 41:14; 49:7; 54:8), then the angel (the messenger or agent) of the BeComingOne is the agent *of* the redeemer and his redemption.

Moses and the Angel

gp207» "And the *angel* of the BeComingOne appeared to him in a flame of fire out of the midst of a bush: and he [Moses] looked, and, behold, the bush burned with fire, and the bush was not consumed" (Ex 3:2). (In other words this "angel of the BeComingOne" looked like he was on fire much like Christ's face appears in the pictures of his glory, Matthew 17:2.) "And when the BeComingOne saw that he turned aside to see, God called to him out of the midst of the bush, and said, Moses, Moses ... Moreover he said, I am the God of your father, the God of Abraham, the God of Isaac, and the God of Jacob. And Moses hid his face, for he was afraid to look upon God" (Ex 3:2, 4, 6). Do not these verses seem to say the *angel* of the BeComingOne appeared in the midst of the bush (V. 2), and he who called out of the bush was God (V. 4). In fact, this angel of the BeComingOne was apparently the God (elohim) of Abraham, Isaac, and Jacob.

There is a very close connection between the angel and the BeComingOne.

gp208» This is confirmed in Acts 7:30-33: "And when forty years had passed, an *Angel* of the Lord [YHWH] appeared to him in the desert of Mount Sinai, in a flame of fire in a bush. And Moses saw and wondered at the sight. And as he was coming near to look, a voice of the Lord came to him: 'I am the God of your fathers, the God of Abraham and the God of Isaac and the God of Jacob.' " Here it says Moses drew near to a bush where an angel of the Lord appeared, and the *voice* of the BeComingOne (YHWH) came out of it, and said that he was the God of Abraham, Isaac, and Jacob. Furthermore in an inspired speech by Stephen we read, "This is he [Moses], that was in the church in the wilderness with the *angel* which spoke to him in mount Sinai, and with our fathers: who received the lively oracles [Ten Commandments] to give unto us" (Acts 7:38). Notice that Stephen had already spoken about an angel of the Lord who appeared in a bush in verse 30, but now he says it was an angel who *spoke* to Moses in mount Sinai when the commandments were given.

Balaam and the Angel

gp209» Notice that the Bible's rendition of the angel and Balaam's ass. In Numbers 22:22, 23, 24, 25, 26, and verse 27 it shows the *angel* of the BeComingOne standing in the way of the ass that Balaam was riding. Then in verse 28: "and the BeComingOne opened the mouth of the ass...." And verse 31, "then the BeComingOne who was there with the ass and Balaam, and he opened the eye of Balaam and the mouth of the ass."

gp210» Notice verse 35, "and the *angel of the BeComingOne* said unto Balaam, Go with the men: but *only the word that I shall speak unto you, that you shall speak.*" And, "and the BeComingOne put a word in Balaam's mouth..." (Num 23:5). And again, "must I not take heed to speak that which the BeComingOne has put in my mouth?" (V. 12) Further, "and the BeComingOne met Balaam, and put a word in his mouth.." (V. 16). Compare Numbers 22:35 with Numbers 23:5, 12, and 16. The "angel of the BeComingOne" and the "BeComingOne" are used interchangeably in these verses because the angel is the agent of the BeComingOne as the messenger of a general is the agent for that general. Also notice that this angel is somehow "for an adversary [Satan] against" Balaam. This cannot be understood without knowing the significance of the cherubs in the holy of holies, which we discuss in later chapters of this work. Again we see the close connection between the angel of the BeComingOne and the very BeComingOne.

Gideon and the Angel

gp211» "And the *angel* of the BeComingOne appeared unto him [Gideon], and said unto him, The BeComingOne is with you" (Judges 6:12). Notice the angel of the BeComingOne appeared. "And the BeComingOne looked upon him" (V. 14). And again, "and the BeComingOne said unto him" (V. 16). The angel of the BeComingOne appeared, but the BeComingOne looked and talked. And further in verse 20 it says: "and the angel of God said unto him." And again, "then the angel of the BeComingOne put forth the end of the staff that was in his hand... Then the angel of the BeComingOne departed out of sight. And when Gideon perceived that he was an angel of the BeComingOne, Gideon said, Alas, O Lord(s) BeComingOne [YHWH — not 'GOD']! for because I have seen the angel of the BeComingOne face to face" (V. 21, 22). Again the angel or messenger of the BeComingOne and the BeComingOne (YHWH) are closely connected.

Manoah and the Angel

gp212» "And an *angel* of the BeComingOne appeared unto the woman ... then the woman came and told her husband, saying, A man of God [thus, the angel of the BeComingOne, looked like a man] came unto me and his countenance was like the countenance of an angel of the God ... the angel of the God came again unto the woman ... her husband was not with her. And the woman ... ran, and showed her husband ... And Manoah [her husband] arose and went after his wife, and came to the man [or angel, V. 6] ... And the angel of the BeComingOne said unto Manoah ... And Manoah said unto the angel of the BeComingOne" (Judges 13:3, 6, 9, 11, 13, 15).

gp213» Remember this angel of the God/BeComingOne looked like a man with an appearance like the angel of God. In verse 15 Manoah asks the "man" to stay and he will fix him something to eat. In verse 16 the angel of the BeComingOne declines the offer, but tells the husband to offer the food to the BeComingOne, "for Manoah knew not that he [the 'man'] was an angel of the BeComingOne" (V. 16). Next the husband asks the "man's" name, but the angel says it is secret (V. 17 & 18). Then Manoah offers his kid of the goats to the BeComingOne, but at that time the angel of the BeComingOne did an amazing work, for he ascended as a flame into heaven (V. 19 & 20). In the Greek translation, the *Septuagint*, of the Old Testament, the Greek words indicate that the angel disunited or separated in form when he did this amazing act of ascending in a flame. This must have amazed Manoah and his wife for they didn't know it was an angel since he looked like a man. "But the angel of the BeComingOne did no more appear to Manoah and his wife. Then Manoah knew that he (the 'man') was an angel of the BeComingOne. And Manoah said unto his wife, We shall surely die, because we have *seen God*. But his wife said unto him, If the BeComingOne were pleased to kill us..." (V. 21-23). Again the angel or messenger of the BeComingOne and the BeComingOne (YHWH) are closely connected.

Jacob and the Angel

gp214» "And Jacob was left alone; and there wrestled a *man* with him until the breaking of the day" (Gen 32:24). In Hosea 12:2, 4 it identifies this "man" Jacob wrestled with as an *angel*. Thus, Jacob was wrestling an angel (in a dream?) who looked like a man. Jacob asks this angel his name, but the angel asks a rhetorical question and then blesses him (Gen 32:29). Jacob then calls the place where he wrestled with the angel/man — *Peniel*, which means "the face of God." Jacob calls this angel/man, *Peniel*, for he saw God face to face: "I have seen God face to face, and my life is preserved" (verse 30, cf Ho 12:3, 4). This event is of a dual significance because it also prophesies of Jacob [Israel, the true Church] until the breaking of the great day of the Lord when Jacob will be redeemed and see God as he is (1 John 3:2). Again the angel or messenger of the BeComingOne and the BeComingOne (YHWH) are closely connected.

Angel of God's Presence

gp215» In Isaiah it speaks of "the angel of his [the BeComingOne's] presence" who saved Israel, and with his love and pity he redeemed them and carried them all the days of old (Isa 63:9, 7-8). Now the angel of the BeComingOne's presence redeemed Israel. Jacob [Israel] spoke of this same angel (Gen 48:16), a redeeming angel who blesses. This same angel is described in Genesis 32:24-30, and is the angel who changed Jacob's name to Israel. As explained before, this angel looked like a man, but Jacob said he saw the face of God. Further we've shown in Judges 13:1-23 that Manoah and his wife had seen an angel of the BeComingOne who looked like a man, who then transformed himself into the flame of fire and ascended into

heaven, *and* they said that seeing this angel of the BeComingOne/God, who looked like a man, was like seeing God. We've also shown that the redeemer is the BeComingOne of the Old Testament who is Jesus Christ's Father: "said the BeComingOne, and your Redeemer, the Holy One of Israel" (Isa 41:14).

gp216» Who is this angel of His presence? The angel of the BeComingOne is the answer. The Hebrew word translated in the King James Version as "presence" (*paniym*, # 6440) is in the plural form, but is used as if it were in the singular form. It means, face(s). It was the angel of the BeComingOne who led Israel out of Egypt and appeared to Moses (Exo 32:34; 33:14, 15; Isa 63:9; etc.). The angel of the BeComingOne's presence is the angel of the BeComingOne/God. He is described again in the New Testament as Gabriel.

Gabriel and the Angel

gp217» Now "Gabriel" means, *man of God*. Notice in Judges 13:1-25, that the angel of the BeComingOne/God that appeared to Samson's parents, was called God (V. 22), but he looked like a man (V. 11). Now Samson's mother, the wife of Manoah, called this angel of the BeComingOne, "a man of God" (V.6). The very word Gabriel means, man of God.

gp218» We have shown that the "angel of his presence" in Isaiah 63:9 is the angel of the BeComingOne. Further the angel Gabriel is the angel of the BeComingOne as we will show. As Satan is called many names in the Bible and as each name helps to describe some characteristic of Satan, so too is the angel of God called many names.

gp219» Note in Luke's first chapter that an angel of the Lord appears to Zechariah, the father of John the Baptist (V. 11, 12). This angel of the Lord in answering a question by Zechariah (V. 18), says to him: "I am Gabriel ["man of God"], that stand in the presence of God..." (V. 19).

gp220» The angel of the Lord (YHWH) says he is Gabriel. And Gabriel means, "man of God." And the angel of the BeComingOne/God of the Old Testament is called "a man of God" (Judges 13:6) because he looks like a man (Gen 32:24; Jud 13:6, 11). Furthermore, it is Gabriel who stands in the presence of God as does the angel of Isaiah 63:9. Let's go back to the Old Testament for a moment to further help connect the BeComingOne of the Old Testament with the angel of the BeComingOne.

Joshua, Satan, and the Angel

gp221» Notice in Zechariah 3:1 that Joshua was standing before the angel of the BeComingOne and Satan was there too. Verse one tells us Joshua and Satan were standing before the angel of the BeComingOne. But in verse two, "and the BeComingOne said to Satan, The BeComingOne rebuke you, O Satan." And in verse three it again tells us they were standing "before the angel." And in verse six it is the angel of the BeComingOne speaking, but in verses seven and nine it says the BeComingOne spoke. Reading this, one has to almost conclude, that the "angel of the BeComingOne" and the "BeComingOne" are one and the same person. The Bible doesn't say the angel of the BeComingOne each time but alternates with either angel of the BeComingOne or BeComingOne. With what has already been shown you, we know that there is a very close connection between the angel of the BeComingOne, and the BeComingOne.

Moses. Satan, and Michael the Archangel

gp222» Notice the BeComingOne or the angel of the BeComingOne says to Satan, "The BeComingOne rebuke you, O Satan" (Zech 3:2). Compare this with: "Yet Michael the archangel, when contending with the devil [Satan] ... said, the Lord rebuke you" (Jude 1:9). Now Satan in Jude is contending over Moses while in Zechariah, Satan was contending about Joshua. But where in the Bible does it say Satan comes before a regular angel of heaven? In Job it says "there is a day when the sons of God came to present themselves before the BeComingOne and Satan came also among them" (Job 1:6; 2:1). Notice Satan is before the BeComingOne in Job, and is contending against Job at that time (V. 1:9, 10). While in Zechariah, it was Joshua he contended against, and in Jude it was Moses. Actually as Gabriel is another name for the angel of the BeComingOne, so too is Michael. The word "Michael" means, "who is God."

gp223» Notice Jude calls Michael the archangel. The word in Greek means, *chief* angel. The prefix of the Greek word translated "archangel" means, *chief*, or *beginning*, or *headship*, or *first* in place or time, or *prince*. Thus, Michael is the head-angel, or first-angel, or beginning-angel, or chief-angel, or prince angel. Notice what Michael is called elsewhere in the Bible, the "great prince" (Dan 12:1), and "your prince" (Dan 10:21). Michael is the great-angel, the prince-angel, the first-angel, the head-angel, the archangel. It is Michael who

"stands for the children of your [Daniel's] people" (Dan 12:1). Who are the children of Daniel's people? By verses 2 and 3 of the 12th chapter of Daniel and the *New Mind Papers*, we can perceive that these children are those who are resurrected at Christ's coming. Michael is the angel who will stand up and deliver the Spiritual Christians at the end of the age of Satan's misrule (See the *New Mind Papers*). The angel of the BeComingOne is Michael, the first-angel. Could another angel be above the angel of the BeComingOne. No, the angel of the BeComingOne is the chief-angel, the archangel.

gp224» In Daniel 12:1 it says that Michael will stand up for the children of God. But in Isaiah 40:10 it says the "BeComingOne will come with strong hand, and his arm shall rule for him: behold, his reward is with him...." Again there is a very close connection between the BeComingOne (YHWH) and the angel of the BeComingOne.

Job, Satan, and the LORD

gp225» A comparative survey of the phrase "sons of God" indicates that this expression is used in the Bible as not only meaning physical sons of God (those who are sons of Adam, for Adam was a "son," so to speak, of God), but also spiritual sons of God, or angels. Notice the following where the LORD spoke to Satan and even in some way directed or gave Satan permission to do certain things to Job:

■ Job 1:6 Now there was a day when the sons of God came to present themselves before the LORD, and Satan also came among them. 7 The LORD said to Satan, "From where do you come?" Then Satan answered the LORD and said, "From roaming about on the earth and walking around on it." 8 The LORD said to Satan, "Have you considered My servant Job? For there is no one like him on the earth, a blameless and upright man, fearing God and turning away from evil." 9 Then Satan answered the LORD, "Does Job fear God for nothing? 10 "Have You not made a hedge about him and his house and all that he has, on every side? You have blessed the work of his hands, and his possessions have increased in the land. 11 "But put forth Your hand now and touch all that he has; he will surely curse You to Your face." 12 Then the LORD said to Satan, "Behold, all that he has is in your power, only do not put forth your hand on him." So Satan departed from the presence of the LORD. (Job 1:6-12)

■ Job 2:1 Again there was a day when the sons of God came to present themselves before the LORD, and Satan also came among them to present himself before the LORD. 2 The LORD said to Satan, "Where have you come from?" Then Satan answered the LORD and said, "From roaming about on the earth and walking around on it." 3 The LORD said to Satan, "Have you considered My servant Job? For there is no one like him on the earth, a blameless and upright man fearing God and turning away from evil. And he still holds fast his integrity, although you incited Me against him to ruin him without cause." 4 Satan answered the LORD and said, "Skin for skin! Yes, all that a man has he will give for his life. 5 "However, put forth Your hand now, and touch his bone and his flesh; he will curse You to Your face." 6 So the LORD said to Satan, "Behold, he is in your power, only spare his life." 7 Then Satan went out from the presence of the LORD and smote Job with sore boils from the sole of his foot to the crown of his head. (Job 2:1-7)

In the book of Job it has the Lord speaking to Satan and directing him or giving him permission to do certain bad things to Job. We can look upon this as the BeComingOne's (YHWH) "predestinating permission" for Satan to do certain things in this age, things predestinated before the cosmos, before law, and before sin. This may be hard to understand, but a thorough reading of this book should help you to understand.

Joshua and the Chief-Angel

gp226» When Joshua was near Jericho he "saw a man" with a sword drawn who said he was the commander of the army of the BeComingOne (YHWH) (Josh 5:13-15). Joshua put his face down to the ground (a sign of worship) and asked, what my lord(s) is the message for his (YHWH) servant.

gp227» Who is this "commander of the army of the LORD" (NIV); or the "captain of the host of the LORD" (KJV); or the "prince of Jehovah's host" (Young's Literal Translation)? The Hebrew word translated commander or captain or prince is *sar* (# 8269) which means prince, head, chief, captain, general, etc. This is the same Hebrew word translated "*prince* of the host" (KJV) in Daniel 8:11, or "*prince* of princes" (KJV) in Daniel 8:25, or "prince" (KJV) in Daniel 11:21 and 12:1. This great prince of Daniel was Michael the angel.

gp228» The Hebrew word translated army or host by some is *Tseba* (Tsaba, # 6635) which means a "mass" of things, people, solders, angels, etc. It is the Hebrew word in such translations as "LORD of *Hosts*" (KJV) in such books of the Bible as Isaiah and Jeremiah.

gp229» Thus, this "man" that Joshua saw was the prince of the host of Yehowah, or the prince of the host of the BeComingOne. With the information shown to you in chapter 3, we see this "man" was the great chief angel of the BeComingOne: Michael.

Review: Angels Close Connection to the BeComingOne

We have seen angels closely associated with the BeComingOne in the cases of Hagar, Abraham, Jacob, Moses, Balaam, Gideon, Manoah, Gabriel ("man of God"), Joshua, Michael, and Job. So close are these associations that it is difficult to see if there is any difference between the angels and the very BeComingOne. But since we know that the true God was, is will be all in all (Rev 1:8; 1Cor 15:28), whose spirit is everywhere (Psa 139:7; Jer 23:24; 1Kings 8:27), and since we know that angels occupy location (a place in the cosmos versus all the cosmos), and are *messengers* of another one, then we know for these reasons and others that the angel of the BeComingOne is not in the truest sense the true God, the "BeComingOne" or the YHWH.

Angel of the BeComingOne (YHWH) is not the BeComingOne (YHWH)

gp230» So far in GP 3 we have seen the very close connection between the BeComingOne (YHWH) and the angel of the BeComingOne. In some scriptures they appear to be the same. But there are scriptures that indicate they are not the same.

(1) Notice that the "word of the BeComingOne" came to Zechariah (Zech 1:1). Now notice the prayer of the *angel* of the BeComingOne:

■ ZEC 1:12: Then the Angel of the BeComingOne answered and said, "O BeComingOne of hosts, how long will You not have mercy on Jerusalem and on the cities of Judah, against which You were angry these seventy years?" 13 And the BeComingOne answered the angel who talked to me, with good and comforting words.

The angel of the BeComingOne prayed to the BeComingOne (YHWH) and asked Him a question, and the BeComingOne answered. This is proof that the angel of the BeComingOne (YHWH) and the BeComingOne (YHWH) are not exactly the same.

(2) Another proof that the angel of God AND God are not exactly the same is that God is Spirit (John 4:24) and God's spirit fills all (Jer 23:24). Angels also are spirits (Heb 1:7), but do not fill all. Angels appear in locations and thus are not the fullness of God. No single angel in himself can be the true God, because the true God fills all. But angels can and do represent God, or speak for God: they are agents of God.

gp231» The very Hebrew word translated into angel in the Old testament, and the very Greek word translated into angel in the New Testament means, "messenger." The angel of the BeComingOne (YHWH) is the agent or ambassador of the BeComingOne (YHWH). A messenger of God brings the words of God. A messenger of a king brings the words of the king. The words that a messenger speaks in the name of a king, are the very words of the king. The message or words of the messenger (angel) of God are the very *words* of God. This is why there is a close connection in the Bible between the angel of God or the angel of the BeComingOne and the BeComingOne, for the angel of the BeComingOne brought the WORD of the BeComingOne.

NAME *Given To the Word/Angel*

gp232» One very important fact we need to know is that the angel of the BeComingOne (YHWH) was given the NAME of the true God:

- "Behold, I send an angel before you, to keep you in the way, and to bring you to the place which I have prepared. Beware of him and obey his voice, provoke him not; for he will not pardon your transgressions: FOR MY *NAME* IS IN HIM" (Exo 23:20-21, KJV).

- "See, I am sending an angel ahead of you to guard you along the way and to bring you to the place I have prepared. Pay attention to him and listen to what he says. Do not rebel against him; he will not forgive your rebellion, since my NAME is in him" (Exo 23:20-21, NIV).

gp233» This angel had the NAME (YHWH) in him. This angel with the NAME in him was the angel of the BeComingOne (YHWH) (compare Isa 63:9 and proof in GP 3). This angel spoke in the NAME of God; he spoke the BeComingOne's WORD.

Remember here that God's very Name is a verb, a verb in the imperfect tense – God very Name tells us that God is <u>Becoming</u> (GP1). This is the great hint that God's Name carries: He is BeComing. He is, He (who) will be. He is the BeComingOne.

Cherubs and the Angel between them

gp234» As we learn in Part 16 of the *New Mind Papers*, Moses' tabernacle was made according to the pattern of the heavenly tabernacle (NM16; Heb 9:1-9, 23-24). In the tabernacle there was a place called the holy of holies. The typical most holy of holies had two cherubs in it, and *between* these cherubs the BeComingOne used to appear and speak to Moses:

- "There I will meet with you; and from above the mercy seat, from <u>between the two cherubim</u> which are upon the ark of the testimony, I will speak to you about all that I will give you in commandment for the sons of Israel. (Exodus 25:22)

- Now when Moses went into the tent of meeting to speak with Him, he heard the voice speaking to him from above the mercy seat that was on the ark of the testimony, from <u>between the two cherubim</u>, so He spoke to him. (Numbers 7:89)

- The LORD said to Moses: "Tell your brother Aaron that he shall not enter at any time into the holy place inside the veil, before the mercy seat which is on the ark, or he will die; for <u>I will appear in the cloud over the mercy seat</u>. (Leviticus 16:2)

Since we know the BeComingOne is Spirit and his spirit fills all (or will fill all), then we know that the BeComingOne <u>himself</u> did not appear between the cherubs, but the angel or messenger or agent of the BeComingOne appeared. How do we know this? As we are seeing in GP3, there is an angel that is closely connected to the BeComingOne. So close, it is difficult sometimes to differentiate between them. This angel went with Israel and gave them the law through Moses (Ex 23:20ff; Acts 7:30, 38; Heb 2:2, 5; see all of GP 3). "The angel the one speaking to him [Moses] in the Mount Sinai" (Acts 7:38, Greek text). This angel was the one that spoke to Moses in the bush and with the fathers of Israel, and this angel is the one who gave the commandments to Moses in Mount Sinai (Acts 7:30,35,38). This angel is the angel with God's Name (YHWH) in him (Ex 23:20-21). The Spirit in Paul told us that in the past God had put mankind under angels (Heb 2:1-2, 5), but now he has put us under his Son (Heb 1:2; 2:5,8-10; 1Cor 15:23-28). God spoke to mankind through his angels in the Old Testament. Once we understand that the Spirits in the prophets, were Spiritual messengers or angels (Heb 1:1), then we further understand how angels ruled before Jesus Christ. This angel who spoke to Moses represented the God who is becoming, for he was the angel who spoke between the cherubs over the mercy seat. The cherubs and the mercy seat represent the BeComingOne (GP8).

Word of God

Word of the BeComingOne

Word: Spoken by an Angel or Spiritual Messenger

gp235» "For if the word [logos] spoken through angels proved steadfast, and every transgression and disobedience received a just reward ... For He has not put the world to come, of which we speak, in subjection to angels" (Heb 2:2, 5 ,NKJ).

gp236» The word or logos was spoken through angels. As Acts 7:38, 53 indicate, the ten commandments were also given through an angel. And this world was in subjection to angels as Hebrew 2:5 indicates. But this angel(s) spoke the very message or word of God, with the power that goes with these words. The words of God have power because all God's words come true (Isa 46:11, etc.).

- In Genesis 15:1, "the *word* of the BeComingOne came to Abram."

- In Exodus 24:3, 4, "And Moses came and told the people all the *words* of the BeComingOne ... And Moses wrote all the *words* of the BeComingOne ..."

- In Deuteronomy 5:4, 5, "The BeComingOne talked with you face to face in the mount out of the midst of the fire (I stood between the BeComingOne and you at that time, to show you the *word* of the BeComingOne ....") But as Exodus 3:2 and Acts 7:30ff say, it was the *angel* who appeared in the bush that appeared on fire. Therefore Moses gave Israel the word of the BeComingOne that Moses received from the *angel* of the BeComingOne.

- In Judges 2:4, "And it came to pass when the angel of the BeComingOne spoke these *words* ...."

- In Isaiah 1:10, "Hear the *word* of the BeComingOne."

- In Jeremiah 1:2, "To whom the *word* of the BeComingOne came ...."

- In Ezekiel 1:3, "The *word* of the BeComingOne came expressly to Ezekiel ...."

- In Hosea 1:1, "The *word* of the BeComingOne came to Hosea ...."

- In Joel 1:1, "The *word* of the BeComingOne came to Joel ..."

- In Jonah 1:1, "Now the *word* of the BeComingOne came to Jonah ...."

- In Micah 1:1, "The *word* of the BeComingOne came to Micah ...."

- In Zephaniah 1:1, "The *word* of the BeComingOne which came to Zephaniah ...."

- In Haggai 1:1, "...came the *word* of the BeComingOne by Haggai ...."

- In Zechariah 1:1, "... came the *word* of the BeComingOne to Zechariah ...."

gp237» In Zechariah 1:1 we see the Hebrew word *debar* translated "word" in English is also *logos* in the Greek translation. The same applies to Haggai 1:1, Zephaniah 1:1, Micah 1:1, Jonah 1:1, Joel 1:1, Hosea 1:1, Ezekiel 1:3, Jeremiah 1:2, and Isaiah 1:10.

gp238» The LOGOS or the WORD of the Old Testament was the *angel* of The BeComingOne (YHWH). **The messenger (or angel) of the BeComingOne (YHWH) is the WORD of the true God.** This Word was also described in the New Testament by John:

- "In the beginning was the WORD, and the WORD was with God, and the WORD was God. 2 He was in the beginning with God. 3 All things came into being through Him, and apart from Him nothing came into being that has come into being. 4 In Him was life, and the life was the Light of men." (John 1:1-4)

This translation is not literal, and it should read:

- John 1:1 In [the] beginning was the WORD, and the WORD was toward the God, and God was the WORD. 2 He was in the beginning toward the God.[1] 3 All things have being [aor] through him, and outside of him have being not one [thing] which have received being. 4 In him was life, and the life was the light of men. (John 1:1-4, BCB; see Notes GP5)

Therefore the WORD, or angel of the BeComingOne, was <u>toward</u> the true God. The angel was the messenger of the great coming one, the BeComingOne, and so his WORD was in reference and toward this great coming God, the God that will be all in all (GP6). Power was given to this WORD, this angel, to create the cosmos.

Can we see Spirits?

gp239» But can spirit ever be seen? Or is spirit absolutely invisible to physical eyes? What is vision anyway? Is it possible that angels or spirits cannot be seen by any kind of eye? Could it be that angels or spirits have no form? Could it be spirits are intelligent modes of energy or power?

Analogous to Burning Flames or Wind

gp240» Now spirits themselves have no flesh and bones (Luke 24:39). Further, spirits or angels are represented by burning flames (Heb 1:7). Burning flames have no particular form, but they do have position — they exist in a certain area. And Job 4:15-16 shows us spirits or angels have no form: "Then a spirit passed before my face ... but I could not discern the form thereof: an image was before mine eyes." He could not comprehend any form, yet there was a spiritual image before his eyes. In verse 8 of John 3 Jesus made an allegory between Spirit and the wind and says as one cannot see the wind so also he cannot see the spirit. Something that is spiritual is something that is not easily detectable by sight, touch, smell, or by our other senses. Spirit is analogous to air: you cannot see air but we see the effect of its wind; spirit is invisible but we see the effects on mankind's behavior: either good or bad.

gp241» Thus, it is impossible to actually physically see the *spiritual* essence. Remember God through Paul said we could figure out the invisible by what appears (Rom 1:20). God used burning flames to represent angels or spirits,

thus from this typical representation we should be able to learn something. And the something we learn, with the verses of Job 4:15-16, is that spirits or angels have no particular form, but are modes of energy or life or power. *Spirits or angels are physically formless modes of energy, or powers, with spatial location, with self-consciousness, and with mental ability.*

Satan a spirit had no Form

gp242» Notice where it describes Satan, in an antitypical way, that it says, "you were perfect *in your ways* from the day you were created, till iniquity was found in you" (Ezek 28:15). Satan was beautiful (Ezek 28:12) in his *ways* before his first sin; he was not beautiful in his physical appearance, for spirits have no physically manifestable form.

Review of GP 3

gp243» The Word (logos) of God before Christ's resurrection was a spirit, the chief-spirit, the chief-angel, the angel of the BeComingOne. Since "angel" is translated from the Hebrew word *malak*, which means messenger, then the angel of the BeComingOne is the messenger of the BeComingOne. A messenger brings words from someone else. Thus, the messenger of the BeComingOne is the "WORD of the BeComingOne," for he brought the very words of the BeComingOne. Since the BeComingOne (YHWH) is the proper NAME of the true God, then the WORD is the WORD of *the* God. The angel of the BeComingOne had the NAME of *the* God in him (Exo 23:20-21). The WORD of God or the angel of the BeComingOne represented and stood in the NAME of *the* God. The age before Jesus Christ was subjected to angels, even the commandments given on Mount Sinai were from an angel (Acts 7:38).

General Review

gp244» Now let's clarify what we have learned so far.

In GP 1 we started our search: who or what is God? From the Bible we learned about the apparent paradoxes of God: "I make peace, and create evil: I the LORD do all these things" (Isa

---

[1] "toward the God" cf John 13:3; "toward the Father" cf John 16:17; 16:28; 20:17, the Father being the God; see Greek text

45:7). God who is Love (1John 4:8) has somehow and for some reason created evil; He has even killed (Deut 32:39). But how can God be Love and also a killer?

We next learned that there are two basic laws and one basic fact we must understand in order to rightly perceive the true nature of God: the Law of Contradiction and the Law of Knowledge plus the fact that the God cannot lie.

We then went on and explained the Law of Contradiction.

We further showed the many attributes and titles of God and put forth that "time" is very important in our understanding of the paradoxes of God.

We also showed you the very NAME of the true God: YHWH, or Jehovah, or Yehowah, or He (who) will-be, or the BeComingOne, or the One who was, who is, and who is coming. God's NAME and its meaning is the real secret in revealing the answer to the Paradoxes of God. God's NAME is an *imperfect* (incomplete) verb and not as would be expected a *perfect* (complete) verb or a noun. Names are very important in the Bible and many times describe some facet of a person. The true NAME of the true God is important for it is the secret in explaining the apparently unexplainable scriptures about God.

In GP 1 we also looked into the meaning of "with God all things are possible," the *"one Yehowah,"* the so-called unchangeableness of God, and other matters concerning the God. What GP 1 does is set the stage in our search for who or what is God.

In GP 2 we learned that Jesus Christ's Father was the BeComingOne (YHWH) of the Old Testament: He was the Jews' God.

In GP 3 we learned that the angel of the BeComingOne and the BeComingOne of the Old Testament were closely connected. Since angels are messengers, this means the angel of the BeComingOne is a messenger of the BeComingOne or this angel is the WORD of the BeComingOne. Therefore, the words that the angel of the BeComingOne spoke belonged to the Great BeComingOne Power — the true God. This angel stood in the NAME of the true God (Exo 23:20-21); he represented the great NAME. This angel was in a sense the very WORD of God. The Word (logos) of God before Christ's resurrection was a spirit, the chief-spirit, the chief-angel, the angel of the BeComingOne. The age before Jesus Christ was subjected to angels, even the commandments given on Mount Sinai were from an angel (Acts 7:38).

=====

GP 4: Jesus Christ the Man

His Name
Promised Seed
God Inside
God Made Flesh
Death of Christ the Man
Jesus' Pre-existence Doctrine
Melchizedek
Genealogy of Christ

Who Was Jesus Christ

gp245» Who or what is Jesus Christ? Is he God, man, or man/God? Was he spirit, or human? Did he pre-exist before he was born? What does the Bible mean by saying he is the Son of God?

His Name

gp246» Let's begin with Christ's name. The Bible uses a name not only to point out a person among others, but also to describe that person. For example "Satan" means the hater, or accuser, or adversary. The name "Christ" tells us also something about him. "Christ" in the New Testament is a translation from a Greek word (*Christos*, # 5547) that is equivalent to "Messiah" of the Old Testament, and the Greek word means "*anointed*" as with oil, and implies to consecrate for an office or religious service. The word "Jesus" is from a Greek word (*Iesous*, # 2424) that means "*Jehovah* ('s or is) *salvation*." The meaning of Jesus Christ's name speaks of Jehovah's [the BeComingOne's] anointed savior which Jesus Christ was and is and will be.

Jesus Not Called "Christ" Openly before his Death

gp247» Although we call Jesus Christ, "Jesus Christ," he was not called "Jesus Christ" by his acquaintances before his death, but he was called "**Immanuel**" and/or "**Jesus**" because that was the name(s) he was given:

- "Behold, a virgin shall be with child, and shall bring forth a son, and they shall call

his name **Immanuel**, which is interpreted, God with us" (Matt 1:23; see Isa 7:14).

- "and he [Joseph] called his name **Jesus**" (Matt 1:25).

- "and, behold, you shall conceive in your womb, and bring forth a son, and shall call his name **Jesus**" (Luke 1:30; see 2:21).

gp248» Before Jesus' resurrection, he was *not* openly called the "Christ" (Matt 16:15-16, 20; Luke 9:20-21). Only after his resurrection was he called Christ by his disciples and by his apostles. By some of his enemies Jesus was called such names as, *impostor* or *deceiver* (Matt 27:63). The New Testament writings were written *after* Jesus's resurrection, and thus the writers used "Jesus **Christ**" because at that time they knew he was the Messiah — the Christ, or the one anointed the YHWH. (See later in this book for details.)

gp249» This is similar to two in marriage. A woman named Mary Jones marries Joseph Smith, so after their marriage she is called Mary Smith. Later in life some may say Mary Smith moved from Seattle to San Jose when she was seven years old, even though Mary Smith was not known as "Smith" when she was seven; she was known as Mary Jones when she was seven. So when the writers of the New Testament say Jesus <u>Christ</u> did this and did that before his death, it did not mean that he was called Jesus <u>Christ</u> at that time. It was only after his death and resurrection (when the New Testament was written) that his followers openly called him, Jesus <u>Christ</u>.

Meaning of Being Anointed

gp250» "Christ" which means *anointed* has a special meaning in context of its usage in the Bible. When one is anointed it represents something, "then Samuel took the horn of oil, and anointed him [David] in the midst of his brethren: and the spirit of the BeComingOne came upon David from the day forward" (1Sam 16:13). And, "but you are an anointing from the Holy One, and you know all things ... But the anointing which you have received of him lives in you, and you need not that any man teach you: but as the same anointing teaching you all things, and is truth, and is no lie, and even as it has taught you, you shall live in him" (1 John 2:20, 27).

gp251» Note that after David was anointed, the spirit came to him. And also notice that with the true anointing, "you know all things," the anointing "lives in you," and the anointing "is truth." Now to be spiritually anointed in the highest sense is to receive God's Spirit and that anointing from the Holy One leads you into truth so you know all things concerning the Spiritual (1 John 2:20, 27).

gp252» To confirm that Spiritual anointing is anointing with the Spirit, compare 1 John 2:20, 27 with: "the Spirit of truth, is come, he will guide you into *all* truth..." (John 16:13). And, "the Comforter, which is the Holy Spirit, whom the Father will send in my name, He shall teach you all things" (John 14:26).

gp253» Thus, Christ's own name tells us he is anointed with the Spirit. "The Spirit of the Lord [= YHWH of OT] is upon me, because he has anointed me [Jesus] to preach the gospel to the poor" (Luke 4:18). Hence, Christ is anointed with the Spirit to preach the gospel. This is one of Christ's commissions from God.

gp254» Remember that the word "Jesus" is translated from a Greek word, *Iesous*, the equivalent of the Hebrew, *Jehoshua* [Yehoshua], which means, The BeComingOne (is) savior, or the BeComingOne's savior. This could be written as, "Yehowah's Savior," or "Jehovah's Savior." Jesus Christ's name has the meaning that is equivalent to the English's, "*The BeComingOne's savior, anointed*" or "*anointed The BeComingOne's savior.*" "For unto you is born this day in the city of David a *Savior*, which is Christ (the) Lord" (Luke 2:11). This could be translated :"... a Savior which is anointed Lord." "The Father sent the Son to be the Savior of the World" (1 John 4:14). "And she shall bring forth a son, and you shall call his name Jesus: for he shall *save* his people from their sins" (Matt 1:21). Jesus Christ is Jehovah's anointed Savior of the world. "Savior" also indicates one who sets free or delivers those captured or enslaved. Jesus Christ will and is setting the world free from the slavery and confusion of this age through the New Mind of the New Spirit (John 8:31-36; Rom 8:2).

Promised Son Of God, Seed Of David, and Eve

Jesus Christ was Prophesied to Come

gp255» In the Old Testament it *foretold* the birth of Jesus Christ: "he shall cry unto Me, you art my Father, my God, and the rock of my salvation. Also I will make him My first born, higher than the kings of the earth" (Psa 89:26-27). "For unto us a child is born, unto us a son is given: and the government [rulership of God]

shall be upon his shoulder: and his name shall be called Wonderful, Counselor, the mighty God, the duration Father, the Prince of Peace" (Isa 9:6). "I will declare the decree: the BeComingOne has said unto me, You art my Son; this day have I begotten you" (Psa 2:7; Heb 1:5).

gp256» "The BeComingOne came unto Nathan, saying, Go and tell my servant David, Thus says the BeComingOne ... and when your days be fulfilled, and you shall sleep with your fathers, I will set up your *seed* after you, which shall proceed out of your bowels, and I will establish his [Christ's] kingdom. He shall build a house for my NAME and I will establish the throne of his kingdom for olam. I will be his [Christ's] father, and he shall be my son" (2Sam 7:4, 5, 12-14).

Fulfillment

gp257» The Bible also shows the fulfillment of these prophesies. "This is my beloved Son, in whom I am well pleased" (Matt 3:17). "And I saw, and bare record that this is the Son of God" (John 1:34). "I come in my Father's name [the son of YHWH, thus Jesus had His NAME]" (John 5:43). "Concerning his [God's] Son Jesus Christ our Lord, which was made of the seed of David according to the flesh" (Rom 1:3; Acts 2:30).

gp258» Thus, Jesus is the Son of God as well as the seed of David as promised (Luke 3:23-38). Further Christ is the seed of Eve that was promised to bruise the head of Satan (Gen 3:15; Rom 16:20; Psa 91:13; see The "Seed Paper" [PR 1]).

Son of Man Through Mary

gp259» Jesus Christ is called Son of God, but also he is called Son of man (Acts 7:56). Now to be a son of mankind, one needs to be human. Jesus was human for he was a son of Mary (Mark 6:3). "And Joseph also went up from Galilee, out of the city of Nazareth, into Judea, unto the city of David, which is called Bethlehem (because he was of the house and lineage of David) to be taxed with Mary his *espoused* wife [notice not wife, but espoused wife; for Joseph hadn't consummated the marriage, see Matt 1:24-25], being large with a child ... And she brought forth her first-born son" (Luke 2:4-5, 7). Thus another prophecy about Christ came true, "but you Bethlehem ... out of you shall he come forth unto me that is to be ruler in Israel" (Micah 5:2). Jesus was born in Bethlehem as prophesied in Micah and confirmed in Luke. "Therefore the BeComingOne himself shall give you a sign; Behold, a virgin shall conceive, and bear a son, and shall call his name Immanuel" (Isa 7:14).

gp260» As we've shown you Joseph hadn't consummated the marriage, and after Gabriel had told Mary about the son that she was to bring forth, "said Mary unto the angel, How shall this be, seeing I know not a man?" (Luke 1:34) Mary was a virgin mother. "But when the fullness of the time was come, God sent forth his Son, *made* of a woman" (Gal 4:4).

gp261» Notice in Galatians 4:4 that God's Son was "made" (KJV) or was "born" (NIV) of a woman. This word "made" is translated from a Greek word *ginomai* (# 1096) which means: "to become, i.e. to come into existence, begin to be, receive being" (*Thayer's Greek-English Lexicon*). From the Greek Gal 4:4 reads:

- "But when had come fullness of the time, the God sent forth the Son of Him, *coming into existence* from [or "out of"] a woman."

It says that God's Son was made out of a woman. This same Greek word, *ginomai*, is also found in two other closely related verses:

- "Concerning his Son Jesus Christ our Lord, which was made [*ginomai*] of ['out of' — Greek] the seed of David according to the flesh" (Rom 1:3; Acts 2:30).

- "And the Word was made [*ginomai*] flesh" (John 1:14).

The Son of God *first* came into existence from a woman, who was flesh ("according to the flesh"), who was also a seed of David.

gp262» Furthermore, these verses are related in meaning to 1John 4:2:

- "By this you know the Spirit of God: Every spirit that confesses that Jesus Christ *has come* in the flesh is of God" (NKJV).

The word translated "has come" is a Greek verb in its participle *perfect* tense (The Analytical Greek Lexicon). Christ *did* come in the flesh as the above verses indicate. In fact, he first came into existence through a woman. He had no pre-existence. The word "pre-existence" is a self contradiction. Scriptures prove that Jesus Christ the man did not exist *before* he was born from a woman (see later).

gp263» Jesus Christ was a Son of mankind, not a product of man through a male and female union, but from mankind through his mother. Mary was Christ's mother, but Joseph was not his real father, for Joseph didn't "know" Mary. At that time Mary was a virgin. Thus, Jesus is a son of man or mankind through the medium of Mary only.

Virgin Birth

gp264» Let's understand this virgin birth, and see how Jesus is a Son of God through being born from a woman:

- "And the angel [Gabriel] said unto her, Fear not, Mary: for you have found favor with God. And, behold, you shall conceive in your womb, and bring forth a son, and shall call his name JESUS ... Then said Mary unto the angel, How shall this be, seeing I know not a man? And the angel answered and said unto her, The Holy Spirit shall come upon you, and the power of the Highest [BeComingOne] shall overshadow you: therefore also that holy thing which shall be born of you shall be called the Son of God" (Luke 1:30-31, 34-35). And, "for that which is conceived in her is of the Holy Spirit" (Matt 1:20).

gp265» This is why the Bible calls Christ the Son of God. God, the highest power, the power of the Holy Spirit, conceived Jesus Christ in the womb of Mary. In this way, Christ is a Son of God. (In another way Christ is the Son of God, as Christians are sons of God, because he had the Spirit of God in him.) Through the power of God, Mary conceived, not through the power of a male's sperm. Basically, what the male's sperm does when it contacts a female's ovum (egg-cell) is to initiate a chemical-biological chain reaction. Through this chemical reaction the egg-cell grows into a child. Jesus Christ began as any other child from an ovum. But this egg-cell was fertilized *not* by a male's sperm, for Mary was a virgin. It was the power of God that fertilized this egg-cell that became Christ. How? God by merely duplicating the chemical code of a male's sperm conceived Jesus. Since God designed the ovum and sperm, he knows how they work and thus is able to initiate the chemical reaction in an ovum needed to produce a child. Through the power of God the Father, Mary conceived. In this way, God is the physical father of Jesus; Joseph is not the physical father of Jesus. Mary was a virgin to man, but not to God, for in a sense God "knew" Mary.

Middle Man: Son of God and Son of Man

gp266» This is why Jesus Christ is called a mediator, "for there is one God, and one mediator between God and men, the MAN Christ Jesus" (1 Tim 2:5). Notice it was the *man* Jesus who was the mediator, *not* the resurrected Christ. Now "mediator" means middle man or middle one. The man Jesus Christ is the middle one between God and mankind. He was "Jesus of Nazareth, a *man*" (Acts 2:22). This offspring of Mary and God was the beginning of a union of man back to God.

Born of the Seed of David

Genealogy of Mary & Joseph

gp267» Matthew 1:1-17 and Luke 3:23-38. Notice that Christ was born "of the seed of David according to the flesh" (Rom 1:3). In order for Jesus to be the Messiah, he must be from David, that is from the seed of King David. Since Mary's husband, Joseph, was not the physical father of Christ (Mary was a virgin), then Joseph's genealogy as listed in Luke 3:23-38 is irrelevant. Jesus' genealogy must come through Mary.

gp268» Matthew 1:1-15 is Jesus Christ's real genealogy because it says specifically in verse one, that it was the "book of the generation of Jesus Christ, Son of David, son of Abraham."

gp269» In most translations Matthew 1:16 reads, "Joseph was the husband of Mary, out of whom was born Jesus." Most translations in English come from Greek texts, and in these texts Joseph is said to be the husband of Mary. But the book of Matthew was first written in Hebrew. In at least two Hebrew texts found recently by a Hebrew scholar named Nehemia Gordon, it says for verse 16 that Joseph was the father of Mary. How can Joseph be the father and also the husband of Mary? It is because both genealogies are not speaking about the same Joseph, but two different Josephs. Those who translated Matthew did not understand this and so they "corrected" the Hebrew text from Joseph being the father of Mary to Joseph being the husband of Mary.

gp270» Notice in Matthew 1:16 that it specifically says Jacob begat Joseph: Jacob was the father of Joseph. But the wording in Luke 3:23 has Heli as the father of Joseph: "Joseph of Heli."(Luke 3:23, 24). There are **two different Josephs** being spoken about: one whose father was Jacob and one whose father was Heli.

gp271» Examine both of these genealogies in Matthew and Luke, they are different. Can Joseph have two genealogies with different fathers? No. The simple answer is that there were two Josephs. The proof is the use of the **format formula of three 14 generations' lists** in the Matthew genealogy:

> "all the generations, therefore, from Abraham to David were **fourteen** generations; and from David until the carrying away to Babylon, **fourteen** generations; and from the carrying away from Babylon unto the Messiah, **fourteen** generations" [Matt 1:17].

The first two lists of 14 generations in the book of Matthew have 14 generations each, but the third list in the Greek/English texts has only 13. This is because these mistaken translations have Joseph as the *husband* of Mary instead of the *father* of Mary. Matthew didn't make a mistake in his original text written in Hebrew, but the translators did make a mistake and thus were left with only 13 generations in their third list instead of 14 generations.

Seed of Nathan and Solomon

gp272» Mary's lineage came from David's son Solomon (2Sam 5:14). Nathan was also a son of David as shown in 2 Samuel 5:14, and Joseph's (Mary's husband) lineage came from Nathan. Notice this in Zechariah 12:12, "and the land shall mourn [in verses 10 and 11 it prophesies of people: first looking on Jesus on the cross; and second looking on Jesus when he returns, Rev 1:7], every family apart ['I am come to set a man at variance against his father, and daughter against her mother,' Matt 10:35]; the family of the house of David apart [Mary's], and their wives apart; the family of the house of Nathan [Joseph's family, Luke 3:31] apart, and their wives apart" (Zech 12:12).

[NOTE: Christ was on the cross, and possibly Christ's brothers were apart from their wives, and Joseph was apart from Mary (Joseph isn't mentioned at the end of the accounts of Matthew, Mark, Luke, and John — he could have been dead or separated from his wife). The Nathan Family was apart because some of them believed in Christ while others didn't. There is also a Spiritual meaning here. Christ had the Spirit of God and was apart from his brethren Spiritually up until the time they received the Spirit.]

gp273» Mary came from Solomon's lineage, but Joseph came from Nathan's, yet both from David since Nathan and Solomon were sons of David. Notice that in Joseph's and Mary's genealogy from David back to Abraham and beyond is the same, although not all of the generations are listed.

Christ's Lineage Review

gp274» We have proved from scriptures that Jesus was a descendant of David as promised David (2Sam 7:4-5, 12-14). Further, we saw that Jesus was the son of God. God (YHWH) begot Jesus the man through the power of the Holy Spirit (Luke 1:30-35). Thus, God is Jesus' physical Father (Psa 2:7; Heb 1:5; Psa 89:26-27; 2Sam 7:4, 12-14). Therefore, Jesus is at once the Son of mankind and the Son of God.

gp275» Jesus is the middle one between man and God, "one mediator between God and men, the MAN Christ Jesus" (1 Tim 2:5). Jesus was flesh and blood, born of woman (Gal 4:4). And his physical birth was initiated through the power of God.

God *inside* Christ The Man

gp276» But Jesus had another facet about him: he had God's Spirit inside him. God's Spirit was not him, but inside him. This is very important:

- ▪ "The BeComingOne has called me [Christ] from the womb; from the bowels of my mother has he made mention of my name [see Luke 1:26-32]. And he [BeComingOne] has made my mouth like a sharp sword [word of God, cf Eph 6:17; Rev 1:16] ... the BeComingOne that formed me [Christ] from the womb His servant, to bring Jacob again to Him ... and my God shall be my strength ... I [the BeComingOne] will also give you [Christ] for a light to the Gentiles" (Isa 49:1-2, 5-6; cf Luke 2:32).

gp277» God gave Jesus a "sharp sword" (word of God) and God was Christ's strength. It was God who gave his strength and words to Christ. Christ didn't speak *his* own words or use *his* own power. Notice,"behold my [the BeComingOne's] servant, whom I uphold; mine elect, in whom my soul delights; *I have put my spirit upon him*: he shall bring forth judgment to the Gentiles" (Isa 42:l). "And the *spirit of the BeComingOne shall rest upon him*, the spirit of wisdom and understanding, the spirit of counsel and might, the spirit of knowledge and of the fear of the BeComingOne" (Isa 11:2). This "spirit of wisdom" is the same "wisdom" pictured in Proverbs 8:22ff. The "spirit of wisdom" is God's Spirit.

gp278» Jesus was prophesied to receive the BeComingOne's spirit of wisdom, knowledge, and so on. Did he receive it? "The words that I speak unto you I speak not of myself: but the Father that lives in me, he does the works" (John 14:10). Notice that the Father, the BeComingOne's Spirit, was inside him and this Spirit, did the works. Even the words of knowledge that Christ spoke were from the BeComingOne as Isaiah 11:2; John 14:10,24; 12:49-50; 8:38 prove. Jesus was set forth distinctively from men, not merely by his resurrection (Rom 1:4), but by his works; Jesus says his Father did the works (John 14:10).

Spirit in Jesus was the Angel of God

gp279» But since angels are spirits (Heb 1:7; compare Heb 1:13 w/1:14), since angels are messengers, and since messengers are word carriers, then the Spirit in Christ was an angel. In fact it was the very angel of the BeComingOne (YHWH). It was this angel who was the WORD of God, who carried the words of the BeComingOne (see GP 3).

gp280» It was the Spirit or the angel of God *in* Christ the man that did the good works. Jesus was sinless (John 8:46) and it was the Spirit in him that did these works. Jesus was a human being, but he had the BeComingOne's Spirit or mind inside him. He was not Spirit himself, for if he were spirit, one could not see him (John 3:8). The man Jesus was flesh and blood, he was human, with the BeComingOne's Spirit inside him. "God was manifest in the flesh" of Christ (1Tim 3:16). "That God was in Christ" (2Cor 5:19).

gp281» "And she [Mary] shall bring forth a son, and you shall call his name Jesus for he shall save his people from their sins. Now all this was done, that it might be fulfilled which was spoken of the Lord [BeComingOne] by the prophet, saying, Behold, a virgin shall be with child, and shall bring forth a son, and they shall call his name Immanuel, which being interpreted is, GOD WITH US" (Matt 1:21-23). Names are used in the Bible to identify characteristics of people. Immanuel, one of Jesus's names, means, "God with us." INSIDE Christ, God was with the world. Christ the man was not God, but God was inside him. Jesus was *man* only, until his resurrection from the dead and his ascension to the Father.

Jesus Christ Came In The Flesh, Anointed by the Spirit

gp282» Notice, "by this you know the Spirit of God: every spirit that acknowledges that Jesus Christ *has* come in the FLESH, is of God" (1 John 4:2, see NKJV & Greek). John gives those reading the Bible a test: IF someone does agree that Christ Jesus came in the flesh, then he is of God's Spirit. But further John says: "and every spirit [in man] that acknowledges that Jesus is not come in the flesh, is not of God: and this is the spirit of antichrist" (1 John 4:3). Jesus was the Christ; He was the Messiah. Jesus was thus the promised anointed one. He was anointed by the Spirit of God. He was anointed by the very Word of God. **If Jesus the man was the very Word of God, then how could he have been anointed by the Spirit?** If Jesus was already Spirit why would he be anointed again with the Spirit again? Was he anointed by himself? No Jesus came in the flesh, as a human being, and then was anointed by the Spirit. The Savior of mankind came in the flesh with the Spirit of God inside him. But this does not mean he saves us in or by his former state as a man. We are saved by his resurrected life, not by his human life or death, but by his new life.

Jesus Christ The Man Was Not God

gp283» Jesus was *not* God before his resurrection. Jesus was a man with God's Spirit inside him as many scriptures prove (John 14:10; 1Tim 3:16; John 10:38; etc.). Jesus Christ before his resurrection was a *man*: "the *man* Christ Jesus" (1 Tim 2:5).

Came From God

gp284» "He [Christ] came from God, and went to God" (John 13:3). The literal translation from the Greek of this verse reads: "and that from God he came *out of* and to the God goes." Jesus Christ came out of God. He was a physical offspring of God, as shown before, through the power of God. And, "I came out from God. I came forth from the Father, and am come into the world; again, I leave the world, and go to the Father" (John 16:27-28). By the fact that God was Christ's "physical" father, by the power of the Father, Christ indeed come *out of* the Father in the same sense that a son comes from his

father's physical seed. (There is a dual sense here, see later.) "Now are we sure that you know all things, and need not that any man should ask you: by this we believe that you came forth *from* God" (John 16:30). If one is *from* or *out of* something, he is not that thing. Christ the man was from God, he was not God.

They were Two

gp285» There is an abundance of scripture that indicates that Jesus Christ the man and His Father were two distinctive individuals or beings at one time:

A. **sender-sent:** The fact that Jesus was sent by his Father means he is someone other than the sender (John 4:34; 6:38-39, 57),

B. **two wills:** the fact that he came not to do his will but the will of him who sent him shows two wills (John 6:38),

C. **two witnesses:** the fact that Jesus Christ the man spoke about the law of *two* witnesses and that he was not alone (*monos*), comparing himself as one witness and his Father as the other witness indicates two (John 8:16-18),

D. **not alone:** the fact that Jesus Christ spoke about himself *not* being alone (*monos* or only) because his Father who sent him was with him indicates two (John 8:29;16:32),

E. **not his words:** the fact that Jesus doesn't say his own words ("not from myself") but his Father's words indicate he was not the Father (John 12:49; 14:24),

F. **"we":** the fact that Jesus spoke of himself and his Father as "we" indicates two (John 14:23),

G. **from-going back:** the fact that Jesus came from the Father and was going back indicates two (John 16:28),

H. **"nor me":** the fact that Jesus said that men didn't know his Father "nor me [Christ]" indicates two (John 16:3),

I. **Father greater:** the fact that Jesus called the Father greater than himself indicates two (John 14:28),

J. **sent by the only God:** the fact that Jesus Christ said he was sent by the only (*monos*) true God indicates two (John 17:3),

K. **prayed to the Father:** the fact that Jesus Christ the man prayed *to* his Father indicates two (John 14:16; 17:1 ff),

L. **resurrected by God:** If he was raised by God (Acts 2:24, 2:32; 13:33-37), at the time he was raised by God, how could he be God? No there were two, the one resurrecting, and the one resurrected,

M. **not yet returned to the Father:** the fact that Jesus said he had yet to return to his Father and called his Father "my Father and your Father"indicate that Jesus Christ the man and God the Father were *two* (John 20:17),

N. **God made him both Lord and Christ:** this speaks of two, the who made Jesus Lord and Christ, and Jesus who was made by God, both Lord and Christ (Acts 2:36).

All the above and other scripture indicate that Jesus Christ the man and his Father were two before Jesus was resurrected and went to the right side of his Father. Yet other scripture indicates somehow they were one (John 10:30) in a way similar to the way Christians are one in God (John 17:21-23, 11; see 1Cor 12:12ff). When Jesus Christ the man was on earth before his resurrection he was separate *from* his Father, yet he was ONE in a Spiritual sense, since Christ the man acted as his Father directed (John 12:49-50).

A Mediator is Not God

gp286» "For there is one God, and one mediator between God and men, the MAN Christ Jesus ... Now a mediator is not of one, but God is one" (1 Tim 2:5; Gal 3:20). Jesus Christ the man as a mediator was not God, but God was inside him, for the Spirit lived inside him.

Angel of the BeComingOne Was in Christ the Man

gp287» *Word of God in Jesus*. Jesus was a Spiritually begotten Son of God through the Spirit inside him. Jesus had inside him the very Spirit or the angel of the BeComingOne. The BeComingOne being the Father because the Power of the BeComingOne predestinated all, thus in this sense the BeComingOne is the Father, while the angel was the messenger and agent of the Father. The archangel (as the messenger of the Father) was actually inside his mind Spiritually leading him. Since Spirit takes up no space (as some imagine space), it can and does live anywhere, including inside a human mind. Since "angel" is a translation from a word meaning "messenger," then the Spiritual Messenger of God the Father was inside Jesus the man's head giving him Spiritual messages that enabled him to fulfill God's will. In a sense, this angel was the WORD of God — the WORD of the BeComingOne.

gp288» And again, here is some proof that this Spirit or angel of God the Father was inside Jesus the man:

- One of Christ's names tells us this: "Immanuel" means *God with us*. God was with man, by the fact His Spirit was *in*side Christ.

- Paul said God was manifested *in* the flesh (1Tim 3:16). This means by the context of the chapter that God was *in* Christ, not as the flesh of Christ. If God was as the flesh of Christ, it means God died with Jesus on the cross. But since God is immortal, then God could not and did not die on the cross (see GP 1). To refute this is to say that God lied when through scripture he said that spiritual beings like angels cannot die (Luke 20:36). How can an immortal being die? How can someone that cannot die, die?

- Jesus said "the son of man shall give you, because God the Father has set His seal on Him" (John 6:27). Jesus was sealed by the God with the Holy Spirit as Christians are sealed: "And do not grieve the Holy Spirit of God, with which you were marked with a **seal** for the day of redemption" (Eph 4:32; see Eph 1:13; 2Cor 1:22).

- "That is, that God was in Christ, reconciling the world unto himself" (2Cor 5:19). God's mark or seal was on Christ the man. He had God's Spirit, God's angel, God's WORD, in him. Again God was *in* Christ.

Christ Suffered, Can God Suffer?

gp289» Jesus Christ the man, from a regular ovum in Mary, grew into a man who was sinless (Heb 4:15) because of the power of the Spirit in him. He was only given what was needed for his commission as the man Jesus Christ. He was given God's Spirit that produced the wisdom of Christ the man (Isa 11:2-4) and the great works (John 14:10). But he was given only just enough to do this. He "who in the days of his flesh, when he had offered up prayer and supplications with strong crying and tears unto Him that was able to save him from death, and was heard in that he feared" (Heb 5:7). Jesus had to cry aloud to his Father for help, "but was in all points tempted like as we are, yet without sin" (Heb 4:15).

Because of this, the resurrected Christ, "in that he himself has suffered being tested, he is able to help them that are tested" (Heb 2:18). Can God suffer or was God inside of Christ, not as Christ?

gp290» We will explain later how Jesus became God, or we should say was infused into the true God, after being resurrected to the right side of the true God. This may be confusing until you understand who or what is the true God. But for now let's go into more details to prove God's Spirit was inside Christ the MAN.

God Not As Christ, But inside Christ

Men as Temples of the Living God

gp291» God was manifested in the flesh of Christ (1Tim 3:16). And one of Jesus Christ's names ("Immanuel") meant "God with us" (Matt 1:23). But we have shown you that Jesus before his death was not God, but God was inside him. Human beings are called temples of God and God lives inside these temples, if these individuals are Spiritual Christians. "For we are the temple of the living God" (2Cor 6:16). Even Jesus the man was the temple of his Father's Spirit: "Jesus answered them. 'Destroy this temple, and I will raise it again in three days.' The Jews replied, 'It has taken forty-six years to build this temple, and you are going to raise it in three days?' But the temple he had spoken of was his body" (John 2:19-21).

gp292» Although God the Father through his Spiritual messenger initiated the chemical process in one of Mary's eggs that produced Christ, God didn't put himself or transform himself *with or as* the flesh of Christ. God in a sense begot Christ, through a chemical process (created sperm) to produce the MAN Christ in the womb of Mary. At that time, God was not man, thus he did not have physical sperm with genes in it. No physical genes from God went into the ovum that produced Christ, for God being Spirit had no human sperm when he produced Christ the man. Nor did a Spiritual "sperm" enter the ovum, or else Jesus would have been part Spirit. But God did put his Spirit *into* the mind of Christ. And it is this Spirit of God that led Christ as it leads all sons of God (Rom 8:14) that are sons in the Spiritual or antitypical sense.

Spirit Did the Works With Jesus

gp293» And it was through this Spirit in him that Jesus did the great works (John 14:10). Yet just because it was the Spirit inside him that did the great works, this doesn't mean the man Jesus did nothing. The Spirit in him worked and produced Christ's good Spiritual fruit as the sap in trees produce the fruit. Yet, as the sap needs the branches and the leaves to produce the fruit, so does the Spirit need a body to produce the good fruit. As the leaves and the branches help with the work, so did Christ the man help the Spirit produce the good fruit. Yet without the Spirit no one can produce good fruit, as no branch can produce fruit without the sap from the trunk and roots of the tree. The Spirit is the energy needed to produce the good fruit, as the sap is the energy needed to produce the fruit. They need each other to produce. The Spirit in Christ worked and led Jesus just enough for Jesus the man to be sinless. God gives enough Spiritual power for a person to do what is asked of him (Rom 8:28ff; see *New Mind Papers*).

Paul is an example.

gp294» Paul was "appointed a preacher, and an apostle, and a teacher of the Gentiles" by God (2Tim 1:11; Acts 9:15). And what did God answer Paul about a problem he had: "my Grace is sufficient for you" (2Cor 12:9). "But by the grace of God I am what I am: and his grace which was bestowed upon me was not in vain; but I labored [he did his appointed work] more abundantly than they all: YET NOT I, but the grace of God which was with me" (1Cor 15:10). The grace of God did Paul's work, as the grace of God, or the Spirit of God, or the New-Mind did the works of Christ (John 14:10).

We Are Saved By the Resurrected Christ

gp295» "And if Christ is not risen, then is our preaching futile, and your faith is also futile ... For if the dead don't rise, then Christ is not raised ... If in this life [our human life] only we have hope in Christ [the MAN], we are of all men most miserable. But now Christ is risen from the dead" (1Cor 15:14, 16, 19-20). Christ must be alive and resurrected or our faith is just plain stupidity. "WE SHALL BE SAVED BY HIS LIFE" (Rom 5:10). The Savior of mankind came in the flesh as a human, but we are saved by his new resurrected life.

When Did Christ Receive the Spirit?

gp296» Jesus the man was unique, he was a physical Son of God (through God's power over the ovum) and a Spiritually begotten Son of God (through the Spiritual power in his mind). Romans 8:9-17 and elsewhere manifests to us that to be a Spiritual son of God, you must have God's Spirit in you.

gp297» In the New Testament, it doesn't say *when* Christ received God's Spirit in his mind, but it was probably from his birth (Isa 49:1-2, 5-6, "from the womb"; cf Luke 2:32; see above), for during his childhood he had the Spirit and the grace of God upon him (Luke 2:40, 46-47). Therefore when Christ was baptized in water by John, the "Spirit of God descending like a dove," was representative of Christ having the Spirit. People saw a dove land on Jesus which fulfilled a sign that Christ was the one who baptized with the Spirit; John merely baptized with water (John 1:33). Those John baptized didn't receive the Spirit (Acts 18:24-25; 19:1-6). Christ received the Spirit at birth. The water baptism by John of Jesus was symbolic only, for Christ already had the Spirit (Luke 2:40). Christ only got water baptized to fulfill prophecy and righteousness and to show a physical example or sign to Israel (Luke 2:40; Matt 3:11-16;Luke 3:15-16; John 1:30-33).

God Made Flesh?

gp298» Now some say that Jesus Christ the man and the BeComingOne (Jehovah) of the Old Testament were one and the same person. They, believing that the *Word* of God, God, and Jesus Christ are one and the same, point out, "In the beginning was the Word ... and the Word was made flesh" (John 1:1,14). Therefore they say God somehow became flesh, but they have a problem since they believe God is immutable or not changeable. Part of the gymnastics that they go through because of their impossible theory is written about in *Trinity v. BeComingOne* paper.[1] But as we have shown with many proofs in GP 3, the WORD of the Old Testament was the *angel* of the BeComingOne. It was through this angel or messenger that God — the BeComingOne — spoke. This angel spoke for the Coming Power: the God all in all; the BeComingOne that will

finally become. The BeComingOne spoke (Isa 52:6) in the Old Testament of the Bible, but he spoke through His agent — the angel, the very WORD of God. The BeComingOne was one and the same being as Christ's Father (GP 2). One of the proofs of this came from Christ's own lips. Since the Jews' God is the God of the Old Testament, then when Christ tells them his Father is the same God, we know that the BeComingOne is the Father (John 8:54).

gp299» Now *if* it was true, that the very God became flesh, and that he became Christ the man, then it means the BeComingOne, the Father, died when Christ died. We ask, if this is true, WHO RESURRECTED CHRIST THE MAN? Scripture plainly shows us God the Father (through His predestinated Power) raised Christ the man from death (Acts 2:32, 24; Rom 8:11; 1Cor 6:14). Thus, they were separate. God inside Christ the man, not God as Christ the man. God was in Jesus Christ because His Spirit, that is His Angel, or His WORD (and the Power of His WORD), was in Jesus Christ. Angels or spirits do not die (Luke 20:36). God is immortal, unable to die (1Tim 1:17). Jesus said scripture could *not* be broken (John 10:35), therefore the BeComingOne or his angels do not die. Before the resurrection God or His angel did not become Christ the man. God in some way was living inside Christ the man. Since the Father is the BeComingOne of the Old Testament, we know that God was with man because he was somehow *in* Christ the man. God was in Jesus through His Spirit, His Angel, His Spiritual messenger.

Meaning of, "the Word Was Made Flesh"

gp300» Let's see what the scripture means by "the Word was made flesh." From the Greek text this sentence reads, "and the Word became Flesh" (John 1:14). The real sense of this verse is that the WORD became flesh *after* the resurrection, for these words were written by John *after* the resurrection (GP 5). But there is another sense to this verse. By other scriptures which we shall show you, we can in light of these make this verse clear as following, "and the Word became [in] flesh." For the antitypical or higher or real meaning of this verse see GP 5.

gp301» "God who at different times and divers manners spoke in time past unto the fathers by the prophets, Has in these last days spoken unto us in his Son" (Heb 1:1-2; cf. Heb 2:2-3). God did not speak in the past (before Christ the man) through His Son because he did not exist

---

[1] See
http://becomingone.org/gp/gp10b.htm

then. God spoke by or inside his Son during his days on earth:

- "The words that I speak unto you I speak not from myself: but the Father that lives in me, he does the works" (John 14:10).

- "For I have not spoken out of myself; but the Father which sent me, he gave a commandment, what I should say, and what I should speak ... whatsoever I speak therefore, even as the Father said unto me, so I speak" (John 12:49-50).

- "The word which you hear is *not* mine, but the Father's which sent me" (John 14:24). And again, "I speak that which I have seen with my Father" (John 8:38).

Notice the prophecy of this, "I [the BeComingOne, Deut 18:15] will raise them up a Prophet from among their brethren, like unto you [Jesus came in the flesh, 1 John 4:2-3, by the seed of mankind not angels, Heb 2:14, 16], and *will put my words in his mouth*; and he shall speak unto them *all* that I shall command him" (Deut 18:18).

gp302» Jesus spoke the Words of his Father because the angel or messenger of the BeComingOne was inside Jesus and was leading him. In this way the Father was *in* Jesus the man leading him with God's Spirit (Isa 42:1; Rom 8:14). The Word was *in* the flesh, the Word was made and manifested *in*side the flesh, but was *not* the same as the flesh before the resurrection. The BeComingOne or the Father didn't become flesh, for if he did then God or His angel died on the cross with and as Christ. But since immortal beings cannot die, then God or His angel was not transformed into Christ in order to die.

Death of Christ the Man

Holy One

gp303» Let's look at Christ's death, for by it we can prove positively what we have put forth herein (For details see CP 4). Notice what is quoted in Acts from the 16th Psalm, "because you [the BeComingOne] will not leave my [Christ's] soul in hell, neither permit your Holy One to see corruption" (Acts 2:27).

gp304» Who is the Holy One? He is the God of the Old Testament. In other words, Christ's Father. "BeComingOne of hosts (is) his name, the Holy One of Israel" (Isa 47:4). "Thus says the BeComingOne, the Holy One of Israel" (Isa 45:11). There are about 40 places in the Old Testament

where the BeComingOne identifies himself as the Holy One. The BeComingOne will not permit the Holy One (the BeComingOne himself) to see corruption. Corruption is used by the Bible to indicate death. Thus God will not die on any cross, as some make Him by having Jesus Christ the man joined to their Trinitarian God.

gp305» Christ the man died according to scripture (1Cor 15:3), and was buried for three days (Matt 12:40). And they buried his body in the grave (John 19:40-42). Now Christ the man was resurrected after exactly three days and three nights in the grave (see CP 4). But notice, "of the resurrection of Christ that his soul was not left in hell [the grave], neither his flesh did see corruption" (Acts 2:31). What?

gp306» If Christ the man died, how could his flesh not see corruption? Scripture as we have shown you said specifically Christ died. Since he was flesh, that means his flesh died. But Acts 2:31 said his flesh didn't see corruption. Notice the verse is speaking of the resurrection, and that his soul was not *left* in the grave. He wasn't left in the grave. He was resurrected, and after this resurrection his flesh didn't and won't see corruption *again*.

gp307» Notice the proof of this rendition, "and as concerning that he [God] raised him [Christ the man] from the dead, now no more to *return* to corruption" (Acts 13:34). Acts 2:31 must be read in context of Acts 13:34. Christ's flesh did see death, but after his resurrection his flesh will not see corruption *again* (NOTE: the resurrected Christ does have a flesh and blood body, Luke 24:39). Thus in context of Acts 13:34, Acts 2:31 means "his flesh did see corruption no more after his resurrection." And Acts 13:37 means in context, "whom God *raised*, saw no [more] corruption."

gp308» "You shall not permit your Holy One to see corruption" (Acts 13:35). Acts repeats this twice in Acts 13:35 and Acts 2:27. The "Holy One" is the BeComingOne or the Lord God of the KJV, an immortal being. He cannot die according to scripture. Those who say Christ is the BeComingOne of the Old Testament or the LORD or Jehovah of the Old Testament, say he became flesh and died. But scripture says God will not permit the Holy One (the BeComingOne [YHWH] himself) to die. Of course not, for the BeComingOne is the Holy One. He is Christ's Father. He lived inside of Christ the man through His angel who carried God's WORD.

gp309» But Acts does use a similar word formation as in Psalms 16:10 and Acts 2:27. Notice Acts 2:31 and Acts 13:37. It seems to say

Christ the man is the Holy One by the word formation. But it does not speak of the Holy One. It uses Christ's name, thus, fooling those who are not discriminating enough to notice the difference. Those who do not know there is a difference between the Holy One and Christ the *man* conclude that Christ didn't really die when they look at Acts 2:31. When they see "neither his flesh did see corruption," they conclude he didn't really die. But as shown, they leave out Acts 13:34 which clears up Acts 2:31.

Spirit Leaves Body At Death

gp310» Therefore Christ died, but the Holy One (the Father's Spirit) inside Christ the man did not die. Notice where the Holy One left Christ the man's body, "Father, into your hands I commend my spirit" (Luke 23:46). At that same time "he gave up the breath." The Bible uses this last expression to indicate dying, for life came at first from the breath of life (Gen 2:7). At the losing of this breath of life, one loses his life. Also, at death, the man Christ gave up his spirit or angel. And since his Spirit was what made the God *in* him, it was at this point that the Holy One's Spirit left Christ's body. This fulfills prophecy that the Holy One would not see corruption, but that Christ the man died.

gp311» Now some will probably point to Acts 3:14 in order to prove (they think) that the *man* Jesus was the Holy One, "but you denied the Holy One and the Just, and desired a murderer to be granted unto you" (Acts 3:14). But HOW did they deny the Holy One? "If you had known me, you should have known my Father also: and from henceforth you know him, and have seen him ... Don't you believe that I am in the Father, and the Father in me? the words that I speak unto you I speak not of myself: but the Father that lives inside me, he does the works" (John 14:7, 10). "It is the Spirit who gives life; the flesh profits nothing. The words that I speak to you are spirit, and are life" (John 6:63). Christ's good works and his words came from the good Spirit inside him. "God was manifest in the flesh" (1Tim 3:16). They denied the Holy One because Christ was a shadow or image of the Holy One (Col 1:15), for Christ did exactly as the Holy One in him led him.

gp312» The angel of the BeComingOne did not die on the cross with Christ the man. Jesus the *man* died. Jesus came in the flesh (1 John 4:2-3). He was made just like his brothers, as flesh and blood, as a human (Heb 2:14, 16). Jesus the man was commissioned before the world began to be sinless, to die for sin, and to be resurrected into God.

Why Christ Died

Predestinated Before the Foundations of the World

gp313» Jesus the man is the Spiritual Lamb of God who *before* the foundations of the world was chosen to be a spotless or sinless "lamb" through the power of God's Spirit inside him (Compare John 1:29; Rev 13:8; Isa 53:7-8; Matt 12:18; 1Pet 2:4; Isa 49:7; John 14:10; Rom 1:4).

gp314» It was by the death of one spotless, sinless *human* being, that God is going to reconcile the world to himself (Rom 5:18; 2Cor 5:19; 1Tim 2:5-6). God who through his Spiritual power begot Jesus physically as well as Spiritually, has given his own physical Son as a sacrifice for all man's sins (1 Tim 2:6). Jesus as a man was a physical son of God as explained before, but a Spiritually begotten son not yet born. After Christ the man was resurrected, he became a Spiritually *born* Son of God as opposed to a Spiritually *begotten* son of God. To understand the difference between being begotten and being born, see the "Begotten, Born Paper" [NM5] in the *New Mind Papers*.

gp315» Notice the proof of Christ's commission, "for as by one man's [Adam's] disobedience many were made sinners, so by the obedience of one [Jesus] shall many be made righteous ... for if, when we were enemies, we were reconciled to God by the death of his Son, much more, being reconciled, we shall be saved by his life" (Rom 5:19, 10). "Christ was without sin, but for our sake God made him share our sin [or share the effects of our sins] in order that in union with him [Jesus] we might share the righteousness of God ... he [Jesus] bore the sin of many, and made intercession for the transgressors" (2Cor 5:21, TEB; Isa 53:12, NIV). "For there is one God, and one mediator between God and men, the *man* Christ Jesus; Who gave himself a ransom for *all*, to be testified in due time" (1 Tim 2:5-6). God the Father gave his physical Son, the man Christ, to be a sacrifice for the transgressions of *all*.

Jesus Christ the Man, the Son, did *not* Pre-exist

gp316» How can someone "pre" exist? It is against the Law of Contradiction to say that Jesus the promised Messiah and Son of God existed before he existed. But since the "pre-existence" of Jesus Christ the man is a popular theory today, we are going to show you the evidence that he *first* came into existence when he was conceived or begotten in the womb of Mary. We will also look at the so-called scriptural evidence used by others to "prove" that Christ the man existed as some kind of "god" or "angel" before he was born.

The following is evidence that Jesus Christ the man, the Son, first came into existence when he was conceived in Mary's womb:

(1) Against the Law of Contradiction

gp317» The biggest proof that Jesus Christ did not exist before he was born, is the evidence given in GP1 of this book: it would be against the very Law of Contradiction for an immortal being (a being not capable of death) to be changed into a being that is mortal in order for that being (Jesus Christ) to die, or for the being to be simultaneously immortal, yet capable of death (see GP 1).

(2) Jesus came into existence in the Flesh

gp318» "By this you know the Spirit of God: every spirit that acknowledges that Jesus Christ *has* come in the FLESH, is of God" (1 John 4:2, see NKJV & Greek). John gives those reading the Bible a test: IF someone does agree that Christ Jesus came in the flesh, then he is of God's Spirit. But further John says: "and every spirit [in man] that acknowledges that Jesus is not come in the flesh, is not of God: and this is the spirit of antichrist" (1 John 4:3). Christ did not come as a transformed spirit, angel, or God, but he "has come in the *flesh*."

gp319» And it was the "man Christ Jesus; who gave himself a ransom for all" (1Tim 2:5-6). It was not the Spirit of Christ that gave himself for mankind, but the *"man* Jesus Christ." The Spirit of Christ the man was inside Christ before his death; thus, Christ the man gave up his Spirit when he died (note Luke 23:46; Matt 27:50; Mark 15:37; John 19:30).

(3) Prophesied Seed Cannot Exist Before He Genetically Passes through the Fathers and then is Born

gp320» "The BeComingOne came unto Nathan, saying, Go and tell my servant David, Thus says the BeComingOne ... and when your days be fulfilled, and you shall sleep with your fathers, I will set up your *seed* after you, which shall proceed out of your bowels, and I will establish his [Christ's] kingdom. He shall build a house for my NAME and I will establish the throne of his kingdom for olam. I will be his [Christ's] father, and he shall be my son" (2Sam 7:4, 5, 12-14).

Look at this carefully. This is a prophecy of the Messiah. He must be a seed or offspring of David, but also the son of the God. How can he be a son of man and a son of God at the same time?

gp321» This prophecy of God's Seed coming out of David who would also be God's son ("my son") pointed to the future when God would make his Son through the means of Mary. It was through Mary that God's one-of-a-kind son was made, not at some previous time:

> ■ "And the angel said to her, Do not be afraid, Mary, for you have found favor with God. And behold, *you will conceive* in your womb and *[will] bring forth* a Son, and *shall call* his name Jesus. He *will be* great, and *will be called* the Son of the Highest" (Luke 1:30-32).

gp322» Notice these emphasized words are in the *future* tense (see Greek text). God's Son did not exist before this time, and was not great before this time (except in the forethought of God). The angel was announcing the coming Son being born by a woman with the help of the Holy Spirit. The Son did not exist before these events.

> ■ "And Joseph also went up from Galilee, out of the city of Nazareth, into Judea, unto the city of David, which is called Bethlehem (because he was of the house and lineage of David) to be taxed with Mary his *espoused* wife [notice not wife, but espoused wife; for Joseph hadn't consummated the marriage, see Matt 1:24-25], being large with a child ... And she brought forth her first-born son" (Luke 2:4-5, 7).

gp323» Joseph hadn't consummated the marriage, for after Gabriel had told Mary about the son that she was to bring forth, "said Mary unto the angel, How shall this be, seeing I know not a man?" (Luke 1:34) Mary was a virgin mother. And with the help of the Holy Spirit Mary conceived a son (Luke 1:35, see above). "But when the fullness of the time was come, God sent forth his Son, made of a woman" (Gal 4:4).

(4) Born of a Woman

gp324» Notice in Galatians 4:4 that God's Son was "made" (KJV) or was "born" (NIV) of a woman. This word "made" is translated from a Greek word *ginomai* (# 1096) which means: "to become, i.e. to come into existence, begin to be, receive being" (*Thayer's Greek-English Lexicon*). From the Greek Gal 4:4 reads: "But when had come fullness of the time, the God sent forth the Son of Him, *coming into existence* from [or 'out of'] a woman." It says that God's Son was made out of a woman. It did not say that before this event God existed with the Son. God's Son was made or came into existence from a woman by the power of the Holy Spirit (note earlier in this chapter).

(5) Son Speaks Now, Not in the Old Testament

gp325» God didn't speak through the Son in "time past," but to the fathers of Israel by the prophets, but in the last days He speaks in his Son (Note Heb 1:1-2). God did not speak by His Son in the Old Testament times because, His Son did not exist at that time.

(6) Proof of Worshiping Angels

gp326» "But again WHEN He [God] brings the firstborn into the world, He says: Let all the ANGELS of God worship Him" (Heb 1:6; Deut 32:43, Greek text). God brought His Son into the world by having him being made by a woman with the power of the Holy Spirit (see above). But *when* the Son came into the world *angels* were present to worship him. Thus, after the Son was born:

- ■ "*Today* in the town of David *a Savior has been born to you*; he is Christ the Lord. This will be a sign to you: You will find a baby wrapped in strips of cloth and lying in a manger. *Suddenly a great company of the heavenly host [angels] appeared with the angel*, praising God and saying, Glory to God in the highest, and on earth peace to men on whom his favor rests. And when the *angels* had left them and gone into heaven..." [Luke 2:11-15, NIV].

gp327» With Heb 1:6 and Luke 2:11-15 we see *when* (at Jesus's physical birth) and *how* (with the angels worshiping him) the first born of God came into the world; He (the man Jesus)

did not exist before Mary conceived him in her womb, except in the forethought of God

(7) Spirit and Flesh

gp328» All the so-called "pre-existence" scriptures can be shown to refer to Christ the man's Spirit or angel existing before Christ the man. The Spirit of Christ the man was the angel of the BeComingOne or the angel of Jehovah (YHWH) as we have so far explained (GP 3). The Spirit of Christ and the fleshly Christ the man are just as different as a man and a woman in marriage even though according to the Bible they are ONE in marriage. That is why Paul said that our fathers passed through the sea, baptized unto Moses in the cloud and in the sea, by the spiritual rock: "and that Rock was Christ" (1Cor 10:1-4). The "spiritual Rock" (1Cor 10:4) was Christ even though Christ himself was not even born yet during Moses' time because:

- ● (1) Paul was talking about the "spiritual Rock that followed them [the fathers]" (1Cor 10:4)

- ● and (2) Paul was talking about Christ in an *ex post facto* manner as one uses a married woman's marital name even when speaking about some event that happened before she was married (see gp313).

(8) Summarized Evidence against Pre-existence Theory

gp329» Jesus Christ the man was/is the *one-of-a-kind* Son of God (John 1:18; 1John 4:9; monogenes=unique). The scriptures we covered above indicate when and how this Son was born: He did not exist before he was begotten; He was not begotten twice for He is the *one-of-a-kind* Son of God. (There are two senses to Jesus being the only one-of-a-kind Son — physical and Spiritual. We speak here of the special physical sense as explained in this chapter.)

gp330» Again, we repeat, there is no way for an immortal being to die. Most, if not all, of those who believe that Jesus Christ existed before he was a man have him existing as an immortal being (an angel or God). There is no way an immortal being can die, for if he ever does die he proves he was never immortal, but mortal.

gp331» Some, but not all, who argue for the so-called "pre-existence" of Christ the man are not thinking through their beliefs and are unknowingly participating in the "big lie":

- "The coming of the lawless one will be in accordance with the work of Satan displayed in all kinds of counterfeit miracles, signs and wonders, and in every sort of evil that deceives those who are perishing. They perish because they refused to love the truth and so be saved. For this reason God sends them a *powerful delusion* so that they will believe the lie and so that all will be condemned who have not believed the truth but have delighted in wickedness" (2Thes 2:9-12).

gp332» It is a powerful delusion to believe in the contradiction of the Trinitarian theory. But since most, if not all of us, at least subconsciously, act in our daily lives as if they know the Law of Contradiction, then we are in a sense knowingly participating in the "big lie" if we believe in the contradictory Trinitarian theory. At one time I "believed" in the Trinitarian theory, but was puzzled by it: on a certain level I believed; on a different level I did not believe. This is not the only delusion we are under. There are just as big contradictions in various fields of "science."

Pre-existence Theory: Refuting their Evidence

Let's look at some of the scripture others use to "prove" their "pre-existence" theory:

(1) In the Beginning was the Word, which beginning?

gp333» "In the beginning was the Word ... the Word was toward the God ... All things were made through him ... In him was life ... And the Word became flesh" (John 1:1-4, 14, see Greek).

But which beginning was John speaking about?:

- (1) the beginning of the Good News of Jesus Christ (Mark 1:1) as witnessed by the disciples of Jesus (Luke 1:2; John 15:27; 1John 1:1); or (2) the beginning of the creation (Gen 1:1) Since John spoke about the "beginning" in his Gospel and letters, it is John's meaning in (1) above that should clarify this verse. Yet the Genesis' "beginning" also applies, being true through the *Spirit* of Jesus, but not true through the flesh of Jesus, for the flesh of Jesus only existed after his birth from Mary.

gp334» The word, "Word," was translated from the Greek word, *logos*. This Greek word means, "something *said* (including the thought)." "Logos" not only indicated the word spoken, but can also indicate the reason or thought behind the word. But a word is *spoken*, that is, a word is spoken by someone. The "Word" in the beginning was spoken by the God through His angel or messenger for it was through angels that God spoke in the Old Testament times (cf Heb 1:1-2 w/ 2:2-5):

- "And God said, let there be light: and there was light" (Gen 1:3).

This is some proof that the Word and the true God are not one and the same in the fullest sense: the Word comes from God, and is spoken ("God said") from God, but is not the God. We have explained already in the Word chapter that the Word was carried by the angel of God. Also in the Greek it says, "the Word was *toward the* God" not the Word was "with" God (John 1:1; cf. John 13:3). And the Greek says, "and God was the Word," not the Word was God. The Word cannot be *toward* the God and at the same time be the true God in the fullest sense. As shown in the Word chapter (GP 3) there is a very close relationship between the Word or angel of God and the God, and as we will show in a later chapter, the Word is indeed *toward* the true God, but is not the God in the fullest sense of God all in all. The Word of God was/is closely related to God, for it carries with it the power of God, for what God says will happen. God does not lie (see GP 5 for more details on John 1:1-18 and the rest of this book).

gp335» The Word became flesh in the truest sense after the death of Christ the man (see GP 5). Another sense of John 1:14 ("And the Word became flesh") is that during Christ the man's life on earth the Word was inside the flesh (see this chapter, GP 4)

(2) Whose going forth was from of Old ...

gp336» This verse is speaking about the place of the birth of Jesus, and is mostly incorrectly translated into something like this:

- "And you Bethlehem Ephratah... out of you shall he come forth unto me to be Ruler in Israel: whose going forth are from of old, from the days of eternity" (Micah 5:2).

It looks like the Messiah comes from past eternity. The problem is that "eternity" should be translated *olam* which as explained in the "Age Paper" (NM7) means an age of unknown length. In one sense Jesus Christ's going forth was from the very old days, for since the beginning God had been prophesying his coming. And the second meaning is that the

Spirit in Christ was from the old days, since the Spirit in Christ existed from the beginning of creation. In fact this very Spirit created the universe. The man, the Messiah born in Bethlehem, did not exist in the old days, for he first came into the world through Mary, but his "going forth" by prophecy was from even before the very beginning for he was predestinated before the cosmos (1Pet 1:19-20) and he was the seed prophesied in Genesis (Gen 3:15).

gp337» Other scriptures such as Proverbs 8:22-23 and John 8:58 can be explained in this way: Christ's Spirit existed before the man Jesus was born. This is the same for us. Our own Spirit has existed since the beginning, but we were born in this age, a long time after the beginning. After we are infused with our Spirit in the resurrection, we can say in a sense that we existed from the beginning since our own Spirit existed from the beginning even though there were many years after the beginning before we were born.

(3) Jesus Christ Existed Before He was Born Only in God's Fore-Thoughts

gp338» "And now, Father, glorify me in your presence with the glory I *had* with you before the world began" (John 17:5). The Greek word translated "had" here is an imperfect Greek verb that means "to possess" or "to hold." Christ possessed in some imperfect or incomplete sense (remember the verb is in the imperfect or incomplete tense) glory *before* the world began.

gp339» Christ was predestinated to possess the glory of God *after* he was born and thereafter was to obtain the Kingdom of God and its glory; in the Old Testament it *foretold* the birth of Jesus Christ:

- "He shall cry unto Me, you art my Father, my God, and the rock of my salvation. Also I will make him My first born, higher than the kings of the earth" (Psa 89:26-27).

- "For unto us a child is born, unto us a son is given: and the government [rulership of God] shall be upon his shoulder: and his name shall be called Wonderful, Counselor, the mighty God, the duration Father, the Prince of Peace" (Isa 9:6).

- "I will declare the decree: the BeComingOne has said unto me, You art my Son; this day have I begotten you" (Psa 2:7; Heb 1:5).

- "The BeComingOne came unto Nathan, saying, Go and tell my servant David, Thus says the BeComingOne ... and when your days be fulfilled, and you shall sleep with your fathers, I will set up your *seed* after you, which shall proceed out of your bowels, and I will establish his [Christ's] kingdom. He shall build a house for my NAME and I will establish the throne of his kingdom for olam. I will be his [Christ's] father, and he shall be my son" (2Sam 7:4, 5, 12-14).

- "And behold, you will conceive in your womb and bring forth a Son, and shall call his name Jesus. And he will reign over the house of Jacob for aeonian, and of his kingdom there will be no end" (Luke 1:31,33).

gp340» Just as Jeremiah was known *before* he was born, and was ordained a prophet *before* he was born (Jer 1:4-5), just as Paul was "appointed" an apostle *before* the world began (2Tim 1:11, 9), just as Christians are chosen *before* the world began (2Tim 1:9; 1 Per 1:2; 2Thes 2:13; see "Predestination Paper" [NM8]), is how Jesus Christ possessed the glory of God before the world began. Before the world began Jesus possessed the glory in an imperfect or incomplete sense: thus the use of the imperfect Greek verb ("I had" — KJV) in John 17:5. It was in an incomplete sense because he at the beginning did not yet have the glory, but was predestinated to have it.

gp341» But when Jesus Christ the man was predestinated before the world began he did not at that time possess the glory because at that time he did not exist as the Son of God, nor did the great glory of God exist before the world began: the great glory of God is coming, not here yet. In the truest sense, the glory of the God will exist when the BeComingOne (YHWH) has come (see GP 6, "Glory of God," and rest of this book). Such verses as Micah 5:2 can be explained in the way we explained John 17:5.

(4) Jesus Christ Existed in Heaven Before His Birth? How?

gp342» The scriptures that apparently say that Jesus Christ the man existed in heaven before his birth can be explained by the fact that Christ the man spoke the words of God, it was the Word of God or the angel of God *inside* Jesus the man that existed before Christ was born (see above under, "God Inside Christ"). Such verses as John 1:30; 3:13, 31; 6:33, 38, 51; 8:23, 42, 58; 1Cor 10:4; Col 1:17 and so forth can be explained in this way. As we will learn in GP 6 we also have a spirit or angel that existed before we were born. Does that mean we existed before we were born just because our spirit existed before we were born?

(5) Jesus Christ Created All Things? How?

gp343» The scriptures that seem to say that Jesus Christ the man created all things can be explained by the fact the Word or Spirit of the God was *inside* Christ the man, and it was this Word of God that created the present universe (Psa 33:6; Gen 1:1, 3 ["and God *said*"]). Another sense of Jesus Christ creating all is explained in the next chapter. Read all this book to understand still more fully the meaning of the verses that seem to indicate that Christ did/will create all. Some of these verses are John 1:1-4, 10; Col 1:16; Heb 1:2.

(6) Personification of Wisdom

gp344» The scripture in Proverbs 8:22-31, concerning the personification of Wisdom, where "The BeComingOne possessed me [Wisdom] in the beginning of his way before the works of old" can be explained by the fact Jesus the man had the Spirit or angel of Wisdom *inside* him (see above under, "God Inside Christ," note Isa 11:2; 42:1).

(7) Jesus Christ Humbled or Emptied Himself?

gp345» The scriptures that seems to say the "pre-existent" Christ "emptied himself" or "humbled himself" of his pre-birth glory or power to be born or transformed or incarnated as Christ the man can be explained by the fact that *inside* Christ the man was the Word or angel with the NAME of God. It was the Spirit *inside* Christ the man that humbled himself by being restricted inside a human being while that being was being humbled by the ignorant around him. And it was the man Jesus Christ the coming king who was also humbled, for he knew he was predestinated by God to be king of the whole world, yet he was treated with irreverence. These scriptures are Phil 2:6-8 and 2 Cor 8:9.

(8) Trinitarians' Bias

gp346» Most, if not all, who point to the scriptures that seem to say that Christ the man existed before he was born, speak of that pre-existent Christ as God (i.e. in the "Trinity" God), or a God, or an angel. But God is immortal and angels do not die (Luke 20:36). Therefore such an immortal being cannot be converted or transformed or incarnated into a being that can die. This is foolishness (See GP1 under, "Law of Contradiction").

Melchizedek

Without Parents, No Beginning of Days?

gp347» Some use the scriptures on Melchizedek to try and prove that Jesus Christ had no beginning of days, that he has always existed, and that Melchizedek (or Melchisedec) and Jesus Christ are one and the same person. They quote from Hebrews:

- "Without father, without mother, without descent, having neither beginning of days, nor end of life; but made like unto the Son of God; abides a priest continually." [Heb 7:3]

They point out that Melchizedek had "neither beginning of days, nor end of life, but made like unto the Son of God." So they conclude wrongly that Melchizedek was Jesus Christ, and that this is proof that Jesus had no beginning of days. But this misunderstands what Paul was saying and misunderstands type and antitype in scripture.

Jesus had a Genealogy

gp348» All one has to do to disprove that Jesus Christ and Melchizedek are not one and the same person is to read Hebrews 7:14, "For *it is* evident that our Lord sprang out of Judah; of which tribe Moses spake nothing concerning priesthood." Jesus, our Lord, was an offspring from Judah. Jesus had a genealogy and it is found in the Bible (Mat 1:1-17). Jesus was born "of the seed of David according to the flesh" (Rom 1:3). All through the Old Testament it predicted Christ's coming in the flesh (GP 4). It is the spirit of the anti-Christ that will not admit that Christ came in the flesh (1John 4:3). Jesus Christ came in the flesh with a clear genealogy. Where Christ differs from others is that his Father was God, not a human father (GP 4). It is clear that Jesus had a father and had a mother: he had a genealogy; he had a beginning. But Paul said that Melchizedek was without father, without mother. So it is clear from this alone that Melchizedek and Jesus are not one and the same person.

Pre-Existence?

gp349» Jesus Christ did not pre-exist as some say, but nevertheless his Spirit did exist before he was born (GP 3), as did my Spirit, as did your Spirit, and as the Spirit of everyone else did exist before they were born (GP 6). Just because

Christ or you or I have a Spirit that existed before we were born doesn't mean we pre-existed. We came into existence only when we were physically begotten.

Mechizedek Prefigured Christ's Perpetual Priesthood

gp350» But why did Paul say that Melchizedek was without father or without mother, "having neither beginning of days, nor end of life." Was Melchizedek an angel who always existed? No, for Paul said Melchizedek was a man (Heb 7:4). And as a man he was born, and he also died, even though the Bible did not record this fact. Look at what a study Bible had to say about this point:

> ■ "*Without father ... or end of life.* Genesis 14:18-20, contrary to the practice elsewhere in the early chapters of Genesis, does not mention Melchizedek's parentage and children, or his birth and death. That he was a real, historical figure is clear, but the author of Hebrews (in accordance with Jewish interpretation) uses the silence of Scripture about Melchizedek's genealogy to portray him as a prefiguration of Christ." [*NIV Study Bible*, footnote for Heb. 7:3]

Just because the Bible doesn't mention Melchizedek's genealogy doesn't mean he had none. Paul shows in the book of Hebrews that the high priest Melchizedek prefigured the priesthood of Christ (Heb 6:20). Christ's high priesthood did not come from him being genealogically linked to the Levites (who were priests by the Law), but through the "order of Melchizedek" (who was without genealogical linkage to the Levites):

> "For it is evident that our Lord was descended from Judah, a tribe with reference to which Moses spoke nothing concerning priests. 15 And this is clearer still, if another priest arises according to the likeness of Melchizedek" (Heb 7:14-15, 99-15).

But just because Melchizedek prefigured Christ in some way doesn't mean he was Christ. The Passover lamb prefigured Christ. Does this mean the lamb was Christ? Of course not. Other real persons in the Bible prefigured Christ, does it mean they were Christ just because they were a type of Christ. To get the "type" mixed up with the "antitype" is to show one does not know what type and antitype mean.

Notes

Genealogy of Jesus

gp351» Notice that there is something strange in Mary's and Joseph's genealogies. During the time of the Babylon captivity there is again a common ancestor to Joseph and Mary's side of the family. "And after they brought to Babylon, Jechoniah begat Shealtiel; and Shealtiel begat Zerubbabel" (Matt 1:12). And, "Zerubbabel, which was the son of Shealtiel, which was the son of Neri" (Luke 3:27). Now in Matthew 1:12 it says Jechoniah begat Shealtiel. This is impossible for Jechoniah had no sons who lived to produce offspring (see 2Kings 24:12, 15; Jer 22:24, 30; note Jechoniah=Coniah=Jehoiachin). But he had wives (2Kings 24:15). The scriptures said Jechoniah begat Shealtiel, yet scriptures say he would have no sons. Thus, the Bible contradicts itself? No! "If brethren [brothers or near of kin] live together, and one of them die, and have not child [Jechoniah had no sons], the wife of the dead [Jechoniah or Jehoiachin died without offspring, see 2Kings 24:12; Jer 52:31-34; Jer 22:24, 30] shall not marry without unto a stranger: her husband's brother [or near of kin] shall go in unto her, and take her to him to wife, and perform the duty of a husband's brother [or near of kin, see Book of Ruth, especially 3:11-13; 4:10, 13-14, this proves one doesn't have to be a brother, but be the nearest of kin willing and able to perform this duty] unto her. And it shall be, that *the first-born which she bears shall succeed in the name of his brother* which is dead, that his name be not put out of Israel" (Deut 25:5-6).

gp352» Therefore, a near of kin to Jechoniah performed his duty as described in Deuteronomy 25:5-10 and married one of Jechoniah's wives. The first-born of this relationship is accounted to the name of Jechoniah *not* the name of the near of kin who married one of Jechoniah's wives. Hence according to the laws of Israel Jechoniah begat Shealtiel even though physically he didn't, but a next of kin did.

gp353» The physical father of Shealtiel was Neri (Luke 3:27), but the *legal* father (for the genealogy of David) was Jechoniah (Matt 1:12). Therefore Neri was the near of kin who married one of Jechoniah's wives to conform to the law of Deuteronomy 25:5-10.

Review of GP 4

gp354» We have shown in this part that *before* Jesus Christ was resurrected, he was a human being because he was born from a woman. Jesus the man was not just any human being. Christ was also a Son of God, both physically (through the Holy Spirit's union with Mary) and Spiritually (through the medium of God's Spirit inside of him). Christ was also a Son of man, for he was born through the means of Mary his mother. Christ was a mediator between man and God; he was the Son of God and the Son of man. Jesus Christ the man actually had the Spirit or Angel of the BeComingOne (YHWH) inside him leading him in the right way. It was because of this Spirit that Christ the man became sinless. God was not Jesus Christ the man, but God was inside of Christ the man. When Jesus Christ the man died, his Spirit was then separated from him. The Spirit or angel of God did not die *as* Christ the man or *with* Christ the man, for Spirit cannot die. The angel of the BeComingOne separated himself from Christ the man when Christ died.

GP 5: Jesus Christ the God

Same Titles and Names
Spiritual Marriage
Image of God: Two into One
Before All
Trinity, Godhead and the Law of Contradiction
John 1:1-18
Sun and Moon
Glory of God

Who He Was/Is/Will Be

Resurrection

gp355» After Jesus Christ the man was dead and in the grave for three days and three nights, he was resurrected by his Father's power, that is, he was brought back to life again.[See *Chronology Papers*]

gp356» But before this resurrection we know that Jesus was *not* God when he lived on earth. He was a son of God, both physically and Spiritually, as just explained in GP 4. The very angel/Spirit of the BeComingOne (YHWH) was inside Christ. This angel/Spirit was in a sense the WORD of God of the Old Testament, for he carried the very words of God (see GP 3). Jesus was anointed by this very angel/Spirit. Jesus was the Messiah because he was anointed by this angel/Spirit. Jesus, you might say, was a shadow of the Spirit or angel inside him. Christ spoke what the Spirit inside him directed him to speak (Isa 59:21; John 14:10). Therefore in a sense he was a reflection of the true light. In fact, a symbol of Jesus Christ the man is the moon (see Notes). To ones on the earth, the moon is the reflection of the light from the sun, so too is Christ the man the reflection of the true Light of the Father to those on the earth. The Spiritual light in Jesus Christ the man came from the Father, the YHWH, the BeComingOne.

BeComingOne's Names And Titles

gp357» The BeComingOne of the Old Testament had such titles as:

A. **Lord of lords** (Psa 136:3);
B. **Lord and God** (Psa 68:19-20; 68:32);
C. **Lord of kings** (Dan 2:47);
D. **King of Glory** (Psa 24:7-10);
E. **Almighty God** (Gen 17:1; 28:3; Exo 6:3);
F. **Creator** (Isa 45:18; 48:13);
G. **Rock** (Deut 32:4; Psa 18:2; 28:1; Isa 26:4);
H. **Father** (Isa 63:16);
I. **Husband of Israel** (Isa 54:5);
J. **Savior** (Isa 45:15; 49:26);
K. **Redeemer** (Isa 47:4; 49:26; 54:5; 60:16);
L. **Word**, the one who speaks through an angel (Isa 52:6, see GP 3 & 4);
M. **True God** (Jer 10:10);
N. **"I am the first, I also the last"** (Isa 44:6);
O. **Only God** (I am YHWH, none else... no God beside me[Isa 45:5])

gp358» But God's proper NAME is YHWH, pronounced Yehowah or traditionally as Jehovah (lately as Yahweh). It means "He-(who)-will-be" or the "BeComingOne" (see GP 1).

gp359» We have shown in GP 4 that the angel of the BeComingOne was not the man Jesus. Jesus the man was born from an ovum as flesh, but born to be Savior of the world (see Matt 1:21; 1John 4:2-3; Luke 9:56; 1John 4:14). Christ the *man* saved no one. The BeComingOne through His Spirit was inside Christ the man, and He (YHWH) is the true Savior of all (Isa 49:26; 45:21).

gp360» The Savior came inside the flesh of Jesus Christ the man. Jesus Christ the man was the real sacrificial Lamb of God who died for mankind's sins. Christ the man lived his commission by revealing his Father's way (John 17:6; 14:10) and then dying as a sacrificial offering for sin (1Pet 1:18-19; Isa 53:10).

Christ's Names And Titles

gp361» But in the New Testament we see that the *resurrected* Christ is called:

A. **Lord of lords** (Rev 19:6);
B. **Lord and God** (John 20:28);
C. **King of kings** (1Tim 6:15; Rev 17:14; 19:16);
D. **King of Glory** (1Cor 2:8);
E. **Almighty** (Rev 1:8);
F. **Creator** (John 1:3; Col 1:16, 17; Eph 3:9);
G. **Rock** (1Cor 10:4);
H. **Father**, through prophecy, (Isa 9:6; 22:21);
I. **Husband of Spiritual Israel** (Rev 21:2; Eph 5:22-23);
J. **Savior** (2Tim 1:10; Titus 2:13; 3:6);
K. **Redeemer** (Gal 3:13; 1Tim 2:6; Titus 2:14; Rev 5:9);
L. **WORD of God** (Rev 19:13);
M. **True God** (1John 5:20);
N. **First and the Last** (Rev 22:13);
O. **Only God** (Jude 1:25; 1Tim 1:17; 1Cor 8:4)

gp362» We see that the resurrected Christ is called by the same names as the BeComingOne of the Old Testament. Christ the man was *not* called by these titles.[1] Only after Christ was resurrected was he called by these titles. The first time Jesus was called God was *after* he was resurrected (John 20:28), not before. But how can this be? Does this mean the BeComingOne gave his titles to Christ after the resurrection?

Glory Not Given to Another

gp363» "I am the BeComingOne: that is my name: and my glory will I *not* give to another, neither my praise to graven images" (Isa 42:8; Isa 48:11). The BeComingOne will not give his glory to another, yet we see the BeComingOne's titles on Christ, and we see that Jesus Christ will come in the glory of his Father (Note Mark 8:38).

Was Jesus YHWH in Old Testament Bible?

gp364» Because many have seen the titles of the BeComingOne on Jesus Christ, they have assumed that Jesus Christ was the BeComingOne of the Old Testament or was inside the God of the Old Testament as some

kind of "Trinity." But at the same time they ignore other scriptures that indicate Jesus Christ the man was not the same as God (see GP 4), or not the same as God the Father (GP 2). You can read books that assert false evidence that Jesus Christ belongs to some type of "Trinity." You can also read books that pretend to show evidence that Jesus Christ was/is only a man. You can also read so-called evidence that seems to show that Jesus Christ existed *before* he was born as a human (his so-called pre-existence). All this can get confusing even for the theologians.

gp365» Why did Christ after the resurrection have the same titles as the BeComingOne? How can this be when the BeComingOne said he would not give his glory to another? (Isa 42:8; 48:11) Can there be two Saviors, or two who are Almighty? Some may try and say that the BeComingOne could have given these titles to the resurrected Christ, but He said he would not give His glory or honor to another. Remember God does not lie; He keeps His Word. There is an answer to this paradox. The answer is all around us. Much of creation shouts to us the answer to this puzzle.

---

[1] Qualification: He was called "king of Israel" in John 1:49 as YHWH was called the king of Israel in the OT (Isa 44:6). But Christ at the time Nathanael called him king of Israel was only king in the sense that Christ was born to be king and born to be savior (John 18:37; Mat 1:2).

Spiritual Marriage: Two into One

Analogy

gp366» God tells us through Paul that we can know "the invisible things of him from the creation of the world." He is "understood by the things that are made" (Rom. 1:20). Through *analogy* of events or persons in the Bible Paul taught some of the New Testament Christians the truth about God. Paul used the analogy between Christ and Adam (1Cor 15:21-22, 45, 47; Rom 5:14, 18). Paul used analogy between Christ and Melchizedek and the high priest (Hebrews chap 7 & 8). Paul used analogy between the seed of Abraham and the true Seed, and between Abraham's faith and the Faith of Christians (Rom chap 4; Gal 3:16; etc.). Paul used analogy between old Israel (physical Jews) and new Israel (Spiritual Jews) [Rom 2:28-29; etc.]. Paul used analogy between a woman in marriage and the Church in marriage (Eph 5:22-33). Paul used analogy between the old physical Jerusalem and the new Spiritual Jerusalem (Gal 4:25-26). John, Paul, James and others from the New Testament used analogy to help explain doctrine. We will also use analogy to help prove doctrine. Paul spoke about the law having a shadow of good things to come, not the very image of the things (Heb 10:1). Paul spoke of events in the Old Testament happening as types or examples for Christians (1Cor 10:11; see "Duality of The Bible" in Intro).

gp367» Through analogy we will be better able to understand the paradox concerning Jesus Christ and the BeComingOne (YHWH) and we will be better able to understand how: "the WORD became flesh" (John 1:14).

Marriage Analogy: Two Become One

gp368» Notice how the physical marriage union was described in scripture? "Therefore a man shall leave his father and his mother and shall cleave to his wife, and they shall become one flesh"(Gen 2:24). And again, "For this cause a man shall leave his father and mother, and the two shall become one flesh; consequently they are no longer two but one flesh" (Mark 10:7-8). According to scripture, when two marry they are no longer two, but they are one. Two become one, not only sexually, but as a union in their work, finance, families, pleasure, hope, life, and name. And through the sexual act of two becoming one a new creation (a child) is born.

From two a newborn is created with the physical traits and genes of both parents. There is a higher meaning to the various aspects of sex especially sexual intercourse. **The sexual act of two becoming one is used in a spiritual sense throughout the Bible**:

- **1)** In a good sense of the Church and God: 2 Cor 11:2; Rev 21:2; Eph 5:23-32; Matt 22:1-14; 25:1-13; or Israel and God, Isa 54:5;

- **2)** or in the evil sense of mankind and Satan: Ezek 16; Exo 34:15-17; Jud 2:11, 17; Jer 3:6-14; Hosea 1:1-3; 2:19-20; 4:10-19; Rev 2:14, 20-22; 17:1-2ff

Therefore, there is a another meaning to sex in the scriptures and when we understand this higher meaning we will understand the mystery of the BeComingOne (YHWH) and Jesus Christ having the same titles even though the BeComingOne said he would not give his glory to another.

God the Husband of Israel

gp369» In the Old Testament the BeComingOne is pictured as the husband of the Israelite people, "for your maker is your husband; the BeComingOne of hosts is his name" (Isa 54:5; see Jer 3:14; Hos 2:19-20; etc.). The BeComingOne is pictured as the husband of Israel. Jesus Christ the man was an Israelite, the true seed of Abraham (Heb 2:16; Gal 3:16). He fulfilled Israel's promise not to sin and thus is the real Israel (Ex 19:5-6). We also see that Christ the man is the true Israel of God by comparing Matthew 2:15 with Hosea 11:1. So following this analogy since God was the husband of Israel, then God is the "husband" of Christ, the true Israel of God.

Spiritual Marriage

gp370» The BeComingOne of the Old Testament was connected closely with an angel as shown in GP 3. In fact, the BeComingOne (YHWH) said his NAME was *in* the angel or messenger (Exo 23:21). Just as in a physical marriage when two become one, so too in a spiritual marriage — two do become one. The angel of the BeComingOne (who had the NAME in him) spiritually married Jesus Christ the man and they, who were two, became one. As Christians have their own angel (Mat 18:10), Christ also had his (Acts 12:8-11; Rev 1:1; 22:16), the very first-angel, the angel that carried the Word of God (GP 3). And as we will see later, in the type and antitype of the Bible, the spiritual are the antitype "males" while the physical are the antitype "females." So the antitype male (the angel of Christ) became one with the antitype female (Jesus the physical Israelite).

Jesus Christ's New Name

gp371» As a woman takes on the family name of her husband, in the fulfillment of this analogy Jesus also took on the NAME of God. The NAME of God is the BeComingOne (see GP 1). God or the BeComingOne is Jesus' *new name* (Rev 3:12, see Isa 62:2; Rev 14:1; Isa 65:15; GP1, "Name in Scripture" at ¶ gp72). Yet as a woman retains her personal name, so does Jesus retain his personal name, but his surname is now the BeComingOne (YHWH), the true God, for he went into the God. In a sense Jesus is now in the Family of YHWH — the BeComingOne.

gp372» The Spirit of the God and Jesus Christ are spiritually married. They are one, thus, share the honors of each other. That is why Jesus is called by the New Testament writers with the same titles as the BeComingOne was called in the Old Testament. Unlike the physical marriage — they are perfectly ONE. It is a perfect and complete relationship. What God does is perfect (Psa 18:30; Deut 32:4).. Two have become one as the commandment demands (Gen 2:24).

Notice how a perfect marriage should be:

> ■ "So ought men to love their wives as their *own* bodies. He that loves his wife loves *himself* ... let every one of you in particular so love his wife even as himself" (Eph 5:28,33).

gp373» Jesus Christ's human body is now God's body. "God is Spirit" (John 4:24). The angel of the BeComingOne (YHWH) is Spirit, for angels are spirits (Heb 1:7). Thomas called Christ, God (John 20:28). In fact, Jesus Christ the man *with* God (His Spirit) through their Spiritual marriage are ONE. Since they are one, the resurrected Jesus Christ is God (John 20:28; Titus 2:13; 2Pet 1:1; 1John 5:20; Jude 1:25; 1Tim 1:17; 4:10; see *The Trinity*, by Bickersteth, chap. 4).

Jesus Christ's Flesh

gp374» But notice *after* Jesus was resurrected and became one with his Spiritual mate, he appeared to his disciples from nowhere and:

> ■ "They were terrified and affrighted, and supposed that they had seen a spirit. And he [Christ] said unto them, why are you

troubled? And why do thoughts arise in your hearts? **Behold my hands and my feet, that it is myself [Greek: 'I am myself']: handle me, and see: for a spirit has not flesh and bones, as you see me have.** And when he had thus spoken, he showed them his hands and his feet ... And they gave him a piece of broiled fish, and of a honey comb. And he took it, and did eat before them."[Luke 24:37-43]

gp375» Scripture tells us of women actually taking hold of him after he ascended to God (Matt 28:9-10). Christ became one with God, thus one with Spirit, but here (Luke 24:37-43) scripture says the resurrected Christ was flesh and blood, he was like a human being. Is this a Biblical contradiction?

Two, Spirit and Flesh, Became One

gp376» No, for as God is spirit, so too the resurrected and ascended Christ as God, is spirit. But further, Christ is flesh and blood. God has Spiritually married and become one with mankind. And now they are one. The God is now in mankind and mankind is in the God through Jesus Christ (2Cor 6:16; John 10:38; 17:21; Rev 21:3). Two, spirit and flesh, have become one as the higher meaning of the commandment of Genesis demands.

Christ in the Image of God

The Image has Something to do with Two in One

gp377» We see that Christ fulfills the image of God, bodily (Col 1:15; 2:9; 2 Cor 4:4; Heb 1:3). What is the image of God? Since God is spirit and spirit is invisible (GP 3), the image or likeness of God is not necessarily the *appearance* of the man, but does include other aspects of the likeness. We see that "God created man in his image, in the image of God created He him; male and female created He them" (Gen 1:27). Notice that God created "him" in the image of God, but also God created "them" male and female. But look, "male and female created He them; and blessed them, and called *their* name Adam" (Gen 5:2). God created "him," meaning the male (Adam), in the image of God. But furthermore, God created *them* male and female and called *their* name Adam or man since the Hebrew word translated "Adam" is the same Hebrew word translated elsewhere as "man." So does this mean that male and female are in the image of God? Yes, because God's word says we can know the Godhead by his creation (Rom 1:19-20), and one of the most obvious aspects of the creation is the male and femaleness of it, and the reproduction aspect of it. Sex and reproduction penetrates the whole creation. So from the Bible we can ascertain that sex and other aspects of man are included in the likeness or the *image of God*:

A. The male (Adam) was created in the image of God (Gen 1:26-27; 5:1) while the female is also the glory of man (1 Cor 11:7)

B. Both male and female are called man, and thus, male and female are man (Adam) and are in the image of God (Gen 1:26-27; 5:1-2), but the female herself is in the reverse sexual image of the male

C. After God created the first man, the man was alone (Gen 2:18)

D. Woman came out of man (Gen 2:21-23), but in turn after the first man, man came out of woman (Gen 4:1; 1 Cor 11:12)

E. Both the man and the women ("them") were to have rulership over the creation (Gen 1:26, 28)

F. Both man and woman became one (Gen 2:24)

G. Both were to be fruitful and multiply and have rulership over the earth (Gen 1:26)

H. Both rested on the seventh day (Gen 2:1-3)

The male (Adam) was created in the image of God, but both male and female are called man. Thus, male and female are man (Adam) and are in the image of God. But as Paul indicated (1Cor 11:7) there is a double meaning here since the male himself is an image of God while the female herself is in the image or glory of man. And both are out of each other, but all things out of God (1Cor 11:12).

Male and Female are One

gp378» From the very beginning of creation God considered male and female as "man." God used the name "man" to include both male and female. In many languages today, "man" has a dual meaning of not only meaning male, but also mankind, which includes both sexes. The very act of sexual intercourse and marriage has the meaning of signifying two (male and female) becoming one (Gen 2:24; Mark 10:6-8). Male and female are two, yet in God's eyes they are one, they are "man" or "Adam."

Spirit and Flesh of Christ Become One

gp379» The antitypical male (Spirit of Christ) has joined with the antitypical female (the flesh of Christ) and they two have become one. Genesis 2:21-25 pictures the woman coming out of the man just as Jesus Christ the man came out of God (see GP 4). And Genesis 2:21-25 also pictures the woman coming back to the man to become one, just as Jesus the man was the first to go back and become one with the Father:

■ "I have come out of the Father, and I have come into the kosmos, again I am leaving the kosmos and I am going toward the Father" (Greek, John 16:28, & see v. 5, 10, 17; and John 13:3; see Notes this Part).

Another Sense of the Image of God

gp380» There is still a higher meaning to the Image of God which we will leave until GP 8 to reveal. Since both male and female are in the image of God, the God also must in some sense be "male" and "female," and two in one. ("And

God said, let <u>us</u> make man...." [Gen 1:26] Hint: there is a right and left side of God. [GP 8].)

Word Became Flesh

gp381» In GP 4 we explain one meaning of John 1:14: "and the WORD became flesh." The typical meaning of this verse is: "and the WORD [in] flesh became," for God came inside Jesus the man. But the Bible is dual (type and antitype) and this verse is of a dual significance. In Isaiah 52:5-6 the Bible shows us the BeComingOne is the one who spoke through the Bible, by the means of his angel (see GP 3). Now in GP 3 of these papers we manifested that the angel (a spiritual being) of the BeComingOne was the one who carried the words for the BeComingOne. He was the messenger for the BeComingOne. Hence the angel of the BeComingOne is the WORD. And in GP 5 we are showing you that the angel of the BeComingOne and Jesus Christ the man are now one. Therefore the antitypical meaning or the higher meaning of John 1:14 is that the WORD *became* flesh, or the angel of the BeComingOne became flesh. Or the angel of God took on the essence of the flesh. The BeComingOne's angel or WORD was infused with the physical Christ: "and the WORD became flesh." But since the angel had the NAME, then in a sense the BeComingOne became flesh.

gp382» Actually, this antitypical meaning of John 1:14 is what God intends us to understand. The book of John was written *after* Christ's resurrection when the angel of the BeComingOne and Jesus the man *had* already become ONE, thus, after the WORD had become flesh, that is, had become ONE with Jesus Christ the man. Read the section on John 1:1-18 later in GP 5 for greater detail.

New Creation: Two into One

gp383» Christ God has two essences: the physical and the Spiritual. This is the new creation (2Cor 5:17). Now the new creation described in 2 Cor 5:17 is speaking about Christians becoming new creations. But since we know that we will be changed into a body like Christ's (Phil 3:21; 1John 3:1-2), then what the Bible says about Christians is also true about Christ. Jesus the man and God were infused together as a new creation. Jesus Christ is the beginning of the creation of the true God (Rev 3:14). Thus, as a new creation, Jesus Christ is able to function as a spirit being (go through matter such as walls, John 20:19), yet he is also able to

function as a physical being (eat, Luke 24:41-43; drink and eat, Acts 10:41).

Two Bodies in One

gp384» As there are two bodies in a physical marriage, so too in the spiritual marriage — "there is a fleshly body, and there is a spiritual body" (1Cor 15:44). "And as we have borne image of the earthly, we shall ALSO bear the image of the heavenly" (1Cor 15:49). "If there is a natural body, there is ALSO a spiritual body" (1Cor 15:44, NIV). The Bible does not say when humans are born of God they get rid of their physical essence.

New Soul

Saving of the Soul

gp385» We know from the "Body, Soul" paper [NM 6] in the *New Mind Papers* that "soul" in the Bible is a living physical body, that is, a physical body with the breath of life in it. And from that paper we know that souls can die. Thus the Bible speaks of the "saving of the soul" (1Pet 1:9; Heb 10:39; James 1:21; 5:20; Luke 21:19). The Bible speaks of the saving and resurrection of the physical body:

- "And if the Spirit of him who raised Jesus from the dead is living in you, he who raised Christ from the dead will also *give life to your MORTAL BODIES* through his Spirit, who lives in you" (Rom 8:11).

- "... the redemption [buying back] of our bodies" (Rom 8:23).

Using the example of Christ and Christians, the soul is saved ultimately by the dead physical body being resurrected from the dead and given the Spirit, and these two become one (GP 6). But the physical living body or soul is living because it has *breath* inside it (see "Body, Soul" paper). There is a dual meaning here. "Breath" in the Bible is translated from words in both Hebrew and Greek that can mean either "breath" *or* "spirit" (see "Body, Soul" paper). The old soul prefigured the new body or New Soul of Jesus Christ, the individual. When a soul dies, the "breath" and the "spirit" leave the body (John 19:30; Matt 27:50; Psa 146:4; cf. Gen 2:7; etc.). But a living physical body has breath and/or spirit in it. Following the analogy we see also that the new living body or the New Soul of Jesus Christ has the Flesh *and* the Spirit.

gp386» The Bible says that our old body will be transformed into a glorious body like Christ's (Phil 3:21; 1John 3:2). "If we have been united with him in his death, we will certainly also be united with him in his resurrection" (Rom 6:5). And the resurrected Christ has both spirit, and flesh and blood. He has two bodies or two essences in one like a marriage union. Did not the resurrected Christ say he was flesh and blood? Yes (Luke 24:37-43). And as God, he must also be spirit.

Angel of the Old Testament Prefigured the New Soul

gp387» As we've shown, in GP 3: God appeared through his angel looking like man *and* like flaming-fire. Flaming-fire is used in the Bible to indicate spirits or angels (see Heb 1:7; Psa 104:4; Ex 3:2). In Judges chapter 13 we've shown you in GP3 how the angel of the BeComingOne appeared as a man to Manoah, but then transformed himself into a flame and ascended into heaven. These manifestations of the angel of the BeComingOne to mankind were a foreshadow of the future Spiritual marriage of God and man. These appearances prefigured the New Soul. From the beginning God was manifesting the future essence of God. Also sex and the laws given to man through Moses concerning marriage and sex have been manifesting God's plan to us, as we will continue to show you. Remember the BeComingOne's Name indicates that he is the BeComingOne, **He will be.** The angel of the BeComingOne of the Old Testament foreshadowed the predestinated future.

Beginning with Moses Christ was Manifested

gp388» "And beginning at Moses [the first 5 books of the Bible] and all the prophets, he [Jesus] expounded unto them in all the scriptures the things concerning himself" (Luke 24:27). This was the resurrected Christ speaking to two disciples. Jesus began at Moses to explain himself. Christ was at that time infused with His Spirit. As Christ began in the books of Moses to show some disciples His essence, so have we begun at the beginning showing you that the God manifested himself in the Old Testament through an angel. But now that angel, the archangel, *and* Jesus Christ the man have become ONE.

Spiritually Married

gp389» Parenthetically, even as the pleasure of becoming one sexually, so too the pleasure of becoming one or being one with the Spirit. The pleasure of the male and female becoming one physically and mentally, foreshadows the pleasure of the human body and Spiritual body becoming one in Spirit and body. This is one reason why in the Kingdom of God there will not be any marriage as we know it on earth (Luke 20:35). To be Spiritually in the kingdom of God you must be complete: you must be one; a new finished creation; you must be spiritually married. Physical marriage has pointed to the Spiritual marriage.

Christ: Second Adam

gp390» In analogy Paul called Christ the second Adam or the last Adam (1Cor 15:47, 45). Let's follow through on this by comparing the first Adam with the Second Adam:

A. **First Adam**: Adam came from God (Gen 2:7); **Second Adam**: Christ came from God (Luke 1:35).

B. **First Adam**: Man was in the image of God, male and female (Gen 1:26-27); **Second Adam**: The physical and Spiritual Christ is/will fulfill the image of God (Col 1:15), as Christ and the Church fulfills the male (husband) and female (wife) roles (Rev 21:2,9; Eph 5:22-32).

C. **First Adam**: Adam started out alone until his wife was taken out of him (Gen 2:18, 21-23); **Second Adam**: Christ started out alone before his "wife" or church came into being from him (Acts 1:4-5, 8; 2:1, 4, 33).

D. **First Adam**: Woman came out of Adam and is under his dominion (Gen 2:21-22; 3:16); **Second Adam**: Christ's wife the church comes out of Him and is under His dominion (John 1:13; Acts 2:33; NM papers; Eph 5:22-25).

E. **First Adam**: Adam and Eve was to have rulership over the world (Gen 1:26, 28); **Second Adam**: Christ will rule the whole world with his "wife" (Rev 4:11; 5:10; Rev 21:2; etc.).

F. **First Adam**: Adam and Eve became one (Gen 2:24); **Second Adam**: Christ's Spirit and

Flesh became one (John 1:14; GP 5) as Christ's Spirit and the Church will become one (Eph 5:31-32).

G. **First Adam**: Adam with his wife was to be fruitful and multiply (Gen 1:26); **Second Adam**: Christ with his "wife" would multiply and fill all (Eph 1:22-23).

H. **First Adam**: Adam with his wife rested on the seventh day (Gen 2:1-3); **Second Adam**: Christ with his "wife" will rest on the 1000 year-seventh-day (Rev 20:3-5; NM papers).

Compare the above (A to H) with the previous section called, "Christ in the Image of God."

Christ the Man Became and is Becoming the God

gp391» Today, because Christ the man became one with the Spirit of God, Christ has become to us:

- "*the* Lord and *the* God" (John 20:28, see Greek).

- Christ as God is the "only God our Savior" (Jude 1:25; 1Tim 1:17; 1Cor 8:4).

- Christ is "*the* great God, our Savior" (Titus 2:13).

- Christ is "*the* true God" (1John 5:20).

- Christ is the "Living God" (1Tim 4:10).

- Christ has been given *all* the power of God (Matt 28:18), he is the Almighty God (Rev 1:8; 15:3).

- Christ thus has the power and wisdom of God (1Cor 1:24).

- Christ was given all things (Luke 10:22; John 3:35).

- But *now* all are not under Christ's power (Heb 2:8).

- But all will be put under Him (1Cor 15:25; Psa 110:1).

- He has "a NAME which is above *every* name" (Phil 2:9).

Christ's NAME is above every name because He is Christ, who has the very NAME of the God, the BeComingOne. The Spiritual being, the angel of the BeComingOne, that Jesus Christ the man

was joined to and became ONE with, had the NAME of the God (YHWH) in him (GP 3). Thus, Jesus Christ is jointed to that NAME and is ONE in that NAME. In a sense, Jesus Christ is now Jesus Christ the God, or **Jesus Christ the BeComingOne** (YHWH).

Christ into the Glory of the Father

gp392» Notice, "and every tongue shall confess that the Lord Jesus Christ [went] *into* the glory of God the Father" (Phil 2:11). What Philippians 2:11 says when correctly translated is what Isaiah 9:6 and 22:21 prophesied about. In Isaiah 9:6 is pictured the Christ child being born, and thereafter receiving the rights of the government of God, and becoming the "duration *father*." In Isaiah 22:20-21 it pictures Eliakim (a typical representation of Christ) becoming "a *father* to the inhabitants of Jerusalem." Also in Jeremiah 3:19 the returning Christ God will be called by Israel, "my *father*." Christ in the glory of God is the Spiritual Father of the New Creation. Christ sitting on God's throne, at the right side of the true God, sends forth the Spiritual Seed, or Holy Spirit (Acts 2:24-33; 7:55-56; see "Seed Paper" [PR 1]).

gp393» Jesus Christ the man went into the God, "For in him dwells all the fullness of the godhead, bodily" (Col 2:9). And, "the light of knowledge of the glory of the God in the person of Jesus Christ" (2Cor 4:6). Note the following scriptures:

gp394» "But Jesus answered them, saying, The time has come for the Son of man to be glorified ... Now is the Son of man glorified, and the God is glorified in him. And the God will glorify him [Christ] in Himself, and at once He will glorify him [Christ]." [John 12:23; 13:31-32]

Thus the glory of the God or the great appearance of the God is manifested in the person of Jesus Christ, and the fullness of the Godhead is (or will be) in him bodily. The great appearance or glory of God has been/will-be fulfilled in the resurrected Christ.

gp395» Christ the man thus went into the Godhead, he became ONE with the good God and the good Father, or God the Father incorporated Jesus the man into Himself. Christ the man was infused into the God, as a new creation, or as a New Soul.

Scriptures Now Make Sense

Better than Angels

gp396» Notice how scripture now makes sense when we know what was shown so far in these papers: "Being so much better than the angels, as he [Christ] has by inheritance obtained a more excellent name than they" (Heb 1:4). Jesus the man and the angel of the BeComingOne together as one have become ever so much better than angels. Both together are in a better state than they were before. The Spirit of God has a physical body; Jesus Christ the physical man has a Spiritual essence. This is much like a man and woman are in a better state when they are married, than when they are alone.

Under Christ, not Angels

gp397» "For if the word spoken by angels [the angel of the BeComingOne was the WORD] was steadfast, and every transgression and disobedience received a just recompense of reward. How shall we escape if we neglect so great salvation; which at the first began to be spoken by the Lord ... For unto angels he has not put in subjection the world to come" (Heb 2:2-3, 5). In the Old Testament, it was the angel(s) of the BeComingOne that ruled through the power that was predestinated to them and through the powerful words that were given them. This will become clearer as you read the rest of this book.

Jesus Christ the BeComingOne

gp398» The kingdom of God will not be under angels, but under the true God. The true God is the BeComingOne, YHWH (GP 1). The truest sense of the BeComingOne is the fullness of the Spiritual Body of Jesus Christ. All will eventually go into this true God (see GP 6, "All into Christ the BeComingOne"). We can call Jesus Christ, "*the* God," only if we know it is at the fullness of the Spiritual Body of Jesus Christ that Jesus Christ, in the truest sense, will be, "*the* God," the God all in all (see GP 6).

He was before All in Two Senses

gp399» Jesus Christ the BeComingOne calls himself "the beginning of the creation of God" (Rev 3:14). And, "he is before ALL things, and through him all things consist. And he is the head of the body: who is the beginning" (Col 1:17-18). Remember Jesus Christ is infused with God though the angel who had the NAME of God (YHWH) in him; thus, Christ shares the titles and honors of his Father because the angel was in his Father's NAME. Christ as the Spiritual wife of the angel of the BeComingOne shares his Spiritual husband's NAME, titles, and honors; as should a physical wife share the name, titles, and honors of her physical husband (1 Pet 3:7; Eph 5:28, 33, etc.).

gp400» Christ the **man** was born, he came in the flesh; he never existed before all things (present cosmos) except in the forethought of God (see GP 4). It was the angel or WORD of the BeComingOne who was before all things in the present or old creation. Christ the man was resurrected and created new with the angel of the Father, as a perfect marriage between Spirit and man. Christ is the first and is before all things in two senses:

- **(1)** Christ God is the beginning of the *new* creation of God. Christ the man is the first of the old creation to go back into God, therefore he is before all things as far as returning to God.

- **(2)** But also with the scriptures shown you previously, we know that the angel (WORD) of the BeComingOne was the first, the very beginning of the first creation of God, for all things came out of God (1Cor 8:6, see Greek), and the first to come out of the Father was the WORD (John 1:1; Psa 33:4; Heb 11:3).

Trinity, Godhead and the Law of Contradiction

Immortality and Death

gp401» According to the Bible God is immortal (1Tim 1:17). Thus, it follows from the Law of Contradiction[1] that God cannot be immortal (not capable of death) and yet at the same time be capable of death. Either God is immortal or he is not. He cannot be both mortal (capable of death) and immortal (not capable of death) at the same time. Furthermore, if at first, God was immortal, then at no time later can he die. If a so-called immortal person ever dies, it merely means, he was never immortal. If one denies this, he either denies the Law of Contradiction or denies the meaning of the word "immortal." To deny the meaning of "immortal" is to play a word game.

"With God Nothing Shall Be Impossible"

gp402» Now some will think that yes God is immortal and yes those who are immortal cannot die, but "with God nothing shall be impossible" (Luke 1:37, KJV). Thus, they reason, in so many words, God (or Jesus Christ) before his "incarnation" was immortal, but nevertheless died on the cross, for without this death no man could be saved. They reason that this contradiction is not contrary because "with God nothing shall be impossible." But this kind of "reasoning" makes a mockery out of reasoning. It is against the Law of Contradiction and against scripture

(They also use the argument that God is timeless, immortal, and immutable, and thus according to their theory, the fleshly body of the Son must have been "taken" because he could not literally be made flesh, for he was not mutable. See more against this argument in the *Trinity v. BeComingOne Comparative Tables* on the web site.)

Incorrect Translation

gp403» There is a problem with the above argument. They use Luke 1:37 of the Kings James Version (KJV), but this is a mistranslation of the Greek text. Luke 1:37 from the Greek language says: "for not shall be impossible with the God any word." The Greek word *rhema* (Strong's #4487) was not translated in Luke 1:37 in the KJV. The Greek word *rhema* means: "word, saying, any thing spoken." Thus in the very next verse, Luke 1:38, we find this same Greek word *rhema* translated as "word": "be it according to thy *word*" (Luke 1:38). What was being said in Luke 1:37, is that God's *word* that Mary would have the child Jesus even though she "know not a man" (Luke 1:34), was not impossible (Luke 1:37). Thus Mary believing the words answered the angel, "be it unto me according to thy *word* [Gk. *rhema*]" (Luke 1:38). The scripture that says in the KJV "with God all things are possible" (Matt 19:26; Mark 10:27; Luke 18:27) is in context merely saying that even the rich men can be saved, for such things are possible with God.

God of Law, Not of Confusion

gp404» To say that God can go against the basis of reason, the Law of Contradiction, is mockery. God is a God of Law (Isa 33:22; etc.), not of confusion (1Cor 14:33). God is a God of His Word: "So shall my word be that goes forth out of my mouth: it shall not return unto me void, but it shall accomplish that which I please, and it shall prosper whereto I sent it" (Isa 55:11). "I have spoken, I will also bring it to pass; I have purposed, I will also do it" (Isa 46:11).

God's Word Not Impossible

gp405» Luke 1:37 is saying that all God's words are possible with God (see Greek text). Nothing God speaks or says is too hard for him for he created heaven and earth by his word (Psa 33:6). "Ah Lord GOD [YHWH] behold; you have made the heaven and the earth by your great power and stretched out arm, and there is not too difficult for you any *word*" (Jer 32:17). Notice in the KJV of Jer 32:17 there is again a mistranslation, "there is nothing too hard for thee," instead of "there is no word too hard for thee." It says in the Hebrew that no word (Heb. *dabar*) of God is too difficult for Him. Why? It is because God does not lie. Therefore any word he speaks is true. "The sum of your word is true," or "chief is your true word," or "truth is the head of your word," or "the head of your word is truth," or "your supreme word is true"

---

[1] See GP1 for definition of the law

(Psa 119:160, see Heb. and various translations). He who is the liar is Satan (John 8:44).

Truth Is

gp406» The truth is: *what is*, at one point in time. What is false is: *what is not*, at a certain point in time. One statement cannot be true and false <u>at the same time</u>. God cannot be immortal and yet be capable of death. God cannot be alive forever, and yet die. God cannot go back on his word (Isa 46:11). It is impossible for God to lie (Heb 6:17-18; 1John 5:18; etc.). Considering everything, it is impossible for God to go against the Law of Contradiction.

It Follows Thus:

gp407» From the above arguments it follows that:

- God did not die on the "cross" (tree "stake" or "post");

- God did not change from immortality to mortality in order to die on the cross (God cannot change what has gone out of his mouth); OR an immortal angel (note Luke 20:36) was not "transformed" from immortality to mortality in order to die on the cross;

- he who died on the cross was not God, <u>when</u> he died;

- he who died on the cross was a mortal, for he died;

- he who <u>died</u> on the cross could not have existed as God or angel (angels do not die, Luke 20:36) before his birth as a mortal, because those who are immortal cannot change from their state to mortality;

- he who was born (or thus came into an existence through a physical birth) and died (through the 'cross') was a man;

- that man was/is Jesus Christ the man who with a resurrection went into his Father (God) and became one with the God and now sits on the right side of the Power of the God (GP 4-8).

Jesus Christ as a mortal man *before* his resurrection was a man, "the man Christ Jesus" (1Tim 2:5). He was a go-between (mediator) between God and man (1Tim 2:5). He was a physical son of God. But he did not exist (the one who died) before his birth (GP 4).

Trinity Belief Impossible

What is the Trinity Belief?

gp408» From the above, especially when you understand the Law of Contradiction, you know that three uniquely different persons cannot exist as one person at the same time (Trinity theory). The Trinitarians have been changing and redefining their theory on God from the beginning of their theory. But what many of them are saying is this:

- There is *one God being* who created all and who is immortal and all-powerful (p. 87, *Systematic Theology*, by L. Berkhof; p 155-156, *The Trinity*, by Bickersteth);

- There are *three persons* (Father, Son, Holy Spirit) in this one God being; these are *not* three different metonymical names for the same essence in the one God being (p. 87, *Systematic Theology*; p. 156, 150ff, *The Trinity*);

- The whole essence of God belongs equally to each of the three persons (p. 88, *Systematic Theology*; p. 155, *The Trinity*).

gp409» From, *Logic and the Nature of God* (1983), by Stephen T. Davis, the trinity doctrine is stated:

- "The Christian doctrine of the Trinity is notoriously easier to state than explain. Augustine states it as follows:

    There are the Father, the Son, and the Holy Spirit, and each is God, and at the same time all are one God; and each of them is a full substance, and at the same time all are one substance. The Father is neither the Son nor the Holy Spirit; the Son is neither the Father nor the Holy Spirit; the Holy Spirit is neither the Father nor the Son. But the Father is the Father uniquely; the Son is the Son uniquely; and the Holy Spirit is the holy Spirit uniquely. All three have the same eternity, the same immutability, the same majesty, and the same power...." [p. 132]

gp410» Davis in his 1983 book further stated:

- "But is the doctrine of the Trinity coherent? Is there any good reason for a Christian to believe it? Let us say that the doctrine consists at heart of five statements:

- (1) The Father is God

- (2). The Son is God

- (3). The Holy Spirit is God

- (4). The Father is not the Son and the Son is not the Holy Spirit and the Holy Spirit is not the Father

- (5). There is one and only one God." [pp. 134-135]

What Mystery?

gp411» "Not surprisingly, Christian theologians almost with one voice have stressed that this doctrine is a great mystery, perhaps the greatest mystery in Christian theology" (p. 132). But some of them use scripture out of context to prove this. They quote Paul speaking about mystery:

- Without any doubt, the **mystery of our religion is great**: [NRS 1 Timothy 3:16]

and misquote Paul:

- That their hearts might be comforted, being knit together in love, and unto all riches of the full assurance of understanding, to the acknowledgment of the **mystery of God, and of the Father, and of Christ**. [KJV, Colossians 2:2]

gp412» Paul wasn't acknowledging any mystery, this last verse is better understood from the NRS version:

- I want their hearts to be encouraged and united in love, so that **they may have** all the riches of assured **understanding** and have the **knowledge of God's mystery, that is, Christ himself,** [NRS Colossians 2:2]

gp413» Paul was praying for others to understand the knowledge of God's mystery like Paul understood it. Paul did understand the mystery and in fact revealed it to the Church:

- he has made known to us the mystery of his will, according to his good pleasure that he set forth in Christ, [NRS, Ephesians 1:9]

- and how the mystery was made known to me by revelation, as I wrote above in a few words, 4 a reading of which

will enable you to perceive my understanding of the mystery of Christ. [NRS, Ephesians 3:3-4]

- **This is a great mystery**, and I am applying it to Christ and the church. [NRS Ephesians 5:32]

- Pray also for me, so that when I speak, a message may be **given to me to make known with boldness the mystery of the gospel,** [NRS Ephesians 6:19]

- I became its servant according to God's commission that was given to me for you, to make the word of God fully known, 26 **the mystery** that has been hidden throughout the ages and generations but has **now been revealed to his saints**. [NRS, Colossians 1:25-26].

Therefore there is no mystery, for Paul was given the revelation to reveal it, and Paul did reveal it, as we will again, in more detail, reveal.

Trinitarian Belief against the Law of Contradiction

gp414» Thus, Trinitarians are saying that three uniquely different beings are a one, single, unique being at the very same time they are three. This is against the Law of Contradiction and against Biblical scripture as we will see. We can understand three in one, or even two in one, or even a million in one, but this "one," in its totality, cannot be exactly the same as *each* of the three, or *either* of the two, or *each* of the million.

One in Number, but Three in Person

gp415» Either there is only one (single/individual) God or there are three (as one, as a unit, in some kind of unity or oneness) at the same time, but both "the only one" and "the three persons" cannot exist as the only one (single/individual) at the same time. When the Trinitarians are speaking of *one* God, they are **not** using one as a synonym for "unity." To them the oneness of God is not the same as the unity of God. When they speak of the "only one" they mean singleness, numerically speaking, of God. Yet to them there are three persons in this *one* (numerically speaking) God. This is the first real contradiction. A drop of water can only appear as either ice, or water, or vapor at any one time: all three states of the same water drop cannot exist at the same moment.

Immortal Person Dies?

gp416» There is another real contradiction in their theory. In the Trinity theory they have an immortal being dying. Someone who is immortal cannot die. Someone with the potential to die is mortal. Anyone who says an immortal being can die either doesn't know the meaning of 'immortal' or is playing a word game. God's essence is not like the god of the Trinity as commonly taught today.

God Not the Trinity

gp417» One reason God is not a trinity is because the Trinity idea is against the Law of Contradiction. For other reasons (documented in this book) that God is not the Trinity, note the following:

- Jesus Christ the man was not God before his death and resurrection: he was the *son* of God.

- Jesus Christ the man did not exist before his birth. Jesus Christ was *foreknown* before his birth and death, but he did not exist before his birth, except in the mind and planning of the God (see Psa 139:16; Jer 1:4-5, NIV).

- But doesn't it seem to say in the Bible that Jesus Christ is God? Yes (see GP 5).

- Don't some titles of God in the Old Testament now appear on Jesus Christ? Yes (see GP 5).

- Doesn't the Bible seem to say that Jesus Christ is the Creator and God? Yes (see GP 5).

- BUT the Bible does not speak of Jesus Christ the **man**, *before* his resurrection and his going to his Father, as *the* God, or a God, or the Creator.

- When Jesus Christ was a man, before going to his Father, he spoke of his Father as the God of the Jews (John 8:54).

- He spoke of his Father as "my God" (John 20:17).

- He spoke of his Father or God as being greater than himself (John 14:28).

- And He prayed to his Father (John 17:1) as if his Father was a separate entity, which the Father was at one point in time.

In this book the problem of God and his Son has been solved. **Time** plays an important role in the answer.

But what about the Holy Spirit?

Fourth Person in the Godhead?

gp418» **The Trinitarians argue:**

- "First, we show that the Bible teaches that there is only one God. Second, we found that the Bible tells us that there are three persons who are called God. Hence, the inescapable conclusion: the three persons are the One God. Theologians have called this the Trinity"

(Of course, whether you know it or not, the theologians also deny that the three "persons" in the Trinity are in fact, persons. They use their own pseudo-word "hypostases" when they speak of the three *persons* in the Trinity, since they know that when we think of persons we think of distinct individuals, and three distinct individuals cannot be one in the sense the Trinitarians want to use "one." The Trinitarians wish to use "one" in a limiting sense that does not even coincide with the historical use and meaning of "one." (See "One in History" gp160)

The Trinitarians tell us that because three "persons" (Father, Son, Holy Spirit) are called God in the Bible that this is proof that they are God and are the three persons in the ONE God. And that this is the Trinity. But –

Four or More Persons in the "Trinity"?

gp419» Wait a moment. With the Trinitarians own logic we can prove that there are four or more "persons" in their "one" God. According to the Trinitarians own special logic the "Holy One" could be the *fourth* person in their "one" God:

- The **Holy One** is God (Psa 71:22; 78:41; Isa 29:23; 43:3; 48:17; 54:5; 55:5; 60:9; Hosea 11:9; Hab 1:12)

- The **Holy One** is YHWH (Isa 30:15; 48:17; 54:5; 55:5; 60:9)

- The **Holy One** is the Savior (Isa 43:3)

- The **Holy One** is the Redeemer (Isa 48:17; 54:5)

- The **Holy One** can be sinned against (Jer 51:5; Hab 1:12)

- The **Holy One** is (in a sense) Jesus (Mark 1:24; Luke 4:34)

- The **Holy One** is a person, a "he" (Isa 49:7; 54:5)

How about Five Persons?

gp420» We just identified a fourth person in the Godhead using the Trinitarians' own logic. How about a fifth person? The "Almighty" in the Bible did the same things the Holy Spirit did.

- The **Almighty** is God (Gen 35:11;49:25; Rev 4:8)

- The **Almighty** is YHWH (Ruth 1:21)

- The **Almighty** is prayed to (Job 8:5; 21:15)

- The **Almighty** answers prayer (Job 31:35)

- The **Almighty** is beyond finding out (Job 11:7; 37:23

- The **Almighty** gives spiritual understanding (Job 32:18)

- The **Almighty** does not sin (Job 34:10)

- The **Almighty** is the Father (2Cor 6:18)

- The **Almighty** is the Lord and God (Rev 11:17)

- The **Almighty** is the was, is, and will be (Rev 1:8; 4:8; 11:17)

- The **Almighty** is king (Rev 15:3)

- The **Almighty** is Jesus (Rev 1:8 cf. w/ 22:12-13, 20)

- The **Almighty** is a person, a "he" (Gen 43:14; Job 37:23; Eze 10:5)

So as the "Holy One" could be one of the "persons" in the Godhead, so could be the "Almighty" according to the logic that the Trinitarians used to put the Holy Spirit into the Godhead.

Metonymy and the Holy Spirit

gp421» The problem with the Trinitarian's logic is that the Holy Spirit, the Holy One, and the Almighty are merely words that describe different aspects of God and are being used in the Bible in a metonymical way. The phrase "Holy Spirit" or "Holy One" or "Almighty" are metonyms or words used in metonymy. Metonymy is a figure of speech. "*Metonymy* is a figure by which one name or noun is used instead of another, to which it stands in a certain relation" (*Figures of Speech Used in the Bible*, by E. W. Bullinger, p. 538). Or metonymy is the use of

the name of one object or concept for that of another to which it is related, or of which it is a part, as "scepter" for "sovereignty," or "the bottle" for "strong drink" (*Random House Dict.*). Another example is the "White House" is used as metonym when we say, "the White House said today...." What we mean is "the President said today...." The White House is not a person or power, but is a name of an object closely related to the President. When we say "the White House said today," we are using language in a metonymical way (a figure of speech) to say "the President said today." When we say the "Holy Spirit dwells in us," (2Tim 1:14) we are using language in a metonymical way to say "God abides in us," (1John 4:12) or "Jesus Christ is in you" (2Cor 13:5; Col 1:27). The "Holy Spirit" is another name for God. God revealed his name of names to Moses in Exodus 3:13-15.

gp422» The Bible uses hundreds of different figures of speech as Bullinger manifested in his book, *Figures of Speech used in the Bible*. One figure of speech the Bible used was the "Holy Spirit" for God, or the "Holy One" for God. The "Holy Spirit" is God's spirit. God possesses spirit because he is made up of spirit, for God is spirit (John 4:24). His spirit is the "Holy Spirit" because He is holy. The phrase "Holy Spirit" is a metonym for God, not a name of a different person of the Godhead, but a name of a different aspect of God. The phrase "Holy One" is a metonym for God because God is Holy. The "Holy One" is not a separate person of the Godhead, but it is just another name for an aspect of God. The word "Almighty" is not a different person of the Godhead, but it is just another name for some aspect of God for God is all mighty. In fact, concerning the "Almighty" the Bible actually says that it was just another name for the God:

- "And I appeared unto Abraham, unto Isaac, and unto Jacob, by the name of God Almighty, but by my name JEHOVAH [YHWH] was I not known to them...." [Ex 6:3]

Father is God; Jesus Christ is God; Holy Spirit is God

gp423» God has many names as we show in chapter one of this book. The Bible used metonymy when referring to God. That is, the Bible used many different names and phrases for God. The Bible used the "Father" as a name for God. The Bible, starting with Christ's resurrection, used "Jesus Christ" as a name for God. The Bible used the "Holy Spirit" as a name for God. The Bible used the "Holy One" as a name for God. The Bible used the "Almighty" as a name for God. But God's most important Name is YHWH. This is why you baptize into the Name of the Father, Son, and Holy Spirit. The Name of the Father, Son, and Holy Spirit is YHWH (Jehovah, or Yehowah, or Yahweh, or). The Father, the Son, the Holy Spirit, the Holy One, and the Almighty are all just other names of YHWH.

Puzzle of the Godhead

gp424» Remember that Jesus did **not** say in Luke 10:22 that no one knew who the Father and who the Holy Spirit were, except the Son. No, Jesus said no one knew who the Father was except the Son, or no one knew who the Son was except the Father. The puzzle of the Godhead had nothing to do with the Holy Spirit, but only the nature of the relationship of the Father and the Son. The "Holy Spirit" is just as much God as the Son, because both the Father and the Son (after the son went to the Father) are Spirit. But the Holy Spirit is not a different person of the Godhead, just as the Holy One or the Almighty are not different persons of the Godhead. The Holy Spirit is a metonym that manifests one aspect of God. It is just another name for God. The name speaks about one aspect of God: He is *Holy* spirit, or He is spirit that is Holy. God has many names and they all teach us something about God. But the main Name for God is YHWH (Yehowah) as revealed to Moses in the original language of Exodus 3:14 ff.

Time and Change Play Major Parts in the Answer

gp425» The immutability theory blinds all who examine the scripture so that they cannot see through the paradoxes and are thus forced to hold on to an obviously contradictory theory, and are forced to call it a mystery even though Paul's writings indicate that the mystery of Christ was solved (Rom 16:25; Eph 1:9; 3:3-4, 9; 5:32; 6:19; Col 1:26-27; Paul's letters did not give *all* the details of the puzzle of the Father and the Son).

gp426» Knowing that time plays a major role, what we will see is this:

- At one time Jesus was a man (GP 4);

- At a later time Jesus was in the God (GP 5).

- At first only the Spirit of Jesus Christ existed (GP 3).

- Before Jesus Christ's physical birth, the flesh of Jesus only existed in the forethought of the God who had predestinated the Seed (through Adam & Eve and the patriarchs), the fleshly Jesus Christ, to come into existence at the chosen time (GP 4 & PR 1).

- Later, when the fleshly Jesus was born, the essence of Jesus Christ was in some way split between his Spirit and his flesh (GP 4).

- At the death of Jesus, his Spirit was given up, thus separating the Spirit from the physical body for three days (GP 4).

- Only after Jesus' resurrection did he go back into the God and became one with the Spirit, and at that time for the first time Jesus was called God and was God (GP 5).

- On earth Jesus was the *Son of God* before his resurrection; he never claimed to be God before his resurrection and his return to his God, only the *son* of God.

Notes for GP 5

John 1:1-18

gp427» Let's go over John 1:1-18 in detail because of the misunderstandings these verses have brought. First of all we must understand that the book of John was written AFTER Christ the man was resurrected, and *after* he became one with God through God's Spirit. They did not go around calling Jesus, "Christ," before he was resurrected (GP 4). But those who wrote about him *after* he went to the Right Hand of the God called him "Christ" (GP 5). Knowing this let's begin.

gp428» We will now give you a literal translation word for word. Then we will explain after each verse(s) about the translation, and point out other scriptures that reiterate the same information.

Translation and Notes

- In [the] beginning was [imperfect tense = imp.] the word, and the word was [imp.] towards the God, and God was [imp.] the word. This [God or the word] was [imp.] in [the] beginning towards the God (John 1:1-2).

gp429» *Notes on Meaning:* These verses say God was the WORD and was in the beginning, as does Psalm 119:160, "your word true the beginning," and Genesis 1:1, "in the beginning God." But notice the WORD was towards the God (cf. Rev 12:5, Greek text). This is movement *towards* the God, the finished essence of God. Notice that in someway, "God was the WORD." The angel of the God [angels are called god(s). See GP 3] in the beginning was the spokesman for the God.

gp430» The "Word" came through the angel of the BeComingOne as Isaiah 52:5-6 indicates and as this book indicates (GP 3). Angels are spiritual messengers. God's angel carried God's message. Thus, it is the angel of the BeComingOne who was the agent and spokesman of the God, and who was in the beginning, and was the beginning, and was the WORD of God (see "Image of God"). But Jesus Christ became the *new BEGINNING* of the new creation of God (Rev 3:14). John 1:1 speaks about the "beginning." What beginning? The following verses indicate that the "beginning" John was talking about was the *new* beginning of Jesus Christ (Compare "beginning" in: John 1:1; 15:27; 1John 1:1; 2:7; 2John 5,6; note Luke 1:1-3).

- All things through him [the WORD] came [aorist] into existence, and apart from him came into existence not even one [thing] which has come into existence (John 1:3).

gp431» *Notes on Meaning:* Genesis 1:3, 6, 9; Rev 21:5 and so on show us God creates through "verbal" (the "WORD") commands. The pre-creation Power creates from himself, everything, for the BeComingOne is everything and everywhere; for he is spirit, and his spirit is everywhere and everything (John 4:24; Psa 139:7-8; 1Kings 8:27; Jer 23:24). Even the visible is from the invisible (his spirit), or a manifestation of his spiritual substance (Heb 11:3). The pre-creation Power, or Life, or Being created from his substance, or from his spirit a new system of things — the present cosmos. And this pre-creation Power predestinated all his power to the BeComingOne (Jer 32:17, 27), who was represented by the angel (Exo 23:21) who was the spokesman of the completed God Being. This BeComingOne (YHWH) is moving towards the completed God, or the BeComingOne who has become (John1:1-2).

- In him was life, and the life was the light of the men (John 1:4).

gp432» *Notes on Meaning:* Jesus the man said he was "the light of the world," and he qualified this by saying this light was "the light of life" (John 8:12). But Jesus the *man* was not "the light of life," but the WORD or God *in*side him spoke the words and did the power that Christ the man spoke and worked (John 12:49; 14:10).

gp433» The BeComingOne, or the WORD of the Father, who was inside Christ the man, was the "light of life," or the way of life as John 8:12 indicates, "I am the light of the world: he that follows me shall not walk in darkness, but shall have the light of life." The WORD, or the BeComingOne was not only the *way*, or the *light* of life, but he was *life* as John 1:4 and John 5:26 indicate.

- And the light [the WORD] in the darkness appears, and the darkness comprehended it not (John 1:5).

gp434» *Notes on Meaning:* The words, "light," and "darkness," are contrasted throughout the Bible with "light" being synonymous with God, or his ways (1 John 1:5; etc), and "darkness" being synonymous with Satan, or Satan's ways (1 Pet 2:9; Col 1:13; Eph 6:12; Rom 13:12; etc). Thus, John 1:5 says God (the light) appeared in the darkness (the world's present system), but those of the darkness comprehended it not. See GP 7 for more understanding on this.

- There was a man sent forth [as a messenger, or agent] from God, his name John. He came for a witness that he might testify [give evidence, to prove] about the light that all might believe through him. He was not the light, but [he was sent so] that he might testify about the light (John 1:6-8).

gp435» *Notes on Meaning:* Now this "John" was the one promised to go forth as a messenger for God. This messenger was to come before God came to earth (Mal 3:1; 4:5). John came "in the spirit and power of Elijah ... to make ready a people prepared for the Lord" (Luke 1:17, 13-17). John was the one prophesied of in Malachi 3:1 and 4:5 as Luke 1:17, 76 and Matthew 11:12-14 indicate. But John was only a typical one to come in the spirit of Elijah to testify of the typical coming of the Lord.

gp436» Because John testified of Christ's first or typical coming, and because the Bible is dual, and because the Bible said John had come in Elijah's spirit, and because John did not fulfill to the fullest extent the prophecy about Elijah; then this means John was a typical representation of the Elijah who is to come before the Lord.

gp437» But an antitypical "Elijah" or "John" shall come before the antitypical coming (2nd physical coming or manifestation) of God to announce His coming. This person will be in the spirit of Elijah, not the Old Testament Elijah himself, but doing in an antitypical way what the Old Testament Elijah did.

- [this light] was the true light which lightens every man coming into the cosmos (John 1:9).

gp438» *Notes on Meaning:* This verse antitypically prophesies of ALL mankind being enlightened by the true light (see "All Saved" paper [NM 13]). This verse typically merely says the sun (the light) shines on every man born.

- In the cosmos was [the light], and the cosmos through him came into existence, and the cosmos of him knew not. Before his own he came, and his own received him not (John 1:10-11).

gp439» *Notes on Meaning:* The Light, or the WORD, or the God (all these names are metonymical descriptive words of the same power) came into the cosmos, or the present order of things, but even though the cosmos came into being through him, and even though he came to his own people, they did not accept him. The Light, or the WORD, or the God was *in*side Christ the man before he died (see GP 4). People of Christ the man's own nation did not receive him, but killed him. After Christ the man was resurrected and became one with God, this Christ appeared to his own disciples and they were afraid of him (Luke 24:36-37, 41 "while they yet believed not"). Only when his disciples received the Spirit on the day of Pentecost did they truly believe. In effect they rejected the "light" inside Christ. As we have seen in this book all mankind are or will belong to the Light. The Light came to His future own and they rejected Him. See *New Mind Papers* for a better understanding on true belief.

- But as many as received him he gave to them ability [or energy, or power] to be children of God [that is] to those that believe into the name of his (John 1:12).

gp440» *Notes on Meaning:* Most of the cosmos or world did not accept the WORD, but a few did, because they were given the power of the Spirit to be children of God, and these believed or will believe into his NAME. God in a sense is the WORD since He gives His full power to His WORD. Thus, those who accepted the WORD of the Light believed into the NAME of God, and became children of God. When the Bible speaks of God's NAME, it means YHWH — the BeComingOne. The angel of the BeComingOne had the NAME, Christ was given that NAME, it is Christ's new NAME (GP 1, 3-5).

■ (these who are children of God are those) who not out of bloods, nor out of the will of flesh, nor out of the will of man, but out of God were begotten (John 1:13).

gp441» *Notes on Translation:* The Greek word translated "born" in the KJV is a Greek word of dual meaning, with a meaning of either: begotten or born. See "Begotten, Born Paper" [NM 5].

gp442» *Notes on Meaning:* Those who are children of God are children of God because from God (not man, not flesh) they are begotten. Since God is Spirit (John 4:24), then the children of God are begotten from God's Spirit (1 Pet 1:23 with Rom 8:14, 16; "Baptism Paper" [NM 4]). They become children of God not through their own will, but the predestinated will of God (see "Predestinated Paper" [NM 8]).

■ And the word [the angel of the Lord] became* [in] flesh, and he lived* among us, (and we discerned* the great appearance of him, an appearance a *one-of-a-kind* from father) completely full of grace and truth (John 1:14).

gp443» *Notes on Translation:* The "*" after the verbs indicate that it is translated from an aorist verb, a verb of action, not time.

gp444» *Notes on Meaning:* Now this verse is dual as is the meaning of "light" [light = God's Spirit] or "John" [John = Elijah]. The WORD, the angel of the Lord, became *in* flesh as we showed in GP 4 by being *begotten* inside Christ the man. And because of this he by being inside Jesus lived among mankind. The great appearance ("glory") of the WORD in the flesh was *as* the one-of-a-kind from the Father. Jesus was

actually the only physically begotten son *from* the Father through a miracle (see GP 4). Since God was in a sense the WORD, and God is spirit, and because spirit can not be seen; then the WORD's glory, or great appearance was *in* Jesus the man.

gp445» But antitypically, the WORD became flesh, or the angel of the Lord did become flesh, or did incorporate flesh (Christ the man) into his essence (see GP 5). And he, as flesh (also spirit), lived among man *after* the resurrection of Christ to God for a short time. And this glory, or great appearance was like the only-BORN of the Father (John 1:18, see some Greed texts). Christ the BeComingOne is the only-BORN son of God until He physically returns again, then 144,000 will also be born of God (Rev 14:1-5; 1Cor 15:23). Later, at the end of creation, the third unit or the rest of mankind will be born of God (see the "All Saved Paper" [NM 13]).

■ John testified about him, and cried, saying, this was he of whom I spoke, he who comes after me he has preference of me, for before [in place or time] me he was. For out of the fullness of him we all receive, and grace upon grace (John 1:15-16).

gp446» *Notes on Meaning:* Christ the man was sinless. Thus, he was before John in *place* or position. The WORD inside Jesus the man did Christ's works (John 14:10). And it was because of this, and by this, that Jesus the man was given the fullness of grace (John 1:14, 16) without measure (John 3:34). Not only was Christ the man given the fullness of the Spirit (grace), but after he was resurrected he was given the power of all the grace (Matt 28:18) to give as God had willed (Acts 2:33; John 5:30) before the cosmos began (Eph 1:4-6).

gp447» In another sense Christ (with God in him) was before John in *time*, because the angel of the Father in Christ was before John in time. See GP 3 and GP 4.

■ For the law through Moses was given: the grace and the truth through Jesus Christ came (comes/will come) (John 1:17).

gp448» *Notes on Translation:* The Greek word translated into "came" is an aorist word that is a verb of *action* only, not a verb of time.

gp449» *Notes on Meaning:* Jesus Christ the man came with truth and grace because he had it *in*side of him since the WORD was inside him. But Jesus Christ resurrected into the Godhead still comes with grace and truth since from him comes the Spirit of God (Acts 2:33). It is through this grace that comes from the Spirit that man is now being freed or saved (Rom 5:10, 15, 21).

■ God, no one has seen at anytime; the one-of-a-kind son who is into the bosom of the father, he revealed [him] (John 1:18).

gp450» *Notes on Meaning:* In GP 2 we showed you that God the Father was the BeComingOne of the Old Testament. Now verse 18 says no one has seen this God, but Christ who went into the Father reveals, or revealed him. Yet we showed you in GP 2 that Moses and others had seen God in visions, and others of God can see God, Spiritually, as John 6:46 and John 14:7 indicate. The reason no one could see God was because God was spirit (John 4:24), and spirit is invisible (Heb 11:3). Hence, the Bible qualifies John 1:18 by saying those of God can "see" God, Spiritually speaking, but not physically speaking. Also we must remember that God is YHWH — the BeComingOne. He exists in his truest self at the end of creation, thus, of course, up to that time no one will have seen Him in the truest sense.

Christ literally did go *into* the bosom of God as John 1:18 says. And Jesus Christ as God, reveals God in himself, or through himself, for he himself is in the God, thus can be called God or even called the BeComingOne, for he has the NAME. But he is God that is moving towards the truest sense of God — the fulfilled God.

Symbolic Meaning of the Sun and Moon

gp451» God at the beginning manifested the fact that the stars, sun, and moon were for signs (symbols) as well as for telling the days and seasons (see Gen 1:14). Further in Romans 1:19-20 it manifests to us that all of God's creation shows us about his essence and thus His plan. His Bible tells us the physical is symbolic of the Spiritual (Heb 10:1; see "Duality of the Bible" [See Intro]). But God cautions us to look to the higher meaning — the Spiritual— and not to the physical symbols (Col 3:2).

gp452» The sun and moon are symbolic of something. The problem is that some look to the physical symbols themselves, instead of the *meaning* of those symbols. This is similar to making a big commotion out of words themselves instead of their meanings. Words are merely symbols of meaning, just as the sun and moon are symbolic of something. What is the Biblical interpretation of the meanings of the sun and moon in God's plan?

Sun

gp453» First of all the sun is merely a star. And the Bible interprets stars as being angels (Rev 1:20). And further the Bible tells us angels are spiritual beings (Heb 1:7). So the sun is symbolic of a spiritual being. But which spiritual being? The sun symbolizes the angel of the BeComingOne or the resurrected Christ the BeComingOne because he now encompasses the angel. By comparing the following verses this is shown:

■ Psalm 19:4-5 equates the sun with the bridegroom coming out of his chamber. Who is *the* Biblical bridegroom? By comparing the parable of the marriage feast (Matt 22:1-13) and the parable of the ten virgins (Matt 25:1-12), with Revelation 19:7-9, we know *the* bridegroom is Christ the BeComingOne at his return. Thus, since Psalm 19:4-5 equates the sun to the bridegroom, and since the Bible equates *the* bridegroom to Christ the BeComingOne, then the sun is symbolic of Christ the BeComingOne.

■ Malachi 4:2 calls the coming Christ the BeComingOne, the "*sun* of righteousness."

■ Christ the BeComingOne calls himself the bright and morning star (Rev 22:16).

- Peter in context tells us Christ the BeComingOne is the day star (2Pet 1:19).

- The book of Revelation tells us Christ the BeComingOne looks like "the sun shines in his strength" (Rev 1:16).

- Deuteronomy 33:14 tells us "the precious fruits [are] brought forth by the sun." Who are the world's precious fruits? They are the Spiritual "first-fruits" (Rev 14:4; 1Cor 15:23). Who brings forth the fruits of the earth? It is God/Christ (Psa 104:30; Rev 14:4). The antitypical meaning of precious fruit is those born of God. The antitypical of the sun is Christ the BeComingOne.

Moon

gp454» What is the moon symbolic of? It is symbolic of Christ the man. As Christ the man was a manifestation of the angel of the BeComingOne in himself (John 14:10; 12:46, 47, 49; 1Tim 3:16; see GP 4), so too is the moon a manifestation of the sun's light. Remember stars or burning heavenly bodies are symbolic of spirits or angels. Now the moon has no light of its self, but is a reflection of the light from the sun. One can say the moon manifests or reveals the light of the sun. The moon manifests a sphere (the sun), as did Christ the man reflect or manifest the true light (of the Father through the Spirit) in himself. From GP 2 we saw that Jesus Christ's father was the BeComingOne (YHWH) of the Old Testament. From the New Testament we see that the God is symbolically or typically represented by light, "God is light" — 1 Jn 1:5. The light of the Sun and the moon represent God — His good Spirit.

gp455» Notice some of the scriptures that indicate how the moon represents symbolically or typically Jesus Christ the man:

- **(1)** Deuteronomy 33:14 tells us the moon (Christ the man) put forth precious things. What precious things did the antitypical moon put forth? They were the words of God; he spoke and put forth to the world the words of his Father (John 12:49; 14:8-10; 1Tim 3:16)

- **(2)** Genesis 3:15 tells us the serpent (Satan) would bruise the heel of the woman's seed, but that this seed would bruise the serpent's head. Now Romans 16:20 tells us this seed of Eve who is to bruise the serpent's head is Christ the BeComingOne. And Revelation shows us that at his coming he will bruise Satan's

head, which is the Beast, Satan's kingdom, with its seven *heads* and the power of the Beast which is the dragon (Satan) (Rev 19:20; 20:10; 13:2; and Psalm 91:13; see "Seed Paper" [PR 1] and Beast-System Paper" [PR 2 & see PR 3]).

- **(3)** But Genesis 3:15 said Satan would bruise the heel of the seed. And since Christ is the seed (Gal 3:16), then Satan has bruised Christ. Now we know Satan is not going to bruise Christ the BeComingOne, but Christ the man was bruised on the cross. The Bible said that Satan put the idea into Judas to betray Christ (Luke 22:3-4; John 13:2).

- **(4)** Now the Bible calls God's Church his body (1Cor 12:12). Christ is the head of the body or his Church (Eph 1:22; 4:15). In other words, in symbolism, all of Christ's body is the Church, but the head of this body is Christ. Further, the Bible calls the Church a wife (Eph 5:22-25; Rev 19:7). Christ the BeComingOne is the head of the Church (his body) as a husband is the head of his wife. Thus, Christ's body (Church) is represented by a woman (wife), and Christ is the head of that woman.

- **(5)** Satan bruised Christ the seed's heel — Christ the man. But Christ's body in Biblical symbolism is a woman (his wife; the Church). Notice Revelation 12:1 where it pictures a woman (the Church; Christ's body) clothed with the sun (the Spirit of God) and the **moon** under her feet (at *her heel*). The **moon** is at the *heels* of the body of Christ (the woman; the Church). It was the heel of the seed that Satan was to bruise. It is the MOON that is at the heels of Christ's body, and it is Christ who was the seed that was promised to come. And it was Christ the man who was bruised on the cross.

Thus, the moon under the woman (Christ's body, Rev 12:1) is symbolic of Christ the man. And remember also that Christ the man died on the Passover, and that on the Passover the moon is full. Christ the man's time was fulfilled on the Passover also.

Glory of God

gp456» What is the "Glory" of God. The word "glory" is translated from the Hebrew word *kabod* (# 3519) or the Greek word *doxa* (#1391). Both of these words had a wide application literally and figuratively in both languages, and have approximately the same meaning as the English word "glory." Thus the glory of God is:

- the great splendor of God

- the great magnificence of God

- the great prosperity of God

- the great honorableness of God

- the great exultation of God

- the great fame of God

- the great grandeur of God

- the great brilliance of God

- the great showing of God

- the great state of God

- the great triumph of God

- the great majesty of God

- the great distinction or honor of God

Great Age and Glory

gp457» The Glory of God will endure for *olam* (Psa 104:31). The Glory of God will be in the Church and in Christ in the great age (Eph 3:21, "age of ages"). The Glory of God in the Church in the great age is the same glory as Christ's agelong glory (2Tim 2:10, see "Age Paper" [NM 7]). This age is the endless age of the Kingdom with its glory (Dan 7:14; 2:44; Luke 1:33, see "Age Paper"). The "Glory of God" and the "Kingdom of God" are different names for the same great age of glory and rulership of harmony. An examination of the scriptures concerning the Glory of God as compared with the Kingdom of God shows the similar qualities and thus points to their agreement and sameness.

gp458» As we learned in this book Jesus Christ went into the BeComingOne [YHWH — Jehovah or Yehowah] and now is in the Name of the BeComingOne with the BeComingOne's titles and powers. Thus we see that what the BeComingOne or Jehovah was said to be in the Old Testament is what Jesus Christ is said to be in the New Testament. Remember this when you compare the following verses:

- Jehovah is king of glory (Psa 24:10)

- Christ is king of glory on the throne (Matt 19:28; 25:31; Rev 5:13; 2Pet 1:11; Rev 12:10; Rev 5:12)

- Jehovah is king for *olam* (Psa 10:16; 29:10; Jer 10:10; Matt 6:13 ["into the ages"])

- Christ is king for *aion*, an *aionios* kingdom (2Pet 1:11; Luke 1:33; Dan 7:13-14)

- This *olam*, this *aion* has no end (Dan 7:14; Luke 1:33; see "Age Paper" [NM 7])

- Christ's Glory is his Father's, who gave ALL to him (Psa 110:1; John 13:31-32)

- "Jesus Christ, Lord, into the glory of God the Father" (Phil 2:11; Matt 16:27; 24:30-31; 25:31)

- "Now is the Son of man glorified, and God is glorified in him [JC]. If God be glorified in him [JC], God shall also glorify him [Christ] in himself [God], and shall straightway glorify him [Christ]." (see Greek, John 13:31-32)

- "to the only wise God be glory *through* Jesus Christ" (Rom 16:27)

- "All things that the Father has are mine" (John 3:35; Luke 10:22; Matt 28:18; John 16:15)

- All nations and peoples are gathered to "see" the GLORY of God (Isa 66:18-19; Psa 97:6; Isa 40:5)

- This is the same gathering of the nations at Christ's coming with his Kingdom to see Christ and be judged (Matt 25:31-32ff; Matt 16:27; Mark 8:38; Matt 24:30-31; Mark 13:26-27; Rev 1:7)

- "EVERY ONE [all, the whole] who is called by my Name and to my glory, I have created him [each of the all who is in His Name], I have formed him, yea I have made him." (Isa 43:7, Hebrew text)

gp459» The fact is that all will *see* (this points to Spiritual sight) the Glory of God (Isa 40:5; Psa 97:6; etc.), and that all will be in God's Name and in God's Glory (Jer 3:17; 4:2; Zeph 3:9, see Hebrew text; Eph 3:15; Phil 2:11). As shown in other papers ALL will also be in God's Kingdom, for ALL will be saved ("All Into Christ," "All Saved," "Does All Mean All" papers).

gp460» Further study by you of the "Glory of God" and the "Kingdom of God" and synonyms for these terms will give you a better understanding that the Glory of God is the Kingdom of God.

Review of GP 5

gp461» We have seen in this part that after Jesus Christ the man's death and his burial for three days and three nights, he was resurrected and became ONE with the Spirit or Angel of the BeComingOne: two became one. The Spiritual marriage of mankind back to God *began* when Jesus Christ the man went into God. Since the beginning of creation, when all went *out of* God, God and man were separate. But through Jesus Christ the man, mankind is beginning to go back into God, or into the BeComingOne (the YHWH). Through the angel with the NAME, the BeComingOne incorporated Jesus Christ the man into Himself: the Spiritual WORD became flesh. The two, the man Jesus Christ and the Angel of the BeComingOne ("Lord" in the New Testament text), as ONE are called: Jesus Christ, God, or in another sense he is/will-be Jesus Christ the God. Jesus Christ the man has become Jesus Christ *the* God, in one sense, because ALL will eventually go into Jesus Christ by the End (see GP 6). But as of now we do not see ALL in Jesus Christ the God (see Heb 2:8). In two senses Jesus Christ is "the beginning of the creation of the [true] God" (Rev 3:14): the Spirit of Jesus was the First, the beginning of the "old" creation of God; but Jesus Christ the man and his Spirit as ONE is the beginning of the "New" creation.

General Review

In GP 1 we started our search: who or what is God? From the Bible we learned about the apparent paradoxes of God: "I make peace, and create evil: I the LORD do all these things" (Isa 45:7). God who is Love (1John 4:8) has somehow and for some reason created evil; He has even killed (Deut 32:39). But how can God be Love and also a killer?

We next learned that there are two basic laws and one basic fact we must understand in order to rightly perceive the true nature of God: the Law of Contradiction and the Law of Knowledge plus the fact that the God cannot lie.

We then went on and explained the Law of Contradiction.

We further showed the many attributes and titles of God and put forth that "time" is very important in our understanding of the paradoxes of God.

We also showed you the very NAME of the true God: YHWH, or Jehovah, or Yehowah, or He (who) will-be, or the BeComingOne, or the One who was, who is, and who is coming. God's NAME and its meaning is the real secret in revealing the answer to the Paradoxes of God. God's NAME is an *imperfect* (incomplete) verb and not as would be expected a *perfect* (complete) verb or a noun. Names are very important in the Bible and many times describe some facet of a person. The true NAME of the true God is important for it is the secret in explaining the apparently unexplainable scriptures about God.

In GP 1 we also looked into the meaning of "with God all things are possible," the "*one* Yehowah," the so-called unchangeableness of God, and other matters concerning the God. What GP 1 does is set the stage in our search for who or what is God.

In GP 2 we learned that Jesus Christ's Father was the BeComingOne (YHWH) of the Old Testament: He was the Jews' God.

In GP 3 we learned that the angel of the BeComingOne and the BeComingOne of the Old Testament were closely connected. Since angels are messengers, this means the angel of the BeComingOne is a messenger of the BeComingOne or this angel is the WORD of the BeComingOne. Therefore, the words that the angel of the BeComingOne spoke belonged to the Great BeComingOne Power — the true God. This angel stood in the NAME of the true God (Exo 23:20-21); he represented the great NAME. This angel was in a sense the very WORD of God. The Word (logos) of God before Christ's resurrection was a spirit, the chief-spirit, the chief-angel, the angel of the BeComingOne. The age before Jesus Christ was subjected to angels, even the commandments given on Mount Sinai were from an angel (Acts 7:38).

We have shown ***in GP 4*** that *before* Jesus Christ was resurrected, he was a human being because he was born from a woman. Thus, Christ before his death and resurrection was a man; he was Jesus Christ the man. He was called Son of man because he was born of mankind through the means of Mary his mother. Jesus the man was not just any human being. Christ was a Son of God, both physically (through the Holy Spirit's union with Mary) and Spiritually (through the medium of God's Spirit inside of him). Christ was a mediator between man and God; he was the Son of God and the Son of man. He is the "one *mediator* between God and men" (1Tim 2:5). Jesus Christ the man actually had the Spirit or Angel of the BeComingOne (YHWH) inside him leading him in the right way. It was because of this Spirit that Christ the man was sinless. God was not Jesus Christ the man, but God (through the power of His Spirit) was inside of Christ the man. When Jesus Christ the man died, his Spirit was then separated from him (Luke 23:46). The Spirit or angel of God did not die *as* Christ the man or *with* Christ the man, for Spirit cannot die. The angel of the BeComingOne separated himself from Christ the man when Christ died.

To say it in slightly a different way, in GP 4 we have shown that Jesus Christ before he was resurrected was a man with the Spirit or angel of the BeComingOne inside him leading him in the way of love. Jesus Christ the man was born from woman and by a miracle of God. Thus, Christ the man was from mankind (through Mary) and from God because of the miraculous conception by the power of God. Jesus Christ the man was a son of man, and a son of the God. He is the "one *mediator* between God and men" (1Tim 2:5). Christ the man wasn't God, but God's WORD and power was inside of him. The angel of the BeComingOne was inside Jesus Christ the man. When Christ died the angel of the BeComingOne separated from Christ the man's body. Thus, Christ the man died, but his angel inside of him stayed alive, for the angel (spirit) separated himself from Christ the man when Christ died (Luke 23:46).

In GP 5, we have seen that after Jesus Christ the man's death and his burial for three days, he was resurrected and became ONE with the Spirit or Angel of the BeComingOne: **two became one**. The Spiritual marriage of mankind back to God *began* when Jesus Christ the man went into the BeComingOne, the God. Since the beginning of creation when all went *out of* God, God and man were separate. But through Jesus Christ the man, mankind is beginning to go back into God, or into the BeComingOne (YHWH). Through the angel with the NAME, the BeComingOne incorporated Jesus Christ the man into Himself: the Spiritual WORD became flesh. The two, the man Jesus Christ and the Angel of the BeComingOne ("LORD"), as ONE are called: Jesus Christ, God, or in another sense he is/will be Jesus Christ the BeComingOne. Jesus Christ the man has become Jesus Christ *the* God, in one sense, because ALL will eventually go into the Spiritual Body of Jesus Christ by the End (see GP 6). But as of now we do not see ALL in

Jesus Christ the God (see Heb 2:8). In two senses Jesus Christ is "the beginning of the creation of the [true] God" (Rev 3:14): the Spirit of Jesus was the First, the beginning of the "old" creation of God; but Jesus Christ the man and his Spirit as ONE is the beginning of the "New" creation.

GP 6: All into Christ

Great Mystery
All Back into God
Angels and Spirits
An Angel for Everyone
BeComingOne Himself the Elohim
Great Cycle
New Body, New Soul

Mankind To Be Like Christ

gp462» "Who will transform our lowly body that it may be conformed to His [Christ's] glorious body, according to the working by which He [Christ] is able even to subdue ALL things to Himself" (Phil 3:21, NKJV — emphasis added). We have just explained to you Christ's body of glory or New Soul. His glorious body is his resurrected body that was joined to the angel of the BeComingOne (YHWH). Thus his body of glory is the new creation: two (the physical and spiritual) are one. But notice carefully that through the power given to Christ, Christians' ("our" — Phil 3:21) bodies will be created like Christ's AND with this same power all will be subdued to Christ. We need to know what it means "to subdue all things to" Christ.

Christ Now Has All The Power

gp463» Jesus Christ was given ALL the power of his Father (John 3:35). All are subjected to Christ (Matt 28:18; 1Pet 3:22) because God gave His NAME and power to Christ (Phil 2:9-11) by infusing Jesus the man into the God through the Spirit of the God (GP 5). But all are not yet under Christ (Heb 2:8). Yet as we see from Phil 3:21, Christ does have the power to subdue all to himself. And as Romans 8:32 indicates, God with Christ gave *us*, meaning Real Christians, all things (note also 1Cor 3:21). Furthermore, not only will Christians receive all things with Christ, but all people ever born will receive all things with Christ. Let's look at the Biblical evidence.

All Saved

gp464» "And the angel said to them, Do not fear. Lo, I bring you good news of great joy, which shall be to *all* people ... and *all* flesh shall see the salvation of God" (Luke 2:10; 3:6; see "All

Saved," and "Does All Mean All" papers). Somehow most have overlooked scriptures concerning *all* of mankind inheriting peace and immortal life like Jesus Christ. Somehow mankind has felt a need to damn others. But while damning others, they project themselves as being superior to the ones they damned.

Jesus Christ's Commission to Save

gp465» Jesus Christ was given a commission to be Savior of the whole world, not just some elect group. Let's see where the Bible says through Jesus Christ that *all* will be saved and go into the God. Jesus Christ the BeComingOne is Savior of the world; he has come to free the world from its madness (1John 4:14; Luke 9:56). God will come to save the world from destroying itself at His physical return to earth (Matt 24:22). Thus, at that time He will have saved the world from destroying itself; and from that point on will begin to set the ground to save the world Spiritually as other scriptures indicate.

Great Mystery – Christ Fulfills All

gp466» In the Book of Ephesians we see that God had revealed to Paul a great mystery,

- "how that by revelation he [God] made known unto me the mystery; (as I wrote afore in a few words, Whereby, when you read, you may understand my knowledge in the mystery of Christ)" (Eph 3:3-4).

All in Heaven and Earth

gp467» What did Paul write about before in a few words that he calls the mystery of Christ? We merely look back from chapter 3 to chapter 1 to see the answer:

- "Wherein he [God] has abounded toward us in all wisdom and prudence; Having made known us the MYSTERY OF HIS WILL, according to his good pleasure which he has purposed in himself: That in the dispensation of the fullness of times he might *gather together in one ALL THINGS* in Christ, both which are in heaven, and which are on earth in him" (Eph 1:8-10).

["Heaven" represents the Spiritual reality as the "earth" represents the physical reality (cf "heaven" and "earth" in Isa 55:8-9; 1Cor 15:44-49; Phil 3:18-19; Col 3:1-2).]

gp468» Thus the answer to the great mystery is that *all* things will be gathered into Jesus Christ by the "fullness of the times." Paul reiterates this in Colossians:

- "and having made peace through the blood of his cross, by him to reconcile *ALL THINGS* unto himself; by him, I say, whether they be things in the earth [physical], or things in heaven [spiritual]" (Col 1:20).

gp469» And again in another book, Philippians, Paul states the answer to the mystery:

- "according to the working whereby he is able even to subdue *all things* unto himself" (Phil 3:21).

All into the Church

gp470» But what does it mean to reconcile all to himself — to Christ the BeComingOne?:

- "This is a *great mystery*: but I speak *into* Christ and *into* the Church" (Eph 5:32).

gp471» This mystery has something to do with reconciling all to Christ the BeComingOne. Now Paul says that this "great mystery" is something about "into Christ and into the Church." Reconciling all into Christ, into the Church, what does that mean? Will all be reconciled to Christ the BeComingOne through the medium of the Church?

Christ's Spiritual "Wife" – the Church

gp472» Notice in Ephesians 5:32 Paul said, "this is a great mystery." He had just finished writing about a subject, and concerning that subject he said: "this is a great mystery." What was he writing about before he made that statement? Read Ephesians 5:22-31 and you will see what he was writing about. He was comparing Christ as the head of the Church with man as the head of his wife. He was comparing and equating in a symbolic way the Church as a wife and Christ as the husband. Thus, in a sense women are symbolic of a church. Hence, in Ephesians 5:22-33 we can transpose Church for wife and Christ for husband and we will get the antitypical or higher meaning for these verses. Let's do that:

- "So ought men [Christ] to love his wife [Church] as his own body. He [Christ] that loves his wife [Church] loves himself. For no man ever yet hated his own flesh; but

nourishes and cherishes it, even as the Lord the Church: For we are members of his body [Christ's Church], of his flesh, and his bones. For this cause shall a man [Christ] leave his father and mother, and shall be joined unto his wife [Church], and they two shall be one flesh. This is a great mystery: but I speak *into* Christ and into the Church" (Eph 5:28-32; see 1Cor 12:12 and Eph 1:22-23; Col 1:18; 2:19; Eph 4:15).

Great Mystery is Going into Christ

gp473» The word Paul used that was translated into "mystery" is a word that actually means *revealed* mystery.

03521 μυστήριον, ου, τό *mystery, secret;* (1) as a relig. t.t. in the cults of the Graeco-Roman world, a relig. *secret* confided only to the initiated, *secret rite,* not used in the NT; (2) in the NT; (a) as what can be known only through revelation mediated fr. God *what was not known before* (MT 13.11); (b) as a supreme redemptive revelation of God through the Gospel of Christ *mystery* (RO 16.25; EP 3.9); (c) as the hidden mng. of a symbol w. metaph. significance *mystery* (EP 5.32).

It is going into Christ, his body, which is the Church. Going into Christ is like going into the Church. Thus, when Paul speaks about the mystery that was revealed to him (Eph 3:3-4; 1:9) and says it is gathering "together in *one all things* in Christ [his body], both which are in heaven [the spiritual] and which are on earth [the physical] in him," we know when Paul says this, he is saying all will come into Christ's Spiritual Body, which is the Church.

All Back Into God through Jesus

gp474» All will be gathered into Christ, his body, his Church:

- "therefore any man be in Christ, he is a new creation: old things are passed away; behold, all things are become new. And *all* things are out of the God, who has reconciled us to himself by Jesus Christ, and has given to us the ministry of reconciliation; to wit, that God was in Christ, reconciling the world unto himself, not imputing their trespasses unto them; and has committed unto us the word of reconciliation. Now then we are ambassadors for Christ, as though God did beseech you by us: we ask you in Christ's behalf, be reconciled to God. For He [God] has made him [Christ] to be sin for us, who knew no sin; that we might be made the righteousness of God in him" (2Cor 5:17-21, see Greek text).

God is reconciling the world back to Himself through Christ. The Church is supposed to be teaching the world this. When one is reconciliated to Christ he becomes a new creation.

gp475» What does it mean reconciling the world back to God: "behold, the man is become as one *out of* us" (Gen 3:22). This is a correct translation from the Hebrew. Man after breaking God's law went out of the way of God. Today, through Christ the BeComingOne, the process of reconciliation is bringing the whole world back to God. How is this being done?

gp476» Notice that Christ was made sin for us so that Christ could make us the righteousness of God by us coming to God through him (2Cor 5:21). Read Isaiah 53:10-12 which shows that God allowed Christ the man to be killed for our transgressions, and that Christ the man "made intercession for the transgressors" (Isa 53:12). Compare Isaiah 53:1-12 with 2 Corinthians 5:19-21. To be an intercessor for something is to go between something — it is to be a mediator. Christ the man was a mediator!

■ "For there is one God, and one mediator between God and men, the *man* Christ Jesus; Who gave himself [his life] a ransom for *all* to be testified in due time." [1 Tim 2:5-6].

gp477» How is God reconciling or harmonizing the world through Christ? God is reconciling the world through Christ by having Christ the man die for our sin as a ransom for ALL. Then Christ the man was reconciliated to God; he actually became God as we have shown previously in this book. He went into the NAME. Christ the man went between man and God, so as to bring God and man back together through Christ's death, a death he did not deserve, for he was blameless.

gp478» "And I, if I be lifted up from the earth, will draw *all* men unto me. This he said, signifying what death he should die" (John 12:32-33). In Christ's own words he said if he was lifted up, he would draw all to himself, he would reconciliate all to God as we've shown you. "For out of him, and through him, and into him, all things" (see Greek text, Rom 11:36). All of us came out of God, all because of the mediator, Christ

the man, will return to God. At the fullness of the time all will be in Christ (Eph 1:10). It is the fullness of Jesus Christ that fills all in all:

■ And He [God] put all under his feet, and gave him [Jesus] to be head over all to the church, which is his body, the fullness of him who fills all in all (Eph 1:22-23).

All things will be put back under the power of God and his system of love so "that may be the God, all things in all" (1 Cor 15:28). When *all* is the God (YHWH) then comes true the word: "I am the BeComingOne [YHWH], and there is none else" (Isa 45:5, 6, 21, 22).

NAME and the *All*

gp479» *All Power.* The true God gave all (including *all* power) to the Son, Jesus Christ (John 3:35; Luke 10:22; Matt 28:18; John 17:2). Since everything belongs to the true God (Deut 10:14), then He can, in some way, give ALL to Jesus Christ after He sets him at His right side of power (Psa 110:1). All will be subjected to Jesus Christ (Heb 2:8; 1Pet 3:22; Phil 2:9-11), because Christ was given *all* the power (Matt 28:18) by being set at God's right hand. That is why Christ is now called the Almighty (Rev 1:8 with 1:17-18).

gp480» *All Spirit.* When Christians are in Christ, they are one with Him, they are one with His Body, they are one with His Church, and they are one with His Spirit (1Cor 8:6; 1Cor 12:4, 11-13). Real Christians are in Christ as Christ is in the true God and as God is in Christ (John 17:22; see 1Cor 12). Thus, as Christ was given ALL, so too are Christians given ALL (Rom 8:32; 1Cor 3:21).

gp481» *All Glory.* The God does not give His glory to others (Isa 42:8; 48:11) because first He brings the others back into Himself, and by so doing gives them some share in *His* glory. Thus when all are back in God, or are in His NAME, then comes true the saying, "I am the BeComingOne [YHWH], and there is none else" (Isa 45:5, 6, 21, 22). As the God gave His glory to Christ (by bringing Christ into His glory), so Christ gives the glory to Christians (John 17:22) by bringing them into his glory.

gp482» *All Name = All BeComingOne.* God's NAME was in the angel who was made one with Christ the man. Thus, Christ the man was given God's NAME through the angel. When Christians are baptized into the NAME of God, they are given the NAME of God (Matt 28:19; Acts 8:16). But ALL will go into the NAME (Jer 3:17; 4:2). All people will be called by God's NAME (Zeph 3:9, see Hebrew text; Eph 3:15). All will be in the NAME, for all will be the BeComingOne. All will be that great Power. There will be nothing but the BeComingOne (Isa 45:5, 6, 21, 22), God, all in all (1Cor 15:28). The NAME of God is for the *new* age — the time of *olam* (Psa 135:13; 102:12; see "Age Paper" [NM 7]).

gp483» When All. Jesus Christ will fulfill all things (Eph 4:10; 1:22-23). When Jesus Christ fulfills all things is when God will be all in all (1Cor 15:28). This will be when all will be the BeComingOne, or when God's Spirit will fill all (Psa 139:7-8; 1Kings 8:27; Jer 23:24; Job 34:14-15; Acts 17:28; etc.). This is when the BeComingOne, will be. This gift of the true God (YHWH) to ALL, is given when all receive God's NAME. Everyone will be born of God by the true end of creation, at the creation of the new heaven and earth. See the "Thousand Years and Beyond" paper [NM 15] and other papers on the end of creation.

Angels and Spirits

gp484» Let's go into the details about "our body" being made like unto Christ the BeComingOne's (note Phil 3:21). As Jesus has his angel, so does each of us have our angel. As Jesus Christ the man was infused with his angel after his resurrection, so too will each of us be infused with our own angel after our resurrection from the dead to the real life. Before we can go into this subject we must remember that the Bible is dual as the creation is dual. There is a type for the antitype, or the shadow for the real. As Romans 1:19-20 indicates, the physical first creation by God is representative and has meaning in the spiritual or invisible dimension. The Bible is one part of that creation, as is marriage, as is sex difference, as is ....

gp485» As we've shown you in GP 5, male and female, and even marriage are representative of Spiritual truth. Even the sun and moon have a meaning in God's plan. The moon represents Christ the man. The light of the sun represents the truth of God as represented by the angel of God or now by the resurrected Christ, who carries the Truth or Light of God (see notes for GP 5).

Stars and Angels

gp486» "And God said, Let there be lights in the firmament of the heaven to divide the day from the night; and let them be for *signs*, and for seasons, and for days, and years" (Gen 1:14). The stars are for signs. What kind of sign?

gp487» What does the word translated "sign" mean. Both in the Hebrew and the Greek, it means — sign. We can look to our dictionary. A good dictionary tells us basically a "sign" represents something else. As a word is a sign of a meaning or a sign for a thing, so too are stars: they are a sign or symbol for something else. The Bible defines its signs or symbols. And in Revelation 1:20 it defines stars as being symbolic of angels. And from other scriptures we know that angels are spiritual beings (Heb 1:7). Angels are classified into two groups, God's and Satan's. As the Bible shows, at the end of the age of Satan, one-third of the angels (stars) will have gone over to Satan's side, thus two-thirds of the created angels are God's angels or God's Spirits since angels are Spirits (Rev 12:4, 7-9). Jesus told us in Luke 24:39 that spirits do not have bodies of flesh and blood, so angels do not have physical bodies.

gp488» God is the Father of spirits (Heb 12:9), or Father of the angels, for angels are spirits. God created these Spirits, therefore he is their "father." The angel of the BeComingOne's father was God the Father — the BeComingOne (YHWH). The BeComingOne is the Father through His *power* of predestination. One-third of these angels or spirits will have been under Satan, after God "allowed" them to separate themselves from his Spirit for a higher purpose; Two-thirds of the rest of the Spirits are God's, for they follow in the true God's way (gp563).

Our Own Angel

gp489» Everyone that becomes a "son of God" must be begotten of the Spirit (Rom 8:9-10, 16). What is this Spirit? What does it mean to be begotten of God's Spirit? Notice that those begotten of the Spirit are led by it (Rom 8:14). The Spirit in them, leads them.

gp490» Notice that "these little ones" have in heaven "their angels" (Matt 18:10). The Greek word translated "their" means "of one's self." These "little ones" are Spiritual children of God (1 John 2:12-13). And these little ones have angels of their own self. Or, thus, since angels are spirits (Heb 1:7), Christians have their own angels or Spirits.

gp491» What do these angels or Spirits do? "For he shall give his angels charge over you, *to keep you in all your ways*" (Psa 91:11). In other words, angels lead them, as the Spirit leads the little ones or sons of God (Rom 8:14). Psalms 91:11 was used in a physical sense concerning Christ (Matt 4:6). But the Bible is dual and speaks in a dual sense, the physical sense and the Spiritual sense. We are to look to the higher sense — the Spiritual (Col 3:1-2). Not only do angels help out physically, but they help out Spiritually. And since Christ is our example, and the forerunner, then what applies to him applies to all others (cf Col 1:18; Rom 8:17; John 14:6).

gp492» Each son of God has his own angel (Matt 18:10). And these angels lead them in the way (Psa 91:11), as the angel of the BeComingOne led Christ (John 14:10, GP 3 & 4), who is our example. Thus, the Spirit of God that leads Christians (Rom 8:14) is an angel of God that is in them. One of God's own Spirits leads each one of them. These Spirits or angels are for the elect humans who are the sons of God (1Pet 1:1-2). These angels or Spirits serve the elect, they are ministers or servants "for them who shall be heirs of salvation" (Heb 1:14). These angels are the "elect angels" (1Tim 5:21).

Two into One

gp493» Christians at their resurrection, like Christ the man, will be Spiritually married to their own Spirit or angel: "to the general assembly and church of the first-born, which are written in heaven [in the Book of Life], and to God the Judge of all, and to the spirits [angels] of just men made complete" (Heb 12:23). The Christians are complete when they are infused with their Spirit.

gp494» Notice what the Bible calls Christ's "completion" with his Spirit: "Though he were a Son, yet learned he obedience by the things he suffered; and being made *complete*, he became the author of aeonian salvation unto all them that obey him" (Heb 5:8-9). And, "that they may be one, even as we are one: I in them, and you in me, that they may be *complete* in one" (John 17:22-23). Jesus Christ is asking in prayer that others ("they") be made complete like he and his Father.

Engaged - Married Metaphor

gp495» The Church is the wife of Christ (Rev 19:7; 21:2,9-10; Eph 5:22-32). But the Church is to be presented to Christ as a chaste virgin (2Cor 11:2). The Church is the five virgins with oil and the bridegroom is Christ (Mat 25:1-13). According to this Spiritual metaphor, before the marriage at Christ's coming, the Church, the betrothed wife of Christ, has not been made <u>one</u> with her future husband, for she is a Spiritual virgin. She enjoys Spiritual interaction (social intercourse) with her future husband, but not Spiritual intercourse (sexual intercourse) with her husband: she is not <u>one</u> with her future husband in the sense that married couples are one.

Begotten - Born Metaphor

gp496» There is a dual meaning here. A Spiritual *begotten* Christian is a typical completed person, but a resurrected Christian (*born* of God) is an antitypical completed person. Those born of God are made complete with their angel or Spirit (see "Begotten, Born Paper" [NM 5] to understand being begotten or born of God; see the "Duality of the Bible" [See Intro] to understand our use of "typical" and "antitypical").

An Angel for Everyone

Stars and Angels

gp497» Not only do the elect have angels but all human beings ever conceived have angels, for all will be born of God eventually by the end of God's Spiritual creation (see "All Saved Paper" [NM 13]). Therefore each person has an angel. This may be how some get their theory that each man born has his own star. We'll now show you *where* they got this belief. Believe it or not they are right, each has his own star. But stars are merely symbolic of angels. Each has his own angel that each will be Spiritually married to. The angels are antitypical males as was the angel of the BeComingOne; mankind are antitypical females. Both the angels and mankind will be spiritually married. Those who think only of the physical while ignoring the spiritual reality are wrong, for they look to the typical creation (stars) instead of to the antitypical meaning (angels). But we are to look to the higher meaning, the antitype.

gp498» From the very beginning God has been saying what we have put forth herein. Notice one of the first promises to Abraham: "And he brought him forth abroad, and said,

Look now toward heaven, and count the *stars*, if you be able to number them: and he said unto him, *so shall your seed be*" (Gen 15:5). Abraham's seed will be numbered as much as the number of stars (Gen 22:17). Further, "and lest you lift up your eyes unto heaven, and when you see the sun, and the moon, and the stars, even all the hosts of heaven, and should be driven to worship them, and serve them, **Which the BeComingOne your God has divided unto all nations under the whole heaven**" (Deut 4:19). Moses is shown in the Bible reminding God of this promise (Ex 32:13). Now remember stars are symbolic of angels. *God in the Spiritual meaning of these verses was saying all people were appointed or given a Spirit or angel.*

gp499» God said *if* they could number the stars, then that number would be how many of Abraham's seed who would be born. In the higher or antitypical meaning of this scripture, an equal number of persons will be born as there are angels (stars). But we know that all who go *into* Christ, or thus are begotten of the Spirit, do become heirs to the promises given to Abraham (Gal 3:16,29). Since it can be proven that all will go into Christ eventually, then all ever conceived will be made complete with their own angel or Spirit and they will become like Christ.

gp500» Now God says he *has* numbered the stars, and further, has given them names (Psa 147:4). Thus, even though *we* cannot count the stars, they are countable. And as many as there are stars, so too, will they receive the promises of Abraham. But the higher meaning of stars is angels. And we are to look to the higher meaning (Col 3:1-2).

gp501» Hence, throughout all the creation, there will be born as many human beings as there are angels (stars). And the higher meaning of Psalms 147:4 is that God has numbered the angels, and named them. God has created stars to represent angels. Furthermore, he knew from the beginning that there would be conceived one man per angel (star). God is in FULL control of the creation.

gp502» Notice in Daniel 12:3, "and they that be wise shall shine as the brightness of the firmament; and they that turn many to righteousness **as the stars** to the age and onward."

gp503» And in Revelation 21:17, "and he measured the wall thereof, a hundred and forty and four cubits, according to the measure of man, that is, angel." **The measure of man, or the number of man is equal to the angels.**

"You Are Gods"?

gp504» Then what does this all mean? What is the God doing? What is the God creating here on earth? "You are Gods" (John 10:34). The whole purpose is to create ALL of mankind into what the Bible calls, Gods. The actual NAME of God is the "BeComingOne" (see GP 1). God is becoming: He will be. God is not only becoming, He is the Gods (Hebrew text, Deut 4:35, 39; 7:9; 1Kings 18:39; etc.). The true translation from the Hebrew of the English translation, "LORD God," is the "YHWH Gods" or "he (who) will be Gods." Most of the places in the Old Testament of the Bible (King James Version) where it has "God" is actually an incorrect translation. It should read Gods since the Hebrew word is *elohim*, which is the plural of *el*. "El" means God, therefore *elohim* means Gods. The true NAME of God is the "BeComingOne" (see GP 1). In the Old Testament the "BeComingOne" (YHWH) was described as the "BeComing Gods" (*Yehowah Elohim*) or the "BeComingOne himself [is] the Gods" (Deut 4:35). The true essence of God is that he is many in one. The BeComingOne, the God who will be, is *many in one*. The BeComingOne (YHWH) is ONE, means He is in unity (see GP 1).

BeComingOne Himself <u>the</u> Gods (Elohim)

gp505» This heading is a literal quote from the Hebrew text (Deut 4:35, 39; 1Kings 18:39). Other scriptures say the same thing (Deut 7:9; Josh 22:34; 1Ki 8:60; 18:24; 2Ki 19:15; 1Ch 17:26; 2Ch 33:13; Ezra 1:3; Neh 9:7; Isa 37:16; 45:18). What does this mean? It has everything to do with everyone in some sense becoming God-like. Notice the following scriptures about people being Gods or children of God:

- "The Jews answered him, saying, We are not stoning you for a good work, but for blasphemy, and because you, being a man, make yourself God. Jesus answered them, Is it not written in your Law? 'I said, you are gods.' If *He called them gods* to whom the word of God came, and scriptures cannot be broken.

- I have said, You are gods; and all of *you are children of the Most High.*

- See what love the Father has given to us, that *we should be called the children of God.*

For this reason, the world does not know us, because it did not know Him. Beloved, now *we are children of God*.

▪ For as many as are led by the Spirit of God, these are the *sons of God*.

▪ For our citizenship is in heaven, from which we also are looking for the Lord Jesus Christ as Savior, who will completely transform our body of humiliation, for it to be *made like His glorious body*.

▪ *you might be partakers of divine* [Greek "godly"] *nature*."

[John 10:33-35; Psa 82:6; 1 John 3:1-2; Rom 8:14; Phil 3:20-21; 2Pet 1:4: see Great Cycle in GP 6]

We see scriptures that say the BeComingOne himself [is] <u>the</u> Gods (Deut 4:35), and we see that people will somehow be Gods (John 10:33), or at least offspring of God. If you are an offspring of a man, are you not a man? If you are an offspring of God, are you not somehow God? But what does this mean?

From One to Many

gp506» The Bible uses the name "Israel" to describe a whole nation. Israel was originally the name of one person. From a man named Israel came a nation of people called Israel. As with "Israel" so too with God. From one God-being (Christ) a whole nation of God-persons will become, they will come into existence and the totality of them will be the Spiritual Body of Jesus Christ. Jesus Christ was the first to go into God. ALL will become like Jesus the man became. Jesus Christ went into God and now is ONE with the God as we have shown in GP 5. After the 1000 years ALL will have followed in Jesus Christ the man's footsteps: All will go into God. All will be God all in all.

Great Cycle

All out of the Father; All Back into the Father

gp507» In the beginning all went out of the Father. But by the end of creation all will be brought back into the Father, the great God. The Great Cycle is the cycle of ALL coming out of the intelligent pre-creation Power (Father) at the beginning of creation, and the returning of the ALL, through Jesus Christ's Spiritual Body, to this great Power by the end of creation. (The

creation is still going on.) For *out of* the One God, the Father, came all things (1 Corinthians 8:6; see Rom 11:36; 1Cor 11:12; 2Cor 5:18). All things will return to the Great God, the Father of ALL, *through* Jesus Christ. "Now all things out of the God, who has reconciled us to himself [Father] through Jesus Christ" (2Cor 5:18). Jesus Christ is the mediator between God and mankind (1Tim 2:5-6). It is through Jesus Christ that all will come back into the Father so that God will again be all in all (everything) (1Cor 15:24-28). The Great God is using Jesus Christ to make all things new:

● "[Jesus Christ] **who is the image of the invisible God** [if Jesus is the "image" of the God, he is not, in the truest sense, the God], **firstborn of all creation** [first born from the dead, the first or beginning of the new creation (Col 1:18; Rev 3:14; James 1:18; 1Cor 15:23)] ; 16 **because in him being created** [*aorist* verb, a verb without a time element] **all things, the things in the heavens and the things upon the earth, the visible and the invisible, whether thrones, or lordships, or principalities, or authorities: all things have been created through him and into him. 17 And he is before all** [the Spirit or angel of Jesus was before all things in the creation (GP 3); after the resurrection, Jesus was first in the new creation (GP 5)], **and all things subsist together in him. 18 And he is the head of the body, the assembly** [Church]; **who is (the) beginning, firstborn from among the dead, that he might have the first place in all things:** [in the creation] 19 **for it pleased** [the Father; pre-creation Power who predestinated all] **that in him** [Jesus Christ] **all the fulness was to dwell**" (Col 1:15-19)

When all are in Jesus Christ, then at that very time will all be back into the Father, so that, God will be all in all (1Cor 15:28). Jesus, "**the son will be subjected to** [the One who gave the] **the subjection to him of all things, in order that, the God, all things in all**" (1Cor 15:28). The One who subjected all to Christ was the Father, YHWH, the BeComingOne. The Father is the source of all. During the creation he has given himself the Name, BeComingOne, to signify that He is becoming something other than what He was before the beginning of the creation.

gp508» ALL went out of the pre-creation Power to learn good and evil so by the end ALL will understand, know, and appreciate the coming utopia (GP 7). The true God is the

BeComingOne (YHWH). The BeComingOne is ALL that will be in the Great Power by the end of creation. The BeComingOne is the goal of creation. When the Great Cycle is complete, when ALL are back in the God Power, then the BeComingOne *will be* in the truest sense and the Great Cycle will be complete. ALL will be back in the Power, but in a different manifestation than before the beginning of creation.

gp509» What the Power will have done by this Great Cycle is create from His pre-creation essence many new individuals with the ability to be joyful and happy without end in the coming Utopia (Kingdom of God or Heaven). The most obvious process in the creation is reproduction — the sexes coming together and creating new life. What the pre-creation Power is doing is reproducing Himself, with the help of the physical dimension, into new individuals with the ability to enjoy life for an endless age. We may never be able to know or understand what was before the beginning of creation or the essence of the God Power before creation. Since there is an analogy of the male and female representing the spiritual and the physical dimension (see "Image of God," GP 5 GP 8), then from Genesis 2:18 — "not good that the man should be alone," we may surmise that the pre-creation Power may have in some way been "alone." Thus, the pre-creation Power is reproducing Himself in some way so as not to be alone.

New Body, New Soul for All

gp510» Now we have shown in GP 5 that the resurrected body or New Soul will be both spirit and flesh. Jesus Christ the BeComingOne has a fleshly body. He was resurrected as a young man (Mark 16:5). Since we will be like Jesus (Phil 3:21), we too will have a body like Christ.

Sexuality

gp511» But what about our sexual organs? will someone resurrected have sexual organs? The answer is yes. Now we'll show you why the answer is yes.

gp512» In Deuteronomy 23:1 it speaks: "he that is wounded in the testes, or has his penis cut off, shall not enter into the congregation of the BeComingOne." It says those without genitals will not enter into the congregation of the BeComingOne. But in Isaiah 56:3-7 it indicates that eunuchs (castrated men) can be in the Church and will be in the kingdom of God.

By putting Deuteronomy 23:1 and Isaiah 56:3-7 together, and knowing that the Bible does not contradict itself in its higher meaning, we know those who are eunuchs *now*, when resurrected *later* will have their genitals. God tells us to look to the higher, heavenly, or Spiritual meanings (Phil 3:19; Col 3:2; John 4:24; 6:63). Thus the "congregation of the BeComingOne" of Deuteronomy 23:1 is the antitypical congregation of God, which is the kingdom of God, or those born of God. Those born of God will have their genitals. Yet those who are eunuchs now, can be in the Church and will be in the kingdom of God, or will be born of God. But these eunuchs will be born of God with genitals. The same with women. Women when born of God will have female genitals.

gp513» Some people take Galatians 3:28 out of context and say men and women will be the *same* when born of God. These people say there will not be sexual differences in the kingdom of God. But the higher sense of Deuteronomy 23:1 says those in the kingdom will have genitals. Galatians 3:28 is speaking about everyone being of the one seed of Abraham's when in Jesus Christ. It is not saying that everyone in Christ is a non-sexual being, but this verse in context is saying that everyone in Jesus Christ is one, because they are of the *one* seed of Abraham's (V. 29, 16).

gp514» Hebrews 4:3 says "the works were finished from the foundation of the world." And "at the beginning made them *male* and *female*" (Matt 19:4). God isn't suddenly going to take the sex differences away. One will be either male or female in appearance when born of God. If one was born male on earth, he will be born male in the kingdom of God. If one was born female on earth, she will be born female in the kingdom of God.

Two Into One

gp515» Remember there are two essences or bodies to each one who is born of God. The spiritual body or essence and the physical body or essence. The physical body is resurrected and infused with a spiritual body. If one had a male body when on earth, this male body will be resurrected and infused with a spiritual body for a new creation. If one had a female body when on earth, this female body will be resurrected and infused with a spiritual body for a new creation.

No Deformities

gp516» In Leviticus 21:16-24 it projects to us that no one who will go into the antitypical sanctuary, or into the holy of holies, or be born of God shall have a physical blemish of any kind. There will be no lame, blind, and so forth among those born of God.

Resurrected Young

gp517» Job 33:25 shows us that those resurrected will have flesh *"fresher than a child's, he shall return to the days of his youth."* Those born of God will be born youngish with skin fresher than a baby. Christ was resurrected looking like a young man (Mark 16:5).

Spiritual Element

gp518» The physical form of those born of God will be perfect. But further since those born of God will have a spiritual essence, they will reap all the benefits of the spiritual dimension. Because of this spiritual dimension, those born of God will be like the angel of the BeComingOne, they will never be faint or weary (Isa 40:28).

Memory of Both the Spirit and the Body

gp519» We are born and die in the physical world. During our life we have a spirit(s) that lives in us. But we do not have full access to his mind, and he does not have full access to our mind. The "other-mind" has some influence over us in this old age, depending on the measure of power given him, and depending on our physical condition. If we have the New Mind he influences us somewhat, depending on the measure of power given to him, and depending on our physical condition. But when we are made one with our own Spirit we will have great influence over him, and he will have great influence over us. This "marriage" of our Spiritual mind to our physical mind will be similar to the intercourse between the left and right sides of our brain; our left and right brains in some ways function differently and independently, but at the same time work together for speech, movement, and behavior. The Spirit we will receive has lived since the beginning of creation and will have memory of a vast period of time, but without any direct memory of physical pain and pleasure.[1] Contrariwise, each of our minds will have the memory of our relatively short live span, but with memory of physical pleasure and pain. When joined both natures will bring together each other's memory. Our future Spirit will have the memory of what we saw, and we will have the memory of what they saw in their life span. If the spiritual world can see our world, then we will be able to see the entire history of the world through the memory of our future Spirit.

Pleasures of Both the Physical and Spiritual Dimensions

gp520» Those born of God will have the "infinite-like" dimension (the spiritual) as well as the finite (the physical). Their bodies will encompass both qualities. They can at the command of their mind change their form into the spiritual essence and travel throughout the earth or universe in an instant if need be. There will be no need for planes, or cars, or any other form of transportation for those born of God. They will enjoy the pleasure of both the physical and spiritual dimensions, as Christ did after he was resurrected and became one with his Spirit. (Gp5 & GP7)

gp521» *In GP 6* we learned how the rest of mankind will follow in Christ the man's footsteps. All of mankind each has their own angel which each will eventually be infused with. As Christ the man Spiritually married his angel, so too will each of mankind Spiritually marry his own angel so that two will become ONE. Since the beginning of creation mankind has been out of God. But through the power of Christ and God all will come back into God so "may be the God, all things in all" (1Cor 15:28).

---

[1] This is reversed in the case of those born during the millennium. See GP7, under "Two-Thirds ..."

GP 7: The Real Reason Why

Law of Knowledge
God has Created Evil? (Isa 45:7)
Why Physical and Spiritual Dimension?
Not New Knowledge
Law of Knowledge Table
More Details on the Law of Knowledge

gp522» Why is there evil in this life? Why are there diseases? Why do children get sick? Why are there natural catastrophes? Why is there war? Why is there death? Why is there hunger? Why is the world the way it is? Why are there male *and* female? Why does God have two becoming one? Why are there good *and* evil? Why has God allowed evil? Why has God split up the creation into the physical and spiritual. Why didn't God just create all his spiritual messengers ("angels") with physical bodies? Why is there a spiritual reality and why is there a physical reality? Why didn't God just make everything complete and perfect at the beginning? Or why has the BeComingOne (YHWH) created evil?:

- "forming light and creating darkness, making peace and creating evil; I, the BeComingOne [YHWH], do all these things" (Isa 45:7, see Hebrew text).

Why evil? If God is all-powerful, why did he allow ("create") the age of confusion, tears, and evil?

To Know Good and Evil?

gp523» "And the LORD said, Behold, the man is become as one *from* us, to know good and evil" (Gen 3:22, see Hebrew; see Greek also). This comment was made right after mankind had broken God's first commandment by the influence of the serpent (see The "Other Mind" paper [NM 20-22] for more details). Thus scripture says that man was getting to know good and evil from the plurality ("us") of God (LORD or YHWH). From the "us" of God man is learning good and evil. There was/is a plurality to God as we are manifesting in this book.

gp524» In the middle of the garden of Eden was "the tree of KNOWLEDGE of good and evil" (Gen 2:17). It was a tree of good and evil, not just a tree of good or not just a tree of evil. It was not just an ordinary tree, but a tree of *knowledge*. After mankind took from the forbidden fruit from the tree of knowledge of good and evil, God said man was getting to KNOW good and evil (Gen 3:22). God then took away the tree of life and placed the cherubs to guard the way to the tree of life (Gen 3:23-24). The Hebrew word translated "*from* us" in Genesis 3:22 can also be translated "*out of* us" or even "*of* us" as it is translated in most English Bibles. Because of Adam and Eve's behavior mankind did at this time go "out of" the God, but also, since the God knows all, including good and evil, then mankind was becoming like ONE *of* the God (of the "us" [His hidden plurality]) by learning good and evil. "One" here can be translated "whole" since in history the word one was more likely to mean "whole" or "unity" rather than just the number one (See this book under "One Yehowah"). Consequently, as events manifested, man was mostly left under the influence of the evil spirit of Satan, who was symbolized by the serpent of Genesis (see "Other Mind" paper [NM 20-22]). In the New Testament Paul said we were and are under the influence of the devil/Satan/evil powers and so forth (Eph 6:12).

gp525» We know from earlier chapters that the true God is ALL MIGHTY. Thus, He has the power to stop the evil, if this is what He wishes. But God has allowed this kind of world because He knows man *must* endure in evil in order to be happy. What are we saying?

Why Know Evil

gp526» Man went out of the Garden to know good *and* evil. Why know good *and* evil? Why know evil? Why live evil to know it? The main difference between a man and any other animal is his higher power to reason and know. So far, it is true he has misused this power, but, nevertheless, greater knowledge is what makes man greater than most other creatures of God. But why know evil at all? Why not just know good? Why good *and* evil? Before we answer this we must know how one knows evil.

Experience Teaches

gp527» In order to know something, to truly know something, you must live it. It takes experience with something to know it. It takes experience with evil to know evil. Our very life today teaches us that. How can you know pain if you had never felt it? How can anyone explain pain to you if you have never felt pain? Just stop and think for a moment. Try to imagine that you have never felt pain. If someone showed you someone else in pain, would you know what it was to be in pain, if you had never *felt* it? As you looked, you would see this person with an expression on his face like he was in pain. But how can you know pain through the face of a person in pain? Remember you have never *felt* pain. Any outward sign of a person in pain is just that, a sign or symbol of pain. Just because you see someone in pain, it doesn't mean you *know* pain for remember you have never *felt* pain, or *experienced* pain. You must *feel* pain to know it.

gp528» The same applies with evil. To truly know evil, one must live it. How would you explain misery to one who never felt or lived misery? How would you explain the pain of losing a loved one to someone who has never felt such a feeling? Now on this latter example, you could compare it with some other form of misery or pain. But, what if the person who you were trying to explain this grief to, had never felt any grief, misery, or pain? You could never compare your grief of losing a loved one with anything that would allow that person to know of your misery. To obtain the knowledge of knowing evil, then, you must *live* it and *feel* it. To obtain the characteristic of knowing evil we must live in such a world as we now live in.

Know Evil To Know Good?

gp529» But this is only a part of the overall picture. We must know evil to know good! Evil and good are inseparable! We must suffer evil to know good. Again, does that shock you? But why should it? Every day you live, you prove the principle that you cannot know good without real knowledge of evil. Every day that you obtain knowledge, you live this principle, and prove this principle. You cannot know good unless you know evil. You cannot separate the knowledge of good and evil. The very Law of Knowledge tells us that. What is that law?

Law of Knowledge

gp530» As mentioned in the first chapter (GP 1), there are three things you must know in order to understand who or what is God. You must know that the God cannot lie. You must know the Law of Contradiction. And you must know the Law of Knowledge. The Law of Knowledge is obvious, almost too obvious. Yet with the cognition of it you will come to understand why God has allowed misery to go on and on.

The Law of Knowledge can be stated:

gp531»

- **Generally.** Knowledge of *A* is dependent upon knowledge of *non-A*. Or to know *A* you must also know *non-A*. Or the knowledge of *A* presupposes knowledge of *non-A*. Or you know what is *A* because you know what is *non-A*. In order to "know" *A* you must compare *A* with *non-A*. Correlatively, the knowledge of *A* is proportional to the knowledge of *non-A*; or the more you know about *non-A* the more you understand the uniqueness of *A*; or the extent of your knowledge of *A* is dependent on the extent of your knowledge of *non-A*; or the more you compare *A* with *non-A* the more you know *A* (For greater detail, see notes in GP 7).

gp532»

- **Opposite qualities**. Particularly, in the case of opposite qualities (light and darkness, etc.) you must know *both* qualities to know either: you must compare each with the other to know either. Thus, in the case of opposite qualities: to know light ("A") you must compare "light" with non-light (non-"A"); "darkness" (the opposite of light) is included in what is non-light (non-"A"); and it is with the knowledge of "darkness" *and* the knowledge of "light" that we are able to know either "light" or "darkness;" but to *know* light ("A") you must compare light with "darkness" (opposite of "A" or opp-"A") and vice versa — you must know *both* qualities to know either (For greater detail, see notes).

gp533» It follows then that since good and evil are opposite qualities, then to know good you must know evil, but also to know evil, you must know good. But in order to "know" either quality you must *compare* both qualities with each other.

In the rest of this paper we will deal with opposite qualities, but see the Notes for this section to understand the Law of Knowledge in a more detailed way.

Knowledge of Opposite Qualities

Blind: Light <u>and</u> Darkness

gp534» To amplify on this law we will use the example of a blind person. We want you to try to empathize with a person that was totally blind from birth. Try to put yourself in such a person's mind. Close your eyes and imagine yourself as being blind. Now such a person has never seen light. Light is the quality that allows one's eyes to see objects. Without light no one would see even if they had perfect eyes. Light is the quality that the totally blind person cannot perceive or comprehend.

gp535» If you had never seen light, how would someone explain light to you? What choice adjectives would describe light to someone who has never seen light? To explain anything to someone who has never seen it, you have to use comparison, and say it is like this or like that. But there is no comparative quality in the universe that compares with light. It would be impossible for someone to explain light to you, let alone sight, if you had never seen light.

Knowledge of Each Presupposes Knowledge of Both

gp536» Yet at the same time one truly doesn't know what *darkness* is until one has seen light. The very definition of dark is: "without light." Darkness means without light as light means "without darkness." Each definition is dependent on its opposite quality. A definition of something is a statement of the knowledge of that thing. To know light or darkness by their very definition presupposes knowledge of each other. A blind person in order to know what darkness is, would have to see light. He knows darkness only if he sees light, for it is only then that he will understand what people were talking about when they spoke of darkness. The only reason that you can close your eyes, and call the result darkness, is because you have *seen* light. One cannot know darkness or light unless one has seen both and compared both qualities with each other.

gp537» Thus, specifically in the case of opposite qualities, your knowledge of light ("A") is dependent upon your knowledge of darkness (opposite-"A"), and vice versa. Because they are opposite qualities, you must know both to know either quality, but in order to know either quality, you must compare each with the other.

gp538» Furthermore, remembering that a blind person is blind because he cannot see light, it also follows that if there was only white light we would also be blind because we would not see or recognize any object, since in order to see anything, we need different shades of light and darkness, or more correctly since most of us see in color, in order to see anything, we need different shades of light and darkness and different hues of color.

Sound <u>and</u> Silence

gp539» The same applies for sound and silence. If you had never heard sound, how would you know what silence was like? Silence and sound are opposite qualities as light and darkness are opposite qualities. You must know both to know either, and you must compare each with the other to know either. Since these two qualities are interrelated, one has to know both to know one. The very basic definition of sound ("without silence") and silence ("without sound") need the opposite quality to define it. To know sound or silence by their very basic definition presupposes knowledge of each other.

Hot <u>and</u> Cold, Good <u>and</u> Evil

gp540» The same can be said about hot and cold. "Hot" and "cold" are relative opposite qualities. One knows something is cold only so far as he has something hot to compare it with. You can place your hand into a container of water that is 90 degrees and it will feel warm to you. But if you place your hand into a container that is 110 degrees and keep it there for a while, and then place it again into the container of water of 90 degrees, the 90 degree water will then feel cool while before it felt warm. Your knowledge of hot or cold is obtained through contrast and comparison of both qualities. Knowledge of hot or cold presupposes knowledge of the other quality. The water of 90 degrees can be compared to a town with 50 murders per year, while the water of 110 degrees may be the same town, but with 500 murders per year. When the town had 50 murders a year, you felt it was bad, but when it became 500 murders a year, you could look back at the 50 murders per year as "the good old days." Here is an example of relative evil. An example could also be given about relative good. So there is relativity to good and evil. But to have real knowledge of either (good or evil) you must have knowledge of both, you must compare one with the other because both are *comparative* qualities. You understand good by comparing it to evil; you understand evil by comparing it to good.

Life *and* Death

gp541» Further, one doesn't know what life is until he has seen death. To have knowledge of life you must have knowledge of death. One is very aware of life only if one has seen or become aware of death. Adam and Eve didn't know death and that is one reason why they chose death in the garden of Eden. Adam had never seen or felt the pain of losing a loved one. All he saw around him was life. This is very difficult for us to perceive today, for around us are the dying and the dead. It is difficult for us to put ourselves into Adam's position.

Right *and* Left & More Examples

gp542» The right side has no meaning unless there be a left side. You do not know what the meaning of right is until you know about left; you do not know what left is until you know what about right. You need knowledge about both to know either. You do not know something is "high" unless you know there is something "lower." You do not know something is "low" unless you know something is "higher." You do not know a "plus" quality until you know its "minus" quality. You do not know a "minus" quality unless you know its "plus" quality. You do not know light if you do not know darkness. But you can know light if you know darkness. You do not know or realize harmony, if you have never known confusion. Think on what is being said. If you had always lived in an environment where there was no confusion, where there was harmony, would you realize the goodness of that harmonic environment? Would harmony mean anything to you in such a harmonic environment? Can you really *appreciate* harmony if you have never lived in confusion?

gp543» If you had good vision for forty years, and then lost your sight, you would truly know the value of sight, as does a blind person who miraculously gains his sight. But how does someone after he loses his sight, come to *appreciate* the sight he once had?

Appreciation

gp544» What does it mean to appreciate something? Webster's Dictionary says that to appreciate something one must: "recognize it gratefully; estimate its worth; estimate it rightly; be fully aware of it; and notice it with discrimination." Thus, when one comes to appreciate something (especially if it is good), one in fact comes to know that thing. To appreciate something is to know it; to know something is to appreciate it.

gp545» When one loses a loved one, one by the loss of the loved one knows the worth of the loved one. The same with good. One comes to know the worth of good only after he has lived in evil.

gp546» How can we know joy, until we have lived sorrow? How can you really become happy unless you have been sad. How can we know good until we know evil? Opposite qualities need to be compared to each other to know either.

God Has Created Evil? ...

gp547» God (YHWH), through his predestination power, before the world began[1] *created* evil (Isa 45:7) so we can know good, to know good's worth, to appreciate good, and to enjoy good. The reason we must suffer the effects of evil is so we can know, to truly know good. To know what is good we must have something to compare good with. God has given man a time for good and a time for evil (Eccl 3:1-8), so as to know each. Thus in this way mankind comes to realize the value of good and harmony. God has given us joy to balance against adversity, so as to know joy (Eccl 7:14). To be able to know goodness, one must know evil. "For in much wisdom is much grief: and he that increases knowledge increases sorrow" (Eccl 1:18). "Sorrow is better than laughter: for by the sadness of the face the heart is made to be good" (Eccl 7:3). When man sinned they went out of God "to *know* good *and* evil" (Gen 3:22).

---

[1] Which was before time, before good, before evil, before law, and before sin

Should We Then Seek Evil?

gp548» Then does this mean we should seek evil? No! Once we come to realize how bad evil is, then evil has served its purpose as the comparative quality to good. But we will not know we live in evil until we see the good. In good is where the happiness lives, not in evil. We in this age are mainly learning evil; we are blind and live in the darkness. There are moments of joy and happiness in this world which allow us to partially perceive just how bad evil is, and at the same time allow us to perceive how precious good is.

Light Brings True Knowledge

gp549» The best and only way to truly perceive good is only with God's Spirit — the New Mind. Through God's Spirit man begins to renew his knowledge and mind to the ways of good (Col 3:10; Rom 12:2). Before man receives God's Spirit, man is like a blind man: he lives in darkness, yet comprehends it not, for the blind do not know light. "And the light shines in the darkness; and the darkness comprehended it not" (John 1:5). Why? Because this world is Spiritually blind, this world or this age cannot perceive their sad state of affairs. This age and most people in it, do not and cannot know how bad this age really is until they receive God's Spirit — the New Mind, which is the Spirit of truth (John 14:17). This age only partially perceives how bad this age is, and this only because there is some joy in this age to compare with the average state of affairs. But those who have received God's Spirit know eversomuch more just how bad this age is (Rev 12:11).

Two Forces

gp550» There are two spiritual forces or extra-physical mental forces in the world today: God's Spirit and the enemy spirit, which we call euphemistically, the other-mind or the Enemy's spirit. God's Spirit is ("A") and the other-mind's spirit is (opposite-"A"). Your knowledge of God's way ("A") is dependent upon your knowledge of the other-mind's way (opposite-"A"); To know the way of God ("A") you must compare it with the way of the Enemy (opposite-"A"). Mankind will only have the knowledge of good and evil after they live under the bondage of the other-mind's rule and under the harmony of God's rule. That is why all, who are eventually born of God, will and **must** live under the other-mind's spiritual law of confusion *and* under God's Spiritual law of harmony.

Sow in Tears, Reap in Joy

gp551» All must suffer evil. So that "they that sow in tears shall reap in joy" (Psa 126:5). The tears come first for man, the joy is the dessert of the creation. We learn unhappiness or the knowledge of sin through the other-mind's way. And it is through this knowledge of sin that we are able to truly know good, for then we have something to compare with God's way and his law of harmony.

Light = Good; Darkness = Evil

It is through God's Spirit and his law in our mind that we see good (the light). And it is because of our former blindness (Spiritually speaking) concerning the good (light) that we are able to comprehend the worth of good. Mankind is like a blind person who has lived in darkness (the other-mind's way) yet really didn't know how bad it was until he gained (or will gain) his sight (through God's Spirit) and was made able to comprehend the light (good), then all became understandable to him.

Time to Love; Time to Hate

gp552» Since the knowledge of God depends upon the knowledge of Satan, then man must have a period under the way of Satan and a period under the way of God in order to understand the goodness and worth of God and His way. "A time to love and a time to hate; a time of war and a time of peace" (Eccl 3:8). "Better is the end of a thing than the beginning thereof" (Eccl 7:8).

Mankind in School

gp553» We are going through a spiritual creation. Mankind is in school. Man is going through a process of discriminating between plus and minus qualities. Mankind is learning to discriminate between good and evil, by living each. Man is living each for it is impossible to teach it through words. How can you know pain through words? How can you teach a blind person what light is by words? No, man must *feel* pain to know pain, and the blind must *see* light to know light. But further, the blind must see light to know darkness, for our very definition of light ("without darkness") and our basic definition of darkness ("without light") projects to us that opposite qualities need each other to *know* either one of the qualities. A totally blind person even though he lives in darkness, doesn't know darkness until he sees light. We only know darkness because we have seen light. To know what is darkness one must have something to compare it with.

- "Except they give a distinction in the sounds, how shall it be known what is piped or harped?" (1 Cor 14:7)

Except that there be a period of time to distinguish between good and evil, how else would mankind learn or understand what is good? We know there is a right only because we know there is a left. We know something is "up" only because we see something below it in position. If everything were of the same height, there would be no "up" or "down."

Harmony Means Nothing without Disharmony

gp554» The same principle, or law of knowledge, holds true for pain and non-pain, or sound and silence, or right and left, or up and down, or big and small, or for that matter clean air and smog. But, what is important for us in this paper is that this principle holds true for good *and* for evil. If you only had lived in an environment of harmony, how would you know it was a good environment? You would have nothing to compare it with. You would be like a person who lived all his life at the top of a hundred story building in a room without any window or way to go downstairs. Even though you have 99 levels below you, you do not know you are at the top, for you do not know there is a down.

Time & Why Did God Do This?

Why did not God just put the knowledge of good and evil in our minds at the beginning and forget the 6000 years of evil?

What is Time?

gp555» The answer to this has to do with the knowledge of *TIME*. "Time" is used in different ways: (1) time is used as if it were *chronology* or *history*, or time is thought of as the passage of sequential events, "the passage of time;" (2) time is used as the *method* of reckoning and measuring events, "time can be reckoned by new moons or seasonal cycles (year) or clocks;" (3) similar to number 2 above, time is used as an *era*, instead of saying the age of communism, one may say the time of communism; (4) time is used as the fourth dimension in mathematical formulas; etc. **When we speak of time, we mean chronology or history; when we speak of _before_ time we mean before history, or before our cosmos.**

Time and Language

gp556» Without time it would be impossible to communicate. Our speech is based on words laid out sequentially in sentences. One word follows another word. Sentences are words one after another spoken in the continuum of time. Words themselves are letters, one letter following another. We could not talk or communicate to each other without time, for all words would blur into one another. We think sequentially. We live in a world of sequential events. In fact we could not know a real contradiction without time, and thus not know anything without time (see GP 1, "Law of Contradiction").

Life, Death and Time

gp557» Without time there could be no life. Death is the opposite to time because it stops the history of each person. Life can only take place in time. We become aware of time because we have historical records which pictures a continuum of events. We become more aware of time because of death which stops time for each individual. Death gives us something to compare time to. Without death we would be less aware of life or time. Death stops time. Life is organized movement within time. It would have been impossible for the God to teach us good and evil without time. We need a period of time with evil in order for us to know the good. See the Chronology Papers for more information on "time."

gp558» Even though we do not enjoy our existence or time in such a world as it is today, we must live in it in order to have *knowledge* of good. If you have never lived sorrow, how would you know what joy is? "Better is the end of a thing than the beginning" (Eccl 7:8). If it is true that mankind will be given immortal life, then this seventy or so years of sorrow on earth will seem little after millions of years of joy. These seventy years seem like a long time, but it is worth it to have the knowledge of good through living in a time period of evil. "Now no discipline for the present seems to be joyous, but grievous: nevertheless afterward it yields peaceable fruit of righteousness unto them which are exercised thereby" (Heb 12:11; see Deut 8:16).

gp559» Mankind has gone out of God to *know* through evil the worth and the fact of good. Man is learning by experience, which is the hard way to learn, but yet the best way to learn. In fact it is the only way that we could have learned.

gp560» Now we know WHY we have the confusion of this age. Today, the true God is not causing the madness of this age. It is Satan, that other-mind, that is now presently in control of this age, and it is the other-mind that is causing the confusion of this age. The BeComingOne in a sense did indeed created evil (Isa 45:7) by predestinating everything good and evil <u>before</u> the cosmos, <u>before</u> good (as we know it), <u>before</u> evil (as we know it), <u>before</u> law (as we know it), <u>before</u> sin (as we know it). And, this is very important, we know that the creation as a whole is not complete until the Spiritual or antitypical "days" are over. There was no way to create the knowledge of good, the knowledge of pleasure, or the knowledge of paradise without a period of time for the creation to learn good and evil and other important knowledge. Only when the entire creation is finished, will God be ONE, and then there will be good in its most perfect sense.

Why Physical and Spiritual Dimensions?

gp561» But why has God created from *two* parts (angel and man) the one New Soul? Why the split creation? Why spiritualkind and physicalkind? Let's look at sex differences for an answer. Why sex? As Romans 1:19-20 and other scripture show, God created the physical world to point to the Spiritual. Adam and Eve were the physical image of the spiritual God (Gen 1:26-27). The sexes come together in complementary ways, both physically and mentally? They fulfill each other. In a relationship a male and female bring a slightly different view point to that relationship. The sexes were meant to complement each other, not only physically, but also mentally. It wasn't good for man to be alone, thus God created a mate fit for him (Gen 2:18). Woman is what man lacks, and man is what woman lacks. Only together are they one (Gen 2:24). They were made for each other. They are complementary to each other. From the example of the sexes we learn why God has split the future one New Soul by first making spirits and mankind as separate entities in the first creation, and later joining them together into the New Soul for the New Creation.

gp562» One needs to live in an incomplete state in order to appreciate a perfect state. For example, if each sex had from birth a continuous state of physical and mental fulfillment, why would contact with the opposite sex be appreciated? Not only would males and females not appreciate that union, but they wouldn't get any pleasure from the union. God has created pleasure between the sexes by making them two. The pleasure is in the coming together mentally and physically. But if all were born unisexual, with the qualities of both sexes in them, a whole dimension of pleasure and good would never have been possible. Because they were apart, the act of being together physically and mentally is good and pleasurable. But if they had always been one, this dimension of pleasure would never have been possible. This same principle also applies to the physical and spiritual beings. Because the spiritual and physical beings were apart, when they are joined their very bodies and very minds will have innate pleasure. This is one reason God has created the spirit half and the physical half apart from each other and later joins them into one.

Another Reason: One-Third / Two Thirds

gp563» There is another reason for the split creation. Remember that one-third of the angels are of Satan, and that two-thirds are of God (Rev 12:4). Now being of God is to be sinless, for those of God do not sin. Sin is behavior that is harmful. Now the Law of Knowledge tells us: one must live good *and* evil to know either good or evil. After man has lived evil he will be "begotten" of his own Godly Spirit or "engaged" to be "married" to his own God-Angel. At that time he begins to know and live the good, while at the same time is able to perceive Spiritually the evil he was living in before. For example, mankind is like a blind man (Spiritually) who has never seen light (the good). But when this blind man (mankind) sees the light (the good), then he knows what darkness was and is. Man sees the light (good) through the medium of God's Spirit which allows man to perceive things Spiritually.

Complementary Knowledge

Man from Satan's Age

gp564» It is one of God's angels who dwells or will live in mankind's mind, and this spirit complements the learned evil in man's mind who lived in the age of Satan. When they are joined in the Spiritual new creation as one (spirit and man), each brings an element to this Spiritual marriage. As a male and female bring elements that the other does not have to their union, so do the Spiritual angels and physical mankind bring different elements to their union. God's angels, who are the antitypical males of the creation, bring the "knowledge" or experience and ways of God to mankind. And mankind, who lived in Satan's age, are the antitypical females of the creation, and they bring the "knowledge" or experience and ways of Satan's influence to God's angels. Thus as one, they complement each other, and this makes it possible for each other to know and appreciate more fully the way of God. As one, they are in God, thus, do not go against the way of happiness any more. But in their minds they have stored in their memory, the way of Satan. Thereby they are able to compare it with their God-life. They have a built-in comparison to always remind them of the other way, thus, giving them appreciation of the ways of love. And since when one appreciates something good, he in essence has pleasure in it, then, because of their time of evil on earth they will be able to live in the New Age with great joy and pleasure. We who have lived and suffered in Satan's age will have great joy in the ways of God because in our memories we will have the way of Satan to always compare to the new great life. Because we have lived the ways of Satan and know its worthlessness, we are able to comprehend and appreciate God's way much more than if we had never lived evil.

"Ought not Christ to have suffered these things"

gp565» Remember if we had never suffered, we could not enjoy harmony and the good. For in order to have joy, one must know sorrow. Man has to suffer as did Christ: "Ought not Christ to have suffered these things, and to enter into his glory" (Luke 24:26). The angel of God, that is, the angel of the BeComingOne, incorporated Christ into his essence so the angel too could know suffering and death, something that the good angel could not know in the sense of the knowledge of someone who has experienced or suffered under the confusion of this evil age. In the book of Revelation God says, "and I became dead, and behold, I am living into the ages of the ages" (Rev 1:18). God, the BeComingOne, through Jesus Christ has actually incorporated physical death into Himself. In this way, God complemented Himself with Christ the man who suffered by the evil age he lived in. The true God knows all things, but remember, God before Christ never lived evil. Also remember that the true God, the BeComingOne, is only manifested at the total fulfilling of the Spiritual Body of Jesus Christ which is fulfilled in the future; there is a difference in "away" knowledge and the knowledge of experiencing. We can know *about* pain, but to suffer pain is *real* knowledge of it.

Man from God's Age

gp566» But what about the spiritual beings of the Enemy? And what about those born of mankind under God's kingdom? God did say that *all* in heaven and earth would be one in Christ (Eph 1:10). That means the Enemy and his angels will also be in Christ the God. After all we were at one time all the enemies of God, but He reconciled us (Rom 5:10). Is God biased or partial? He asked us to love and forgive our enemies (Luke 6:27-37). He won't forgive His enemies?

gp567» Now those of mankind physically born under Satan's kingdom experience evil because they have lived in evil before they come into God's kingdom. But those physically born under God's kingdom will not experience evil because they never will have lived in evil. How can those born under God's kingdom obtain the experience of evil and thus be able to obtain the knowledge of good? One must simply reverse what we have put forth so far. Those born in God's kingdom will be joined to a former-enemy spirit: good thus complements the evil and allows both to understand good.

gp568» If God is to draw *all* to him, as explained before in these papers, the Power/God must change Satan and his spiritual power of evil to good. But when? After Satan's 1000 year judgment (Rev 20:1-3), then he will be loosed to Spiritual atonement for the short period after the millennium. The period after the millennium is a Great Day of Atonement, or union with God (see "Thousand Years and Beyond" paper [NM 15]).

Young Ones

gp569» So the angels who once belonged to Satan's power will be joined in the Great Day of Atonement to those born after the beginning of God's Government on earth, and those angels will also be joined to the young ones who died in early childhood or possibly through abortion during Satan's age. And after the 1000 year refinement period for Satan's angels, they in totality will be changed from evil to good, and will be begotten or joined to those of mankind born during God's millennium or those young ones who died during Satan's age. In other words, Satan and his angels because of their changed nature will then be at that time God's angels. Thus, after the millennium those of mankind born during the millennium will have a repentant or changed spirit in their minds that will at that time have the qualities of God's Spirit. But there is one exception. Satan's angels lived evil, thus in their spiritual minds they will have the experience of evil. With this experience of evil, Satan's repentant angels will complement mankind born in God's kingdom who didn't known evil. Do you see?

Two-Thirds and One-Third of Mankind

gp570» About two-thirds of mankind will be born under Satan's kingdom, and will have learned evil with Satan's angels or messengers of evil. Then at or before the end of the spiritual creation these two-thirds of mankind will be joined to the two-thirds of God's angels. Another approximately one-third of mankind will not have lived under the evil of Satan's age, for they were born in God's age or kingdom, or they died in Satan's kingdom as young ones with none or very little knowledge of evil. This one-third of mankind will be joined with the repentant angels of Satan, who at that time will have become God's angels. Thus one-third of the total number of angels, who belonged to Satan's way during Satan's age, will be joined to the one-third of the total number of mankind, who knew no or little evil, and will complement each other's former experience in order to form the knowledge of good and evil. Then *all* will become somewhat equal in their knowledge of good and evil. Typically, this equality was foreshadowed by equalizing designs of the Jubilee and other examples (Lev 25:15-16; Ex 16:18; 2Cor 8:14-15). Everyone will be more or less equal in knowledge and ability when the creation is finished, except that each will be unique individuals with individual experiences.

Reincarnation?

gp571» Each person in the present age of confusion has at least one angel or messenger of evil inside their mind feeding negative and confusing thoughts. When a human being dies in this age, their messenger of evil is released and *may* be allowed to enter another person's mind, and in most cases this is an infant who has just been born. Remember that two thirds of mankind live in evil, but only one third of the angels live in evil. Thus, the evil angels on average live in two humans during the first 6000 years. As each person dies, the evil angel may enter another person. As we can see this is a form of reincarnation. But it is the evil spirit that is reincarnated. You and I, that is, our physical bodies, are not reincarnated. It is the evil spirit in us that is reincarnated. The idea of reincarnation comes from the evil spirit that lives in man's mind. He is projecting his own experience to us through his thoughts that he feeds us from time to time. It is through the evil spirits that evil is spread throughout the ages. It is also through this evil that the evil spirits come to know evil. It is this experience with evil which helps to create the real knowledge of good and evil, for evil is the comparative quality that helps to create good.

Mankind From The 1000 Years

gp572» Soon a 1000 year utopia will be created on the earth. Those born in this age will not see nor learn evil. Because they will not know evil, they will not know good, in its truest sense. They will not appreciate the good in the 1000 year age. Real Christians who before their resurrection lived and learned evil in Satan's age, will be amazed how mankind during the 1000 years will not really grasp the greatness of the 1000 years. These Christians will then understand why the God had to create evil through the age and way of Satan.

Not New Knowledge

gp573» The idea that good and evil are inseparable to the knowledge of either in not a new idea. In C.K. Barrett's *The New Testament Background* (1961, Harper Torchbooks, he has a fragment from the Stoic writer Chrysippus (about 280-205 B.C.), we read:

gp574»

- "There can be nothing more inept than the people who suppose that good could have existed without the existence of evil. Good and evil being antithetical, both must needs subsist in opposition, each serving, as it were, by its contrary pressure as a prop to the other. No contrary, in fact can exist, without its correlative contrary. How could there be any meaning in 'justice,' unless there were such things as wrongs? What *is* justice but the prevention of injustice? What could anyone understand by 'courage,' but the antithesis of cowardice? Or by 'continence,' but for that of self-indulgence? What room for prudence, unless there was imprudence? Why do not such men in their folly go on to ask that there should be such a thing as truth, and not such a thing as falsehood? The same may be said of good and evil, felicity and inconvenience, pleasure and pain. There things are tied, Plato puts it, each to the other, by their heads: if you take away one, you take away the other." [*Chrysippus, Fragment* 1169. On the problem of evil. Barrett, p. 64]

Law of Knowledge

(Pertaining to Opposite Qualities)

Both sides complement the other and give meaning to each other;

you must know both qualities to know either: you must compare each with the other to know either

One Side	Opposite Side
love	hate
light	darkness
right	left
front	back
up	down
affection	contempt
good	evil
grace	ungracefulness
peace	war
kind	unkind
helpful	troublemaker
forgiving	unforgiving
thankful	unthankful
reconciliatory	revengeful
lawful	lawless
hope	hopeless
truthful	liar
fairness (impartial)	unfairness (partiality)
brave	coward
temperance	overindulgence
honorable	dishonorable
unpretentious	pretentious
elegant	crude
patient	impatient
sympathetic	unsympathetic

More Details on the General Law of Knowledge

Specifically, Knowledge of Non-Opposite Qualities

gp575» In this chapter we mostly talked about so-called opposite qualities such as light and darkness or good and evil. But the Law of Knowledge not only explains knowledge of opposite qualities, but also knowledge of everything capable of being known. The General Law of Knowledge Is:

> **gp576» *Generally.*** Knowledge of *A* is dependent upon knowledge of *non-A*. Or to know *A* you must also know *non-A*. Or the knowledge of *A* presupposes knowledge of *non-A*. Or you know what is *A* because you know what is *non-A*. In order to "know" *A* you must compare *A* with *non-A*. Correlatively, the knowledge of *A* is proportional to the knowledge of *non-A*; or the more you know about *non-A* the more you understand the uniqueness of *A*; or the extent of your knowledge of *A* is dependent on the extent of your knowledge of *non-A*; or the more you compare *A* with *non-A* the more you know *A*.

Basic Definition of the Law of Knowledge can <u>also</u> be stated as:

Knowledge of **A** *is equal to and dependent on the knowledge of* **non-A**.

> Where **A** can be any particular object, technique or belief;
> n**on-A** is anything but that particular object, technique or belief.

It follows —

> The depth of one's knowledge of **A** (and it truthfulness) is contingent upon the depth of one's knowledge of **non-A**; particularly, in the case of opposite qualities (light and darkness), you must know both qualities to know either; you must compare each with the other to know either.

In other words —

- To know **A** you must also know something to everything about **non-A**;
- The knowledge of **A** presupposes at least some knowledge of **non-A**;
- In order to know **A** you must compare **A** with **non-A**;

- the knowledge of **A** (and its truthfulness) is proportional to the knowledge of n**on-A**.

True Knowledge through the law of knowledge:

The continuum from incorrect knowledge —> to absolute true knowledge

- The less one knows about **non-A**, the less one knows about the truthfulness of **A** and the more likely one's knowledge is incorrect.
- The more one knows about **non-A**, the more certain one knows the truthfulness of **A**.
- If one knows all that is **non-A**, one knows absolutely the truthfulness of **A**.

   (An omniscient being would know the full truth; less than omniscient beings would not know the full truth.)

How Children Learn

gp577» One way to understand the Law of Knowledge is to understand how a child learns. Children's simple generalizations reflect lack of differentiation. That is, a child's wrong generalization about *A* (cow) reflects lack of knowledge of the difference between a cow and all that is not a cow (*non-A*) such as other four legged animals.

gp578» A child when he is first learning about four legged animals sometimes may mix up a cow and a horse, or a cow and a deer, or even a cow and a dog. This is because the child does not know what a cow is not. When parents first begin telling their child what a cow is, they point to a cow and say, "that is a cow." The child with the aid of other knowledge in his memory and his senses "sees" this living animal with four legs. Depending on how many other four legged animals are pointed out to him, he may mix the cow up with any or all other four legged animals.

gp579» After a cow is pointed out to him he may call a horse a cow, after all, to the child a horse is a four legged living animal (not a two legged animal or a toy animal or stuffed animal) just like the one pointed out earlier by his parents. But the child is wrong. This four legged animal is a horse, not a cow. The child fails to differentiate between a cow and a horse. How does the parent correct the child? The parent says, "no, it is not a cow, it is a horse." The parent is telling the child what a cow is not. The parent by telling the child what is not a cow is helping the child to learn what is a cow. Normally, after the child learns that a horse is not a cow, he doesn't call a horse a cow again. But the child may call a deer or other four legged animals a cow. When the child does this he is again corrected, "no, it is not a cow, it is a deer." The child has learned something else is not a cow (*A*); he has learned one more of the *non-A's* (all else besides cows). The more the child learns about other four legged animals not being cows, the better he is able to understand what a cow is. A cow is a four legged animal of a certain size (a cow is not a dog because for one thing a cow is bigger than a dog, etc.), but it is not any other four legged animal: it is not a dog, it is not a horse, it is not a deer, it is not an elephant, it is not a bear, etc.

gp580» But further the child from other knowledge knows a living cow is not a mountain, it is not dead (not a dead toy, not a dead stuffed animal, etc.), it is not a rock, it is not the sky, it is not a two legged animal, it is not an ant, it is not a fish, it is not fog, it is not a color, it is not a quality like "good," it is not a plant, it is not water, etc. The child knows more what a cow is, by the more he knows what a cow is not. Thus, the knowledge of a cow (*A*) is dependent on the knowledge of what a cow is not (*non-A*); or the child knows more about what is a cow (*A*), by the more he knows what is a cow is not (*non-A*).

The Color Green

gp581» Let's take another example, the color green. The more we know what the color green is *not* the more we know the uniqueness of the color green. The only way to point green out is to show what green is *not*. Since most of us know what the color green is (because we know what green is not), we will again try to understand how a child learns about the color green.

GREEN a color is "A"

gp582» The knowledge of GREEN (A) is dependent upon the knowledge of all that is not green (non-A).

- First "green" is a subdivision of color. Before a child can learn what the color "green" is, he must know what is color. In order for a child to understand "color" his parents tell him, "that thing is the *color* red, that thing is the *color* blue, that thing is the *color* orange, that thing is the *color* green, that thing is the *color* ...." Along the line of learning "color" the child comes to understand (through comparison) what "color" is *not*: the color blue on a wall is not the wall, it is not the *material* that makes up the wall such as wall board, or wood studs, or nails, etc., but the quality on the wall that we call "color" is the *color* of the wall. A child learns what color is by understanding what color is not. So before a parent can make a child understand what the "color" green is, the child has to understand what "color" is, by understanding what "color" is not.

Now assuming that the child knows what "color" is we will continue:

- We know GREEN by knowing what is *not* green (non-A). Thus the child comes to know GREEN by knowing what is not green.

What the color green is not (non-A)

gp583» Green Is Not:

- **More generally green is not**: a tree, a bush, a rock, an animal, a fish, a man, the universe, the sun, the moon, our parents, a car, a road, atoms, space, form or shape, relative position in space, time, a dimension, or any other thing or quality except for a quality we call "color."

- **More specifically green is not**: red, blue, orange, purple, or any other color, but the color we call green.

To summarize, *GREEN* is A; *GREEN* is not non-A. We know *GREEN* (A) because we know what *GREEN* is not (it is not non-A).

Review of GP 7

gp584» In this part we learned the reason evil was created (Isa 45:7) by the God. It was "allowed" because man needs a time of evil in order to have something to compare with the coming utopia that God will soon create. If God just put mankind into the coming utopia without a time of evil, man would not comprehend the worth of the utopia, and thus, man would not be able to enjoy the utopia. We showed in GP 7 that the very Law of Knowledge tells us man must have a period of evil in order to understand and enjoy the good. In order to know the good, we must know evil because good and evil are opposite qualities that need to be compared against each other in order to know either. Also in this part we learned the reason God created the creation in two: the spiritual part and the physical part. It was done this way basically so that both the spiritual dimension and the physical dimension can comprehend the knowledge of good and evil and thus be able to enjoy the coming utopia.

General Review

gp585» In GP 1 we started our search: who or what is God? From the Bible we learned about the apparent paradoxes of God: "I make peace, and create evil: I the LORD do all these things" (Isa 45:7). God who is Love (1John 4:8) has somehow and for some reason created evil; He has even killed (Deut 32:39). But how can God be Love and also a killer?

We next learned that there are two basic laws and one basic fact we must understand in order to rightly perceive the true nature of God: the Law of Contradiction and the Law of Knowledge plus the fact that the God cannot lie.

We then went on and explained the Law of Contradiction.

We further showed the many attributes and titles of God and put forth that "time" is very important in our understanding of the paradoxes of God.

We also showed you the very NAME of the true God: YHWH, or Jehovah, or Yehowah, or He (who) will-be, or the BeComingOne, or the One who was, who is, and who is coming. God's NAME and its meaning is the real secret in revealing the answer to the Paradoxes of God. God's NAME is an *imperfect* (incomplete) verb and not as would be expected a *perfect* (complete) verb or a noun. Names are very important in the Bible and many times describe some facet of a person. The true NAME of the true God is important for it is the secret in explaining the apparently unexplainable scriptures about God.

In GP 1 we also looked into the meaning of "with God all things are possible," the "*one* Yehowah," the so-called unchangeableness of God, and other matters concerning the God. What GP 1 does is set the stage in our search for who or what is God.

In GP 2 we learned that Jesus Christ's Father was the BeComingOne (YHWH) of the Old Testament: He was the Jews' God.

In GP 3 we learned that the angel of the BeComingOne and the BeComingOne of the Old Testament were closely connected. Since angels are messengers, this means the angel of the BeComingOne is a messenger of the BeComingOne or this angel is the WORD of the BeComingOne. Therefore, the words that the angel of the BeComingOne spoke belonged to the Great BeComingOne Power — the true God. This angel stood in the NAME of the true God (Exo 23:20-21); he represented the great NAME. This angel was in a sense the very WORD of God. The Word (logos) of God before Christ's resurrection was a spirit, the chief-spirit, the chief-angel, the angel of the BeComingOne. The age before Jesus Christ was subjected to angels, even the commandments given on Mount Sinai were from an angel (Acts 7:38).

We have shown *in GP 4* that *before* Jesus Christ was resurrected, he was a human being because he was born from a woman. Thus, Christ before his death and resurrection was a man; he was Jesus Christ the man. He was called Son of man because he was born of mankind through the means of Mary his mother. Jesus the man was not just any human being. Christ was a Son of God, both physically (through the Holy Spirit's union with Mary) and Spiritually (through the medium of God's Spirit inside of him). Christ was a mediator between man and God; he was the Son of God and the Son of man. He is the "one *mediator* between God and men" (1Tim 2:5). Jesus Christ the man actually had the Spirit or Angel of the BeComingOne (YHWH) inside him leading him in the right way. It was because of this Spirit that Christ the man was sinless. God was not Jesus Christ the man, but God (through the power of His Spirit) was inside of Christ the man. When Jesus Christ the man died, his Spirit was then separated from him (Luke 23:46). The

Spirit or angel of God did not die *as* Christ the man or *with* Christ the man, for Spirit cannot die. The angel of the BeComingOne separated himself from Christ the man when Christ died.

To say it in slightly a different way, in GP 4 we have shown that Jesus Christ before he was resurrected was a man with the Spirit or angel of the BeComingOne inside him leading him in the way of love. Jesus Christ the man was born from woman and by a miracle of God. Thus, Christ the man was from mankind (through Mary) and from God because of the miraculous conception by the power of God. Jesus Christ the man was a son of man, and a son of the God. He is the "one *mediator* between God and men" (1 Tim 2:5). Christ the man wasn't God, but God's WORD and power were inside of him. The angel of the BeComingOne was inside Jesus Christ the man. When Christ died the angel of the BeComingOne separated from Christ the man's body. Thus, Christ the man died, but his angel inside of him stayed alive, for the angel (spirit) separated himself from Christ the man when Christ died (Luke 23:46).

In GP 5, we have seen that after Jesus Christ the man's death and his burial for three days, he was resurrected and became ONE with the Spirit or Angel of the BeComingOne: **two became one**. The Spiritual marriage of mankind back to God *began* when Jesus Christ the man went into the BeComingOne, the God. Since the beginning of creation when all went *out of* God, God and man were separate. But through Jesus Christ the man, mankind is beginning to go back into God, or into the BeComingOne (YHWH). Through the angel with the NAME, the BeComingOne incorporated Jesus Christ the man into Himself: the Spiritual WORD became flesh. The two, the man Jesus Christ and the Angel of the BeComingOne ("LORD"), as ONE are called: Jesus Christ, God, or in another sense he is/will be Jesus Christ the BeComingOne. Jesus Christ the man has become Jesus Christ *the* God, in one sense, because ALL will eventually go into the Spiritual Body of Jesus Christ by the End (see GP 6). But as of now we do not see ALL in Jesus Christ the God (see Heb 2:8). In two senses Jesus Christ is "the beginning of the creation of the [true] God" (Rev 3:14): the Spirit of Jesus was the First, the beginning of the "old" creation of God; but Jesus Christ the man and his Spirit as ONE is the beginning of the "New" creation.

In GP 6 we learned how the rest of mankind will follow in Christ the man's footsteps. All of mankind each has their own angel which each will eventually be infused with. As Christ the man Spiritually married his angel, so too will each of mankind Spiritually marry his own angel so that two will become ONE. Since the beginning of creation mankind has been out of God. But through the power of Christ and God all will come back into God so "may be the God, all things in all" (1Cor 15:28).

In GP 7 we learned the reason why God created evil (Isa 45:7) in this age. We saw that in order to understand the good and to enjoy the coming utopia, mankind and angelkind must have a time of evil. Good and evil are opposite qualities that need to be compared against each other in order to understand either. A person who grew up with all the good things that the earth has to offer doesn't know the worth of these things until he actually loses them. After he loses the good things he then begins to understand the worth of the good things. If God just automatically at first put the creation into an everlasting utopia, the creation itself would not have the understanding of the worth or value of the utopia. In fact as we learned in GP 7 mankind would not have any understanding of good, if they never had learned evil. But by God putting the creation through an evil period God is actually teaching the creation the value of good and the evil of evil. The creation now is actually learning about good and evil. It is through the evil spirits in the minds of mankind that evil is spread on earth. It will be through the good spirits that mankind will live in good.

GP 8: Right & Left Hand of God

gp586» "But he, being full of the Holy Spirit, looked up steadfastly into heaven, and saw the glory of God, and Jesus standing on the right hand of God" (Acts 7:55, KJV) or "**right hand of the power of God**" (Luke 22:69; Mat 26:64; Mark 14:62; see Psa 110:1; Heb 1:3). If Christ the man became God, then why does the Bible say he is on or at the right side or *right hand* of the power of God. Surely this means there is a God besides Christ, and Christ is on his right side? (Remember we are speaking of the resurrected Christ who went into the Father. See GP5.) What does it mean to be on the right side of God? To these questions we'll add: What is the mercy seat, with those two cherubs on each side, one on the left, and one on the right? The angel of the BeComingOne (YHWH) usually appeared between these two cherubs (Exo 25:22; Lev 16:2).

Paul and the Cherubs

gp587» What did Paul have to say about the mercy seat and the cherubs?

- "And over it the cherubs of glory overshadowing the mercy seat; of which we cannot speak particularly" (Heb 9:5).

Paul could not speak particularly on that subject at that time. Does this mean he did not know the symbolic meaning of it? This is a possibility, yet, also, he may have had doctrine on it, but God did not see fit for it to be placed in the Bible. Thus, no one really knows the symbolic meaning of the cherubim and mercy seat?

Knowledge Not to Stay Hidden

gp588» But,

- "fear them not therefore: for there is nothing covered, that shall not be revealed; and hid, that shall not be known. What I tell you in darkness, that speak you in light: and what you hear in the ear, that preach you upon the house tops" (Matt 10:26-27).

- "For there is nothing hid, which shall not be manifested; neither was any thing kept secret, but that it should come abroad. If any man have [Spiritual] ears to hear, let him hear" (Mark 4:22-23).

- "Behold, the former things are come to pass, and new things do I declare: before they spring forth I tell you of them [through his Spiritual power, 1Cor 2:10]" (Isa 42:9).

- "Surely the BeComingOne will do *nothing* ["no word"], unless he reveals his secret unto his servants the prophets" (Amos 3:7).

- "But you, O Daniel, shut up the words, and seal the book, even to the time of the end: many shall run to and fro, and *knowledge shall be increased*" (Dan 12:4).

Knowledge *shall* be increased, not reiteration of old knowledge.

Christ on the Right Side of The God

gp589» Jesus Christ being on the right hand side of the power of the God has something to do with the mercy seat and cherubs. Let's correct something first before we go on. In the Greek text of the Bible, it doesn't say "on" or "by" or "at" the right hand of the God as translated in the King James Version. It says in Greek:

- "*Out of*" (Acts 7:55, 56; 2:25, 34; Heb 1:13; Matt 26:64; Mark 14:62);

- "*In*" (1 Pet 3:22; Col 3:l; Heb 1:3; Rom 8:34);

- "*To the*" (Acts 5:31) right hand of the God

The majority of these scriptures say that Christ is now out of, or in, or to the right hand of *the* God (see Greek text). Furthermore the "right hand" can be correctly translated "right side." "This one [Christ], Prince and Savior, the God exalted to the right side of Him" (Acts 5:31). Jesus Christ was made the right side of *the* of the power of the God, and as the right side, Christ is God, but not all *the* God in His truest sense. Christ was given all the power of the *right* side of the God, but not all the power has been taken (from the left side) by Jesus Christ yet (Heb 2:8). At the End is when all power is taken by the God so at that time all will be in the God, God all in all (1Cor 15:24-28).

First and Last

gp590» The NAME of God (YHWH) was explained in GP 1, and it means the *BeComingOne*, or *He-(who)-will-be.*

- I the BeComingOne, the **first**, and with the **last ones**; I *am* he (Isa 41:4; 44:6; 48:12; see Hebrew text).

What does this mean? The BeComingOne is the first and with the last ones?

First

gp591» The great God, the great He-will-be ONE (YHWH), is now somehow not yet complete:

- "I the BeComingOne, the first" (Isa 41:4).

- "Before me there was no God [*el*] *formed*, neither shall there be after me" (Isa 43:10).

Not only does the God say in Isaiah 43:10 that no God was formed before him, but "neither shall there be *after* me," or "neither shall there be formed a God after me." Remember that Christ now has the very NAME, and he is the only God (1Tim 1:17), because he is the only one now truly in the God or the only one born of God (John 1:18; some Greek texts). Jesus the man became God when the first-angel of the BeComingOne (who had the NAME) and Jesus Christ the man became one or were both formed as one (GP 5). He was the first and there will be no other God formed after him. Not only is the BeComingOne the first, but he is the last, "I am the first, and I am the last" (Isa 44:6). Or as the Hebrew reads, "I the BeComingOne [YHWH] the first and with the last ones" (Isa 41:4).

First or Beginning of the Creation of the God

gp592» It was the resurrected Jesus Christ who was given through the spiritual marriage the NAME and titles of the BeComingOne (see GP 5). It is through this Christ with the NAME that ALL will be saved, and the Spiritual Body of Christ will be filled so that ALL will be in Christ (see GP 5 and 6). *It is this Christ who was the first,*

the **alpha**, the **beginning**, "the **beginning** of the *creation of the God*" (Rev 3:14; see John 6:62, Greek text; Acts 26:23; etc.).

Last and with the Last Ones

gp593» Jesus the man was the *first* one to go back into the God (see GP 5). He is the beginning of the new creation. But not only is he the beginning, but he is the last, the end, and with the last oneS (Rev 1:11, 8; Isa 41:4, remember Christ now has the power and the NAME).

Christ is the BeComingOne
All Not Yet Under Him

gp594» But not only is this Christ the first and the last he has been given ALL the power; thus, this Christ is or will be the Almighty (Jude 1:25; Rev 1:8; 19:6; John 3:35; Matt 28:18; 1Pet 3:21-22; Heb 2:8, etc.). This Christ is the first, the last, the almighty, and He is: "Lord God Almighty, *which was, and is, and is to come*" (Rev 4:8; 1:8). In other words, this Christ is the very YHWH; he has the NAME; He is Yehowah; he is the BeComingOne. "*But* now we see not yet ALL things under him" (Heb 2:8) because he is waiting for all his enemies to put under his control (Psa 110:1; GP 6). In the highest sense, the Spiritual sense, the Biblical words are meant in their literal sense except if they are clearly figures of speech (see "Duality of the Bible" [see Intro]). "All" being under him, not only includes all that is good, but all that is evil. All will be saved; all will be in the NAME (see "All Saved," "Does All Mean All," and "All Into Christ" papers; Jer 3:17) through the two coming resurrections to life.

Not Yet Complete

gp595» We know that the two resurrections when mankind will be born of God have not happened yet. We also know that this Christ has "the fullness of the Godhead, bodily" (Col 2:9). After these two resurrections, then all will be in the Spiritual Body of Christ. But as of now the fullness is not manifested in this Christ (Heb 2:8). Therefore the God cannot be totally formed yet; He cannot be totally complete yet. For one thing, His NAME indicates He is incomplete: the BeComingOne (see GP 1). Jesus Christ who is now in the God, and in His NAME, cannot be totally complete yet, because after he is totally formed no others will be formed (Isa 43:10). Yet the Bible indicates other offspring of God will be formed. ALL will go into the true God. Thus in order for other offspring of the God to be formed, Christ

God must be "formed" *again*. The individual and the Spiritual Body of Christ are not a finished creation; he is not the completed God: He is Becoming. Between now and Christ's total completion other sons and daughters will be born, for God says: "I am the last; apart from me there is no Gods [*elohim*]" (Isa 44:6, see Hebrew text).

Other Offspring in the BeComingOne

gp596» We know there will be formed other offspring of the God besides Christ (GP 6). We know the Bible calls men "Gods" (John 10:34-35), and that mankind will be born of God (see "Begotten, Born Paper" [NM 5]). Remember scripture cannot be broken (John 10:35). Thus, mankind will become sons or daughters of God (in a sense a part of the God's family, nation, or Name):

- Mankind will go into in the NAME of the God.(Jer 3:17; 4:2; GP1, under "Great Significance of the Name")

gp597» As scripture shows *all* the offspring of the God now in the process of being created, are being created by and through the Godhead, Christ the God (Col 1:16-20). Thus none will be created *apart from* God as Isaiah 44:6 manifests. "Before me there was no God formed [he was the first], neither shall there be after me [he will be the last formed]" (Isa 43:10). It is between God's first formation and God's last formation that the rest of the offspring of the God will be created.

Christ the Individual and the Spiritual Body of Christ

gp598» Yet this should be qualified. The BeComingOne, Christ the individual, will be the last formed, but he will not be lastly formed *alone*. "I the BeComingOne, the first, and *with the last ones*; I am he" (Isa 41:4). At the moment of the creation of the totally NEW heaven and earth billions of mankind and angelkind will be fused into the Godkind (GP 6), and at that moment the BeComingOne (Christ) also will be lastly formed with the last ones of creation. Thus all formed at that time are the last ones formed. But there is more to Christ than the individual Christ. Paul's letters in the Bible spoke of the Body of Christ, and we are members of his Body (1Cor 12:12 ff). Christ, the individual and Christ's Spiritual body, will in some way be the last formed "Gods."

Many in ONE Body

gp599» Remember, from GP 1, that God's true NAME is YHWH, the BeComingOne. This NAME was used interchangeably with *elohim* or Gods. The God's NAME is *BeComingOne of Gods*, or *BeComing Gods* or *He (who) will be Gods*. There can be and is plurality in the ONE true God (see GP 1). There can be and are more than a single individual in the BeComingOne. **A definition of the BeComingOne: "The BeComingOne [YHWH] you yourself the Gods"** (2Sam 7:28, Hebrew text). Remember here that Christ's Body has many members (1Cor 12:12). Remember he called them Gods and scripture cannot be broken (John 10:34-35; Psa 82:6). Jesus our God told us this, not the preacher down the road. This was said to the congregation or church of the God (Psa 82:1). Also remember that the church of God is the Body of Christ, which will fill all (Eph 1:22-23).

Besides those of the second resurrection what aspect is missing from Christ, the individual and the Spiritual Body of Christ, that will be in the completed God when God is all in all?

Cherubs

In The Tabernacle

gp600» Now the mercy seat and the cherubs are in the tabernacle (Heb 9:1-5). Behind the first veil was the candlesticks, etc (Heb 9:2). The room behind the first veil as a whole is called the sanctuary or holy place. All the things in the sanctuary have to do with those in the Church of God who lived in the age of Satan and received the new Spirit in the age of Satan. For example, the candlestick in the sanctuary is symbolic of the Church itself (Rev 1:20), and the altar of incense is symbolic of Christian's prayer (Rev 8:3-4).

Most Holies

gp601» But behind the second veil is the room called the holy of holies, or the most holy place (Heb 9:3). And it was in this room where the mercy seat and cherubs sat. What was this room symbolic of? The things of the sanctuary or holy place had to do with the Church, that is, those in the Church who receive the Spirit before the end of Satan's age (NM16). But what about the most holy place? The "holy of holies" means that this place was set apart much more than the sanctuary or holy place. In other words, the "holy of holies" was more holy or set apart than the holy place or what the holy place represented. When Christ died the veil that separated the holy of holies was ripped in half signifying that it was time for man to enter the most holy place (Matt 27:51; Heb 6:19; 9:3; 10:20). But what was the "holy of holies" or "most holy place" symbolic of?

Sanctuary a Type of the Real

gp602» Scripture shows us that the tabernacle is a *type* or pattern of God's plan (cf Heb 8:5; 10:1; 9:22-24; Rom 1:20; NM16). "For Christ did not enter a man-made sanctuary that was only a copy of the true one; he entered heaven itself" (Heb 9:24, NIV). Christ didn't go into a physical sanctuary or holy place, he went into heaven or the Spiritual essence itself. Now up to this verse in Hebrews chapter 9, it had been talking about Christ's sacrifice of himself for the sins of mankind, and how this was a much better sacrifice than the sacrifice of animals as physical Israel used to do.

High Priest a Type of the Real

gp603» Physical Israel used to have a high priest who offered once a year: "the high priest entered into the holy place [most holies — cf. Lev 16:2, 29; Heb 9:7] every year with blood of others" (Heb 9:25). But unlike these high priests of Israel, Christ has offered himself *once* for the sins of all mankind (Heb 9:26, 15; 1Tim 2:6). Christ is the antitypical high priest (Heb 3:l; 7:21; 10:21). Christ is the heavenly high priest. After his sacrifice he entered into heaven or the Spiritual dimension, for any time "heaven" is used we know that it is speaking of the Spiritual dimension (compare the usage of "heaven" in Isaiah 55:8-9; 1Cor 15:44-49; Phil 3:18-19; Col 3:1-2). Christ the man after his physical sacrifice entered himself into the Spiritual, he became God as explained previously. **The most holy place or holy of holies is symbolic of Jesus Christ the BeComingOne.** Going into the most holy place is symbolic of Him being *born* of God (see "Begotten, Born Paper" [NM 5]).

Christ: Right Cherub

gp604» Christ fulfilled the prophecy of Psalms 110:1:

- "The BeComingOne said unto my Lord (Jesus), You site at my right hand..."

- "This one, after he had offered one sacrifice for sins into the perpetuity, sat down in the **right** side of the God" (Heb 10:12).

Christ went into the antitypical most holy place and sat down in the right hand side of God. Now the typical most holy place had two cherubs in it, and *between* these cherubs in the old days the BeComingOne used to appear (Ex 25:22; Lev 16:2). But it was the *angel* of the BeComingOne who appeared on the mercy seat in a cloud between the two cherubs with one cherub on his right, and one cherub on his left (Exo 25:22; Lev 16:2; the angel went with Israel in the OT: Exo 23:20ff; Acts 7:30, 38; Heb 2:2, 5; see GP 3).

- "The angel the one speaking to him [Moses] in the Mount Sinai" (Acts 7:38, Greek text).

This angel was the one that spoke to Moses in the bush and with the fathers of Israel, and this angel is the one who gave the commandments to Moses in Mount Sinai (Acts 7:30,35,38). This angel is the angel with God's Name in him (Ex 23:20-21). This angel represented the God who is becoming, for he was the angel who spoke between the cherubs, and the cherubs represent the BeComingOne (2Kings 19:15; Psa 80:1). As quoted previously Christ is *out of* and *in* the right side of God. Or as other scriptures show, Christ is *the* right side of the power of the God. Thus, when Christ went into the Most Holies He sat down in the place of the cherub on the *right* side of the BeComingOne, the true God. The Spirit of Christ is now the real right cherub, that is, he is the right side (hand) of the power of the God. He is the (Right) arm of God (Ex 6:6; Deut 9:29; Psa 44:3; 77:15; 98:1; Isa 40:10; 51:9-10). He is the right side waiting for all to come under his becoming one power (Psa 110:1).

Two Cherubs: One with Mercy Seat

gp605» Notice how the two cherubs were constructed at first,

- "One cherub on the one end, and one cherub on the other end; *of one piece* with the mercy seat he made the cherubs on its two ends" (Ex 37:8, RSV).

The two cherubs were of one piece with the mercy seat. Hence, the two cherubs and the mercy seat were of one piece physically. As God's festivals, described in the Old Testament, are foreshadows of what is to come (Col 2:16-17), so too are the two cherubs. These two cherubs represent different qualities of the finished God. We have just seen how Christ is in the place of the right cherub. Now we see that *physically* the two cherubs and the mercy seat were of *one* piece. Thus, since we know the Bible is type and antitype, whatever the physical or typical is like, is what the Spiritual or antitypical *will be* like. Whatever the cherubs and mercy seat represent Spiritually, eventually we know they will be of one.

Christ The Man: Mercy Seat

gp606» Notice that Jesus the man, the one who died, is the mercy seat, "the redemption that is in *Christ Jesus. Who God has set forth to be a mercy seat* through faith in his blood" (Rom 3:24, 25). [The word translated into "mercy seat" in Hebrews 9:5 is the same Greek word translated in the King James Version as "propitiation" in Romans 3:25.] So of the symbolic items in the holy of holies, Christ now fulfills the mercy seat *and* the right cherub.

Cherubs Face Each Other

gp607» Notice further that the faces of the cherubs are toward each other and both are facing toward the middle of the mercy seat (Ex 37:9). It was between these cherubs on the mercy seat that the angel of the BeComingOne appeared (Ex 25:22; Lev 16:2). The angel of the BeComingOne represented the goal of creation, the completed God. He had God's NAME in him (Exo 23:21). The two cherubs thus faced toward the goal, the BeComingOne. But since the *angel* of the BeComingOne was not the true complete God, but a Spiritual messenger for Him, then actually the *angel* prefigured the true God — The BeComingOne who will become.

Left Side of God?

gp608» Christ after his resurrection became one with the right cherub. Christ the man is also the true mercy seat as we just showed you. He is the one who covers all sin. Thus the only part that remains to be incorporated into the Godhead is the left cherub. So far the God is only completed in the sense that Christ, the individual, has the new body, both the physical and the Spiritual essence (typical sense of Col 2:9). But the Spiritual Body of Christ will incorporate those of the next two resurrections so that "in him [Christ] dwells all the fullness of the Godhead, Bodily" (Col 2:9; see Eph 1:22-23, 10; Col 1:19; see "All Saved" paper [NM 13]; etc.).

Left Cherub?

gp609» What about the left side of God? What about the left cherub? Who is it, or what is it? There are only two cherubs, one on the right, and one on the left of the mercy seat. If we can ascertain what is the left side of God, then we will know the secret of the cherubs and how God will be completed.

Three Things to Understand

gp610» We must understand three things:

- **(1) God is the BeComingOne:** From GP 6 we learn that the true God will bring ALL back into Himself through Christ, so that "God will be ALL in ALL." In other words, God is the BeComingOne (Elijah = "God is the BeComingOne").

- **(2) Law of Contradiction:** Remember, from GP 1, that the true God not only is good, but He somehow kills and He somehow destroys. But we know from the Law of Contradiction that for the God to be love, to be good, He cannot *at the same time* when He is good also be evil or kill or destroy. A Christian can be "good" now, but in past could have done evil or even killed. We can in the English language call this Christian a killer, for in the past he has killed. A person in prison who has killed is called a "killer" even though he may never kill again, for he may have repented of his past behavior.

- **(3) Law of Knowledge:** And remember from GP 7 that because of the very Law of Knowledge, those who know good must also know evil. From the first chapter we know that the God knows *all* things. In order for the true God to know ALL things (that includes evil), then the God Himself must know evil. GP 7 clearly slows that one can only know evil by knowing good AND evil. The only way to learn good or evil is to live in both.

Missing Part of the God are the Last Ones

gp611» In order for the true God (YHWH), who is good, to be somehow a killer (Deut 32:39), He must somehow incorporate such evil into Himself before the End of Creation. Since we know that the "all" that will go into the Spiritual Body of Christ, means all (see "Does All Mean All" paper [NM 14]), then this "all" *must* include those who have been evil. The missing parts of the completed God — the true God, YHWH — are the evil ones. Not only is YHWH the first He is also "with the last ones" (Isa 41:4). The "last ones" can mean either the last ones in time or the last ones in position. To the good God sinners are definitely the last in position or rank, and most who have sinned will be completed last in time. They are the third unit and last in time to go into the God (see "All Saved" paper [NM 13]). The missing part of the God is the evil ones, or last ones, going into the God. Hence, the *left* side of God, when incorporated into the God will be repentant "sin"[1] infused into the God. Notice that "death" in one sense has already been incorporated into God. "I am He who lives, and *I became dead*; and, behold, I am alive into the ages of ages" (Rev 1:18, Greek text). Christ the man, who physically died was physically resurrected, was then incorporated into the God (Parts 4 & 5). In this sense death and the knowledge of death came into God because Christ the man went into the God.

One Side Good; One Side Evil

Right and Left Side

gp612» Notice that the right side in the Bible is the good side (Matt 25:32-34, 41; Isa 41:10; 48:13; 62:8; Psa 78:54; 80:15; 89:13; 91:7; 98:1; etc.). Christ as the right side of God, is the right side of the totally completed God; he is the good side of God. Christ has not sinned. The left side then must be the evil side. Scripture projects that spiritually speaking the left side is the evil side (Matt 25:32-34, 41).

Who is the Left Side of The God?

Left Cherub

gp613» Who or what is the left side of the God or the left cherub? "You art the *anointed cherub* that covers" (Ezek 28:14). This *cherub* was the antitype of the King of Tyrus (Exek 28:12), who was in the garden of eden (v. 13), but who became corrupt and puffed-up (Ezek 28:17), and because of this will be sent down to a fire destruction (Ezek 28:18). This describes an evil cherub, not a good one. Since there are only two cherubs in the most holy place, and Christ now

fulfills the good right cherub, and because the right side is good in Biblical symbolism, then this cherub described in Ezekiel 28:14 is the left cherub, and is the evil side. If one reads all the 28th chapter of Ezekiel in its higher sense, he knows the cherub described here is Satan (see "Other Mind" paper [NM 21]). God has an evil side since God created evil by predestinating evil (Isa 45:7; see GP 1). Thus, *Satan is the left cherub or the left side of God*, for everything in the most holy place foreshadowed the true God, the BeComingOne. The two cherubs are in the place called the most holy place, thus the two cherubs somehow represent the BeComingOne, the Will-be-One, the Most Holy. In fact the Bible says that the BeComingOne dwells [as] the cherubs (1kings 18:39 [Heb.]; 2Kings 19:15), not "sits on the" cherubs (cf. phrase usage in 1Ki 1:48; 3:6; 8:25). But also in the Old Testament the BeComingOne also spoke to Moses *between* the two cherubs in a cloud (Ex 25:22; Num 7:89; Lev 16:2). The one who spoke to Moses was an angel (GP3), so this angel was an angel or messenger of the BeComingOne; he spoke for the future One who would fulfill both cherubs and the mercy seat.

Right and left Side Join to Create Knowledge

Right and Left Hemispheres of the Brain

gp614» The greatest evil (left side) will be fused with the greatest good (right side) so that "the God all things in all" (1Cor 15:28). In such an infusion of two qualities, the good and evil will contrast each other, and thereby make it possible for all to know good and evil. Genesis 3:22 can also be translated from the Hebrew:

- "the man has become like ONE of us, to know good and evil."

The BeComingOne here speaks of the *one* of *us*. And it is the *one* or *unity* of the BeComingOne who is the "us" that will bring the knowledge of good and evil. The right and left side of God are similar to the right and left hemispheres of the brain. Both sides process information in different ways, but both together create our thoughts and knowledge.

Mercy Seat and Cherubs Prefigured God

gp615» The mercy seat and the cherubs prefigured the coming true God. They were of *one* piece, thus pointed, in its higher meaning, to the *oneness* of the Right *and* Left side of the God. It is the *oneness* of the Right and Left side of the God that will produce the knowledge of good

---

[1] Past "sin," not one still sinning

and evil. Further when the former Enemy is infused into the Godhead, then the two antitypical cherubs and the mercy seat will be one piece like the physical cherubs and mercy seat. The angel of the BeComingOne was represented by the right cherub; Christ the man, the Lamb of God, was represented by the mercy seat; and the former Enemy was/is represented by the left cherub. The angel of the BeComingOne and Christ the man have become one (GP5). But repentant evil, the left side of the God, has yet to somehow be joined to God (see Heb 2:8).

Image of God

Male & Female are One

gp616» From GP 5 we see that from the very beginning of creation God considered male and female as man ("Adam") and as one. God used the name "man" to include both male and female. In many languages today, "man" has a dual meaning of not only meaning male, but also mankind, which includes both sexes. The very act of sexual intercourse and marriage has the meaning of signifying two (male and female) becoming one (Gen 2:24; Mark 10:6-8). Male and female are two, yet in God's eyes they are one, they are "man" or "Adam."

Two in One

gp617» From GP 5 we see that man is in the image of God, and that "man" consists of *both* male and female. Thus, the very image of God must in some way consist of two in one. There are and were more than one entity in the God. That is why the Bible says in the Hebrew text:

- "In the beginning *Gods* created the Heavens and the Earth" (Gen 1:1). "And *Gods* said, Let *us* make man in *our* image, after *our* likeness" (Gen I:26). "So *Gods* created man in his image, in the image of *Gods*" (Gen 1:27). [*Gods = elohim*] Also note the "creators" of Eccl 12:1, the "makers" of Psa 149:2, the "us" of Gen 3:22, and the "us" of Gen 11:7, the "themselves" of the Hebrew of Gen 35:7 : "for there *the* Gods [elohim] revealed themselves."

There are more than one in the BeComingOne, and the very *image of God* has to do with two in one.

Two Opposite yet Complementary Qualities

gp618» Adam is the name of man. Adam or man are two physical sexual opposites, male and female. BeComingOne is the NAME of the true God. In the BeComingOne there are/were two opposite spirits; one good and one evil (see GP 7, 8, 9). From the beginning there were *two* opposite spiritual beings. And as we prove in this book, one was good and the other was evil. As scripture manifests, the evil god rules now, but the good God will rule in the New Age and beyond – the fruit of knowing good and evil will create happiness for all (GP7). See Happiness and Knowledge paper.

From an Imperfect One

gp619» Since the very beginning, when Satan manifested himself as evil, the true God and the other god (Satan) were not one in the truest sense, much like the male and female were never one in the truest sense:

- Male and female were one in a *typical* sense, through physical intercourse. In another sense male and female become one — they have a child with the combined elements or traits of each other. Both male and female complement each other: you cannot have one or the other without each other (1Cor 11:11-12).

- The right and left side of God were one also in a *typical* sense, because both sides were/are/will be the BeComingOne. They were one in the typical sense in that they spiritually "worked" together in creating the knowledge of good and evil and even the universe (Gen 1:1, Hebrew Text). Both good and evil complement each other: you cannot have one or the other without each other (GP 7).

Male and female as two were in the *physical* image of the right and left sides of the God. Since the beginning until now, the right and left side of God were not one in the truest sense, and neither were the male and female one in the truest sense. Both man's oneness and God's oneness pointed towards the oneness of the physical perfect harmony in the Spiritual Body of Jesus Christ.

Christ Fulfills Image of God

Different Senses to the Fulfillment

gp620» When God appeared in the Old Testament times, he always appeared in visions, or dreams as an angel. But when God shall appear again he will be in the form of Christ's body, for Christ fulfills the image of God, bodily (Col 1:15; 2:9; 2 Cor 4:4; Heb 1:3). There are different senses to this fulfillment:

- Physically now Jesus is in the image of God (GP 5), but Christ the individual will also fulfill all aspects of the image of God by fulfilling both cherubs and the mercy seat: He becomes the cherubs and the mercy seat.

- Jesus Christ's Spiritual Body which is the Church fulfills the image of God *when* the Church is totally fulfilled: when *all* go into the God.

Fulfilled within Three Days

gp621» When he walked on earth before his death Jesus in a sense fulfilled the image in the holy of holies because the physical Jesus was the Mercy Seat, the Spirit of God in him was the Right Side of God; the spirit of Satan in him (tempted him, Luke 4:2) was the Left Side of God. When Jesus Christ predicted that he would raise from the dead in three days, he was predicting when he would again and perfectly fulfill the Mercy Seat and both sides of God. Notice the time aspect to Jesus' prophecy:

- John 2:19 Jesus answered and said unto them, Destroy this temple, and **in three days I will raise it up**. 20 Then said the Jews, Forty and six years was this temple in building, and wilt thou rear it up in three days? 21 But he spake of the temple of his body.

Remember the Power of Christ came from the Spirit of his Father, his Father did the works (GP 4). So it is the Power of the Father that will raise Jesus up in all his aspects in three days. And Remember that there is a type and antitype to scripture. So there are 24 hour physical days and there are three 1000-year spiritual days, since a day to God is like a thousand years (2Pet 3:8; Psa 90:4; see NM).

Fulfillment of the Image of God in Three Orders in Three Days

gp622» Christ is different from all others, he, or his Spiritual Body, fulfills all things (Eph 1:23; 1Cor 15:28):

- **Christ the Individual – First Order of two becoming one:** three 24 hour days after Christ's death and burial, the angel of the right side of the power of God was united with Christ our mercy seat (Rom 3:25) as the beginning of the new creation (Rev 3:14; Col 1:15,18). In the case of Christ the individual, there were two becoming one: the physical being infused with the Spiritual (GP 6).

- **Christ's Spiritual Body – Second Order of two becoming one:** two 1000 year-days (Hosea 6:2) after Christ's death and resurrection, in the beginning of the seventh spiritual day, will occur the infusion of the physical Christians into the Spiritual Body of Christ through a Spiritual "marriage" of each Christian to their own Spirit or angel of Christ. Again, two (physical and spiritual) are becoming one, as well as the good and evil are joined to create the knowledge of good and evil (NM16, "Pentecost").

- **Both Christ the Individual and Spiritual Body – Third Order of two becoming one:** three 1000 year-days after Christ's death and resurrection, in the beginning of the eighth spiritual day, the repented spirits of Satan will be brought into the Spiritual Body of Christ so that God will be all in all (GP 7). The *individual* Christ will be infused with his repentant spirit as explained in GP 8, and the rest of mankind and angel kind will be joined so that God will be all in all when all go into the Spiritual Body of Christ, and at the same time will occur the joining of good and evil to create the knowledge of good and evil (GP 6 & GP 7).

Evil: Who Is To Blame?

gp623» At the end of creation the evil force will be fused with the good force thus giving the result of the true knowledge of good and evil. Now since the "works were finished from the foundation of the world" (Heb 4:3), or as good as finished, for all that was planned will happen; then the angel of the BeComingOne as the representative or messenger of the completed God could say before his true completion, "I kill, and I make alive; I wound, and I heal" (Deut 32:39). In a sense God does kill and wound, for a part or one side of the totally finished God is the power of this kind of work; and that part or side was/is Satan — who after he is changed will be the future left side of God. But now the right side of God, he himself, is not doing any evil. It is Satan, the left side of the power of the God (who will in the future become one with God) who is now leading the madness. Thus, as of now, Christ is in himself *not* to blame for today's madness. Satan is doing it. Satan is to blame, but not God, if one wants to find fault. For example compare the following parallel verses in two different books of the Bible:

- "And again the **anger of the LORD** was kindled against Israel, an he moved David against them to say, Go, number Israel and Judah." [2Sam 24:1]

- "**Satan** stood up against Israel, and provoked David to number Israel." [1Chron 21:1]

The *anger of the BeComingOne* that provoked David was Satan (the left side of God), not the good right side of God.

But remember all things have been predestinated to happen as they will happen (see "Predestination Paper" [NM 8]) before the cosmos, and consequently before sin (as we know it). Our God is the BeComingOne, who is good in totality when He becomes One (Deut 6:4; Matt 19:17). The evil that was predestinated was predestinated before sin. Therefore, predestination is outside sin, even before sin. One cannot find fault for the creator's predestination.

Blame to All

gp624» In the true God, the completed God, or that is, God in its truest sense, there will be no one who can stand up and say he is better than any other, for all will have contributed to the madness, and soon to come, all will contribute to the harmony. Thus, at the end no one can stand up and accuse another, for all will have killed, and all will have healed in a sense. The BeComingOne said, "I kill, I make alive" (Deut 32:39). At the End there is nothing, but the BeComingOne, "I the BeComingOne [YHWH] and none else" (Isa 45:18, 5, 22). The truest sense of the

NAME, the BeComingOne, comes true only at the End (1Cor 15:24-28; etc.). At that time God — the BeComingOne — will be all in all, or that is, God will be everything, or everything will be the God: "I the BeComingOne [YHWH] and none else" (Isa 45:18, 5, 22).

Repentant Evil plus Good are <u>One</u> in Harmony

gp625» As we saw in GP 7 the two-thirds of mankind who will have lived under Satan, along with Satan and his angels, are the killers and destroyers. But the others of mankind along with God's angels are the healers. One killer will be fused with one healer to make ONE completed person. Or more technically, the one-half (one-third of the angels, two-thirds of mankind) of the completed Godship who had been destroyers will be fused with the one-half (one-third mankind, two-thirds of the angels) that had been the healers by the End of Creation. And thus the same with Christ the individual. He is the healer, the right half of the completed God (Angel of God, 1/3) with the mercy seat (Jesus Christ, 1/3), who is to be infused with the converted former destroyer (Satan, 1/3), the left side of God, by the end of the spiritual creation, so that $1/3 + 1/3 + 1/3 = $ ONE.

Cherubs Protecting the Garden

Why the Contrast Between the God of the Old Testament and the God of the New Testament

gp626» In Genesis 3:24 it says that the BeComingOne (YHWH) put the cherubs in the garden to guard the way to the tree of life. Now that we know what the true meaning of the cherubs is, we can now know what Genesis 3:24 says.

- So He drove the man out; and at the east of the garden of Eden He stationed the cherubs [right & left] and the flaming sword which turned every direction to guard the way to the tree of life. (Gen 3:24)

What this verse is really saying in context with what we now know about the cherubs is this:

- Mankind finds it difficult to get to paradise because the left cherub keeps misleading and contradicting the right cherub.

gp627» When two people (one who is a liar and one who is not) tell you two different things, how do you tell who is telling the truth, if you do not have any experience with either person? You must get to know each for a while until by experience you are able to ascertain who is the liar. Both cherubs guarded the garden. One was a liar, the greatest liar; one was the truth teller. Since the left cherub misled through the other-mind (see "Other Mind" paper [NM 21]), and since God did not give many the New Mind, mankind as a whole was misled (Rev 12:9). It was only after Satan's spirit was cast out by the death of Jesus that the way to life was/is/will be made clear to ALL (John 12:31-33; Heb 2:14-15).

Two in One and the Duality of *Elohim*

"In the beginning **God** created the heavens and the earth." When you see "God" in most English translations of the Old Testament it is translated from its plural form (*elohim*). The Hebrew meaning for *elohim* is "powers" in the plural or dual depending on context. See Gesenius §87(a) &88(a)(e) to see how the plural and dual forms work. Also study GP8.

Many think that Hebrew *elohim* ("God" with the capital G) in the Old Testament of most English Bibles is speaking of the true *one* God, as they understand Him. So they think of Him in a singular way, not a plural or dual way. (Of course the three in one Trinity god confuses all of this.) Because of this they say the plural *elohim* is a "*pluralis excellentiae.*" (Gesentius' Heb. Gram. § 132*g,h*) But this is mistake.

> **First** the word is NOT in a singular form in the Hebrew language. It is in a plural form by the addition of "im" as its suffix. There is nothing special about this that it should be given a special name – *pluralis excellentiae.*

> **Second** because of other biblical scriptures we know there are **two** spiritual gods: one being the True God and one being Satan. And each god has it own kingdom: Kingdom of God and a kingdom of Satan.

> **Third** we know that mankind was made in the **image** of God (*elohim*). (Gen 1:26-27) Therefore from mankind being in the "image" or likeness of God, we can ascertain what the real essence of God must be.

Literally, "And said (the) Gods, Let **US** make man in **Our** image and according to **Our** likeness... And (the) Gods created <u>the</u> man in His image and in the image of Gods He created him [Adam]; He created them – male and female." (Gen 1:26-27) Both of them, male and female, were made corresponding and complementary to one another as well as opposite to one another [Gen 2:18, 20, Heb.], yet – these **two were one** [Gen 2:24]. Both together are called Adam or mankind [Gen 5:1-2] *and* they both were made in the image of *Elohim* [Gen 1:27]. There is a *two in one* aspect to mankind, and since they are in the image of God, there MUST be a *two in one* aspect to the essence of God. This is why when speaking about God, the Bible uses the plural for the word God. Study GP6 - GP9 & Ex. 25:17-22.

Following this logic, what is the two in one aspect of the real God?

Notice in the Holy of holies (Ex 25:17-22ff) where the *Becoming-One* appeared during Moses' time, there were two Cherubs representing two angels. The two cherubs/angels were made from ONE piece with the mercy seat. The faces of the cherubs faced toward each other and toward the center of the mercy seat as the faces of male and female face each other in sexual intercourse. We show in the *God Papers* that the two cherubs represent two spirits or gods who were different in behavior: one was good; one was evil. The two behaviors were different, but complemented each other, for without the two no one could know about good and evil. If you never learn about good and evil, you would never know what good was or feel pleasure or know how wonderful life was or any of the other aspects of knowing good and evil. (See GP7; NM19-21) The construction of the two cherubs and the mercy seat were made of one piece thus foretelling the coming together of what the two cherubs and the mercy seat represented because Moses' tabernacle and its Holy of Holies were a pattern of what was in the heavenlies. (Ex 25:8-9; Heb 8:5; 9:24) Read all of the *God Papers* for fuller details and see the *Male and Female He Created Them* book.

Way to the Tree of Life

One scripture says that the two cherubs and its flaming sword were put in place to hid the way to the tree of life:

> Gen 3:24 So he drove out the man; and he placed at the east of the garden of Eden Cherubs,[1] and a flaming sword which turned every way, to guard the way of the tree of life.

---

[1] *Cherubim*, See Holy of holies & GP8

But it was through Christ (our Messiah, our King of kings, the Right Cherub and Mercy Seat), who defeated the left cherub and its dysfunction, and made the path to the Tree of life possible through His Spirit and Truth. He took away the flaming sword that guarded the way to the Tree of Life.

Remember Moses wrote the first five books of the Bible which included the books of Genesis and Exodus. He used the plural (*elohim*) Hebrew word for God instead of the singular Hebrew word (*el* or *eloh*) for God because he was inspired to write that and because he was told that the Holy of Holies was the physical representation of the True God – it was the pattern of the Heavenly/Spiritual essence. (Ex 25:17-22)

Satan Cast out of Man

gp628» Christians in the old age had the "other-mind" tempting them (NM20). Formerly, Christ had an "other-mind" tempting him (Mark 1:13) and that was Satan or a spirit of Satan (NM 21). Christ of course had the pure mind of the angel of God inside him when he was a man, but he also had the other-mind inside his mind that was testing and trying to confuse him (note Mark 1:13; Heb 2:18). It was this other-mind of Satan that was cast out of Jesus at his death (John 12:31-33; Heb 2:14-15), and this was the beginning of Satan's power being put out of mankind.

God of the New Testament

gp629» The New Testament was written after Christ sat at the right side of the power of the God. Thus, the God of the New Testament times, Christ the BeComingOne, is pictured as the Right Side of the true God — the good God of the *Gods* of the BeComingOne (YHWH). Remember the BeComingOne is of the *Gods* or is the *Gods* (Deut 4:35, 39). The BeComingOne is the good God and in a sense also the bad god, since *all* power comes from the BeComingOne. The bad god is Satan (2Cor 4:4; 11:14). When Christ died Satan was put out of him, and in a sense, out of the BeComingOne (note John 12:31-32; Heb 2:14). This is one reason in the New Testament that the word God is written in the singular case and the not in the plural case as in the Old Testament. In the Old Testament the BeComingOne was of *Gods*. An examination of the Hebrew Old Testament shows us that it wasn't just a singular "BeComingOne" who did things, but it was the "LORD God" or "YHWH Elohim" or the "He (who) will be Gods" that did things in the Old Testament. Jesus Christ is now the BeComingOne, but He is *not* now of gods in the Old Testament sense because the evil god was cast out. Christ has nothing to do with evil, for He is the only good God — the true God. The BeComingOne will be of *Gods* in a different sense because of the next two resurrections to life.

Scriptures on Satan

gp630» The scriptures concerning Satan and his angels can be found throughout the Bible. Most of these scriptures speak in a typical way or physical way about Egypt, or Babylon, or the Pharaoh, or the evil one(s), or Elam, or Edom, or the prince of Tyrus, or the king of Tyrus (Ezek 28:12), or the king of Assyria (Isa 10:12), or king of Babylon (Isa 14), or Lucifer (Isa 14), or Leviathan (king of the children of pride — Job 41), and so forth. But these verses point to the antitype or to the spiritual or higher meaning. For example, the scriptures of the physical Pharaoh point to the spiritual evil Pharaoh. God through the Bible is speaking to us in a spiritual language (see "Duality of the Bible" [See Intro]).

gp631» See the "Last War and God's Wrath" paper [PR 5], the "Other Mind" paper [NM 21], and other papers such as the "Thousand Years and Beyond" paper [NM 15] to help you understand the higher meaning of scripture concerning Satan, his angels, and their fate.

Satan Does Not Understand Now

gp632» But why doesn't Satan understand his future now? He does not understand now nor does he wish to understand, the same way as those without God's Spirit do not understand now nor do they wish to understand. These are the ones with physical ears that can only hear physically. Only those of the Spiritual ears, hear Spiritually. Jesus said, "every one of the truth hears my voice" (John 18:37). And those of the Truth, have the Spirit of Truth (John 16:13).

gp633» This is why Satan and those belonging to Satan cannot understand that Satan and his angels are also to become one with God:

- "But we speak the wisdom of God in a mystery, even the hidden wisdom, which God ordained before the world unto our glory: which *none* of the princes of this world [the spirits of Satan are the power behind the physical princes] knew: for had they known it, they would not have crucified the Lord of glory. But as it is written, Eye has not seen, nor ear heard, neither have entered into the heart of man, the things which God has prepared for them that love him. But God has revealed them unto us by his Spirit: for the Spirit searches all things, yes, the deep things of God" (1 Cor 2:7-10).

Satan in Darkness

gp634» "And angels who had not kept their own original state, but had abandoned their beginning, for [the] judgment of [the] great day; chained perpetually[1] under gloomy darkness, he keeps [them]. (Jude 1:6) This "day" begins when Christ returns and continues for an antitypical Sabbath, the seventh 1000 year Spiritual day. Thus, during the millennium of joy for God's Kingdom, it will be the millennium of judgment for Satan's kingdom. (Satan did not keep the physical seventh day for 6000 years, so his punishment is for him to keep quiet during the seventh 1000 year period [Lev 26:34-35].) But *after* Satan's judgment (they are in perpetual chains during that time — Rev 20:2, 7) he will be atoned to the true God (see "Thousand Years and Beyond" paper [NM 15]).

gp635» Satan is in the darkness concerning this fact, "he believes *not* that he shall return out of darkness" (Job 15:22). This verse in the typical reading speaks of a "wicked one" (V. 20). Yet since a wicked man is merely a shadow of what is in him (a spirit of Satan with its law of confusion), then Satan himself is in the dark. He believes he will not return from the bottomless pit (Rev 20:1-3). **That makes Satan extremely dangerous: that is why Satan goes completely mad at the end of the age** (Rev 12:12; and see God's Wrath papers [PR 6]).

Review of GP 8

gp636» In this part we learned the meaning of the cherubs. The *right* cherub represents the right hand or right side of the Power of God — the good side of God. The *left* cherub represents the left hand or left side of God — the evil side of God. The BeComingOne is actually

foreshadowed by the cherubs and the mercy seat in the most holy place. The right side of God is Jesus Christ, while the left side of God is that one called Satan. Both Jesus Christ and Satan are represented by the cherubs and the mercy seat. Jesus Christ the man was represented by the mercy seat. The cherubs and the mercy seat represent the BeComingOne, and the truest sense of the BeComingOne is the finished God — the true God — who is the good God. The true God will change Satan and then put this *reformed* spirit back into Himself. Satan when he is put back into the Power will then be a good spirit or power with the memory or knowledge of his former evil ways. The knowledge of evil was needed so as to give the true knowledge of good to ALL, since good and evil must be compared in order for us to know either. In the beginning all went *out of* God. This included Satan. But at or before the end all will be brought back into the God, so that God will be all in all. The physical cherubs and the mercy seat were one piece, thus pointing to the future oneness of the Real cherubs and mercy seat. Satan will thus be brought back into the God. The meaning of the cherubs indicates this to us.

---

[1] Strong's # 126 (from #104) = always, continual, perpetual, not necessarily "forever." See NM24.

God has all power

("I form the light, and create darkness; I make peace, and create evil, I the BeComingOne do all these things" [Isa 45:7])

All went out of the all powerful God to learn good and evil

Right Side of God	Left Side of God
system of love (harmony) – light	system of hate (disharmony) – darkness
affection (of good: "good is good")	affection (of evil: "evil is good")
anger (against evil)	anger (uncontrolled)
hate (evil – Ps 97:10; Prov 8:13; Amos 5:15)	hate (good – Micah 3:2-4)
good (all in harmony with God's laws)	evil (all in disharmony with God's laws)
graciousness (respectful, polite)	ungraciousness (rude, vulgar)
peaceful (calming)	unpeaceful (agitating, reckless)
kind	unkind (harsh, malevolent)
helpful	troublemaker
forgiving	unforgiving
thankful (grateful)	unthankful (ungrateful)
reconciliatory	revengeful
lawful	lawless
rebuke (for edification)	rebuke (for berating or belittling)
truthful	liar
fairness (impartial)	unfairness (partiality)
brave (gives life for truth, justice)	coward (fearful, hides)
temperance (life in balance)	overindulgence (life of extremes)
honorable	dishonorable
unpretentious (reserved, humble)	pretentious (arrogant, boastful)
elegant (graceful)	crude (vulgar)
patient	impatient (rash, reckless)
light	darkness
sympathetic	unsympathetic (callous, indifferent)

Both sides create the knowledge of good and evil

"like God [*elohim*] knowing good and evil"

"God all in all" – 1 Cor 15:28

GP 9: God's Symbolic Throne – Power of God

Outline of Scriptures
What is the Meaning of the Throne?

gp637» Those who have read the scripture of God's symbolic throne have more than likely come away mystified. The description of this symbolic throne is found in Ezekiel chapters 1 and 10, Isaiah chapter 6, and Revelation chapters 4 and 5. Various other places in the Bible have further information on this mysterious throne of God. The description of this throne is truly "out of this world." We will in this paper tie-in the scriptures of Ezekiel, Isaiah, and Revelation. By this we will manifest that all these scriptures are describing the same throne, but each set of scripture is qualifying and amplifying each other set of scripture.

gp638» The synthesizing of Biblical scripture is a key to understanding the Bible. We must synthesize the Biblical verses of God's symbolic throne in order to understand it. By synthesizing these verses we are conforming to one of the major principles in Biblical study (see "Premises for Belief" paper [BP 5] in the *Beginning Papers*). This principle is the one of "here a little, and there a little" (Isa 28:10). After we synthesize the verses on God's symbolic throne, we will give the meaning of this throne and its specific elements. By doing this *we will prove that God's symbolic throne represents the God's full power, authority, and control of everything.*

gp639» We will now begin to synthesize scripture on this symbolic throne. We will use the outline form mostly because of the complexity of synthesizing these scriptures by any other method. The outline form is easy to check and to review which is another reason we will use it.

Outline of Scriptures

Throne's Setting

gp640»

- The appearance of the symbolic throne of God occurs with a whirlwind, great clouds, great noise, great fire, and brightness of light. [Ezek 1:4, 13, 14, 24; 3:13; Dan 7:9-10; Rev 4:5; Psa 97:2-4]

  [This pictures the Last War with its atomic weapons, its clouds, and its lightning. See God's Wrath papers.]

- As in the midst of this fire appears the throne. [Ezek 1:4, 5; Rev 4:5]

- In the midst of the throne are *four living creatures*. [Ezek 1:5; Rev 4:6 (KJV "beasts")]

Biblical Description of the Four Living Creatures

gp641»

- They have a likeness of a man. [Ezek 1:5]

- Each one has *four faces* on each of the four sides of the head. [Ezek 1:6, 10]

- One face was like a *man*; one face was like a *lion*; one face was like an *ox* (calf); one face was like an *eagle*. [Ezek 1:10; Rev 4:7 (note Jer 11:19)]

- Every living creature had *six* wings. [Rev 4:8]

gp642» [[Notice that Ezekiel 1:6, 8 seems to say they had only *four* wings. Let's straighten out this apparent contradiction: "And every one [living creature] had four faces, and every one had four wings" (Ezek 1:6). Yet these four wings are the four wings on each creature's four sides: "their wings on their four sides; and they four [sides] had their [four] faces and their [four] wings" (Ezek 1:8). But these four wings, one on each of the four sides, are *two* pairs of wings, for these wings are in pairs: "two wings on every one [set or pair] joined one to another, and two [pair, thus 4 wings] covered their bodies," (Ezek 1:11) and "every one [set of wings] had two wings, which covered on this side, and every one had two [or each set of wings had two wings, thus these *two* pair of wings made 4 wings covering the four sides of their body], which covered on that side, their bodies" (Ezek 1:23).

> Besides these two pairs of wings (4 wings) that covered their bodies, there were two wings that were stretched over their head: "and their *wings* were stretched upward" (Ezek 1:11). Remember, the other four wings (two pairs) "covered their bodies," but this set was "stretched upward." Hence these living creatures had 6 wings: two pairs of wings (four) — one wing per each of the four sides of the body — and a pair of wings stretched overhead.]]

■ Each creature had four hands under each of the four wings that covered their bodies. [Ezek 1:8]

■ The general appearance of these four living creatures were like: "the color of burnished brass;" or "burning coals of fire;" or "the appearance of flames" (KJV, "lamps"); or "as the appearance of a flash of lightning." [Ezek 1:7, 13, 14]

> [In other words, these living creatures appeared as a burning flame with sparks of lightning coming out at times. Another way to describe such an object is to say it looked like the sun or a star. The stars are burning flames and are symbolic of angels (Rev 1:20). Angels are spirits that are each represented as a "flame of fire" (Heb 1:7).]

Cherubs

gp643» Notice in Ezekiel 10, where it also describes the symbolic throne, that the cherubs in this chapter are one and the same with the four living creatures: "and the cherubs were lifted up. This is the living creature that I saw by the river of Chebar [Ezek 1:1, 4-5] ... This is the living creature that I saw under the God of Israel by the river Chebar; and I knew that they were the cherubs" (Ezek 10:15, 20).

gp644» *The cherubs that Ezekiel are describing in chapter 10 are similar to the four living creatures* he described in chapter one. Notice the similarities between the cherubs of Ezekiel 10 and the four living creatures of Ezekiel 1 and Revelation:

■ Each cherub of the cherubs has *four faces*. [Ezek 10:21]

■ One face like a *cherub*; one face like a *man*; one face like a *lion*; one face like an *eagle*. [Ezek 10:14]

■ "The likeness of their faces was *the same* faces which I saw by the river of Chebar" (Ezek 10:22; note Ezek 1:10). But if you compare these faces with Ezekiel 1:10 you see they are not exactly the *same* faces. A Biblical contradiction? No, for the word "same" comes from a Hebrew word meaning "they," and not "same." Corrected it should read: "And the form of their faces, they (are) the faces ..." Thus, these four faces were similar, but not exactly the same (Check this with the Hebrew text, and *Young's Literal Translation of the Holy Bible*). [Ezek 10:22]

[Remember what we are doing now is tying-in the scriptures of the symbolic throne. Each description in the Bible adds information to our general overall view of this throne.]

■ Each cherub had four wings, one for each side or face. Each wing had a man's hand under each wing. This is the same as each living creature's wing (Ezek 1:6, 8). [Ezek 10:21, 8]

gp645» Thus, we see the similarities of the four living creatures and the cherubs. Yet everywhere in the Bible where it describes the cherubs, it says there are two cherubs, not four. Thus, there is no reason to say that there were four cherubs in this throne like there were four living creatures. There are only two cherubs in Ezekiel's cherubs, but the cherubs are the living creature. Each living creature is like each cherub in appearance.

gp646» We know cherubs are angels, for when it describes the angel Satan in Ezekiel 28, it calls him an "anointed cherub" (Ezek 28:14). Further we have just shown you how the general appearance of the four living creatures are like fiery flames — thus symbolic of angels (GP 3). The four living creatures were symbolic of angels as the cherubs are symbolic of angels.

Seraphs

gp647» Now Isaiah 6 also describes God's symbolic throne. In Isaiah 6:2 it speaks of the seraphs with their six wings. The word "seraphs" means, "burning nobles." The seraphs looked like a burning creature of noble status. As we manifested before, a burning flame in the

Bible is symbolic of an angel. Thus the seraphs were spirit beings. Isaiah 6 doesn't say how many Seraphs there were. But the seraphs are somewhat like the four living creatures ("beasts") of Revelation chapter 4:

■ The seraphs and the beasts both have six wings. [Isa 6:2; Rev 4:8]

■ Both speak about the same words around the symbolic throne. [Isa 6:3; Rev 4:8]

gp648» Thus, we can reasonably conclude that the seraphs, and the four living creatures of the book of Revelation are one and the same. Further the four living creatures are the same four living creatures as the ones in Ezekiel one. And the four living creatures of Ezekiel one are the same as the cherubs of Ezekiel 10 as shown previously.

Throne of God

gp649» *The seraphs, four living creatures, and the cherubs are all metonymical terms which help to describe the same symbolic throne of God.*

Now let us continue to synthesize these scriptures.

■ The four living creatures and the cherubs had four wheels. [Ezek 1:15-16; 10:9]

■ Each wheel looked like they were in each other. [Ezek 1:16; 10:10]

■ Eyes were in the wheels. Notice the four living creatures ("beasts") were filled with eyes (Rev 4:8). [Ezek 1:18; 10:12]

■ Everywhere the cherubs, or the wheels, or the four living creatures went, they all went *together* as a unit "for the spirit of the living creature was in them." They all went together as a unit, for all the descriptions of this throne are metonymical descriptions of the same thing. [Ezek 1:19-21; 10:16-17]

gp650» Since the general appearance of this throne is like a burning flame (Dan 7:9; Ezek 1:7, 13, 14) which is symbolic of angels or spirits, and since the throne is a unit that goes and moves together because of the spirit in it; *then this symbolic throne is of a spiritual dimension.*

■ The sound or noise of the cherubs and living creatures were "as the voice of the Almighty God when he speaks." [Ezek 10:5; 1:24]

■ Above the four living creatures, or the cherubs, or the seraphs was the throne. [Ezek 1:26; 10:1; Isa 6:1-2; Rev 4:2, 6]

■ Above the whole throne and immediate area was a rainbow. [Ezek 1:28; Rev 4:3]

■ Similar things happen around the seraph's throne, and the cherub's throne. [Ezek 10:6-7, 2 with Isa 6:6]

■ SITTING ON THE THRONE was one with "the appearance of a MAN." [Ezek 1:26]

■ "The BeComingOne (YHWH) sitting upon a throne." [Isa 6:1 (6:5); 2Chron 18:18; note Ezek 10:19]

■ The God sits on the throne. [Rev 7:10; 19:4; 12:5; 22:1, 3; Psalm 47:8: Heb 1:8 Psa 45:6]

■ The MOST HIGH sits on the throne. [Psalm 9:2, 4]

■ The BRANCH on the throne. [Zech 6:12-13 (Isa 11:1-4)]

■ JESUS CHRIST on the throne. [Acts 2:30; Rev 3:21 (Rev 1:4 with 4:5)]

[Compare Revelation 1:4 with 4:5 to see that each throne named has the seven Spirits of God before it because they are the same throne. As Revelation 1:4 says it is the throne of "him which is, and which was, and which is to come." Comparing this with Revelation 1:8 we see this throne is Christ's.]

■ The LAMB on the throne. [Rev 5:6; 7:17; 22:3]

■ The SON OF MAN on the throne. [Matt 19:28; 25:31]

■ The ANCIENT OF DAYS on the throne. [Dan 7:9]

■ An ANGEL on the throne. [Rev 10:1 with 4:3 and Ezek 1:27]

gp651» *Christ sitting on the throne.* All these Biblical names of those sitting on the throne are metonymical names of the ONE Christ the BeComingOne. All the above scripture quoted are in context and in a sense concerned with the same throne of God at the very instant of the Last War and the Day of the Lord (note Ezek 1:4; Rev 1:10 & 4:1; with 11:15; etc.). *This throne is describing Christ the BeComingOne's throne with all its spiritual manifestations.*

More Detail on the Throne

gp652» The F*our Living Creatures* of God's throne picture the four beasts of Daniel 7, for the word "beasts" of Daniel 7 in the King James Version (KJV) should have been translated as "living creatures." The four living creatures of Daniel 7 are the kingdoms of Satan. Thus the kingdoms of Satan belong to the total power and authority of God.

gp653» The *seraphs* of the throne are just another way of describing the four living creatures and cherubs (see outline above).

gp654» The *wheels* of the throne are chariot wheels. These wheels can be synthesized with the four chariots of Zechariah 6:2-3. These are symbolic again to the four beasts of Daniel 7 (see "Last War and God's Wrath" paper [PR 5]).

gp655» The *eyes* of the throne (Rev 4:8; Ezek 1:18; 10:12) are the seven eyes of Christ the Lamb (Rev 5:6), or the seven eyes of Christ the Stone or Rock (Zech 3:9), or the seven eyes of Christ the Plummet (Zech 4:10). And these eyes are "the seven Spirits of God sent forth into all the earth" (Rev 5:6; note Zech 4:10). And the seven Spirits of God are the "seven lamps of fire burning before the throne" (Rev 4:5; note 1:4). Spirits are angels (Heb 1:7), and lamps or flames of fire are symbolic of angels or spirits (Heb 1:7). Thus the seven Spirits of God are the seven angels of God (note Rev 1:20). And these seven angels are the seven angel-Spirits of the Church (Rev 1:20). Thus, the eyes of God's throne are the Spirits or angels of God's Church. God's throne or power and authority includes the seven angels of God's Church. In other words, God's power includes his Church and the angels thereof.

gp656» The *faces* of the throne are symbols of the various sides or faces or facets of God. Christ the BeComingOne is a "son of man," thus the face of a man on the cherubs or living creatures. Christ the BeComingOne is an angel, or spirit, thus the face of a cherub. Christ the BeComingOne is symbolized as a Lamb, thus the face of the calf or ox. Christ the BeComingOne is symbolized as a lion (Rev 5:5, "the lion of the tribe of Judah"), thus the face of a lion. Christ the BeComingOne is symbolized as an eagle (Ezek 17:3, with 17:22, its higher meaning), thus the face of an eagle.

gp657» Also Satan, who is part of the throne, is symbolized by an eagle (Ezek 17:7, its higher meaning; Obadiah 1:4; Luke 17:37). And he is symbolized as a lion (1Pet 5:8). And further Satan is a cherub (Ezek 28:14). Thus, these faces also represent Satan. Of course, Christ *the* BeComingOne is/will-be the fulfilled completed God and thus the faces of Satan will be in Christ the BeComingOne at the End, in the sense and way we have explained previously.

Putting Scripture Together

gp658» By studying Ezekiel chapter 1 and 10, and Isaiah 6, and Revelation 4 and 5, and then reading this outline and looking up all the verses quoted (studying and comparing) one will see the great similarities and that indeed these scriptures do pertain to the same throne of God.

gp659» The simplest explanation for the given facts is usually the right answer. These scriptures are not speaking of different thrones, but the same one. These scriptures fit together as the scriptures of Daniel 7 and Revelation 13 & 17 on the "Beast" fit together (see the Beast-System Paper [PR 2]).

gp660» The Bible is made plain only when one synthesizes all scripture on any one topic. By putting the scripture together that pertains to God's throne we get all the facts on it together, and thereafter are ready to understand it. But if we conclude that these scriptures are speaking of different thrones, then we are confused. When we understand that the four living creatures, the cherubs, and the seraphs are merely metonymical names of the overall throne of God, we are on the way to understanding God's mysterious throne.

Throne and the Glory of God

gp661» Notice how Ezekiel generalized on his vision of this throne:

- "I saw the visions of God [*elohim*] ... this was the appearance of the likeness of the glory of the BeComingOne" (Ezek 1:1, 28; Rev 4:2-3).

This whole vision Ezekiel had was of the glory of God or visions of God. In the truest sense, the Glory of God, is the totally fulfilled God, the BeComingOne who has become. The details of the vision are merely particular manifestations of God. Yet since these visions were mainly of the true God's throne, and its immediate surroundings, and because thrones are symbols of power or authority; then these visions were of the true God's power and authority.

Satan's Throne versus God's Throne

gp662» A parallel would be the scripture on the Beast of Revelation with its 10 horns and 7 heads, etc. The "Beast" represents Satan's spiritual "power, and his throne [KJV, 'seat'], and great authority" (Rev 13:2). Just as the "Beast" represents Satan's power and authority, so too does God's symbolic throne represent God's power and authority. It represents his SPIRITUAL power and authority, for the throne is of a spiritual dimension as we previously indicated in this paper.

What is the Meaning of the Throne?

gp663» As we just noted this throne is symbolic of the BeComingOne's power and authority. The various details of this throne add to our knowledge of God's power and authority.

Power: All

gp664» Just how much power and authority does the BeComingOne hold? Christ the BeComingOne has been *given* ALL the power in heaven and earth (Matt 28:18). As of now Christ does not hold all the power for Satan now has the power of death and destruction (Heb 2:14). **"But now we see not yet all things put under him"** (Heb 2:8). The true and complete God is the "Almighty" (Rev 1:8). This God is All-powerful, He holds ALL the power and authority. The true God, which is the fulfilled Spiritual Body of Christ, will be complete, the "God, all in all," only at the End (1Cor 15:24-28).

gp665» This *all-mighty* Power includes authority over the kingdoms of men: "to the intent that the living may know that *the most High rules in the kingdom of men*, and gives it to whomever he will" (Dan 4:17). And again, "by me kings reign, and princes degree justice. By me princes rule, and nobles, even all the judges of the earth" (Prov 8:15-16). So the BeComingOne somehow has power and authority over the kingdoms of the world.

Predestinated Power

gp666» The kingdoms of the world in this age *are* directly under the authority of Satan as Revelation 13:2 shows (see the "Beast-System Paper" [PR 2]). But these kingdoms under Satan are under the true God's predestinated overall power and authority. Before the beginning of the present universe, God predestinated everything which has happened, which will happen, and which is happening (see the "Predestination Paper" [NM 8]). Then *out of* the pre-creation God Power came everything including the angel or messenger of God who spoke for the BeComingOne. All came out of the God Power, but *all* were at that time predestinated to return to that God Power and become the BeComingOne Power. This is known as the Great Cycle (see GP 6). Since the pre-creation God Power predestinated all to return to Him, He in effect has not lost control of the cosmos. In this old age the God has given the right to rule this age to Satan and his kingdom and the rulers therein: "you [Pilate] could have no power at all against me, except *it were GIVEN you from above*" (John 19:11). Pilate and all physical and spiritual rulers of this evil age were given their power by the predestination of the God before the cosmos began. Even Christ the man's death was predetermined by God (Acts 4:27-28). But remember, all was predestinated before the cosmos, before law (as we know it), and before sin (as we know it). We know it was not possible to create good without in some way creating evil because good is a comparative quality to evil (GP7). Therefore how can anyone call predestinating evil, evil, when it was done before evil and before sin?

To Review

gp667» Hence we are perceiving that the throne, or the power and authority of the God includes both parts or sides or hands of the God. At *this* time, or in this age the left side of God rules through Satan. In the next age the right hand or side of God will rule in the Kingdom of God. But in this age of Satan's direct rulership of man, the God (the BeComingOne) is the over-all ruler for He has predestinated ALL through His PERPETUAL power (note Rom 1:20, in Greek). He is the Almighty God. The BeComingOne is in control of the situation, but has in this age predestinated some of his overall power to Satan for a purpose. This purpose being that man needs a period of madness in order to have future happiness (see GP 7).

Review of GP 9

gp668» We have seen basically that the throne of God represents the power of the true God — the BeComingOne (YHWH) — in its totality. The symbolic throne of God represents all of God's power: both the good and the bad. But let us not forget that the true and completed God is good, but in Him there will be the memory of the evil that was "allowed" by God and predestinated by the pre-creation Power. In the beginning all went *out of* the pre-creation Power to learn about evil so that all would be able to know and enjoy the good. The true God is the finished and completed God, the God all in all.

General Review

gp669» In GP 1 we started our search: who or what is God? From the Bible we learned about the apparent paradoxes of God: "I make peace, and create evil: I the LORD do all these things" (Isa 45:7). God who is Love (1John 4:8) has somehow and for some reason created evil; He has even killed (Deut 32:39). But how can God be Love and also a killer?

We next learned that there are two basic laws and one basic fact we must understand in order to rightly perceive the true nature of God: the Law of Contradiction and the Law of Knowledge plus the fact that the God cannot lie.

We then went on and explained the Law of Contradiction.

We further showed the many attributes and titles of God and put forth that "time" is very important in our understanding of the paradoxes of God.

We also showed you the very NAME of the true God: YHWH, or Jehovah, or Yehowah, or He (who) will-be, or the BeComingOne, or the One who was, who is, and who is coming. God's NAME and its meaning is the real secret in revealing the answer to the Paradoxes of God. God's NAME is an *imperfect* (incomplete) verb and not as would be expected a *perfect* (complete) verb or a noun. Names are very important in the Bible and many times describe some facet of a person. The true NAME of the true God is important for it

is the secret in explaining the apparently unexplainable scriptures about God.

In GP 1 we also looked into the meaning of "with God all things are possible," the "*one* Yehowah," the so-called unchangeableness of God, and other matters concerning the God. What GP 1 does is set the stage in our search for who or what is God.

In GP 2 we learned that Jesus Christ's Father was the BeComingOne (YHWH) of the Old Testament: He was the Jews' God.

In GP 3 we learned that the angel of the BeComingOne and the BeComingOne of the Old Testament were closely connected. Since angels are messengers, this means the angel of the BeComingOne is a messenger of the BeComingOne or this angel is the WORD of the BeComingOne. Therefore, the words that the angel of the BeComingOne spoke belonged to the Great BeComingOne Power — the true God. This angel stood in the NAME of the true God (Exo 23:20-21); he represented the great NAME. This angel was in a sense the very WORD of God. The Word (logos) of God before Christ's resurrection was a spirit, the chief-spirit, the chief-angel, the angel of the BeComingOne. The age before Jesus Christ was subjected to angels, even the commandments given on Mount Sinai were from an angel (Acts 7:38).

We have shown *in GP 4* that *before* Jesus Christ was resurrected, he was a human being because he was born from a woman. Thus, Christ before his death and resurrection was a man; he was Jesus Christ the man. He was called Son of man because he was born of mankind through the means of Mary his mother. Jesus the man was not just any human being. Christ was a Son of God, both physically (through the Holy Spirit's union with Mary) and Spiritually (through the medium of God's Spirit inside of him). Christ was a mediator between man and God; he was the Son of God and the Son of man. He is the "one *mediator* between God and men" (1Tim 2:5). Jesus Christ the man actually had the Spirit or Angel of the BeComingOne (YHWH) inside him leading him in the right way. It was because of this Spirit that Christ the man was sinless. God was not Jesus Christ the man, but God (through the power of His Spirit) was inside of Christ the man. When Jesus Christ the man died, his Spirit was then separated from him (Luke 23:46). The Spirit or angel of God did not die *as* Christ the man or *with* Christ the man, for Spirit cannot die. The angel of the BeComingOne separated himself from Christ the man when Christ died.

To say it in slightly a different way, in GP 4 we have shown that Jesus Christ before he was resurrected was a man with the Spirit or angel of the BeComingOne inside him leading him in the way of love. Jesus Christ the man was born from woman and by a miracle of God. Thus, Christ the man was from mankind (through Mary) and from God because of the miraculous conception by the power of God. Jesus Christ the man was a son of man, and a son of the God. He is the "one *mediator* between God and men" (1 Tim 2:5). Christ the man wasn't God, but God's WORD and power were inside of him. The angel of the BeComingOne was inside Jesus Christ the man. When Christ died the angel of the BeComingOne separated from Christ the man's body. Thus, Christ the man died, but his angel inside of him stayed alive, for the angel (spirit) separated himself from Christ the man when Christ died (Luke 23:46).

In GP 5, we have seen that after Jesus Christ the man's death and his burial for three days, he was resurrected and became ONE with the Spirit or Angel of the BeComingOne: **two became one**. The Spiritual marriage of mankind back to God *began* when Jesus Christ the man went into the BeComingOne, the God. Since the beginning of creation when all went *out of* God, God and man were separate. But through Jesus Christ the man, mankind is beginning to go back into God, or into the BeComingOne (YHWH). Through the angel with the NAME, the BeComingOne incorporated Jesus Christ the man into Himself: the Spiritual WORD became flesh. The two, the man Jesus Christ and the Angel of the BeComingOne ("LORD"), as ONE are called: Jesus Christ, God, or in another sense he is/will be Jesus Christ the BeComingOne. Jesus Christ the man has become Jesus Christ *the* God, in one sense, because ALL will eventually go into the Spiritual Body of Jesus Christ by the End (see GP 6). But as of now we do not see ALL in Jesus Christ the God (see Heb 2:8). In two senses Jesus Christ is "the beginning of the creation of the [true] God" (Rev 3:14): the Spirit of Jesus was the First, the beginning of the "old" creation of God; but Jesus Christ the man and his Spirit as ONE is the beginning of the "New" creation.

In GP 6 we learned how the rest of mankind will follow in Christ the man's footsteps. All of mankind each has their own angel which each will eventually be infused with. As Christ the man Spiritually married his angel, so too will each of mankind Spiritually marry his own angel so that two will become ONE. Since the beginning of creation mankind has been out of

God. But through the power of Christ and God all will come back into God so "may be the God, all things in all" (1Cor 15:28).

In GP 7 we learned the reason why God created evil (Isa 45:7) in this age. We saw that in order to understand the good and to enjoy the coming utopia, mankind and angelkind must have a time of evil. Good and evil are opposite qualities that need to be compared against each other in order to understand either. A person who grew up with all the good things that the earth has to offer doesn't know the worth of these things until he actually loses them. After he loses the good things he then begins to understand the worth of the good things. If God just automatically at first put the creation into an everlasting utopia, the creation itself would not have the understanding of the worth or value of the utopia. In fact as we learned in GP 7 mankind would not have any understanding of good, if they never had learned evil. But by God putting the creation through an evil period God is actually teaching the creation the value of good and the evil of evil. The creation now is actually learning about good and evil. It is through the evil spirits in the minds of mankind that evil is spread on earth. It will be through the good spirits that mankind will live in good.

In GP 8 we learned the meaning of the cherubs. The *right* cherub represents the right hand or right side of God — the good side of God. The *left* cherub represents the left hand or left side of God — the evil side of God. The BeComingOne is actually foreshadowed by the cherubs and the mercy seat in the most holy place. The right side of God is Jesus Christ, God, while the left side of God is that one called Satan. Both Jesus Christ and Satan are represented by the cherubs and the mercy seat. Jesus Christ the man is represented by the mercy seat. The cherubs and the mercy seat represent the BeComingOne, and the truest sense of the BeComingOne is the finished God — the true God — who is the good God. The true God will change Satan and then put this *reformed* spirit back into Himself. Satan when he is put back into the God will then be a good spirit or power with the memory or knowledge of his former evil ways. The knowledge of evil was needed so as to give the true knowledge of good to ALL, since good and evil must be compared in order for us to know either. In the beginning all went *out of* God. This included Satan. But at or before the end all will be brought back into the God, so that God will be all in all. The physical cherubs and the mercy seat were one piece, thus pointing to the future

oneness of the real cherubs and mercy seat.
Satan will thus be brought back into the God.
The meaning of the cherubs indicates this to us.

We have seen **in GP 9** that the throne of God
represented the power of the true God — the
BeComingOne (YHWH) — in its totality. The
symbolic throne of God represents *all* of the
true God's power: both the good and the bad.
But let us not forget that the true and completed
God is good, but in Him there will be the
memory of the evil that was "allowed" by God
and predestinated by the pre-creation Power. In
the beginning all went *out of* the pre-creation
Power to learn about evil so that all would be
able to know and enjoy the good. The true God
is the finished and completed God, the God all in
all.

GP 10: God Is ...

God is Love
God, Predestination, and His Essence
God is Omnipresent
Who or What is God

gp670» It is difficult to put into a few words who God *is* because the Bible does not say who God is, but who God *was*, who God *is*, and who God *will be*. It is the ultimate-goal of God that all will live in happiness. The ultimate state is what is becoming and will become by the end of the creation. The creation is still going on. The Almighty by giving us a time to learn good and evil is creating in us happiness, pleasure, goodness, love, and all else that will serve us well in our endless future.

God Is Love

gp671» Ultimately the true God in his totality[1] will be love for "God is love" (1John 4:8, 16). And what is love? "Love is patient; love is kind and envies no one. Love is never boastful, nor conceited, nor rude; never selfish, nor quick to take offense. Love keeps no score of wrongs; does not gloat over other men's sins, but delights in truth. There is nothing love cannot face; there is no limit to its faith, its hope, and its endurance. Love will never come to an end" (1Cor 13:4-8). God's nature is as these verses describe.

gp672» Further, all the verses that describe how a real Christian should behave are more qualifications on love. A person who can follow all these qualifications on love, is love or a living being of love.

- Thus, love (God is love) does *not* give evil for evil (Rom 12:17). Love does not bless *and* curse at the same time (James 3:10).

- But love is "peaceable, and easy to be entreated, full of mercy and good fruits, without partiality, and without hypocrisy" (James 3:17).

---

[1] Of course his "totality" also includes his other qualities

- Love does *not* worry or fear (Matt 6:25-29).

- Love, loves its enemies yet *not* their ways (Matt 5:44; Prov 8:13).

- Love does *not* hate his brother (the person himself), yet if his brother's ways are of evil he hates his brother's ways, yet not the person himself (Prov 8:13 & Luke 14:26).

- Love *does* in deeds what it utters in tongue (1John 3:18 & James 1:22).

- Love does not test others with objects that might lead them away from the truth, to a reasonable degree (Rom 14:20-21),

- yet Love knows nothing in itself is bad (Rom 14:14 & Titus 1:15).

See the "Freedom and Law" paper [NM 17] for more information on "love" and the way of love.

God, Predestination, and his Essence

God is the Creator, We are the Clay

gp673»

- You turn *things* around! Shall the potter be considered as equal with the clay, That what is made would say to its maker, 'He did not make me'; Or what is formed say to him who formed it, 'He has no understanding'? (Isa 29:16)

- I form the light, and create darkness: I make peace, and create evil: I the LORD do all these *things*. 8 Drip down, O heavens, from above, And let the clouds pour down righteousness; Let the earth open up and salvation bear fruit, And righteousness spring up with it. I, the LORD, have created it. 9 Woe to *the one* who quarrels with his Maker-- An earthenware vessel among the vessels of earth! Will the clay say to the potter, 'What are you doing?' Or the thing you are making *say*, 'He has no hands'? 10 Woe to him who says to a father, 'What are you begetting?' Or to a woman, 'To what are you giving birth?' 11 Thus says the LORD, the Holy One of Israel, and his Maker: Ask Me about the things to come concerning My sons, And you shall commit to Me the work of My hands. 12 It is I who made the earth, and created man upon it. I stretched out the heavens with My hands And I ordained all their host. (Isa 45:7-12)

gp674» God has made everything. He predestinated everything before the cosmos was created by him (NM8 & NM9). In GP1 we list some of the verses where God said that he even predestinated some to evil. Now this is very difficult to understand. How can God be love yet predestinate some to evil? How can this be? Yet the Bible clearly speaks about predestination. Not only does God predestinate good, but also evil (NM 8, GP 1).

Free Will?

gp675» There is no easy way of explaining predestination if you believe in free will. But once you understand that free will only exists under the all-powerful free and ultimate will of God, you will understand that your free will is limited. What most think of as free will is just another philosophical belief, not a Biblical teaching. We have some free will, but only under and after the ultimate free will of our God. God is not a fool as to give real free will to each in his creation. This knowledge should not be used as an excuse for our sins, since we do have relative free will under the free will of God. See the Predestination Paper (NM8) for more information.

Paul Explained Predestination

gp676» Paul explained predestination in this way:

- "It does not, therefore, depend on man's desire or effort, but on God's mercy. For the Scripture says to Pharaoh: 'I raised you up for this very purpose, that I might display my power in you and that my name might be proclaimed in all the earth.' Therefore God has mercy on whom he wants to have mercy, and he hardens whom he wants to harden. One of you will say to me: 'Then why does God still blame us? For who resists his will?' " (Rom 9:16-19, NIV)

- "Does not the potter have power over the clay, from the same lump to make one vessel for honor and another for dishonor? 22 What if God, wanting to show His wrath and to make His power known, endured with much longsuffering the *vessels of wrath prepared for destruction*, 23 and that He might make known the riches of His glory on the vessels of mercy, which He had

prepared beforehand for glory" (Rom 9:21-23, NKJV).

Let me explain predestination in another more detailed way. You must read the following in context of all this book.

God and Predestination

gp677» There is One God, who is all-powerful and His Name is the *BeComingOne* (Ex 3:14 ft, God, chap. 1). Before the creation of the universe, God was the highly intelligent pre-creation Power who set up his creation to do exactly as he planned (Eph 1:4; 1Peter 1:20; John 17:24). He thus predestinated everything <u>before</u> the cosmos, <u>before</u> good (as we know it), <u>before</u> evil (as we know it), <u>before</u> law (as we know it), and <u>before</u> sin (as we know it), thus one cannot find sin in God's predestination because it was done <u>before</u> sin. Yet the real God is all-powerful and is responsible for all that takes place in the creation (2Chron 20:6; Isa 44:24; Rom 9:19) and his power is perpetual (Rom 1:20).

Cosmos Created out of Spirit

Ultimate Building "Blocks"

gp678» God was in someway "alone" before the creation (Adam as a type for God, Gen 2:18). God's very essence is spirit, that is, God is spirit (John 4:24). Because spirit is invisible God is invisible (Col 1:15; John 1:18). From God's own invisible essence he created all things because all came out of Him (Rom 11:36; 1Cor 8:6; 15:28). The Spirit of God is the ultimate building block of the present universe and for the coming new universe, when God, in some sense, will be the all in all (1Cor 15:28). The reason we cannot see "spirit" is because it is our ultimate building block. If God, whose essence is spirit was/is/will be everywhere (Jer 23:24; Psa 139:7; Acts 17:27), then he must, he has to be, in some sense everything that was/is/will be. God did not create the universe out of nothing; God created it out of Himself, that is, out of His spirit. In other words, all came *out of* him as Romans 11:36 and other scriptures read in the Greek. By the end of creation God will have created all things new (Rev 21:5).

Word and Jesus Christ

gp679» God created everything by His Word (John 1:1). The Word was the first-angel or Spiritual messenger who spoke the powerful words belonging to the God (GP 3 & GP 4). Since God's essence or spirit is everywhere (Jer 23:24; Psa 139:7; Acts 17:27) and in fact there is nothing else but God (Isa 45:6), in order for God to speak (in the way we understand speech) he must speak through someone who is an individual with spatial position within the ONE who fills all space. Since the Bible calls those who speak or carry messages *angels*, the Word must have been an angel. In fact the Word was the very first-angel or archangel or angel of God's Presence (see GP 3). This archangel who spoke in the Old Testament was not only the first-angel, he was the angel who carried God's very Name (Ex 23:20-21). The Word was made flesh by incorporating Jesus Christ the man into himself as explained in this book (GP 4 & 5). The Word will make all things ever created or ever to be created through him (John 1:3). Apart from him nothing will be created (John 1:3). Christ revealed the invisible Spirit through his behavior (John 14:9-10). Christ reveals God now, for God and Christ the man were fused together as a new creation (GP 5). Christ reveals God now by revealing himself (John 1:18; 14:7, 9).

Father and Jesus Christ

gp680» God who created and will create everything through his Word, is also called the Father since he fathered all things through his great power; he has life in himself, for he is life itself (John 5:26; Acts 17:28). The Father gave all his glory and power to Jesus Christ after Jesus Christ died and rose and was infused with the Spiritual Word (Matt 28:18; John 13:31-32; 17:4-5; GP 5). Jesus Christ in a sense is now the Father (Isa 9:6; GP 5) because he went into the glory of the Father (GP 5). Jesus Christ is now our God, the BeComingOne, since he was infused into the angel with the Name (YHWH), and since all will come back into Jesus Christ's Spiritual Body (GP 6) so that God will be all in all (1Cor 15:28). **The ONE who is the Father of everything and in a sense is everything, now speaks and works through Jesus Christ:**

- Philip said to Him, Lord, show us the Father, and it is enough for us. 9 Jesus said to him, Have I been so long with you, and *yet* you have not come to know Me, Philip? He who has seen Me has seen the Father;

how *can* you say, 'Show us the Father'? (John 14:8-9)

Evil Predestinated?

gp681» We know from scripture that God, the super intelligent pre-creation Power, chose Jesus Christ to be the Savior of the world before the foundation of the creation (1Pet 1:19-20; Rev 13:8), and thus before sin. The Bible by saying this projects to us that God knew beforehand that his creation would need saving. The question is, why didn't God create a universe that did not need saving? Some think mistakenly that God could have created a universe without death, without pain. Why didn't he? Because he knew that in order to create good, happiness, and pleasure he must also create evil, unhappiness, and pain (NM 19). The fact that you cannot have pleasure without knowing about pain and the fact that you cannot know happiness without first knowing unhappiness is understood when you understand the Law of Knowledge. The simple fact is that God could not create good, without in someway creating evil. God created evil through predestination before good (as we know it), before evil (as we know it), before law (as we know it), and thus <u>before</u> sin (as we know it), and therefore God did not sin in his predestination.

Should we then be angry with God?

gp682» Remember all will come back into God (1Cor 15:28; GP 6). Therefore, at the end of creation no one can stand up and say he is better than anyone else, for all will have contributed to the madness, and soon to come, all will contribute to the harmony (GP 7, GP 8, GP 9). No, we should not be angry with God because would we want to live forever without the understanding of immortality, or pleasure, or happiness?

Time and the God

gp683» One secret to unlocking the paradoxes in the Bible pertaining to God is to know the Name of the God, and the significance of that Name. God's Name is the *BeComingOne* or *He (who) will-be*. The God who is everywhere and thus in a sense is everything, or at least his Power is everywhere, and since he is all-powerful, then his power is responsible for everything that was, is, and will be. When the BeComingOne speaks it pertains to the time that **was**, or that **is**, or that **is to come**, since he is becoming and his essence and works are in what was, what is, and what will be. Although God did create evil, he created evil through predestination <u>before</u> time or history (as we

know it). To better understand this read and study the following chart.

gp684» **Chart of Father and Time Below:**

Time	Father and Time				
Before Creation	Pre-Creation Power **Father of All Things** predestinates everything before the cosmos				
Beginning	Beginning of Creation				
Creation Period [1 thru 7th day] **Father gives all Power to his Right and Left Side,** but all future power given to the Word, and through the Word to Jesus and his Church	[Power to the Right Side for all that is good]		Word of the BeComingOne Creates all things	[Power to the Left Side for all that is evil]	
	God is Light Right Cherub 2/3 of all angels		*God separates all things between light and darkness during a 6000 Year Period*	**Satan is Darkness** Left Cherub 1/3 of all angels	
	First-Angel	God's Angels		angel of Satan	Satan's angels
	Angel of God acts throughout OT in the affairs of the patriarchs of the Messiah	Influences and guides God's Prophets and OT Saints	**All Father's power exists as everything diffused throughout the Creation**	Tempts Adam and Eve to sin, is the father of lies and sin; misleads all mankind	Evil minds of Satan become the "other-minds" in first 6,000 years
	Angel Begets Jesus supernaturally		All power diffused between good and evil	Tempts Jesus throughout his life	Tempts all mankind
	Jesus is the **First Resurrection**		All power predestinated to return to the Father by the end of creation	Satan thrown out of the Godhead	
	Establishes his Church	Churches of Jesus Christ	Father still in control through predestination	Satan dwells in the Anti-Christ	Angels of Satan destroy world
	Jesus as King of Kings rules with the Saints	Christians Born of God in **Second Resurrection**	**1000 Year Sabbath** 7th Spiritual Day Kingdom of God rules on earth Earth is renewed	Satan punished in hell-fire for 1000 years. Mankind deceived by the evil angels during the first 6000 years are dead during the 1000 year Kingdom	
100 Years [8th Day] **Father is All Typically**	100 year Great Last Day 8th Spiritual period **Third Resurrection** All Become One in Good Spirit and in the System of Love; all evil behavior taken out of the Creation; heaven and earth still belong to first creation				
Endless Age [9th Day] **Father exists as ONE with New Creation**	New Heaven and Earth Created **Father is all in all** All return to the Father's Power through Jesus Christ's Spirit except at that time **billions of individuals** will have been created with the knowledge of good and evil Endless Age of God's Kingdom Continues forever				
Note: See *New Mind Papers* [NM 16] for explanation of spiritual days (1st through 9th day)					

God is Omnipresent: Everywhere

gp685» Besides the typical sense of God's omnipresence, which is that the God's spirit is everywhere in the sense that everything came out of God's spirit when the universe was created, and everything is a manifestation of that pre-creation spirit, there is an antitypical sense of God's omnipresence. As shown in this book, the true God is the BeComingOne. It is the BeComingOne who will incorporate ALL into Himself, so that will be "the God, all in all" (1Cor 15:28). It is the true God, which has ALL in Him, and it is this God that is omnipresent in the truest sense. At the true end of creation, God is everywhere because God is everything, "the God, all in all" (1 Cor 15:28). "I am the BeComingOne [YHWH] and there is none else" (Isa 45:6).

God is Omnipotent: Almighty

gp686» The typical sense of God's omnipotence (all-powerfulness) is that God predestinated everything through his power. But at the end of the true creation God is omnipotent because all will be in the God, thus, this true God will have *all* the power. At that time all people and powers will be in the One Spirit — the Spirit of LOVE. All at that time will be in the God: God will be many in ONE. All the powers will be in this true God.

God is Omniscient: Knows All

gp687» The typical sense of God's omniscience (all knowledge) is that God knows all because He predestinated all, so of course, He knows all because He predestinated all that happens before the true end of creation. But at the true end of creation ALL will be in the true God, therefore, all the individuals, with all of their total knowledge, will be in the true God. Thus, the true God has *all* knowledge because *all* those with knowledge will be in Him.

Who or What is God?

gp688» The following short summary is based on the Bible and all of this book:

- The **God (YHWH) is unlike any thing else**; there is none like him (Isa 46:9), and that was why Israel was not allowed to make any graven image or any likeness to depict Him (Exodus 20:4).

- **All came out of the pre-creation Power, the Father of all** (1Cor 8:6; Rom 11:36). This means that all have come out of the pre-creation Power, all were made out of the Father's Power and His "substance" or "essence."

- During the creation period (from day one to the End) the Father gave himself the Name, BeComingOne [YHWH] (Ex 3:14-15), for He is all that **was**; all that **is**, and all that **Will Come**, the **Almighty** (Rev 1:8); there is **none else but the BeComingOne** (Isa 45:18); there is no life outside of Him (John 5:26; Acts 17:28).

- **All came out of Him through predestination to learn about good and evil**, about time, about death, and about life, thus to learn about good and be able to appreciate it forever (GP 7). God predestinated everything that will be <u>before</u> the creation, <u>before</u> law, and <u>before</u> sin; God's predestination is without sin.

- **Because the true God is the BecomingOne**, because the God is all-powerful, because the God wishes all to be saved (1Tim 2:3-4; 2Pet 3:9), because the God will bring all back into himself through Jesus Christ (GP 6), this means that:

- all of creation, both physical and spiritual (Col 1:20; Eph 1:8-10,23; Phil 2:10; 3:21), will become one as individuals within the Spiritual power of Christ by the end (Eph 1:23; 1Cor 15:24-28), so that all will be in the true God, **God all in all:** all will be in subjection to the ONE ALL (1Cor 15:28) under the system of love, for the true God is Love (1John 4:8).

- **Adam prefigured:** From one Adam came all of the individuals of mankind; from one mind-substance Power and essence is coming the **many in the ONE God**. God is reproducing Himself as Adam reproduced himself.

- **"BeComingOne, himself, <u>the</u> Gods."** [Deut 4:35, 39]

The true God is the BeComingOne that has become the ONE-ALL.

GP: Appendix

More Details

Yehowah / Yahweh / Jehovah / Lord
Massoretic Text
More Language Details on God's NAME
More on "I will be"

Hebrew Words Written Without Vowels

gp689» At first the Hebrew language, as with other Semitic languages, was written only with consonants and was written from right to left. When the Hebrews read, they added the vowels in their mind to the words. In Moses' time there was apparently no method of writing vowels in Hebrew. Two thousand years after Moses a system of vowel points was developed that was added below, between, and sometimes on top of the letters:

- "The present pronunciation of this consonantal text, its vocalization and accentuation, rest on the tradition of the Jewish schools, as it was finally fixed by the system of punctuation (§ 7 h) introduced by Jewish scholars about the seventh century A. D." [*Gesenius' Hebrew Grammar*, p. 12]

Therefore when Moses wrote down God's NAME he did not write any vowels.

Is the Correct Pronunciation of the NAME Possible?

gp690» Moses did not write down the vowels for God's NAME, since in his time there was no method to write vowels. But it is said that the correct vowels for God's NAME were passed down orally through the years and are preserved in today's vowel point system. But it may be unlikely that the exact sound of the Biblical Hebrew has been preserved for us today because there were different schools

with different methods and interpretations, and there were Jews with different ways of pronouncing the Hebrew words (*Gesenius' Grammar*, p. 38, footnote 2; see § 7 *i*; § 8 "Preliminary Remark"; p. 42 footnote 3; etc.). And of course there were no tape recorders in Moses' time. But we will examine this question anyway to help us better understand the problem over the last three millenniums.

Yehowah or Yahweh or Jehovah or LORD

gp691» In the King James Version of the Bible, we see the word "LORD" was used throughout the Old Testament for the NAME of God. As we have indicated in GP 1, LORD is a mistranslation. "LORD" was translated from a Hebrew word "YHWH," which means — the BeComingOne, or he (who) will-be. In square-shaped letters of the Hebrew language the NAME looked like this: יהוה (read right to left). The square-shaped letters are the ones we see in today's copies of the Hebrew Old Testament. But the more ancient Hebrew letters looked somewhat like the ancient Phoenician or ancient Greek letters. Because of different scribal styles or schools, the ancient Hebrew alphabet varied slightly through the ages. In one style of the old-Hebrew alphabet God's NAME, YHWH, looked something like this:

$$\exists Y \exists \dagger$$

gp692» *Consonants Only*. Hebrew is read from right to left vis-a-vis English's left to right. Originally, the Old Testament was written with only the consonants. "As the Hebrew writing on monuments and coins mentioned in [2] *d* [dated

c. 850 B.C. to c. 138 A.D.] consists only of consonants, so also the writers of the Old Testament books used merely the consonant-signs (§ 1 *k*), and even now the written scrolls of the Law used in the synagogues must not, according to ancient custom, contain anything more. The present pronunciation of this consonantal text, its vocalization and accentuation, rest on the pronunciation of the Jewish schools, as it was finally fixed by the system of punctuation (§ 7 *h*) introduced by Jewish scholars about the seventh century A.D.; cf. § 3 *b*" (§ 2 *i*, pp. 11-12, *Gesenius' Hebrew Grammar*, Oxford, 1910 [1980 reprint]). The Hebrews' written language was thus a "shorthand" language. The vowels were dropped to shorten the space and the time needed to write documents. Other ancient languages were also written only with their consonants. Yet today when we look at the Hebrew texts of the Bible, we see square-lettered Hebrew with *vowel-points* under them. Vowel-points are little dots or lines written under, inside, and over the consonants.

gp693» *No Paragraphs, No Verses, No Spaces*. Up to the finding of the Dead Sea Scrolls sometime around 1945 there were no vowels in these older manuscripts, there were no verses, and there were no paragraphs. But one copy of Isaiah of the Dead Sea Scrolls had "paragraph divisions correspond almost exactly to those in the modern Hebrew Bible" (St. Mark' Monastery Isaiah Scroll, IQIsᵃ, Willaim Sanford LaSor, *The Dead Sea Scrolls*, 1972, pp 29-30). In most of the older manuscripts there were not any separations or spaces between words, all the consonants ran together (Ginsburg, *Introduction to the Massoretico-Critical Edition of the Hebrew Bible*, pp. 158ff).

gp694» *Meticulous Transcription*. The Scribes did not have printing presses or computers; they copied the Bible by hand. In order to preserve the original words as best as possible, the Scribes were very meticulous, they counted words on each page (C.D. Ginsburg, *Introduction to the Massoretico-Critical Edition of the Hebrew Bible*, p. 109) and numbered the letters (Ginsburg, p. 113) and made lists in order to check each manuscript for error (see later).

Spelling of the NAME of God

gp695» *Vowel Letters, Vowel Signs*. The Scribes and readers of the Bible learned from each other the correct pronunciation of each word. But the Masoretes, sometime near 600-700 A.D., began to place graphic-signs for vowels, which led to different systems of vowel-points seen in different Hebrew texts (Ginsburg, pp. 449ff; *Gesenius' Gram.* § 3*b*, 7*h,i*). Today most scholars from the West only study one system of vowel-points. Much earlier than this some scribes made use of vowel-letters (see later), although there seemed to be no uniform tradition (Ginsburg, p.299ff; *Gesenius' Gram.*, §7*a-g*).

Yehowah or Yahweh

gp696» Today (1989) we have only two Hebrew-Greek-English Interlinear Bibles. One Interlinear Bible (Pub. 1976/1986) was edited by Jay P. Green, and uses the so-called *Letteris Bible* (published by the British and Foreign Bible Society in 1866); the other Interlinear Bible (Pub. 1979/1985) was edited by John R. Kohlenberger III and uses the *Biblia Hebraica Stuttgartensia* (BHS) text (published 1967 / 77 by the German Bible Society in cooperation with the United Bible Societies, which reproduces the Leningrad Codex B19*a* [L]) with only a few deviations and is but a version of the *Biblia Hebraica* (BHK), edited by Kittel-Kahle (1905/1947). The Leningrad Codex was previously known as the St. Petersburg Codex B19*a*. This Codex is recognized as the oldest complete Hebrew Old Testament text of the Bible; it has vowel signs and is dated about 1009 A.D.

Yehowah

gp697» The *Letteris Bible* has vowel signs, and the NAME of God is spelled: YEhOwAh. The vowels are **e**, **o**, and **a**. The vowel points for Yehowah looked as follows:

- **e** The short or half vowel "e" was called the Sewa and was two dots, one above the other: . It was placed under the consonant ֯ (Yod), together they looked like this: ֯ . The Sewa sounds like an "eh," or the "e" in emit, or no sound at all when used as a syllable divider.

- **o** The "o" was a dot " " called the Holem and was placed above the " ֯ " (Waw) The Holem sounds like the "o" in roll or mold. The Waw and Holem together looked like this ֯ .

- **a** The "a" was the vowel "ָ" called the Qames or Kamets and was placed under the ו (Waw) just before the ה (He). It looked something like a small compressed capital "T" and was placed under its letter. It sounds like the "a" in father. Waw used to be more commonly called Vav. This is one reason Jehovah had the "v" in it instead of the "w."

gp698» In the square-shaped Hebrew alphabet Yehowah looked like this:

יְהֹוָה

This spelling of God's NAME is also common in major Jewish-Hebrew texts. It is found in *The Pentateuch and Haftorahs*, edited by J.H. Hertz, Chief Rabbi, and published by the Soncino Press, 1956; the spelling is found in the *Interlinear Hebrew-English Old Testament* (Genesis-Exodus), by George R. Berry; the spelling is found in the C.D. Ginsburg's Hebrew Bible; the spelling is also found in some verses of the *Biblia Hebraica Stuttgartensia* (BHS), such as Gen 3:14; 9:26; Ex 3:2; 13:3,9,15; 14:1,8; etc.

Yehwah

gp699» In the *Biblia Hebraica Stuttgartensia* (BHS) God's NAME is spelled: YEhwAh. The vowels are **e** and **a**. The vowel mark called the Holem is missing. In the square-shaped Hebrew alphabet Yehwah looked like this: יְהוָה . Notice this is not Yahweh, but Yehwah. But as noted above Yehowah does appear in some verses in the BHS text. Yahweh does not.

Theory of Yahweh

gp700» But in many Biblical dictionaries, encyclopedias, and some translations of the Bible we see: **Yahweh**. This spelling has the same consonants, but with the vowels **a** and **e** instead of **e**, **o**, and **a**, or **e** and **a**. The spelling, Yahweh, does not appear in any Hebrew text. I repeat, Yahweh does not appear in any Massoretic text, or any ancient manuscript, or papyri, or on any coin. The same consonants, YHWH, appear in ancient writings, but not the vowels.

gp701» There is a popular theory that says that the Hebrew word "Yehowah" does not have its original vowel points, and that the original vowel points would make YHWH to be "Yahweh" instead of Yehowah. But there is no real proof of this spelling as we will show. This is a very popular theory. But it is only a theory. Just because a theory is popular doesn't make it a correct theory. It started out as a theory of a few, most notable was Gesenius, the great grammarian. Some of his students embraced this theory, and helped to make it dogma.

gp702» The reason many think that Yehowah is not the correct rendering of the Hebrew word is because over 2000 years ago, according to some, some of the Jews began substituting another word that meant "Lord" (the Hebrew, *'adhonay*, or Greek, *kurios*) when they read the Hebrew NAME for God in public. It is said that some of the Jews began doing this because they became very cautious about misusing the NAME of God due to a superstitious misunderstanding of the commandment given to Moses: "You shall not take the name of the Yehowah your God in vain" (Exodus 20:7). These Jews were extremely careful about taking the NAME of Yehowah in vain — they didn't use it at all, for they substituted the word "Lord" for "Yehowah." Therefore we see that one version of the Greek translation of the Old Testament (the *Septuagint*, LXX) had the Greek word, *Kurios* ("Lord") translated in place of Yehowah (YHWH).

gp703» According to tradition, this Greek translation was completed in Egypt in about the third century BC. F.F. Bruce in his, *The Canon of Scripture*, states that the original Greek text probably only contained the Law or the first five books of the Bible (p. 43). There are also copies of the Greek text of the Old Testament that have the ancient Hebrew letters for God's NAME instead of the Greek, *Kurios* (see below, *God's NAME in Greek ...*).

Gesenius and Yahweh

gp704» Gesenius, the famous 19th century expert in Oriental literature, popularized this theory:

- "Whenever, therefore, this *nomen tetragrammaton* [the four letter NAME of God] occurred in the sacred text, they were accustomed to substitute for it *'adhonay*, and thus the vowels of the noun *'adhonay* are in the Masoretic text placed under the four letters יהוה, but with this difference, that the initial Yod [י in יהוה] receives a simple and not a compound Sh'va [Sheva v. Hateph Patah or the vowel e v. a].... As it is thus evident that the word יְהֹוָה does not stand with its own vowels, ..." (see *Gesenius' Hebrew and Chaldee Lexicon*, Translated by Tregelles, Eerdmans Pub., 1974 printing, p. 337 under YHWH, from 1857 Eng Ed.).

gp705» Notice carefully that the vowels, that were according to this theory, transposed from *'adhonay* to Yehowah, were not '**e**' (Sheva) '**o**' and '**a**,' but were '**a**' (Hateph Patah) '**o**' and '**a**.' Right here you should stop and think. From the beginning of their theory they use a sleight-of-hand to set this theory on its way. They say that the vowels from *'adhonay* were substituted for the real vowels in YHWH. (This is very suspicious because this is exactly opposite to the Written-Read or the *Kethib-Qere* method.) Yet Yehowah does not even have the vowels from *'adhonay*, for Yehowah does not read Y**a**howah, but Y**e**howah. After this sleight-of-hand they go on and make up a word, Yahweh, and say this is the true pronunciation. Yahweh, with its vowels, does not, I repeat, does not appear in any ancient document; only the constants YHWH appear. Arrogantly, a theory is made up that the NAME for God, יְהֹוָה, has the wrong vowels, and that the vowels were taken from the Hebrew,*'adhonay*, a word that meant lord or "my lords." (Read the discussion in *Gesenius' Hebrew-Chaldee Lexicon*, and see GP 1.) In reality Yahweh (or Yehwah [BHS] or Yehowah [Letteris]) does not have the vowels for *'adhonay*. The vowels in each word are different, thus the theory is nonsense. If you change one vowel in a word, you most often change the meaning of the word. If you change one vowel in the Hebrew אַל ('*l*) you get either the meaning of "not" (*'al*) or "God"(*'el*) or "these" (*'el*) or "towards" (*'el*), which correspond to Strong's numbers 408, 410, 411, and 413. Notice 410, 411, and 413 are spelled the same, but have very different meanings; these different meanings are ascertained by context (Ginsburg, p. 451). There is a lot of craftiness going on here by the advocates for the spelling of Yahweh.

Gesenius admits the spelling "Yehowah" fits the evidence

gp706» But at the same time Gesenius made this argument for the spelling, Yahweh, he wrote, "**Also those who consider that Yehowah was the actual pronunciation, are not altogether without ground on which to defend their opinion. In this way can the abbreviated syllables Yeho and Yo, with**

which many proper names begin, be more satisfactorily explained." As the editor of said, "This last argument goes a long way to prove the vowels Y**e**howah to be the true ones" (*Gesenius' Hebrew and Chaldee Lexicon*, Tran. by S. P. Tregelles, 1949, Eerdmans Pub p. 337 [1857 Eng Ed.]).

gp707» To repeat, Gesenius said that those who hold that Yehowah is the actual pronunciation, "are not altogether without ground on which to defend their opinion. In this way can the abbreviated syllables **Yeho** and **Yo**, with which many proper names begin, be more satisfactorily explained."

Ginsburg lists some evidence for the use of "Yehowah"

gp708» From Ginsburg *Introduction to the Massoretico-Critical Edition of the Hebrew Bible* we quote:

- "There are, however, a number of compound names in the Bible into the composition of which three out of the four letters of the Incommunicable Name have entered. Moreover, these letters which begin the names in question are actually pointed Jeho [Yeho], as the Tetragrammaton itself and hence in a pause at the reading of the first part of the name it sounded as if the reader was pronouncing the Ineffable Name. To guard against it [according to a theory] an attempt was made by a certain School of redactors [editors] of the text to omit the letter *He* so that the first part of the names in question has been altered from *Jeho* into *Jo*." [P. 369ff]

gp709» Ginsburg then lists proper names which have the first three consonants of the Tetragrammaton (YHW [יהו]) which are mistranslated with their vowels into English as **Jeho** but should have been translated as: **Yehow**. Notice the third letter is left out of the English translations. Ginsburg first lists, the names with *Jeho*, then the same name altered by using *Jo* instead of *Jeho*.:

- Jehoachaz (Yehoachaz) appears 20 times in the Bible; Joachaz 4 times
- Jehoash (Yehoash) 17 times; Joash 47 times
- Jehozabad (Yehozabad) 4 times; Jozabad 9 times
- Jehohanan (Yehohanan) 9 times; Johanan 24 times
- Jehoiada (Yehoiada) 42 times; Joiada 5 times

- Jehoiachin (Yehoiachin) 10 times; Joiachin 1Time
- Jehoiakim (Yehoiakim) 37 times; Joiakim 4 times
- Jehoiarib (Yehoiarib) 2 times; Joiarib 5 times
- Jehonadab (Yehonadab) 8 times; Jonadab 7 times
- Jehonathan (Yehonathan) 79 times; Jonathan 42 times
- Jehoseph (Yehoseph) 1Time; is found as Joseph in all other passages
- Jehozadak (Yehozadak) 8 times; Jozadak 5 times, no distinction in the KJV
- Jehoram (Yehoram) 29 times; Joram 20 times
- Jehoshaphat (Yehoshaphat) 83 times; Joshaphat 2 times

So there were about 349 times where proper names were written in the Bible, using the first three consonants and first two vowel of God's NAME , which most agree is referring to Yehowah (Ginsburg, pp. 370-75). The consonant and vowel left out was the last letter and last vowel of God's NAME. That is the *a* and the *h*. It so happens that when an *ah* is used at the end of a verbal word, in many instances, in the Biblical Hebrew, it adds *emphasis* to the meaning of the verb.

Fourth Letter in God's NAME

Cohortative Verb?

gp710» The fourth consonant-letter of God's NAME, **Yehowah**, is the last **h**. In Hebrew when a word takes on a new letter (sometimes with a vowel), either as a prefix or suffix, the new letter adds a secondary meaning to the root word. Notice the suffix in God's NAME: the "ah" in Yehow**ah**. This is important. God's NAME has the suffix "ah" because God's NAME may be in the *cohortative* or is like a cohortative. Words in the Hebrew cohortative are imperfect verbal words with the suffix "ah" which has the effect of emphasizing the word (*Gesenius' Hebrew Grammar*, Oxford 2nd English Edition, § 48c, d, e, & i; Driver *Hebrew Tenses*, Chap IV). As explained previously in this book, God's NAME is an imperfect verbal word that may be called a proper noun because of the way it is used and explained in the Bible (Exo 3:12-16). And it was repeated twice by God for emphasis because when words are repeated twice in the Bible it was for emphasis to show the importance of the of the idea.

God's NAME is Emphasized – He will be!

gp711» Remember that when God first revealed his NAME He repeated it twice: "**I will be** that **I will be**." It is known that when words are repeated in Hebrew it has the effect of **emphasizing** the word (see Introduction in the *Emphasized Bible*, and *Gesenius' Hebrew Grammar*, § 133 *k,l*). For example in Genesis 2:17, the Hebrew word for "death" is repeated twice, and can be literally translated, "dying, you shall die." But when translated into English it becomes "you shall *surely* die." Or in Exodus 26:33 in Hebrew it has, "holy of the holies," and is translated as "the most holy" or "the most holy place." Therefore when God repeated his NAME twice (**I will be** that **I will be**), He was giving *emphasis* to his NAME.

gp712» God repeated his NAME twice for emphasis, "**I will be** that **I will be**." He again says that his NAME is **I will be**. He then changes it to **He will be** or **Yehowah** only because this is the only grammatically correct way for Moses or anyone else to address God. Moses couldn't grammatically say, "**I will be** has sent me," but he could correctly say, "**He (who) will be** has sent me."

What is a Cohortative Verb?

- The Hebrew cohortative "**lays stress on the determination underlying the action, and the personal interest in it**" (*Gesenius' Hebrew Grammar*, Oxford's 2nd English Edition, § 108a and § 48k, § 110c).

- "The cohortative, then, marks the presence of a strongly-felt inclination or impulse: in cases where this is accompanied by the ability to carry the wished-for action into execution, we may, if we please, employ *I, we will* ... in translating" (Driver, *Hebrew Tenses*, p. 53; "..." are in text).

- It is similar to the Arabic *energetic*, "which expressed a strongly-felt purpose or desire," "an emphatic command," or was used "to add a general emphasis to the assertion of a future fact" (*Hebrew Tenses*, Driver, p. 241).

- Notice Exodus 34:14 correct translation: "For you shall worship no other god: for the BeComingOne [YHWH], whose **name** is Zealous, is a zealous GOD." Zealous is from a Hebrew verb (*qanah*) that describes action that is intent on reaching a desired outcome. His name is zealous because He is intent on becoming, "I will be that I will be" and thus it is in the cohortative form. "The verb expresses a very strong emotion whereby some quality or possession of the object is desired by the subject [GOD]." (See p. 802 in the *Theological Wordbook of the Old Testament*, Vol. 2; p. 896 # 2127 in Jeff A. Benner, *The Torah: A Mechanical Translation*, 2019) This same Hebrew word is sometimes translated as "jealous" because, depending on context, can mean zeal for one's own property.

gp713» Grammarians have found a pattern or "rule" — the Hebrews added "ah" to the end of imperfect verbs to add emphasis to these verbs (*Gesenius' Hebrew Grammar*, § 48c, d, e, i, k; and § 46). We emphasize a word in writing by italicizing it or underlining it; in speech we emphasize a word by the way we stress the word. Names like "cohortative" or "imperative" are arbitrarily chosen by grammarians to explain apparently slight variations of the emphatic use of the "ah" suffix on imperfect verbs in the Hebrew language.

gp714» In our books, in order to write something, we have picked the spelling of **Yehowah**, which is the spelling found in major Jewish-Hebrew texts of the Old Testament (See "More Details" below). Nehemia Gordon in the last ten year or so has make some good arguments that indeed **Yehovah may be the correct spelling.** (see www.nehemiaswall.com/nehemia-gordon-name-god)

The above gives some evidence that Yehowah or Yehovah may have the original vowels.

Yah

gp715» There are 149 proper names in the Hebrew Bible which according to the Massoretic text end with **Yah** (Jah) (Ginsburg, pp. 387-96). For example, Abi**jah**, Uri**jah**, Hezeki**jah**, etc. The **Yah** in the 149 proper names (Jah in most English Bibles) are at the end of the words. Yeho**wah** has "ah" at the end of it. Hebrew has certain ways of ending words. Because Yehowah is similar to a cohortative verb, it ended with, "ah". Yehowah fits the cohortative verbal rules, it fits the rule for verbs being used as nouns, and it also fits the Biblical text (Ex 3:9-16; gp92; see all GP 1 & GP: Appendix). Yah is merely an abbreviation for Yeho**wah**: **Yah**.

gp716» **Yehowah or Yehwah has the vowel-points written by "the" Massoretes.**

The spelling of Yahweh is found nowhere in any Massoretic text. But there is more against this theory.

Hebrew New Testament

gp717» The Jews of Christ's time were Hebrew and spoke and read in Hebrew or Aramaic (*He Walked Among Us*, pp.234ff). "The Israelites never wrote their sacred literature in any language but Aramaic and Hebrew, which are sister languages. The Septuagint was made in the 3rd century, B.C., for the Alexandrian Jews. This version was never officially read by the Jews in Palestine who spoke Aramaic and read Hebrew. Instead, the Jewish authorities condemned the work and declared a period of mourning because of the defects in the version.... Greek was never the language of Palestine. Josephus' book on the Jewish Wars was written in Aramaic. Josephus states that even though a number of Jews had tried to learn the language of the Greeks, hardly any of them succeeded.... Indeed, the teaching of Greek was forbidden by Jewish rabbis. It was said that it was better for a man to give his child meat of swine than to teach him the language of the Greeks" (*Holy Bible From the Ancient Eastern Text*, George M. Lamsa, Translator, Aramaic text, pp. ix & x; see Josephus' *Antiquities of the Jews*, Book 20, Chapter 11, Paragraph 2; *The Life and Times of Jesus The Messiah*, Alfred Edersheim, Bk. 1, Chap. 1, footnote #34; Ginsburg, p. 306). And there is some proof that at least some of the New Testament was originally written in Hebrew. In the fourth century, Jerome in his *Concerning Illustrious Men*, wrote:

- "Matthew, who is also Levi, and who from a publican came to be an apostle, first of all composed a Gospel of Christ in Judaea in the Hebrew language and characters for the benefit of those of the circumcision who had believed. Who translated it after that in Greek is not sufficiently ascertained. Moreover, the Hebrew itself is preserved to this day in the library at Caesarea, which the martyr Pamphilus so diligently collected. I also was allowed by the Nazarenes who use this volume in the Syrian city of Beroea to copy it" (Translated from Latin for the series "Texte und Untersuchungen zur Geschichte der altchristlichen Literatur," Vol. 14, Leipzig, 1896, edited by E.C. Richardson).

Also, recently a fragment of Mark was found written in Hebrew (*Bible Review*, 1989?). Nehemia Gordon has also found other fragments of NT Hebrew texts.

God's NAME in Greek Text Written in Hebrew

gp718» There is also some evidence today that there were Greek versions of the Old Testament that used the Hebrew word for God (YHWH) everywhere it should have been translated (*Bible Review*, "Glossary: New Testament Manuscripts," Feb. 1990, p. 9 top picture and inset text; *The Dead Sea Scrolls and the New Testament*, 1972, chapter 2, p. 30; Foreword, pp. 10ff, *The Kingdom Interlinear Translation of the Greek Scriptures*, 1969; see Appendix 1A, 1C, & 1D, pp. 1561ff of the *New World Translation of the Holy Scriptures — with References*, 1984 revised ed.). It is not only a possibility, but a probability that some used a Greek text that had the equivalent to the Hebrew YHWH in it instead of the Greek *Kurios* ("Lord"). But for some reason, either by historical accident or conspiracy to rid the church of Jewish tradition, this version did not prevail and thus we see many of today's translations are influenced by an Egyptian Greek version (*Septuagint*) of the Old Testament. Just how much this Greek version influenced theology can be seen by the following quote from Augustine in about the fourth century A.D.:

- There have, of course, been other translations of the Old Testament from Hebrew into Greek. We have versions by Aquila, Symmachus, Theodotion, and an anonymous translation which is known simply as the 'fifth edition.' Nevertheless, the Church [Catholic] has adopted the Septuagint as if it were the only translation. [*City of God*, by "Saint" Augustine, book 18, chapter 43]

Jehovah

gp719» The reason you see "Jehovah" used by some today is because it is a common translation of Yehowah. Even such names as "Jehoachaz," "Jehozabad," and "Jehohanan" should be rendered as "Yehoachaz," "Yehozabad," and "Yehohanan." Most English translations have a "J" in these words instead of a "Y."

gp720» The first known use of "Jehovah" is found in the book, *Pugeo Fidei*, on page 559, where it is spelled "Jehova" and where the square-lettered YHWH is found next to "Jehova." This was written or published by a Spanish monk, Raymundus Martini, in 1270 A.D (see photographic copy of the page in *Aid to Bible Understanding*, p. 885).

gp721» The reason "J" is found in Jehovah instead of "Y" is the same reason "J" was written in the King James Version instead of "Y" for such words as Jehoachaz, Jehozabad, and Jehohanan. The translators at that time felt that this translation was correct. Comparative studies with other related languages in the last two centuries has refined the art of translation. "Y" is now used to transliterate the Hebrew **Yod** (**·**) instead of "**J**." Because the Jewish race was dispersed, either usage may be right, depending on local Jewish pronunciation norms.

gp722» The reason "**v**" is found in Jehovah instead of "**w**," is because in the past, at least some of the linguists believed that the Hebrew **waw** (**·**) should be pronounced as the "**v**" and because some of the Jews pronounced it that way. In older Grammars and Biblical works ***waw*** was called ***vav*** (see "A Comparative Table of Ancient Alphabets," just before page 1 in *Gesenius's Hebrew and Chaldee Lexicon to the Old Testament Scriptures*, Wm. B. Eerdmans Pub, reprint of 1857 edition, reprinted 1974; *Gesenius' Grammar*, see §6a). *Gesenius' Hebrew Grammar* in the German language (1817-1909) had "**w**" for the Hebrew "**w**" (which is pronounced as a "**v**" in German), so the "**w**" in Gesenius' Grammar should be pronounced as a "**v**" in English.

Yehowah

gp723» From the above evidence and from the rest of GP 1 we see that **Yehowah** is the most likely correct transliteration from the Hebrew, and the "BeComingOne" is a correct translation of the true meaning of the Hebrew word into English. There is no good reason to use Yahweh instead of Yehowah. The spelling of Yahweh comes from an arrogant-intellectual mindset.

Massoretic Text

gp724» *Note*: **The quotes in the following section** were published in 1965 by Harry M. Orlinsky, Professor of the Bible Hebrew Union College, and were included in the "Prolegomenon" of Ginsburg's *Introduction to the Massoretico-Critical Edition of the Hebrew Bible* (the KTAV Publishing House's 1966 printing).

gp725» First "the" Massoretic Text is not one text. It is a collated and compiled text from many different texts. On the whole the variations between the texts are minor. The variations being mostly spelling, order of words, and a few additions by the scribes in order to clarify. After the invention of the printing press, there were no less than twenty-two printed texts of the Hebrew Bible printed between 1477 and 1521, eight of these containing the entire Bible (Ginsburg, p. X). Since then there have been the following editions of the Bible:

- Bomberg Rabbinic Bible (1524-26), edited by J. ben Chayim Bibles of Johannes Buxtorf (1611 & 1618-19)
- Joseph ben Abraham Athias's Bible (1661)
- Daniel Ernest Jablonski's Bible (1699)
- Johann Heinrich Michaelis's Bible (1720)
- Everard van der Hooght's Bible (1705)
- Benjamin Kennicott's Edition of the Bible (1776, 1780) August Hahn's Bible (1831)
- Meir Halevi Letteris' Bible (1852)
- The Letteris Bible (1866, British and Foreign Bible Society)
- Kittel's Biblia Hebraica, 1905-6, 1912/36 2nd & 3rd Ed (BHK).
- *Biblia Hebraica Stuttgartensia* (BHS), 1967/77
- and others...

gp726» The *Biblia Hebraica* (BHK) appeared with much fanfare because "it was supposed to represent the pure text achieved by Aaron ben Moses ben Asher, the great Masorete of the tenth century" (p.XIII).

gp727» "We are now ready to deal with the crux of the whole matter, something that the numerous editors of 'masoretic' editions of the Bible have overlooked, namely: **There never was, and there never can be, a single fixed masoretic text of the Bible!** It is utter futility and pursuit of a mirage to go seeking to recover what never was" (XVIII).

The Massoretic Text?

gp728» "There never was and there can never be 'the masoretic text' or 'the text of the Masoretes.' All that, at best, we might hope to achieve, in theory, is 'a masoretic text,' or a text of the Masoretes,' that is to say, a text worked up by Ben Asher, or by Ben Naftali, or by someone in the Babylonian tradition, or a text worked up with the aid of the masoretic notes of an individual scribe or of a school of scribes. But as matters stand, we cannot even achieve a clear-cut text of the Ben Asher school, or of the Ben Naftali school, or of a Babylonian school, or a text based on a single masoretic list; indeed, it is not at all certain that any such ever existed.... At the same time, it cannot be emphasized too

strongly that none of these manuscripts or of the printed editions based on them has any greater merit or 'masoretic' authority than most of the many other editions of the Bible, than, say, the van der Hooght, Hahn, Letteris, Baer, Rabbinic and Ginsburg Bibles" (pp. XXIII-XXIV).

Written-Read, Kethib-Qere

gp729» In the margins of the Massoretic text(s) they have notations about certain word variations. The written text (Kethib or Kethiv) was how the text was received; the notes in the margin were how some believed it should be read (Qere or Keri). "It is now scarcely possible to deny that the system of Kethib-Qere readings had its origin in variant readings; by the same token, the theory that the Qere readings are but corrections (really a euphemism for 'emendations') of the Kethib readings has no real justification" (p. XXIV). Examples by Orlinsky followed to page XXIX. "It is now admitted by the best textual critics that in many instances the reading exhibited in the text is preferable to the marginal variant, inasmuch as it sometimes preserves the archaic orthography [spelling] and sometimes gives the original reading" (Ginsburg, p. 184).

gp730» There is no single manuscript that contains all of the Kethib-Qere variations: "In order to exhibit, therefore, all the Keris [marginal readings] irrespective of the different Schools, it is absolutely necessary to collate all the existing MSS. which at present is almost an impossible task" (Ginsburg words, p. 185-86).

- "In summary: none of the 'masoretic' editions of the Bible published to date has genuinely masoretic authority for hundreds of Kethib-Qere that they offer the reader" (p. XXIX).

Ben Asher V. Ben Naftali

gp731» "The vast majority of the scholars who have attempted to work up 'the' masoretic text of the Bible have scarcely bothered with the system of Ben Naftali.... A few scholars, e.g., Ginsburg and Baer, did pay attention to Ben Naftali, even if they usually preferred Ben Asher's readings... But the question asks itself: What is there inherently in the masoretic work of Ben Asher school that gives it greater authority than that of the Ben Naftali school?" (p. XXIX-XXX)

gp732» "All the Masoretes, from first to last, were essentially preservers and recorders of the pronunciation of Hebrew as they heard it" (p. XXXII).

Tiberian Massoretes

gp733» "Due to the efforts of the Tiberian Massoretes their system of punctuation had displaced all the others by the end of the 9th century. But by this no absolutely uniform text of the Bible was yet established. These Tiberian Massoretes among themselves continued to hold different views on many issues" (p. XXXV).

gp734» *Note.* The following quotes (gp735, gp737-741) are from C.D. Ginsburg's *Introduction to the Massoretico-Critical Edition of the Hebrew Bible*:

Vowel-Letters Theory

gp735» "To facilitate still further the study of the unpointed consonants on the part of the laity, the Scribes gradually introduced into the text the *matres lectionis* which also served as vowel-letters. But in this branch of their labours as is the case in the other branches, the different Schools which were the depositories of the traditions themselves were not uniform."(p. 299) It should also be noted that vowel-letters when used were used before there were vowel-points. Vowel-points superseded the system of vowel-letters.

gp736» According to the *Gesenius's Hebrew Grammar*,

- "the partial expression of the vowels by certain consonants (א י ו ה), which sufficed during the lifetime of the language, and for a still longer period afterwards...."(§7*b*)

- "When the language had died out, the ambiguity of such a writing [using vowel-letters] must have been found continually more troublesome; and as there was thus a danger that the correct pronunciation might be finally lost, the vowel signs or vowel points were invented in order to fix it.... To complete the historical vocalization of the consonantal text a phonetic system was devised, so exact as to show all vowel-changes occasioned by lengthening of words, by the tone, by gutturals, &c.... The pronunciation followed is in the main that the Palestinian Jews of about the sixth century A.D."(§7*h,i*)

From §7*b* of *Gesenius' Grammar* we see that the consonant:

- י = ê and î,

- ו = ô and û,

- ה = "in the inflection of the verbs לָהֹ the long vowels **a, e,** and **è.**

Thus, even using the theory of the vowel-letter system, God's NAME reads, Yehowah.

Children Reading the Bible

gp737» Just before the time of Christ, schools were or had been established and "at the age of five, moreover, every boy had to learn to read the Bible. As a consequence it was strictly enacted that the greatest care was to be taken that the copies of the sacred books from which the Sopherim imparted instruction should be accurately written. It is to these facts that Josephus refers when he declares 'our principal care of all is to educate our children.' "(p. 304-05)

Josephus, Titus, Vespasian, and Severus to the Massorah

gp738» "Josephus tells us that Titus presented him with Codices of the Sacred Scriptures from the spoils of the Temple, and we know that there were others [MSS.] in the possession of distinguished doctors of the Law, which exhibited readings at variance with the present textus receptus.... Josephus records that among the trophies which Vespasian brought from the Temple to Rome was the Law of the Jews. This he ordered to be deposited in the royal palace circa 70 A.D. About 220 A.D. the emperor Severus who built a synagogue at Rome which was called after his name, handed over this MS. to the Jewish community, and though both the synagogue and the MS have perished, a List of variations from this ancient Codex has been preserved. This List I [Ginsburg] printed in my Massorah from the able article by the learned Mr. Epstein. Since then I have found a duplicate of this List in a MS of the Bible in the Paris National Library No. 31 (folio 399a) where it is appended as a Massoretic Rubric. The List in this Codex, though consisting of the same number of variations and enumerated almost in the same order, differs materially from the one preserved in the Midrash as will be seen from the following analysis of the two records, exhibits the primitive Rubric. The heading of the Paris List is as follows: These verses which were written in the Pentateuch Codex found in Rome and carefully preserved and locked up in the Synagogue of Severus, differs as regards letters and words" (pp. 409-411). Examples of differences followed this quote (pp. 411-20).

Massorites

gp739» "We thus see that the registration of anomalous forms began during the period of the second Temple. The words of the text, especially of the Pentateuch were now finally settled, and passed over from the Sopherim or the redactors to the safe keeping of the Massorites. Henceforth the Massorites became the authoritative custodians of the traditionally transmitted text. Their functions were entirely different from those of their predecessors the Sopherim. The Sopherim as we have seen, were the authorised revisers and redactors of the text according to certain principles [This is a popular theory; the Bible was a Holy book, and thus was not allowed to be tampered with; any revisions or editing was at most minor.], the Massorites were precluded from developing the principles and altering the text in harmony with these canons. Their province was to safeguard the text delivered to them by 'building a hedge around it,' to protect it against alterations or the adoption of any readings which still survived in MSS. or were exhibited in the ancient Versions. For this reason they marked in the margin of every page in the Codices every unique form, every peculiarity in the orthography, every variation in ordinary phraseologies, every deviation in dittographs &c. &c.

gp740» "In the case of the Pentateuch, the Massoretic work was comparatively easy since its text, as we have seen, was as a whole substantially the same during the period of the second Temple as it is now.... The present text, therefore, is not what the Massorites have compiled or redacted, but what they themselves have received from their predecessors and conscientiously guarded and transmitted with the marvelous checks and counter checks which they have devised for its safe preservation" (pp. 421-22). Examples are then given of the care the Massorites took (pp. 423ff).

gp741» Ginsburg gives information on the vowel-points (pp. 451-68).

More Language Details

Yehowah, Similar to Participles

gp742» God's NAME is a verb used as a noun. The English language has verbals that act as adjectives or nouns. The English present participle is the *ing* form of verbs used as adjectives; the gerund is the *ing* form of verbs used as a noun. A Greek participle is a verb or verbal used as an adjective or noun, and is thus a verbal adjective or is a verbal "noun" or verbal substantive when it is used with the article. A Greek participle partakes of both the noun and verb. In Matthew 11:3, John the Baptist sent two of his disciples to ask Jesus, "Are you The **Coming One**, or do we look for another." John wanted to know if Jesus was the Messiah, The Coming One. John was expecting the Messiah (Matt 3:11). The Greek word with its definite article in Matthew 11:3 is ὁ ἐρχόμενος, or Strong's # 2064. It is classified as a verb, participle, present tense, masculine, and singular (*Analytical Greek New Testament*, Friberg, p. 33). The Greek participle partakes "of both noun and verb" (A. T. Robertson's Grammar, p. 372; see Friberg, p. 810). Robertson classifies this participle as a "future participle" (p. 1118). The same word is in Revelation 1:8 but is translated as, "who is to come" in "Lord the God, who is, who was, and **who is to come**, the almighty" (see Rev 1:4; 4:8; 11:17). This can also be correctly translated:

- "Lord, the God, the is, the was, and **the Coming-One**, the almighty."

Is is in the present tense, **was** is in the imperfect tense, and **Coming-One** is a verb in its present participle tense, but A.T. Robertson classifies this particular participle as a "future participle" (p. 1118). The Hebrew word **Yehowah** is a verb used as a noun, while the Greek participle **coming** (#2064) or **Coming-One** is a verb used as a noun; it is a verbal substantive. In Matthew 11:3 the "coming one" is synonymous for the Messiah.

Hebrew Participle

gp743» It should be noted that Hebrew has different shades of the participle: some act in a more verbal character, some more as adjectives, and some as nouns depending on context or syntax (*Gesenius' Gram.* §50 & §116*a,g,f*). The Hebrew participles "occupy a middle place between the noun and the verb." A *participle active* is dissimilar from an imperfect verb: participle active expresses simple duration of an activity; an imperfect expresses progressive duration (*Gesenius' Gram.* §116c; Driver, p. 35ff).

Yehowah, The NAME, is a Verb

gp744» After Moses asked God His NAME, He answered with **I will be** repeating it twice, then He told Moses to tell Israel that His NAME was **I will be**, and right after this He told Moses to tell Israel that His NAME was **Yehowah** [יְהֹוָה]. As we saw above "I will be" was in the imperfect tense. Also "Yehowah" is in the imperfect tense. The Hebrew YHWH is a verb. God's NAME comes from a verb. The stem or root of God's NAME is, HWH, which is a *to be* verb. By looking at *Gesenius' Hebrew Grammar* § 40 c, we see that the normal method of converting the *to be* verb (HWH) into the imperfect, 3rd person, masculine gender form, is to add the Y [י] to the front of the verb. This makes God's NAME mean *he will be* according to the Hebrew grammar rules. But we see in Genesis 27:29 and Ecc. 11:3 (*Gesenius' Hebrew Grammar* § 75 s), YHW [יהו], does mean *will be* or even, *he will be*. Notice in these two scriptures they don't have the last letter, H [ה], of God's NAME as written by Moses. Most experts affirm that YHWH is from Hebrew verb [HWH] and that it is in the imperfect, 3rd person, masculine gender, and when used as a noun means *He who will be* (*Brown, Driver, Briggs, Gesenius Hebrew and English Lexicon*, p. 218 Col. 1).

NAME: An Imperfect Verb

gp745» **Yehowah** is an imperfect verb in the third person singular pronoun form of the verb HWH. The Hebrew HWH [הוה] is Strong's # 1933 and means "to be, become, or come to pass" (*Hebrew and English Lexicon*, Brown, Driver, Briggs, & Gesenius, under הוה). It is felt by some to be a more ancient form of the verb HYH, and is found in Genesis 27:29 (*Analytical Hebrew and Chaldee Lexicon*, Zondervan, p.171; and other Hebrew Lexicons). **Yehowah** is the correct form for an imperfect verb in its third person, singular, masculine of the verb, *hwh*, according to the table in *Gesenius' Grammar*, §40.

gp746» When Moses wrote God's NAME he used a less common form of the verb *to be*. The common form was, *hyh*. If Moses used the *to be* verb "hyh," then God's NAME would have been expressed as, *'hyh* when spoken by God, or *yhyh* when spoken by us. For God's NAME Moses used the less common form of the verb *to be*; Moses may have used *hwh* instead of *hyh* in his books in order to differentiate God's NAME from the more common, *hyh*. The meaning of either *yhyh* or *YHWH*, is **He Will Be**.

NAME: Imperfect Verb, Not Future Tense

gp747» Some call the Hebrew imperfect verb a future tense word, but this is not correct. From *Gesenius' Hebrew Grammar* (Oxford, 1980 reprint) we see that:

- "The Hebrew (Semitic) *Perfect* denotes in general that which is *concluded*, *completed*, and *past*, that which is *represented* as accomplished, even though it is continued into present time or even be actually still future. The *Imperfect* denotes, on the other hand, the *beginning*, the *unfinished*, and the *continuing*, that which is just happening, which is conceived as in process of coming to pass, and hence, also, that which is yet future; likewise also that which occurs repeatedly or in a continuous sequence in the past (Latin Imperfect)." [§ 47.1, note 1].

gp748» More on the Hebrew Imperfect verb from S.R. Driver's *Hebrew Tenses*,

- In marked antithesis to the tense [perfect] we have just discussed, the imperfect in Hebrew, as in the other Semitic languages, indicates action as *nascent* [beginning], as evolving itself actively from its subject, as developing. The imperfect does not imply *mere* continuance as such (which is the function of the participle), though, inasmuch as it emphasizes the process introducing and leading to completion, it expresses what may be termed *progressive* continuance." [p. 27]

More on "I will be"

gp749» From *Aid to Bible Understanding*, a 1971 Jehovah Witnesses' book, we see under "Jehovah":

- "God's reply in Hebrew was "'*Ehyeh asher 'ehyeh*." While some translations render this as '**I am that I am**,' the Hebrew verb (*hayah*) from which the word *'ehyeh* is drawn does not mean simply *to exist*. Rather, it means *to come into existence, to happen, occur, become*, ... Thus, the footnote of the *Revised Standard* version gives as one reading '**I will be what I will be**' (similar to Isaac Leeser's translation 'I will be that I will be') while the *New World Translation*, reads '**I shall prove to be what I shall prove to be.**' " [p. 888, col. 2]

In the Jehovah Witnesses' translation of the Hebrew verb, *'ehyeh* (from Strong's # 1961), in their *New World Translation*, they add "prove to" to their "I shall be" by way of extending the meaning of the Hebrew word, not by way of its most common usage of the verb in the Bible. This extending of the meaning is not necessarily wrong, for God will prove to be all that He says he will be.

New Mind and Christianity

Love is the New Mind, the New Law is Love

_____

Love is patient, kind, forgiving, full of joy and goodness, faithful, hopeful, gentle, not jealous, not arrogant, not unbecoming: love seeks good and shuns evil. It's the new law.

_____

by

Walter R. Dolen

Becoming-One Publications
B1Publ.com

2012 Revised Edition

This edition supersedes all previous editions

This work is a corrected and enlarged version of a 1970-71 non-published work and the 1977, 1989 and 2000 published works.

(Since the 2000 edition an Introduction was added and changes made to the Old Mind/Other Mind chapters, the Reason Why chapter and other minor changes; this book was previously called, *New Minds Papers*)

Printed and Published in the USA

April 2023 Printing

(Preface added)

This book:

ISBN13: 978-1-61918-000-0

(Trade Paper)

Other formats:

ISBN13: 978-1-61918-009-3

(iBook)

also in other e-book formats (Kindle, etc.)

Becoming-One Publications

Pennsylvania

b1publ.com

Preface

This book pertains to the **first-century beliefs** or 'doctrines' of the followers of Yehoshua, otherwise known today as Jesus Christ and thus are called "Christians." Yehoshua is Jesus's Hebrew name. The followers of Yehoshua were the believers who existed before the bureaucrats took over the Church. The bureaucrats went out and lied, killed and sinned in Christ's name: thus maligning the name Christianity and the name Jesus Christ. Therefore in this book when we speak of Christians we are referring to those who follow the real Christ, not the imposters who took over the Church in the decades following His resurrection.

We are also not going to refer to Christ – the Messiah – in this book by his Hebrew name, since today most know him by Jesus Christ, and most popular Bibles use this name. In my opinion it would be too confusing and counterproductive if we used his Hebrew name. Names of other famous people today are also misspelled and mispronounced from how they were spelled and pronounced in their own times.

What this book attempts to do is to simplify the Christian beliefs found in the Bible, not by studying the so-called fathers of the Church, but by analyzing the very words of the Bible. If the Bible was inspired by God, then the truth will be found there, not in theological essays written by the so-called 'fathers' of the Church. If the Bible was not inspired, then how can anyone ascertain anything relevant to Christianity? The only father of the Church is Yehoshua not Augustine or others. Paul was an apostle, not a father. "Call no man your father on earth, for you have one Father, who is in heaven" (Matt. 23:9). "There is but one God, the Father, from whom all things came and for whom we live; and there is but one Lord, Jesus Christ, through whom all things came and through whom we live (1 Cor 8:6).

What do real Christians believe in? Who are Christians? What is the Church? How can we tell if we are real Christians? What 'reward' do Christians receive? Why be a Follower? Can anyone be a Follower? Who is saved, or is everyone saved? Is there a hell, a heaven? Immortality? Is there an end to the world? What hope do we have? What is the meaning of life? Is there evil? What is evil?

This book is an accumulation of over 50+ years of study by one man interested in finding the truth, as of 2019 printing.

May Grace Abound to All,

Walter R. Dolen

Introduction

New Law

Love is patient, kind, forgiving, full of joy and goodness, faithful, hopeful, gentle, not jealous, not arrogant, not unbecoming; love shuns evil and seeks good. Love is the new law and is what Christianity must be in order to be Christ's church. In this book we go into great detail about this. However, others only see the negativity of religion. Mark Twain[1] was disillusioned with Christianity and religion because he only saw the paradoxes and the hell-damnation of religiosity. So he wrote the following in a book not published until after his death:

> "A God who could make good children as easily as bad, yet preferred to make bad ones; who could have made every one of them happy, yet never made a single happy one; who made them prize their bitter life, yet stingily cut it short; who gave his angels eternal happiness unearned, yet required his other children to earn it; who gave his angels painless lives, yet cursed his other children with biting miseries and maladies of mind and body; who mouths justice and invented hell – mouths mercy and invented hell – mouths Golden Rules, and forgiveness multiplied by seventy times seven, and invented hell." [Mark Twain, *The Mysterious Stranger*, Chap. 11]

This perception of the inexplicable paradoxes and negativity found in religion, or the emphasis upon such, is one-sided and unfair, for such negativity was superseded by Christ's teaching on Love.

Jesus Christ, for whom Christianity is named, changed the way some perceived God. Unfortunately, Jesus' teaching was taken over by those who didn't understand and they changed Christ's teachings of forgiveness and love into the teachings of hell and damnation. Because of this, we are forced to review in detail the doctrines of Christianity because the negativity of the world has been interjected into religion, not only Christianity, but all religion. This projects something about man's mind in

---

[1] A pen name for Samuel Clemens, one of America's best know writers

this age, which we call the old mind. But Christ announced a new mind, a new spirit, and a new commandment – the commandment of love. Originally this book was called the *New Mind Papers* because the new mind was the mind of love, not hate. We think our new title for this book more reflects and projects the real essence of Christianity, as taught by Jesus Christ. This book is comprehensive: we cover all the important doctrines found in the Bible about Christianity and attempt to negate the misguided teachings of religiosity.

Before we start examining Christianity, let me give you some of the premises for my belief.

I believe God did create the universe and here are a few reasons why I do

Law. The evolutionary theory always starts with, and assumes, the eternal existence of laws like those of mass, energy, motion, gravity, conservation, chemical bonding and so forth. Laws, in and of themselves, *are* systematic order and project intelligence and power outside of the law itself. The genetic code of life found in DNA also projects high intelligence and power. *How can* the *code of DNA evolve* or any law such as gravity or chemical bonding evolve? How can any code or law itself have any power? What gives a code power? I am speaking about the code itself, the order of the elements within the code. How can the *arrangement* of the code itself have power? The apparent connection between the code and its effect on a body or plant projects, or strongly suggests some kind of force or power *behind* the law. The code itself doesn't do anything, just as the letters in this book don't do anything by themselves. If you change the arrangement of the letters of the code or a word, it has a different result or may not have any. A seed grows into a certain kind of flower, not because of the code per se, but because of the power behind the code. The basic laws of the universe must have come from somewhere and the power behind these laws must have some connection to the law. Evolution has yet to explain the source of the power behind the universal laws. Science can only *describe* gravity (through mathematical formulas) and partially *describe* the code of life, but it has no idea how the power of gravity works or how or where the code of DNA gets its power. I believe that God, as described in the *God Papers*, is the creator and power behind all universal laws. And I believe it is more naive to believe in a cosmic soup theory (evolution) than in a powerful God, although I agree that common descriptions of God are naive and do not explain the paradoxes pertaining to God.

Beginning. Radioactivity and laws of thermodynamics indicate there was no eternity of matter and it corollary: there was a beginning of matter. If matter always existed, without a starting point, then the "life" period of the radioactive elements would have long ago run its course and the whole universe would be the same temperature (thermodynamic laws). The radioactive elements would have run down and there would not be any radioactive elements left; the whole universe should be the same temperature. Thus, there was a *beginning* of matter, and it wasn't that long ago, since there are still radioactive elements. The "science" of evolution cannot explain energy or matter or its source nor will it ever because it has no witnesses and has no real explanation for their beginning. A mathematical description of energy doesn't explain it, it only describes what it does in a quantitative manner in *our* solar system. God created matter and energy and in some way God is matter and God is energy as we attempt to explain in our book pertaining to God (*God: God is the Becoming-One* aka *God Papers* or *My God is the BeComing-one*).

Life. The relative harmonic-symbiosis of the ecosystems, from the biochemical cell to the earth-sea-heavens, projects design. There is a co-operation, interaction and mutual dependence among life forms; one species cannot live well, or at all, without mutual-beneficial interaction of the whole: the flowers need the birds and insects for pollination in order to continue to exist and vis versa; the seed needs its DNA, the dirt with its nutriments, water and the power behind the DNA for it to grow. Our bodies need a heart, lungs, liver, intestines and so forth in order to exist: we need our whole factory of body parts and a compatible earth in order to live. **The whole cannot live without the parts; the parts cannot exist without the whole.** The theory of evolution maintains that life is arbitrary, for life came from a hit and miss adventure ("natural selection" or "mutation," etc.). If life is arbitrary, then the universe would be filled with the inferior products of this evolutionary process, and the inferior and half-made life-forms would greatly outnumber the surviving species. There should be fossils of the inferior products of the evolutionary process in all strata, in the rocks everywhere. In other words, the rejections of

the evolutionary process should be polluting the universe. Where are the fossils of these inferior life-forms? For that matter, where are the masses of missing links in the evolutionary process? Where? Life came from God, not from the mindless soup of evolution.

The Proof. The big bang theory and other theories need to explain where the material and energy for the big bang theory came from. God, the all powerful Being, by definition, must have always been there, or else there is nothing and we are nothing and so this dialogue doesn't exist. Either the all powerful god of Evolution (mindless soup) was there at the beginning or the all powerful Being was there. Of course we cannot prove God by definition, but there is a way to settle this disagreement:

- The evolutionists can prove the universe came into existence through evolution by physically demonstrating evolution. For example, a new species being spontaneously 'created' before our eyes, or at very least finding the massive amount of missing links in the fossils record and logically explaining where laws get their power;

- The believers in the God can prove to others that there is an all powerful God by people seeing God create a new heaven and earth or by seeing God resurrect the dead back to life. Such is the prophecy recorded in the Bible: all will see the resurrection of the dead and the creation of the new heaven and earth, as apparently the angels witnessed the creation of the present universe at the beginning of the present heaven and earth.

"Love is patient, love is kind. It does not envy, it does not boast, it is not proud. It is not rude, it is not self-seeking, it is not easily angered, it keeps no record of wrongs. Love does not delight in evil but rejoices with the truth" [1Cor 13:4-7].

===

NM 1: What is a Christian?

Christian Doctrine
Physical v. Spiritual Meaning
What is a Christian?
How does one know he is a Christian?
What can one expect as a Christian?
Other names for Christians?

NM1 Abstract

In this book we put forth the doctrines of the Bible as we found them, taking into consideration the type and antitype found throughout scripture. What we are doing is attempting to simplify various doctrines of Christianity. In this paper we give short renditions as to what is a Christian, how one becomes one, what one can expect as a Christian, and so forth, thus setting the stage for the rest of this work.

Christian Doctrine

nm1 » In the New Testament of the Bible you can read about many important subjects. Significant subjects such as heaven, hell, sin, law, freedom, miracles, death, resurrection, immortality, predestination, the kingdom of God, and so forth are spoken about throughout the New Testament of the Bible. These subjects and others have to do with our very life, our souls, and our future. These subjects are very important and we cannot permit tradition to dictate to us concerning these subjects. We must test and analyze our views to see if they are correct. If they are not we must correct them. We must take charge of our storehouse of beliefs; we must correct our false beliefs; we must become sound in our knowledge.

Many Opinions

nm2 » But there are many opinions on all of these subjects, and it is difficult to find the truth. There are so many who claim to hold the truth. There are so many traditions, so many teachers, so many differing beliefs. There is confusion on how one is baptized, on the soul, and on the other doctrines of Christianity. There are the liberal Christians, the conservative Christians, and many other classifications.

Doctrines as Found in the Bible

nm3 » In this book we will put forth the real doctrines of the Bible as we found them. It is up to you the reader to prove or disprove what is presented in these papers. Only you can make the decision for yourself. That is, only you *should* make the decision for yourself. Do not let tradition or the authorities of your church or your science prevent *you* from making your own decision.

Physical Meaning versus Spiritual Meaning

nm4 » The mistakes in Christian doctrine were made because of the lack of knowledge of the pattern manifested in the Bible and the inability to see these patterns. In the "Duality Paper" we have spoken briefly about this Biblical pattern. Those making mistakes are only looking at the physical meaning or typical meaning of scripture instead of the higher meaning or Spiritual meaning. We are to worship God in Spirit (John 4:24). We are to look away from the physical to the Spiritual for the true meaning of God's word (see the "Duality Paper"). If we take or understand Jesus Christ's words only in a typical or physical manner we will not understand what he was trying to tell us. Not only this, but we will dramatically misunderstand him. When Jesus spoke about eating his body (John 6:53-56) he was speaking in a Spiritual way (see John 6:63). But if we only take Jesus' words in their literal, or simple, or physical meaning, we will drastically misunderstand him like many of his disciples did at that time (John 6:60-61). When some of Christ's disciples heard him, they mistakenly thought he was advocating cannibalism, a hideous crime against mankind, instead of encouraging the *spiritual* eating of his body. To eat or drink Jesus Christ in a Spiritual way is to *Spiritually* eat and drink his Spiritual body, or that is, "eat" his Spirit or "drink" his Spirit. Being baptized with God's Spirit, or eating Christ's Spirit, or drinking Christ's Spirit, and so forth are all signifying one thing – having God's Spirit or the New Mind. To have God's Spirit is to have the power to do good works. But if we only take the physical or typical meaning of scripture we will not understand the Spiritual words. We must look to the higher or Spiritual meaning of scripture or we will not understand.

What is a Christian?

nm5 » There is a direct relationship between Christianity and the New Mind. We call the Spirit of God the New Mind. You are a Christian when and only when you have the New Mind. Christianity is not Christianity without the New Mind. If you have the New Mind you are a Christian. But what is a Christian? Is a Christian someone who only goes to a Christian church? Can a real Christian not go to church? Are all Christian churches in reality, Christian? Are all who call themselves Christians in reality Christians?

nm6 » A Christian is a believer in Jesus Christ. Yet not only is he a believer, he is also a doer of what Christ did (James 1:22). Christians follow in Christ's way (1Peter 2:21). When Christians follow Christ, they are following God because God was manifested in Christ (1Tim. 3:16). God cannot sin (1John 3:9). Christ didn't sin (2Cor. 5:21). God's behavior was manifested in Christ. God and Christ the man behaved the same. Therefore God was manifested in Christ the man's behavior. God is love (1John 4:8). And Paul said: "love does not work any ill to its neighbor, so love is the fulfilling of the law" (Romans 13:10). To follow Christ is to follow God. Since God is love, then to follow God is to follow love. "Love is patient, love is kind. It does not envy, it does not boast, it is not proud. It is not rude, it is not self-seeking, it is not easily angered, it keeps no record of wrongs. Love does not delight in evil but rejoices in the truth. It always protects, always trusts, always hopes, always perseveres. Love never fails" (1Cor. 13:4-8, NIV).

New Man versus Old Man

nm7 » The main difference between a real Christian and others is that Christians follow after love to the degree of power they were given to follow after love. All Christians in the old age never get close to the level of Jesus Christ's love because Christ was given the full power. Love is a system of behavior that is quite different than the system of behavior we observe around us today. Real Christians belong to the New Age with its New Mind. Today, Christians are New Age People who live in the old age. The old age is the present age with its confusion and hate. The New Age is the Kingdom of God with its system of love.

How Does One Become a Christian?

nm8 » To become a real Christian you must have the Spirit or Mind of God, that is, the New Mind. To receive this New Mind, you must be Spiritually baptized into the Name of the Father, Son, and Holy Spirit (NM 4).

How Does One Know He Is a Christian?

nm9 » You know you are a Christian if you have the New Mind. There is another mind, the old mind, that works in the old age (see NM 21). We can see the power of the old mind working in the old age every day. The confusion of this world comes from the old mind. But when we receive the New Mind we begin to see the difference in our thinking. Instead of getting flash-thoughts concerning evil things, we begin to get flash-thoughts concerning the beautiful and good things. With the New Mind we get flash-thoughts that help us to begin to be patient, kind, truthful, hopeful, trustful, etc. You know you are a real Christian when you have the New Mind. And you can *prove* you have the New Mind by your new behavior. If you have a new behavior that is more in keeping with the way of love, then you can be sure you have been given the New Mind (see "Proof Paper" [NM 10]).

What Can One Expect As a Christian?

nm10 » When you are a Christian you have the Spirit of God, the New Mind. You see matters from a different viewpoint. You have put on the New Mind which is being renewed in knowledge, and thus you are a new person in God (Col 3:10; Rom 12:2). You begin to understand that no one thing is bad in itself (Rom 14:14; 1Tim 4:4), but only wrong activity is bad (Prov 8:13). As a Christian you begin to do to others as you would like them to do to you (Rom 13:8-10). A new Christian begins to change and to do things differently, for he has a New Mind with a new attitude.

nm11 » But because Christians are changed, others around them will notice this transformation (1Peter 4:4). Because mankind as a whole feels threatened when others do not believe as they do, a Christian can expect to be disliked by people, even those of his own family (John 15:18-19; Mat 10:34-37). But a real Christian is to be peaceful and try as much as possible to keep the peace. But at times because true Christians do not run after twisted things as much as others, the people belonging to this age will not like real Christians.

nm12 » Besides receiving the New Mind (Spirit) a Christian will also receive the life in and throughout the first 1000 years of the coming kingdom of God. Read the paper entitled, "Reward for Christians" [NM 11] for more details.

What Are Other Names For Christians?

nm13 » In the Bible Christians are called:

- the Israel of God (Gal 6:16);

- the sheep (John 10);

- the holy temple (Eph 2:21);

- Jews, meaning Spiritual Jews (Rom 2:29);

- the Lamb's wife (Rev 21:9);

- Christ's wife (Rev 19:7);

- the 144,000 (Rev 14:3; 5:9-10);

- virgins (Rev 14:4);

- Zion (Heb 12:22);

- New Jerusalem (Rev 21:2);

- The body of Christ (Rom 12:4-5; 1Cor. 12:27);

- the church of the first born (Heb 12:23);

- the first fruits (Rev 14:4);

- saints (Eph 1:1);

- little children (1John 2:13; 5:21);

- living sacrifices (Rom 12:1);

- the holy nation (1Peter 2:9); etc.

Simplify Doctrine

nm14 » What we will be doing in the rest of this book is to simplify doctrines by examining in detail the scriptures on the doctrines. The main aspect of what we will be doing is showing you about Jesus Christ's Spirit, the New Mind,

and the fruits or effects of the New Mind. The Bible speaks on many different subjects such as repentance, baptism, grace, and so forth. We will see what these subjects have to do with the Spirit of God and the effects of having this Spirit.

NM 2: On The Church of God

What is the Church?
Church Separate from the World?
Church is in Christ's Spiritual Body
How is the Church One?
How can Christians be of Christ's Flesh?
What is the Church Founded Upon?
Can the Church make any Law?
Behavior of those in the Church
Physical Organization for the Spiritual Church?
Is the Physical Church the Spiritual Church?
Doctrinal Errors?
False-Shepherds over the Church?

NM2 Abstract

In this paper we give Biblical definitions as to what the Church is, who or what it is founded upon, whether or not Churches can make just any church law or ruling, what should be the behavior of Church members, and consider the problem of doctrinal errors or false-shepherds over the Church.

What Is The Church?

nm15 » The word "church" comes from a Greek word that means, "called out." Those of the Church are called out from the world, or the way of the world: "Come out from among them, and be you separate" (2Cor 6:17). "Come out of her [Babylon] my people" (Rev 18:4).

nm16 » The word "virgin," which the Church is Spiritually called (Rev 14:4), is translated from the Greek word *parthenos*, which literally means, "one put aside." That is, the Church, the Spiritual wife of God (Rev 19:7), is put aside from the way of the world. Also the word "saint," which is used to describe those in the Church of the New Testament, is translated from the Greek word *hagios*, which means, "set apart." Those in the Church are set apart from the world. They are set-apart by God (see "Predestination Paper"[NM 8]). It is God who puts people in the Church (see Acts 2:47).

nm17 » Christ when he was praying to his Spiritual Father spoke about those of the Church: "Holy Father, keep them in your own NAME, which you have given me, that they may be one, as we are ... I have given them your word, and the world hates them, because they are not of the world, even as I am not of the world. I pray not that you should take them out of the world, but that you should keep them from evil. They are not of the world even as I am not of the world" (John 17:11, 14-16). Therefore those of the Church are Spiritually called out of the world, and are separate from it even though they are physically still in it. How are they apart from the world?

Church Separate From The World?

nm18 » They are apart from the world because they have the set-apart Spirit, or as some translations have it, the Holy Spirit. This set-apart Spirit, or Holy Spirit is the Spirit of God, or of Christ because Christ is God (John 20:28). The Spirit of God is the New Mind (cf Rom 12:2; Eph 4:23; Rom 6:4; 7:6). And those of the Church have the Spirit of God in them (Rom 8:9-11, 14-16).

nm19 » The one main thing that makes you a member of the body or Church of Christ is that you must have the Spirit of God (1Cor 12:12-13). "There is one body, and one Spirit ... One God and Father of all, who is above all, and through all, and in all" (Eph 4:4, 6). "For through him [Christ] we both have access by one Spirit unto the Father" (Eph 2:18). This one kind of Spirit is the Spirit in the body of Christ, which is the Church (1Cor 12:4, 13).

Church is in Christ's Spiritual Body

nm20 » Those in the Church are in the collective body of Christ, and are the collective body of Christians. "For as the body is one, and has many members, and all the members of that one body, being many, are one body; so also is Christ. For in one Spirit are we all baptized into one body, whether we be Jews or Gentiles, whether we be bond or free; and have been all made to drink into one Spirit ... Now you are the body of Christ, and members in particular" (1Cor 12:12-13, 27).

nm21 » Therefore, you are in the Church, or in the body of Christ, when you are in the Spirit of God (you have the New Mind), or when you have the Spirit of God in you (Rom 8:14-16).

nm22 » Those in the Church are the members of Christ's body, and Christ is "the *head* over all things to the church, which is his body" (Eph 1:22, 23).

- Christ is the head of the body, and we are the members of his body (1Cor 12:27).

- "Christ is the head of the church: and he is the savior of the body," as "the husband is the head of the wife" (Eph 5:23).

- We are the wife of Christ, and the bride of the lamb (Rev 19:7). Christ is the lamb of God (John 1:29).

- Those in the Church are the sheep, and Christ is the shepherd (John chap 10).

- Those in the Church are the branches, Christ is the root of the vine (John 15).

- Those in the Church are the stones of the building, Christ is the chief cornerstone (1Pet 2:5; Eph 2:19-22).

- Those in the Church are of the kingdom of priests, and Christ is the High Priest (Rev 1:6; 1Pet 2:5-9; Heb 3:1; 5:1-10).

How Is The Church One?

nm23 » There is only one body of Christ, one Lord, one Faith, one Baptism (Eph 4:4-5). There is only one Church, with one baptism, one faith, one Lord. The Church is not many different groups. The Church is ONE group, ONE body with ONE Spirit. Those with the Spirit of God, that New Mind, are the people of the ONE true Church with the one baptism, one faith, and one Lord.

nm24 » The Church is made up of people who have the Spirit of God inside them leading them into the way of harmony. It is the Spirit that sets people apart from the world. It is the Spirit that makes people one. "For in one Spirit are we all baptized into one body" (1Cor 12:13). With this same Spirit we receive the gifts of the Spirit (1Cor 12:4). These gifts of the Spirit are "love, joy, peace, long-suffering, gentleness, goodness, faith" (Gal 5:22). When we have this one Spirit, we have the same gifts or fruits from this Spirit, "according to the measure of the gift of Christ" (Eph 4:7). These gifts are given "for the edifying of the body of Christ: till we all come into the unity of the faith" (Eph 4:11-13).

How Can Christians Be of Christ's Flesh?

nm25 » When we have the Spirit we are in the Church, and "we are members of his body, of his flesh, and of his bones" (Eph 5:30).

nm26 » When we are Spiritually baptized into the body of Christ, we are baptized into it by the one Spirit of God (1Cor 12:13). Those who have been baptized into the body of Christ, "have been baptized into Christ have put on Christ" (Gal 3:27). When one is Spiritually baptized into Christ, "there is neither Jew nor Greek, there is neither bond nor free, there is neither male nor female: for you are one in Christ Jesus" (Gal 3:28). These are one in Christ because they "have been all made to drink into one Spirit" (1Cor 12:13). They are one in the sense that they have the one Spirit. Males are still males in the physical sense. Females are still females in the physical sense. But they are one in the Spiritual sense because they have the one true Spirit of God. "And if you are Christ's, then you are Abraham's seed, and heirs according to the promise" (Gal 3:29). Christ was a descendant of Abraham (Mat 1:1-17), he was of the seed of Abraham. Therefore when we are in Christ's body (his Church) we too are the descendant of Abraham, we are of the promised seed of Abraham because we have "that holy Spirit of promise, which is the evidence of our inheritance" (Eph 1:13, 14). Because we are of the seed of Abraham, we are also of the Flesh of Christ (Eph 5:30), we are of Israel, we are of the Spiritual "Israel of God" (Gal 6:16).

Who Or What Is The Church Founded Upon?

nm27 » Christ said, "and I say also unto thee, that you are Peter, and upon this the rock I will build my church" (Mat 16:18). Now the Roman Catholic Church has used this verse incorrectly to say that Christ built his Church on the foundation of Peter. The word "Peter" comes here from a Greek word that means, "stone" or "rock." But the sentence reads, "you are Peter and upon this *the* rock I will build my Church." Who is *the* rock Christ was speaking about here? Was it Peter, or was it THE ROCK? Which rock is the Church founded upon?

> "For through him we both have access by one Spirit unto the Father. Now therefore you are no more strangers and foreigners, but fellow citizens with the saints, and of the household of God; And are built upon the foundation of the apostles and prophets, Jesus Christ himself being the chief corner *stone*" (Eph 2:18-20).

nm28 » It is Christ who is the chief foundational stone, or rock. Christ is "the stone which the builders disallowed, the same is made the head of the corner, and a stone of stumbling, and a rock of offense" (1Pet 2:7-8). Christ is God (John 20:28, see *God Papers*). And God is the ROCK (Deut 32:4; Psa 18:2, 31). It is God, or Christ who is the head Rock of the Church, He is the foundation, the chief foundation, not Peter. Peter is just one of the "living stones" of the Church (1Pet 2:5; Eph 2:20), he is not the main foundation. Therefore the "rock" spoken about in Matthew 16:18 is God, not Peter.

Can Churches Make Any Law They Wish?

nm29 » Note Matthew 16:19. Many churches tell their flock that this verse gives them the right to make laws on their own, and that such laws are binding. This is wrong! Notice, "and I will give unto you the keys of the kingdom of heaven: and whatsoever you shall bind on earth shall be bound in heaven" (Mat 16:19). And from the *Twentieth Century New Testament* translation: "whatever you forbid on earth will be held in Heaven to be forbidden." Now IF this translation is correct, it is saying the earthly can tell the heavenly what is right or what is wrong. But this is contrary to all the Bible. It is the heavenly that shows the earthly what is right, not vice versa. In the prayer which Jesus asked us to pray in like manner, Jesus said to ask our Father: "your wish be done in the earth, as it is in heaven" (Mat 6:10). The Church is only to bind on earth things that have already been bound in the heavenly sense. *The Church can only bind things on earth that reflect the heavenly or spiritual sense or spiritual dimension*.

What Was The Bible In Christ's Day?

nm30 » The only Bible they had during the time of Christ and shortly thereafter was the Old Testament scripture. Paul's letters to the Church of God were merely letters explaining the Old Testament promises in light of the things of Jesus Christ. When the Bereans searched the scriptures, they searched the Old Testament (Acts 17:10-11). But today the Christian Bible is the Old and New Testaments of the Bible. The Christian Bible includes the inspired material of the apostles.

How Many Are To Become Members of The Church?

Few Saved Now; All Later

nm31 » Everyone will become members of the Church before the plan of God is completed, but up to the Messiah's return only the few will be saved (see "All Saved Paper" [NM 13]). "Wide is the gate, and broad is the way, that leads to destruction, and many there be which go in thereat: because narrow is the gate, and narrow is the way, which leads unto life, and *few* there be that find it" (Mat 7:13-14). Many will go into aeonian destruction, for it is the many who are misled. "And Jesus answered and said unto them, Take heed that no man mislead you. For many shall come in my name, saying, I am Christ, and shall mislead many." It is the confusion of Satan that has misled the whole world (Rev 12:9). Most will be misled. In fact all have been, for those who learn of the Way were themselves at one time deceived. It is only the *few* who will be given the Spirit which leads them into all the truth (Mat 7:14 & John 16:13). But as we will see in these papers all will receive the Spirit in the Great Last Spiritual Day of Creation, thus all will be saved eventually.

What Is The Behavior of Those In The Church?

nm32 » Since those in the Church are Christians in the truest sense, then they behave as Christians should behave, not like most so-called Christians behave in this age. This book, *New Mind and Christianity*, manifests that true Christians follow the law or system of love. They follow it according to the degree of Spiritual power given them. All Christians produce much Spiritual fruit (see paper, "Prove Paper" [NM 10] and the "Freedom and Law" [NM 17]).

Physical Organization for the Spiritual Church?

nm33 » As the Church was shown organized physically in the book of Acts, so too the Church may be at times physically organized since the days of the book of Acts. Christ said that the Church would be scattered: "I will smite the shepherd, and the sheep shall be scattered" (Mark 14:27). Christ said since they persecuted him, they would persecute the Church (John 15:20). And in Acts 8:1 it reads, "and at that time there was a great persecution against the Church which was at Jerusalem; and they were all scattered abroad throughout the regions of Judah and Samaria, except the apostles." So the Church was scattered from Jerusalem and throughout the known world as Paul's letters indicated.

nm34 » But even though Paul's epistles showed they were scattered, some came nevertheless together in certain cities throughout the world. The Church of Acts had a center at Jerusalem, and was physically organized and did send out teachers from Jerusalem throughout the world. But Jerusalem was destroyed by the invading Roman troops, and tradition has it that the apostles were killed about 40 years after Christ died. Thus, according to the records available today, at that time at least a center of the physically organized Church ceased, yet the Spiritual Church didn't cease since the Church is Spiritually organized through the medium of the Spirit. As long as there is one person on earth with God's Spirit, the New Mind, there is the Church of God.

nm35 » After the center in Jerusalem of the Church was destroyed about 40 years after Christ was killed, more than likely the Church was physically organized in some way many times again over the years. But there does not need to be a physically organized Church on earth in order for there to be a Church of God on earth. The Church of God is made up of those who have the New Mind or the Spirit of God, irrespective of whether there is or isn't a physically organized Church. We can never say with any certainty that any physically organized church, after the Apostles died, was indeed the Church. Even in the physical churches organized by the Apostles, real Christians were put-out (3John 1:9-10). To repeat: the Church is Spiritual and various physical churches cannot in this age be identified with any certainty as being the Church.

Is a Physical Church the Spiritual Church?

nm36 » To be in the Church is to have the Spirit. Yet Jude wrote of "certain men who crept in unawares, who were before of old ordained to this the judgement, ungodly men, turning the grace of our God into loose conduct, and denying the only Lord God, and our Lord Jesus Christ" (Jude 4). "These are they who separate themselves, sensual, having not the Spirit" (Jude 19). This shows there were some physically in the Church who didn't have the Spirit of God (see Gal 2:4).

nm37 » Also 1John 2:18-19 shows that some left the Church because they weren't of the Church. But in some cases the true Christians were forced out of the physical churches in certain areas of the world (see 3John 1:9-10). This was prophesied by Christ (see Luke 6:22). There will even be false teachers among the physical church (2Pet 2:1). Therefore not all in a physically organized Christian Church are of the Spirit of God.

Find the True Physically Organized Church?

nm38 » The best indication that you have found it is by the behavior of the people in the church. Are the people following Spiritually the law of love? Or is the Church merely serving a social function? Whatever, we must be careful of how we judge. We must remember that there may be more non-Christians in a physical church than true Christians with the New Mind. Therefore be careful how we judge.

What About Doctrinal Errors of Churches?

nm39 » When errors of doctrine are brought forward, the Church will admit the error, as Peter admitted one of his (Gal 2:11-14), and as all those with the New Mind should admit their errors when reproved (Job 33:27; Prov 28:13; 1John 1:8-9).

nm40 » Some reasons for error in doctrine may be mistranslations of the Bible, or non-Spiritual leaders, or lack of the Bible or parts of the Bible. There will be errors until Elijah comes (Mat 17:10-11).

nm41 » In Ezekiel 34 it shows that the shepherds of Israel are false shepherds. This is dual: (1) it speaks of false shepherds over physical Israel; and (2) it speaks of false shepherds over Spiritual Israel. But God says, "I will deliver my flock from their mouth" (Ezek 34:10). Therefore it is possible for false shepherds to actually mislead some or many in the true Church, and this could be a reason for doctrinal error. In the truest sense it indicates the other-mind that remains in the minds of Christians in the old age. This evil mind continues to attempt to mislead Christians and sometimes succeeds.

False Shepherds Over The True Church?

nm42 » We will now discuss the scripture about false shepherds 'over' the True Church. Peter wrote about false teachers among Christians (2Pet 2:1). But in order to understand this further we must know that the scripture is dual, with a typical and antitypical fulfillment.

Now, or near the end of the age all prophecy will be fulfilled that has seemed to fail previously:

> "Son of man, what is that proverb that you have in the land of Israel, saying, The days are prolonged, and every vision fails? Tell them therefore, Thus says the Lord GOD; I will make this proverb to cease, and they shall no more use it as a proverb in Israel; but say unto them, The days are at hand, and the effect of every vision" (Ezek 12:22-23).

Now the time is close at hand (see, the "End of the Age"[PR7]). Therefore all prophecy will be fulfilled soon.

nm43 » In the scriptures it indicates that over physical Israel false shepherds would rule at various times. And throughout the history of Israel there were evil leaders over Israel and its congregation. Since the New Testament Church is the antitypical Israel ("Israel of God," Gal 6:16), then Spiritual Israel will have false shepherds misleading it at different times.

nm44 » In Ezekiel 34, it pictures "the shepherds of Israel that do feed themselves" (v. 2). God continues to the false shepherds,

> "You eat the fat, and you clothe you with the wool, you kill them that are fed: but you feed not the flock. The diseased have you not strengthened, neither have you healed that which was sick, neither have you bound up that which was broken, neither have you brought again that which was driven away, neither have you sought that which was lost; but with force and with cruelty have you ruled them" (Ezek 34:3-4).

nm45 » Among these false shepherds is the idolatrous or worthless shepherd. Zechariah 11:16-17 tells of this worthless shepherd:

> "For, lo, I will raise up a shepherd in the land, which shall not visit those that be cut off; neither shall seek the young one, nor heal that which is broken, nor feed that which stands still: but he shall eat the flesh of the fat, and tear their claws in pieces. Woe to the idolatrous shepherd that leaves the flock! the sword shall be upon his arm, and upon his right eye: his arm shall be clean dried up, and his right eye shall be utterly darkened."

Spiritually speaking, this means the worthless shepherd will not have the right (good) side or good eye; he will not have the Spirit of God. In the truest sense this scripture and others like it point to Satan and his spirit of evil (the other-mind).

nm46 » Ezekiel 44 projects to us some more details on these false shepherds:

> "And you shall say to the rebellious, even to the house of Israel, Thus says the Lord GOD; O you house of Israel, let it suffice you of all your abominations, In that you have brought into my sanctuary strangers, uncircumcised in heart [without God's Spirit], and uncircumcised in flesh [without the flesh of Christ, Eph 5:30], to be in my sanctuary [Church], to pollute it, even my house, when you offer my bread [Spiritual], the fat and the blood, and they have broken my covenant because of all your abominations. And you have not kept the charge of mine holy things: but you have set keepers of my charge in my sanctuary for themselves ... And the Levites that have gone away far from me, when Israel went astray, which went astray away from me after their idols; they shall even bear their iniquity. Yet they [the non-Spiritual Levites] shall be ministers in my sanctuary [Church], having charge at the gates of the house, and ministering to the house ... Because they ministered unto them before their idols ["idols in their heart," Ezek 14:3], and caused the house of Israel [Church] to fall into iniquity, therefore have I lifted my hand against them, says the Lord GOD, and they shall bear their iniquity" (Ezek 44:6-12).

nm47 » In verse 13 it says these uncircumcised shall not come near to minister, but verse 14 says they *do*, yet verse 15 speaks of the "sons of righteous" ("Zakok") that keep charge in God's Church. Thus, both the Spiritual and non-Spiritual ministers have "ruled" over the Church. Yet Ezekiel 34 and Zechariah speak of the idolatrous shepherd misruling the sheep with other false shepherds. But God said in Ezekiel 34:10 that he will deliver his flock from the false shepherds' mouth. And Ezekiel 13:23 says the same thing: "for I will deliver my people out of your hand: and you shall know that I am the LORD." It is/was Jesus Christ that delivers his people, the Spiritual Israel, out of the hands of the false shepherds, by overcoming Satan and giving the New Mind to the New

Israel. It is with the New Mind that the New or Spiritual Israel is defeating and will defeat the other-mind and its evil power.

nm48 » In Ezekiel 13 and 14 it describes antitypically some things about these false ministers. They "prophesy out of their *own* hearts," but they call it the word of God: "hear you the word of the LORD" (Ezek 13:2). They see visions of peace for the Church, Spiritual Jerusalem (Ezek 13:16). But Christ prophesied of trouble within the Church (Luke 21:12, 16; Mark 13:9, 12; Mat 24:9-10). And Revelation 2:10 says the Church will have tribulation.

nm49 » In Ezekiel 9:6 it says God will begin to destroy the abomination of Israel at his sanctuary (see verses 4-11). He will begin with the elders of the Church (last part of Ezek 9:6). Three shepherds will be cut off in a month or new moon (Zech 11:8). Again this speaks of the destruction of the spiritual abomination that was in old Israel – the old mind or other-mind – through the power of the New Mind. The old evil spiritual shepherds who ruled over Israel were and are being destroyed by the New Mind that was given to Jesus Christ and Jesus Christ is now giving the New Mind or Spirit to the New Israel (see the "Seed Paper" [PR1] and the *God Papers*).

NM 3: Repentance

NM3 Abstract

In this paper we examine what real repentance is, how many will repent, and who it is that gives repentance. Do we repent on our own or does God give us the power to repent?

nm50 » The New Testament of the Bible speaks of repentance. People are warned to repent for the kingdom of heaven is near (Mat 3:2). People are asked to repent and be baptized or converted (Acts 2:38; 3:19). The repentance the Bible is talking about here is a changed mind. To repent is to change your mind. The Greek word which was translated into our English "repent" is *metanoeo*, which means "to have another mind," or "to change your mind," or to think anew. When the Bible says to repent it speaks to those with the old mind, that evil mind of the old age. To repent is to change your mind. The way those with the old mind change their mind is to receive the New Mind. *To repent then is to change minds.* To repent is to change from the old mind with the evil spirit to the New Mind with the good Spirit.

All To Repent

nm51 » God wants all to come to repentance (2Peter 3:9). In other words, God wants all to have a changed mind. In the paper entitled, "All Saved" [NM 13], we show how all the creation will be freed from the old mind and old cosmos and given life in the new cosmos. The life in the new cosmos will be with the New Mind, the changed mind from the old evil mind.

Repentance Is a Gift From God

nm52 » It is God who *gives* us a changed mind (Acts 5:31; 11:18; 2Tim 2:25). Paul confirms this, "Or do you despise the riches of his goodness and forbearance and long-suffering, not knowing that the goodness of God leads you to repentance" (Romans 2:4). Paul's own changed mind came through God's power, not Paul's power (Acts 9:1-18).

Result of Repentance

nm53 » The result of a changed mind is a new attitude towards God. Paul taught that people "should repent and turn to God, doing works worthy of repentance" (Acts 26:20). Paul did earnestly testify "both to Jews and Greeks repentance toward God and Faith toward our Lord Jesus Christ" (Acts 20:21). Jesus taught, "repent and believe in the gospel" (Mark 1:15). Those with a changed mind believe in the good news of Christ (the gospel), they have turned to God, they do works worthy of their changed mind, they have faith toward Christ, etc.

nm54 » Repentance is also for the sending away ("remission") of sins (Luke 24:47). When you repent you are converted toward the blotting out of sins (Acts 3:19). And it is this repentance that leads to salvation (2Cor. 7:10).

How Does God Give Repentance?

nm55 » As Romans 8:7-9 shows us, it is through God's Spirit that mankind's attitude changes from the way of death towards the way of life. This Spirit of God is a gift from God (1Thes. 4:8; 2Cor. 1:22; 5:5; Romans 5:5; Acts 2:38). With this gift of the Spirit we receive the fruits or effects of this Spirit. "But the fruit of the Spirit is love, joy, peace..." (Gal. 5:22). It is through God's free gift of the Spirit of God that mankind's mind changes from the old mind to the New Mind. In this age the gift of this New Mind is given to those who were predestinated before the world began to receive the New Mind in the old age (see "Predestination Paper" [NM 8]).

NM 4: Baptism: Physical & Spiritual

Real Christian Baptism
Water Baptism v. Spiritual Baptism
Spiritual Baptism
Baptized in the Name
In the Name of Christ?

NM4 Abstract

In this paper we learn that there are two kinds of baptisms. One is the physical and is only symbolic. The other is Spiritual baptism in which we are put into the Name of the true God. John the Baptist who only baptized with water predicted the Spiritual baptism.

Real Christian Baptism

nm56 » The baptism of Christians is the baptism of the Spirit of God. Baptism means to *dip*, *immerge*, or *submerge*. To be baptized with the Spirit is to be submerged into the Spirit. To be baptized with the Spirit is another way of saying you are sealed with the Spirit or that you have received the Spirit, or that you have put on the New Mind, or that you have received the Promise, or that you have the New Life, etc. When you are baptized with the Spirit, you have the New Mind of Love that thinks the positive thoughts of the Spirit of the True God. When you are baptized with the Spirit you are in the body or assembly of Jesus Christ, you are a part of the BeComingOne, which is the True Oneness. When we are baptized with the Spirit our old life is put to death and we are raised up into the New Life (see Rom 6:4). Baptism with the Spirit is different from baptism with water. Baptism with the Spirit is a gift from the True God given to those in this age who were predestinated before the world began to receive the Spirit or New Mind in this age (see "Predestination Paper" [NM 8]). All will eventually be Spiritually baptized (see "All Saved Paper" [NM 13]).

Baptism With Water versus Baptism With Spirit

nm57 » It was John the Baptist who baptized with water. In John's own words: "I indeed baptize you in water to repentance. But He [Jesus] who is coming after me is mightier than I ... He shall baptize you in the Holy Spirit and in fire" (Mat 3:11). Again John's words: "I baptize with water" (John 1:26). John baptized with water, but it was He who was coming after John who would baptize with the Spirit (John 1:26-33). And in Jesus Christ's own words after He was resurrected from the dead: "And gathering them together, He commanded them not to leave Jerusalem, but said, Wait for the promise of the Father which you heard from Me. For John indeed baptized with water, but you shall be baptized in the Holy Spirit not many days after this" (Acts 1:4-5). And while Jesus Christ's followers were waiting in Jerusalem the Promise did come in the form of the Spirit (Acts 2:1-47). Now the Promise is the Holy Spirit: "the Promise of the Holy Spirit from the Father" (Acts 2:33). This Promise was not only to Christ but to "as many as the Lord our God shall call" (Acts 2:39). It is the baptism with the Spirit that counts, not baptism with water. "Water" is merely a symbolic representation of Spirit. When Christ spoke of water he meant the Spirit (see John 7:38-39). Remember here and remember always: look at the higher or Spiritual meaning.

nm58 » John the Baptist's baptism with water prefigured Christ's baptism with the Spirit. As water cleans the body, so does the Spirit clean the body. Water cleans in a physical way, but the Spirit cleans in a Spiritual way. When we are cleansed with the Spirit our minds are cleansed from the dirt of the other-mind, that old twisted mind of the old age. The baptism of John was the baptism with water. This water baptism does not bring with it the Holy Spirit, or the New-Mind (note Acts 18:24-19:6). The Spirit of God comes with the baptism of the Spirit. When one is submerged into the Spirit, he takes on the Spiritual reality. When one is baptized in physical water, he is only cleansed physically. We are to look to the higher and Spiritual meaning in the Bible in order to learn the Truth. Water can only clean physically, but the Spirit cleans Spiritually.

nm59 » Now we see in certain verses in the New Testament of the Bible where some of the early disciples used water to baptize (Acts 8:36-38). Even after some received the Holy Spirit Peter had some baptized with physical water (Acts 10:44-48). The reason for this was because at that time they did not understand fully the power of God and that it is the Spiritual reality that counts not the physical types of the Spiritual reality. The early leaders gradually learned that physical rituals such as water baptism and circumcision were not the important things (see "Freedom and Law Paper"[NM 17]). The old laws of the Old Testament were done away with. They were merely types of the True Reality (Heb 10:1; see the "Freedom and Law Paper" [NM 17]).

Spiritual Baptism

nm60 » Real Christians are baptized with the Spirit (Acts 1:5; 1Cor 12:13). "For by one Spirit also we were baptized into one body" (1Cor 12:13). Christians are in the body of Christ. On the Pentecost the first Christians were filled with the Holy Spirit "suddenly" (Acts 2:4,2). And again later, "Even while Peter was speaking these words, the Holy Spirit fell on all those hearing the word" (Acts 10:44). "And as I began to speak, the Holy Spirit fell upon them also, even as on us in the beginning. And I remembered the word of the Lord saying, John indeed baptized with water, but you shall be baptized with the Holy Spirit" (Acts 11:15-16). God saves people through "the washing of regeneration and renewing of the Holy Spirit, which He poured out on us richly through Jesus Christ our Savior" (Titus 3:5-6). It is the Spiritual washing of the Spirit of God that Spiritually cleans people, not the physical water. *In the True Church of God there is no need or requirement for water baptism.* Water baptism is merely a physical ritual that represents a Spiritual truth. As water baptism cleans the physical body, so does Spiritual baptism clean the Spiritual body in a Spiritual way. Physical ritual does not free anyone from the mad cosmos we live in. It is the Spiritual gift from God, the Spirit of God, that gives us the freedom. There is no certain set of words (magic) that gives us True Life (such as, "I baptize you in the name of the ... "). There are no physical rituals that give us Life. True baptism is Spiritual baptism not water baptism.

Baptized *into* Christ's Name & Spiritual Body

nm61 » When real Christians are Spiritually baptized they are baptized *into* Christ's body (1Cor 12:13). They are baptized into Christ (Rom 6:3; Gal 3:27). They are baptized into his Name (Mat 28:19-20; Acts 2:38; 8:16; 10:48; 19:5).

Baptized Into The Name of Christ?

nm62 » What does it mean to be Spiritually baptized into the NAME of Christ? In the Bible many times a name signifies something about that person. Thus, "Jesus" signifies that Christ is the Savior, for "Jesus" means, *savior*. When one is baptized *into* the NAME of Christ he takes on the NAME of Christ. Since a name describes characteristics of someone, then if one is put into a name, he is actually being put into the characteristics of that person. Allegorically, when a woman marries, she is married into a name. She becomes a part of the family. If the family is rich she shares in the riches. A person baptized into Christ's NAME takes on the NAME and characteristics of Christ. The person is married into Christ's NAME in a sense. In fact women are allegorical to the Church (Eph 5:21-32). Christ is going to marry this woman (Church) allegorically at his physical return (Rev 19:7). Those baptized into the NAME of Christ take on his NAME and some of his characteristics. They become Christians; they receive his Spirit. You are a Christian only when you have the Spirit of God (Rom 8:9, 14).

Baptized into the Name of the Father, Son, and Holy Spirit

nm63 » "Baptize them into the Name of the Father, and of the Son, and of the Holy Spirit" (Mat 28:19). When you are baptized are you baptized into three names or one Name? When you are in Christ, you are in his Father because Christ is in his Father (John 14:11). Since Christ's Father is God, when you are in the Father you are in God – you are a child of God. When you are in Christ's NAME you are in the NAME of God because Christ came in the NAME of God (John 5:43; 10:25; 12:13; Luke 13:35; 19:38; Mat 21:9; John 17:11). When you are Spiritually baptized into Christ, you are baptized into the Name of God the Father, for Christ is in God the Father (John 14:11). You thus become a part of the Coming Oneness (see, *God Papers*).

nm64 » The Spirit of God and the Spirit of Christ are the same thing, "for in one Spirit are we all baptized into one body ... and have been made to drink into one Spirit" (1Cor 12:13). This one Spirit is the Spirit of God. But when you are baptized you receive the Holy Spirit (Acts 2:38; 10:47; 11:16; Mark 1:8; Acts 1:5). The Holy Sprit is the Spirit of God, the Spirit of Christ. Everyone who is a Christian has the Holy Spirit of God. When one is baptized into the NAME of Christ, he receives the Spirit of God, that is, the Spirit of Christ, that is, the Holy Spirit, for Christ is God (John 20:28; Jude 25) and has God's Name (John 5:43; 10:25; 12:13; Luke 13:35; 19:38; Mat 21:9; John 17:11; see *God Papers*). "The Father, the Word, and the Holy Spirit are one" (1John 5:7) because they are the same One Spirit, the Holy Spirit of God the Father. You are not baptized into three names, but into the very NAME of God.

In the Name of Christ?

nm65 » Acts 2:38 is one place where it speaks of being baptized in the NAME of Christ. The word translated "in" is a Greek word that can mean, *in* or *among*. Thus it could as easily be translated, "and be baptized every one of you *among* the name of Jesus Christ." When one is baptized into the NAME of Christ he is also baptized *among* those belonging to Christ.

nm66 » Notice that those "in the name" or "among the name shall cast out demons" (Mark 16:17). Those who have the Spirit of power, God's Spirit, will be in the body of Christ, will be among the others who are in Christ. And it is those in Christ or among his body members that will cast out the demons, or the other-minds. (*Some* had that power; others will do it at the Messiah's return when all demons will be cast out of mankind's mind.) Paul speaking to Christians said, "you are the body of Christ, and members in particular" (1Cor 12:27). When people do things "in the name of Christ," they do these things while they are among or in Christ. That is, they do these things because they have the Spirit which puts them in the NAME of Christ.

nm67 » When you are in the NAME of Christ:

- you are saved (Acts 4:12);

- you have life (John 20:31);

- you are justified (1Cor 6:11);

- you preach boldly (Acts 9:27, 29);

- you may do signs and wonders (Acts 4:30);

- devils ['other minds'] are subjected to you (Luke 10:17); etc.

NM 5: Begotten, Born – the Difference

Difference between Born & Begotten
Pregnant Woman Metaphor
Born of Flesh, Born of Spirit
Fleshly Body, Spiritual Body
Greek Word, *gennao*, and its ambiguity

NM5 Abstract

In this paper we show you the difference between being born of God and being begotten of God. There are Biblical verses that use the metaphor of the pregnant woman (Church) giving birth at the coming of God to set up the kingdom on earth. Before Christ's coming Christians are like babies in the Church's womb waiting to be born. A Greek word that can mean either begotten or born has added to the misinformation on this subject.

Difference Between Born & Begotten

nm68 » Christ spoke about being born again (John 3:5-6). When one is baptized in the Spirit that person receives, or is sealed with the Spirit of God. At that time he becomes a child of God (Rom 8:14). One aspect of this has been overlooked by most. That is, there is a difference between being *born* of God and being *begotten* of God. We need to know the difference. The Bible speaks of both. In short we will find that when one is Spiritually baptized, he is begotten of God. But when one is resurrected to God, he is born of God. Let's explain this.

Pregnant Woman Metaphor

nm69 » The Church is pictured allegorically as a woman or wife of Christ (Eph 5:22-32). But further it is pictured as a "mother of us all" (Gal 4:26; cf "Jerusalem" in Rev 21:2, 9). Yet it is pictured as a pregnant mother (Rev 12:1-2, 4). And in Isaiah 66:6-8 it pictures this woman in labor pains ready to bring forth. The time setting here is when the Lord will recompense or repay his enemies which is the day of the Lord or on the day of God's wrath (Isa 66:6; see "God's Wrath Paper" [PR4]). This woman, or Church, brings forth a whole nation at once (Isa 66:8). This whole nation is a holy nation (1Pet 2:9) of *born* children of God as we will see.

nm70 » This pregnant woman is allegorical to the Church. Inside her womb are her children, a whole nation of children (Isa 66:8). But before one is born of a woman, he is begotten or conceived inside her womb. And in Revelation 12:1-2, 4 it pictures the Church allegorically as a pregnant woman ready to deliver (Rev 12:2). This pictures the Church ready to be born of God with the dragon (who is Satan, Rev 12:9) waiting "for to devour her child as soon as it was born." This is the "day of trouble" for the Church. See the papers on God's Wrath to understand this "day of trouble."

nm71 » In the Bible it speaks about people being born of God or begotten of God (1John 2:29; 3:9; 5:1; 5:4; 5:18). Some people teach being born of God as some heartfelt feeling. They do not take it literally; thus they do not understand God's plan. Being born of God is not just a feeling in the heart.

Born of Flesh, Born of Spirit

nm72 » Christ the man said during his ministry that, "except a man be born of water and of the Spirit, he cannot enter into the kingdom of God. That which is born of the flesh is flesh; and that which is born of the Spirit is Spirit" (John 3:5-6). In verse 8 of John 3 he makes an allegory between Spirit and the wind and says as one cannot see the wind so also he cannot see the spirit. As we just quoted Christ, "except a man be born of water and of the Spirit, he cannot enter into the kingdom of God." Thus since Spirit is invisible, when one is born of God he is invisible. Yet there are verses where the resurrected Christ as God was also flesh and blood (Luke 24:39). Yes, Christ was born of flesh, he was a son of mankind (Gal 4:4). Thus, "that which is born of flesh is flesh" (John 3:6). But Christ is a son of God also (Rom 1:4). Thus when Christ was resurrected he was born of God and became Spirit. Therefore as spirit he could become invisible as he did after he was born of God (Luke 24:31).

Fleshly Body, Spiritual Body

nm73 » The resurrected Christ is a son of God, and a son of man. Once he was born of flesh; those born of flesh are flesh (John 3:6). Once he was born of God by a resurrection (Rom 1:4); those born of Spirit are Spirit (John 3:6). Christ the God is flesh and Spirit. He has two essences, he has two bodies. Christ the God as a son of man has a fleshly body, "there is a fleshly body" (1Cor 15:44). But as a son of God he has a Spiritual body, "there is also a spiritual body" (1Cor 15:44). The resurrected Christ has two bodies, or two essences – a spiritual and a fleshly essence. Scripture does not say that when one is born of God he loses his fleshly body. No it says they are made immortal (1Cor 15:52-55). And as Christ is, so shall all born of God be, for He is the first born of many brethren (Rom 8:29). Scripture indicates those born of God will be like Christ, "we shall be like him" (1John 3:2). "The Lord Jesus Christ who shall change our vile body, that it may be fashioned like unto his glorious body" (Phil 3:20, 21).

Greek Word, gennao, Ambiguity of

nm74 » But not only was Christ born of the flesh and born of the Spirit, he was also begotten of the flesh (inside Mary's womb) and begotten of the Spirit (while he was in his first fleshly state). Christ not only was born of flesh, he was also begotten of flesh. Christ not only was born of Spirit, he was also begotten of the Spirit. But because of the vagueness of a Greek word many do not understand this. The word translated born in many English translations comes from the Greek word, *gennao*. In contrast, this Greek word can mean either to *beget* or to be *born*. In the English language we have two separate words for the process of being begotten and being born. But the Greek word *gennao* can be used to mean either being begotten or being born. Because of this there is ambiguity when translating *gennao* into English.

nm75 » Being begotten is the same as being conceived, or fertilized, or impregnated. To be begotten is to be conceived. An egg-cell is begotten by a sperm cell. This is being begotten. Once begotten an egg-cell grows inside the womb of its mother. Allegorically, a Christian is begotten by the Spirit of God and grows Spiritually inside the womb of the Church, their heavenly mother, or Spiritual mother.

nm76 » But after the egg-cell has grown inside the mother's womb it is born of mankind. Allegorically, after a Christian is begotten, he grows Spiritually in the Church's womb until he is born of God. Isaiah 66:6-8 pictures the Christians all at once being born of God. This will happen at the last trumpet (1Cor 15:52-55; Rev 11:15). This is the time of Christ's physical return (1Thes 4:15-18).

nm77 » Therefore because the New Testament was written in Greek, and because in Greek there is a word that can express two different processes or stages of birth, and because this Greek word (*gennao*) was used in verses to express either "begotten" and/or "born," and because many of those who translated the Bible didn't understand God's plan; then the translators sometimes mistranslated "born" where they should have translated "begotten" and vice versa. And because of this vagueness of the Greek word *gennao* many people today do not understand what it means to be begotten or born of God. (Many places, if not all places, where *gennao* is used, can be and should be understood in the sense of begotten and/or born.) See "Last War and God's Wrath" PR5, in its Notes for more information or details on "begotten" and "born."

NM 6: Body, Soul, Spirit, and Immortality

Soul
Immortal Soul?
Spirit

NM6 Abstract

In this paper we look at what scripture says about the body, soul, and spirit. There is a difference between all three. The scriptures indicate that the soul is not the spirit of man, but tradition mixes these two different things. When some today speak of the soul, what they mean is the spirit. This causes confusion since the soul can die (it is mortal), but the spiritual element cannot die (it is immortal).

Soul

nm78 » *To help clarify*: in this paper we are not referring to the "soul" in the sense that it is used today in music or art ("he has soul"), or as a synonym for emotion or passion or feelings or spiritual depth or mind or psyche, or in any other way except as the word soul is used in the Bible (old and new testaments). In context with the Bible, what does the soul have to do with the body or with the spirit? What is the body? What is the soul? What is the spirit? And is the soul immortal? There is much confusion about what a soul is and if it is immortal. The view of Catholics, many Protestants, and some Jews is that the soul is immortal. From *This is the Faith* [3rd ed.], written by Canon Francis J. Ripley, we see the Catholic view on the soul:

> Man ... has a body and a soul ... it is a spirit, immortal, and endowed with intelligence and free will. Soul is not just another word for spirit. Animals have souls, but their souls are not spirits. Only man's soul is a spirit; in man is the only kind of spirit that is a soul.... There is an obvious difference between a living human body and a corpse. That difference is the soul." (3rd edition, p. 8; earlier ed. pp. 21-22)

As we will see from scripture there are several assertions here that are wrong. First the soul is not a spirit, second it is not immortal, and third its free will is limited under the absolute free will of God. You will not find immortality connected to the soul in scripture, but you will find Satan telling Eve that she is immortal (Gen 3:4; serpent = Satan, Rev 12:9). So how do the Catholics "prove" that the soul is immortal?

> "Scripture is full of proof that the soul of man is spiritual and immortal. 'The Lord God formed man of the slime of the earth and breathed into his face the breath of life; and man became a living soul' (Gen 2:7)

> The souls of the just are in the hand of God, and torment of death shall not touch them. In the sight of the unwise they seemed to die .... but they are in peace .... their hope is full of immortality." (Wis 3:1-4)

These two scripture quotes are from *This is the Faith* under the heading, "What Does Scripture Say?," and are the two main verses used by the author to prove his assertion. The first quote of Genesis 2:7 says nothing about the soul being immortal. The second, even though it is not from the Bible, does not say the soul is immortal, but that man has hope of immortality. Of course they have hope of immortality because of the resurrection. The other scriptures quoted by the author speak of the hope of immortality because of the resurrection or the immortality given through the Spirit. The author substitutes the *hope* of immortality for mankind with mankind's mortality, and mixes the soul with the spirit. The hope of immortality is not the same as immortality; soul and spirit are not one and the same. Let's see what the scripture actually says about the body, soul, spirit, and immortality.

Man is a Living Soul

nm79 » In the Bible the word soul is translated from the Hebrew word *nephesh* (נֶפֶשׁ) and from the Greek word *psuche* (ψυχῆς). From Genesis we see what a soul is:

> ■ "And Jehovah God formed the man, dust from the ground, and breathed into his nostrils breath of life; and man became a living soul." (Gen 2:7)

With the help of God man became a living soul. Man was formed from the dirt of the ground.

Man is earthly. Then Jehovah [YHWH] breathed into man the breath of life and he became a living soul. In context with the other scriptures in Genesis, chapters 1 and 2, we see that Jehovah first made man's body, and second he made man a living soul when and because Jehovah breathed into him. It took the breath of Jehovah to make man a living soul. The living soul is not just a body: it is a body with God's breath in it.

God has a Soul

nm80 » Although theologians call it anthropopathy even God has a soul according to Bible:

- DBY Leviticus 26:11 And I will set my habitation among you; and **my soul** shall not abhor you;

- DBY Leviticus 26:30 And I will lay waste your high places, and cut down your sun-pillars, and cast your carcases upon the carcases of your idols; and **my soul** shall abhor you.

- DBY Isaiah 42:1 Behold my servant whom I uphold, mine elect {in whom} **my soul** delighteth! I will put my Spirit upon him; he shall bring forth judgment to the nations. [Compare with Mat 12:18.]

- DBY Jeremiah 5:9 Shall I not visit for these things? saith Jehovah, and shall not **my soul** be avenged on such a nation as this?

- DBY Jeremiah 6:8 Be thou instructed, Jerusalem, lest **my soul** be alienated from thee; lest I make thee a desolation, a land not inhabited.

- DBY Ezekiel 23:18 And she discovered her whoredoms, and discovered her nakedness; and **my soul** was alienated from her, like as **my soul** was alienated from her sister.

- DBY Zechariah 11:8 And I destroyed three shepherds in one month; and **my soul** was vexed with them, and their soul also loathed me.

- DBY Matthew 12:18 Behold my servant, whom I have chosen, my beloved, in whom **my soul** has found its delight. I will put my Spirit upon him, and he shall shew forth judgment to the nations. [Compare with Isa 42:1.]

- DBY Hebrews 10:38 But the just shall live by faith; and, if he draw back, **my soul** does not take pleasure in him.

(Lev 26:11, 30; Isa 42:1; Jer 5:9; 6:8; Ezek 23:18; Zech 11:8; Mat 12:18; Heb 10:38)

Jesus became or was the soul of God (Mat 12:18; see *God Papers*).

Animals also have Souls

nm81 » Not only does mankind, and God, have souls, so do animals. The following verses in Hebrew show that animals have a soul (*nephesh* or *psuche*) also:

- DBY Genesis 1:20 And God said, Let the waters swarm with swarms of living **souls**, and let fowl fly above the earth in the expanse of the heavens.

- DBY Genesis 1:21 And God created the great sea monsters, and every living **soul** that moves with which the waters swarm, after their kind, and every winged fowl after its kind. And God saw that it was good.

- DBY Genesis 1:24 And God said, Let the earth bring forth living **souls** after their kind, cattle, and creeping thing, and beast of the earth, after their kind. And it was so.

- DBY Genesis 1:30 and to every animal of the earth, and to every fowl of the heavens, and to everything that creepeth on the earth, in which is a living **soul**, every green herb for food. And it was so.

- DBY Genesis 2:19 And out of the ground Jehovah Elohim had formed every animal of the field and all fowl of the heavens, and brought {them} to Man, to see what he would call them; and whatever Man called each living **soul**, that was its name.

- DBY Genesis 9:10 and with every living **soul** which is with you, fowl as well as cattle, and all the animals of the earth with you, of all that has gone out of the ark -- every animal of the earth.

- DBY Genesis 9:12 And God said, This is the sign of the covenant that I set between me and you and every living **soul** that is with you, for everlasting generations:

- DBY Genesis 9:15 and I will remember my covenant which is between me and you and every living **soul** of all flesh; and the waters shall not henceforth become a flood to destroy all flesh. 16 And the bow shall be in the cloud; and I will look upon it, that I may remember the everlasting covenant between God and every living **soul** of all flesh that is upon the earth.

- DBY Leviticus 11:10 but all that have not fins and scales in seas and in rivers, of all that swarm in the waters, and of every living **soul** which is in the waters -- they shall be an abomination unto you.

- DBY Revelation 8:9 and the third part of the creatures which were in the sea which had life [**soul**] died; and the third part of the ships were destroyed.

(Gen 1:20, 21, 24, 30; 2:19; 9:10; 9:12, 15, 16; Lev 11:10; Rev 8:9; etc.)

Many English Bibles have translated the word "creature" or "life" for soul or *nephesh* or *psuche*. In the *BeComingOne Bible* we have translated *nephesh* or *psuche* into soul consistently.

Souls Can Die

nm82 » According to the official Catholic view the soul is immortal:

Lateran Council of 1513

"Whereas some have dared to assert concerning the nature of the reasonable soul that it is mortal, we, with the approbation of the sacred council do condemn and reprobate all those who assert that the intellectual soul is mortal, seeing, according to the canon of Pope Clement V, that the soul is [...] immortal [...] and we decree that all who adhere to like erroneous assertions shall be shunned and punished as heretics."

But contrary to the Catholic view, scripture indicates that souls can die. The following verses indicate that souls are destructible:

- DBY Genesis 17:14 And the uncircumcised male who hath not been circumcised in the flesh of his foreskin, that **soul** shall be cut off from his peoples: he hath broken my covenant.

- DBY Genesis 37:21 And Reuben heard {it}, and delivered him out of their hand, and said, Let us not take his life [**soul**].

- DBY Exodus 12:15 Seven days shall ye eat unleavened bread: on the very first day ye shall put away leaven out of your houses; for whoever eateth leavened bread from the first day until the seventh day -- that **soul** shall be cut off from Israel.

- DBY Leviticus 7:20 But the **soul** that eateth the flesh of the sacrifice of peace-offering which is for Jehovah, having his uncleanness upon him, that **soul** shall be cut off from his peoples.

- DBY Leviticus 24:17 And if any one smiteth any man mortally [kills any **soul**], he shall certainly be put to death.

- DBY Numbers 23:10 Who can count the dust of Jacob, and the number of the fourth part of Israel? Let my **soul** die the death of the righteous, and let my end be like his!

- DBY Numbers 31:19 And encamp outside the camp seven days; whoever hath killed a person [**soul**], and whoever hath touched any slain; ye shall purify yourselves on the third day, and on the seventh day, you and your captives.

- DBY Numbers 35:30 Whoever shall smite a person mortally, at the mouth of witnesses shall the murderer be put to death; but one witness shall not testify against a person [**soul**] to cause him to die.

- DBY Deuteronomy 19:6 lest the avenger of blood pursue the manslayer, while his heart is hot, and overtake him, because the way is long, and smite him mortally [slay his **soul**]; whereas he was not worthy of death, since he hated him not previously.

- DBY Joshua 2:13 that ye will let my father live, and my mother, and my brethren, and my sisters, and all that belong to them, and deliver our **souls** from death.

- DBY Judges 5:18 Zebulun is a people {that} jeoparded their lives [**souls**] unto death, Naphtali also, on the high places of the field.

- DBY 1Kings 19:4 And he himself went a day's journey into the wilderness, and came and sat down under a certain broom-bush, and requested for himself that he might die; and said, It is enough: now, Jehovah, take my life [**soul**]; for I am not better than my fathers.

- BY Job 36:14 Their **soul** dieth in youth, and their life is among the unclean.

- DBY Psalm 22:29 All the fat ones of the earth shall eat and worship; all they that go down to the dust shall bow before him, and he that cannot keep alive his own **soul**.

- DBY Psalm 78:50 He made a way for his anger; he spared not their **soul** from death, but gave their life over to the pestilence;

- DBY Isaiah 55:3 Incline your ear, and come unto me; hear, and your **soul** shall live; and I will make an everlasting covenant with you, the sure mercies of David.

- DBY Jeremiah 4:10 And I said, Alas, Lord Jehovah! surely thou hast greatly deceived this people and Jerusalem, saying, Ye shall have peace; whereas the sword reacheth unto the **soul**.

- DBY Ezekiel 13:19 And will ye profane me among my people for handfuls of barley and for morsels of bread, to slay the **souls** that should not die, and to save the **souls** alive that should not live, by your lying to my people that listen to lying?

- DBY Ezekiel 22:27 Her princes in the midst of her are like wolves ravening the prey, to shed blood, to destroy **souls,** to get dishonest gain.

- DBY Matthew 2:20 Arise, take to {thee} the little child and its mother, and go into the land of Israel: for they who sought the life [**soul**] of the little child are dead.

- DBY Matthew 10:28 And be not afraid of those who kill the body, but cannot kill the **soul**; but fear rather him who is able to destroy both **soul** and body in hell.

- DBY Matthew 26:38 Then he says to them, My **soul** is very sorrowful even unto death; remain here and watch with me.

- DBY Mark 3:4 And he says to them, Is it lawful on the sabbath to do good or to do evil, to save life [**soul**] or to kill? But they were silent.

- DBY Mark 14:34 And he says to them, My **soul** is full of grief even unto death; abide here and watch.

- DBY Luke 6:9 Jesus therefore said to them, I will ask you if it is lawful on the Sabbath to do good, or to do evil? to save life [a **soul**], or to destroy {it}?

- DBY Luke 17:33 Whosoever shall seek to save his life [**soul**] shall lose it, and whosoever shall lose it shall preserve it.

- DBY John 10:15 as the Father knows me and I know the Father; and I lay down my life [**soul**] for the sheep.

- DBY John 12:25 He that loves his life [**soul**] shall lose it, and he that hates his life [**soul**] in this world shall keep it to life eternal [aeonian].

- DBY Acts 3:23 And it shall be that whatsoever **soul** shall not hear that prophet shall be destroyed from among the people.

- DBY Romans 11:3 Lord, they have killed thy prophets, they have dug down thine altars; and I have been left alone, and they seek my life [**soul**].

- DBY Hebrews 10:39 But we are not drawers back to perdition, but of faith to saving {the} **soul**.

- DBY James 5:20 let him know that he that brings back a sinner from {the} error of his way shall save a **soul** from death and shall cover a multitude of sins.

- DBY Revelation 8:9 and the third part of the creatures which were in the sea which had life [**soul**] died; and the third part of the ships were destroyed.

- DBY Revelation 12:11 and they have overcome him by reason of the blood of the Lamb, and by reason of the word of their testimony, and have not loved their life [**souls**] even unto death.

- DBY Revelation 16:3 And the second poured out his bowl on the sea; and it became blood, as of a dead man; and every living **soul** died in the sea.

(Gen 17:14; 37:21; Ex 12:15; Lev 7:20; 24:17; Num 23:10; 31:19; 35:30; Deut 19:6; Joshua 2:13; Jud 5:18; 1Kings 19:4; Job 36:14; Psalms 22:29; 78:50; Isa 55:3; Jer 4:10; Ezek 13:19; 22:27; Mat 2:20; 10:28; 26:38; Mark 3:4; 14:34; Luke 6:9; 17:33; John 10:15; 12:25; Acts 3:23; Rom 11:3; Heb 10:39; James 5:20; Rev 8:9; 12:11; 16:3; etc.)

nm83 » That souls can die is absolutely clear in the Hebrew or Greek as well as the *BeComingOne Bible* and other more literal Bibles. Sometimes English translations leave out soul from the translation as in Judges 16:30 where it should read: "And Samson said, Let my soul die...." Sometimes English translations have life instead of soul as in John 10:15.

nm84 » The following verses indicate that there can be dead souls:

■ DBY Leviticus 21:11 Neither shall he come near any person [**soul**] dead, nor make himself unclean for his father and for his mother;

■ DBY Leviticus 22:4 Whatsoever man of the seed of Aaron is a leper, or hath a flux, he shall not eat of the holy things, until he is clean. And he that toucheth any one that is unclean by a dead person [**soul**], or a man whose seed of copulation hath passed from him;

■ DBY Numbers 5:2 Command the children of Israel, that they put out of the camp every leper, and every one that hath an issue, and whosoever is defiled by a dead person [**soul**]:

■ DBY Numbers 6:11 And the priest shall offer one for a sin-offering, and the other for a burnt-offering, and make an atonement for him, for that he sinned by the dead person [**soul**]; and he shall hallow his head that same day.

(Lev 21:11; 22:4; Num 5:2; 6:11; etc.)

[In some translations "body" or "dead" = soul; see Hebrew text.]

Two Meanings of a Dead Soul

Because there is a type and antitype to the Bible, there are two meanings to dead souls, the physical and the spiritual:

■ Those who lose the breath of life become dead

■ Those with the breath (or spirit) of Satan are dead

Immortal Soul: Satan's Lie

nm85 » Even though mankind will in the future become immortal, the above scripture indicates that the soul is not now immortal. The idea that the soul is now immortal did not come from the Bible. The false idea that humans in this age have immortal souls came from Satan's first lie. Satan's lie occurred right after God said that man would die if they ate from the tree of knowledge of good and evil (Gen 2:16-17):

■ And the serpent said unto the woman, Ye shall not surely die. [Gen 3:4]

This idea of the immortality of the soul continued through the Greeks, Babylonians, Egyptians, Romans, and so forth. Of course these cultures got this idea of the immortality of the soul from the power of Satan which feeds mankind false and destructive information (NM 21). From this theologians down through the years have interjected this false idea into the doctrines of Christianity and other religions.

Salvation of the Soul

nm86 » From Genesis 2:7 we know that a body can be a soul, if it has breath in it. From other scripture mentioned above we see that any living and breathing body, even an animal's body, is called a soul. Also we know from the scripture referred to above that a soul can be destroyed and can die and so can be called a dead soul. Thus in this age mankind's soul is *not* immortal. This is why the Bible talks about the *salvation* of the soul (1Pet 1:9; Heb 10:39; 1Thes 5:23; Luke 21:19; James 5:20) and talks about going from mortality to immortality (1Cor 15:53-54; Rom 8:11; 2Cor 5:4). The idea of the immortal soul came from the influence of Satan, the Greeks, and other ancient peoples. But although the soul is *not* immortal, there is one aspect of man that is immortal. That immortal aspect is the spirit.

Spirit

nm87 » As shown in this book and in the *God Papers* spirits are immortal, and there is a particular spirit for each and every person. For clarification you must read the books: *New Mind and Christianity* (aka: *New Mind Papers*) and the *God Papers* [aka *God*]. For now let it be said that there is an immortal aspect to mankind, and it has to do with man's own spirit. Spirits are immortal; souls can and do die. What most religions do is mix-up the nature of the spirit with the nature of the soul. Our soul is not our spirit; our spirit is not our soul. This mix-up began at the time of Adam and Eve (Gen 3:4) and still causes confusion today. If the translators had only consistently translated the Hebrew *nephesh* and the Greek *psuche* into soul we would not have as much confusion. Of course the reason they didn't translate these words consistently is because of their mindset which made them twist the scriptures to force their incorrect ideas into the text of the Bible.

Note: Remember that in Genesis 2:7 Jehovah "breathed" into man the "breath" of life. In the English version of the Old Testament the word "breath" is translated from either the Hebrew word *neshamah* or *ruah*. These words differ slightly in meaning, both signifying sometimes "wind" or sometimes "breath." The word translated into "breath" in Genesis 2:7 is the Hebrew word *neshamah*. Both Hebrew words, *neshamah* and *ruah* are translated as "spirit" in various places in the English translations of the Bible. This means that the book of Genesis could just as well have been translated showing God breathing into man a spirit. This is just another example of duality (type and antitype) of the Bible. As it turned out both senses are true: God breathed a breath of air into man; God also breathed a spirit into man. See *New Mind* 22 for more information on this. But that spirit was not the soul of man.

NM 7: Age Paper

NM7 Abstract

One of the biggest mistakes in traditional Christianity is the mistranslation of two words into "eternal" that actually mean aeon or age. From this mistake came the eternal hell and punishment for those who never had the chance to learn about Christ or who simply did not believe in him. Not only this, but from this mistake we get such nonsense as, sacrifices for eternity, slaves for eternity, time before eternity, more than one eternity, and other such impossibilities. In this paper we refute the best arguments from Augustine and others. This paper is a key to unlocking the truth that has been hidden behind this mistranslation.

Eternal & Forever in the Bible

nm88 » Do you know that *eternal* and *forever* in most translations of the Bible are incorrectly translated from words that mean **age?** Since the early fifth century AD and probably long before, this major inaccuracy in translation has filtered and shaded most doctrines of the Bible.

Olam & Aionios

nm89 » In the Old Testament the Hebrew *olam* [עוֹלָם] is the most common word translated to English as 'forever.' In the New Testament the Greek *aionios* [αἰώνιός] is the most common word translated to English as 'forever' or 'eternal.' From Young's *Analytical Concordance to the Bible* and Strong's *Exhaustive Concordance of the Bible* we see the proof that the words 'everlasting' and 'forever'

were most often translated to English (KJV) from the Hebrew *olam* or the Greek *aionios*.

nm90 » Even though most translations of the Bible incorrectly translate the Hebrew *olam* or the Greek *aionios* in scripture, there are translations that correctly use them. Translations such as Young's *Literal Translation of the Holy Bible* and Rotherham's *The Emphasized Bible* do use 'age' or 'age-abiding' instead of 'forever.' You will find that in our papers we use the words 'age' or 'agelasting' or 'aeonian' instead of the inaccurate 'forever' or 'everlasting' or 'eternal.' Why do most translations use *forever, everlasting*, and *eternal* (or comparable words in other languages) while we use 'age' or 'agelasting' or 'aeonian'?

Vague Time Period

nm91 » We will show in this paper that the Hebrew *olam* means age or agelong or an eon of indefinite length, and that the Greek *aion* and its adjective *aionios* mean an age of unknown length or agelong or aeonian. The main and only real meaning of these words is an *age or eon of unknown length*. The words in and of themselves tell us nothing about duration, or the beginning or end of the age. In context they *may* indicate "an age (of foreverness)" when it is speaking of an age that will not end, an endless age. But here they only indicate "an age (of foreverness)" by auxiliary words that clarify their normal vague meaning: the word 'age' by itself never tells us the length of the age or the beginning or end of the age. Without auxiliary words that specify its length, the word 'age' or 'eon' is always unclear as to its length.

Damning, Unforgiving Mindset

nm92 » But there is a desire in mankind that wants and needs to believe that the Hebrew *olam* and the Greek *aionios* mean forever or eternal, and because of this they ignore the doctrine of forgiveness and the great all powerfulness of God. The desire has turned into a mindset.

nm93 » This mindset traps mankind into paradoxes that make God contradictory and impotent. Mankind has its 'hell' theories where a supposedly good, forgiving, and almighty God puts humans in a hell-fire that somehow burns their fleshly body for ever and ever: their god does not terminate their life, he tortures them forever. They say, "those who don't believe or commit themselves to Christ, are damned forever." But all one has to do is to translate *olam* and *aionios* into 'agelong' or 'aeonian,' as we did in the *BeComingOne Bible*, and many of the great paradoxes of Biblical doctrine will end. But this isn't easy for many. They *insist* on holding on to their tangled doctrines.

nm94 » I once believed in this distortion. But once I learned of the mistranslation in 1969 it was obvious that major doctrines of the Bible were being taught incorrectly. There is a large difference between the word age and the word forever. Age normally indicates limits (most ages have beginnings and/or ends), but forever and eternal always indicate no limits and no end. In this paper we will go into much more detail on the mistranslated words *olam* and *anionios*, which are incorrectly translated into 'forever,' 'everlasting,' and 'eternal.' This is very important.

Contradictions

nm95 » First let's look at some paradoxical translations caused by the incorrect translation of *olam* and *aionios*. If we translate these words correctly there are no paradoxes.

Sacrifices Forever?

nm96 » Some sacrifices, offerings, and rituals of the Old Testament were *olam* or *aionios* sacrifices. (Note: Lev 3:17; 6:18; 7:36; 10:9, 15; 16:29; 17:17; 23:14; 24:3; Num 10:8; 15:15; 18:8; 19:10; etc.) If *olam* or *aionios* mean forever or eternal why are these sacrifices, offerings, and rituals not now being performed? They are not still being performed because they were for an age, not forever. Christ abolished them by his perfect sacrifice (Heb 10:10-14).

Circumcision Forever?

nm97 » If *olam* or *aionios* means forever or eternal why are not Christians following this *olam* or *aionios* covenant of circumcision:

- "My covenant shall be in your flesh for an *olam* [or *aionios*] covenant" (Gen 17:13).

If *olam* or *aionios* mean forever or eternal, how can anyone unbound such a regulation? Of course *olam* and *aionios* only mean age, thus the reason Christians are no longer bound by physical circumcision (note Acts 15:5-29; 1Cor 7:18-19; Gal 5:1-4, 6; Col 3:11).

Slaves Forever?

nm98 » If *olam* or *aionios* mean forever or eternal, then according to the law of the Old testament, some can be made slaves forever or for eternity (Lev 25:46; Deut 15:17). Of course there are no forever slaves: *olam* and *aionios* do not mean forever, they speak of an age.

Before Eternity?

nm99 » If *aionios* means forever or eternal how could there have been any time *before* eternity, "before the times of eternities [*aionion*, plural of *aionios*]" (Greek text, 2Tim 1:9)?

More Than One Eternity?

nm100 » Is there more than one eternity? *If* the Greek *aionios* or the Hebrew *olam* mean eternal, then according to 2Tim 1:9 there was a time before *eternitieS*. How can there be more than one eternity. There are at least two other places in the Greek New Testament that has *aionios* in its plural form:

- "But the things invisible are *aionia*" (2Cor 4:16). [This Greek *aionia* is the plural form of aionios (*Analytical Greek Lex.*, Zondervan)..]

- "Which God, who cannot lie, promised before times of *aionion*" (Titus 1:2). [According to the Lexicon *aionion* here is in its plural form.]

nm101 » The Hebrew word *olam* is also translated forever or everlasting. *Olam* is also found in its plural form in such verses as in Isaiah 26:4; 45:17; Psa 77:6; 145:13. Is there more than one eternity? Of course not.

nm102 » The Greek *aionios* and the Hebrew *olam* are speaking about an age or ages of secret or hidden or unknown time lengths. We can only ascertain the time periods of each *olam* and *aionios* by other words in context that explain to us what age the Bible is speaking about. The words *olam* or *aionios* in and of themselves tell us nothing about the duration, or the beginning or the ending of the age. We only know that the age of Satan will end because of scripture. We only know the new and coming age of the True God will *not* end because of scripture. And we can know through scripture that there are ages within the great age of God just as there are ages within Satan's age.

Never Die?

nm103 » In John 8:51 and verse 52 we see contradictions: "If a man keep my saying, he shall never see death. Then said the Jews unto him [unto Jesus], Now we know that you are possessed of a demon. Abraham is dead, and the prophets; and you say, If a man keep my saying, he shall never taste of death." And in John 11:26 we see a similar contradiction: "And whosoever lives and believes in me shall never die."

nm104 » In this translation Christ seems to say that if one kept his words and believed in him that such a person would *never* die. But by reading the New Testament we know that those who do keep his word and that do believe in him do die. A contradiction? No! It is a mistranslation.

nm105 » The English word "never" in these verses was mistranslated. It should have been translated as follows: "no not death should he behold into the age" – John 8:51; "no not should he taste of death into the age" – John 8:52; "no not [anyone believing] should die into the age" – John 11:26. Double negatives in Greek adds emphasis to the negative. "No not" can be translated "absolutely not." Thus "absolutely not should anyone who believes in Christ die into the age."

nm106 » These scriptures speak about an *age* and that into that age or in that age those who believe in Christ (those who keep his word) will not, absolutely not, die. This great age begins with the 1000 year age as other scripture indicates (see *Reward for Christians* [NM 11]). When we translate *aion* literally these scriptures make sense. But when we translate it as some *think* it ought to be translated we come up with contradictions. In the above three scriptures "never" was mistranslated for the Greek words that literally meant "no not" and "age."

David's Throne Forever?

nm107 » If *olam* or *aionios* mean forever or eternal, then David and Solomon's thrones should have lasted forever (1Kings 9:5; 2Sam 7:12-13, 16). Of course this kingdom of David and Solomon lasted only for an age. It is the Spiritual Seed of David that will establish the kingdom in the endless age or endless *olam* or *aionios*.

Cities and Land Destroyed Forever?

nm108 » If *olam* or *aionios* mean forever or eternal, then there are cities and lands that forever or for eternity will be in ruins (Isa 25:2; 32:14; Ezek 26:21; 27:36; 28:19; Jer 18:15; 25:9, 12; 49:13, 33; 51:26, 62; Ezek 35:9; etc.). But this cannot be true because when Jesus Christ returns to the earth, the earth will be renewed and eventually created totally new (Rev 21:5; Psa 104:30; Isa 61:4; Ezek 36:10-11; Amos 9:14; etc.). Of course, since *olam* or *aionios* only indicate an age, these cities and lands will not be forever ruined, but will be renewed and the people of these former cities will be resurrected.

Present Earth Forever?

nm109 » In such places as Eccl 1:4 and Psa 104:5 it speaks about the earth standing for *olam* or *aionios*. But Christ said that heaven and earth would pass away (Mat 24:35; Mark 13:31; Luke 21:33). The present earth stands not forever, but for an *olam* or *aionios*, that is, it stands for an age. But after that age it will be totally created new (Rev 21:5).

Land Forever?

nm110 » If *olam* or *aionios* means forever or eternal, then Israel would have continually and forever possessed the land (Note Gen 13:15; see Greek trans.). But if *olam* or *aionios* means agelong, the Genesis 13:15 promise means that Israel would possess the land during an age. This is what happened: physical Israel did possess the land for an age – not forever. The true higher meaning of this scripture is that the Spiritual Israel will possess the land during the age (*olam*) of the True God.

Aeonian or Agelasting Meaning in Harmony With Scripture

nm111 » When the Greek *aion*, its adjective *aionios*, and the Hebrew *olam* are translated anything but age or aeonian or agelasting, contradictions occur in scripture. But when these words are correctly translated a clear meaning is projected to the reader. In all our papers on doctrine we always use the correct translation of these words because it is the best way to translate these words. And because throughout our papers and in the BeComingOne Bible we have translated "aeonian" as it should be translated, we project to you the mercy of God and the great plan of the God. The True God is not a damning forever God. Our God is a God of love and forgiveness. He punishes, but He also will eventually save all. See our paper "All Saved" [NM 13] for there are many scriptures in the Bible where it says that *all* will eventually be saved. There is/was a purpose for evil. There is hope for all.

Hebrew Meaning of Olam

nm112 » In Hebrew, the language of the Old Testament, the Hebrew word most often translated "forever" or "everlasting" is *olam*.

[or *alam*, Strong's # 5956 & # 5957; *'owlan*, #5769; *'eylowm*, # 5865; see also *'ad*, # 5703. Note that the word is spelled in Hebrew differently at different times because of prefixes and suffixes attached to it.]

From the *Hebrew and Chaldee Lexicon* by Gesenius, it shows that the Hebrew word *olam* (Strong's # 5956) has the meaning of a hidden age or hidden time specifically "hidden time, long."

nm113 » From the *Analytical Hebrew and Chaldee Lexicon* by Benjamin Davidson (Pub., by Zondervan, l970), it shows that the Hebrew word *olam* means a hidden time or secret time or age.

nm114 » This word was first used in the Bible to describe the hidden or secret age that Adam and Eve missed because of their sin: "and live into *olam*" (Gen 3:22). Its basic meaning concerns a hidden or secret age, or simply an age of unknown length. At the time Genesis 3:22 was spoken, Adam and Eve were only alive a short while. At that time Adam and Eve did not and could not understand time. Time is something one learns to understand through living in time. See "Reason Why" paper [NM 19] to understand how one learns.

Greek Meaning of Aion

nm115 » In Greek, the language of the New Testament, the Greek words most often translated "forever" or "everlasting" or "eternal" are *aion* or *aionios*. (Note: both of these words are spelled somewhat differently in the New Testament text depending on the usage in the sentence.)

nm116 » From Thayer's *Greek-English Lexicon of the New Testament*, the Greek word *aion* is said to mean age or a human lifetime.

nm117 » From *The Analytical Greek Lexicon* (Pub. Zondervan), the Greek word *aion* is said to mean "a period of time of significant character; life; an era; an age."

nm118 » From William F. Arndt and F. Wilbur Gingrich's *A Greek-English Lexicon of the New Testament*, the meaning of *aion* is "time, age."

nm119 » From the Lexicons in Young's and Strong's concordances, the Greek word *aion* also is indicated as meaning an age or time.

nm120 » From Wuest's *Word Studies*, Volume 8, *Studies in the Vocabulary*, under "world," we see concerning *aion* the following:

> "*Aion* which comes from *aio* [it is debatable whether *aion* came from *aio*] 'to breathe,' means 'a space or period of time,' especially 'a lifetime, life.' It is used of one's time of life, age, the age of man, an age, a generation." And in the same place, "as to *aion*, the papyri speak of a person led off to death, the literal Greek being 'led off from *aion* life.' A report of a public meeting speaks of a cry that was uttered by the crowd, namely, 'The Emperors forever' (*aion*). It is also found in the sense of 'a period of life.'"

nm121 » In the Greek translation of the Old Testament, *aion* was used for *olam* in such verses as Psa 90:2: "from everlasting to everlasting ... "; Hebrew has it: 'from olam to olam; and Greek has it: "from the aion until the aion;" the literal text of the *Emphasized Bible* has it: "from age unto age";

Meaning of aionios

nm122 » The Greek word *aionios* is merely an adjective that comes from the root *aion*. While *aion* means "age," *aionios*, being an adjective, means "agelasting," or "aeonian," or "agelong." When the *Septuagint* was translated, the translators used in many cases the Greek word *aionios* for the Hebrew word *olam*.

nm123 » For example the Greek *aionios* was used for the Hebrew *olam* in Genesis 13:15: "For all the land which you see I will give to you, and to your seed during *olam*."

nm124 » Or again in Genesis 3:22: "And now, lest he put forth his hand and also take from the tree of life and eat and live into *olam* [Greek *aiona*]."

nm125 » And again, "Every man child among you shall be circumcised ... my covenant shall be in your flesh for an *olam* [or *aionion*] covenant" (Gen 17:10, 13).

nm126 » Thus, from the above scripture and many others, we see the Greek *aionios* must be a synonym for the Hebrew word *olam*. To ascertain the meaning of *aionios* we can look to the meaning of *olam*. As shown above *olam* means basically a hidden or secret age or time. Thus *aionios* also must mean a hidden age or time or an unknown age or time. And so it is, the basic meaning of *aionios* is an indeterminate time or age. The word *aionios* can be translated as "agelasting" or "aeonian" or "agelong." It speaks of an age which has an unknown length and which begins and ends at an unknown time.

nm127 » From William Barclay's *New Testament Words* we note the following concerning the Greek word *aionios*. Barclay says *aionios* is an adjective formed from the noun *aion*. "In classical Greek this word *aion* has three main meanings. It means a *life-time* . . .. Then it comes to mean an *age*, a *generation*, or an *epoch* . . . But then the word comes to mean a *very long space of time*." Then Barclay's goes on and tries to say that the "strange" word *aionios* somehow means *eternal* and gives some examples from Plato to back up his contention.

Magic Word

nm128 » One question will do here. How can an adjective that came from a word that means age, come to mean eternal? It would be comparable to the word "some" (i.e. some of time) coming to mean "all" (i.e. all of time). You cannot correctly use the adjective of *age*, which is agelasting or aeonian, as if it meant forever. The whole idea that *aionios* means eternal or everlasting is ludicrous. It is a lie. It is magical in an evil way. That lie has twisted scripture, and has made true doctrine in the Bible almost impossible to see. It has put a blindfold over peoples' eyes.

Aionios in Context

nm129 » Barclay gives a few examples of Plato using *aionios* as if it meant eternal.

> "The most significant of all the Platonic passages is in the *Timaeus* 37d. There he speaks about the Creator and the universe which he has created, 'the created glory of the eternal [aionios] gods.'"

To Barclay, for some reason, because Plato used *aionios* in connection with "gods," it is some kind of proof that *aionios* means eternal. According to this reasoning since God ("gods") is eternal, then *aionios* must mean eternal. But this overlooks the fact that, yes, God in someway was/is perpetual (His power, Rom 1:20), but also God may in some sense also be or relate to an aeonian (aionios) time. Just because the word *aionios* is used in connection with "gods" does not give it the meaning of eternal. Furthermore, what do Plato's writings have to do with our God and the definition of His "eternalness"? Plato was speaking in the above example about *gods*, not God. Plato's writings were not inspired by God: they are full of myth and faulty thinking.

nm130 » Others like Barclay try to make the adjective *aionios* ("agelasting"), which comes from the noun *aion* ("age"), mean "forever," "everlasting," and "eternal." They say *aionios* means "eternal" because in context of its usage it is used as if it literally means "everlasting." They call the Greek word *aionios*, "strange." And to me it is strange that a word that is derived from a word that means "age" should somehow mean "everlasting."

nm131 » We will give you hereafter two of their "best" arguments in favor of the idea that *aionios* means "everlasting," and then we will refute their wrong reasoning. These same arguments were used by Augustine in the fourth-fifth century AD (see below).

Context Argument One

Aionios God; Olam God

nm132 » This argument deals with the usage of *aionios* in connection with God. We will only examine this argument of context by referring to the relevant Biblical text usage. To try and say *aionios* means eternal because Plato or Aristotle seems to use it that way is off the mark. We are only interested in how the Bible uses the word *aionios*, not how some Greek philosopher seems to use it.

nm133 » Romans 16:26: "the *aionios* God." Romans 16:26 speaks of the aeonian God or the God of the aeonian time. Notice that Genesis 21:33 speaks of the God of *olam* and Isaiah 40:28 speaks of the God of *olam*. These verses were translated by the Greek *aionios* in the Greek text. Somehow this usage of *aionios* (or *olam*) is proof positive to many that *aionios* means "everlasting." But the Greek word *aionios* is simply an adjective that comes from the noun *aion*, which means age. The literal meaning of *aionios* is "aeonian" or "agelong." The book of Romans is speaking about one aspect of God. Somehow God is "aeonian." Of course since God is Spirit (John 4:24), and since spirits or angels do not die (Luke 20:36), then God will not die. God is immortal (1Tim 1:17). God's power and Godship was/is/will-be continuous (Rom 1:20, Greek *aidios*). But Romans 16:26 tells us that in some way God is aeonian.

nm134 » In one aspect God *is* aeonian. The true God is aeonian in the respect that He rules through Jesus Christ as King of kings beginning in the age of 1000 years (Rev 19:16; 20:4). In this present age in which I write, the god of the world is a false one, the god of this age is the one called Satan by the Bible (2Cor 4:4). Satan is an agelasting god ruling in the old age. Satan is the power of death. The true God (His Good Spirit) does not rule this old age and that is the reason this age is so twisted. The true God through Jesus Christ rules beginning in the aeonian time, the 1000 year age. This is an aeonian aspect of God. Of course, since God's agelong (*olam* or *aionios*) kingdom will not end (Dan 2:44; 7:14; Luke 1:33), then this new aeonian kingdom (see Greek text, 2Peter 1:11) and rule will never end. It is an endless age, but it does have a beginning at the coming of Jesus Christ (Rev 11:15). Therefore it is an aeonian rulership. The word *aionios* (or *olam*) by itself means agelong, but the new and coming age belonging to God will not end like most ages because other clarifying words tell us that this special age, unlike ages before it, will not end (Luke 1:33; Dan 2:44; 7:14; Isa 9:6-7). It also can be said that the new age began in one sense at Christ's first presence or coming.

Greek Words that Mean Forever

nm135 » If Paul wanted in Romans 16:26 to describe God as the everlasting God or the endless God, Paul had many other Greek words to use to say this. Paul could have used *akatalutos*, which means "indissoluble." Paul could have used *atelestos*, which means "endlessness." Or Paul could have used *aperantos*, which also means "endless," as he used this word in 1Tim 1:4. But Paul did not use these words or other Greek words or phrases because he did not want to use them, for he was simply mentioning in Romans 16:26 that some aspect of God is "aeonian." God is a lot of things, and one of these is that he is in a certain way

"aeonian." God belongs to the new never ending age – God's age or God's *olam* or *aion*. God is God of *olam* or God of *aionios*, He is the God of the new, great, never ending age. We know it is a new great and never ending age, not by the word *olam or aionios*, but by **other** words and sentences because *olam* and *aionios* speak only of a vague or undefined age, not a forever age.

nm136 » In some translations of the Old Testament, it has "eternal God" in Deuteronomy 33:27, but this should be translated "ancient God" or "God of old." This mistranslated word is strong's # 6924, *qedem*. In some translations of Genesis 21:33 and Isaiah 40:28 it has "everlasting God," but should be translated "*olam* God" or "God of *olam*." This means that God belongs or pertains to *olam* or the hidden age of the future first mentioned in Genesis 3:22.

Context Argument Two

Aionios Life and Punishment

nm137 » The second argument of context which Augustine cleverly articulated back in the early fifth century AD (413-26), in his *City of God* (trans. W.M. Green, The Loeb Classical Library, 1972), deals with two items: one is punishment, and the other is life. In Augustine's time some were teaching that the *aionios* punishment would end. Augustine in book 21 of his *City of God* was in part arguing against this position.

nm138 » Augustine translates the word *aionios* from scripture into the Latin, *aeternus* and *aeternitas*. These Latin words are related to the Latin *aevum*, which in turn is related to the Greek *aion* (cf. *Oxford Latin Dict.*, 1968, p. 74, col. 2, under *aeternus*, "[aevvm + -ternvs]").. Augustine knows that these words can mean longlasting, with the possibility of an end. Thus Augustine must emphasize in his writings that he is not speaking of the Latin, *aeternus*, in the sense of a long period, but in the sense of eternal, a period without end. Notice Augustine's own words, translated from Latin:

> The term "eternal," as applied here, does not refer to a long period of time (*aetas*) lasting through many ages, but still at some time bound to end.[1] Rather, as it stands written in the gospel, "of his kingdom there shall be no end.[Luke 1:33]" (*City of God*, book 22, part 1; page 173)

[1] The words "eternal" and "eternity," from Latin *aeternus, aeternitas*, are related to

aevum, which means both "unending time" and "a period of time"; for the second meaning the commoner word is *aetas*. Augustine seeks to make it clear that the "eternal" happiness of the saints is unending happiness, that is, an unending immortality for each individual (text of editorial footnote 1 on p. 173).

And from the *Oxford Latin Dictionary*, 1968, we see that:

aetas means

> 1. The number of years one has lived, one's age. 2. Period or time of life. 3. A person or person of a particular age or period of life, an age group. 4. a. youth. b. old or advancing age; greater age. 5. The whole period of a man's life, the mortal span, one's lifetime, an age. 6. Human life and all that goes with it. 7. The passage or lapse of time. 8. The time or period to which a person or thing belongs, an era, age; the duration of this as a unit of time, a generation.

aevum means

> 1. time of life. 2 a generation. 3. age.

nm139 » Thus, even in his time, Augustine (354-430 AD) knew that *aionios* meant agelong or longlasting, thus he reasoned by context:

> Then what sort of reasoning is it, to take the eternal [*aeternum*] punishment of the wicked as a fire of long duration and believe that eternal [*aeternam*] life is without end? For Christ said in the very same place, including both in one and the same sentence: "So these will go into eternal [Lat., *aeternum*; Greek *aionios*] punishment, but the righteous into eternal [Lat., *aeternam*; Greek, *aionios*] life." [Mat 25:46] If both are eternal, then surely both must be understood as "long," but having an end, or else as "everlasting," without an end. For they are matched with each other: in one clause eternal punishment, in the other eternal life. But to say in one and the same sentence: "Eternal life shall be without end, eternal punishment will have an end," is utterly absurd. Hence, since the eternal life of the saints will be without end, eternal punishment also will surely have no end, for these whose lot it is. [book 21, part 23; p. 113]

nm140 » Augustine finds a place in the Bible where the Greek *aionios* is used to speak both of the punishment of sinners and the reward of the saints (Mat 25:46). But because all Christians think and know that their 'reward' is immortality, an everlasting life, and since in one sentence *aionios* describes the life for the righteous, and the punishment for evil, then according to this argument, *aionios* must mean forever, at least in context. This may seem logical, but in context of other scriptures it is not logical.

All Made Alive

nm141 » Notice one sentence where it shows that since all die, all will be saved:

From Paul's resurrection chapter we read:

> For as in Adam ALL die, even so in Christ shall ALL be made alive (1Cor 15:22; see Rom 5:14-18; Psa 82:7-8).

In this one sentence we have "all" repeated twice. Most believe the first "all," that is, because of Adam's sin *all* die. Most do *not* believe the second "all," that is, because of Christ *all* will be made alive.

nm142 » What is meant in 1Corinthians 15:22 by "be made alive"? Does it mean be made alive (resurrected) and thereafter be killed in some kind of hell-fire? In context what does it mean to "be made alive"?

> ■ "For since by a man came death, by a man also came the resurrection of the dead. For as in Adam all die, so also in Christ *all* shall be made alive. But each in his own order" (1Cor 15:21-23, NASB).

nm143 » Here it speaks of the resurrection of the dead. All shall be made *alive*, but not at the same time. There is an *order* to the "all shall be made alive." In context this resurrection has something to do with "the resurrection of the dead."

nm144 » As we clearly show in our papers, 'All Saved' and 'Does All Mean All?,' there are three orders or ranks of resurrection. The first was Jesus Christ. The second will be the real Christians at Christ's second coming. The third and final resurrection will be at the *end* of creation when the universal resurrection of all others occur. This will not be a resurrection to death, but one to life. Please read these two papers so as to begin to see the truth about the universal resurrection to life. Also read the "Reason Why" [GP 7] paper in the *God Papers* to understand the need for the present age of confusion. There is hope for all. All will be given the good Spirit and the good mind.

But they forget...

nm145 » According to those who say *aionios* means "everlasting," those who do evil in this age will go away to serve an everlasting punishment because they did not bring God into their lives. And also according to these same people, those who do good in this age and/or those who accept God in this age will be given everlasting life. But they forget that it is God who gives the power for people to be good, to repent ("change one's mind"), to receive God's Spirit, and so forth (see the rest of this book, *New Mind and Christianity*, for documentation). Thus, simply, those who will be damned for everlasting punishment, as some assert, will be damned merely because God did not give them the power to save themselves. What these people are saying is that God is discriminating wrongly against some: he saves some forever through his grace; he damns others forever by not giving them his grace.

Paradox: All Eventually Saved; Evil Damned Forever

nm146 » The answer is not as Augustine argues, that all does not mean all (Book 21, part 24). Augustine needlessly throws in more confusion by alleging that 'all' does not mean 'all.' But the real answer negates the confusion. The answer to the paradox is that *aionios*, an adjective of *aion* (which is a word that means *age*), means aeonian, and that 'all' means 'all.' The *aionios* punishment the Bible speaks about is *age*lasting. There is an agelasting or aeonian punishment for many, not an everlasting punishment. Notice the following scriptures concerning the *aionios* (aeonian) punishment or judgment.

Aeonian or Agelasting Punishment.

nm147 » There is an agelasting fire: "And if your hand or your foot offend you, cut them off, and cast them from you: it is better for you to enter into life lame and crippled, rather than having two hands or two feet to be cast into the *aionios* fire" (Mat 18:8).

nm148 » At the physical return of the Messiah some people will burn in the fire (caused by the Last War) which is to last for an age. This fire was meant for the other-mind, our spiritual adversary, and his spiritual friends: "Then shall he say also unto them on the left hand, Depart from me, you cursed, into *aionios* fire, prepared for the devil and his angels" (Mat 25:41).

nm149 » This *aionios* fire is an *aionios* punishment: "And these shall go away into *aionios* punishment" (Mat 25:46). As shown in the "Thousand Years and Beyond Paper" [NM 15] this fire is mainly for the twisted spirits of Satan because people can't live in fire. People die in fire. But with this fire many will die. Their punishment is death from the fire. But the twisted spirits will be punished in the fire because they can't die in it.

nm150 » This *aionios* fire, this *aionios* punishment is an agelasting destruction away from the glory of the Lord in his 1000 year rule: "The Lord Jesus shall be revealed from heaven with his mighty angels, In flaming fire taking vengeance on them that know not God, and that obey not the good-news of our Lord Jesus Christ: Who [those not knowing God] shall be punished with *aionios* destruction from the presence of the Lord, and from the glory of his power" (2Thes 1:7-9).

Aeonian or Agelasting Life versus Immortal Life

nm151 » Now the Bible speaks of an *aionios* life, thus an *age*lasting or aeonian life. As shown in the "Reward for Christians" paper [NM 11] this age-life begins at the beginning of the 1000 year rule of Christ the God. Of course, since this coming new age will never end, the *aionios* life continues after the 1000 years. There are at least 44 scriptures mentioning the *aionios* life and other scripture mentioning the age-life for those who are in God (see the Englishman's Greek Concordance under *aionios*; See "Reward for Christians Paper" [NM 11] and other papers for more detail on this *age*lasting "life."). When one is resurrected in the resurrection at Christ's coming, he/she will receive immortal life *and* he/she will live during an *aionios* period of 1000 years, *and* will also live after that 1000 year age into the next age or endless period of time since he/she will be immortal. It is possible for someone to live in the aeonian life and be mortal for those physically born during the 1000 years will be mortal. There is a difference between immortal life and aeonian life as explained in the *Reward for Christians* paper [NM 11].

Two Ages: One Ends; One Does Not End

Old Age

nm152 » There are two main ages. There is Satan's age of confusion with its spirit of confusion. There is the True God's age of harmony with its Spirit of harmony. Matthew 12:32 speaks about the present age (*aion*) and about the coming age. The KJV translates *aion* as "world." In Matthew 13:22 it speaks about, "the care of this *aion* [KJV, "world"], and the deceitfulness of riches." In Matthew 13:39-40 it speaks about "the end of the world," that is the end of the age (*aion*). There is a certain "wisdom" of this age or aion (1Cor 2:6). There are certain children of this age (Luke 16:8; 20:34 - KJV "world"). These children are of the devil (Mat 13:38-40). There is a god of this age or *aion* (2Cor 4:4; KJV, "world"). People fight "against the rulers of darkness of this age [*aion*]" (Eph 6:12).

New Age

nm153 » Scripture indicates that the end of this wicked age comes at the beginning of the new age – at Jesus Christ's coming. (Mat 13:38-40; 24:3-31; 2Thes 2:1, 8; Rev 11:15; 12:10-11; 20:1-5; Dan 2:44; 7:17-18, 25-27; see "Beast Paper" [PR2, PR3], "God's Wrath Paper" [PR4], etc.)

nm154 » There is a coming age at the end of the old wicked age. Such scriptures as Mark 10:30; Luke 18:30; 20:35; Eph 1:21 in the Greek text indicate this. This New Age (*aion*) or agelasting (*aionios*) period under God's Spirit will not end as Luke 1:33, Dan 2:44, and Dan 7:14 indicate. But during this endless age their will be the 1000 year age in which some will be punished. After this age there is another short age called the Great Last Day (see "Thousand Years and Beyond Paper" [NM 15] and others).

Why Misusage of these Words?

nm155 » There could easily be a book written on the story behind the misusage of *olam* and *aionios*. One reason was some of the early fathers of the Catholic Church such as Augustine relied too heavily on the Greek literature especially Plato's to obtain doctrine instead of relying on Biblical scripture. Augustine used the faulty Greek text of the Old Testament instead of the inspired Hebrew text (*City of God*, book 18, chapter 43). Some of the very arguments used by Augustine to "prove" that *aionios* means eternal (*City of God*, book 21, chapters 23, 10-22, 9; etc.) are used

today by theologians and preachers to "prove" that *aionios* means eternal (Berkhof's *Systematic Theology*, "The Duration of their Punishment," p. 736; etc.).

nm156 » We should no longer take our doctrine on the God and His ages from Plato or the other Greeks. The truth is found in the Bible, not Greek literature. Do read all our papers on Christianity to better understand the age plan of God.

Review And Further Arguments for the Use of *Age*lasting or Aeonian

nm157 » The Issue:

- The misuse of the words "forever," "everlasting," and "eternal" instead of the correct usage of "age" or "aeonian."

nm158 » The Question:

- Why do Christians teach of an "everlasting" damnation (death or punishment) when the very word "everlasting" from which these Christians obtain their doctrine is a mistranslation according to reliable Biblical aids?

Explain to me why Christians use the traditional mistranslation of the Hebrew word *olam* and the Greek word *aionios* instead of the inspired meaning of these words?

Significance:

nm159 »

- The great importance of using the correct translation is that it opens up highly important truths of the Bible. If one takes the literal meaning of these words every place they appear in the Bible, an "age" plan of God is projected.

Further Response to the Arguments of Context

nm160 »

1. *The use of the context argument in the English language.*

   One way to disprove the context argument is to try and use the English word "age" to mean "everlasting" or "eternal." This can *not* be reasonably done.

2. *The quantity of misusage of the Hebrew and Greek words for age.*

If the context argument is correct then why should the great majority of the original words, that mean "age" in Hebrew and Greek be translated to mean "forever"? It would be more reasonable if most of the original words were translated with their literal meaning instead of their evolved "context" meaning.

3. *Why didn't the writers of the Bible use other words or phrases in Greek that meant everlasting or eternal?*

If the context argument is correct, in that the Greek word that means "age" should be read "forever" because of the context, then *why* didn't the original writers use other Greek words or phrases that literally meant "forever" when they were writing? No, they used words that meant literally *age*lasting because they were speaking of an *age*lasting time not an everlasting time. The writers of the Bible were referring to the time of *olam* that Adam and Eve missed by their sin (Gen 3:22).

4. *The dubiousness of writers using two different languages using words that literally mean "age" which they intended to be understood by their readers to mean "everlasting."*

The fact that the two main original languages of the Bible have words that mean "agelasting," which are said to mean "everlasting" because of their usage in context, helps to rule out the argument of context. Maybe, just maybe, *one* language could use a word that literally means "agelasting" in context so as to mean "everlasting." But both of the languages used words that literally meant "*age*lasting" to describe the agelasting reward and agelasting punishment of Christians and non-Christians because they were speaking of an *age*-time reward and punishment not an everlasting reward and punishment.

5. *The dubiousness of most of the 40 or so writers of the Bible using words that literally meant "age" which they meant to be understood by their readers to mean "everlasting."*

Maybe some of the writers would use words that mean "agelasting" in context to mean "everlasting," but surely not most of them.

6. *On the passages such as Romans 16:26 where if taken literally God would be an agelasting God in some aspect.*

Now there are many other places where it indicates that God is in someway an eternal God. God is in a sense eternal – His Power. But in another sense he is also an *age*lasting God. He is an agelasting God because His age begins in the coming 1000 years of the Kingdom of God. This is the age when God is King of kings, this is the age wherein God will rule all. In the present age God is not the God of the world, for Satan is now that god (2Cor. 4:4). The new age of the true God will not end (Luke 1:33; Dan 2:44; etc.). The new age is different from the age of Satan, for Satan's age will end when God's age begins.

7. *On the reward for Christians – agelasting life.*

Christians live the agelasting life as servant-rulers under Christ, but since they also are made immortal at the beginning of the 1000 year age (1Cor 15:52-54), then they live on forever after that 1000 year age (see "Reward for Christians" paper [NM 11]). This 1000 year age is an age within an age or within the great age or great ages. The great age is the *olam* or *aion* of God, and this new age of ages will not end (Luke 1:33; Dan 2:44; 7:14; Isa 9:6-7).

———————————————

Everliving? or Ageliving

Rosetta Stone: Egyptian Holy or Hieroglyphic Script.

nm161 » The error of turning words that mean age into everlasting is not confined to the Bible. Of some interest is the translation of the Egyptian hieroglyphic signs into "everliving." On the Rosetta Stone certain hieroglyphs were translated into, "everliving." On the Rosetta Stone there are two languages: Egyptian and Greek. The Egyptian language is cut into the stone in two different kinds of characters: (1) Hieroglyphic characters were used for state and ceremonial documents that were intended to be seen by the public; and (2) Demotic characters, were "the conventional, abbreviated and modified form of the Hieratic character, or cursive form of hieroglyphic writing, which was in use in the Ptolemaic Period" (E.A. Wallis Budge, *The Rosetta Stone*, Ares Publishers; Chicago:1980, reprint of 1922 work, p. 2). The Greek portion of the inscription was cut into the stone in ordinary uncials. "The inscription on the Rosetta Stone is a copy of the Decree passed by the General Council of Egyptian priests assembled at Memphis to celebrate the first commemoration of the coronation of Ptolemy V. Epiphanes, king of all Egypt" (p. 7). This coronation of Ptolemy to king of Egypt took place about 196 BC. "The original form of the Decree is given by the Greek section, and the Hieroglyphic and Demotic versions were made from it" (p. 7).

Greek/English Translation of "Everliving"

nm162 » In Budge's *The Rosetta Stone*, he has the translation of the Greek into English. On lines 4, 8, 37, 49, and 54 of the translation he has the Greek *aionobioy* translated into "everliving." But this Greek word is made up of two parts. The Greek *aion* is the word for "age" or "era." The Greek *bioy* is the genitive singular for *bios*, which means "life" or "living." Thus this Greek word means "age-living" or "era-living" or in a sense "long-living." This word does not mean everliving.

Copic Translation of "Eternity"

nm163 » Quoting Budge from page 6,

- it was therefore guessed that the next sign [the next part of the hieroglyphic signs translated into "everliving"] meant

"ever." Coptic again showed that one of the old Egyptian words for "ever, age, eternity," was Djet, and as we already know that the phonetic value of the second sign in the word is T, we may assume that the value of [sign] is DJ.

Budge attempts to show in another way through a Coptic word that a certain hieroglyphic sign means "everliving." But notice this Coptic word means, "ever, age, eternity." Here it is again, the mixing of the word age with eternity. But as we see using the Greek translation of the Rosetta Stone, this sign reads "age-living" or "long-living" or "era-living." Remember the original was written in Greek and the Egyptian translation was taken from the Greek.

Egyptian's "Everlastingness," "Eternity," or "Millions of Years"

nm164 » In Budge's, *The Gods of the Egyptians*, this **same** hieroglyphic sign mentioned on page 6 of Budge's, *The Rosetta Stone*, is translated as either "ever" or "everlastingness" (Vol 1, pp 54-55, line 521). But also note that the hieroglyphic sign, *heh*, is translated "eternity" in line 520 of this same book. Yet,

- "according to Dr. Brugsch the name Heh is connected with the word which indicates an undefined and unlimited number, *i.e.*, *heh*, when applied to time the idea suggested is "millions of years," and Heh is equivalent to the Greek αιων [*aion*]" (Budge, *The Gods of the Egyptians*, Vol 1, p. 285).

Thus, the sign, *heh*, is wrongly translated as "eternity" and sometimes translated "millions of years," but is equivalent to the Greek *aion*, which we have seen in this paper means an age of undefined time.

To Conclude

nm165 » Not only was the Bible mistranslated, but what we are manifesting here is that one should be very careful when reading translations in general. There is a bias against translating words or signs that mean "age" correctly. They prefer the hyperbolic mistranslations of "forever," "everlasting," or "eternity." Be careful.

The Saying, "To the Age"

nm166 » A saying that goes back to ancient time is, "long live the king." Or in the Bible it has it:

- "let live my lord king David to *olam*" (1Kings 1:31)

- "the king to *olam* live" (Neh 2:3)

- "king to *alam* live" (Dan 2:4; 3:9; 5:10; 6:6, 21)

nm167 » As we have learned in the "Age Paper" [NM 7] *olam* or *alam* means a hidden age or time. It in fact is that great age of the kingdom of God promised since the garden of Eden after mankind lost the right to "live to *olam*" (Gen 3:22, see Hebrew). The Hebrew preposition *el*, which means to or towards or into, is connected in the above verses with the Hebrew *olam* or *alam*, thus our translation of "to *olam*" or "to (the) age." Throughout my life I have heard the saying "til kingdom come" or "to kingdom come." These sayings are versions of the Bible's 'to *olam*,' that is, to (the) age of the kingdom.

Examples in the NT

KJV Revelation 20:10 And the devil that deceived them was cast into the lake of fire and brimstone, where the beast and the false prophet *are*, and shall be tormented day and night for **ever and ever**.

YLT Revelation 20:10 and the Devil, who is leading them astray, was cast into the lake of fire and brimstone, where {are} the beast and the false prophet, and they shall be tormented day and night--to the **ages of the ages**.

GNT Revelation 20:10 καὶ ὁ διάβολος ὁ πλανῶν αὐτοὺς ἐβλήθη εἰς τὴν λίμνην τοῦ πυρὸς καὶ θείου ὅπου καὶ τὸ θηρίον καὶ ὁ ψευδοπροφήτης, καὶ βασανισθήσονται ἡμέρας καὶ νυκτὸς εἰς τοὺς αἰῶνας τῶν αἰώνων.

NM 8: Predestinated: Called and Chosen

Many are Called, Few are Chosen
Christians are Predestinated
God Chooses
Predestinated to Destruction?
Two Groups
Losing God's Spirit?
Chosen are Called
Are you Predestinated?

NM8 Abstract

Why are many called, but few chosen? Why does the Bible say that Christians are predestinated if there is no predestination? And if there is predestination, how can God damn anyone forever who was predestinated to do evil? Predestination is a difficult subject for anyone who believes in the false translation we corrected in NM7. When you understand the mistranslation of the Bible corrected in NM7, you will understand how there can be a logical doctrine of predestination.

nm168 » The doctrine of Predestination has been twisted and turned by those who do not understand the great power of God. What is the Biblical definition of "predestination," being "called," and being "chosen"? There is a difference between being just called or invited to the kingdom of God than being called, chosen, and predestinated. Let's define these three words found in the King James Version. These words were translated from Greek words which were inspired to be written through the power of God (2Pet 1:20-21).

■ *Predestination* (#4309, from 4253 & 3724) means, to be marked off beforehand.

■ *Chosen* (#1588 & 1586; O.T. #977) means, to be laid out or chosen or select.

■ *Called* (#2821, 2822, 2564) means, to be invited.

Many are Called but Few are Chosen

nm169 » There is a difference between being just called and being chosen, "for many are called but few are chosen" (Mat 22:14; 20:16). Many are called or invited to God's kingdom, but few are chosen for that kingdom. Many will be invited to be in the new reality of God, but few will be in it at its beginning. In fact near the end of the age of Satan's kingdom (the old age) all will be invited to the kingdom of God (see Mat 24:14; Mark 16:15; Psa 19:4; Col 1:23), but few are chosen. We notice that at Christ's coming "they that are with him are called, and chosen, and faithful" (Rev 17:14). We need to know what the Bible means by being chosen, and being predestinated.

Father Chose Before the Foundation of the World

nm170 » "According as He [the Father] has *chosen us* in him [Christ] *before the foundation of the world* [cosmos], that we should be holy and without blame before him in love: Having *predestinated* us unto the adoption of children by Jesus Christ to Himself, according to the good pleasure of His [the Father's] will" (Eph 1:4-5). "In whom also we have obtained an inheritance, being *predestinated according to the purpose of Him* who works all things after the counsel of His own will" (Eph 1:11).

Chosen <u>Before</u> the Cosmos

When scripture says that God chose or predestinated before the foundation of the world (Eph1:4-5; 1Pet 1:19-20), since the word "world" is from the Greek *kosmos*, this means God predestinated <u>before</u> the creation (cosmos), <u>before</u> good (as we know it), <u>before</u> evil (as we know it), <u>before</u> law (as we know it), and consequently <u>before</u> sin (as we know it).

nm171 » The ones chosen, thus, were chosen by the Father "before the foundation of the world," and these chosen were predestinated to "the adoption of children [of God] through Jesus Christ." And these predestinated to be children of God were predestinated "according to the good pleasure of His [God's] will."

Foreknown, Predestinated, Called, Justified & Glorified

nm172 » "And we know that all things work together for good to them that love God, to them who are called according to his purpose. For those he did *foreknow*, he also did *predestinate* to be conformed to the image of his Son, that he

[His son] might be the first-born among many brethren. Moreover those he did *predestinate*, them he also *called*: and those he called, them he also *justified*: and those he justified, them he also *glorified*" (Rom 8:28-30).

nm173 » We see that those predestinated, God also did foreknow. Now from Ephesians 1:4-5 we see that the chosen are predestinated and that they were predestinated before the world began. And here in Romans 8:29 we see God foreknew those who are "predestinated to be conformed to the image of his Son," or to be a child of God. And we see (Rom 8:28) that those called according to God's purpose are the ones who were foreknown and predestinated.

nm174 » Then in Romans 8:30 we see those predestinated are called or invited. Called for what? They are called "according to his [God's] purpose" (v. 28). Now those called (who were predestinated) are justified (v. 30). As we see in the paper on justification [NM 18] those justified are those who have God's Spirit or the new mind.

Christians Are Predestinated

nm175 » Thus, those *chosen*, were chosen before the world began (Eph 1:4). And these are predestinated to be children of God (Eph 1:5). And those chosen and predestinated are called according to God's purpose (Rom 8:28). Also these were foreknown and predestinated to conform to the image of Christ (Rom 8:29). We also see those chosen, predestinated, foreknown, and called are justified (Rom 8:30). From the paper on justification (see "Other Papers" section [NM 18]) we know that those justified have become real Spiritual Christians. And in Romans 8:30 we see that those justified "he also glorified." Hence, real Christians are the ones chosen, predestinated, and called. This is confirmed in God's word:

- "*Elect* [chosen] according to the *foreknowledge* of God the Father, through sanctification of the Spirit, unto obedience and sprinkling of blood of Jesus Christ: Grace unto you, and peace, be multiplied" (1Pet 1:2).

- "But we are bound to give thanks always to God for you, brethren beloved of the Lord, because *God has from the beginning chosen you* to salvation through sanctification of the Spirit and belief of the truth" (2Thes 2:13).

- "Who has saved us, and called us with a *holy calling*, not according to our works, but according to his own purpose and grace, *which was given us in Christ Jesus before the world began*" (2Tim 1:9).

- "But *you are a chosen generation*, a royal priesthood, a holy nation, a peculiar people; that you should show forth the praises of him who has *called you* out of darkness into his marvelous light" (1Pet 2:9).

nm176 » Furthermore, the word *elect* (#1588 &1589) used to describe Christians in the New Testament means "chosen;" and *Saint* (#6918) means "set apart" or "sacred." The real Christians were set apart, chosen, and predestinated before the world began as the above scriptures clearly indicate. And now the chosen are being called or invited.

Christians Predestinated for Good

nm177 » Notice the real Christians were "created in Christ Jesus unto good works, which God has before ordained that we [Christians] should walk in them" (Eph 2:10). Christians were not predestinated to be free to do anything, but to do good works. The very proof that one is a real Christian is that he does good works (see "Proof Paper" [NM 10]).

God Chooses

nm178 » Hear what Christ says: "you have not chosen me, but I have chosen you, and ordained you" (John 15:16). No one can choose to follow Christ, unless Christ has chosen them. Yet Christ only does the will of his Father (John 6:38). Christ chooses what his Father chose for him. Is this confirmed anywhere else in the Bible? "No man can come to me, except the Father which has sent me draw him" (John 6:44).

nm179 » The only way to follow Christ is Spiritually with the aid of God's Spirit (John 4:23-24). "Jesus says unto him, I am the way, the truth, the life: no man comes unto the Father, but by me" (John 14:6). No one comes to the Father's Spirit (see verses 5-10) except through Christ. One comes to the Father's Spirit through Christ, but it is God who chooses (Eph 1:4-5). Although Christ also chooses, he chooses only what his Father wills (John 6:44). One can only come to the Father's Spirit if the Father draws that one to himself through Christ.

God Draws To Himself Through Grace

nm180 » And how does the Father draw one to himself through Christ? "But when it pleased God ... called me [Paul] by his grace" (Gal 1:15). Now grace is merely a free spiritual gift of God. One form of grace is Faith (1Cor 12:9). And it is through the medium of this Faith of God's Spirit (Gal 5:22; Eph 2:8) that God baptizes his chosen.

nm181 » No man can come to Christ or choose Christ unless God the Father draws him through his grace. "All that the Father gives me shall come to me" (John 6:37). And, "Holy Father, keep them in your own name which you have given me, that they may be one, as we are" (John 17:11). It is God the Father that gives the chosen to Christ (Eph 1:3-4; John 17:11; John 6:44), and no one can come to Christ unless the Father has chosen them (John 6:44; Eph 1:4-5). What does this mean?

nm182 » These scriptures mean exactly what they say! None can come to Christ to be a Christian unless they were chosen before the foundation of the world. One, if he was not predestinated to be a real Christian, cannot come to Christ's body or Church.

Marriage as a type

nm183 » God has made the physical to be a type of the Spiritual or the true reality (Rom 1:20). The physical marriage between man and woman prefigured the true marriage – the marriage of Christ to his wife, the Church (Rev 19:7). Now, who traditionally has chosen the bride? – the man. Yet it was traditional that the man only chose a bride that was acceptable to his parents. In fact the parents usually chose the bride for him. It was the parents who made the arrangements for their children (Gen 21:21; 24:3-4). God the Father was the parent of Christ. And in God's Word it says that God the Father has chosen those who will be in the Spiritual marriage to Christ. The physical arrangements for marriage prefigured the true arrangement that God the Father made before the beginning of the cosmos, the Spiritual wedding of his Son to his Son's wife – the Church. It is the real Christians who are chosen and predestinated to be the Spiritual wife of Christ. Only those chosen, will be married to Christ. Is this confirmed yet again in God's word?

Predestinated to Destruction?

nm184 » "Has not the potter power over the clay, of the same lump to make one vessel unto honor, and another unto dishonor? What if God, willing to show his wrath, and to make his power known, endured with much long-suffering the vessels of wrath fitted to destruction. And that he might make known the riches of his glory on the vessels of mercy, which he had before prepared unto glory" (Rom 9:21-23). The potter is God as such verses as Jer 18:6 and Isa 45:9 show. But what is being said here?

nm185 » God has prepared from the same lump of clay: vessels of destruction and vessels of mercy, and the latter were before prepared unto glory? Has God prepared some for destruction? In Jude 4 it speaks of those "who were before of old ordained to this condemnation, ungodly men." And, "the LORD (Jehovah) has made all things for himself; yea, even the wicked for the day of evil" (Prov 16:4). "Even to them which stumble at the word, being disobedient: whereunto also they were *appointed*" (1Pet 2:8).

nm186 » We have already shown you that real Christians were chosen and predestinated before the foundations of the world to become Christians unto good works. Now we see that the others were prepared for destruction. But remember, God predestinated everything <u>before</u> the creation, thus before time (as we know it), before good (as we know it), before evil (as we know it), before law (as we know it), and thus before sin (as we know it).

nm187 » The true Christians are "not appointed as to wrath, but to obtain salvation by our Lord Jesus Christ" (1Thes 5:9). "I make peace, and create evil: I the LORD [Jehovah] do all these things" (Isa 45:7). Are we to take these words literally? If you do not, you have nothing to base your faith on. One cannot read and believe one part of the Bible and then pass over another part. Let's look closer at this apparent problem.

Two Groups

nm188 » The Bible clearly speaks of two groups of people: the "dead" and the "living," the children of the devil and the children of God, the children of wrath and the children of light, and so forth. One group is identified with God and his way. These belong to the "living." The other group follows the way of destruction. These belong to the "dead." The group that follows God's way are the vessels of mercy. The other group are the vessels of wrath. One was appointed to salvation; the other group was appointed to wrath (1Thes 5:9). They were appointed before the foundation of the world as Ephesians 1:4 and Jude 4 tells us. Now those chosen to be Christians are created "unto good works" (Eph 2:10). In fact it can be proven that doing good works is a proof of one's own Christianity (see "Proof Paper" [NM 10]). The chosen group of God does good works as opposed to the other group which follows in the ways of confusion. Now is this fair that one group is set up one way and the other group another way?

Is God fair?

nm189 » Is God fair? Can God predestinate one group to mercy and another group to wrath and still be fair?

- "So then He has mercy on whom He desires, and He hardens whom He desires. 19 You will say to me then, 'Why does He still find fault? For who resists His will?' 20 On the contrary, who are you, O man, who answers back to God? The thing molded will not say to the molder, 'Why did you make me like this,' will it?" (Rom 9:18-20)

- "Shall the clay say to him that fashions it, What are you doing? or to your work, He has no hands? Woe unto him that says unto his father, What are you doing? or to the woman, What have you brought forth?" (Isa 45:9-10)

- "Yet the children of your people say, The way of the Lord is not equal. O you house of Israel, I will judge you every one after his ways" (Ezek 33:17, 20; see 18:30).

nm190 » God judges man according to man's ways, according to his works, according to his behavior. This means that those who do evil will destroy the world by their own behavior (See PR4 to PR6) and their own destruction will judge them (NM24). Each group was prepared before the cosmos began either to do good works, or to do the works of confusion. This means they were predestinated <u>before</u> good (as we know it), <u>before</u> evil (as we know it), <u>before</u> law (as we know it), and <u>before</u> sin (as we know it). You cannot accuse God of sin for something done before there was law or even a cosmos.

Seven Days to Finish the Creation

nm191 » Also the type and antitype nature of the Bible teaches us that we are now in the midst of a creation process. God is now creating, and has yet to finish. Look at the typical creation mentioned in the book of Genesis. First God created the universe in seven days: actively working for six days (Gen 1:1-31), but also in some way finishing the creation on the seventh day (Gen 2:2). Accordingly, it will only be on the antitypical seventh day that God's real work will be complete (NM15; NM16; NM26). As we learn in other parts of this book, there are seven 1000 year periods or days wherein God is creating the knowledge of good and evil, the knowledge of time, the knowledge of peace, the knowledge of paradise and other important spiritual things. Therefore it will not be until the end of the seventh 1000 year period that mankind will be atoned, and consequently be brought back into the great God. All at the beginning went out of God, but all by the end of creation will be brought back into God. This is called the Great Cycle (NM13, "Three Divisions"). We are now in the midst of a creation process, and therefore anything being created in it is not yet done. The creation will only be good at the end of the creation process, when all are back into God (1Cor 15:28), when only the One God is in the truest sense, good (Mat 19:17, see Greek text). God is only One, when all come back into the God (1Cor 15:28; GP6). Even the scriptures about Jesus Christ being now at the right side of God (Psa 110:1; Acts 2:32-35; Heb 2:7-10) tell us that all things are not yet under Christ, but all will be: "But now we see not yet all things put under him" (Heb 2:8). The creation is not yet finished in the truest sense. All must be under Christ, then, and only then, will all be back in God, so that God will be all in all, so that, God will be ONE. When the antitypical creation is finished, no one can accuse God of being partial, for as we see in NM13 and GP6, God will bring all back into God, so that God will be all in all (1Cor 15:28). The creation is similar to a clay pot, and God is the potter forming the pot. When the potter is making the pot there is waste (vessels of wrath), but God is perfect and will not waste any material in the process of making the pot. God will pick up the clay discarded on the ground and reintroduce it back into the clay pot, so all will be in the pot (new creation). As we cannot blame the potter for any spillage, until he has totally completed his work, we also cannot blame God for any temporary waste, until he has finished all in all.

nm192 » Read the "Reason Why" paper [NM 19] to understand *why* God did such a thing. Also, after you read the *God Papers* you will understand that God the Father gave all his power to his Son, Jesus Christ. Jesus Christ is our God and he has done no wrong. Jesus Christ is now the right side of God's power. If Jesus Christ is the right side, who is the left side? In the *God Papers* we prove that the left side of God is the evil facet of God (Isa 45:7) that was predestinated by God before law and before sin, and thus we cannot find fault with God over this matter. God who is *all* powerful even has power over evil, for he somehow has created evil (Isa 45:7). We cannot speak of this in detail here. We cover the left side of God in the *God Papers*. It should be noted here that predestination was done <u>before</u> the creation, thus before law and sin.

Both Groups Exercise Each Other

nm193 » It is true that there are two groups on the earth today. One group is appointed to God and one group is appointed to Satan. Each group is against the other as the law of confusion is against the law of love. Each is exercising the other. Satan's way and his group are allowing the world to learn of evil. God's group has come out of the other group, and they have a changed attitude towards the world's way. And God's group is now learning of good through God's Spirit given to them. God has granted one group mercy. These are the vessels of mercy (Rom 9:23). It is through the vessels of mercy that all will eventually learn of good through the kingdom of God or the rulership of God. All will eventually be saved (see "All Saved Paper" [NM 13]). Each group is now using each other in order to produce the knowledge of good and evil (see the *God Papers*).

nm194 » All people in this old age are, or were, in Satan's group, but to the predestinated vessels of mercy, God has given mercy by allowing them to see the true light. The vessels of Satan's group are vessels from the same lump of clay but are fitted to destruction – aeonian punishment (Mat 25:46). The vessels of God's group are the vessels of mercy who will have aeonian life while the other group has aeonian death away from the glory of the millennium (2Thess 1:9; see *Last Judgment*).

nm195 » Can any of Satan's group become through any effort of their own a part of God's group? "No man can come to me, except the Father which has sent me draw him," answers Christ (John 6:44). "Consider the work of God: for

who can make straight, which he made crooked? ... that which is crooked cannot be made straight" (Eccl 7:13; 1:15). Spiritually, only God can make the crooked things straight (Isa 42:16).

Losing God's Spirit?

nm196 » Can a real Christian return completely to Satan's group after he has turned to God? "And grieve not the holy Spirit of God whereby you are sealed unto the day of redemption" (Eph 4:30; see Isa 59:21).

nm197 » Can one lose the New Mind or Spirit of God after he has received it? The following scriptures tell us no!

- "Of them which you gave me I have lost none" (John 18:9).

- "All that the Father gives me shall come to me; and him that comes to me I will in no way cast out" (John 6:37).

- "And this is the Father's will which has sent me, that all which he has given me I should lose nothing, but should raise it up again at the last day" (John 6:39).

- "No man can come to me, except the Father has sent me draw him: and I will raise him up at the last day. It is written in the prophets, and they shall be all taught of God. Every man therefore that has heard, and has learned of the Father, comes unto me" (John 6:44-45). That is, all that learns Spiritually or hears Spiritually will come to Christ.

- "And when he puts forth his own sheep he goes before them, and the sheep follow him: for they know his voice" (John 10:4).

- "My sheep hear my voice, and I know them, and they follow me: and I give unto them aeonian life; and they shall *not* be destroyed for the age, neither shall any man pluck them out of my hand. My Father, which gave them to me, is greater than all; and no one is able to pluck them out of my Father's hand" (John 10:27-29).

nm198 » Notice that Jesus said no one could take those chosen out of his Father's hand – that means a chosen one himself has no power to change his predestination:

> ■ "While I was with them in the world, I kept them in your NAME, which you have given me, and I have kept, and none of them is lost, but the son of perdition; that the scripture might be fulfilled" (John 17:12).

nm199 » If Christ was not lying when he said he lost none, then Judas was not a chosen one of God the Father. That is, Judas was not chosen to salvation, but to wrath.

Chosen Are Called

nm200 » Those chosen, predestinated before the world began are called: "and the sheep follow him: for they know his voice" (John 10:4). Further, "all that the Father gives me shall come to me" (John 6:37).

nm201 » Why are there *many* called, but *few* chosen? (Mat 22:14) For the whole world is invited or called to the kingdom of God (Mat 24:14; Mark 16:15; Psa 19:4; Col 1:23), but only a few of these were predestinated and chosen to begin that kingdom. It is those who are of the chosen who hear the words of the calling and understand (John 10:4, 27).

Predestinated: Are You to Live in The 1000 Years?

nm202 » Are you one of the called, predestinated, and chosen to live in and throughout the New Age during the first 1000 years? Your proof is in your good works. If one is not doing works of the Spirit then they do not have the New Mind or Spirit of God (see "Proof Paper" [NM 10]).

Why Have Some Been Chosen?

nm203 » "I returned, and saw under the sun, that the race is not to the swift, nor the battle to the strong, neither yet bread to the wise, nor riches to men of understanding, nor yet favor to men of skill; but *time and chance happens to them all*" (Eccl 9:11). "For you see your calling, brethren, how that not many wise men according to the flesh, not many mighty, not many noble, are called: But God has chosen the foolish things of the world to confuse the wise; and God has chosen the weak things of the world to confuse the things which are mighty; And the base things of the world, and things which are despised, has God chosen, yea, and things which are not, to bring to nought things that are: *That no flesh should boast in his presence*" (1Cor 1:26-29).

nm204 » Those chosen were chosen by time and chance, and were mostly from the unwise in the ways of the world. And these where chosen for good works (Eph 2:10).

NM 9: Free Will versus Predestination

God Predestinates All Things
Free Will
Free Will v. Happiness
Self-righteousness and Free Will
Job Seemed Upright
God Answers Job
Way to Wisdom and Knowledge
God Does All

NM9 Abstract

If there is predestination, how can there be free will? Or, if there is free will, how can there be predestination? We learned about predestination in NM8. There is ample scripture that tells us that God in some way is all powerful and is our creator. Knowing this, in this paper we will learn that true free will is impossible when you have a creator who made us and is making us into what he wishes us to be. We will also study the book of Job in order to better understand this subject.

Many Predestination Scriptures: Most Still Believe in Free Will

nm205 » Many refuse to believe in predestination even though the scripture clearly teaches that the God has predestinated everything to be as it was, as it is, and as it will be. They believe in "free will." In this paper we will contrast the "free will" scriptures and arguments against the predestination scriptures.

nm206 » In the paper, "Predestination: Called and Chosen" [NM 8], we saw many of the predestination scriptures. Some are chosen to mercy; some are chosen even to wrath. Christians were chosen and predestinated for good while others were chosen for wrath and evil. There is a reason for this.

God Predestinates All Things

nm207 » God in someway has predestinated, chosen, elected, ordained, set apart beforehand, or set in motion before the cosmos, before good, before evil, before law, and thus before sin:

- The nation of Israel [Deut 7:7-8; 10:15; 1Sam 12:22; Psa 135:4]

- Jacob versus Esau [Mal 1:2-3; Rom 9:11-13]

- The Christians [Eph 1:4-5,11; Acts 10:41; Rom 8:28-29; 1Thes 5:9; 2Thes 2:13; 2Tim 1:9; 1Pet 5:10]

- The Church [Acts 2:47; 1Pet 1:2; 2John 1:2]

- The Christ [1Pet 1:19-20; 2:6; Isa 42:1; Luke 24:26-27; see *All the Messianic Prophecies of the Bible*, by Lockyer; etc.]

- Christ's death [Acts 4:27-28; 2:23; 3:18; 1Pet 1:19-20]

- The results of sin [Gen 2:7; 3:16-19; Rom 5:12; 6:23]

- Nations and their leaders [Jer 18:7, 9; 1:10; Acts 17:26; Job 12:23-25; Dan 4:28-35; 2:44-45; 7:14]

- Individuals (and nations from some of these individuals) [Paul, 2Tim 1:1,11; Gal 1:15-16]

- Esau [Mal 1:2-3]

- Jacob [Mal 1:2-3]

- Pharaoh [Rom 9:17]

- Samson [Judges 13:3-5]

- Solomon [2Sam 7:12-13; 1Chron 22:6-19]

- Josiah [1Kings 13:2]

- Jeremiah [Jer 1:5]

- Cyrus [Isa 45:1]

- John the Baptist [Luke 1:13-17]

- Judas Iscariot [Acts 1:16-17]

- Jesus, see "Seed Paper" [PR1]

- Elijah [Mal 4:5; Mat 11:14; Luke 1:17; Mark 9:12]

- Noah, Abram or Abraham, Isaac, Pharaoh's butler, Joseph, Aaron, Angel of Yehowah, Korah, Dathan & Abiram, Moses, Judah, Simeon, Levi, Reuben, Zebulun, Issachar, Naphtali, Dan, Benjamin, Gideon, Manoah's wife, Ahab,

Elisha, Jonah, etc (see *Encyclopedia of Biblical Prophecy*, by Payne, "Summary A & B").

- Vessels of wrath and of mercy [Rom 9:21-23]

- All those appointed to wrath, evil, condemnation, wickedness, etc. [Rom 9:21-23; Jude 1:4; Prov 16:4; 1Pet 2:8; see "Predestination Paper" (NM 8)]

- All generations [Isa 41:4]

- The future [Isa 41:4, 22, 26; 44:7; 46:9-11]

God Predestinates: Summary

- "Known unto God are all his works from the beginning of the world." [Acts 15:18]

- "God makes all things." [Eccl 11:5]

nm208 » God can and does predestinate all things because He is all knowing (1John 3:20; Psa 147:5), because he is all powerful (Gen 17:1; Rev 1:8; 4:8; 15:3), and because he creates all (Eccl 11:5; 2Chron 20:6; Eccl 11:5). God has even in some sense created evil (Isa 45:7; see *God Papers*). But God has promised that before the end he will make all that is crooked or dark – straight or light (Isa 42:16; 1Cor 15:24-28; see "All Saved Paper" [NM 13]). God will make the lamb live with the wolf (Isa 11:6, spiritual meaning, see *God Papers*). The answer to the paradox of the God creating evil, yet being Good, is explained in the *God Papers*. There is an answer to this.

Free Will

nm209 » But what about "free will"? Much of the "free will" doctrine comes from the Greeks and other ancients. In the truest sense of the word, no one is *free*, except God. It is God's will that will be done over and above all others' will (Isa 46:10-11; Acts 15:18). Since from the beginning God has through his power created all things with the laws concerning these things (Eccl 11:5; James 4:12; Isa 33:22), then all are limited by these laws. We have physical limitations. We can only run so fast, climb so high, or live so long. We have mental limitations. We can only think or concentrate on one thing at a time, while God can think on a million, a billion, a trillion ... things at one time. His mind is not limited like ours. There is some freedom within these laws. God could have limited our minds and ability so as to make it impossible for us to sin. But God gave us the apparent "freedom" to choose to sin (Gen 2:16-17; Deut 30:19-20). But if we choose sin,

then there are evil results for this sin (Gen 2:17; Deut 28:15ff). And if we "choose" not to sin then there will be certain "rewards" (Deut 28:1-2ff). God judges solely by our ways or behavior (Ezek 18:20, 25, 30). If mankind kills those of mankind, then some of mankind will be killed (Rev 13:10; Gen 9:6).

Do we have Free Will?

No Power or Knowledge to Choose Good

nm210 » But do we have free-will? Did Israel have free-will? Do we have the power to choose not to sin? Did Israel have the power or freedom to choose not to sin?:

- "Moses summoned all the Israelites and said to them: 'Your eyes have seen all that the LORD did in Egypt to Pharaoh, to all his officials and to all his land. With your own eyes you saw those great trials, those miraculous signs and great wonders. **But to this day the LORD has not given you a mind that understands or eyes that see or ears that hear.**'" (Deut 29:2-4, NIV)

Israel, in fact, did not have the mind to understand. Mankind cannot please God (Rom 8:7-8). They cannot understand or please God because they do not have the power of the Spirit to see, and thus do not truly have the freedom to choose. Israel and mankind are blind. They have been limited. They cannot see the truth.

nm211 » Does mankind in this old age have the freewill to do good?

- "Because the carnal mind is enmity against God; for it is not subject to the law of God, nor indeed can be. **So then, those who are in the flesh cannot please God**" (Rom 8:7-8, NKJV).

Good Spirit Has the Power to Choose Good

nm212 » It is only when mankind gets the New Mind (the Spirit) that he understands and is able to choose the good:

- "And the LORD your God will circumcise your heart and the heart of your descendants ... that you may live" (Deut 30:6, NKJV).

- "In him you were also circumcised with the circumcision made without hands, by putting off the body of the sins of the flesh, by the circumcision of Christ" (Col 2:11).

- "For we are the circumcision, who worship God in Spirit . . ." (Phil 3:3).

- "These things we also speak, not in words which man's wisdom teaches but which the Holy Spirit teaches, comparing spiritual things with spiritual. But the natural man does not receive the things of the Spirit of God, for they are foolishness to him; nor can he know them, because they are spiritually discerned" (1Cor 2:13-14).

Those with the circumcised heart, the true circumcision of Christ, that is, the Spirit, can see and choose the good things of the Spirit of God.

Reasons For No Real Freedom to Choose Good

nm213 » There are some reasons for mankind and angelkind not having the power or freedom to choose good over evil:

nm214 »

- **Lack of Spirit**. Adam and Eve did not have the New Mind and therefore did not have the ability of obeying God (see "Other Mind Paper" [NM 21]). It is through the Spirit of God (the New Mind) that mankind is given the gifts of goodness (Gal 5:22-23). It is through the first sinner, Satan & his lie (Gen 3:4; John 8:44), that sin entered the world through the willingness of man (Gen 3:4-12; Rom 5:12).

nm215 »

- **Lack of Knowledge**. Mankind and angelkind at the beginning did not have the *knowledge* of good and evil. It was only after mankind ate from the tree of *knowledge* of good and evil that mankind and angelkind began to know about evil, and thus were, also in a manner, learning about good (see "Other Mind Paper" [NM 20-21], "Reason Why" paper [NM 19], and the *God Papers* [GP 7]). So in part and in one sense, the ignorance of both Satan and mankind led to the first sin.

nm216 »

- **God Created With Limitations**. But it was God's power that made all, with its limitations and abilities. And all that we are, or all that we have, comes from outside of us: "For

what makes you differ from another? And what do you have that you did not receive? Now if you did indeed receive it, why do you glory as if you had not received it?" (1Cor 4:7) The lack of positive goodness on Satan and mankind's part was because God did not *give* Satan or mankind his own Good Spirit.

Freewill v. Happiness

nm217 » But if God at first gave mankind and angelkind (specifically Satan) his goodness, then how would we come to appreciate the goodness, how would we understand good, if we had never lived in an evil place? *If God had given us goodness and the environment of paradise (peace and harmony) at first with immortal life, we would never have been happy.* In order to be happy, in order to know good, in order to know peace, in order to know harmony, in order to know pleasure, in order to know life, we MUST first know unhappiness, evil, war, disharmony, pain and death. The very basic Law of Knowledge tells us that (see "Reason Why" paper [NM 19] and the *God Papers* [GP 7]). God has no pleasure in our present evil (Heb 10:8), He, of course, did not want or desire the evil period (sacrifice), but he knew that we first must suffer in order to know the Good and to be able to enjoy paradise. God by creating evil (Isa 45:7), in a sense through Satan, was creating good, because according to the Law of Knowledge:

> "Particularly, in the case of opposite qualities (good and evil) you must know *both* qualities to know either: you must compare each with the other to know either." (NM19)

Evil a Mistake?

nm218 » Since God is all powerful, He is in the last analysis responsible for evil. God did not make a mistake when he allowed evil through Satan. God created all and He knew what would happen when he created the universe and all that was in it. God at some time did create the angel Satan, although at first Satan did not appear as the evil Satan until God stated a law (see "Other Mind Paper" [NM 21]). Satan has not gone beyond what he was allowed to do (note Job 1:12; 2:6; see below). Satan's evil came about because he lacked the knowledge of good and evil and because he did not have the good Spirit. See and read all of the *God Papers* carefully to understand the paradox of God's goodness and his creating of evil through Satan. In this age for certain we can say that Satan is the power behind evil, and that God is the power behind good, but in a sense it is the God who has created evil (Isa 45:7), by predestinating evil before creation, before law (as we know it), and before sin (as we know it).

Self-righteousness and Freewill

nm219 » Those who do not believe in the Biblical doctrine of Predestination are in reality saying that their "good" behavior is because *they* are somehow doing this "good" through their *own* striving and thus are "qualifying" for or "earning" a reward – the kingdom of God (heaven, paradise, eternity, etc.). *They* are being good, thus will reap *their* just reward. They say or imply that their witnessing, or tithing, or going to church, or giving to the poor, or eating the right foods, or believing in the right doctrines are their means to their reward – God's Kingdom or paradise. They think (at least subconsciously) that they *deserve* God's good reward because of their apparent good behavior. They are being righteous through their own efforts: they are *self*-righteous. Let's study Job and his *self*-righteousness to understand why self-righteousness is wrong and to understand its connection to the freewill doctrine.

Job

Job Seemed Upright

nm220 » Job was a rich man from the land of Uz. "And this man was perfect and upright, and fearing God, and turning away from evil" (Job 1:1).

Satan against Job

nm221 » But Satan went against Job to test him with trials to see if he would still love God (1:6-12). After a great loss to his sheep, camels, servants, and sons and daughters (1:13-19), "Job rose up and tore his robe, and shaved his head. And he fell on the ground and worshiped. And he said, I came naked out of my mother's womb, and naked I shall return. Jehovah gave, and Jehovah has taken away. Blessed be the name of Jehovah. *In all this Job did not sin, nor charge wrong to God*" (1:20-21).

nm222 » To continue from the book of Job:

■ Again a day came when the sons of God came to present themselves before Jehovah. And Satan also came among them to present himself before Jehovah. And Jehovah said to Satan, From where have you come? And Satan answered Jehovah and said, "From going to and fro in the earth, and walking up and down in it." And Jehovah said to Satan, "have you set your heart on My servant Job, that there is none like him in the earth, a perfect and upright man, fearing God, and turning away from evil? And he is still holding to his integrity, although you incited Me against him, to destroy him for nothing." And Satan answered Jehovah and said, "Skin for skin. Yea, all that a man has he will give for his life. Put out Your hand now and touch his bone and his flesh, and he will curse You to Your face." And Jehovah said to Satan, "Behold, He is in your hand; but save his life." And Satan went out from the presence of Jehovah. And he struck Job with bad burning ulcers from the sole of his foot to the top of his head. And he [Job] took a broken piece of pottery with which to scrape himself. And he sat down among the ashes. And his wife said to him, "Are you still holding fast to your integrity? Curse God and die!" But he said to her, "You speak as one of the foolish women speaks. Indeed shall we receive good at the hand of God, and shall we not receive evil?" In all this Job did not sin with his lips (Job 2:1-10).

nm223 » Then three of Job's friends came to him after hearing about his troubles: "They sat on the ground with him for seven days and seven nights. No one said a word to him, because they saw how great his suffering was" (Job 2:11-13).

nm224 » But after this Job "cursed the day of his birth" (Job 3:1). "why did I not perish at birth, and die as I came from the womb? ... I have no peace, no quietness; I have no rest, but only turmoil" (Job 3:11, 26; see 3:1-26).

Charge: Job Suffered Because He Sinned

nm225 » In chapter 4 to 36 Job's friends Eliphaz, Bildad, and Zophar gave discourses that suggested that Job's suffering was *only* because Job had sinned against God:

- Consider now: who, being innocent, has ever perished? Where were the upright ever destroyed? As I have observed, those who plow evil and those who sow trouble reap it (Job 4:7-8).

- Blessed is the man whom God corrects; so do not despise the discipline of the Almighty (Job 5:17).

- When your [Job's] children sinned against him [God], he gave them over to the penalty of their sin. [Job's children were destroyed by a natural disaster (Job 1:5, 18)]

- If you are pure and upright, even now he [God] will rouse himself on your behalf (Job 8:4,6).

- You [Job] say to God, My beliefs are flawless and I am pure in your sight ... If you put away the sin that is in your hand and allow no evil to dwell in your tent ... [then] you will surely forget your trouble ... Life will be brighter than noonday (Job 11:4, 14, 16, 17).

- All his days the wicked man suffers torment (Job 15:20).

- The lamp of the wicked is snuffed out (Job 18:5).

- A flood will carry off his house, rushing waters on the day of God's wrath. Such is the fate God allots the wicked (Job 20:28-29).

- Is it for your piety that he rebukes you and brings charges against you? Is not your [Job] wickedness great? Are not your sins endless? (Job 22:4-5)

- Submit to God and be at peace with him in this way prosperity will come to you (Job 22:21).

- Evil, a Mistake? How then can a man be righteous before God? (Job 25:4)

Job answered these charges: 'I am blameless'

nm226 »

- Teach me, and I will be quiet; show me where I have been wrong (Job 6:24).

- If I have sinned, what have I done to you, O Watcher of men? Why have you made me your target? (Job 7:20)

- How then can I dispute with him [God]? ... though I were innocent, I could not answer him (Job 9:14-15).

- Although I am blameless ... (Job 9:21)

- I loathe my very life ... I will say to God: Do not condemn me, but tell me what charges you have against me. Does it please you to oppress me, to spurn the work of your hands, while you smile on the schemes of the wicked? (Job 10:1-3)

- Though you know that I am not guilty . . . (Job 10:7)..

- If I sinned ... If I am guilty – woe to me (Job 10:14, 15).

- Now that I have prepared my case, I know I will be vindicated. Can anyone bring charges against me? If so, I will be silent and die (Job 13:18-19).

- How many wrongs and sins have I committed? Show me my offense and my sin (Job 13:23).

- My face is red with weeping, deep shadows ring my eyes; yet my hands have been free of violence and my prayer is pure (Job 16:16-17).

- Though I cry, 'I've been wronged!' I get no response (Job 19:7).

- I have kept to his way without turning aside (Job 23:11).

- As surely as God lives, who has denied me justice, the Almighty, who has made me taste bitterness of soul (Job 27:1). [Here Job accuses God of denying him justice.]

- I will never admit you [his friends] are in the right; til I die, I will no way deny my integrity. I will maintain my righteousness ... (Job 27:5-6).

- ...He will know that I am blameless (Job 31:6).

- Job runs through his list of his righteousness (see Job 31:1-40).

- But the text says, "So these three men stopped answering Job, because *he was righteous in his own eyes*" (Job 32:1).

Charge: Job Is Saying God Is Unjust

nm227 » Elihu, who had listened to the exchanges between Job and his three friends, answered correctly Job's cries:

- Job says, "I am innocent, but God denies me justice. Although I am right, I am considered a liar; although I am guiltless, his arrow inflects an incurable wound" (Job 34:5-6).

- For he says, "It *profits* a man nothing when he tries to please God" (Job 34:9).

- Therefore, O man of heart, listen to me; far be it from God to commit iniquity; and the Almighty, to do wrong. For He repays man's work to him; and according to a man's way. Surely God will not do wickedly, nor will the Almighty pervert justice (Job 34:10-12).

- He punishes them for their wickedness ... . They cause the cry of the poor to come before him ... But if he [God] remains silent, who can condemn him (Job 34:26, 28-29).

- Job speaks without knowledge, his words lack insight, Oh, that Job might be tested to the utmost for answering like a wicked man. To his sins he adds rebellion (Job 34:35-37).

- Yet you ask him. "What *profit* it to me, and what do I gain by not sinning?" (Job 35:3)

- He [God] does not answer when men cry out because of the arrogance of the wicked (35:12).

- So Job opens his mouth with empty talk; without knowledge he [Job] multiplies words (Job 36:16).

- God is mighty, but does not despise men; he is mighty, and firm in his purpose (Job 36:5).

- The godless in heart harbors resentment (Job 36:13).

- How great is God – beyond our understanding (Job 36:26).

- In his justice and great righteousness, he does not oppress (Job 37:23).

God Answers Job

nm228 » After this Jehovah answered Job "out of a whirlwind" stating some of his great power (see Job 38:1-40:7; and see Job 40:9-42:1) and then adding:

- Would you discredit my justice? Would you condemn me to justify yourself? (Job 40:8)

Job Repents

nm229 » Job's reply to God:

- I know that you [God] can do all things; no plan of yours can be thwarted. You asked, "Who is this [Job] that obscures my counsel without knowledge?" Surely I spoke of things I did not understand, things too wonderful for me to know (Job 42:2-3).

- My ears had heard of you but now my eyes have seen you. Therefore I despise myself and repent in dust and ashes (Job 42:5-6).

Job's Self-Righteousness

nm230 » Job was declared righteous *before* his hard testing by Satan (1:1). Sometime after Job's friends mourned with him, and after Satan tested Job the second time, Job did not recognize God's all mightiness – that it is God who, in the truest sense, gives and takes away (Although, *immediately* after the second testing by Satan, Job did verbally recognize this. – Job 2:10). According to the text, Job did not do any wrong until sometime after Satan tested him for the second time. And because Job had done no wrong before Satan's testing, Job thought that he was being punished unjustly and was challenging God to prove or show his sin. Job incorrectly thought like his friends, that *only* those who did wrong suffered in this present life ("Only those who do wrong do not have expensive cars, houses, goods. Those who are righteous have physical rewards in this life."). But suffering comes in this life to the righteous as well as the unrighteous (see below). Therefore Job in declaring his own righteousness was projecting his own *self-righteousness*, and thus denying God's *all* mightiness. Everything that man is comes from God. And everything that man gains or suffers comes from God:

■ "Shall we receive good at the hand of God, and shall we not receive evil? In all this Job did not sin with his lips" (Job 2:10).

Job a Prostitute?

nm231 » Furthermore, besides Job's self-righteousness, he acted like a prostitute:

■ What *profit* is it to me, and what do I gain by not sinning? (Job 35:3; 34:9)

nm232 » If you are righteous, if you are of the good, you do not behave honorably for rewards – for profit or gain – you behave honorably because you are righteous, because you are of the good and *hate* all evil. If you hate evil you do not want any part of it even if you are apparently not physically rewarded for good. (Christ is an example: He never sinned, but died because of His good behavior, without any apparent reward before His death.) In the truest sense of the word, Job was *not* righteous, when he acted *apparently* righteously before his testing by Satan, because afterward he showed his real color – his prostitute mind: "Where is my reward for being good" (cf. Job 34:9; Job 35:3). Satan's testing merely brought out the evil in Job.

nm233 » But Job finally *saw* God and understood that God does *all* things (Job 42:2-6). There is a reason for evil and suffering even though it might not be understood by most men in this age (Job 36:26). God does all things (Job 12:10-25; 42:2). God is just (Job 34:10; 37:23). If God allows evil it is for a just and noble purpose (Job 4:8-9; 11:7; 37:23; Rom 8:28; see "Reason Why" paper [NM 19] and the *God Papers*). Satan does not interfere with God's purpose, for Satan can only do what God allows (Job 42:2; 1:12; 2:6; Isa 46:9-11; 55:11; Psa 115:3).

Way to Wisdom and Knowledge

nm234 » Although the book of Job does not reveal the purpose of God allowing evil and suffering it says something that implies the answer:

■ There is a mine for silver and a place where gold is refined .... But where can wisdom be found? Where does understanding dwell? *Man does not comprehend its worth* .... It cannot be bought with the finest gold, nor can its price be weighted in silver .... Where then does wisdom come from? Where does understanding dwell? .... God understands the way to it and he alone knows where it dwells .... He [God] looked at wisdom and appraised it; he confirmed it and tested it. And he said to man, the *fear* of the Lord – that is wisdom, and to shun evil is understanding (Job 28:1,12-13,15, 20,23,27-28, see Hebrew for verse 28 "Lord" = *Adonay* # 136, which is in a plural form)..

nm235 » Notice there is wisdom in the *fear* of the Lord (Job 28:28) or the fear of the LORD (Prov 1:7) and to shun evil is understanding (see Prov 1:7). When you *fear* the Lord, who does *all* things, even in someway creating and predestinating evil, then there is wisdom. And when you shun evil or hate evil there is understanding. But you must know evil to shun it or hate it. The only way you can know evil is to learn about it. The only way to learn about evil is to live in a time of evil. Thus, God saw the great *worth* of wisdom (wisdom is the fear of the Lord & to shun or hate evil – Prov 8:13), thus He created evil (Isa 45:7) by predestination before creation and sin through Satan (Job 1:7-12; 2:2-7) so that mankind could learn to hate evil and thus obtain wisdom and understanding. Therefore, God allowed man to take from the tree of *knowledge* of good and evil for a higher purpose (see "Reason Why" paper [NM 19] and the *God Papers* [GP 7]).

nm236 » *False common thinking.* Job and his friend projected the common thinking that *only* the evil ones are supposed to suffer, and that if you do good you will be rewarded in this life or age. Their thinking says that there is *profit* for being good in this age.

nm237 » *Both the good and the evil suffer.* But scripture clearly shows those with good behavior also suffer (note Christ's human life; see Hebrews chapter 11; Rom 8:17; 1Pet 3:14, 17; 4:1; etc.). Of course, the evil ones are also suffering for their sins and the spiritual evil ones will suffer in the 1000 years (Job 15:20; 31:3; note "Thousand Years and Beyond Paper" [NM 15]). Those who do good suffer because they live in an evil environment. The world is under the influence of the evil mind – the other-mind. Therefore and thereby, the whole creation is now suffering (Rom 8:22).

nm238 » But the common mistaken thinking of Job and his friends, is the kind of thinking that ignores the many predestination scriptures *and* ignores the many scriptures that clearly indicate that the God does all and gives all. We have looked at some of the predestination scriptures, now let us look at the scriptures that indicate that the God does all.

God Does All

nm239 »

- First, God is Almighty – He has ALL the power. "I am Almighty God." (Gen 17:1; Rev 15:3; etc.)

- In His All mightiness God creates ALL. "God who makes all things." (Eccl 11:5; Gen 1:1; Jer 10:16)

- In God's creating, God creates good. "I make alive ... I heal ... I make peace ... the LORD [YHWH] is good to all: and his mercies are over all his WORKS." (Deut 32:39; Isa 45:7; Psa 145:9; Gen 1:31)

- God even predestinates some to good. (Rom 9:21-23; Eph 1:4-5; etc.; see "Predestination Paper" [NM 8])

- But somehow in God's all powerfulness, He creates evil. "I kill ... I wound ... and create evil." (Deut 32:39; Isa 45:7)

- God even predestinates some to evil. (See Romans 9:21-23; Jude 1:4; Prov 16:4; 1Pet 2:8; see "Predestination Paper" [NM 8])

- Yet God will in the future make ALL THINGS NEW. "I make all things new." (Rev 21:5; Isa 65:17)

- And then ALL WILL BE IN GOD. "Then the end ... the last enemy that shall be destroyed is death ... that God may be all in all." (1Cor 15:24-28; "All Saved Paper" [NM 13]; etc)

- God even gives:

  the Spirit (Gal 4:6; 1Cor 12:1 ff)

  repentance (Acts 5:31; 11:18; 2Tim 2:25)

  grace (Rom 11:5-6; 15:15)

  salvation (Tit 3:5-7)

nm240 » *In summary*, anything that you are (whether you are good or evil) is from outside of you:

- "For who makes you different from anyone else? What do you have that you did not receive? And if you did receive it, why do you boast as though you did not?" (1Cor 4:7)

nm241 » We receive all from God, even our good works are from God's predestination – he *gave* us our ability to do good (the Spirit), he gave us the physical body in order to do this good, he gave us a place (the earth) to do good, he gave us the physical energy to do good, and he gave us a time of evil so that there could be good, for without evil there could never have been good because evil and good are comparative qualities (see "Reason Why" paper [NM 19] and the *God Papers* [GP 7]). What, then, is there to boast about?

nm242 » See the "Proof Paper" [NM 10] to understand why Christians run the race, why they continue to try to do good even though they were predestinated. When you are predestinated to evil, you do not know it. When you are predestinated to good, your good works are your *proof*, but God gives you the power. When you are predestinated for good, you do good essentially because you hate evil and love the good, and because you have no desire for evil, not for some kind of reward you may receive for good behavior.

See also "Reward for Christians Paper" [NM 11] and "According to Works Paper" [NM 12].

NM 10: Proof Paper

NM10 Abstract

There is a physical faith; there is a Spiritual faith. Can you prove what kind of faith your have? How do Christians "keep the commandments"? Can Christian lose salvation? Give yourself the test.

Proof of Being a Christian: Prove Yourself

nm243 » How does one know he is a Christian? We know that one is a Christian when one has a changed attitude away from the ways of this age and towards the ways of the coming age. We know that a true Christian has the New Mind, which is the Spirit of God, and that this New Mind makes the Christian behave in a more positive manner. The main difference between a Christian and non-Christian is the fact that a Christian has the Spirit of God in himself. How does one know he really has the New Mind or Spirit of God? How do you know? How can we tell if we have the New Mind or not?

nm244 » The word of God, the Bible, tells us to "prove all things."(1Thes 5:21) Is there a way to prove one's Christianity, to prove to oneself, whether one is a real Christian?

nm245 » "Prove yourselves, whether you are in the faith; prove your own selves" (2Cor 13:5). God says through Paul's words, "prove yourselves"! Prove what? – whether you are in the faith. What faith? What kind of faith?

What Kind of Faith?

nm246 » "What does it profit, my brethren, though a man say he has faith, and have not works? can faith save him? ... show me your faith without your works, and I will show you my faith by my works ... But don't you know, O vain man, that faith without works is dead?" (James 2:14, 18, 20) Faith without good works is dead. "For as the body without the Spirit is dead, so faith without works is dead also" (James 2:26). Thus, we see there is a faith that is dead because there are no good works with this dead faith.

nm247 » Is this dead faith, the faith spoken about when, through Paul, God says?: "Prove yourselves, whether you are in the faith; prove your own selves" (2Cor 13:5). No, God is not talking about dead faith, for to continue: "Know you not your own selves, how that Jesus Christ is in you, if not you are reprobates?" Those not in the faith, who have not Christ in them are reprobates. What are reprobates according to the Bible?

Reprobates

nm248 » People who are reprobates must have a reprobate's mind. Paul defines a reprobate mind:

- "And just as they did not see fit to acknowledge God any longer, God gave them over to a depraved mind, to do those things which are not proper, 29 being filled with all unrighteousness, wickedness, greed, evil; full of envy, murder, strife, deceit, malice; *they are gossips.*" (Rom 1:28-29)

A reprobate mind does the opposite of good works. When God tells us to prove if we are in the faith, he is not speaking of dead faith belonging to the class of people called reprobates by God. By the way, the word translated "reprobate" is from a Greek word which means worthless.

nm249 » Those of the dead faith are those who "profess that they know God; but in works [they do works of reprobates, Rom 1:28-29] they deny him, being abominable, and disobedient, and unto every good work reprobate" (Titus 1:16).

nm250 » Those of the dead faith, do reprobate works; "they profess that they know God; but in works they deny him" (Titus 1:16). Those reprobates say (profess) that they know God, but they have the dead faith that does not do good works (James 2:20).

Law Keeper: Truth in Them

nm251 » "He that says he abides in him [Christ] ought himself also to walk, even as he [Christ] walked" (1John 2:6). How did Jesus walk? Jesus said, "I have kept my Father's commandments" (John 15:10).

What are his Father's commandments?

nm252 » What are Jesus Christ's Father's commandments? When Jesus Christ was on earth before his death, he was under the laws and commandments of the Old Testament. Christ kept these laws perfectly (not the add-on laws of the Rabbis). But now the law or system of love is in effect and this is the law or commandment that Christians keep. Please see the "Freedom and Law Paper" [NM 17] for more information on the system of love and God's laws.

Truth in them

nm253 » To continue, "He that says I know him, and keeps not his commandments, is a liar, and the truth is not in him" (1John 2:4). What is this "truth" that is not in him (the reprobate) who does not keep God's commandments? – "the *Spirit* of Truth," "the *Spirit* is Truth" (John 14:17; 1John 5:6). Hence, those who say they know Christ (1John 2:4), and those who say they abide in Christ (1John 2:6) ought to walk in God's commandments (1John 2:6; John 15:10). If they do not keep the commandments, then those who say they are in Christ and know Christ are liars, if they do not keep the commandments (1John 2:4). These reprobates do not have the Spirit of truth in them. By putting these verses together we can see that those who say they know and are in Christ (they call themselves Christians), but who do not keep God's commandments of love are liars. They are not Christians according to God's Word. And they are not Christians, for the *Spirit* of God or the *Mind* of God isn't in them.

nm254 » "Whosoever transgresses, and abides not in the doctrine of Christ [words of Christ, which are the words of God, John 17:14, 17], has not God. He that abides in the doctrine of Christ, he has both the Father and the Son" (2John 9). Whoever sins and does not keep the doctrines of God has not God, but he that keeps God's Word does have both God the Father and Christ in them. What does it mean to have the Father and the Son?

God's Spirit in All Christians

nm255 » "You Father, art in me, and I in you ... I am in my Father, and you [speaking of Christians] in me, and I in you" (John 17:21; 14:20). "One God and Father of all, who is above all, and through all, and in you [Christians] all" (Eph 4:6). What does it mean here about God in Jesus, Jesus in God, Christians in Christ, Jesus in Christians, and God the Father in all?

nm256 » "For as the body [the Church – 1Cor 12:27; Eph 4:4; 5:23-32; Col 1:24; Rom 12:4-5] is one, and has many members, and all the members of that one body [Church], being many, are one body: so also Christ. For by one Spirit are we all baptized into one body [Church] ... and have been all made to drink into one Spirit" (1Cor 12:12-13). Here it pictures the Church of God, which has one Spirit that all spiritually drink from. The Spirit or New Mind is in all the body. When Jesus says his Father is in him and he in his Father, that Christians are in him and he in Christians, and that God is in all, he is merely saying that God the Father's Spirit or Mind is in all – "all made to drink into one Spirit." Romans 8:9, 10-11, 14-16 reiterates this: to be the son of God one must have God's Spirit dwelling in him. Thus, "the body [Church] without the Spirit [of God] is dead" (James 2:26).

nm257 » Now we can better understand the verse: "whosoever transgresses, and abides not in the doctrine of Christ, has not God" (2John 9). Thus, those who go against the doctrines of love do not have God's Spirit or God's Mind. But "he that abides in the doctrine of Christ, he has the Father and the Son" (2John 9). He that has God's Spirit that is common to the Father and Son (John 14:10, 20) does abide in the doctrine of Christ. "He that says, I know Him ["I'm a Christian, I believe in him"], and keeps not his commandments, is a liar, and the truth [God's Spirit, 1John 5:6] is not in him" (1John 2:4). Thus, those who do not keep God's commandments are not Christians, they are liars because they do not have God's Spirit. Conversely, those who follow God's commandments are Christians and have God's Spirit (end of 2John 9). Those who are Christians will walk as Christ walked (1John 2:6). Yes, "and he that keeps His commandments dwells in Him, and He in him" (1John 3:24). The Spirit or Mind of God makes you in Christ, and Christ in you.

Christ The Vine

nm258 » In John 15:1-8 it pictures the body of Christ or the Church of God as a vine, with the branches as the members of the Church, and the root as Christ (See also Rom 11:13-24, "root" is Christ – Rev 22:16). Jesus says, "every branch in me that bears not fruit he [God] takes away" (John 15:2). Let's stop here. Now since this vine is metonymical for the Church, we know that the fruit this branch should bear is the fruit of God's Spirit: "love, joy, peace," etc (Gal 5:22). Thus, God takes away the branch that does not produce love, joy, peace, and so on. We know from 1Cor 13:4-7 that love is patient, kind, envies not, is not puffed up, rejoices in truth, and so on. Hence, the branch that does not produce fruit, is the one who does not produce good works of God's commandments, for "love is the fulfilling of the law" (Rom 13:10).

nm259 » Notice that Jesus says this branch that does not do good works or that does not produce fruit is in Him, but notice what Christ does not say: that Christ is in this branch. This is very significant! Why?: for "he that says he abides in Him ought himself so to walk, even as he [Christ] walked" (1John 2:6). He that says he is in his vine [his body, God's Church] should walk as Christ walked. He that says he is in Christ ought to produce fruit. "He that says he is in the light [God is light, 1John 1:5], and hates his brother, is in darkness even until now" (1John 2:9). Those who say they are in Christ but who hate their Spiritual brother, are in Satan (darkness) even until now. These are in Satan's church, not in God's if they hate the Spiritual brother: "The man that wanders out of the way of understanding shall remain in the congregation of the dead" (Prov 21:16). Notice they remain or rest (Hebrew) in the church of the dead, they do not go back to it when they wander, but remain or rest in it – "is in darkness even until now" (1John 2:9).

nm260 » "He that says, I know Him, and keeps not His commandments, is a liar, and the truth [God's Spirit, 1John 5:6] is not in him" (1John 2:4). If the Mind of truth is not in them, they are not Christians.

nm261 » "But whoso keeps his word, in him verily is the love of God perfected: hereby know we that we are in Him" (1John 2:5). Those in Christ know they are in Him when they keep his word.

nm262 » Hence, the branch taken away for not bearing fruit (John 15:2) was never in Christ (in the vine), for the proof that you are in Him is that you produce fruit [keep his Word – 1John 2:5; keep his commandments –1John 2:4; walk like Christ – 1John 2:6; not hate your brother –1John 2:9]. You know you are in Christ when you produce the fruit of the Spirit (Gal 5:22-23). Those in Christ, and Christ in them, have what is in Christ. They have God's Spirit and Mind. And those in Christ and Christ in them do produce the fruit of love.

nm263 » "And every branch that bears fruit, he purges ["cleans"] it, that it may bring forth more fruit ... abide in me, and I in you. As the branch cannot bear fruit of itself, except it abide in the vine; no more can you, except you abide in me. I am the vine, you are the branches: He that abides in me, and I in him, the same brings forth much fruit: for apart from me you can do nothing" (John 15:2, 4-5). Notice it says abide in me, and I in you. As explained previously, those in Christ and with Christ in them are in the Spirit of God, are in the body of Christ and drink in the same Spirit (1Cor 12:12-13; John 14:10, 20). The branch that was in Christ (John 15:2), did not produce fruit, because Christ was not in the branch. Verse 2 only says that it was in Christ, the verse did not say Christ was in the branch. This branch, in other words, *says* it is in Christ or in the Church, but the fact that it does not produce fruit means it was never in Christ.

nm264 » Now we see those in Christ, and Christ in them do produce much fruit (John 15:5). And it adds, that one can't do anything if it is apart from Christ, that is, apart from the Church because it does not have God's Spirit. One does produce much fruit if the Spirit is in him. "If you keep my commandments, you shall abide in my love" (John 15:10). "God is love" (1John 4:8, 16). "God" can be used metonymically here for love: "If you keep my commandments, you shall abide in my God." When one keeps the commandments he is in God, and God is in him. "And he that keeps His commandments dwells in Him, and He in him" (1John 3:24). Those in God, and God in them, produce much fruit of the Spirit (John 15:5, 8, 16). "The Head [Christ is the Head – Eph 1:22], from which all the body [Church] by joints and bands having nourishment ministered, and knit together, increases with the increase of God" (Col 2:19).

nm265 » Thus, we see that you are in Christ or in the Church, if God's Spirit is in you. And that you will keep the commandments if God is in you as opposed to those who <u>say</u> they are in

Christ, but have not God's Spirit and do not keep the commandments.

Keep the Commandments?

nm266 » What does the Bible mean, keep the commandments? We know that only one person who ever lived on earth was sinless, and he was Christ (Heb 4:15; John 8:46; John 15:10; 1Pet 1:19). And we know that all others have transgressed or sinned (Rom 3:9; 1John 1:8, 10). What does God mean when through John, he says, "And hereby we do know that we know him, if we keep his commandments" (1John 2:3). Doesn't this verse, and other similar ones, mean to keep the law as Christ the man did while he was living? Doesn't this mean keep the law of love perfectly?

nm267 » When we keep God's law of love in this age we keep it Spiritually. That is, we keep it in our attitude or mind. But because we are in the age of the other-mind, the mind of confusion, the mind of Satan that misleads us, then the other thoughts from the other-mind will sometimes confuse us, yet our true desire is toward the way of love. Now the Bible does say that those born of God do not sin: "We know that whosoever is born of God sins not" (1John 5:18). This should be taken in the Spiritual sense. That is, those with the Spirit of God in their minds will desire the good. But sometimes, in physical weakness, they may do what is evil because of the other-mind and outside pressures. Yet in their Spiritual mind they will hate their act. They will never enjoy their wrong act like most of this age which do enjoy their evil. It is only when one is BORN of God that one won't sin at all (see "Begotten, Born Paper" [NM 5]). See the "Freedom and Law Paper" [NM 17] to understand what sin is.

Christians Do Not Sin Willfully

nm268 » Paul is speaking,

- "I do not know what I am doing. For what I want to do I do not do, but what I hate I do. And if I do what I do not want to do, I confirm [by my actions] that the law is good. As it is, it is no longer I myself who do it, but it is sin living in me. I know that nothing good lives in me, that is, in my sinful flesh. For I have the desire to do what is good, but I cannot carry it out. For what I do is not the good I want to do; no, the evil I do not want to do – this I keep on doing. Now if I do what I do not want to do, it is no longer I who do it, but it is sin living in me that does it. So I find this law at work: When I want to do good, evil is right there with me. For in my inner being I delight in God's law; but I see another law at work in the members of my body, waging war against the law of my mind and making me a prisoner of the law of sin at work within my members" (Rom 7:15-23).

Paul does not of his true self want to sin, it is the law of sin that dwells in him that wishes to sin. When Christians sin, it is because of this law of sin (the other-mind), that satanic spirit in their minds or that "spirit that dwells in us lusts to envy" (James 4:5). See the "Other Mind" paper [NM 21] for more details.

nm269 » Therefore Christians may sometimes physically sin, but not willfully or deliberately or intentionally. True Christians want to follow in the way of love. They do not willfully sin because it proves something:

- "For if we sin willfully after we received the knowledge of the truth, there remains no more sacrifice for sins, but a terrifying expectation of judgment and the fury of a fire which will consume the adversaries" (Heb 10:26).

If we sin willfully, we go to judgment. This judgment as we learn in NM24 will be with fire for the angels of sin, and death for mankind sent to the judgement. But notice that it says, "if we sin willfully . . ." It does not say after we have received the Spirit of God, but after we receive the knowledge of truth. And it did say IF. "If" is a hypothetical word. The hypothetical word "if" is used to begin a train of thought "based on, involving, or having the nature of a hypothesis" (Webster's Dictionary). A hypothesis is a supposition. Another word for "if" is suppose. So this train of thought beginning in Hebrews 10:26 is a hypothetical situation: IF Christians do this, then they can expect to go to the judgment. Notice verse 39: "but we [Christians] are not of them who draw back into perdition; but of them that believe to the saving of the soul." Now that is no hypothetical statement. It is a positive statement. Real Christians do not fall back. Those that sin willfully are those who follow their "other-mind" and sin willingly. When you sin willfully you are on the way to proving that you may not be a real Christian.

nm270 » Notice the positive statement: "and grieve not the holy Spirit of God, whereby you are sealed unto the day of redemption" (Eph 4:30; see Isa 59:21). This isn't a statement of hope! If one was to say this was only positive thinking on Paul's part, then what about the statement about the Messiah's return? How would we distinguish between "positive thinking" statements and fact? Who would decide? No, Ephesians 4:30 and Hebrews 10:39 are to be taken as factual statements. Christians do not sin willfully. They will sin, but they will not be proud of their sins and will repent of them quickly. Real Christians are sealed with the Spirit to the day of redemption and their belief will save their souls.

Tests to Prove Yourself

Hypothetical Word Formations

nm271 » We'll explain why Paul used the hypothetical word formation in Hebrews 10:26ff. This is not the only place Paul uses this type of word formation. For example, Hebrews 6:6 is the beginning of another hypothetical train of thought, yet notice once again the "but" sentence which is somewhat like Hebrews 10:39, "but, beloved, we are persuaded better things of you, and things that accompany salvation, though we thus speak" (Heb 6:9).

nm272 » These word constructions are much like those in the First Letter of John. For example, "and hereby we do know that we know him, IF we keep his commandments ... IF any man love the world, the love of the Father is not in him" (1John 2:3, 15). These are tests of proof of one thing or another.

nm273 » In 1John 2:3 what are we testing for? If one passes this test in this verse, what does he prove? "If we keep his commandments," if we pass this test what do we prove? "And hereby we do know that we know him." If we keep his commandment we know God. If we pass the test we prove that we know him.

nm274 » In 1John 2:15, what are we testing for? If one passes this test what does he prove? "If any man love the world," If we pass this test (if we love the world's way), what do we prove? We prove that the love of the Father is not in us. If we love the world's way the Father's love is not in us.

nm275 » In Hebrews 10:26-39, what are we testing for? If we pass this test in these verses, what does it prove? "If we sin willfully after that

we have received the knowledge of truth," if we pass this test (if we sin willfully), what does it prove? "There remains no more sacrifice for sins, but a certain fearful looking for judgment." If we sin willfully, then we must look fearfully for judgment which shall devour the adversaries of which we would be one. If we pass the test (by sinning willfully), we prove that we shall be devoured. But this word formation adds, after further amplification in verses 28-38, that "we [Christians] are *not* of them who draw back into perdition; but of them that believe to the saving of the soul" (Heb 10:39).

Lose The Spirit, Lose Salvation?

nm276 » Do you see the pattern? In Hebrews 10:26-39 it says that those who sin willfully will receive judgement of being devoured for such a sin. If any that are called Christian sin willfully, then they prove that they belong to those who are going to be devoured. But in verse 39 it qualifies the "we" in verse 26 by stating a positive fact that we, the real Christians, are to the salvation of the soul. And verse 39 is confirmed by Ephesians 4:30 where it says those sealed with the Spirit are sealed to redemption. And Ephesians 4:30 can be confirmed by the fact there is not a verse in the Bible that states as fact that one who receives the Spirit of God can lose it.

nm277 » Notice: "And grieve not the holy Spirit of God whereby you are sealed unto the day of redemption" (Eph 4:30). Christians are sealed with the Spirit to the day of redemption. This is so because in Christ's words, "My Father, who has given them [Christians] to me, is greater than all; no one can snatch them [Christians] out of my Father's hand" (John 10:29).

nm278 » Notice further proof that once the Spirit is given it is sealed to the day of God. "For whatsoever is begotten of God overcomes the world: and this is the victory that overcomes the world: our faith" (1John 5:4). Faith is the power that helps the Christian to overcome the world. Now Faith is the fruit of the Spirit (Gal 5:22). "For by grace are you saved through faith; and that [faith] not of yourselves: it is the gift of God" (Eph 2:8). Why does Faith help us to overcome and be saved? Or better yet, why does the Spirit, for Faith comes from the Spirit, help Christians to overcome and be saved? "You are of God, little children, and have overcome them: because greater is he [God] that is in you, than he [Satan] that is in the world" (1John 4:4).

With The Spirit You Are Saved

nm279 » Do you see? Once one has the Spirit of God he is sealed to the day of redemption. Once one becomes a Christian, he will be saved (Heb 10:39). There is no way that anyone once begotten of God's New Mind can lose out on salvation. But once begotten, can you do anything? No, "for we are his workmanship, created in Christ Jesus unto good works" (Eph 2:10). If one wants to do anything but good works, how can he be a Christian? Christians are created for good works (Eph 2:10). A Christian wants to do good works; he has a good attitude.

nm280 » "If any man love the world, the love of the Father is not in him" (1John 2:15). The "love of God" can only come from the Spirit of God, for the love of God is a fruit of God's Spirit (Gal 5:22). With this knowledge what is the test being given here to man by God through John? "If any man love the world, the Spirit of the Father is not in him." If you pass the test of loving the world, you prove that you have not the Spirit or Mind of God, thus, you are not a Christian. Remember to be a Christian one must have a changed attitude away from the world's way. If one is convinced that the world's way is wrong, why would he go back after he had received God's Spirit which is greater than the other-mind or other spirit. "Prove yourselves, whether you be in the Faith; prove your own selves. Know you not your own selves, how that Jesus Christ is in you, except you be reprobates" (2Cor 13:5). Prove yourselves to see if you have God's Spirit that does produce much fruit, and helps you to have a good attitude. If you do not find this proof, then you are worthless (a reprobate) with a worthless mind concerning the way to love, peace, and harmony (see Rom 1:28-31).

More Tests To Give Yourself

nm281 » What are some tests you can give yourself in order to prove to yourself that you are a true Christian? We have given you many tests already in this paper. Do you love God's law of love? Why not?, unless you are a reprobate. If you sin willfully, doesn't that mean you love sin? And if you love sin, how can you love God's law? For "love is the fulfilling of the law" (Rom 13:10). "But he that sins against me [God] wrongs his own soul: all they that hate me love death" (Prov 8:36). If one loves to watch violence, isn't he projecting the fact that he

loves forms of death-producing activity? He loves death! (note Rom 1:32)

nm282 » Look at the test in Romans 1:28-31, if you pass that test you prove yourself a reprobate or one with a worthless mind. If you love the world's way, then you prove you do not have the Spirit (1John 2:15).

nm283 » Which of the following tests do you pass. "He that loves his life [in this world] shall lose it; and he that hates his life in this world shall keep it unto life of aeonian" (John 12:25). If the world is as bad as God tells us through his Bible, why, if you have God's New Mind would you like your life in this world or this old age?

nm284 » "They [the ones who call themselves Christians] went out from us, but they were not of us; for if they had been of us, they would have continued with us" (1John 2:19). Those who physically go out of the Church prove they were never a part of the Church. But this does not mean that those who were in the Church physically and then went out, can't at some future time come Spiritually into the Church.

nm285 » "In this the children of God are manifest, and the children of the devil: whosoever does not righteousness is not of God, neither he that loves not his brother" (1John 3:10). A person proves he is not a child of God by not doing righteousness, and by not loving his Spiritual brother.

nm286 » "Whosoever hates his brother is a murderer: and you know that no murderer has aeonian life abiding in him. Hereby perceive we the love of God, because he laid down his life for us: and we ought to lay down our lives for the brethren" (1John 3:15-16). If one hates his Spiritual brother he is as good as a murderer, and he has not the aeonian life in him, or he has not the Spirit that brings this aeonian life in him; he is not a true Christian (1John 3:10). Then verse 16 shows that those who have the love of God, prove they have God's Spirit if they do lay down their life for their Spiritual brother like Christ did. "He that finds his life shall lose it [in the seventh millennium]: and he that loses his life for my sake shall find it [in the seventh millennium]" (Mat 10:39). If one is willing to give his life for Jesus Christ's sake, this proves God's Spirit is in him. A real Christian if called upon must give up his life, if not, he proves he is not a Christian. There are two senses of giving up one's life: (1) physical sense – the giving up of one's physical life; (2) the Spiritual sense – the giving up of one's former spiritual life of confusion.

nm287 » "My little children, let us not love in word, neither in tongue; but in deed and in truth. And hereby we know that we are of the truth, and shall assure our hearts before him" (1John 3:18, 19). If one is a hypocrite and only does God's law through words and not deeds, he is not of the truth. God's truth comes from his Spirit (John 14:17). One does not have the Spirit if he is a hypocrite.

nm288 » "And he that keeps His commandments dwells in Him, and He [Christ] in him. And hereby we know that he abides in us, by the Spirit [New Mind] which he has given us" (1John 3:24).

nm289 » "We are of God: he that knows God hears us; he that is not of God hears us not. Hereby know we the Spirit of truth, and the spirit of error" (1John 4:6). The one that hears has the Spirit of truth; the one that does not hear has the spirit of error. "God has given them the spirit of slumber, eyes that they should not see, and ears that they should not hear" (Rom 11:8). But those of the Spirit of truth (the New Mind) will hear Spiritually. "And when he puts forth his own sheep he goes before them, and the sheep follow him: for they know his voice" (John 10:4).

nm290 » "We glory in trial also; knowing that trial works patience; and patience proof [good works like patience are proof one has the New Mind]; and proof, hope [hope for a better resurrection, Heb 11:35]: and hope makes not ashamed [shame of aeonian contempt, Dan 12:2]; because the love of God is shed abroad in our hearts through the Holy Spirit which is given to us" (Rom 5:3-5).

Prove Yourself

nm291 » "Prove yourselves, whether you be in faith; prove your own selves, how that Jesus Christ is in you; if not you be reprobates?" (2Cor 13:5)

nm292 » Test yourselves to see if you have the Spirit or Mind of God, for he who does have it does produce much fruit. Those who do not have the Spirit are reprobates or worthless and are the ones of the "dead," "for as the body without the Spirit is dead" (James 2:26).

nm293 » You will be resurrected, if the Spirit is in you. "Now if man have not the Spirit of Christ, he is none of his ... But if the Spirit of

him [God] that raised up Jesus from the dead dwell in you, he that raised up Christ from the dead shall also quicken your mortal bodies by his Spirit that dwells in you" (Rom 8:9, 11). That is a statement of *fact*, not hope, for as we have shown, once one has received the Spirit he can't lose it. It is the Spirit of God, or the New Mind which brings salvation.

nm294 » *To Review*, the Bible tells us those begotten of God are sealed with the Spirit to the day of redemption. Further, the Bible gives its readers many tests to take. If one passes one kind of test, then this proves he has the Spirit or New Mind and will be saved. If one passes the other kind of test he proves himself a reprobate, and manifests that he does not have God's Spirit, thus, is not a real Christian. All the tests that prove one a Christian have one thing in common: you prove you have the Spirit of God when you keep the law of love (see "Freedom & Law Paper" [NM 17]). All the tests that prove one is a reprobate also have one thing in common: you prove you do not have the Spirit of God when you do not follow the law of love.

nm295 » A Christian's good works are his proof that he is a real Christian. We do not prove anything to others when we do good works. We do not try to show-up others when we do good works. When a Christian does good works they come from his heart. A Christian has the New Mind with the new attitude. This New Mind gives real Christians the power to do good works. We keep God's law of love because it comes from the heart, not because we are trying to prove anything to others.

Why Paul Ran the Race

nm296 » When we do good works it is our proof that we do have the New Mind – the Spirit of God. That is why Paul ran the race, to prove to himself he was a real Christian. If Paul let up on his good works, it would mean that he was a reprobate. But it was God's Spirit in Paul that gave Paul the power to overcome the world (see 1Cor 9:24-27; 1John 5:4 & 1Cor 15:10). Paul ran to prove to himself his salvation.

What about Doubts?

nm297 » You run the race to prove to yourself about your salvation. But what about doubts? Some admit that they have doubts about their Christianity. This is not unusual. Every Christian in the old age will have doubts about their own Christianity. This is because of the other-mind (NM 21). The other-mind puts doubts, fear, and other uncomfortable ideas into our minds. Look at Elijah. He doubted himself and felt he was no better than anyone else (1Kings 19:4), yet God chose him to do his will. But we are the overcomers; we will overcome to reach our salvation because we have the Spirit and our Spirit is greater than the one who rules the evil age (1John 4:4).

"But let every man prove his own work, and then shall he have rejoicing in himself alone, and not in another." (Gal 6:4)

Caution must be had here. Each Christian is given the Spirit by measure (Rom 12:3; Eph 4:7). There may be some very weak Christians. We not only need to be careful how we judge others (NM23), but also how we judge ourselves.

NM 11: "Reward" for Christians

Eternal Life: Our Reward?
Aeonian Life?
Immortality Promised to each Christian
Aeonian Life: What is it?
Aeonian Life: When does it Begin?
Aeonian Life: Where will this Age be?
Aeonian Life: How Long is this Age?
Aeonian Life: An Age within an Endless Age
Aeonian Punishment
Resurrection from Aeonian Punishment

NM11 Abstract

In this paper we will learn the very important distinction between aeonian life and immortality. We will again analyze and examine the words mistranslated into eternal and forever. Because these words were mistranslated in the Bible the difference between aeonian life and immortality has been hidden from almost everyone.

Eternal Life: Our Reward?

nm298 » What is the "reward" for being a Christian? That is, what is the result or effect of being a Christian besides having the New Mind of love, joy, and peace. "The gift of God is *eternal* life through Jesus Christ our Lord ... the righteous into life *eternal*" (Rom 6:23; Mat 25:46).

nm299 » Many believe that the reward for being a Christian is eternal life. (This is not to say that a Christian has *earned* the reward by himself. See the "Predestination Paper" [NM 8].). What they mean is that at the resurrection they will be made alive forever, because they can't mean eternal life, for life eternal means: life without beginning or end.

Key Word Means Aeonian, not Eternal

nm300 » The Bible was inspired mainly in two languages, Hebrew and Greek. In Romans 6:23 and Matthew 25:46 the Greek word that was translated into English as "eternal," is *aionios*. This Greek word at the time it was inspired to be used in the New Testament meant, agelasting, aeonian, or a period of time of

significant character, or an indefinite period of time, or a indeterminate period of time, or simply an age. Thus *aionios* means an agelasting period of time of unknown length. This Greek word has been mistranslated into such English words in the New Testament as: eternal, everlasting, and forever. But the word does not mean forever. At the time it was written by the New Testament writers it meant – an agelasting period of unknown length. One can only ascertain the length of time by the context in which the word is used, or from other scriptures related to the subject. The inspired word means: *agelasting* or aeonian (see "Age Paper" [NM 7]).

nm301 » ***Meaning of Words***. Just because a word has evolved through misuse to mean something, are we to take its evolved meaning, or its original meaning? Words are only symbols of meaning. The important quality about a word is the meaning that was meant to be conveyed by the speaker or writer. When the New Testament writers were inspired to write, they were inspired to write down meanings through the medium of symbolic words. A word is only as good as the meaning it stands for. If we want God's inspired meaning, we should take the word for what it meant when it was written, and not the evolved meaning.

Agelasting or Aeonian Life?

nm302 » After we correct the Greek word we get, "the gift of God is *aeonian* life through Jesus Christ our Lord ... the righteous into life *aeonian*" (Rom 6:23; Mat 25:46). Now these verses are in context telling us about one result of being a Christian is aeonian life. What does that mean? It means what it says, life for an aeonian period of unknown time.

nm303 » But we have a problem here. Does this mean the result of being a Christian is an aeonian period of time? Yes. But does it mean, then, that Christians born of God will die after this aeonian life, for an aeonian period of time to be aeonian or agelasting indicates or implies the possibility of an end to the age at some time? (see below, *An Endless Age?*)

Immortality Promised to each Christian

nm304 » Will Christians die after this period of time? No, for those resurrected will be born of God (1John 5:18). God is a spiritual being. Those born of spirit are spirit (John 3:6). We know that spiritual beings do not die (Luke 20:36). In Christ's own words, those resurrected, "neither can they die any more ... and are the children of God, being the children of the resurrection" (Luke 20:36). This is confirmed in 1Corinthians 15:52, "for the trumpet shall sound [Rev 11:15], and the dead shall be raised incorruptible [immortal], and we shall be changed." Thus, those resurrected will have life forever.

Paradox

nm305 » After we correct the translation for Romans 6:23 and Matthew 25:46 we have a paradox: the gift, or the "reward," or the result of being a Christian is life aeonian not life forever. But because they will be resurrected or born into an immortal state as 1Cor 15:52 and other verses say, they, of course, will not die. But immortal life is not the promise made by such verses as Romans 6:23 or Matthew 25:46. These verses only promise aeonian life, or life for an age. We need to know what this *aeonian life* means and why it was used. There is a logical answer to this as you shall see shortly.

nm306 » We have shown that Christians when born of God in the resurrection will not die. We will now show you several more verses to confirm that one benefit for being a Christian is aeonian life. In other words, immortal life is but one result of being a Christian. But immortal life will also be given to those who will not become Christians in the age of Satan's spiritual rule (NM13). Therefore, Christians from the present age will receive *another gift* besides immortality, and that gift is called aeonian life by the Bible (Rom 6:23). What is this aeonian life?

Aeonian Life What is it?

nm307 » "There is no man that has left house, or parents, or brethren, or wife, or children, for the kingdom of God's sake, who shall not receive manifold more in this present time, and in the age ["world"] to come life *aeonian*" (Luke 18:29-30).

- "Labor not for the meat which perishes, but for that meat which endures unto *aeonian* life" (John 6:27).

- "That whosoever believes in him should not perish, but have *aeonian* life" (John 3:15).

- "And he that reaps receives wages, and gathered fruit unto life *aeonian*" (John 4:36).

- "Search the scriptures; for in them you think you have *aeonian* life" (John 5:39).

Thus, the effect of being a Christian, as projected by the above five verses and Romans 6:23 and Matthew 25:46, is *aeonian* life. But we still haven't any real knowledge about this aeonian life. What is it?

- "Whereas you have been forsaken and hated, so that no man went through you, I will make you an *aeonian* excellency, a joy of many generations" (Isa 60:15). Even in the Old Testament such words as eternal, everlasting, and forever are mistranslated. Over 400 places in the Old Testament the Hebrew word "olam" is mistranslated into eternal, everlasting, and forever, while it should have been translated, *aeonian*. Isaiah 60:15 is speaking about Zion. Zion in the Spiritual sense represents Christians. God says he will make them an aeonian excellency.

- "And every one that has forsaken houses, or brethren, or sisters, or father, or mother, or wife, or children, or lands, for my name's sake, shall receive a hundredfold, and shall inherit *aeonian* life" (Mat 19:29).

- "And if children, then heirs; heirs of God, and joint-heirs with Christ; if so be that we suffer with him, that we may be also glorified together" (Rom 8:17).

Those who will receive an aeonian excellency (Isa 60:15) and inherit aeonian life (Mat 19:29), will be glorified together (Rom 8:17). When? Where?

Aeonian Life: When Does it Begin?

nm308 » "Therefore I [Paul] endure, all things for the elect's sake, that they [the elect] may also obtain the salvation which is in Christ Jesus with *aeonian* glory" (2Tim 2:10). "He [Christ] became the author of *aeonian* salvation unto all them that obey him" (Heb 5:9). Here it is speaking of aeonian salvation (Heb 5:9), and that with this salvation will be aeonian glory (2Tim 2:10). When and where is this aeonian excellency, aeonian salvation, aeonian glory, and aeonian inheritance of life?

nm309 » "And this is the will of him that sent me, that everyone which sees [Spiritually] the Son, and believes [Spiritually] into him, may have *aeonian* life: [when?] and **I will raise him up at the last day**" (John 6:40). "This is that bread which came down from heaven: not as your fathers did eat manna, and are dead: he that eats of this bread shall live in the *age*" (John 6:58).

nm310 » Those Christians who will have aeonian life will be raised or resurrected up at the last day (the 1000 year Sabbath). As Revelation 20:4-6 indicates, this is the resurrection at the beginning of the millennium. And they shall live in the age. This aeonian life begins at the start of the 7th millennium when Christ physically returns (see "Thousand Years and Beyond" and "Last Judgment" papers). At that same time Christians will be made immortal (1Cor 15:52-55).

nm311 » Thus, the aeonian excellency, glory, and salvation begins at Christ's physical return. And Christians will live in that age. What age? We know when this age begins (at Christ's return), but when does it end, and what happens in it, and where will this age period of glory be located at?

Aeonian Life: Where will this Age be?

nm312 » Romans 8:17 tells us Christians will be glorified *together* with Christ. Isaiah 60:15 speaks of an aeonian excellency, and 2Timothy 2:10 speaks of an aeonian glory. And all these begin at Christ's return ("Thousand Years and Beyond" paper [NM 15]). What else begins at Christ's return?

- "The *aeonian* kingdom of our Lord and Savior Jesus Christ" (2Pet 1:11). The word, "everlasting," in this verse in the KJV is translated from the Greek word *aionion* which means, aeonian. What does this mean? It means that the ruling kingdom of God under Christ will be an aeonian one.

- "The most High [Christ], whose kingdom is an *aeonian* kingdom" (Dan 7:27). "How great are his signs and how mighty are his wonders! his [Christ's] kingdom is an *aeonian* kingdom" (Dan 4:3).

- In Daniel 7:14 it is speaking about the return of Jesus, "and there was given him dominion, and glory, and a kingdom, that all people, nations, and languages, should serve him: his [Christ's] dominion is an *aeonian* dominion, which shall not pass away, and his kingdom that which shall not be destroyed."

- Even the promised land that was promised to Abraham and his seed, was promised for an aeonian period of time *not* forever as mistranslated in Genesis 13:15.

- Christians will be glorified with Christ in the kingdom of God as rulers (Rev 20:4) in this age-Kingdom that has been put under the authority of Christ the God, who will be King of kings. This Kingdom will be located *on* the earth (Rev 5:10). It is the government of God under the aeonian dominion of Christ the God (Dan 7:14; 1Cor 15:27-28, 24).

nm313 » **Therefore the effect for being Christians** (besides immortality, which all others will eventually obtain) is aeonian existence and rulership, in the aeonian glory of the kingdom of God under Christ the God, until the typical end of the Spiritual creation when all of the God family will typically take over the kingdom in equality. The Family of God will continue forever since they are immortal, but the government under Christ the God will be given over to the God, who at that time will be all in all (1Cor 15:28; *God Papers*).

Aeonian Life: How Long is this Age?

nm314 » The *aionios* time or aeonian time begins when Christ returns (Rev 11:15) and exists during the 1000 years (Rev 20:4-6). After this age there will be a short age before the creation of the New Heaven and Earth, wherein *every*one will be alive (Isa 65:20; John 7:37; see "Thousand Years and Beyond"). Thus, the only special "reward" of being a Christian is rulership for 1000 years, an aeonian time. After this aeonian kingdom of God under Christ, the kingdom of God will continue into the next age of unity until the God is all in all (see the *God Papers*).

Aeonian Life: An Age within an Endless Age

nm315 » There are ages within the great new age of the True god. The 1000 years is an age, as well as the short age after the 1000 years. These are ages within the greater age. As shown in the "Age Paper" [NM 7] there is an age of Satan and an age of God. Satan's age *ends* at Jesus Christ coming back to the earth. But God's age does not end: (1) "His dominion is an aeonian dominion *which shall not pass away*, and his kingdom that which shall not be destroyed" (Dan 7:14, see Hebrew text).; (2) "And he will reign over the house of Jacob into the ages, and of his kingdom there will be no end" (Luke 1:33, see Greek).

nm316 » To repeat. This aeonian kingdom has a beginning: it starts at Christ's return. (In a sense it started with Christ first appearance.) But this aeonian kingdom has no end, 'which shall not pass away, and his kingdom that which shall not be destroyed.' It is an aeonian system because it has a beginning, but it has no end. Contrariwise, Satan's system has a beginning, and it has an end.

nm317 » Thus, the Bible by definition tells us that God's age will not end. This does not mean *aionios* means forever. It still means age or eon, but this new age or eon of harmony will never end. God won't allow it to end like other ages. Christians are "rewarded" with life throughout the new age – from its very beginning at the start of the 1000 years. The others miss the full 1000 years.

Aeonian Punishment

nm318 » Conversely, the effect for the others, besides the Christians who lived during the age of the other-mind (Satan), will be death (Rom 6:23) as punishment (Mat 25:46; Ezek 33:8; see Last Judgment paper [NM 24]). This punishment will be an "aeonian punishment" (Mat 25:46). These are those "who shall be punished with aeonian destruction [death, Rom 6:23] from the presence of the Lord, and from the glory of his power" (2Thes 1:9). How long will this aeonian punishment last? As a whole the non-Christians will be punished with death for 1000 years during the seventh millennium. Then after this 1000 year death the "dead" will be resurrected as humans in the resurrection of the dead (1Cor 15:21; Acts 24:15; Rev 20:5, 13). They will live after the 1000 year age in the atonement period (Great Last Day) as humans begotten of God's Spirit. Then after this atonement period they will be born of God at the antitypical or true end of creation (1Cor 15:24-28) much like those real Christians who are still alive at Christ's physical return (1Cor 15:52-55).

Resurrection from Aeonian Judgment

nm319 » John 5:29 which reads, "the resurrection of judgment," is one and the same with the "resurrection of the dead." This is so because the people called the "dead" by the Bible will be dead during the 1000 year judgment or punishment. The resurrection of judgment is the resurrection of those of the judgment of the dead. Therefore the "dead" will be raised from their judgment in the "resurrection of the judgment" (See Last Judgment paper [NM 24]).

NM 12: According to Works

God Awards Every Man According to his Work
Negative Rewards for Negative Behavior
God Gives Good for Good
God Rewards According to Works
God Gives the Power
Grace Given According to Measure
Paul as an Example
Notes: Parable of Pounds and Talents
Usury

NM12 Abstract

What does the Bible mean when it talks about mankind being awarded "according to his works"? Does this mean we can actually earn our way into paradise by working at it, or is paradise a gift given by the grace of God? What is grace? We will examine scripture pertaining to Paul to learn more about these things.

God Shall Award Every Man According to his Work

nm320 » "For the Son of man shall come in the glory of his Father with his angels; and then [at that time] he shall reward every man according to his work" (Mat 16:27). God through his word tells us he rewards men according to their works. "I will judge you every one after his ways" (Ezek 33:20). What does he reward for men's work?

By Grace you are Saved

nm321 » Before we show you what is meant by according to works, we need to know the fact that "by grace you are saved ... for by grace are you saved through faith; and that [Faith] not of yourself: it is the gift of God: Not of works, lest any man should boast" (Eph 2:5, 8-9). People are saved (freed) by the free gift of God. People aren't freed or saved through something they do. It is a free gift. But what does God reward *according to works*?

Negative Reward for Negative Behavior

nm322 » "Alexander the coppersmith did me [Paul] much evil: the Lord reward him according to his works" (2Tim 4:14). "Therefore it is no great thing if his [Satan's] ministers also be transformed as the ministers of righteousness; whose end shall be according to their works" (2Cor 1:15). "And the dead were judged out of those things which were written in the books [Bible], according to their works" (Rev 20:12). Thus, those doing bad works are judged according to their works and given the reward suitable for their ways. What is the "reward" for wrong?

Wages of Sin is Death

nm323 » "For the wages of sin is death" (Rom 6:23). The reward or wages for bad works is death. And this death is for an age: "these shall go away into aeonian punishment" (Mat 25:46).

Punished by Aeonian Destruction

nm324 » And these, "shall be punished with aeonian destruction from the presence of the Lord, and from the glory of his power" (2Thes 1:9). As we show in the paper, "Reward for Christians" [NM 11], the punishment for those who do evil will be a 1000 year death away from the coming utopia. The words "forever" and "eternity" in the Bible are wrong translations of words that mean either *age* or *agelasting (aeonian)* (see "Age Paper" [NM 7]). Those rewarded for wrong behavior will be given their "reward" of death in the aeonian judgement that lasts for 1000 years. They are dead and in the ground ("hell") while the utopia goes on above them on earth. This "reward" begins at the Messiah's coming (Rev 22:12; Mat 16:27). But what is the reward for good works?

God Gives Good For The Good

nm325 » "I will give unto every one of you according to your works" (Rev 2:23). "And he that overcomes, and keeps my works unto the end, to him will I give power over the nations: and he shall rule them with a rod of iron" (Rev 2:26-27). "To him that overcomes will I grant to sit with me in my throne, even as I also overcame, and am set down with my Father in his throne" (Rev 3:21). Those doing good works to the end will become rulers in the New Age, the kingdom of God, beginning on the first day of the 1000 years (Rev 20:4c). "And has made us unto our God kings and priests: and we shall reign *on* the earth" (Rev 5:10). Those who do good works will become rulers on earth as the kings and priests of God's Kingdom on earth beginning at his coming (Mat 16:27). But God says he will reward *according* to their works. What does that mean?

God Rewards According to Works

nm326 » "They thought that the kingdom of God should immediately appear. He said therefore, A certain nobleman [J.C.] went into a far country [heaven] to receive for himself a kingdom [of God], and to return. And he called his ten servants, and delivered them ten pounds, and said unto them, Occupy till I come" (John 14:2-3). "And it came to pass, that when he had returned, having received the kingdom [of God], then he commanded these servants to be called unto him to whom he had given the money, that he might know how much every man had gained by trading. Then came the first, saying, Lord, your pound has gained ten pounds, and he said unto him, Well, you good servant: because you have been faithful in a very little, have you authority over ten cities." This servant gained ten times what he was given. According to this work he was given rulership over ten cities. "And the second came, saying, Lord, your pound has gained five pounds. And he said likewise to him, Be you also over five cities" (Luke 19:11-19). The second servant who had increased what was given to him five fold, according to his work, he received rulership over five cities. When God rewards good works *according* to them, he is rewarding according to the amount of good works. The more good works in this age, the more service one will do in the new age (note Mat 20:25-28).

Not for Reward

nm327 » From this we know what the Bible means by "reward according to works." Or do we? Just whose works are these that are performed? Did the person who will receive rulership over ten cities earn these rewards through his own will and power? Is the degree of good works achieved through a person's *own* will and power? Is the degree of good works from, and of, the person who performs them? Or, do they come from another source? Could it be that one is *given* the power to perform these good works? Notice the following: "I have raised him up in righteousness, and I will direct all his ways: he shall build my city, and he shall let go my captives, not for price nor reward, says the LORD [YHWH] of hosts" (Isa 45:13).

God Gives the Power to do Good Works

nm328 » "Are you so foolish? having begun in the Spirit, are you now made complete by the flesh?" (Gal 3:3) Are those who have the New Mind (Spirit of God) made complete through the flesh? Notice Paul is speaking to those who are supposed to be Christians. He asks now after you have the Spirit are you made complete through the flesh or by the flesh?

nm329 » Paul reveals something when he says: "But when it pleased God, who separated me from my mother's womb, and called me by his grace, to reveal his Son in me" (Gal 1:15-16). Paul was called by grace to reveal Jesus in himself. What does this mean? How does Paul reveal Christ in himself?

nm330 » "But by the grace of God I am what I am" (1Cor 15:10). Paul is what he is through God's grace. "And his grace which was bestowed upon me was not in vain; but I labored more abundantly than they all: Yet not I" (1Cor 15:10). It was Paul who labored more than the other apostle's (1Cor 15:9), yet he says it wasn't him who worked. "But the grace of God which was with me" (1Cor 15:10). It wasn't Paul who worked (the fleshly Paul – Gal 3:3), but it was the grace *given* to Paul. Isn't that what is being said here? Or was Paul being falsely modest? For if what Paul said is not true, then Paul is being falsely modest! This verse, 1Cor 15:10, is to be taken literally. Paul does not lie in these inspired scriptures! What is this grace Paul is speaking of that allows him to work so abundantly?

Grace Given According to Measure

nm331 » "But unto everyone of us is given grace according to the measure of the gift of Christ" (Eph 4:7). "Now, concerning spiritual gifts, brethren, I would not have you ignorant ... But the manifestation of the Spirit is given to every man to profit everybody. For to one is given by the Spirit the word of wisdom; to another the word of knowledge by the same Spirit; to another faith by the same Spirit; to another the gifts of healing by the same Spirit ... But all these works that one and the selfsame Spirit, dividing to every man separately as He desires" (1Cor 12:1, 7-9, 11). These free gifts of Spiritual power that work the works described in this chapter of First Corinthians, are *given* by God's will. "And there are diversities of operations, but it is the same God which works all in all" (1Cor 12:6).

nm332 » When Paul told us that it wasn't him that worked or labored but the grace within him (1Cor 15:10), he meant, by grace, the free Spiritual gifts and the power of these gifts (1Cor 12:11, 6 & all of 1Cor 12). Is this confirmed elsewhere in the Bible?

nm333 » "Whereunto I also labor, striving according to his workings, which works in me mightily" (Col 1:29). Paul works or strives according to God's workings inside him through the power of God's Spirit or New Mind.

nm334 » "According to the grace of God which is given unto me, as a wise master builder, I have laid the foundation, and another builds thereon" (1Cor 3:10). According to Paul's grace, he works.

nm335 » "I speak not of myself: but the Father that dwells in me, he does the works" (John 14:10). Even Christ didn't do his great works; God's Spirit in him did the works!

nm336 » "Now the God of hope fill you with all joy and peace in believing, that you may abound in hope, through the power of the Holy Spirit" (Rom 15:13). Even hope comes through the power of the Holy Spirit.

nm337 » Again Paul reiterates God's message that it's the power of God's Spirit that works: "Whereof I was made a minister, according to the gift of the grace of God given unto me by the effectual working of his power" (Eph 3:7).

Paul As An Example

nm338 » Paul epitomizes what God is doing on earth. It is through Paul that we can understand how God is working his wonder on earth! (1Tim 1:16, 12-16) Paul had been the antithesis of a Christian, for he had persecuted the Church of God before his conversion (1Cor 15:9 & Acts 8:3). How was Paul converted to a real Christian? Read yourself about it in Acts 9:1-18. Paul was not seeking to be a Christian. It was Christ through the power of God's Spirit that changed Paul's attitude towards Jesus Christ. Even a changed mind (repentance) is *given* to people by God (2Tim 2:25; Acts 11:18). Paul was given repentance (a changed mind) by God.

nm339 » Paul was "appointed a preacher, and an apostle, and a teacher of the Gentiles" (2Tim 1:11) in spite of what Paul was up to the time God brought him to repentance. Paul received his Spiritual gifts of being a preacher, apostle,

and teacher through the will of God (1Cor 12:28, 11). Hear Paul's inspired words: "Who has saved us, and called us with a holy calling, not according to our works, but according to his own purpose and grace, which was given us in Christ Jesus before the world began" (2Tim 1:9).

nm340 » Good works come through the power of God's Spirit (grace), and God's Spirit is given to man because of God's purpose. It does not matter what a man was before God gives him his Spirit. As in the case of Paul, God gives it to one in spite of what that one has done before; he gives his Spirit of power to those who were set aside before the world began. See the paper on Predestination [NM 8] to understand this last point.

God Gives The Power for Good

nm341 » God gives us Spiritual power to do good works; He makes us good through his Spiritual power. God is the source of energy for good works:

- "But when he sees his children, the work of mine [God's] hand" (Isa 29:23).

- Speaking about Zion (God's church): "I will make you an aeonian excellency" (Isa 60:15).

- "But now, O LORD, you art our Father; we are the clay, and you our potter; and we are the work of your hand" (Isa 64:8). "We will not boast of things apart from our measure, but according to the measure of the rule which God has distributed to us, a measure to reach even unto you" (2Cor 10:13). What is this measure spoken of here? "But unto every one of us is given grace according to the measure of the gift of Christ" (Eph 4:7).

- Faith, a grace of God, for example, is given according to measure (Rom 12:3). Thus, Christians are given a measure of grace "according to the measure of the rule which God has distributed." "Not boasting of things apart from our measure, that is, of other's labors: but having hope [Hope comes from the Spirit, Rom 15:13] when your faith [Faith is a gift of God, Eph 2:8; Rom 12:3] is increased [God gives the increase, 1Cor 3:6; Col 2:19], that we shall be enlarged by you according to our rule [measure of rule is from God, 2Cor 10:13] abundantly" (2Cor 10:15). Are you beginning to see?

nm342 » Paul adds, "but he that glories, let him glory in the Lord" (2Cor 10:17). "And whatsoever you do in word or deed, do all in the name of the Lord Jesus, giving thanks to God and the Father by him" (Col 3:17). Why?: for God does all through his Spiritual power! It's God's Spirit that does the good works (Isa 26:12). All men are clay in God's molding Spirit. But how does God create through his Spirit?

nm343 » Paul again gives us the answer, Paul had a thorn in his flesh – the angel (KJV "messenger") of Satan that troubled him (2Cor 12:7). This angel of Satan (the other-mind) was the other spirit in Paul's mind that was warring against God's Spirit (Eph 6:12; Rom 7:22-23, see "Other Mind Paper" [NM 21]). Paul had prayed to God that God would take this spirit from him (2Cor 12:8). God had answered, "My grace [Spiritual power] is sufficient for thee." Paul concludes, "for my strength is made perfect in weakness." What does Paul mean that weakness makes for strength?

From Trials to Strength

nm344 » "No trial has taken you except what is common to man. But God is faithful, who will not permit you to be tested above what you are able. But with the trial, he will make a way of escape, so that you may be able to bear it" (1Cor 10:13). Thus so far we see God's grace to Christians is sufficient, and that God will not permit any Christian to be tested in a Spiritual trial above what he is able to endure. God further tells us that for Christians "all things work together for good" (Rom 8:28). This includes trials.

nm345 » "And not only so, but we glory in trial also: knowing that trial works patience; and patience, proof; and proof, hope" (Rom 5:3-4). "My brethren, count it all joy when you fall into divers trials; knowing this, that the trying of your faith works patience. But let patience have her complete work, that you may be perfect and entire, lacking nothing" (James 1:2-4). It is through trials that God works good fruit in Christians.

nm346 » "Now no chastening for the present seems to be joyous, but grievous: nevertheless afterward it yields that peaceable fruit of righteousness unto them which are exercised thereby" (Heb 12:11). Those exercised by these trials produce good fruit, if they have the grace that allows them endurance and overcoming of these trials (2Cor 12:9 & 1Cor 10:13). The whole

world is now in a trial because of the "other-mind." See New Mind 20.

nm347 » Of course some fruits of the Spirit are not produced through trials such as those described in Mark 13:11, "but when they shall lead you, and deliver you up, take no thought beforehand what you shall speak, neither do you premeditate: but whatsoever shall be given you in that hour, that speak you: for it is not you that speak, but the Holy Spirit."

nm348 » *Review*. It is God's Spirit that produces good works, and God calls and chooses people to fulfill positions in the Church according to his will (2Cor 10:13; 2Tim 1:9). God gives to Christians Spiritual power by measure (Eph 4:7). God gives roles in the aeonian Kingdom of God under Christ, by the fact that God through his Spirit and by the measure of power given, does the good works. Why?

No Reason To Boast: God Gives The Power

nm349 » "Not of works, lest any man should boast" (Eph 2:9). Although in this verse it is referring in context to being saved through Faith, and is not speaking about reward according to works, the same conclusion is still valid. The scriptures quoted in this paper clearly indicate it is God who does the good works, not man. Even Christ the man said it was his Father that did his works (John 14:10). Shall we take Christ's word as true? Further Christ the man did his works according to the power given him: "And declared to be the Son of God with power, according to the Spirit of holiness" (Rom 1:4). But unlike regular Christians, Christ the man was given the Spirit without measure (John 3:34). He was a man given enough power to be the only sinless human in this age (see *God Papers*).

But We Work

nm350 » Let's clarify something here through an allegory. We are like the branches of a tree (Rom 11:17), and God is like the root and trunk. God as the root and trunk supports us, the branches (Rom 11:18). God provides the sap (Spirit), for us, the branches, to produce the good fruit (John 15:5). Without the sap and support from the root and trunk, we would produce nothing. God does the works by providing the support and sap for us to produce the fruit. Yet we, the branches, do perform works, but only because of the support of the sap from the trunk and root. We, the branches, produce fruit according to the amount and quality of sap from the roots reaching us.

nm351 » Hear what Paul says, "I have in a figure transferred to myself and to Apollos for your sakes; that you might learn in us not to think of men above that which is written [God gives the increase, 1Cor 3:7], that no one of you be puffed up for one against another. For who makes you to differ from another? and what have you that you didn't receive? now if you did receive it, why do you glory, as if you hadn't received it?" (1Cor 4:6-7) "And base things of the world, and things which are despised, has God chosen, yes, and things which are not, to bring to nought things that are: that no flesh should glory in his presence ... That, according as it is written, he that glories, let him glory in the Lord" (1Cor 1:28, 29, 31).

nm352 » It is God who does the good works; it is God who creates Spiritual men according to the measure of Spiritual power given them (Deut 8:17-18). God gives his Spiritual power; one cannot earn Spiritual power; one cannot qualify for Spiritual power, it is given to those it is given to (see "Predestination Paper" [NM 8]). Spiritual power is given as a non-earned gift, so people won't become puffed up and have glory in their own selves.

nm353 » Let's summarize this topic through a few scriptures. Note the following verse in a Spiritual way:

- "Surely your turning of things upside down shall be esteemed as the potter's clay: for shall the work say of him that made it, He made me not? or shall the thing framed say of him that framed it, He had no understanding?" (Isa 29:16)

nm354 » God is the potter; we are the clay (Isa 45:9-10; 64:8). Notice the rhetorical question in Isaiah 29:16. Shall real Christians, the clay, the Spiritual work of God's hand, say God made them not? The answer, "we all are the work of your [God's] hand" (Isa 64:8).

––––––––––––––––––––––––––––––––––––

–

Notes for NM 12

Parables of Pounds And Talents

nm355 » We have just seen in the paper, "According to Works" [NM 12], that God gives Spiritual power by measure. As explained, God gave Paul just enough Spiritual power to do the task he was commissioned to do, "my grace is sufficient for you" (2Cor 12:9).

nm356 » We saw in that paper that the parable of pounds shows us that Spiritual Christians are given rulership (serviceship) over cities according to how much they produce from the one "pound" that was given to each at the start. One gained ten times as much as what he was originally given, and he received rulership over ten cities. Another gained five times as much, and he was given rulership over five cities in the kingdom of God. Remember rulership in the kingdom of God will be different from how it is now done in this age (see Mark 10:42-44). But the one who gained nothing, his only pound was taken away from him because he gained nothing.

nm357 » Now the "pounds" of this parable can be looked at as being good Spiritual fruit. The more good fruit (pounds) one produces, the more responsibility that person will have in the kingdom of God. Look up the following verses and see how those who will rule in the kingdom of God are those who produce much good Spiritual fruit: Mat 13:23; Luke 8:8, 15; John 15:5, 8, 16; Rev 2:26. Thus, those who produce much Spiritual good fruit receive rulership (serviceship) according to their production of good fruit (pounds).

nm358 » Notice that all the servants in the parable of pounds received just one pound to begin with. This should be looked upon as what any one person has in good fruit without God's Spirit. Someone without God's Spirit is like one who has one pound. But with God's Spirit, one produces much good Spiritual fruit (pounds).

Now let's examine the parable of talents:

- "Again, it will be like a man going on a journey, who called his servants and entrusted his property to them. To one he gave five talents [a talent was worth about a thousand dollars] of money, to another two talents, and to another one talent, each according to his ability. Then he went on his journey. The man who had received the five talents went at once and put his money to work and gained five more. So also, the one with the two talents gained two more. But the man who had received the one talent went off, dug a hole in the ground and hid his master's money. After a long time the master of those servants returned and settled accounts with them. The man who had received the five talents brought the other five. 'Master,' he said, 'you entrusted me with five talents. See, I have gained five more.'

■ His master replied, 'Well done, good and faithful servant! You have been faithful with a few things; I will put you in charge of many things. Come and share your master's happiness!'

■ The man with two talents also came. 'Master,' he said, 'you entrusted me with two talents; see, I have gained two more.'

■ His master replied, 'Well done, good and faithful servant! You have been faithful with a few things; I will put you in charge of many things. Come and share your master's happiness!'

■ Then the man who had received the one talent came. 'Master,' he said, 'I knew that you are a hard man, harvesting where you have not sown and gathering where you have not scattered seed. So I was afraid and went out and hid your talent in the ground. See, here is what belongs to you.'

■ His master replied, 'You wicked, lazy servant! So you knew that I harvest where I have not sown and gather where I have not scattered seed? Well then, you should have put my money on deposit with the bankers, so that when I returned I would have received it back with interest [KJV, "usury"].

■ 'Take the talent from him and give it to the one who has the ten talents. For everyone who has will be given more, and he will have an abundance. Whoever does not have, even what he has will be taken from him. And throw that worthless servant outside, into the darkness, where there will be weeping and grinding of teeth.'" (Mat 25:14-30, NIV)

nm359 » Notice the servants in this parable were apparently given talents according to their ability. Since we are to perceive Spiritual lessons out of the Bible (John 4:24), then the "ability" spoken about is Spiritual ability. These servants were given "talents" according to their Spiritual ability. And we showed in the paper, "According to Works" [NM 12] that Spiritual ability is a gift. Thus, the "talents" given correspond exactly to the given Spiritual ability. Hence, these "talents" can be looked upon as being one and the same as the degree of Spiritual ability that is given to each Christian. The person receiving five talents can be looked upon as one who is given five times the Spiritual ability as the person with one talent; he thus will grow five times, so to speak, in Spiritual ability.

nm360 » You can see in this parable that as each was given, so did they gain, except the person given one talent. The person who was given five talents brought back to Christ an increase in good Spiritual works according to what was given to him in Spiritual ability. And the one given two talents brought back two more talents. But the one given one talent brought back just one talent – the one talent given to him. Each (except the one) brought back an increase according to what was given to him. And each was given talents according to their ability, Spiritually speaking. Therefore, each produced according to what was given to him, except the person given the one talent: he produced no increase.

Both Parables Together

nm361 » The servant in the parable of pounds who gained ten pounds is merely a servant who was given ten talents of Spiritual ability (cf. Mat 25:28 with Luke 19:24 in context noting in Mat 25:15 that the parable didn't mention someone receiving ten talents). Not only do these parables need to be looked upon as going together, they have to be looked upon that way in order to understand them. The servant who gained five "pounds" is merely the one who was given five "talents" of Spiritual ability. The servant who gained two "pounds" is the servant who was given two "talents" of Spiritual ability.

nm362 » Now the one who gained nothing, who just kept the one "pound" was the one given the one "talent" of ability. Yet since we know that those who are real Christians do produce much fruit (John 15:5, 8, 16; and paper called

"Prove Paper" [NM 10]), then we know that the one who produced no pounds above and beyond the original one given him, is a non-Christian. A non-Christian is one without the New Mind, he is one with the old mind. The one "talent" can be looked upon as the normal Spiritual ability of anyone without God's Spirit. And the one "pound" can be looked upon as the degree of good fruits one has without God's Spirit.

Usury

nm363 » For another way to prove that the person who had but one talent and one pound was a non-Christian please note Luke 19:23 and Mat 25:27. Here it says God was requiring usury (interest) from this person. But notice it is against the law of God in the Old Testament to take interest from a member of Israel, yet it is alright to take interest from a stranger or Gentile (Deut 23:20). Now Spiritual Christians are Spiritual Israelites (Gal 6:16). Therefore those without God's Spirit are spiritual strangers or spiritual Gentiles. Since God is a brethren of the Israelites, he is their Father, he cannot require interest from them, but he can require interest from strangers. Since God does not break his own laws in an antitypical way, the person from whom God requires usury or interest is a spiritual stranger, a non-Christian. The persons with one talent and one pound God required usury from. Thus, they are spiritual strangers, or non-Christians.

NM 13: All Saved

Scriptural Proof
God's Will
As One Goes All Must Go
Lazarus and the Rich Man
Unpardonable Sin?
Sin to Death?

NM13 Abstract

Why are we calling this paper, "all saved"? Can all be saved when in our translations of the Bible it says some will be damned or judged forever? Again, in this paper, we are dealing with the mistranslation of words that mean aeonian or age but have been mistranslated into the words forever or eternal. We will examine some scripture that says that God will save all and we will thus begin to see that God will save all by three orders or ranks or at three different times. We will also look at the parable pertaining to Lazarus and the rich man and the scripture about the unpardonable sin.

nm364 » Why do we call this paper, "all saved" when so many believe that some will be damned forever? There is a good reason for many believing in an *ever*lasting punishment for when you read most of today's translations of the Bible you see scripture that says some will be damned forever. But these scriptures have been mistranslated. In the Bible words that mean aeonian or age in the original languages were mistranslated into everlasting: resulting in the confusion and wrong doctrine of the majority who call themselves Christians.

nm365 » There are many verses in the Bible that indicate that *all* will be saved. Because of the mistranslation of words that mean aeonian into words like "forever," many are blinded to the fact that all will be saved. In this paper we will show you some of the scripture, and amplify on the scripture.

God: Good to All & Mercy on All.

nm366 » Notice the following scripture: "The LORD is good to *all*, and His mercies are over *all* His works. *All* Thy works shall give thanks to

Thee, O LORD" (Psalm 145:9-10, NASB). This scripture is speaking about ALL. God's mercies are over *all* His works. In the truest sense, all means all (see, "Does All Mean All Paper" [NM 14]). In the higher meaning, or Spiritual meaning of the Bible we take this for what it means. All will receive mercy (note Romans 11:32). As Romans 9:23-24 indicates, Christians (the ones with the New Mind) are given mercy. As with the Christians, so too with the rest of mankind. All will obtain mercy. All will receive the Spirit (New Mind) that brings with it true freedom and true salvation because God is not partial. Each receives salvation in his own order or appointed time (1Cor 15:23-24).

All Saved: Scriptural Proof

nm367 » **Everyone into the Kingdom**. Let's look at other scriptural proof that all will be saved, eventually. "The law and the prophets were unti1 John: since that time the kingdom of God is preached, and *everyone* presses into it" (Luke 16:16). This says that EVERYONE presses into the kingdom of God. It does not say some, or most, but it does say everyone presses into it. This scripture means what it says.

nm368 » **All Israel Saved**. "And so *all* Israel shall be saved" (Rom 11:26). Now either this statement is true or it is wrong. It says that ALL Israel will be saved. It does say *all* Israel, therefore it means all who ever lived in Israel. But what about the Gentiles? The Bible uses the word "Gentile" to mean all who are not part of Israel. Now God's word says that God is not a respecter of persons (Deut 10:17; Job 34:19; Acts 10:34; Eph 6:9; Rom 2:11). God is not biased. God has no partiality. Now if it is true that God is not partial, why would He save *all* of Israel and not save all of the Gentiles? Thus, since God is not partial, then all of Israel *and* all of the Gentiles will be saved. We just showed you (Luke 16:16) that God says all will be pressed into the Kingdom. Now through a logical construction, we have shown you the same thing. God meant what He said when He said: "in him [Christ] shall the Gentiles trust" (Rom 15:12).

nm369 » **All to Worship God**. "All people whom You have made shall come and worship before You, O Lord; and shall glorify Your name" (Psalm 86:9). If *all* will, in the truest sense of the word, worship God and glorify Him, then doesn't that mean all will turn to God and be saved thereby?

nm370 » *Mercy on All*. "For God has concluded them all in unbelief, that He might have mercy upon *all*" (Rom 11:32). Now in the 11th chapter of Romans, Paul is speaking to true Christians. Paul tells them that they have obtained mercy (v. 30). God gave mercy on all of them and made them part of the vessels of mercy (Rom 9:23). Now since we have proved in this book that all true Christians will be saved, and since in Romans 11:30 it shows Christians being allowed to be Christians through mercy, then when God shows mercy on the rest of mankind, the rest of mankind will be saved. Did it or did it not say, "mercy upon *all*" (Rom 11:32).

nm371 » *Lost Saved*. "But if our gospel be hid, it is hid to them that are lost" (2Cor 4:3). God, through Paul, says the gospel has been *hidden* from some who are called the "lost." Other scripture clearly indicates that God has hidden from many the good news of God for a purpose (Rom 9:18; 11:8-10; Mat 13:10-17; etc.). That purpose is for man to have a time, or an age with wrong so he will be able to know good (see "Reason Why" paper [NM 19] and the *God Paper* [GP 7]). "For the Son of man is come to seek and to save that which was lost" (Luke 19:10). Jesus is not only the savior of the Christians, but to those the gospel has been hidden from – the lost. "And we have seen and do testify that the Father sent the Son to be the Savior of the world" (1John 4:14). Now the word "world" is used to mean those of the worldly ways (1John 2:15-16). Jesus Christ was sent to save the "lost." He was sent to save the whole world, not just Christians.

nm372 » *Those that Erred will Understand*. "They also that erred in spirit shall come to understanding, and they that murmured shall learn doctrine" (Isa 29:24). Who are those who erred in spirit? In 1John 4:1-6 it gives a test so as to ascertain who are real Christians, and who are not. Those who do not pass the test are in the spirit of error (1John 4:6). Those who erred in spirit, are those who are being led by a satanic spirit of error (the other-mind), and are not being led by the Spirit of God (the New Mind). Thus, according to Isaiah 29:24, those who were led by the spirit of error shall understand, they will learn doctrine. Other scripture shows that they will learn after they are resurrected in the resurrection of the dead, which occurs after the 1000 years (Rev 20:5; 1Cor 15:21; NM24).

nm373 » *All to Bow to and Agree with God*. "I have sworn by myself, the word is gone out of my mouth in righteousness, and shall not return, that unto Me *every* knee shall bow, *every* tongue shall swear" (Isa 45:23). What does the Bible mean "every knee shall bow"? What does it mean to bow?

nm374 » Now when a person bows before an important person, he is in essence saying he respects the person he bows down to. Today many bow to a god, and think they bow to the God. Yet these people bow falsely to *their* god. If they were true in their bowing, they would show their respect by obeying their gods. To bow before something or someone is to show respect, or to humble one's self.

nm375 » When God says that *all* will bow before God, He means to bow truly to God. The world already bows hypocritically to their gods, while thinking these gods are the true God. Why would God say *all* would bow to God, or to Christ (He is in the God), when most do already, but hypocritically? God is not talking about hypocritical bowing to God. There is enough of that now. That kind of bowing is wrong. God desires truth.

nm376 » But notice further, "every tongue shall swear" (Isa 45:23). This does not mean swear against God. No, all will swear with God. All will pledge or bow with God, by God, and for God.

nm377 » Let's look at another very similar verse to Isaiah 45:23: "As I live, says the Lord, every knee shall bow to me, and every tongue shall confess to God" (Rom 14:11). Let's look at the phrase, "every tongue shall confess to God." What does the word "confess" mean? The word translated "confess" here comes from a Greek word that means, to acknowledge; to agree fully. Thus a better translation would read: "and every tongue shall agree with God." *All will acknowledge God, and agree with God*. "That at the name of Jesus *every* knee should bow, of things in heaven, and things in the earth, and things under the earth; and that *every* tongue should confess that Jesus Christ is Lord, to the glory of God the Father" (Phil 2:10-11).

nm378 » *Jesus Died for All*. "For there is one God, and one mediator between God and man, the man Christ Jesus; Who gave himself a ransom for *ALL*, to be testified in due time" (1Tim 2:5-6). "And He Himself is the forgiveness [propitiation] for our sins; and not for ours only, but also for *those of* the whole world" (1John 2:2). Now it is due time that everyone knows that Christ died for *all*, not just Christians. Jesus came to save *all* those of this age (1John 4:14).

God's Will: All To Be Saved

nm379 » Let's look at God's will. What does God wish or will for mankind? Can God do what He wishes? Can anyone prevent God from doing His will? Let's see what God has to say about this subject. "For this is good and acceptable in the sight of God our Savior; Who wishes *all* to be saved, and to come unto the knowledge of the truth" (1Tim 2:3-4). That is His wish or will. What else is His will? The Lord is "not willing that any should perish, but that ALL should come to repentance" (2Peter 3:9). Now the Lord does his Father's will (Heb 10:7; John 6:38). Thus, God the Father's will is for all to come to repentance, and for all to be saved and come to the knowledge of the Truth (2Pet 3:9; 1Tim 2:4)..

nm380 » *All to Repent*. How does one come to repentance? How does one come to turn away from the confusion of the world's ways? As 2Timothy 2:25 and Acts 11:18 show, it is God who *gives* or grants repentance (see "Repentance Paper" [NM 3]). Thus, this clearly shows that God can have His wish that all will come to repentance. All God has to do is give mankind the power to repent. Why hasn't God done this? God is allowing most people in this age to do wrong for a purpose (see "Reason Why" paper [NM 19]).

nm381 » *Christ Does God's Will*. Now Christ had come to do just one overall thing. He came to do the will of his Father (Heb 10:7; John 6:38). "My food is to do the will of Him [the Father] that sent me, and to FINISH HIS WORK" (John 4:34). Christ is the savior of the world (1John 4:14). Jesus prayed not too long before his death "that the world may believe that You have sent me" (John 17:21). How does one pray? "If we ask any thing according to His will, He hears us" (1John 5:14). Jesus was asking according to his Father's will when he prayed asking that the world believe. It is through Faith that one overcomes (1John 5:4-5). And those who overcome shall be saved (Rev 2:7). Jesus said he came to do his Father's will. Will Jesus do his Father's will? Or will Christ fail? Christ will do his Father's will, and the whole world will believe. They will come to believe when they are raised-up to life in the Great Last Day after the Thousand Years (see "Thousand Years and Beyond Paper" [NM 15]).

God's Will Shall Be Done

nm382 » "So shall my word be that goes forth out of my mouth: it shall not return unto me void, but it shall accomplish that which I please [will or wish], and it shall prosper in the thing whereto I sent it" (Isa 55:11). "Declaring the end from the beginning, and from the ancient times the things that are not yet done, saying, My counsel shall stand, and I will do all my pleasure [will] ... I have spoken it, I will also bring it to pass; I have purposed it, I will also do it" (Isa 46:10-11). God said He would do *all* his pleasure (his will), and that He will bring it to pass. God's will is to have all to be saved (1Tim 2:4). **God does his will, God is God, who can stop God?**

As One Goes All Must Go

nm383 » **Everything that a person is, has been given to him either by his parents, his environment, or his spiritual father.** All the great minds of the world are what they are from what they received physically or mentally from their parents and environment. All the great Spiritual people are what they are through the gifts of the Spirit (1Cor 12:1-31). Thus, if what we are, is because we have received it from outside of us, then as some go (saved at Christ's return, 1Cor 15:23), then *all* must go, or God is being partial to some in the creation. Since God says He is not partial (Rom 2:11), then what He *gives* to one (Christ the man), or to some (the Christians), He must give to ALL. Doesn't God's word say you are what you are because of what you have received? (Prov 22:6; 1Cor 4:7; etc.) Does not common sense tell us this also? Did not God's word say God wasn't partial? So as one goes, ALL must go. Is this confirmed elsewhere in God's word?

nm384 » Compare the following verses. "Therefore as by the offense of one [Adam's] judgment came upon all men to condemnation; even so by the righteousness of one [Christ] the free gift came upon *all* into justification of life" (Rom 5:18). And, "for as in Adam all die, even so in Christ shall *all* be made alive" (1Cor 15:22). In verses 45-47 of 1Corinthians, chapter 15, it tells us that Adam is the first man and Christ is the second man (the antitypical man). By comparing Romans 5:18 (first part) and 1Corinthians 15:22 (first part) we see that Adam died for his offense – the first man died for his sins. But by comparing the last parts of these two verses we see that through the medium of Christ, "ALL men into justification of life," and "in Christ shall ALL be made alive." Thus, as Adam died, all died. And as Christ went, all will go eventually (see, "Does All Mean All Paper" [NM 14]).

nm385 » ***Potter and the Clay***. To prove the point we are making, "has not the potter power over the clay, of the same lump to make one vessel unto honor, and another unto dishonor?" (Rom 9:21) Yet Romans 11:16 shows us that as the first fruit (Jesus) goes, so will the whole lump. And Romans 9:21 says from the same lump God has made two groups of people – the vessels of wrath, and the vessels of mercy. In the "Predestination Paper" [NM 8] we proved that the vessels of mercy were the real Christians, while the vessels of wrath were the rest who have only the "other-mind." In the "Predestination Paper" [NM 8] we proved that *all* the vessels of mercy will be saved. Since we know that God is not partial (Rom 2:11), then if the first fruit (Christ) is saved, and the vessels of mercy (real Christians) are saved, then *all* of the lump of mankind will be saved.

Three Divisions/Orders/Groups

nm386 » There are three groups or divisions being saved. One, Jesus Christ the man, was saved from death at his resurrection from the dead. The next to be saved is at Christ's (the Messiah's) coming (1Cor 15:23). This group is the called and chosen *few* (Mat 22:14) who are allowed to see the way "which leads unto life" (Mat 7:14). As Matthew 7:14 says, *"few* there be that find it." This group is also called the "vessels of mercy" (Rom 9:23).

nm387 » The last group or division is called the "vessels of wrath," the "dead," the "children of Satan," and so forth. The first book of Corinthians 15:24 shows us **when** this last group will be saved – at "the end." This "end" is the end of the Spiritual creation. It is at this "end" of creation, which is also the beginning of the totally NEW cosmos, that all will be in God, for at that time all will be "the God, all in all" (1Cor 15:28). As we have shown in the *God Papers*, all went out of God in the beginning of creation, but all will go back into God so that may be "the God, all in all" (1Cor 15:28; see *God Papers*, GP 6). This is the great cycle. All went out of God, but all will go back into God. God tells us through scripture that all will be in God (see Greek text: John 17:23; 2Cor 5:17-21; Eph 1:9-10; John 12:32; Col 1:20; 1Cor 15:28; etc.). Thus, all will be saved.

nm388 » ***Three Measures***. "Another parable spake he unto them; the kingdom of heaven is like unto leaven, which a woman took and hid in THREE measures of meal, till the WHOLE was leavened" (Mat 13:33). These three measures of meal are shown in 1Corinthians 15:23-24. The FIRST measure is Christ (the first fruit). The SECOND measure is the Christians (the first fruits). The THIRD measure is the rest of mankind (the vessels of wrath). There are three measures or groups from the same lump (mankind). They will all be saved, but at different times (see "God's Appointed Times Paper" [16]). Here Christ uses "leaven" to teach us a truth even though leaven is used in a negative way in other scripture. This is good leaven (not sin) because the parable teaches us about the "kingdom of heaven," and how the *whole* lump of bread is saved – in three measures.

Lazarus and the Rich Man

nm389 » Many "prove" their "hell" and "lake of fire" theories from the parable of Lazarus and the Rich Man. Let's look at this parable. Turn to Luke 16:19-31 and read verses 19-21. "And it came to pass, that the beggar [Lazarus] died, and was carried by the angels into Abraham's bosom." Now this parable has no time element. Jesus does not say *when* this will take place or if it already happened. Of course, remember this is a *parable*. Notice that the beggar died. The Bible tells us that those who die have no consciousness after death (Eccl 9:5; Psalm 146:4). Those who die are dead. Man is not immortal. Man is mortal, he has no immortal soul, but he will have an immortal Spirit (see "Body, Soul Paper" [NM6]).

nm390 » Next it says that Lazarus was carried to Abraham's bosom by the angels. Now scripture tells us that Abraham is also dead now, but he still is "waiting" dead in his grave for the promise (read Acts 7:1-5; Heb 11:8-13). The promise given to Abraham was a kingdom in the land of Palestine (Gen 12:5-7; 13:15,18). This land was to Abraham and his seed for an aeonian period of time. As other scriptures show this land for an aeonian time is the same Kingdom of God that is promised to Christians (see "Seed Paper" [PR1]). Jesus said that Abraham would be in the kingdom of God (Luke 13:28). Notice that Lazarus was carried by angels into the bosom of Abraham. Now, we know the beggar will be carried into "Abraham's" bosom by angels at Christ's return (Mat 16:27; 24:31; 25:31). And that they will meet Christ in the air at cloud level (after being resurrected) to bring Jesus down to earth (1Thes 4:17; Rom 10:6) to rule *on* earth (Rev 5:10).

nm391 » We see that Lazarus will be carried to Abraham's bosom. Turn to Isaiah 40:11. Here God will care for His people as a shepherd does for his sheep, which He will carry "in His bosom." Jesus was "in the bosom" of his Father (John 1:18), enjoying the Father's Spiritual power. Moses carried the children of Israel in his bosom. To be in one's bosom is to have that one's love and protection, and share in his inheritance. Therefore at the resurrection to the kingdom of God at Christ's coming, Lazarus will be carried into the care of "Abraham" and his seed.

■ *Rich Man in Hell*. In the last part of verse 22 of the parable we see that the rich man dies and is buried. "And in hell [the grave] he lifts up his eyes, being in torments, and sees Abraham afar off, and Lazarus in his bosom" (v. 23). The word "hell" comes from the inspired word "hades" in the Greek language. It simply means "grave." But look, the rich man lifts up his eyes. He is in the grave, and he lifts up his eyes. Now it did say he died, that he was in the grave, and that he lifted up his eyes. But this isn't all. After the rich man cried to Abraham to have Lazarus come and cool his tongue with a few drops of water, the rich man said: "I am tormented *in* this flame" (v. 24). He said he was tormented IN a flame. It didn't say he was beside the flame, it said he was *in* the flame. In the inspired Greek it says he was *in* the flame. Does this mean there is a hell as the world knows it? No, but it does mean this "rich man" was in the flame, and that he was being tormented in the flame. It also says he was in the grave ("hell"), that he lifted up his eyes in this grave, and that he was *in* a fire. Thus, his grave must have a fire in it. The Bible clearly describes a grave with fire in it – the lake of fire (Rev 20:10).

nm392 » How can a man be dead, and in a grave with fire, while being tormented in it, when the Bible clearly tells us a dead man is dead and unconscious? The Bible does say the rich man was in a flame, not near it, or about to go into it, but in a flame which is also called the rich man's grave. Have we disproved the reliability of the Bible? No, for is the rich man, in reality a human being?

Satan and the Rich Man

nm393 » Turn to Ezekiel chapter 28. "Son of man, say unto the prince of Tyrus, Thus says the Lord GOD; Because your heart is lifted up, and you have said, I am a God, I sit in the seat of God, in the midst of the seas, yet you art a man, and not God" (v. 2). Now continue to read on until verse 13, "You have been in Eden the garden of God; every precious stone was your covering, the sardius, topaz, and the diamond..." (v. 13). Now this physical prince of Tyrus was not in the Garden of Eden. Further these precious stones mentioned in verse 13 are used throughout the Bible to describe spiritual beings (see Mal 3:17; Isa 54:11-12; Rev 4:3). Man is not a spiritual being. Now look at verse 14. "You art the anointed cherub that covers . . ." By noting what we have shown so far and by examining the whole chapter we can see it is dual. The typical version speaks about the prince of Tyrus as a man. The antitypical version speaks about the prince of Tyrus as Satan, for it was Satan who was in the Garden of Eden, who is a spiritual being, and who is an anointed cherub. In fact, not only is chapter 28 of Ezekiel dual, but all the Bible (see the "Duality Paper"). **The parable of Lazarus and the rich man is only a parable, but it is a parable with a higher meaning**.

nm394 » Notice verse 5, "By your great wisdom and by your traffic have you increased your *riches*, and your heart is lifted up because of your *riches*." Verse 16, "By the *multitude of your merchandise* they have filled the midst of you with violence" (see also Rev 18:11-16). Thus, both the prince of Tyrus and the spirit of Tyrus (Satan) are called RICH. "I will cast you to the ground ... by the iniquity of your traffic; therefore will I bring forth a fire from the midst of you, it shall devour you..." (v. 17, 18).

nm395 » God is describing the coming fate of Satan, he is to be cast into a pit of fire. The prince of Tyrus in Ezekiel 28 is used metonymically to represent Satan as is the Pharaoh and the tree in Ezekiel 31. Note the same fate for Satan in Ezekiel 31:18, 14 – cast into a pit.

nm396 » In Isaiah 14 it also describes Satan in the antitypical, and Babylon in the typical meaning of this chapter. What does God, through the medium of Isaiah, say about Satan?: "Yet you shall be brought down to hell [the grave], to the sides of the pit" (v. 15). Therefore, from the above verses we see that Satan is rich, has a multitude of merchandise, and will go down to hell (a pit in the ground with fire in it).

nm397 » We are beginning to see that the rich man in the *parable* is much like the prince of Tyrus. This prince in Ezekiel 28 is called a man, but some of the things attributed to him could only be attributed to Satan. Ezekiel 28 is dual, as the parable of Lazarus and the Rich Man is dual. The rich man of this parable not only represents a rich *man*, but also the spiritual Satan. In Ezekiel 28 the man prince could not have been in the garden of Eden, thus we know it was talking only of Satan at that point in the chapter. The same with the parable of the rich man – it was Satan who lifted up his eyes in the flame, not a man. For a man cannot live in a burning flame, but a spiritual being can. Satan is a spiritual being. **Transpose "Satan" for the rich man in the parable and the parable comes alive**.

nm398 » "And the devil that deceived them was cast into the lake of fire and brimstone, where the Beast and the false prophet are [the Beast and the false prophet are physical, thus, they are burned up and their ashes are left in the pit, but the spiritual Beast (Satan) and the spiritual false prophet (Satan) are in the pit of fire], and shall be tormented day and night into the ages of ages" (Rev 20:10). Revelation 20:1-3 describes the place of Satan's "*death*" (he is as good as dead since he can't come up out of the pit of death, Rev 20:1-3), and Satan's *trying* (the Greek word translated "tormented" means – trying as by fire).

nm399 » How is Satan being tormented? (Luke 16:23-25) A few verses in the Bible, project that Satan and his spiritual demons *need* to dwell inside animal beings, if not, they are tormented thereby (see Luke 11:24; Acts 8:7; Mark 5:12). Thus, since Satan will be sealed in a burning pit where there can be no animal (mortal) life because of the flame, Satan will be tormented because he cannot dwell inside animals.

nm400 » For 1000 years Satan and his demons (the "other-mind") must live in a state of torment (Rev 20:3, 7). But the torment of not dwelling in another physical being is not the only factor that will torment Satan. "You believe that there is one God? You do well, the demons also believe, and tremble" (James 2:19). Satan and his demons will be "in perpetual chains sealed under darkness into the judgment of the great day" (Jude 6). Satan is in the dark Spiritually, will

be in a sealed bottomless pit that burns with fire, and will be tormented mentally thereby.

Great Gulf

nm401 » The "great gulf" that is fixed between Abraham and Lazarus, and the Rich Man (Satan) is: (1) the seal that will be set upon Satan (Rev 20:1-3) for the 1000 years (Rev 20:7).; (2) in the case of the typical rendition of the parable (a rich man belonging to the vessels of wrath), it speaks of the great gulf between the two appointed parts of mankind, that is, between the vessels of mercy and the vessels of wrath. After the vessels of mercy are saved, it will be 1000 years before the vessels of wrath are saved.

nm402 » The "problem" of the parable of Lazarus is solved when one sees the higher meaning of the parable. Remember this parable is just that – a parable. Parables are meant to teach us something; they are aids to understanding God and his plan.

Unpardonable Sin?

nm403 » Is there an unpardonable sin? Is there a sin that God will not pardon?:

- "'Assuredly, I say to you, all sins will be forgiven the sons of men, and whatever blasphemies they may utter; but he who blasphemes against the holy Spirit never has forgiveness, but is subject to eternal condemnation' – because they said, 'He has an unclean spirit.'" (Mark 3:28-30, NKJV)

- "Therefore I say to you, every sin and blasphemy will be forgiven men, but the blasphemy against the Spirit will not be forgiven men. Anyone who speaks a word against the Son of Man, it will be forgiven him; but whoever speaks against the Holy Spirit; it will not be forgiven him, either in this age or in the (age) to come" (Mat 12:31-32, NKJV).

nm404 » IF there is an unpardonable sin – a sin that can *never* be forgiven, then God cannot keep his word. There are many scriptures that speak of ALL being saved (see *Does All Mean All*, GP 6 of *God Papers*, etc). God tells us through the Bible to forgive and not to repay evil for evil (Rom 12:17); overcome evil with good (Rom 12:21); do not render evil with evil or insult with insult but with blessing (1Pet 3:9); to forgive your brother time after time (Mat 18:21-35).

nm405 » From Mark 3:28-30 and Matthew 12:31-32, some think they prove that there is a forever judgement or damnation for sinners. BUT they are using a mistranslation of the words *aion* and *aionios*. There are many scriptures that speak about ALL being saved (see "All Saved" and "Does All Mean All" papers, and *God Papers*, GP 6). *If* there is an unpardonable sin, then God will not keep his Word – all will not be saved. But God keeps all his words, thus, all will be saved and all will go into the God (see *God Papers*). Notice what the correct translation does to the so-called unpardonable sin:

- "Verily I say to you, that all the sins shall be forgiven to the sons of men, and evil speakings with which they might speak evil, but whoever may speak evil in regard to the Holy Spirit hath not forgiveness – to the *age*, but is in danger of *age-during* judgment; because they said, 'he hath an unclean spirit.'" (Mark 3:28-29, Young's *Literal Translation of the Holy Bible*)

- "And whoever may speak a word against the Son of Man it shall be forgiven to him, but whoever may speak against the Holy Spirit, it shall not be forgiven him, neither in this *age*, nor in that [age] which is coming" (Mat 12:32, Young's *Literal Translation of the Holy Bible*).

nm406 » Because they were speaking against Jesus Christ's Holy Spirit ("He said this because they were saying, 'He has an evil spirit.' " – Mark 3:30, NIV), Christ said their sin is not forgiven "neither in this age [age of Satan] nor in that (age) which is coming" (Mat 12:32). That age that is coming is the age of the true God. Further their sin would not be forgiven "into the age but is in danger of agelasting [*aionios*] judgment" (Mark 3:28-29, Young's translation and see Greek text – the Greek word for "judgment," Strong's #2920, is in the older Greek texts instead of the word for "sin," Strong's #265; see Greek text, George Ricker Berry, Zondervan, 1969 printing).

nm407 » Those who speak against Jesus, also speak against His Holy Spirit, and they are in danger of the aeonian judgment *because* they do not have the Spirit that enables them to see the truth. "Therefore I tell you that no one who is speaking by the Spirit of God says, 'Jesus be cursed.'" If you have the Spirit of God you cannot speak against the real Jesus Christ. It is impossible because you have the Spirit of Truth that knows the truth when it sees it. That is why Christ said their sin is not forgiven, 'neither in this age [Satan's] nor in that (age) which is coming'– because they did not have the Spirit to see the Truth of Christ. And those without the Spirit of Truth will not be forgiven in this age of evil or in the coming 1000 years. But after the 1000 years they will be forgiven (see "Thousand Years and Beyond" paper [NM 15]). These scriptures about the so-called unpardonable sin are tests that one can take to prove to one's self if he is or is not a real Christian. If one blasphemes the Spirit, or those with the Spirit, he is in danger of the aeonian punishment if he does not repent, because he is proving he does not have the Spirit that sees the Truth and brings life during the 1000 years instead of the judgment during the 1000 years.

Sin to Death?

nm408 » Now we can understand the sin to death (1John 5:16). Those who sin to death (aeonian death as judgment for their sins) are those not in the Spirit. Those who sin not to death (aeonian death) are those begotten of the Spirit. Remember there are sins among real Christians (1John 1:8-10), but not willful or intentional sinning. If they do sin, it is because of weakness, not because they wish to sin or want to sin. Those in the Spirit sin in this age of Satan, but not to the aeonian death and judgment.

Let us look at 1John 5:16:

- **"If any man see his brother sin a sin which is not unto death** [he has the Spirit, he is not sinning willfully], **he shall ask, and he shall give him life** [aeonian life] **for them that sin not unto death. There is a sin unto death** [those without the Spirit]: **I do not say that he shall pray for it."**

nm409 » How should we pray? "If we ask anything according to his will, he hears us" (1John 5:14). Now the will of God is to have two groups of people on earth (the vessels of mercy and the vessels of wrath – Rom 9:22-23). The vessels of mercy are for aeonian life and rulership (but they are to suffer like Christ in order to rule – Rom 8:17), and the vessels of wrath are for aeonian punishment away from the glory of the 1000 years (2Thes 1:9). The vessels of mercy were predestinated to be what they are to be. They were set out before the world began for their job. The person predestinated as a vessel of mercy cannot become a vessel of wrath, and contrariwise. Thus, when we pray, we cannot ask God to save a person for aeonian rule, if God had not predestinated that person to be a vessel of mercy. Hence, we should pray thusly: "If you have chosen him to be a vessel of mercy, please give him aeonian life." If he is a vessel of mercy God will give him that life and forgive him his sins. If that person is not a vessel of mercy, then God will not answer and give him mercy in this age.

All Saved: Review

nm410 » We see in this paper that all will be saved. That is, *all* will be freed from the confusion and tears of the old age, and all will be made NEW in the New Age. The tears and confusion of the old age are for a purpose (see *Reason Why* NM 19). Eventually ALL will be made into immortal beings, and all will live in freedom and harmony forever. Many will be punished, for there is a time when people will be punished, but it does not last forever. As we have shown in this book, this age of punishment is for 1000 years. Those who do not receive the New Mind (Spirit of God) in the old age will be dead during the 1000 year age. Their death is their punishment. They will miss 1000 years of life. And as we have shown in the paper called, the "Thousand Years and Beyond Paper" [NM 15], the evil spirits (the invisible evil powers of the old age) will be punished during the 1000 year age: this is the evil minds' (spirits') aeonian punishment for misleading mankind during the old age. Even the evil spirits know that they will be punished at a certain time. About 2000 years ago demons who possessed two individuals said the following to Jesus:

■ "And they [demons] cried out, saying, What business do we have with each other, Son of God? Have You come here to torment us before the appointed time?" (Mat 8:29)

nm411 » See "Predestination," "Proof," and "According to Works" papers to better understand the vessels of mercy and wrath; see the "Prayer" [NM 18] to understand praying.

NM 14: Does All Mean All?

NM14 Abstract

This is a follow-up paper to NM13 ("All Saved") in which we examine more scripture about all being saved. We will, in so doing, consider whether or not the Bible really meant to say "all." Does all mean all?

"All" in Context

nm412 » This paper is concerned with the word "all" as it appears in scripture. When God, through the writers of the Bible, uses the word "all," does He mean *all*, or does he exaggerate? Does God stretch the truth like humans? This is very important. There are scriptures in which, if we take the word "all" to mean *all*, we see a different view of scripture than those who think God makes overstatements or exaggerates. There are psychological and spiritual reasons for some not understanding "all" as meaning *all* in scripture.

nm413 » Let us first examine some of the "all" scriptures. Then we will discuss some of the alleged impossibilities about taking "all" as truly meaning ALL.

From Paul's resurrection chapter we read:

> For as in Adam ALL die, even so in Christ shall ALL be made alive (1Cor 15:22; see Rom 5:14-18; Psa 82:7-8).

nm414 » In this one sentence we have "all" repeated twice. Most believe the first "all," that is, because of Adam's sin *all* die. Most do *not* believe the second "all," that is, because of Christ *all* will be made alive. Does God, through Paul's writing, use "all" in two different ways in the same sentence? Is the second "all" ("all be made alive") a hyperbole or just a positive

hope? Is the *all* powerful God just exaggerating here?

All Made Alive in Christ

nm415 » What is meant in 1Corinthians 15:22 by "be made alive"? Does it mean be made alive (resurrected) and thereafter be killed in some kind of hell fire? In context what does it mean to "be made alive"?

> ■ "For since by a man came death, by a man also came the resurrection out of the dead. For as in Adam all die, so also in Christ *all* shall be made alive. But each in his own order" (1Cor 15:21-23, NASB).

Christ is the First

nm416 » Here it speaks of the resurrection of the dead. All shall be made *alive*, but not at the same time. There is an *order* to the "all shall be made alive." In context this resurrection has something to do with "the resurrection of the dead.":

> ■ "But each in his own order: Christ the first fruit, after that those who are Christ's at His coming" (1Cor 15:23, NASB).

nm417 » Who was the first one in this "order" of being made alive or thus being resurrected? Christ was the first one resurrected from the dead to life (1Cor 15:12, 19; Rom 1:4; Rev 1:18; etc.). This resurrection of Christ was not a resurrection like Lazarus or like other resurrections reported in the Bible (John 11:23-44; Mat 27:52-53). Lazarus was merely resurrected back to physical and mortal life. But Christ the first of the "order" of the "all shall be made alive," was resurrected to permanent life – to immortal life. Paul in this very important chapter of 1Corinthians chapter 15, explains what kind of resurrection he had in mind:

> ■ "So also is the resurrection of the dead. It is sown a perishable body, it is raised an imperishable body ... for this perishable must put on the imperishable, and this mortal must put on immortality" (1Cor 15:42, 53).

nm418 » So in context of this chapter in Paul's writings, Christ the first of the *order* was resurrected to immortal life.

> ■ "For as in Adam *all* die, so also in Christ *all* shall be made alive. But each in his own order: Christ the first fruit, after

that those who are Christ's at His coming."

nm419 » After Christ, the first of the order, comes "those who are Christ's at his coming." After Christ's resurrection to immortal life there is another resurrection at Christ's coming.

Two More Resurrections

nm420 » Scripture speaks of two more great resurrections. Acts 24:15 speaks of the resurrection of the JUST and UNJUST. John 5:29 speaks about the good having a "resurrection of life" and those who have done evil having a "resurrection of judgment." Those doing good, those that come to Christ and eat of the Spiritual manna will be resurrected "at the last day" and live "into the age" (John 6:39-44, 58, Greek text). These are those who are "worthy to obtain that age" – and who "are the children of God, being children of the resurrection" (Luke 20:35, 36).

nm421 » This "last day," this "age" can be ascertained by other scripture and Biblical patterns. This last day is the 1000 year day (2Pet 3:8; Rev 20:2; see "Thousand Years and Beyond Paper" [NM 15]). Thus, "they lived and reigned with Christ a thousand years. (But the rest of the dead lived not until the thousand years were finished.) This is the first resurrection" (Rev 20:4, 5). There are two resurrections pictured here. The first one being those who will reign with Christ for 1000 years (Christians); the others are not resurrected until *after* the 1000 years.

nm422 » This first resurrection of Revelation 20:4-5 is also the "resurrection of the just," or "the resurrection of life" (Acts 24:15; John 5:29). These are resurrected to "that age," the "last day," or that is, the 1000 year antitypical Sabbath day – the last day. The next resurrection after the one at Christ's coming or at the beginning of the "last day," is the resurrection *after* the 1000 year day (Rev 20:5a). This is the "resurrection of the unjust," or the "resurrection of judgment," that is, the resurrection of the evil bunch who are to be judged-down for the 1000 years.

> "For as in Adam all die, so also *in Christ all shall be made alive*. But each in his own order: Christ the first fruit, after that those who are Christ's at his coming." (1Cor 15:22-23)

nm423 » The first in the "order" to the new life was Christ who was the first to be resurrected from death to immortality. The second order in the resurrections mentioned in 1Cor 15:22-24 is also the "first resurrection" of Revelation 20:4-5, and is the resurrection of the real Christians who will be resurrected to immortality at Christ's coming and will rule with Christ for the 1000 years.

nm424 » Scripture says "in Christ ALL shall be made alive. But each in his own order." If ALL are to be made alive, then most will be resurrected in the third of the "order." Christ was one person. Christians are few in number compared to the billions who died without Christ's Spirit. Thus, the resurrection with the greatest number of persons is the third order of resurrections, which is the resurrection after the 1000 years (Rev 20:4-5).

nm425 » *Pattern*. As we have seen, these resurrections are resurrections to immortality. There is a pattern here. Christ, the first in the order, was the first to be resurrected to immortality. Christians, the second in the order, will be resurrected to immortality. The third in the order must *also* be resurrected to immortality. Thus, "in Christ ALL shall be made alive" (1Cor 15:22). They will truly be made alive; they will be given immortality. By taking these scriptures in context, by taking "all" to mean ALL, then "as in Adam all die, so also in Christ *all* shall be made alive." Thus, "so then as through one transgression there resulted condemnation to ALL men, even so through one act of righteousness there resulted justification of life to ALL men" (Rom 5:18) Or, "for since by man came death, by man came also the resurrection of the dead" (1Cor 15:21).

Three orders to immortality

nm426 »

- "And again he said, Whereunto shall I liken the kingdom of God? It is like leaven, which a woman took and hid in THREE measure of meal, till the whole was leavened" (Luke 13:20-21).

- "For the earth brings forth fruit of herself; first the blade [Christ], then the ear [Christians], after that the full corn in the ear" (Mark 4:28).

- "Three times in a year shall your males appear before the LORD your God in the place which he shall choose; in the feast of unleavened bread, and in the feast of weeks, and in the feast of tabernacles" (Deut 16:16).

nm427 » Christ's resurrection was represented by the waving of the sheaf of the first fruits of Israel's first harvest of the year. Christians' resurrection was represented by Israel's spring harvest and its feast of weeks. The resurrection of the final harvest is represented by the final harvest of Israel in its feast of tabernacles (see "God's Appointed Times or Seasons Paper" [NM 16]). Another antitypical pattern is the Biblical Joseph (Christ), Manasseh (Christians), and Ephraim (the resurrection of the multitude of nations) (see "Seed Paper" [PR1]).

Notice:

> ■ "For as in Adam ALL die, so also in Christ ALL shall be made alive. But each in his own order: Christ the first fruit, after that those who are Christ's at his coming, *then comes the end*" (1Cor 15:22-24).

nm428 » The third order is resurrected at the *end*. The "end" Paul is speaking about here is amplified on in verses 24 to 28. The end comes when:

> ■ Christ has abolished all rule and all authority and power of his enemies (v. 24b & 25).

> ■ The last enemy abolished is death.

> ■ At that time ALL things have been put under Christ's feet (The exception to "all" is indicated in verse 27b).

> ■ At that time God will be ALL in ALL.

nm429 » At the *end* there will be no death. This chapter in 1Corinthians chapter 15 is speaking about resurrections to life, about immortality. Thus, after the third resurrection to immortality, then death is abolished, then the God is ALL in ALL. Satan and his evil influence will be abolished. Death, Satan's greatest power (Heb 2:14), is abolished at the END. Since God is ALL in ALL at the "end," and since God is love (1John 4:8), then Biblical love will be in ALL at the "end." There will be NO evil. In Christ all shall be made alive. This is, *alive* in the Spiritual sense of alive – being immortal and being inside the true life, being inside of God and His true Biblical love.

nm430 » After death is abolished, then comes true the saying, "death is swallowed up in victory. O death, where is thy sting? O grave where is thy victory?" (1Cor 15:54, 55; see Hosea 13:14)

More Scriptural Proof that All Will Be Saved

nm431 » Here is some further proof that *all* will be saved:

> ■ **(1)** "For God has concluded them ALL in unbelief, that he might have mercy upon ALL" (Rom 11:32).

> ■ **(2)** "For out of Him, and through him, and to Him, are ALL things" (Rom 11:36, see Greek text).

> ■ **(3)** "The LORD is good to ALL, and His mercies are over ALL his works. ALL thy works shall give thanks to thee, O LORD" (Psa 145:9-10, NASB).

> ■ **(4)** "And so ALL Israel shall be saved" (Rom 11:26).

> ■ **(5)** "In the LORD shall ALL the seed of Israel be justified, and shall glory" (Isa 45:25).

> ■ **(6)** "ALL the ends of the world shall remember and turn unto the LORD: and ALL the kindreds of the nations shall worship before thee" (Psa 22:27). [Here, in the antitype, it speaks of *real* worship; the only way you can really worship God is with the Spirit of God (Rom 8:8-9; John 4:24).]

> ■ **(7)** "Behold, I am the LORD, the God of ALL flesh: is there any word too hard for me?" (Jer 32:27)

>> God is the God of ALL flesh. But now in this age Satan is the god of the flesh belonging to this present evil age (2Cor 4:4). Thus at some future time God will truly be God of all flesh.

>> After Abraham, Isaac, and Jacob were dead, God spoke to Moses, "I am the God of thy fathers, the God of Abraham, the God of Isaac, and the God of Jacob" (Exo 3:6). To give proof for the resurrection of the dead Christ said, "now that the dead are raised, even Moses showed at the bush, when he called the Lord the God of Abraham, and the God of Isaac, and the God of Jacob. For he is not a God of the dead, but of the living:

for ALL live unto him" (Luke 20:37-38).

By God calling Himself the God of Abraham, Isaac, and Jacob *when* they were dead, He was projecting their future resurrection to life. **By God saying He is "God of ALL flesh," He is projecting the resurrection of** *all* **flesh to life in the future.**

■ **(8)** "Praise ye the LORD: praise ye the LORD from heavens: praise him in the heights. Praise ye him, ALL HIS ANGELS ... Praise the LORD from the earth ... kings of the earth, and ALL people, princes, and ALL judges of the earth: both young men, and maidens; old men, children ... Let EVERYTHING that has breath praise the LORD" (Psa 148:1, 2, 7, 11-12; 150:6).

> These Psalms were not placed in the Bible for mere words of hope; they are words of prophecy like Psalms 22, 110, and all others. All will praise God in the truest sense (note Phil 2:10-11; Rom 14:11; Isa 45:23).

In order for ALL to praise God, then ALL must be resurrected to life. Thus,

> ■ **(9)** "For the grave cannot praise thee, death cannot celebrate thee: they that go down into the pit cannot hope for thy truth. The living, the living, he shall praise thee" (Isa 38:18, 19).

> As this scripture shows, for number (8) to come true, then all must be resurrected to life - Spiritual life.

■ **(10)** "The Father loves the Son, and has given ALL things into his hand. He that believes on the Son has aeonian life: and he that believes not the Son shall not see life [in the aeonian life – 1000 years]; but the wrath of God abides on him" (John 3:35-36).

> In order to understand the wrath of God, when it is, and what it is; and in order to understand the aeonian life, you must read the "God's Wrath Paper" [PR4], the "Thousand Years and Beyond Paper" [NM 15], and the "Reward for Christians Paper" [NM 11]. Those that do not believe the Son

are those resurrected with the unjust after the 1000 years (see above).

■ **(11)** "For the love of Christ controls us, having concluded this, that one died for ALL, therefore ALL died; and he died for ALL, that they who live should no longer live for themselves, but for Him who died and rose again on their behalf" (2Cor 5:14-15).

> He died for ALL (1John 2:2); we live for Him and to Him.

■ **(12)** "The man Christ Jesus; Who gave himself a ransom for ALL to be testified in due time" (1Tim 2:5, 6).

> Now is that time.

■ **(13)** "And the bread that I will give is my flesh, which I will give for the life of the world [kosmos]" (John 6:51).

■ **(14)** "And he [Christ] is the propitiation [mercy seat] for our sins: *and not for ours only*, but *also* for the sins of the whole world [kosmos]" (1John 2:2)

■ **(15)** "And we have seen and do testify that the Father sent the Son to be *savior of the world* [kosmos]" (1John 4:14).

■ **(16)** "And I, if I be lifted up from the earth, will draw ALL unto me" (John 12:32).

■ **(17)** "And the nations shall bless themselves in him [God], and in him shall they glory" (Jer 4:2).

■ **(18)** "And ALL the nations shall be gathered unto it, to the NAME of the LORD" (Jer 3:17).

■ **(19)** "And teach ALL nations, baptizing them [all nations] into the NAME of the Father..." (Mat 28:19).

■ **(20)** "For the Son of man is not come to destroy men's lives, but to save" (Luke 9:56).

■ **(21)** "For I came not to judge the world [kosmos], but to save the world [kosmos]" (John 12:47).

■ **(22)** "I bring you good tidings of great joy, which shall be to ALL people" (Luke 2:10).

- **(23)** "And ALL flesh shall see the salvation of God" (Luke 3:6).

- **(24)** "With a view to an administration suitable to the *fullness of the times*, that is, the summing up of ALL things in Christ, things in the heavens [angels] and things upon the earth [mankind]" (Eph 1:10).

- **(25)** "For it was the Father's good pleasure for ALL the fullness to dwell in Him, and through Him to reconcile ALL things to Himself ... whether things on earth or things in heaven" (Col 1:19-20).

- **(26)** "And has put ALL things under his feet, and gave him to be the head over ALL things to the church, which is his body [the Church is his Body], the fullness of him [his Body, his Church] that fills ALL in ALL" (Eph 1:23).

- **(27)** "He who ascended far above all the heavens, that he might fill ALL things" (Eph 4:10).

- **(28)** "Who will transform the body of our humble state into conformity with the body of his glory, by the exertion of the power that he has even to subject ALL things to himself" (Phil 3:21).

Remember the True God is ALL powerful; He can "even subject ALL things to himself."

- **(29)** "Therefore if any man is in Christ, he is a new creation; the old things passed away; behold, new things have come. Now ALL things are out of God, who reconciled us to Himself through Christ, *and gave us the ministry of reconciliation, namely, that God was in Christ reconciling the world [kosmos] to Himself, not counting their trespasses against them*, and He has committed to us the word of reconciliation. Therefore, we are ambassadors for Christ" (2Cor 5:17-20).

> We are ambassadors for Christ, and our word, our message, is the message of reconciliation, that is, the reconciliation of the whole cosmos to Christ. All are to go into the Church of Christ, that is the body of Christ (Eph 1:23; and other above scripture).

What Is Meant By *all*?

nm432 » "Therefore also God highly exalted him, and bestowed on him the name which is above every name, that at the name of Jesus EVERY knee should bow, of those who are in heaven [angels], and on earth [mankind], and *under* the earth [the dead]" (Phil 2:9-10).

nm433 » By putting the above scriptures together, how can anyone with the Spirit of God say anything else but that Christ is the Savior of the whole cosmos – all the angels and all the flesh. The ALL powerful God has the Spirit of Biblical love. This love, this Spirit is one of MERCY:

- "For God has concluded them ALL in unbelief, that he might have mercy upon ALL" (Rom 11:32).

God's Spirit is one of forgiveness:

- "If your enemy be hungry, give him bread to eat; and if he be thirsty, give him water to drink" (Prov 25:21, 22).

nm434 » The higher meaning here is giving your enemy *Spiritual* bread to eat, and *Spiritual* water to drink. The truest enemy of God is Satan. Since Satan is a spiritual angel, and since angels cannot die (Luke 20:36), then the only way God can be ALL in ALL at the "end," is for Satan to be given repentance and then given the Spiritual bread and Spiritual water. Remember God is ALL powerful, and it is God who *gives* repentance (Acts 5:31; 11:18; 2Tim 2:25; Rom 11:29; 2:4; 2Peter 3:9).

nm435 » If you look up these scriptures just listed you will see God does *give* repentance. If you read other papers we have like, "Predestination Paper" [NM 8], the "Proof Paper" [NM 10], the "Reward for Christians Paper" [NM 11], the "According to Works Paper" [NM 12], the "All Saved Paper" [NM 13], and other papers, then you will begin to understand that the ALL powerful God is in total control of the cosmos and is leading it to the final state where God will be in the truest sense – ALL in ALL.

How Much Does The Spirit of Love Forgive?

nm436 »

- "Then came Peter to him, and said, Lord, how often shall my brother sin against me, and I forgive him? till seven times? Jesus said unto him, I say not unto you, until seven times: but seventy times seven" (Mat 18:21-22).

- But love covers ALL sins (Prov 10:12).

- "It [love] covers ALL, it believes ALL, it hopes ALL, it endures ALL" (1Cor 13:7, Greek text).

- Christ came to save ALL (John 4:42; 1John 4:14). Christ saves ALL, by forgiving ALL sin or by covering ALL sin by his sacrifice.

- "And the bread that I will give is my flesh, which I will give for the life of the world" (John 6:51).

- "But this one [Christ] after he had offered one sacrifice for sins for all time..." (Heb 10:12).

- "And he is the propitiation for our sins: and not for ours only, but also for the sins of the whole world" (1John 2:2).

nm437 » The ALL powerful God will be ALL in ALL (1Cor 15:28) because God's Spirit of love will fill the whole universe at the true END or at the true fullness of the ages (Jer 23:24 – truest sense of this scripture, see *God Papers*).

What About *forever* Punishment?

nm438 » If we are to believe that the ALL powerful God is in full control, and will be the Savior of the whole cosmos through his Son, and will fill ALL in ALL with his Spirit, then what about the scriptures that apparently say that those who do evil will be punished with *forever* punishment?

nm439 » First it is true that God through his built-in laws (cause and effect) does punish (Exo 34:7; etc). Satan – the main cause of evil (see "Other Mind Paper" [NM 21]) – will be punished for 1000 years (Rev 20:2).

God's Wrath and Punishment

nm440 » But how does God judge? How does God's wrath work? How does God make war against his enemies?

- **(1)** "And if any man hear my words, and believe not, I judge him not: for I came not to judge the world, but to save the world. He that rejects me, and receives not my words, has one that judges him: the **word** that I have spoken, the same shall judge him in the **last day**" (John 12:47-48; see Hosea 6:5; Rev 20:12-13).

  Notice they will be judged in the last day – that 1000 year day (see above, see "Thousand Years and Beyond Paper" [NM 15]).

- **(2)** "In RIGHTEOUSNESS he does judge and make war" (Rev 19:11).

- **(3)** "The LORD: for he comes, for he comes to judge the earth: he shall judge the world with RIGHTEOUSNESS" (Psa 96:13; 98:9; 1Chron 16:33).

- **(4)** "The LORD is known by the judgment which he executes. The wicked is snared in the work of his own hands" (Psa 9:16).

nm441 » God's wrath is not like man's wrath, "but the wrath of man works *not* the righteousness of God" (James 1:20). God's wrath is to let evil destroy evil (Psa 9:15; 10:2; Prov 11:6; 12:13; Isa 3:11; 59:18; Joel 3:4; Obad 1:15; Jer 25:32; Isa 19:2; Mat 24:7; Ezek 38:21; 32:12; Hag 2:22; Zech 14:13; Rev 17:16; see "God's Wrath Paper" [PR4]).

- Psa 10:2 (NKJV): The wicked in his pride persecutes the poor; Let them be caught in the plots which they have devised.

- Pro 11:6 (NKJV): The righteousness of the upright will deliver them, but the unfaithful will be taken by their own lust.

- Pro 12:13 (NKJV): The wicked is ensnared by the transgression of his lips, but the righteous will come through trouble.

- Isa 3:11 (NKJV): Woe to the wicked! It shall be ill with him, for the reward of his hands shall be given him.

- Isa 59:18 (NKJV): According to their deeds, accordingly He will repay, fury to His adversaries, recompense to His enemies; the coastlands He will fully repay.

- Joe 3:4 (NKJV): Indeed, what have you to do with Me, O Tyre and Sidon, and all the coasts of Philistia? Will you retaliate against Me? But if you retaliate against Me, swiftly and speedily I will return your retaliation upon your own head;

- Oba 1:15 (NKJV): For the day of the Lord upon all the nations is near; as you have done, it shall be done to you; your reprisal shall return upon your own head.

- Jer 25:32 (NKJV): Thus says the Lord of hosts: Behold, disaster shall go forth from nation to nation, and a great whirlwind shall be raised up from the farthest parts of the earth.

- Isa 19:2 (NKJV): I will set Egyptians against Egyptians; everyone will fight against his brother, and everyone against his neighbor, city against city, kingdom against kingdom.

- Mat 24:7 (NKJV): For nation will rise against nation, and kingdom against kingdom. And there will be famines, pestilences, and earthquakes in various places.

- Eze 38:21 (NKJV): I will call for a sword against Gog throughout all My mountains, says the Lord God. Every man's sword will be against his brother.

- Eze 32:12 (NKJV): 'By the swords of the mighty warriors, all of them the most terrible of the nations, I will cause your multitude to fall.' They shall plunder the pomp of Egypt, and all its multitude shall be destroyed.

- Hag 2:22 (NKJV): I will overthrow the throne of kingdoms; I will destroy the strength of the Gentile kingdoms. I will overthrow the chariots and those who ride in them; the horses and their riders shall come down, every one by the sword of his brother.

- Zec 14:13 (NKJV): It shall come to pass in that day that a great panic from the Lord will be among them. Everyone will seize the hand of his neighbor, and raise his hand against his neighbor's hand;

- Rev 17:16 (NKJV): And the ten horns which you saw on the beast, these will hate the harlot, make her desolate and naked, eat her flesh and burn her with fire.

nm442 » There is nothing righteous about man's wrath, but God's wrath is fair – mankind punishes themselves. Satan also punishes himself. He will start the Last War, through his evil influences. This Last War's bombs will be flying everywhere at once. The earth will actually begin to burn up (2Pet 3:10; Isa 64:1-3; Psa 97:1-5; Rev 18:8; Mal 4:1; etc.). But Christ is the Savior, thus he will cut this day of wrath short (Mat 24:22). But Satan and his angels will dwell in the fire they made for 1000 years (Rev 20:2; Mat 25:41).

nm443 » But what about the *forever* punishment? There are scriptures in many English translations that speak of "forever" or "everlasting" punishment. How can God be ALL in ALL at the "end," if at the same time there are those being punished with forever punishment? How can God have mercy on ALL if some are being punished forever? How can God be the Savior of the whole cosmos, if some are being punished forever? How can ALL praise and worship God, even those *under* the earth (in graves - Phil 2:9-10), if some are being punished forever? Isn't this a great contradiction? You bet it is!

The Contradiction

nm444 » We know that all the words that go out of God will happen (Isa 55:11; 46:10-11; etc.). We also have intuitive knowledge about the Law of Contradiction: One thing or event cannot *at the same time* be and not be (see *God Papers*). We know that God, who created wisdom and logic, cannot be illogical. God does not think against the very basic Law of Contradiction. It would be pure foolishness to reason against the Law of Contradiction as Aristotle once showed (*GP1; and Metaphysics*). We have in this paper gone over in some detail the scriptures on ALL being saved. It would be very difficult for someone with God's Spirit to deny these scriptures and their ordinary meaning. Thus, we come to the word translated "forever" in many (but not all) translations of the Bible. The apparent contradiction has to do with a gross mistranslation.

nm445 » The mistranslation of "forever" appalled me when I first found it out. The mistranslation projects the length some will go to feed *fear*, *hate*, and *unforgiveness* to their sheep. This mistranslation also forces some not to take God's word for what it says – all will be saved – God is in control. Satan is foolish and does not know the mercy and kindness of God. Satan was allowed his evil power for a purpose much like the Pharaoh (a type of Satan):

> ■ "For scripture said unto Pharaoh, even for this same purpose have I raised thee up, that I might show my power in thee, and that my NAME might be declared throughout all the earth" (Rom 9:17; see the "Predestination Paper" [NM 8]).

nm446 » But because Satan does not understand God's mercy, he is *very* dangerous. So dangerous is he, that he will influence mankind to begin to destroy themselves in the Last War (see "God's Wrath Paper" [PR4]).

nm447 » The mistranslation of "forever" has helped to cause many to ignore the predestination scriptures. But what is "free" choice? Did Israel have *free* choice? They did not have God's Spirit, thus could not "please God" (Rom 8:7-9). Because of this, physical Israel did not have the power to choose good (note Deut 30:15-20). Adam and Eve also did not have any "free" choice (see "Free Will Versus Predestination" paper [NM 9]).

nm448 » Not all English translations have mistranslated "forever." These translations correctly translate this word(s):

> ■ **(1)** *The Emphasized Bible,*
>
> 　　by Joseph Bryant Rotherham
>
> ■ **(2)** *Young's Literal Translation of the Holy Bible,*
>
> 　　by Robert Young
>
> 　　Author of the *Analytical Concordance to the Bible*
>
> ■ **(3)** *The Holy Scriptures: A New Translation from the Original Languages,* by J. N. Darby [not in all verses]

nm449 » Many Interlinear Bibles in Greek also translate these words correctly (the Zondervan Parallel New Testament in Greek and English). Some English translations from time to time correctly translate these words. (Note "for ever" in the KJV versus "lasting" in NIV, Exo 28:43; in Gen 17:13 note "everlasting covenant" in the KJV versus "a covenant to time indefinite" in the New World Translation; etc.)

The Correct Translation

Aeonian Punishment not Forever Punishment

nm450 » In many English translations "forever," "everlasting," and "eternal" have been mistranslated from words from Hebrew (*olam*) and words from Greek (*aion, aionios*) that mean age, ages, agelong or aeonian, hidden time, life-long, life-age. We have covered this subject in our "Age Paper" [NM 7]. We constantly use the correct translation throughout our study of the Bible. Thus, those of evil go to *agelong* punishment, not "forever" punishment. That age is the 1000 year age (Rev 20:2). Thus, an agelong or aeonian plan of God is projected in scripture once the correct and proper translation is used. Therefore, it is not only possible for God to save ALL, but it will be. God will be ALL in ALL. God will prove his ALL powerfulness. God is God. Who can stop Him?

NM 15: Thousand Years and Beyond

NM15 Abstract

In the Bible there is a reoccurring cycle of seven units of time or activity followed by an eighth one in which atonement is given. In many of these seven unit patterns there is a rest period during the seventh unit. In this paper we are going to learn the higher meaning of these patterns and in so doing we will understand more about the meaning of the millennium rest period and the time period after it when atonement will be given. This paper should be read in conjunction with NM16, which adds to the meaning of this paper.

Patterns and the Thousand Years

nm451 » There are patterns in the Bible. There is a physical meaning and there is a spiritual meaning to scripture throughout the Bible. We learned something about these patterns in the Premises of this book (see also NM16) and in the previous parts of this book. Moses was told to make a tabernacle according to the pattern (Ex 25:8-9; NM16). Paul wrote about scripture foreshadowing the Spiritual things in his letters.

nm452 » The Bible shows us something about a millennium period of time in the book of Revelation. An utopian society will be set up at that time and continue for 1000 years (Rev 20:4 ff). Concerning these 1000 years we need some more information. Also after these 1000 years we need to know what will happen. We will find some not so obvious things about the millennium by examining various patterns in the Bible. For example, we have learned in earlier parts of this book that before the

millennium Christians receive the good Spirit that enables them to eventually be resurrected and live in the millennium, but will people born during the 1000 years receive the good Spirit or is there some reason for them not to receive it then? By looking at the various patterns in the Bible we will find the answer to this question.

Sabbath Patterns

nm453 » There is a great pattern in the Bible. This pattern is a cycle of seven units of time, or seven units of action. But in many of these cycles, *the seventh unit of time or action is different from the other six units*. For example, six days were for working and the seventh day was for resting. Six years were for sowing the fields and the seventh was for resting the land. These and other cycles of seven were not mentioned in the Bible for no reason. Rituals and ceremonies in the Bible have a higher meaning. The pattern of the cycles of seven units also has a higher meaning.

Sabbath of Rest

nm454 » "For if Jesus had given them rest [physical or Spiritual Israel], then would he not afterward have spoken of another day [7th millennium]. There remains therefore a rest to the people of God. For he that is entered into his rest [7th millennium] he also has ceased from his own works; as God did from his" (Heb 4:8-10; see, Heb 4:4-5).

nm455 » The seven day week with its Sabbath prefigured the true week or the seven millenniums with its millennial Sabbath. As there is a rest day at the end of the week, on the seventh day, so too with the antitypical week, there is a rest period in the seventh millennium.

nm456 » "And God blessed the seventh day, and sanctified it [set it apart]: because that in it he had rested from all his work which God created and made" (Gen 2:3).

nm457 » "Six days may work be done; but in the seventh is the Sabbath of rest, holy to the LORD: whosoever does any work in the Sabbath day, he shall surely be put to death" (Ex 31:15).

nm458 » The Bible is dual, the prefigure and the true figure (type and antitype) or physical and spiritual. There are seven physical days and seven spiritual days or seven millennia (2Pet 3:8). Satan, the spiritual enemy, has been busy sowing his spirit and word in the world for six spiritual millenniums; while God has sowed his Spiritual seed only in a few called/chosen by God the Father, who were chosen at the foundations of the world (see Mat 13:38-39). The seventh millennium day of rest is closing in upon us, when Satan will have to be bound and chained from any spiritual work during the seventh millennium (Rev 20:1-3).

nm459 » During the seventh millennium Satan will be prevented from spiritually sowing souls with his spiritual seed (note Mat 13:38-39; NM16), *and* God during this seventh millennium will not sow souls with His Spiritual seed. The cycles of seven in the Bible indicate in their higher meaning that the seventh millennium, the Spiritual Sabbath, will be different from the previous six millenniums. There will be spiritual rest for mankind.

nm460 » *Weekly Sabbath.* "Six days you shall work, but on the seventh day you shall rest; in plowing time and in harvest you shall rest" (Ex 34:21). As there is no harvest on a physical Sabbath so too will it be on the Spiritual Sabbath. God says the physical is a type or shadow of the real in God's plan (Rom 1:20; Heb 8:5; 9:23-24; 10:1). God tells us to look at the higher meaning (Phil 3:19; Col 3:2; John 6:63; 4:24; see "Duality Paper").

nm461 » *Land's Sabbath.* "Six years you shall sow your field and six years you shall prune your vineyard, and gather the fruit thereof; but in the seventh year shall be a Sabbath of rest unto the land, a Sabbath for the LORD: you shall neither sow your field nor prune your vineyard. That which grows of its own accord of your harvest you shall not reap, neither gather the grapes of your vine undressed; for it is a year of rest unto the land" (Lev 25:3-5). There will not be a Spiritual harvest during the Spiritual Sabbath, for none will be sowed with the Spirit during this period.

nm462 » *Manna.* Concerning the manna given to the children of Israel in the wilderness, Moses relayed the law of the LORD to the people: "six days you shall gather it; but on the seventh day, which is the Sabbath, in it there shall be none" (Ex 16:26). As there was no manna given on the typical seventh day, so too on the antitypical seventh day, the Spirit will not be given to man (note John 6:32, 35, 63).

nm463 » *Dual Meaning.* When one understands that the Old Testament laws apply to both the physical and Spiritual units of time, then he will understand what is wrong with saying God will sow his Spirit in the seventh

millennium. The land is dealt with in a seven-unit cycle. In Matthew 13:38-39, we see the land or field being compared with the world, and the good seed that is planted is God's Spirit as opposed to the evil seed of the devil.

nm464 » "And six years [or millennia] you shall sow your land [sowing of the good and evil spirits], and shall gather in the fruits thereof [the evil seed produced the tares in Matthew 13:40 which 'are gathered and burned in the fire' at the end of this age as opposed to the good seed which are the first fruits in Revelation 14:4 and in Matthew 13:30]: But the seventh year you shall let it rest and lie still; that the poor of your people may eat" (Ex 23:10-11). The "poor" are those that are to inherit the kingdom of God (Mat 5:3). Thus the Spiritual poor must be those with the New Mind or the Spirit of God.

Sabbath of Purifying

nm465 » Not only is the seventh unit of time for resting from work, and resting the land, and so forth; but also the seventh unit of time or activity is for purifying and cleansing:

- "And you shall wash your clothes on the seventh day, and you shall be clean, and afterward you shall come into the camp" (Num 31:24).

- "And he shall look on the plague on the seventh day ... then the priest shall command that they wash the thing wherein the plague is, and he shall shut it up seven days more" (Lev 13:51, 54). On the seventh day the priest came, and if they were still not clean of the plague, then on that seventh day he ordered them to wash.

- "This shall be the law of the leper in the day of his cleansing: He shall be brought unto the priest ... But it shall be on the seventh day, that he shall shave all his hair off his head and his beard and his eye-brows, even all his hair he shall shave off: and he shall wash his clothes, also he shall wash his flesh in water, and he shall be clean" (Lev 14:2, 9).

- "Then he shall shave his head in the day of his cleansing, on the seventh day shall he shave it" (Num 6:9). Note: It is true that there was cleansing on a daily basis (Lev 15:16, 18), but what we are speaking about herein are ceremonies with a cycle of seven units of time or activity.

Baptism Patterns

nm466 » These physical cleansing ceremonies were foreshadows of the Christians Spiritual baptism (Eph 5:26-27; Titus 3:5; 1John 1:7, 9) which makes possible the Spiritual washing away of sin instead of filth in the typical ceremonies. These old ceremonies happened on the seventh unit of time. Therefore baptism can be said to be a seventh unit ceremony.

nm467 » But as we know baptism is representative of a death (Rom 6:4; Col 2:12; 3:3). And as with the old ceremonies so too with their higher meaning. The seventh 1000 year period will be for purifying or cleansing: the fire baptism for the enemy angels (see later in this paper); and the death for mankind who lived in the first six millenniums without the Spirit. This is man's aeonian death. Thus, this 1000 year period is a baptism in a sense, for the earth will be cleansed of sin because the transgressors will be dead, or as good as dead in the case of the enemy angels. The Spiritual Sabbath or the seventh Spiritual day does both the things that baptism does: (1) it cleanses away sin; and (2) it destroys sin because sin and the sinner are dead. Hence, we can call the seventh millennium a baptismal period.

nm468 » During this time period while the transgressors are being purified either by death or fire, the kingdom of God will be renewing the face of the earth (Psa 104:29-30). The earth itself will be cleansed after 6,000 years of misuse by man and Satan.

nm469 » With the exception of the set-apart people, everyone during the seventh millennium period will be made ready for the atonement or unity with God. Ever since the garden of Eden when it was said, "man is become as one out of us, to know good and evil" (Gen 3:22), man has been away from the way of love, that is, away from the good God's way.

Sin And Satan Put Away

nm470 » During the seventh millennium all human beings[1] living will be without either God's Spirit or Satan's spiritual influence.

---

[1] Of course we are not talking about the resurrected saints here, for they will be new creations with the new mind infused into their new nature.

Satan's spiritual influence will be locked up or sent away (Zech 13:2; Rev 20:1-3). This is the Spiritual Sabbath, a time of spiritual rest. No working of spirit or spirits will be allowed to form man's mind during the Spiritual Sabbath. The proof of this is the physical Sabbath which is a type or shadow of the Spiritual and real Sabbath – the seventh millennium. Since it can be proven that the other-mind's influence (satanic influence) is the cause of sin, then when this influence is sent away to the pit we know there will be no willful sin.

nm471 » Now when one is baptized it is "for the remission of sins" (Acts 2:38). This word translated "remission" in the KJV comes from a Greek word that means: *sending away*. When one is baptized, sin is sent away. This same Greek word is translated in the KJV as "remission" (Acts 2:38); "forgiven" (Col 1:14; Acts 13:38; etc). The same is true during the world's baptism – the seventh millennium. Since sin and Satan are metonymical terms in that Satan is the cause ("father," John 8:44) of sin, and in that a satanic spirit in man causes sin (see, the "Other Mind Paper" [NM 21]); then in the Spiritual baptism Satan and sin are also *sent away* into the bottomless pit (Zech 13:2; Rev 20:1-3, 10), for their fire baptism.

Fire Baptism

nm472 » At Christ's return to the physical dimension there will be the Last War. And because of this war many people alive at that time will die (see "Last War and God's Wrath" paper [PR5]). At the same time one-third of the angels will also "die" (note Rev 12:4). This death will be their fire baptism of 1000 years in the lake of fire. This fire baptism will help to purify the evil spirits. This is the fire baptism mentioned in the Bible (Mat 3:9-12; Luke 3:16, 17). Notice carefully "every tree which brings not forth good fruit is hewn down, and cast into the fire" (Mat 3:10). And again "he will burn up the chaff with unquenchable fire" (v. 12). Study the context carefully. Now compare it with Malachi 4:1 and Matthew 13:30, 39-43 which proves this fire baptism begins at Christ's return. This is an aeonian fire "prepared for the devil and his angels" (Mat 25:41). ("Everlasting" should be "aeonian" in this verse since "everlasting" is translated from a Greek word that means aeonian.) The fire lasts for 1000 years, or the time period that the angels will be in this pit of fire. While the peace goes on above, in this pit Satan and his demons will be kept in a fire for

1000 years. This is the fire baptism Christ spoke about.

nm473 » Now baptism is a Biblical type which indicates washing away of the dirt of sin from the person (Eph 5:26-27; Titus 3:5). In other words, baptism represents the washing away of sins or of purifying one's self. What baptism does for humans, so does fire baptism do for spirits like Satan.

nm474 » Notice: "Only the gold, and the silver, the brass, the iron, the tin, and the lead, Every thing that may abide the fire, you shall make it go through the fire, and it shall be clean: nevertheless it shall be purified with the water of separation: and all that abides not the fire you shall make go through the water" (Num 31:22-23). Notice that these two verses were an "ordinance of the law which the LORD commanded Moses" (Num 31:21). As you are seeing all the apparently non-important ceremonies are representative of higher meanings. The basic patterns of these ceremonies are of a great significance in finding out God's plan.

nm475 » Notice that anything that can abide in fire shall be made to go through this fire to be cleansed. Now flesh and blood can't abide in fire, for it burns up. But spirit can abide in fire, for spirit can't die (Luke 20:36). Further, the Bible makes comparisons between spiritual things and the metals described in Numbers 31:22. For example, see the symbolic image of Satan's spiritual kingdoms in Daniel 2:31-40. The metals described in Numbers 31:22 are used to describe the make-up of the image in Daniel 2:31-40. During the 1000 years this fire baptism will purify the satanic spirits of their spiritual impurity. This is their judgement.

nm476 » The satanic spirits are in the dark Spiritually about God's plan (Jude 6). These spirit beings do not know their fate. They think God will come to permanently destroy them (Mark 1:23-24; Luke 4:33-34). Jude 6, 2Peter 2:4, and 1Corinthians 2:7-8 prove they do not know their true and final end. They too will be freed from their evil ways like all of mankind (see "All Saved Paper" [NM 13]). They do not know that this trial of fire is a fire baptism that will purify them. They do not understand God gives repentance or the changed mind (2Tim 2:25). They do not understand that they are tools of the spiritual creation. They are helping to build knowledge in man (Rom 9:17, the Pharaoh as a shadow of Satan, cf Isa 14:12, 17-18; Ezek 28:17-18; 31:18). Thus, during the 1000 year utopia Satan and his

angels will be refined in the lake of fire, in their fire baptism.

Great Last Day Patterns

nm477 » What happens after one goes through baptism? As shown in the "Baptism Paper" [NM 4] after one is baptized, one receives the Spirit of God, which is the New Mind. The same is true after the seventh millennium baptism. After the seventh millennium the world as a whole will receive the Spirit (John 7:37-39). And that age after the 1000 years is called the Great Last Day.

nm478 » God has set aside a great day of Spiritual atonement for mankind which we call The Great Last Day. This is a Spiritual day when all will be in Spiritual unity with God. This great last day is the first day *after* seven units of time, or *after* the seven millenniums. This Spiritual Great Last Day has been pictured in the Bible as the eighth day of the Feast of Tabernacles (Lev 23:34-42; John 7:37-39). The first seven days of this festival represents the first seven millenniums (see God's Appointed Times). The eighth day of the festival represents the eighth Spiritual day of the Spiritual creation.

nm479 » At the beginning of this eighth spiritual day, which is after the 1000 years, all the dead will be resurrected to human life (Rev 20:4-5, 13; Ezek 37:1-13). This will fulfill the verse "through man came also the resurrection of the dead" (1Cor 15:21). After mankind is resurrected, then they will receive the Spirit (Ezek 37:14). After mankind has gone through a cycle of seven 1000 years, then they are atoned to God by the medium of God's Spirit. At the beginning of the eighth spiritual day man is atoned to God by the Spirit of God. When mankind receives the Spirit in the Great Last Day, it will fulfill the typical day of atonement which always came on the eighth day. In the Old Testament, atonement was always after seven units of time or activity. This atonement represents or typifies the true atonement of man to God in the Great Last Day.

nm480 » Hereafter, in an outline form, we will show you that atonement comes on the eighth unit of time or after seven units of time or activity. We will show you this, for we want to reconcile what the Bible calls atonement with the Great Last Day and the giving of the Spirit of God to ALL of mankind on that day.

Atonement Patterns

Atonement Is on The 8th Day or After 7 Units of Time:

nm481 »

- cleansing on the 7th day [Num 6:9]

- 8th day is for offering [Num 6:10] and 8th day for atonement Num 6:11]

- 7th day is to prepare offerings [Ezek 43:25]

- 7 days for purifying the altar [Ezek 43:26]

- after these days (on the 8th) God will accept, God will atone [Ezek 43:27]

- 7th day is to cleanse [Lev 15:28]

- 8th day offering is made [Lev 15:29] for atonement [Lev 15:30]

- 7 days for cleansing [Lev 15:13]

- 8th day is for offering to [Lev 15:14] make atonement [Lev 15:15]

- 7th day is for cleansing [Lev 14:9]

- 8th day is for offering for [Lev 14:10-17] atonement [Lev 14:18]

- 7 days for consecration of priests [Lev 8:33]

- 8th day is for offering for [Lev 9:1-6] atonement [Lev 9:7]

- 7 days the sheep are with their mother [Ex 22:30]

- 8th day they are given to God [Ex 22:30; see Lev 22:27]

- at the *end* of 7 years a release [Deut 15:1 (31:10) is called the LORD's release [Deut 15:2] a release from debt [Deut 15:2]

["Debt" indicates sin, compare Matthew 6:12 with Luke 11:4 and with Matthew 18:21-35. Thus, this is a release from sin or the cause of sin, which is Satan.]

The set-apart people are released from Satan after six units of time, at the beginning of the seventh millennium (Deut 15:12, 18).

Atonement Is After 7 Units of Activity

nm482 »

- 7 units of activity (sprinkling blood 7 times) [Lev 4:17]
- *after* this comes atonement [Lev 4:20]
- 7 times sprinkling blood [Lev 8:11] for reconciliation (atonement) [Lev 8:15]
- 7 times sprinkling blood [Lev 14:16]
- *after* this comes atonement [Lev 14:18]
- 7 times sprinkling blood [Lev 14:27]
- *after* this comes atonement [Lev 14:29]
- 7 times sprinkling of blood [Lev 14:51]
- *after* this comes atonement [Lev 14:53]
- 7 times sprinkling of blood [Lev 16:14]
- *after* this comes atonement [Lev 16:16]
- 7 times sprinkling of blood [Lev 16:19]
- *after* this comes reconciliation (atonement) [Lev 16:20]

nm483 » All these atonement rituals in their higher meaning represent or prefigure the future when all mankind is to be atoned to God on the Great Last Spiritual Day of Creation.

Eighth Day of Festivals or Rituals

nm484 » *Eighth Day of Festival*. Another proof that the Spirit will be given on the eighth day to the rest of mankind is John 7:37-39. Christ is pictured on the great last day of the feast of tabernacles. The great last day of this Biblical festival was the *eighth* and last day of the festival. On that day, Christ said, "if any man thirst, let him come unto me, and drink. He that believes on me, as the scripture has said, out of his belly shall flow rivers of living water. But this he spoke of the *Spirit*" (John 7:37-39). By the living waters, Christ meant his Spirit would be given. What day will he give his Spirit? The Great Last Day. This is the day *after* seven days of the Biblical feast of tabernacles (Lev 23:34-42). This feast is symbolic of the whole creation. The first seven days represents the first seven Spiritual days of creation. The last day represents the unity of man to God through the medium of God's Spirit – the living water of Christ (John 7:38-39; Rev 21:6).

Pentecost

nm485 » *Pentecost*. Still another proof that God's Spirit will be given out in the period after the seventh millennium is the feast of Pentecost. The Pentecost is a Biblical feast or appointed time after seven units of time (seven weeks of weeks, Deut 16:9; Lev 23:15-16). The day after these seven units of time is the Pentecost. The Pentecost is at the beginning of the eighth cycle of weeks. It was on the Pentecost that the New Testament Church first received its Spirit (Acts 2:1-41). The pattern of the Pentecost proves further that the Spirit will be given on the eighth Spiritual day or time period of creation.

Circumcision

nm486 » *Circumcision*. Also circumcision points to the 8th Spiritual day of creation. Circumcision was done always on the 8th day after birth (Gen 21:4; Lev 12:3; Luke 2:21). A circumcised person represents the Spiritual person (Col 2:11-13; Phil 3:3). A Spiritual person is a person with a circumcised heart as opposed to an uncircumcised heart which one can have even if he is physically circumcised (Jer 4:4; 9:26; Rom 2:28-29). Since the physical circumcision happens on the 8th physical day, then the Spiritual circumcision will happen on the 8th Spiritual day.

nm487 » Therefore the 8th Spiritual day of creation will be a period where man will be with God in Spirit. It will come after a 1000 year death of sin, a fire baptism. It is the great last day of the Spiritual creation, just before the creation of the new heaven and earth.

Jubilee

nm488 » *Jubilee.* The year of the jubilee is a type of this 8th Spiritual day of creation. The jubilee year was every fiftieth year after seven weeks of years (Lev 25:8, 10), or thus after seven seven of years (49 years). Notice the jubilee came at the beginning of the 8th cycle of seven years, or after seven units of time (7 units of 7 years).

nm489 » "That fiftieth year shall be a jubilee year for you ... and in the year of jubilee each man of you shall go back to his own property" (Lev 25:11, 13 – *Moffatt*). The year of jubilee for Israel was a year when everyone was released from slavery, debts, and so forth, and "when every man of you" (Israel) goes back to his own property and family (Lev 25:10 – *Moffatt*). Chapter 25 of Leviticus explains the year of jubilee.

nm490 » The jubilee thus represents the Great Last Day, wherein everyone will be released from Satan's misrule, and released from the 1000 year baptismal death. The Great Last Day is the age when everyone will return to life and to his own property: "O my people, I will open your graves, and cause you to come up out of your graves, and bring you into the land of Israel" (Ezek 37:12). Ezekiel 16:53-55 also indicates that at this same time when Israel will be resurrected and brought back to their land of Israel, so too will the other nations be returned to their property: "I will restore their fortunes, the fortunes of Sodom and her daughters, and your fortunes along with theirs ... When Sodom and Samaria, your sisters, and their daughters, regain their former state [at the resurrection of the dead, after the millennium, in the beginning of the atonement age], you and your daughters also shall regain your former state" (Ezek 16:53-55 – *Moffatt*).

nm491 » If God's Spirit, which is the New Mind, is one of joy, happiness, and so forth, then the Great Last Day will prove this absolutely. The Great Last Day will be even more joyous than the millennium, for during the millennium no one except those born of God at, or by the time of, the physical return of Christ, will have God's Spirit. The millennium will be great because of the lack of Satan's ways. Man during the millennium will be led physically by the resurrected who will rule this age period. But the Great Last Day will be an exceptionally beautiful age of joy much better than the great last day in the feast of tabernacles. Now let's show the references of Satan coming into the atonement period. Satan will be atoned to God in the Great Last Day. But first we must know something else about the 7th millennium.

1000 Year Judgment of God

nm492 » Not only is the 7th millennium like a baptismal, a cleansing, and/or a death period for mankind and/or the satanic spirits who lived during the first six thousand years; but, also, it is the judgment period. "And as it is appointed unto men once to die, and with [#3326] this the judgment" (Heb 9:27).

nm493 » The judgment for sin is death (Ezek 18:11-13; Gen 2:17; Rom 6:23). This judgment is an aeonian punishment or death (Mat 25:46). It is a death judgment of an "aeonian destruction from the presence of the Lord, and from the glory of his power [in the seventh millennium]" (2Thes 1:9).

nm494 » This 1000 year judgment begins at Christ's return (see Mat 25:31-33, 41, 46; Rom 2:5; Joel 3:12, 14; 2Pet 3:7; Rev 11:18; Dan 7:10, 26). At that time the saints are resurrected to judge-down the way of man/Satan (Rev 11:18; 1Cor 6:2-3).

nm495 » Satan and his angels are now in Spiritual darkness because of their rebellion, and are being reserved for the day of judgment (2Pet 2:4; Jude 6). The day of judgment is the day of wrath when Christ returns (Acts 17:31). Also the day of judgment is the antitypical 1000 year day of judgment, the aeonian judgment, or the aeonian death of the way of man/Satan.

nm496 » This judgment is the sending away of sin. The sentence for this judgment is given at Christ's return. The judicial sentence is for the 1000 years during the millennium. This is pictured in Revelation 20:1-3 where Satan's power is put in chains (prison) for 1000 years.

nm497 » This is also pictured typically by baptism which is the Christians' judicial sentence (note 2Cor 1:9). And because of this "death," sin is sent away (Acts 2:38).

Azazel – Removal of Sin And Satan

nm498 » That the real judgment of God is the sending away or destruction of sin is further proven by the once-a-year ritual on the Day of Atonement mentioned in Leviticus 16. Here the second goat was brought before Aaron (v. 21) who confesses the sins of Israel over the goat's head. In other words, he put the sins of the people on the head of the goat, and then sends it away "by the hand of a fit man" (v. 21). This pictures the sending away of sin, or the remission of sin by Aaron (a type of Christ). This goat is let go "for the entire removal" (Lev 16:26). The Hebrew word "Azazel" translated "scapegoat" in the KJV means according to *Brown Driver Briggs Gesenius Hebrew Lexicon* (p. 736) – "entire removal." And in the Greek text it means – "for the dismissal." Thus, the sending away of the goat pictures the entire removal of sin because the goat had all the sins on its head.

nm499 » There were two "twin"[1] goats in this ritual, so they cast lots to find out which goat was for the LORD and which goat was for Azazel (Lev 16:8). The first goat was killed or offered and had its blood sprinkled on the mercy seat for a sin offering for the people and the Holy Place without the priest laying his hands on its

head (Lev 16:9,15-16,20). After the first goat was killed and its blood sprinkled on the mercy seat, the second goat had the sins of Israel confessed over his head and was let go or sent away for Azazel or for an "entire removal" into the wilderness (Lev 16:21). The first goat and the offering of that goat signified Christ's sacrifice (Isa 53:10; Heb 9:11-14,23-25). The second goat (sent away *after* the first goat was killed) signifies Satan being sent away with his sins into the 1000 years of separation (Rev 20:1-4; 2Thes 1:9). As the second goat was sent away *after* the first one died, other scriptures show that Satan will be sent away two millenniums *after* Christ died (see NM16 & *Prophecy Papers*). Even though Christ died for sin (Rom 5:6; 6:10; 8:3,34; 1Cor 15:3; 1Pet 3:18), and even though he was accounted with the transgressors and bore their sins (Isa 53:11-12), he had no sin (2Cor 5:21; 1Pet 2:21-22), thus sin in the truest sense could not be placed on his head, but had to be placed on the head of the second goat. It is the true or antitypical scapegoat (Azazel) that bears the real blame for sin, because it is Satan and his "other-mind" that put sin in the world (see NM20). In order to understand more on this subject, you must read and understand the scriptures on the right and left side of God. In the *God Papers* we discuss this.

Satan Let Loose

nm500 » Satan and his angels are judged for 1000 years (Rev 20:1-3; Isa 24:22). And after this 1000 year period they will be let loose out of their prison (Rev 20:7; Isa 24:22). They will be let loose for atonement to God's way, not to wage war again as Revelation 20:8 *seems* to indicate.

nm501 » First Revelation is in parts generally sequential, but it has insert or qualifier verses and chapters throughout the book. Second, Revelation reiterates in many cases the same events in sightly different words (see "Last War and God's Wrath" paper [PR5]). And third, always remember that there is no authority as to where the punctuation ought to be in the Bible. The original text had no punctuation. Translators have added punctuation where they felt the punctuation should be placed in the sentence structure. After Revelation 20:7 should be a period; and verses 8 to 10 should be in parentheses for they indicate what Satan did *before* he was cast into the pit. Verses 8 to 10 are simply telling the reader that this one that is to be let loose, will go out at some time and gather the nations together. But it does not say

---

[1] "The two goats, however, must be altogether alike in look, size, and value." (*The Temple*, by Alfred Edersheim, p. 312)

when. It does not say that after he is let loose out of the prison that he will gather the nations. It merely tells us he is the one who will gather the nations, but it does not say *when*. We find out when he gathers the nations from other verses in Revelation and other books in the Bible.

nm502 » Let's prove that Satan will not gather the nations again once the kingdom of God is set up on earth. Remember if Satan does gather the nations to make war after he is let loose from the pit, he would be doing wrong in God's kingdom. But once God's kingdom is set up there will be no evil.

nm503 » If we can prove that there will be no more war or wrong behavior after the kingdom of God is set up, then we prove that verses 8 to 10 are not speaking about Satan misleading the nations *after* God's kingdom is set up, but *before* it is set up. Verses 8 to 10 do *not* say that after he is released he will go out to gather the nations, for these verses are vague in their original language as to *when* he will gather the nations to fight.

nm504 » Revelation 20:8-10 is at the time of Christ's physical return, not after Satan is released. The proof is in Isaiah 2:4. The verses around this one describe Christ at his return. Note in verse 4 it speaks of nations *not* learning war again. Micah 4:3 says the same thing! And in speaking about God's return to earth: "He makes wars to cease unto the end of the earth" (Psa 46:9). And again speaking about his return: "but with an overrunning flood he will make an utter end of the place thereof, and darkness shall pursue his enemies. What do you imagine against the LORD? he will make an utter end: AFFLICTION SHALL NOT RISE UP THE SECOND TIME" (Nah 1:8-9).

nm505 » So Satan can *not* bring Gog and Ma Gog against the saints camp a second time. By comparing scriptures we can see that Revelation 20:8-10 is speaking about the gathering against the Saint's camp at Christ's return. Compare Revelation 20:8, 9 with Revelation 16:16; Isaiah 54:15; Ezekiel 38:8, 16; Joel 3:11-12, 14. Now for Revelation 20:10 compare it with Ezekiel 28:12-19 (especially Ezek28:17, 19); Isaiah 27:1, 4; Isaiah 14:11-12, 18-20; Ezekiel 31:18, 14 (Pharaoh as a type of Satan; the "tree of Eden" as a type of Satan's kingdom). You can now see that verse 10 is an amplification of Revelation 19:20-21. And these latter verses can be compared with Ezekiel 38:22; 39:11; 32:26; etc, which are all

amplification of Revelation 16:21 (see "God's Wrath, An Outline" PR6).

nm506 » Thus, Satan is set loose on the Great Last Day of atonement and will not make war. The Great Last Day of creation will prove once and for all that God's way of Love is the best and only way. Everyone will be resurrected to human life into this Great Last Day except of course those who are already born of God. Those resurrected in the Great Last Day are those of the "resurrection of the dead" (Acts 24:15; 1Cor 15:21). Those resurrected into the Great Last Day as humans will then be begotten of God's Spirit and will live as Spiritually begotten children of God. But then comes the true end (1Cor 15:24) of creation after the atonement period and then all will be *born* of God (see "Begotten, Born Paper" [NM 5]; see "God's Appointed Times Paper" [NM 16]).

Seven Times of Nebuchadnezzar

nm507 » In Daniel, chapter 4, Nebuchadnezzar, the king of Babylon had a dream, "thus were the visions of mine head in my bed; I saw, and behold a *tree* in the midst of the earth, and the height thereof was great. The tree grew, and was strong, and the height thereof reached unto heaven, and the sight thereof to the end of earth" (Dan 4:10-11).

nm508 » But then a watcher and a holy one came down from heaven; He cried aloud, and said:

- cut the tree down and destroy it (V. 23), scatter the fruit, and cut down the branches of the tree (Dan 4:14).

- but leave the stump of the roots with a band of iron and brass (Dan 4:15)

  - change the tree's heart from a man's to a beast's heart, (Dan 4:16)

  - and let SEVEN TIMES pass over him.. Why?

  - to show who rules in the kingdom of men (Dan 4:17) God gave Daniel power through the spirit in him (Dan 4:9) to interpret dreams (Dan 1:17). Daniel interprets the dream (Dan 4:19)

  - the TREE represents the KING (Dan 4:20-22)

- and SEVEN TIMES will pass over the KING until he knows that the most High rules in the kingdom of men (Dan 4:25).

- the king's kingdom will be sure to the king after he learns who rules in the kingdom of men. Thus, it would be sure to him after the SEVEN TIMES (Verse 25) (Dan 4:26).

- a warning is given to the king (Dan 4:27-28)

- But after 12 months the king lifted up his heart and spoke proud words about himself alone building the great kingdom which he controlled (Dan 4:28-30).

- While these words were in his heart, a voice out of heaven spoke: "The kingdom is departed from you" (Dan 4:31).

Seven Times as a Mad Man

nm509 » The kingdom departed from the king, and the king was to dwell with the beasts of the field until SEVEN TIMES should pass over the king so that the king would learn who really rules in the kingdom of men (Dan 4:32).

nm510 » Nebuchadnezzar "was driven from men, and did eat grass as oxen, and his body was wet with the dew of heaven, *till his hairs were grown like EAGLES' feathers*, and his nails like birds' claws" (Dan 4:33). But at the end of the seven times, the king's understanding or reasoning was returned to him (Dan 4:34, 36).

nm511 » Now for SEVEN TIMES the king was given the mind or reasoning of a beast, "let his heart be changed from man's, and let a beast's heart be given unto him" (Dan 4:16). The king for SEVEN TIMES or seven years was given a beast's mind, he acted like a beast, like an animal. But after the *seven times* a man's mind was again given to him.

Now what is the meaning of these seven times?

Duality

nm512 » The Bible is dual. this is the consistent pattern of the Bible. Duality: *type* and *antitype*. God works in twos. He created male and female throughout His creation. He created the spiritual and physical dimensions. But further He created the Bible through inspiration to be dual in meaning (note Heb, chaps. 8, 9, 10). The old and new covenant, and the physical

Passover and the Spiritual Passover are examples. If you read the "Duality Paper" you may see duality throughout God's plan, and the duality throughout the Bible. Prophecy is also dual. Genesis has a dual or antitypical meaning. Revelation has a dual meaning. Matthew, Mark, Luke, and John have a dual meaning. Not every word has a dual meaning, but every complete thought in the Bible, or every event pictured has a dual meaning. Everything prophesied about will be fulfilled in duality. The antitypical fulfillment is its truest sense.

nm513 » The duality of the Bible can't be overemphasized. It is because the Jews did not comprehend the duality of the Bible that they did not accept Christ. They did not understand that the Messiah was to come twice: once physically to die and be ridiculed, and the second time to rule the earth for 1000 years.

nm514 » Moses was warned to make the earthly tabernacle according to the pattern shown him in mount Sinai (Heb 8:5). Hebrews, chapters 8, 9, and 10 project that this pattern is one where the physical or type is a shadow of the heavenly or Spiritual or antitype.

nm515 » With this understanding of duality, is the only way one can comprehend the seven times. If one does not understand the duality of the Bible we suggest he read our "Duality Paper" or other papers we have written to begin to see the duality of the Bible. In fact if you do not understand the duality, or believe in the duality of the Bible, then you can't possibly believe the content of this paper. For the premise of this paper is the fact of duality. From now on we will take it for granted that the content of Daniel 4 has a dual significance.

Seven Times Higher Meaning

nm516 » Now the king was punished for SEVEN TIMES (years) for his false pride (Dan 4:30-31; 5:18-21). Why?, "a high look, and a proud heart ... is sin" (Prov 21:4). It is sin because everything a man has, or is, was given to him (1Cor 4:7). To say that one has glory because of himself, like Nebuchadnezzar did (Dan 4:30), is a lie, for it is the LORD who grants everything (Dan 5:18-19). And to lie, is to sin, for to lie is to transgress the law of Truth.

nm517 » So because of the king's sin, he was punished SEVEN TIMES or years. During these seven years he was given a mind of a beast (Dan 4:16). And after these seven years he was given back the mind or reasoning of a man (Dan 4:34,

36). But since the Bible is dual, and the king also represents an antitypical or Spiritual truth.

nm518 » Note Nebuchadnezzar during the seven years was given a beast's mind, and "his hairs were grown like EAGLES' feathers" (Dan 4:33). Now notice Daniel 7:4, the first beast. This first beast represents Babylon, Nebuchadnezzar's kingdom (see "Beast-Man Paper" [PR2]). Notice what Daniel 7:4 says about the king's KINGDOM, "the first was like a lion, and had EAGLE's wings [eagle wings are made of feathers]: I beheld till the wings thereof were plucked, and it [the kingdom] was lifted up from the earth, and made to stand upon the feet as a man, and a man's heart was given to it."

nm519 » Notice the parallel between the KING Nebuchadnezzar and his KINGDOM (1). Both had eagle wings or feathers (Dan 4:33; 7:4). (2). Both had their reasoning returned to them so that they could think like a man instead of a beast (Dan 4:34, 36; 7:4). Further we know in Daniel he uses "king" and "kingdom" interchangeably (Dan 7:17, 23). Because the *king* was punished SEVEN TIMES, the *kingdom* also would be punished seven times.

nm520 » Also in speaking about the tree, which signified the king *and* the kingdom, the scripture said that the kingdom would be cut down, and *seven times* would pass over it before the kingdom would *know* who ruled in the kingdom of men (Dan 4:23-25). Now the *king* was cut down and made to dwell among the beasts (Dan 4:32-33), and after seven times (years) he knew who ruled on earth (Dan 4:34-35). The seven times that affected the physical king and kingdom of Babylon was seven *years*. **The seven times that affect the spiritual king (Satan) and kingdom (Satan's kingdom) is seven spiritual days, or *seven thousand years*.** The book of Genesis showed the start of the kingdom of Satan. Ever since the beginning of this kingdom, its leaders have acted beast-like. After 7000 years a clear mind will be given to Satan and his kingdom, they will be resurrected and given the New Mind in the Great Last Day, the short period after the 1000 year period.

Notes

More on when the Spirit will be Given by God

nm521 » Now we have shown that the millennium will be a rest from spiritual work. Let's note other verses about the Spirit not being given out in the seventh millennium but in the period after it. In Isaiah 55:1, 3, 6 we read:

nm522 » "Every one that thirsts, come you to the waters ... incline your ear, and come to me: hear, and your soul shall live; and I will make an aeonian covenant with you ... Seek you the LORD while he may be found, call upon him while he is near."

nm523 » Compare the above with John 4:13: "whosoever drinks of the water that I shall give shall never thirst; but the water that I shall give him shall be in him a well of water springing up into aeonian life."

nm524 » Notice the similarities in these verses. By studying the context of these verses from Isaiah and John we see it is talking about a future salvation, in an aeonian covenant or life, through the waters that are now being poured out through the Church. These running waters represents God's Spirit being poured out (John 7:37-39).

nm525 » Also Revelation 21:6 denotes the physical waters that will pour out of Jerusalem at Christ's coming (Zech 13:1; Ezek 47:2, 8-9), as well as denoting the *Spiritual* water being poured out in the Great Last Spiritual Day.

nm526 » Note that Isaiah 32:15 has no time element. We ask the question *when* will the Spirit be poured down? This verse does not say, but we've shown you when it will be poured down.

nm527 » We ask the same question about Isaiah 44:3. *When* will it be poured down? Now Zechariah 12:10 is dual. It speaks about people looking on Jesus "whom they *have* pierced." This happened while Jesus was on the cross. It will also happen when Christ returns (Rev 1:7). The world will look upon Christ who *was* pierced by the old age. (This is not to say that Christ at His return will have pierced marks on Him, for Christ was resurrected with a new perfect body.) This verse also says God will pour down his Spirit on the house of David and the inhabitants of Jerusalem. This happened typically on the Pentecost (Acts 2:1-41).

Antitypically this will happen at God's return, when he sends his Spirit (his angels) to resurrect New Jerusalem, the true Christians. See the paper on New Jerusalem [NM 18] where it equates real Christians, when born of God, with the New Jerusalem described in Revelation. Thus, God sends his Spirit, his angels, to the inhabitants of the antitypical Jerusalem, which is the New Jerusalem.

nm528 » Now turn to Joel 2:27-28: "and my people shall not aeonian be ashamed" (v. 27). Notice the correct translation is: "not aeonian" as against "never." The translation of "never" is incorrect. From the Hebrew it should read, "not aeonian." In the Greek translation (Septuagint) it should read, "not into the age" instead of "never." This aeonian period is proven to be the seventh millennium (see the "Reward Paper" [NM 11]). Thus Joel 2:27 speaks about God's people ("my people") not being ashamed during the seventh millennium. Notice Joel 2:28: "and it shall come to pass *afterward* [after the millennium, v. 27], that I will pour out my Spirit upon all flesh; and your sons and daughters shall prophesy" (Joel 2:28). This is one more proof that the Spirit will be given after the millennium.

NM 16: God's Appointed Times: Tabernacle, Festivals, and Sacrifices Foreshadowed Christ

NM16 Abstract

Appointed times were festivals and holy days such as the Sabbath, Passover, and Pentecost. During these appointed times there were various rituals and sacrifices that were performed. Moses constructed a Tabernacle in which to perform these rituals. What we show in this paper is that all these things foreshadowed the appointed times of God's plan and the Coming Christ and his Spiritual Body (Church). This paper again manifests the type and antitype of the Bible.

Patterns in the Bible Foreshadow the Future

nm529 » When attempting to understand the higher or Spiritual meaning in the Bible, the patterns of the Bible *must be* observed, marked and understood. If our beliefs go against the patterns in scripture we are in error. Moses wrote the first five books of the Bible. When you read these books you see patterns occurring again and again. For example, again and again, you see the pattern of six periods of work, one of rest, and one of atonement (see NM15). What we do in the *BeComingOne Papers* is to point out these patterns, and the type and antitype aspect of these patterns. Not only in the books of Moses do you see these patterns, but in all the books of the Bible. But why did Moses and others write down these patterns? It is because God directed this to be done. In Moses' case, Moses was directed to make the tabernacle and all its furniture by the pattern shown to him on the mountain:

■ Let them construct a sanctuary for Me, that I may dwell among them... According to all that I am going to show you, *as* the **pattern [type or image] of the tabernacle** and the **pattern [type or image] of all its furniture**, just so you shall construct *it*. (Ex 25:8-9)

■ See that you make *them* **after the pattern** [type or image] for them, which was shown to you on the mountain. (Exodus 25:40)

■ Then you shall erect the tabernacle **according to its plan** which you have been shown in the mountain. (Exodus 26:30)

■ Now this was the workmanship of the lampstand [candlestick], hammered work of gold; from its base to its flowers it was hammered work; **according to the pattern** [type or image] which the LORD had showed Moses, so he made the lampstand. (Numbers 8:4)

■ Our fathers had the tabernacle of testimony in the wilderness, just as He who spoke to Moses directed *him* to make it **according to the pattern** which he had seen. (Acts 7:44)

■ [Priests] who serve a **copy and shadow of the heavenly things**, just as Moses was warned *by God* when he was about to erect the tabernacle; for, see, He says, that you make all things **according to the pattern** which was shown you on the mountain. (Hebrews 8:5)

■ For the Law, since it has *only* a **shadow of the good things to come** *and* not the very form of things, can never, by the same sacrifices which they offer continually year by year, make perfect those who draw near. (Hebrews 10:1)

■ Now **these things** [in O.T.] **happened as examples** for us, so that we would not crave evil things as they also craved. (1 Corinthians 10:6)

nm530 » Even the way the sacrifices were performed was done by the direction of God. The book of Leviticus was primarily devoted to the ministry and ceremonies of the priests. In this book, over fifty-times Moses wrote that God spoke, or, that is, God told Moses what to write pertaining to the sacrifices and ceremonies. In this book there was

■ the **law of** the grain offering (Lev 6:14)

■ the **law of** the sin offering (Lev 6:25)

■ the **law of** the guilt offering (Lev 7:1)

■ the **law of** the sacrifice **of** peace offerings (Lev 7:11)

In fact there was a "law of the burnt offering, the grain offering and the sin offering and the guilt offering and the ordination offering and the sacrifice of peace offerings, which the LORD commanded Moses at Mount Sinai in the day that He commanded the sons of Israel to present their offerings to the LORD in the wilderness of Sinai." (Lev 7:37-38) These laws were a shadow of things to come (Heb 10:1).

nm531 » Not only were there patterns to the tabernacle and the rituals of the Old Testament, but as Paul indicated also other laws, even laws that dealt with animals were a shadow of future things that pertain to Christians:

■ The elders who rule well are to be considered worthy of double honor, especially those who work hard at preaching and teaching. For the Scripture says, "you shall not muzzle the ox while he is threshing," and "The laborer is worthy of his wages." (1Timothy 5:17-18)

Notice how Paul used a scriptural law meant for an ox to teach Christians about how to best treat their Spiritual elders.

Appointed Times

nm532 » In Leviticus, chapter 23, it describes the appointed times ("feasts") of God: "And the LORD spoke to Moses, saying, Speak unto the children of Israel, and say to them, Concerning **the appointed times of the LORD**, which you shall proclaim to be holy convocations, even these are my appointed times" (Lev 23:2).

Festivals Foreshadowed the Future

nm533 » Most of the so-called Christians say these appointed times or seasons were done away with. But we will show that these appointed times or seasons have a higher and Spiritual meaning. The New Testament Church, unlike the physical Israelites, perceived these appointed times in a Spiritual way because of Christ's own Spirit and these words: "But the hour comes, and now is, when the true worshipers shall worship the Father in Spirit and in truth: for the Father seeks such to worship him. *God is Spirit: and they that worship him must worship him in Spirit and in truth*" (John 4:23-24). "Therefore no one is to act as your judge in regard to food or drink or in respect to a festival or a new moon or a Sabbath day – things which are a *mere* shadow of what is to come; but the substance belongs to Christ" (Col 2:16-17). These Old Testament festivals were a shadow of things to come.

nm534 » Throughout Paul's scripture, he manifests Spiritual worship, or Spiritual thoughts toward God. Paul's writings show the type and antitype of the Bible. Paul, speaking about the Old Testament scripture, says:

■ "These things were our **types** . . .. Now all these things happened **typically** unto them: and they are written for our admonition, upon whom the ends of the ages have come" (1Cor 10:6, 11).

■ The things described in the Old Testament were typical events just as Moses' tabernacle was a shadow of the heavenly things to come (Heb 8:5).

■ The tabernacle "which *is* a type for the present time. Accordingly both gifts and sacrifices are offered which cannot make the worshiper perfect in conscience." (Hebrews 9:9)

- "For the Law, since it has *only* a shadow of the good things to come *and* not the very form of things, can never, by the same sacrifices which they offer continually year by year, make perfect those who draw near." (Hebrews 10:1)

nm535 » The Old Testament festivals and Sabbaths were typical representations of things to come afterward. We will explain these appointed times and show their higher meanings. In short, we will see that the appointed times or seasons pictured the appointed times of God's plan and Jesus Christ and his Spiritual Body (Church).

Sabbath

Six Days of Work; One Day of Rest

nm536 » The Sabbath is one of God's appointed times. Notice in Leviticus 23:2, "these are my appointed times." Then in the next verse, "Six days shall work be done: but the seventh day is the Sabbath of rest, a holy convocation; you shall do no work therein: it is the Sabbath of the LORD in all your dwellings."

nm537 » The Sabbath is the seventh day of the week. The Jews of today celebrate the Sabbath. The Sabbath is what we call Saturday. Most so-called Christians keep Sunday, but there is no scripture that says we should keep Sunday. God instituted the Sabbath as the seventh day, the day of rest:

- "Thus the heavens and the earth were being [imperfect verb] finished, and all the host of them. And on the seventh day God will end [imp. verb] his work which he had made; and he will rest [imp. verb] on the seventh day from all his work which he had made. And God blessed the seventh day, and set it apart: because that in it he had rested from all his work which God created and made" (Gen 2:1-3).

So God blessed and set the seventh day apart from the rest of the days.

nm538 » Before Moses was given the precepts of Jehovah, the Sabbath was pointed out to the children of Israel (Ex 16:23, 26, 29-30, 1-31). In the words of the ten commandments:

- "Remember the Sabbath day, to keep it set apart. Six says shall you labor, and do all your work: but the seventh day is the Sabbath of the LORD your God: in it you shall not do any work, you, nor your servant, nor your maidservant, nor your cattle, nor your stranger that is within your gates: for in six days the LORD made heaven and earth, the sea, and all that is in them, and rested [imp. verb] the seventh day: wherefore the LORD blessed the Sabbath day, and set it apart" (Ex 20:8-10).

nm539 » The Sabbath is a memorial. The Sabbath has us remember the six days of creation when God created the universe. On the seventh day of this creation God rested from his work. But further, the Sabbath is a memorial to the Israelite people coming out of Egypt:

- "And remember that you were a servant in the land of Egypt, and that the LORD your God brought you out of there through a mighty hand and by a stretched out arm: therefore the LORD your God directed you to keep the Sabbath day" (Deut 5:15).

nm540 » Now as Paul projected to us in his letters, the Old Testament was the type of the antitype, or the Shadow of the true or real. The Sabbath has an antitypical meaning. It foreshadows something (Col 2:16-17).

If God had given them Rest

nm541 » In the fourth chapter of Hebrews Paul shows us the antitypical meaning of the seventh day rest, the Sabbath:

- "For he spoke in a certain place of the seventh day this way, 'and God did rest [aorist verb] the seventh day from all his works.' And in this place again, 'If they shall enter into my rest [Sabbath].' Seeing therefore it remains that some must enter therein, and they to whom it was first preached entered not in because of unbelief . . .. For if Jesus had given them rest [aorist verb], then would he not afterward have spoken of another day. There remains therefore a rest [Sabbath] to the people of God. For he that is entered into his rest, he also has ceased from his own works, as God did from his. Let us labor therefore to enter into that rest [Sabbath] lest any man fall after the same example of unbelief" (Heb 4:4-11).

Now this other rest that God's people shall enter is the 1000 year rest (Rev 20:4-5), the millennium Sabbath of peace. This is the 1000 year Sabbath when the Spiritual "Israel of God" (Gal 6:16) shall have fled out of spiritual Egypt and entered God's promised land. "The Son of man [Christ] is Lord also of the Sabbath" (Mark

2:28). It is Christ who shall rule as King of kings in the 1000 year Sabbath wherein the Israel of God will cease from its own works.

Physical Sabbath	Spiritual Sabbath
Sabbath Begins When? **nm542 »** *Physical Sabbath* The Sabbath begins on Friday at the setting of the sun as it begins to darken, for the days of the Bible begin at evening, the very last part of sunlight. The physical Sabbath lasts the whole day, from sunset on Friday to sunset on Saturday (Neh 13:19; Lev 23:32).	*Spiritual Meaning* The Spiritual Sabbath begins at the return and physical manifestation of the Messiah, Jesus Christ, and will last for 1000 years because a day to God is like 1000 years (2Peter 3:8).
Sabbath Preparation **nm543 »** *Physical Sabbath* The Sabbath must be prepared for on the sixth day of the week by: preparing & cooking the food for the Sabbath (Ex 16:5, 22-23, 29), for there should be no cooking on the Sabbath. Further one was to prepare in any other way so there wouldn't have to be any unnecessary work on the Sabbath (Ex 35:2-3).	*Spiritual Meaning* The Spiritual meaning of this is that all the preparing of the Spiritual food for the 1000 year Sabbath will happen in the sixth millennium and before.
Sabbath a Delight **nm544 »** *Physical Sabbath* The Sabbath is to be a delight (Isa 58:13; Ex 20:8-10; Neh 13:15-21; 10:31).	*Spiritual Meaning* The higher meaning here is that the New Mind, will delight in the new ways of the New Age. Those of the New Mind and those of mankind will cease from their old way during the Spiritual Sabbath.
Sabbath: an Assembly **nm545 »** *Physical Sabbath* The Sabbath is kept by assembling with others (Lev 23:3; Luke 4:16).	*Spiritual Meaning* This pictures the higher meaning of those of the Spirit of God who will be gathered by God at the end of the old age to live in the New Age, the 1000 year Sabbath, together with the others who are gathered. The gathering of physical Israel in their physical Sabbath pictures the gathering of Spiritual Israel (the Church) in the Spiritual Sabbath which will last for 1000 years.
Sabbath for Good Works **nm546 »** *Physical Sabbath* Christ taught that it was right to do good works on the Sabbath (Luke 13:14-16; 14:3-5; Mark 3:1-6; John 5:8-16; Mat 12:11-12).	*Spiritual Meaning* This pictures the good works that will be done on the Spiritual Sabbath by Spiritual Israel.

Physical Sabbath	Spiritual Sabbath
Sabbath Fulfillment: A Rest Period from Satan **nm547 »** *Physical Sabbath* The physical Sabbath was a day of rest.	*Spiritual Meaning* The Spiritual Sabbath will be a day of rest also, that is, rest from the evil spiritual work of the spiritual enemy of mankind. Only the good works of the true and good God will be performed during the 1000 year Sabbath. It will be an utopia for those in the Spirit because the enemy, that spiritual enemy of mankind, will be put away during the 1000 year rest (Rev 20:2-3).

Three Harvests of the Land

Agricultural Metaphor

nm548 » In the past, and until recent times, most people in the world were involved intimately in agriculture. The Bible appropriately uses agriculture metaphorically to manifest God's plan of salvation. Notice the metaphor of sowing seed and harvesting the crop from the seed:

- Jesus presented another parable to them, saying, The kingdom of heaven may be compared to a man who **sowed good seed** in his field. 25 "But while his men were sleeping, his enemy came and **sowed tares** among the wheat, and went away. (Matthew 13:24-25)

- and the field is the world; and *as for* the **good seed, these are the sons of the kingdom**; and the **tares are the sons of the evil** *one*; (Matthew 13:38)

- And another angel came out of the temple, crying out with a loud voice to Him who sat on the cloud, "Put in your sickle and reap, for the hour to reap has come, because the **harvest of the earth is ripe**." (Revelation 14:15)

There were three harvest festivals in the year. One was at the beginning of the grain harvest at the Passover feast, one at the end of the grain harvest at the Pentecost, and one at the end of the last and final harvest at the Feast of Tabernacles or Booths.

Barley, Wheat, Fruit Harvests

nm549 » To many of us, harvest time is of little concern, because in our complex life we are far removed from the actual production of our food supplies, but for the Hebrew people, as for those in any agricultural district today, the harvest was a most important season (Gen 8:22; 45:6). Events were reckoned from harvests (Gen 30:14; Josh 3:15; Jdg 15:1; Ruth 1:22; 2:23; 1Sam 6:13; 2Sam 21:9; 23:13). The three principal feasts of the Jews corresponded to the three harvest seasons (Ex 23:16; 34:21,22):

- **(1)** the feast of the Passover in April at the time of the barley harvest (compare Ruth 1:22);

- **(2)** the feast of Pentecost (7 weeks later) at the wheat harvest (Ex 34:22),

- and **(3)** the feast of Tabernacles at the end of the year (October) during the fruit harvest.

The seasons have not changed since that time. Between the reaping of the barley in April and the wheat in June, most of the other cereals are reaped. The grapes begin to ripen in August, but the gathering of crops for making wine and molasses (*dibs*), and the storing of the dried figs and raisins, is at the end of September. [Paragraph taken from *ISBE* (1915), under "Harvest"]

Three Harvests Correspond to the Three Times Before YHWH

nm550 » Corresponding to the three harvests, the males of Israel were to stand before God three times in a year:

> **Three times in a year all your males shall appear before the LORD** your God in the place which

He chooses, at the **Feast of Unleavened Bread** [Passover] and at the **Feast of Weeks** [Pentecost] and at the **Feast of Booths** [Tabernacles], and they shall not appear before the LORD empty-handed. (Deuteronomy 16:16)

Three times a year you shall celebrate a feast to Me. 15 "You shall observe the **Feast of Unleavened Bread**; for seven days you are to eat unleavened bread, as I commanded you, at the appointed time in the month Abib, for in it you came out of Egypt. And none shall appear before Me empty-handed. 16 "Also *you shall observe* the **Feast of the Harvest** *of* **the first fruits** of your labors *from* what you sow in the field; also the **Feast of the Ingathering** in the produce of the year when you gather in *the fruit of* your labors from the field. 17 Three times a year all your males shall appear before the Lord GOD. (Exodus 23:14-17)

nm551 » Notice that all three times the males were to appear before God was at the time of the three harvests. The second main festival which was the feast of the harvest of the first fruits is also known as the feast of the wheat harvest, and the last of the three main festival was also known as the Feast of Ingathering:

You shall celebrate the **Feast of Weeks**, *that is*, **the first fruits of the wheat harvest**, and the **Feast of Ingathering** at the turn of the year. (Exodus 34:22)

There are many patterns in the Bible, but the pattern of the three harvests corresponded to the three times males were to stand before God and as we will see this in turn pointed to the harvest of souls at the appointed times.

Three Pilgrim-Feasts for Males & Angels

nm552 » The three main feasts were pilgrim-feasts, which the males were required to attend. "In Hebrew two terms are employed – the one, *Moed*, or appointed meeting, applied to all festive seasons, including Sabbaths and New Moons; the other *Chag*, from a root which means 'to dance,' or 'to be joyous,' applying exclusively to the three festivals of Easter [Passover], Pentecost, and Tabernacles, in

which all males were to appear before the Lord in His sanctuary" (*The Temple*, by Alfred Edersheim, p. 196) This word [*Chag*] is closely related to an Arabic word that means pilgrimage (BDBG Lexicon, p. 290). These three main feasts were *pilgrim-feasts*, a time of great joy. When you understand the type and antitype of the Bible you understand that males represent the angels, while females represent mankind (*God Papers*). Notice that Job wrote about how Satan also came before God when the sons of man came to present themselves:

■ Now there was a day when the sons of God came to present themselves before the LORD, and Satan also came among them. (Job 1:6)

■ Again there was a day when the sons of God came to present themselves before the LORD, and Satan also came among them to present himself before the LORD. (Job 2:1)

The higher meaning of "all males of Israel standing and presenting themselves before God" (Deut 16:16) points to the fact that all angels ("sons of God") will and must stand before God and present themselves to God, and "all" means all.

Not Empty Handed

nm553 » The males were to stand before God not empty handed (Deut 16:16; 23:15). They were to bring the fruits of their labor with them.

■ "Three times in a year all your males shall appear before the LORD your God in the place which He chooses, at the Feast of Unleavened Bread and at the Feast of Weeks and at the Feast of Booths, and they shall not appear before the LORD empty-handed. 17 Every man shall give as he is able, according to the blessing of the LORD your God which He has given you." (Deut 16:16-17)

Angels are the antitypical meaning of "males" in the Bible. Therefore in the higher meaning, all angels were to appear before God three times with the blessings that God had given to them. These three times were at harvests, and at these harvests the males or angels were to bring their blessings. These blessings were usually from the harvest. So if it was the Pentecost or the harvest of the wheat, then the males brought their first fruits or blessings from that harvest. Notice that at the end of the age it is the angels who are the ones to do the harvesting:

■ And the field is the world; and *as for* the good seed, these are the sons of the kingdom;

and the tares are the sons of the evil *one*; and the enemy who sowed them [tares or weeds] is the devil, and the harvest is the end of the age; and **the reapers are angels**. (Mat 13:38-39)

This "the end of the age harvest" is the harvest of the bodies of Christians who died in the old age, therefore the blessings that the angels bring to God are the physical bodies of Christians who are united with their own angels and become one with God thereby.

nm554 » From the *God Papers 6*, we take the following which shows that Christians do have their own angel:

Our Own Angel

Now everyone that becomes a "son of God" must be begotten of the Spirit (Rom 8:9-10, 16). What is this Spirit? What does it mean to be begotten of God's Spirit? Notice that those begotten of the Spirit are led by it (Rom 8:14). The Spirit in them, leads them.

Notice that "these little ones" have in heaven "their angels" (Mat 18:10). Now the Greek word translated "their" means "of one's self." These "little ones" are Spiritual children of God (1 John 2:12-13). And these little ones have angels of their own self. Or, thus, since angels are spirits (Heb 1:7), Christians have their own angels or Spirits.

What do these angels or Spirits do? "For he shall give his angels charge over you, *to keep you in all your ways*" (Psa 91:11). In other words, angels lead them, as the Spirit leads the little ones or sons of God (Rom 8:14). Now Psalms 91:11 was used in a physical sense concerning Christ (Mat 4:6). But the Bible is dual and speaks in a dual sense, the physical sense and the Spiritual sense. We are to look to the higher sense — the Spiritual (Col 3:1-2). Not only do angels help out physically, but they help out Spiritually. And since Christ is our example, and the forerunner, then what applies to him applies to all others (cf Col 1:18; Rom 8:17; John 14:6).

Each son of God has his own angel (Mat 18:10). And these angels lead them in the way (Psa 91:11), as the angel of the BeComingOne led Christ (John 14:10,

GP 3 & 4), who is our example. Thus, the Spirit of God that leads Christians (Rom 8:14) is an angel of God that is in them. One of God's own Spirits leads each one of them. These Spirits or angels are for the elect humans who are the sons of God (1Pet 1:1-2). These angels or Spirits serve the elect, they are ministers or servants "for them who shall be heirs of salvation" (Heb 1:14). These angels are the "elect angels" (1Tim 5:21).

nm555 » It is the angels (Mat 13:38-39) who will be sent by God to resurrect the dead in Christ (Christians):

> "For this we say to you by the word of the Lord, that we who are alive and remain until the coming of the Lord, will not precede those who have fallen asleep. 16 For the Lord Himself will descend from heaven with a shout, with the voice of *the* archangel and with the trumpet of God, and the dead in Christ will rise first." (1Thes 4:15-16)

> "Behold, I tell you a mystery; we will not all sleep, but we will all be changed, 52 in a moment, in the twinkling of an eye, at the last trumpet; for the trumpet will sound, and the dead will be raised imperishable, and we will be changed." (1Cor 15:51-52)

Remember it is the angels (who are spirits of God) who will resurrect (Mat 13:39; Rev 14:15; Mat 24:31; Mark 13:27; Mat 16:27; 2Thes 1:7)

Three Harvests Point to Three Orders of Resurrection to Immortality

nm556 » God directed that all males must stand before the LORD during three of the annual harvest-festivals: "Three times in a year shall your males appear before the LORD your God in the place which he shall choose; in the feast of unleavened bread, and in the feast of weeks (Pentecost), and in the feast of tabernacles" (Deut 16:16). And they were supposed to offer gifts from the harvest, "not empty handed." We project in this book that the words in the Bible are shadows or types of the real and true. There is a duality of meaning in the Bible: a physical meaning and a spiritual meaning. Thus, the three harvests and three main festivals have a higher meaning. Through comparison of scripture we can see the connection between the three harvests in Israel and the three resurrections mentioned by Paul.

nm557 » Paul wrote about three resurrections in 1Corinthians, chapter fifteen:

> "In Christ shall *all* be made alive. But every man in his own order:
>
> > **[1]** Christ the first fruit;
> >
> > **[2]** afterward they that are Christ's at his coming [Rev 14:1-4].
> >
> > **[3]** Then the end...." (1Corinthians 15:22-24)

Paul called these three resurrections, orders, "each man in his own order." This word translated "order" is *tagma* in the Greek text, which means rank or division or proper order. Paul manifested the three orders or ranks of resurrections. Christ fulfilled the first festival ceremony (sheaf of 1st fruit) by being resurrected and then going to his Father on the exact day that the sheaf of first fruits was waved (see below). Christians will fulfill the second festival by becoming the first fruits of the creation (1Cor 15:23; Rev 14:4).

nm558 » Other allusions to these three resurrections are found in the Bible:

■ **(1)** Mark 4:28: "For the earth brings forth fruit of herself; first the **blade** [Christ], then the **head** [first-products], after that the **full gain** in the head."

■ **(2)** Luke 13:20-21: "And again he said, Whereunto shall I liken the kingdom of God? It is like leaven, which a woman took and hid in **three measures** of meal, till the whole [*all* of mankind] was leavened." [Leavened here is used in a different way than when it is used to signify sin or the way of sin in the Feast of Unleavened Bread.]

■ **(3) Noah's ark with three stories** was also a foreshadow of the three orders of salvation of mortal mankind to immortality. All life forms were saved in the ark that was built with three stories: "You shall make a window for the **ark**, and finish it to a cubit from the top; and set the door of the **ark** in the side of it; **you shall make it with lower, second, and third decks**. (Genesis 6:16)

■ **(4)** Three sections of the tabernacle of Israel indicated and pointed to the three orders of resurrection. The **Holy of Holies** definitely points to the first resurrection to immortality – Jesus Christ. The **Holy Place** points to the Church and represents the second resurrection

to immortality. Both of these were called the inner court or the upper court. The **outer court** was for the gentiles and pointed to the last resurrection to immortality. The inner court, with its Holy of Holies and Holy Place, was holy and set apart from the outer court. In the Bible the inner court was counted or measured, but the outer court was not (Rev 11:1-2). We know the count of the inner court (Holy of Holies & Holy Place) represents Jesus Christ and the first fruits of 144,000 (Rev 14:3-4), but we do not know the count or number of the outer court (Rev 11:1-2; Rev 7:9).

> Then there was given me a measuring rod like a staff; and someone said, "Get up and measure the temple of God and the altar, and those who worship in it. 2 **"Leave out the court which is outside the temple and do not measure it**, for it has been given to the nations; and they will tread under foot the holy city for forty-two months. (Rev 11:1-2)

> After these things I looked, and behold, **a great multitude which no one could count**, from every nation and *all* tribes and peoples and tongues, standing before the throne and before the Lamb, clothed in white robes, and <u>palm branches</u> *were* in their hands; (Revelation 7:9; Notice the palm branches, see below under Feast of Tabernacles)

■ **(5)** The *three orders* or divisions of mankind can be seen typically in Joseph and his two sons, Manasseh and Ephraim: [1] *Joseph*, the one set apart from his brethren (Deut 33:16), represents Jesus Christ, who was the first to be born of God (1Cor 15:22-28). [2] *Manasseh*, the one that was to become a great nation, represents the second group of first fruits to be presented to God, which are the Christians who lived in the old age. [3] *Ephraim*, the one that was to become a multitude of nations, represents the third group to be presented to God, which are the multitudes of peoples who will be born of God at the end of creation. (See PR1)

■ Also see NM24 for more details on these three resurrections or harvest of souls.

Harvest of Souls

nm559 » The Bible speaks about the saving of souls (1Pet 1:9; Heb 10:39; James 1:21; 5:20; Luke 21:19). In a sense, souls are sown with a seed, either a good seed or a bad one.

- **"I will sow her for Myself in the land**. I will also have compassion on her who had not obtained compassion, And I will say to those who were not My people, 'You are My people!' And they will say, '*You are* my God!'" (Hosea 2:23)

- "Behold, days are coming," declares the LORD, "when **I will sow the house of Israel and the house of Judah** with the seed of man and with the seed of beast. (Jeremiah 31:27)

- Jesus presented another parable to them, saying, "The kingdom of heaven may be compared to **a man who sowed good seed in his field**. 25 "But while his men were sleeping, **his enemy came and sowed tares among the wheat**, and went away. (Matthew 13:24)

- Then He left the crowds and went into the house. And His disciples came to Him and said, "Explain to us the parable of the tares of the field." 37 And He said, **"The one who sows the good seed is the Son of Man, 38 and the field is the world; and *as for* the good seed, these are the sons of the kingdom; and the tares are the sons of the evil *one*.** (Matthew 13:36-38)

nm560 » This harvest of the antitypical wheat occurs at the coming of Christ (1Cor 15:23) when the first fruits (the "wheat") are redeemed from among men (Rev 14:4).

- "His winnowing fork is in His hand, and He will thoroughly clear His threshing floor; and **He will gather His wheat** into the barn, but He will burn up the chaff with unquenchable fire." (Mat 3:12)

- Allow both to grow together until the harvest; and in the time of the harvest I will say to the reapers, First gather up the tares and bind them in bundles to burn them up; but **gather the wheat into my barn.** (Mat 13:30)

- Then He left the crowds and went into the house. And His disciples came to Him and said, "Explain to us the parable of the tares of the field." 37 And He said, "The one who sows the good seed is the Son of Man, 38 and the field is the world; and *as for* the good seed, these are the sons of the kingdom; and the tares are the sons of the evil *one*; 39 and the enemy who sowed them is the devil, and the harvest is the end of the age; and the reapers are angels. 40 "So just as the tares are gathered up and burned with fire, so shall it be at the end of the age. 41 "The Son of Man will send forth His angels, and they will gather out of His kingdom all stumbling blocks, and those who commit lawlessness, 42 and will throw them into the furnace of fire; in that place there will be weeping and gnashing of teeth. (Mat 13:36-42)

nm561 » Therefore at the end of the evil age the evil ones are taken out of the world and burned, while God through his angels gathers the wheat into the barn (Mat 13:30).

When was the harvest of wheat in Israel? It occurred typically just before the Pentecost, which was the festival that celebrated the harvest of wheat and other first grain products (fruits). Is this telling us something about *when* the coming of Christ is and when the harvest of Spiritual first fruit happens?

We will now study the Feast of the Passover. From this study we will see that what the Passover typified was fulfilled perfectly by Christ, our Spiritual Passover.

Feast of the Passover							
Represents first Harvest (Barley)							
Beginning of the Harvest (Rev 3:14; Col 1:15, 18)							
First in Rank (1Cor 15:23)							
Preparation	1st day	2nd day	3rd day	4th day	5th day	6th day	7th day
Passover 14th Nisan Death & burial	**Sabbath** 15th [1st day in grave]	**Friday** 16th [2nd day in grave]	**Weekly Sabbath** 17th [3rd day in grave] Resurrection>	**Sunday** 18th Sheaf-wave offering			**Sabbath** 21st Nisan
Signifies predestination of Christ's suffering	< **Signifies** First seven millenniums of creation > < Seven Days of unleavened bread: signifying purity of Jesus Christ						

First Harvest: Passover

Passover Foreshadowed Christ

nm562 » As we will see in this section, the Passover Feast foreshadowed Christ's death and resurrection. The Passover was one of God's appointed times:

■ "These are the appointed times of the LORD, even holy convocations; which you shall proclaim in their seasons. In the fourteenth day of the first month between the two evenings is the LORD's Passover" (Lev 23:4-5).

Sacred Months

nm563 » First of all what does the Bible mean by, "in the fourteenth day of the first month"? The Old Testament uses the Hebrew's Sacred Calendar, which some call the Jewish Sacred Calendar. The first month is also called Nisan or Abib, and occurs in March to April on today's calendar. Hence on the fourteenth day of Nisan is the Passover. But the Passover occurs "between the two evenings" on the 14th day as correctly translated in the following verses: Lev 23:5 (Hebrew and Greek); Num 9:3,5; Ex 12:6. In the King James Version "between the two evenings" is translated wrongly as "even" or "evening."

Between the Two Evenings

nm564 » There is error concerning the meaning of "between the two evenings." First of all note that the Passover is in the fourteenth day of the first month between the two evenings. The Passover is between the two evenings. This merely means that the Passover happens on the 14th between the evening of the 13th and the evening of the 14th.

Morning and Evening Time

nm565 » The proof of this is shown in the appointed time for Atonement: "Also on the tenth day of this seventh month there shall be a day of atonement . . .. It shall be unto you a Sabbath of rest, and you shall afflict your souls: in the ninth day of the month at even, from even until even, shall you celebrate your Sabbath" (Lev 23:27, 32). The day of Atonement is on the tenth of the seventh month, but it is celebrated from the evening of the ninth to the evening of the tenth, thus between these two evenings. And the word "evening" is defined by the Bible: "at even, at the going down of the sun" (Deut 16:6; note Neh 13:19; Lev 22:6). Evening is the latter part of the 24 hour day. In the Bible "evening" is used in two ways: (1), the later hour(s) before sunset; (2) the time or moments just before sunset, or at sunset. Morning is "when the sun rises" (2 Sam 23:4)

Hence, the Passover is on the fourteenth of the first month in the Hebrew Sacred Calendar, between the two evenings, or thus from the evening of the 13th to the evening of the 14th.

Festival of Unleavened Bread

nm566 » Now with this day of the Passover there are seven other days that are called the festival of unleavened bread or the appointed time for unleavened bread:

- "And they killed the Passover on the fourteenth day of the first month . . .. And the children of Israel that were present kept the Passover at that time, and the feast of unleavened bread" (2Chron 35:1, 17).

- "In the fourteenth of the first month is the Passover of the LORD. And in the fifteenth of this month is the festival [appointed time]: seven days shall unleavened bread be eaten. In the first day [15th] shall be a holy convocation; you shall do no manner of servile work therein . . .. And on the seventh day [the 21st, the 7th day of the festival] you shall have a holy convocation; you shall do no servile work" (Num 28:16-18,25).

- "In the first month, on the fourteenth day of the month at even [sunset], you shall eat unleavened bread, until the one and twentieth [21st] day of the month at even [sunset]. Seven days shall there be no leaven found in your houses . . .. Seven days shall you eat unleavened bread; even the first day you shall put away leaven out of your houses: for whoever eats leavened bread from the first day until the seventh day, that soul shall be cut off from Israel" (Ex 12:18-19, 15).

nm567 » By putting together the above, we see that the fourteenth is the Passover, from the evening of the 13th to the evening of the 14th. After this begins the festival of unleavened bread from the evening of the 14th until the evening of the 21st, which is exactly seven days of eating unleavened bread.

Passover's Various Names

nm568 » The killing of the Passover occurred on the 14th (Lev 23:5; Num 28:16; 2Chron 35:1; Ezra 6:19); the festival of unleavened bread began on the 15th and ended on the 21st at sunset (Lev 23:6-7; Num 28:17-18, 25; Ex 12:18-19; Note: Referring to Ex 12:18, remember the 15th day begins at evening, sunset, of the 14th). Yet sometimes the "Passover" refers to the whole event from the 14th to the 21st, or thus both the Passover day and the festival of unleavened bread: "Now the feast [appointed time] of unleavened bread drew near, which is called the Passover" (Luke 22:1). "Then were the days of unleavened bread ... intending after the Passover to bring him forth to the people" (Acts 12:3-4). Therefore technically, the Passover is *on* the 14th (evening of the 13th to the evening of the 14th), and the festival of unleavened bread is from the 15th (at the end of the evening of the 14th) to the 21st at evening. Yet the words "Passover" or "feast," or "days of unleavened bread" are used interchangeably in the Bible. Thus, we should be careful when we read about this event, so that we do not misunderstand the descriptions of it.

Sabbaths: 15th and 21st of Nisan

nm569 » Now the scripture indicates that the 15th is a day of "holy convocation" or assembly, and the 21st is another assembly or holy convocation (Num 28:17-18, 25; Deut 16:8; Ex 1:18). A holy convocation or assembly is a coming together of set-apart people or holy people. Actually these appointed days of assembly are called Sabbaths (Lev 23:24, 27, 32, 39). Therefore the 15th is an annual Sabbath, and the 21st of the first month is an annual Sabbath.

Passover's Meaning

nm570 » Now we know *when* the Passover and the Festival of Unleavened Bread occur. But we must find out what is the meaning of this event. We will explain the event as described by the Bible, and at the same time explain the higher or Spiritual meaning. Because of the way we will present this subject, we suggest that the reader go over this paper at least twice before forming an opinion. Many of the things presented are different, therefore they may seem strange, yet they are strange only because they are different or new to you.

Physical Passover	Spiritual Passover
Tenth of Nisan (Abib) 　nm571 » *Physical Passover* A lamb was picked out on the 10th of the first month, "a lamb for a house" (Ex 12:3).	*Higher Meaning* Jesus Christ is the lamb of God (John 1:29, 36). Jesus Christ was betrayed by Judas to the chief priests on the 10th of the first month: "And he promised, and sought opportunity to betray him unto them in absence of the multitude" (Luke 22:3-6; see notes in back of this paper).
Passover Without Blemish 　nm572 » *Physical Passover* This lamb was to be without blemish.(Ex 12:5) The lamb without blemish is called the Passover (Ex 12:21).	*Higher Meaning* Christ is the lamb without blemish, or spot (1Pet 1:19; Heb 9:14). This means Christ is sinless (1Pet 2:21-22). Jesus is the Passover lamb (1Cor 5:7). He is the true Passover lamb that was set forth or predestinated to be the Passover lamb before the world began (1Pet 1:19-20; Rev 13:8).
Fourteenth: Preparation Day 　nm573 » *Physical Passover* The 14th day was the day of preparation for the Passover lamb, for other ceremonies of the day, and for the Passover meal (2Chron 35:1-6, 10-13, 16).	*Higher Meaning* The 14th was the day Christ the true Passover was taken and prepared before his slaughter: it was the day of preparing Christ for his death, just before the high Sabbath, or the 15th day of Nisan which is a Sabbath day for the Feast of Unleavened Bread (John 19:14, 31, 42).
Passover Killed on 14th of Nisan 　nm574 » *Physical Passover* The lamb was killed on the 14th of the first month, between the evening of the 13th and 14th towards the evening of the 14th (Ex 12:6; 2Chron 35:1; Deut 16:6).	*Higher Meaning* Christ the true Passover was killed on the 14th of Nisan (John 18:28; and the rest of the scripture on the death of Christ, see notes).
Passover Killed Outside Gates 　nm575 » *Physical Passover* The lamb was killed outside the gates of the city (2Chron 35:11; Ex 12:6, 21; Deut 16:5).	*Higher Meaning* The lamb of God died also outside the camp, or city of Jerusalem (Heb 13:11-12). Christ had not yet ascended into the New Jerusalem; He died outside or before the kingdom of God (New Jerusalem) was set-up on earth.
Passover's Blood on House 　nm576 » *Physical Passover* The blood of the lamb was sprinkled on the door posts of the house wherein the people had gathered to eat one Passover lamb, for only *one* lamb per gathering house was allowed (Ex 12:6-7, 3-4, 46; Exo 12:21-22; 2Chron 35:11).	*Higher Meaning* Christians are of the house of God (1Pet 2:5; 4:17; etc); they have the blood of Christ (the Passover) sprinkled on them (1 John 1:7; Heb 10:22; 12:24; 13:12, 20). Christ is the *one* lamb of God for all the house of God (John 1:36; 1Pet 1:19; 1Pet 2:3-5; Heb 10:4; Heb 3:6).

Physical Passover	Spiritual Passover
Passover Roasted As One nm577 » *Physical Passover* Thereafter, since one could not eat the flesh raw, they roasted it as a unit (head, body, legs together), which takes some time (Ex 12:9; Deut 16:7; 2Chron 35:13).	*Higher Meaning* Christ the Passover was "roasted" out in the sun light for hours (Mark 15:25, 33-34).
All Passover's Blood Spilled Out nm578 » *Physical Passover* No blood could be eaten with the sacrificed lamb; his blood had to be spilled completely out (Lev 7:27; 17:12-14).	*Higher Meaning* All of Christ's blood was poured out, until water came out instead of blood (John 19:34).
Passover Had No Bone Broken nm579 » *Physical Passover* No bone could be broken on the Passover lamb (Ex 12:46; Num 9:12).	*Higher Meaning* No bone of Christ was broken (John 19:33, 36).
Passover Eaten in One House nm580 » *Physical Meaning* The Passover was eaten in one house; no flesh was to be carried outside the gathering place (Ex 12:46), or outside the house on which the blood was sprinkled.	*Higher Meaning* Only those Spiritually in the house of God (those with the "blood" of Christ sprinkled on them) can eat the Passover, if we attempt to eat it without being in the Church, we are guilty (1 Cor 11:26-27).
No Stranger May Eat the Passover nm581 » *Physical Passover* No stranger (non-Israelite) could eat the Passover unless he was circumcised, for no uncircumcised person could eat the Passover lamb (Ex 12:43-44, 48).	*Higher Meaning* No stranger outside of the Spiritual Israel of God, can eat the Passover Christ (Eph 2:12, 19; 1Cor 11:28-29). That is, eat his Spiritual bread (see John 6:56-63; see "Spiritual Bread" below).
Unleavened Bread nm582 » *Physical Passover* No leavened bread was to be eaten with the Passover (Deut 16:3; Ex 23:18; 12:8).	*Higher Meaning* No "old leaven" or "the leaven of the Pharisees," or the doctrines of man can be eaten with the Spiritual Passover (Christ). When you have Christ (the Spiritual Passover) you have his Spirit or the New Mind with the doctrine of the good God, not the doctrines and ideas of mankind belonging to this age (Mat 16:6, 11-12; 1Cor 5:7-8; Mark 8:15; Luke 12:1).

Physical Passover	Spiritual Passover
Passover Eaten in the Night nm583 » *Physical Passover* The lamb was eaten in the night of the 15th; the flesh had to be roasted, and eaten with unleavened bread (Ex 12:8, see *Septuagint*).	*Higher Meaning* In a physical sense, the Passover Christ was consumed by the tomb in the late evening of the 14th as it became night (Mark 15:42-47). But, in real sense, since Paul called Christians "unleavened" (1Cor 5:7), and since unleavened bread is that of truth (1Cor 5:8); then Christ and his truth are the unleavened bread consumed by Spiritual Israelites during the spiritual night – the darkness of Satan's kingdom.
Passover Eaten in Haste in the Night nm584 » *Physical Passover* The Passover lamb was eaten in *haste* in the *night* with shoes on their feet, fully clothed, and ready to leave Egypt in a moment's notice (Ex 12:11, 33, 39; 2Chron 35:13).	*Higher Meaning* Spiritually, the Passover Christ is eaten in the night of Satan in trepidation or tribulation as we are fleeing spiritual Egypt.
Passover Saves Us from the Destroyer nm585 » *Physical Passover* The houses wherein the Passover was eaten in the night, and where the blood of the Passover was put on, therein the destroyer would not smite the first born of the house as the destroyer struck down the first born of the Egyptians (Ex 12:7, 13, 22-23; Heb 11:28; Ex 12:29; 13:15: 11:5).	*Higher Meaning* Because Christians so to speak have Christ's blood sprinkled on them, they will become the first born of mankind to immortality (Heb 12:23; Rev 14:4; James 1:18). The destroyer (Satan) does not spiritually destroy them in the "night." But the first born of the Pharaoh is destroyed in the "night." The "night" is the darkness of Satan's kingdom. The "destroyer" was and is Satan (Psa 78:49; Isa 14:17, 12-17).

Physical Passover	Spiritual Passover
Passover and the Seven Days **nm586 »** *Physical Passover* Now after Israel went out of Egypt, or as they were going out, they ate unleavened bread for seven days from the 15th to the 21st day, from the evening of the 14th to the evening of the 21st day. The main reason they baked unleavened bread seven days was "because they were thrust out of Egypt, and could not tarry, neither had they prepared for themselves any victual" (Ex 12:39). The eating of unleavened bread for the seven days was for a memorial or remembrance of Israel's delivery from the bondage of Egypt (Deut 16:3; Ex 13:3-16; Exo 12:14-19).	*Higher Meaning* Now the True unleavened bread, is Truth (1Cor 5:8). And God's word or Bible is the Truth (John 17:17; James 1:18). And since Christ spoke the word of God, which is the Truth (John 12:49), which comes from the Spirit of Truth (John 14:17; 15:26; 16:13), then the real unleavened bread is Christ's Spiritual Word, his Truth, his Living Bread (see "Spiritual Bread" below). Moreover, since the True seven days are the seven 1000 year days (2Pet 3:8), then the True picture of the Passover Festival is that Christ the Passover Lamb of God was predestinated before the world began to die for all mankind's sin, and because he was slain, it makes it possible for Spiritual Israel (Christians) to come out of spiritual Egypt (Satan's kingdom). Now in the typical Passover festival the unleavened bread was eaten as a remembrance of Israel's deliverance out of Egypt. We eat the Spiritual unleavened bread (The Spiritual Christ is the bread of life) in remembrance of Christ's sacrifice (1Cor 11:24-26) which makes it possible for us to come out of spiritual Egypt.

Sacrifices Fulfilled in Christ's One Sacrifice

nm587 » As Paul showed in Hebrews and as Daniel 9:27 indicated all ritual sacrifices for sin have been fulfilled in Christ's great sacrifice for sin, with the one important qualification, that where the remission of sin is, there is no more offering for sin (Heb 10:18). But this remission or forgiveness of sin only occurs when one has the laws of God in their heart (Heb 10:16-17). And you only have the laws in your heart or mind when you have the Spirit of God inside you leading you (2Cor 3:3-6). Therefore, although Christ died as the Passover for all sin (1John 2:2), all have not been forgiven yet for "we see not yet all things put under him" (Heb 2:8b). But when the true end comes:

- 1Cor 15:24: then *comes* the end, when He hands over the kingdom to the God and Father, when He has abolished all rule and all authority and power. 25 For He must reign until He has put all His enemies under His feet. 26 The last enemy that will be abolished is death. 27 For he has put all things in subjection under his feet. But when He says, "All things are put in subjection," it is evident that He is excepted who put all things in subjection to Him. 28 When all things are subjected to Him, then the Son Himself also will be subjected to the One who subjected all things to Him, **so that God may be all in all.** (1Cor 15:24-28)

When our God is all in all, then all will be atoned to God because when one receives atonement or reconciliation, his sins are covered, and he is brought back to God (Rom 5:9-11).

nm588 » Paul wrote to us about Christ's sacrifice as the real Passover (1 Cor 5:7) who takes away the sins of the world in due time:

- And not through the blood of goats and calves, but through His own blood, He entered the holy place once for all, having obtained eternal redemption. 13 For if the blood of goats and bulls and the ashes of a heifer sprinkling those who have been defiled sanctify for the cleansing of the flesh, 14 how much more will the blood of Christ, who through the eternal Spirit offered Himself without blemish to God, cleanse your conscience from dead works to serve the living God? (Heb 9:12-14)

- 22 And according to the Law, *one may almost say*, all things are cleansed with blood, and without shedding of blood there is no forgiveness. 23 Therefore it was necessary for the copies of the things in the heavens to be cleansed with these, but the heavenly things themselves with better sacrifices than these. 24 For Christ did not enter a holy place made with hands, a *mere* copy of the true one, but into heaven itself, now to appear in the presence of God for us; 25 nor was it that He would offer Himself often, as the high priest enters the holy place year by year with blood that is not his own. 26 Otherwise, He would have needed to suffer often since the foundation of the world; but now once at the consummation of the ages He has been manifested to put away sin by the sacrifice of Himself. 27 And inasmuch as it is appointed for men to die once and after this *comes* judgment, 28 so Christ also, having been offered once to bear the sins of many, will appear a second time for salvation without *reference to* sin, to those who eagerly await Him. (Heb 9:22-28)

- For the Law, since it has *only* a shadow of the good things to come *and* not the very form of things, can never, by the same sacrifices which they offer continually year by year, make perfect those who draw near. 2 Otherwise, would they not have ceased to be offered, because the worshipers, having once been cleansed, would no longer have had consciousness of sins? 3 But in those *sacrifices* there is a reminder of sins year by year. 4 For it is impossible for the blood of bulls and goats to take away sins. (Heb 10:1-4)

- 9 then He said, "behold, I have come to do your will." He takes away the first in order to establish the second. 10 By this will we have been sanctified through the offering of the body of Jesus Christ once for all. 11 Every priest stands daily ministering and offering time after time the same sacrifices, which can never take away sins; 12 but He, having offered one sacrifice for sins for all time, **Sat down at the right hand of God**, waiting from that time onward **until his enemies be made a footstool for his feet**. (Heb 10:9-13)

- Now where there is forgiveness of these things, there is no longer *any* offering for sin. (Heb 10:18)

- "And he will make a firm covenant with the many for one week, but in the middle of the week **he will put a stop to sacrifice and offering**; and on the wing of abominations *will* come one who makes desolate, even until a complete destruction, one that is decreed, is poured out on the one who makes desolate." (Dan 9:27; there are two different interpretations here and both may be partly correct)

What these scriptures are saying is that Christ's great Passover sacrifice of himself perfectly fulfilled all ritual sacrifices for sin. Thus, in the middle of the week (Christ died on a Wednesday in the middle of the seven year period: he fulfilled 3 ½ years of it) Christ's death cut off the need of sacrifice for the forgiveness of sin. But men will continue to suffer because of sin until they stop sinning, for sin is wrong behavior that causes suffering.

Spiritual Unleavened Bread as Memorial

nm589 » Now in the typical Passover festival the unleavened bread was eaten as a remembrance of Israel's deliverance out of Egypt. We eat the Spiritual unleavened bread (The Spiritual Christ is the bread of life) in remembrance of Christ's sacrifice which makes it possible for us to come out of spiritual Egypt.

See Spiritual keeping of the Passover in the last part of this paper for more information on other Spiritual aspects of the Passover rituals.

Sheaf of the First Fruits: Time of First Harvest

Foreshadowed First Resurrection

Christ is the Sheaf of First Fruits

nm590 » Now within the festival of the Passover there was another ritual performed, and that was the waving of a sheaf of the first fruits of the first harvest:

- "And the LORD spoke to Moses, saying, Speak unto the children of Israel, and say unto them, When you come into the land which I give unto you, and shall reap the harvest thereof, then you shall bring a sheaf of the first fruits of your harvest unto the priest: And he **shall wave the sheaf before the** [faces or presence of the] **LORD**, to be accepted for you: on the day *after* the Sabbath the priest shall wave it ... And you shall eat neither bread, nor parched corn, nor green ears, until the selfsame day until you have brought an offering [Sheaf] unto your God: a statute aeonian throughout your generations in all your dwellings" (Lev 23:9-11, 14).

Higher Meaning

nm591 » This festival of the Passover and the ritual of the sheaf of first fruits had an important higher meaning. The man Jesus Christ died on the Passover (14th of Nisan), was buried at the very end of the 14th of Nisan, was resurrected three days later, and went back into his Father on the very day the Jews used to wave the sheaf of barley "before the LORD." Paul shows us that Christ was that sheaf of first fruits:

- 21 For since by a man *came* death, by a man also *came* the resurrection of the dead. 22 For as in Adam all die, so also in Christ all will be made alive. 23 But each in his own order: **Christ the first fruits**, after that those who are Christ's at His coming (1Corinthians 15:21)

nm592 » Jesus and Paul explained how even nature projects to us a reason for Christ's death:

- Truly, truly, I say to you, unless a grain falls into the earth and dies, it remains alone; but if it dies, it bears much fruit [produce]. (John 12:24)

- You fool! That which you sow does not come to life unless it dies; and that which you sow, you do not sow the body which is to be, but a bare grain, perhaps of wheat or of something else. (1Cor 15:36-37)

So from this very first fruit (Jesus Christ) will come "much fruit."

nm593 » Notice Leviticus 23:9-11 that the sheaf was waved "before the LORD." Compare this to the order for all males to stand before the LORD three times a year in the three main harvest festivals (Deut 16:16):

- He shall **wave the sheaf before the LORD** for you to be accepted; on the day after the Sabbath the priest shall wave it. (Lev 23:11)

- Three times in a year all your males shall appear **before the LORD** your God in the place which He chooses, at the Feast of Unleavened Bread and at the Feast of Weeks and at the Feast of Booths, and they shall not appear **before the LORD** empty-handed. Every man shall give as he is able, according to the blessing of the LORD your God which He has given you. (Deut 16:16-17)

Jesus Christ was the first of the first fruits (Rev 3:14; 1Cor 15:23; Col 1:15,18) who fulfilled the sheaf of first fruits ritual, perfectly. Only through Christ can anyone be acceptable to God, because you must have the Spirit of God to be acceptable to God as this book prove through Biblical scripture. It is Christ who gives this Spirit (Acts 2:33). So it was by the death of Christ, the Seed [PR1], that "much fruit" will come (John 12:24).

Sheaf of First Fruits Perfectly Fulfilled

nm594 » This ceremony of waving the sheaf of first fruits towards heaven was perfectly fulfilled by Jesus Christ when he ascended to his Father at the beginning of the new 24 hour day *after* the Sabbath, after sunset (see the *Chronology Papers,* CP4). The sheaf of barley, the first of the first fruits, was cut down at the very end of *the Sabbath*, "just as the sun went down" and was waved on the day after the Sabbath (*Unger's Bible Dict.*, p. 355). This Sabbath being the regular weekly Sabbath as the Sadducees and others interpreted Lev 23:9-11 (*Unger's Bible Dict.*, p. 356). As we see when we study the scripture (CP4) Jesus Christ was resurrected just before or at sunset on Saturday (the Sabbath) exactly three days after he was buried. After sunset in the beginning of the first day of the week, he was made into one with his Spirit, and at that time became the first-fruit of the new creation (1Cor 15:23a; Rev 3:14). Christ was foreshadowed by the "sheaf of the first fruits" in this ceremony; He is the first of the first fruits. Christ is the first product, or the first born of the new creation (note Rev 3:14; 1 Cor 15:23; Rom 8:29; Col 1:15, 18; Rev 1:5; see "All Saved Paper" [NM 13]).

Fulfillment on Exact day

nm595 » This ceremony happened typically on the day *after* the weekly Sabbath, a Sunday, within the seven days of unleavened bread (Lev 23:15-16,9-11, 14). Christ fulfilled this perfectly by going back to his Father on that Sunday (cf John 20:17; Mat 28:9; see the *God Papers*). As of now Jesus is the only born of God, or only born God as John 1:18 says in certain Greek texts ("only born God").

Also see *Prophecy Papers 7* [PR7] under "Sheaf of the First Fruits," and *Chronology Papers 4* [CP4] under "Ascension."

Feast of Pentecost Represents second Harvest (Wheat) Harvest of First Fruits 144,000 (Rev 14:3-4; Rev 7:4) Second Order or Rank (1Cor 15:23)							
< Grain Harvest Period >							Pentecost
1st Sabbath after sheaf waved	2nd Sabbath after sheaf waved	3rd Sabbath after sheaf waved	4th Sabbath after sheaf waved	5st Sabbath after sheaf waved	6st Sabbath after sheaf waved	7st Sabbath after sheaf waved	Sunday Celebrated after grain harvest
< **Signifies** the seven millenniums of creation >							Two leavened loaves waved
							Atonement

Second Harvest: <u>Pentecost</u>

Foreshadowed Second Resurrection

Feast of First Fruits; Feast of Weeks

nm596 » This festival or appointed time is sometimes called in the Bible, the Feast of Weeks, or the Feast of First Fruits (Exo 34:22).

- "And you shall count unto you from the day after the Sabbath [thus the 1st day of the week, the day Christ went back to his Father], from the day that you brought the sheaf of the wave offering; seven Sabbaths shall be complete: Even unto the day after the seventh Sabbath shall you number fifty days ... And you shall proclaim on the selfsame day that it may be a holy convocation unto you: You shall do no servile work therein: it shall be a statute aeonian in all your dwellings throughout your generations" (Lev 23:15-16, 21).

Counting <u>weeks</u> from the Sheaf of First Fruits

nm597 » "Seven weeks shall you number unto you: begin to number the seven weeks **from such time as you begin to put the sickle to the grain [barley]**. And you shall keep the feast of weeks unto the LORD your God with a tribute of freewill offering of your hand, which you shall give unto the LORD your God, according as the LORD your God has blessed you" (Deut 16:9-10).

nm598 » One counts seven weeks "from such a time as you begin to put the sickle to the crop." Or as it said in Leviticus 23:15-16: "you shall count unto you from the day after the Sabbath, from the day that you brought the **sheaf of the wave offering**; seven Sabbaths shall be complete: Even unto the day after the seventh Sabbath **shall you number fifty days**."

Counting of the 50 Days

nm599 » Now since Christ is the antitypical or Spiritual sheaf, and since Christ the man went to his or ascended to his Father (see John 20:17) after sunset on a Sunday (the day *after* the weekly Sabbath), and since this was the beginning of the time the sickle was put to the harvest, and since seven weeks or seven Sabbaths from the beginning of the Sunday is complete at the end of the Sabbath seven weeks later; then the 50th full day is the Sunday after the seventh Sabbath. The word *from* means: "used to specify a starting point in spatial movement: a train running west from New York City." Or "used to specify a starting point in an expression of limits." Or "used to indicate source or origin" (from the Random House Dictionary). Since we count from a point, the origin point, and since the Pentecost is *the* 50th full day of a limited period; then we count seven weeks from the starting point or origin day to the end of a Sabbath seven weeks thereafter. The 50th full day is the Sunday, the day after the seventh Sabbath (Lev 23:15-16).

Pentecost of the Church not Perfect Fulfillment

nm600 » On the 50th day (after seven Sabbaths [49 days]) from when the sheaf of barley was waved by the Jews at the Festival of the Passover (which was the 50th day after Christ went to the Father), the apostles gathered together to celebrate the Pentecost

(Acts 2). On that day the Spirit was first given to the Church. Some therefore think (as I once mistakenly believed) that this is the antitypical or higher meaning of the Pentecost – the giving of the Spirit to the Church. But this is mistaken. We see in this paper that Christ **fulfilled** the Passover festival, perfectly (see also Luke 22:15-16). So what the Pentecost represented typically, the antitypical Pentecost must fulfill it perfectly, not partially. If the physical Pentecost was the day Israel celebrated the <u>harvest</u> of the first fruits of the land, then the antitypical Pentecost <u>must</u> perfectly fulfill the *harvest* of the first fruits of the new creation (1Cor 15:23, 2nd part; James 1:18; Rev 14:4; See "All Saved Paper" [NM 13]). The Spirit on the first Church Pentecost was given to only a few of the total "first fruits" – only about 3,000 (Acts 2:41), not all of them.

nm601 » **Thus, not all of the first fruits mentioned in Rev 14:3-4 received the Spirit on this first Pentecost of the Church. But in order to fulfill the Pentecost perfectly, all things pertaining to the physical Pentecost must be fulfilled Spiritually and perfectly, as Christ fulfilled the physical Passover.**

Trumpets Blown on the Pentecost

nm602 » We need to know something about the blowing of trumpets, as in all other holy days, trumpets were blown on the Pentecost:

- "Also in the day of your gladness **and in your appointed feasts**, and on the first *days* of your months, **you shall blow the trumpets** over your burnt offerings, and over the sacrifices of your peace offerings; and they shall be as a reminder of you before your God. I am the LORD your God." (Numbers 10:10)

Blowing of Trumpet versus Blowing an Alarm
Trumpet for Assembly

nm603 » There were two different types of trumpet blowing. One way was for assembly and one was for an alarm of war. The trumpet for assembly, "the calling of the assembly"(Num 10:2-3, 7; Judges 3:27), was blown in "the day gladness, and your solemn days you shall blow with trumpets" (Num 10:10).

Trumpet for War

nm604 » There was a difference between blowing the trumpets for the assembly or gathering of Israel, and the blowing of trumpets to sound an alarm for war: "If you go to war in your land against the enemy that oppresses you, then you shall blow an alarm with the trumpets; and you shall be remembered before the LORD your God, and you shall be saved from your enemies" (Num 10:9; Neh 4:20), "the sound of the **trumpet**, the alarm of war" (Jer 4:19). "Declares the LORD, That I will cause a **trumpet** blast of war" (Jer 49:2). Therefore when Israel "is to be gathered together, you shall blow, but you shall not sound an alarm" (Num 10:9); one kind of trumpet blowing was for gathering Israel for a holy assembly, and one was for a gathering for war.

Trumpets at end of World

nm605 » There are also the blowing of trumpets associated with the end of the old age and the beginning of the Kingdom of God.

- **Blow a trumpet in Zion**, And sound an alarm on My holy mountain! Let all the inhabitants of the land tremble, For the day of the LORD is coming; Surely it is near, A day of darkness and gloom, A day of clouds and thick darkness. As the dawn is spread over the mountains, *So* there is a great and mighty people; There has never been *anything* like it, Nor will there be again after it To the years of many generations. (Joel 2:1-2)

- 'They have blown the **trumpet** and made everything ready, but no one is going to the battle, for My wrath is against all their multitude. (Ezekiel 7:14)

- "And He will send forth His angels with a great **trumpet** and they will gather together His elect from the four winds, from one end of the sky to the other. (Matthew 24:31)

- In a moment, in the twinkling of an eye, at the last **trumpet**; for the **trumpet** will sound, and the dead will be raised imperishable, and we will be changed. (1Corinthians 15:52)

- For the Lord Himself will descend from heaven with a shout, with the voice of *the* archangel and with the **trumpet** of God, and the dead in Christ will rise first. (1Thessalonians 4:16)

■ I was in the Spirit on the Lord's day, and I heard behind me a loud voice like *the sound* of a **trumpet**, (Revelation 1:10)

■ "And the seventh angel sounded [his *trumpet*, cf Rev 8:6, 7, 8, 10, 12; 9:1, 14]; and there were great voices in heaven, saying, the kingdoms of this world are become the kingdoms of our Lord, and of his Christ" (Rev 11:15).

Trumpets at the Presence of the King

The blowing of trumpets also represented other things such as to introduce a new king (1Kings 1:34-41), and were used in the Bible to indicate the presence of God (Ex 19:16, 19; 2Kings 9:13). This is fulfilled on the antitypical Pentecost when Christ comes with his saints.

Perfect Fulfillment of Pentecost

Pentecost is the Second Order of Persons to Immortality

nm606 » The Pentecost was a day of assembly for the Israelites, but it foreshadowed the first harvest of the earth's people to become immortal beings, the first fruits or products of God (Rev 14:4). The typical or physical Pentecost occurred after the grain harvest in Palestine. The higher meaning of this indicates the <u>harvest</u> of the first fruits of mankind to God (Rev 14:14-20, 4; Mat 13:38-43; James 1:18).

Because at the harvest of first fruit there will be the Last War (see PR 4-6), then the blowing of trumpets on the Feast of Pentecost will perfectly fulfill this, and any other blowing of trumpets at future feasts will not represent war. The blowing of trumpets on the antitypical Pentecost represents the (1) gathering; (2) Last War; (3) presence of the King of Kings.

Two Loaves of <u>Leavened</u> Bread Waved Before Yehowah

nm607 » Leviticus 23:17 You shall bring in from your assembly places **two loaves of bread for a wave offering**, made of two-tenths *of an ephah*; they shall be of a fine flour, **baked with leaven** as first fruits to the LORD ...19 'You shall also offer **one male goat for a sin offering** and two male lambs one year old for a sacrifice of peace offerings. 20 The priest shall then **wave them with the bread of the first fruits for a wave offering with two lambs**

[peace offering] **before the LORD**; they are to be holy to the LORD [and] to the priest. (Lev 23:17-20)

Notice a few things here:

■ The two *baked* leavened loaves of bread were waved before the LORD.

■ Two *baked* leavened loaves were taken out of Israel's assembly.

■ The two *baked* leavened loaves of bread were made of *leaven*.

■ The two *baked* leavened loaves of the first fruits were waved with two lambs for the peace offering (Lev 23:19-20).

■ Leavened bread could be used for peace offerings (Lev 7:13).

■ Leaven was not to be used in any offering made by fire (Lev 2:11).

■ The offerings of first fruits were not to be burnt on the altar (Lev 2:12).

■ Therefore the two leavened loaves were not placed in the fire because they were *baked* with leaven (Lev 23:17).

■ All offerings *baked* in an oven are for the Priest (Lev 7:9, 13-14) and therefore, of course, not burnt on the altar.

■ The two leavened loaves were waved with one male goat being offered for a sin offering (Lev 23:19), while the waving of the <u>sheaf</u> of first fruits (barley) did not require a sin offering since it was without leaven (cf. Lev 23:12-14)

■ The two leavened loaves were holy in reference to the LORD and to the priest (Lev 23:20).

What does this all mean?

Leavened Bread

nm608 » Israel was forbidden from eating leavened bread during seven days of the Passover Festival beginning on the 15th of Nisan and ending on the 21st of Nisan (Ex 12:18-19; 13:6-7; Lev 23:6-7; Num 28:16-8,25; Deut 16:3). This rule was a memorial of Israel coming out of Egypt because they were driven out before their bread was leavened (Ex 12:39). Spiritually, the leavened during these seven days represent the doctrines and hypocrisy of the Pharisees, Sadducees, and those like Herod (Mat 16:12; Mark 8:15; Luke 12:1; 1Cor 5:8). This festival foreshadowed Christ's unleavened sacrifice. And thus, blood could not be sacrificed with leavened bread (Ex 23:18; 34:25) because this blood foreshadowed Christ's blood, which was sacrificed without sin (leaven). As we see from the evidence above leaven could be used in a few offerings as long as it wasn't burnt on the altar (Lev 7:13; 23:17-20; 2:11). Although generally, leaven indicated sin or the doctrines of sin, in at least two places leaven and leavening of bread is used allegorically to represent a positive activity:

- The **kingdom of heaven is like leaven**, which a woman took and hid in three pecks of flour until it was all leavened (Mat 13:33; Luke 13:21)

The meaning here has nothing to do with sin (leaven), but with the three separate pecks of flour that were hidden (unknown to most) until the bread was ready to eat (when it had fully risen).

Leavened Bread & Sin Offering

nm609 » Since there was a sin offering (Lev 23:19) associated with the waving of the two loaves, then sin is somehow connected with the two baked loaves. Remember, there was no sin offering associated with the waving of the sheaf of first fruits (Lev 23:10-14). Yet does this mean that the two leavened loaves themselves represented sin? No, because they were waved before God to be accepted by God. And they were for the Priest to eat after being waved or heaved (Lev 7:9, 13-14). Since the Priest could not touch or eat anything defiled (Lev 7:21; 10:10), then these two baked loaves are not defiled, for they were "holy to the LORD" (Lev 23:20). The sin offering made it possible for the loafs to be accepted. Considering the relevant scriptures, whatever the loaves represented, we know that the leaven in the bread somehow was associated with sin, but the sin offering made it possible for them to be waved and accepted by God.

Time of Wave Offering at a Harvest

nm610 » We must consider in our analysis the time of the loaves being waved. It was done at the time of the harvest of wheat. In the case of the waving of the sheaf of first fruits, which was the beginning of the harvest of barley grain, it was waved before the LORD. As we saw previously this represented Jesus Christ the man going back into his Father and becoming one with his Father. Thus, to follow the metaphor, the two loaves must also become one with God as Christ prayed in John 17:21-23 that Christians would become one with God. The two loaves represented the grain harvest since they were taken from the grain of the harvest belonging to the assemblies or households of Israel (Lev 23:17).

Two Loaves Baked with Leaven

nm611 » The two loaves of bread were baked with leaven. Since this was not the Feast of the Passover, but an offering of baked bread (Lev 23:17) which was for the priest (Lev 7:9) and thus could be baked with leaven (Lev 7:13-14), then leavened-baked bread was perfectly correct for this ritual. Baked leavened bread is edible bread, edible in the way most people eat bread. It is a finished product while the sheaf of first fruits was not baked, but was an unbaked sheaf of grain.[1] This wave offering of two loaves of baked bread at the Pentecost was a finished product, so to speak. It was baked; it was baked with leaven. It was a finished product that used leaven (or sin, since leaven represents sin) to make it. If we take the leaven in these two loaves as somehow representing sin or the effects of sin, then sin played a part in the making (baking) of the two loaves. As leaven was used to make the finished product (bread), so must sin be used in some way to produce the finished or harvested man. To understand how sin could play any part, you must understand the law of knowledge and the reason for good and evil (see NM 19 and GP7). Remember, we learn from John that Christians are forgiven through Christ, for Christ's sacrifice (blood) made us acceptable to God, yet we are not without sin (leaven):

- but if we walk in the Light as He Himself is in the Light, we have fellowship with one another, and the blood of Jesus His Son cleanses us from all sin. 8 If we say that we have no sin, we are deceiving ourselves and the truth is not in us. 9 If we confess our sins, He is faithful and righteous to forgive us our sins and to cleanse us from all unrighteousness. 10 If we say that we have not sinned, we make Him a liar and His word is not in us. (1John 1:7-10)

So Christians, even though Christ cleansed them from all sin (forgave all their sins), still are

---

[1] We learn more about Jesus Christ in the God Papers, so we will not mention the significance of the sheaf wave offering in the Passover festival. But there is meaning to this. As there is meaning to the fact that the unleavened bread was eaten for seven days and not eight days.

sinners and have sin and the results of sin in them: they still have leaven in them; none are perfect like Jesus Christ. The leaven in the two loaves represented the sin or age of sin that gives us the knowledge of evil, which in turn allows us to understand and appreciate good (study NM19).

Two Loaves Represent ...

nm612 » So what do the two loaves of baked leavened bread mean in the Feast of Pentecost?

■ Knowing, first that this festival represents the second order to salvation as explained by Paul (1Cor 15:20-28);

■ knowing, second that it occurs at the time of the coming of Christ and the setting up of his kingdom on earth (1Cor 15:23);

■ knowing, third that there will be two witnesses representing Christ 3 ½ years before Christ returns (PR8; Rev 11);

■ knowing, that two churches are associated with the two witnesses ("two candlesticks," Rev 11:4 cf. Rev 1:20);

■ then we know that the two loaves represent, all in the second order to salvation and also the two witnesses with their associated two churches at the end of the world.

nm613 » But also since we know there is only ONE church of Christ (which includes both the living and the dead in Christ [1Cor 15:52-56]), then these two loaves or churches also represent metaphorically the ONE church. The two loaves only add another point of information as do the seven churches in Revelation add to our information of the state of the Church at the end of the age. It is the 144,000 that will be infused into God at Christ's coming and it is one Church with one Spirit.

■ But each in his own order: Christ the first of the first fruits, **after that those who are Christ's at His coming**, (1Cor 15:23)

■ Then I looked, and behold, the Lamb *was* standing on Mount Zion, and **with Him one hundred and forty-four thousand**, having His name and the name of His Father written on their foreheads.... 14:4 These are the ones who have not been defiled with women, for they have kept themselves chaste. These *are* the ones who follow the Lamb wherever He goes. These have been purchased from among men **as first fruits to God and to the Lamb**. (Rev 14:1,4)

■ For even as the body is **one and *yet* has many members**, and all the members of the body, though they are many, are **one** body, so also is Christ. For **by one Spirit** we were all baptized **into one body**, whether Jews or Greeks, whether slaves or free, and we were **all made to drink of one Spirit**. (1Cor 12:12-13)

Feast of Trumpets: Great Assembly

Trumpets for Resurrection

nm614 » On the first day of the seventh month of Israel's Sacred Calendar (Sept-Oct) Israel had a festival of trumpets, which we call the Feast of Trumpets and others call the "Day of the Awakening Blast," which started a repentance period ("days of repentance") of nine days before the Day of Atonement (*The Temple*, by Alfred Edersheim, p. 297). Moses in the Bible writes about the Feast of Trumpets as follows:

- Speak to the sons of Israel, saying, 'In the seventh month on the first of the month you shall have a rest, **a memorial by blowing *of trumpets***, a holy convocation. (Leviticus 23:24)

- Now in the seventh month, on the first day of the month, you shall also have a holy convocation; you shall do no laborious work. It will be to you **a day for blowing trumpets**. (Numbers 29:1)

Therefore the feast of trumpets was another annual Sabbath for the Hebrews. It wasn't a festival where all males had to appear before God in Jerusalem, nor was it a harvest, but it was a holy convocation, a Sabbath. It happened on the first day of the seventh month. The main aspect of this holy day was that trumpets were blown on it, some accounts tell us that trumpets were sounded all day long:

Feast of Trumpets not for War

The time when the Kingdom of God is set up on earth is after Jesus Christ comes, *after* the great tribulation (Mat 24:29-30), on the antitypical *Pentecost*. And since after the kingdom of God is set up there will never be war again (PR6, pr447, Sec [13]), then the fulfillment of the Feast of Trumpets has to do with something else besides war.

Trumpet blowing also represented other things such as to introduce a new king (1Kings 1:34-41), and was used in the Bible to indicate the presence of God (Ex 19:16, 19; 2Kings 9:13). But these two aspects of trumpet blowing were fulfilled by the trumpet blowing on the antitypical Feast of Pentecost.

There will be trumpets blown at the end of the world (old evil age), but just because the trumpets are blown at the end of the world and just because the feast of trumpets is a day of blowing of trumpets, does not necessarily mean that the feast represents the end of the world. Originally, the trumpet blowing on the Feast of Trumpets was not for war, but to assemble Israel for a holy day where no work was to be done (Num 29:1; Lev 23:24-25). Remember that *every* festival had horns blown on them (Num 10:10).

Perfect Fulfillment always occurs on Same Day

nm615 » The various meanings of blowing the trumpets are telling us something. As we have seen by our study of the Passover, events were fulfilled anti-typically on the exact day on which they happened in the Old Testament feast. The typical Passover was killed on the 14th of the first month; the antitypical Passover (Christ) was killed on the 14th also. The taking of the sheaf of the first fruits typically was performed at the beginning of the day after the Sabbath; antitypically this was fulfilled when Christ ascended to his Father on the beginning of the day, after sunset, after the weekly Sabbath. The Pentecost typically happened after the harvest of first fruits; antitypically the Pentecost indicates the first harvest of the world's people to immortality and this harvest will happen on the antitypical Pentecost. The first giving of the Spirit that makes this harvest of people to immortality possible also happened on the exact day of the Pentecost (Acts 2).

Feast of Trumpets Fulfillment

nm616 » The Feast of Trumpets will be the resurrection of the third order or rank of all that died in the first seven 1000-year periods. Trumpets were used to gather Israel together. As we see in the "Seed Paper" [PR1], physical Israel represents allegorically all of mankind. The Feast of Trumpets was the grand day of trumpet blowing for trumpets were blown all day long. The Pentecost foreshadowed the resurrection at the beginning of the 1000 year-day of rest (Rev 20:4-5), while the Feast of Trumpets foreshadows the far greater (in number) resurrection of all the rest of the dead from their graves after the 1000 year-day (Rev 20:4-5). This is the resurrection of judgment, or those from the 1000 year judgment (John 5:29; see NM24). All the dead will be resurrected and taught from the Bible the meaning of life and the plan of God as foreshadowed by such verses as Nehemiah 7:73; 8:1-13. Remember all resurrected in this great resurrection did not understand who or what God was or the purpose of life while they lived on earth previously. So at the time of their resurrection from the dead and from their judgment they will learn the truth and they will weep and repent when they start to understand how grand God was/is/will be, and how evil they had been in their past behavior.

- "For the people wept when they heard the words of the law" (Neh 8:9)

But their weeping will turn to joy as they learn of the great good news of God's great plan of creation. As in the book of Nehemiah, the people on the first day of the seventh month (Neh 8:1-3) and on the second day of that month gathered "to understand the words of the law" (Neh 8:13). From the first day when they will be resurrected on the Feast of Trumpets until the Day of Atonement, they will study and learn from the Bible all that they overlooked while on earth. The Feast of Trumpets gathers the people of the earth from their judgment and starts to teach them the truth until the Day of Atonement when something even greater happens.

Day of Atonement

Atonement for All
Physical and Spiritual Join

nm617 » "And the LORD spoke unto Moses, saying, also on the tenth day of the seventh month there shall be **a day of atonement**: it shall be a holy convocation unto you; and you shall afflict your <u>souls</u> and offer an offering made by fire unto the LORD. And you shall do no work in that same day: for it is a day of atonement, to make an atonement for you, before the LORD your God. For whatsoever <u>soul</u> it be that shall not be afflicted in that same day, he shall be cut off from among his people. And whatsoever <u>soul</u> it be that does any work in that same day, the same <u>soul</u> will I destroy from among his people. You shall do no manner of work: a statute aeonian throughout your generations in all your dwellings. It shall be unto you a Sabbath of rest, and you shall afflict your <u>souls</u>: in the ninth day of the month at evening, from evening unto evening, shall you celebrate your Sabbath" (Lev 23:27-32).

nm618 » "*This* shall be a permanent statute for you: in the seventh month, on the tenth day of the month, you shall humble your <u>souls</u> and not do any work, whether the native, or the alien who sojourns among you; 30 for it

is **on this day that atonement shall be made for you to cleanse you; you will be clean from all your sins before the LORD**. 31 "It is to be a sabbath of solemn rest for you, that you may humble your <u>souls</u>; it is a permanent statute. 32 "So the priest who is anointed and ordained to serve as priest in his father's place shall make atonement: he shall thus put on the linen garments, the holy garments, 33 and **make atonement for the holy sanctuary**, and he shall make **atonement for the tent of meeting and for the altar**. He shall also make **atonement for the priests and for all the people of the assembly**. 34 "Now you shall have this as a permanent statute, to make **atonement for the sons of Israel for all their sins** once every year." And just as the LORD had commanded Moses, *so* he did. (Lev 16:29-34)

The key word here is <u>soul</u> as we will see below after we see what is the antitypical meaning of "fast."

Physical Fast and Soul

nm619 » Therefore the day of Atonement was a day of rest, and a day to "afflict your souls." The day of Atonement was an annual Sabbath for the Hebrews wherein they came together; it was a day of rest from labor. Now what does it mean to afflict their souls? First of all, "soul" here is translated from a Hebrew word that means, *breathing animal* (NM6). To afflict one's soul is to afflict one's living body.

Real Fast is Behavior without Evil

nm620 » The word "afflict" is translated from a Hebrew word that means, to humble, or to lower. Therefore on the day of Atonement we humble or lower our bodies. Now this means to the physical Jews to *fast* on that day, and thus to humble their bodies. Isaiah 58:3 seems to use afflicting one's soul and fasting interchangeably. Therefore to afflict one's souls is to fast. But we are to worship God Spiritually (John 4:23-24). What is the Spiritual or higher meaning for fasting?:

- "Is not this what I require of you as a fast: to loose the fetters of injustice, to untie the knots of the yoke, to snap every yoke and set free those who have been crushed? Is it not sharing your food with the hungry, taking the homeless poor into your house, clothing the naked when you meet them, and never evading a duty to your kinfolk?" (Isa 58:6-7, NEB)

nm621 » In other words, Spiritual lowering of our bodies, or fasting, is the way or system of love. Therefore since we are to worship God Spiritually, then to Spiritually fast is to follow in the way of love. Since Christians are supposed to have the Spirit, which brings them atonement to God, then Spiritual Christians always Spiritually fast – they always follow in the way of love, at least in their inner minds, according to the power given them.

New Soul

Also notice that the fast on the Day of Atonement is a fast of the soul. As we learned in the *Body, Soul, and Spirit Paper* [NM6], there are two meanings to soul. The physical meaning is that a soul is a physical body with breath in it; while the higher meaning of soul is a physical body with God's Spirit in it. At the antitypical Feast of Atonement, the Spiritual meaning of "fast" will apply and the Spiritual meaning of soul will apply. At the antitypical Atonement all will be in their real soul: a physical body with a good Spirit of God in them (Rom 5:11).

Atonement to God

nm622 » The Hebrew's Day of Atonement Spiritually represents to us the atonement of man to God. In the paper "Thousand Years and Beyond" we explained that the time of Atonement for mankind to God through the medium of the Spirit of God is in the eighth Spiritual time period, the Great Day of Atonement. This Spiritual day of Atonement lasts for 100 years, and occurs after the seventh millennium as shown in the paper just mentioned. The day of Atonement pictures the snapping of all the yokes of bondage from mankind and angelkind, and the practicing of the way of love for ALL. We receive atonement through Christ (Romans 5:11). And we receive this atonement when we have the Spirit of God in us.

Trumpets on the Day of Atonement on the Jubilee

nm623 » In Leviticus, chapter 25, it explains the year of Jubilee. This was a year when all returned to their homeland and families. This Jubilee pictures the Great Day of Atonement to God when everyone ever born of mankind has been resurrected and will return to their families and live in that age. It is significant that on the day of atonement, just at the beginning of the Jubilee that there was to sound "the trumpet of Jubilee" throughout the land (Lev 25:9). This trumpet does not antitypically represent a resurrection (1Cor 15:52; 1Thes 4:16), since that happened earlier on the Feast of Trumpets. The trumpet here does not

signify war because at the beginning of the 7th millennium, all war will be stopped and will never happen again (see "Last War and God's Wrath" paper [PR5]). On the Great Day of Atonement after the thousand years, all will have been resurrected back to life (Ezek 37:1-13; 16:55). Thereafter on the Day of Atonement the Spirit of God will be given them (Ezek 37:13-14). The trumpets on the Day of Atonement signify the gathering of the physical and Spiritual coming together into atonement.

Day of Atonement Fulfillment: Release From Sin

nm624 » At the end of seven units of time ("years"), which represents the Spiritual seven units of time (millenniums), is the release from debt (Deut 15:1-2). Debt and sin are used interchangeably in the Bible (Luke 11:4). Therefore, according to the pattern, after the seven millenniums, all will have been released from sin. The Spiritual law will be imputed to all and thereafter they shall follow all the ways of love perfectly because on the eighth Spiritual day they will receive the Spirit or New Mind in the full measure of power as Christ had when he was on earth.

nm625 » The festival of Atonement pictures the atonement of all mankind in the eighth Spiritual day and receiving the Spirit in the eighth unit of Spiritual time. This is pictured in John 7:37-39:

- Now on the last day, the great *day* of the feast, Jesus stood and cried out, saying, "If anyone is thirsty, let him come to Me and drink. 38 "He who believes in Me, as the Scripture said, 'From his innermost being will flow rivers of living water.'" 39 But this He spoke of the Spirit, whom those who believed in Him were to receive; for the Spirit was not yet *given*, because Jesus was not yet glorified. (John 7:37-39)

nm626 » The Jubilee also started on the day of Atonement and the Jubilee represented atonement of all the people to God.

- You shall then sound a ram's horn abroad on the tenth day of the seventh month; on the day of atonement you shall sound a horn all through your land. (Lev 25:9)

Everyone was to return to his family and possessions, and the poor and servants were able to redeem themselves and their property. It was a year of liberty and freedom: You shall "proclaim liberty throughout the land and unto all the inhabitants." (Lev 25:10) But the only true way to get true freedom is through the Spirit of God (2Cor 3:17; Gal 5:1). "And not only *so*, but we also joy in God through our Lord Jesus Christ, by whom we have now received the **atonement**" (Rom 5:11). The real atonement is through Christ and his Spirit of God. When the last group or order receives the Spirit, at that time, *all* will be atoned to God through the New Mind or new Spirit of God. But all will not yet be immortal as we see by studying how the seed was sown in the Jubilee. See Below. Spiritual atonement must happen on the <u>eighth</u> period of time after the seven millenniums as shown in NM15.

Ritual of the "twin" goats fulfilled on the antitypical Day of Atonement

The ritual of "twin" goats on the Day of Atonement was partially fulfilled when Christ died for our sins. Christ fulfilled the ritual of the first goat mentioned in Leviticus 16:7-9,15-19. But the fulfillment of the second goat's ritual will not be fulfilled until the antitypical Day of Atonement, when all sin is sent away for ever. The antitypical Day of Atonement will fulfill perfectly atonement to God because all mankind on that day will receive the Spirit, and because of this, **sin will on that day be sent away** in the perfect sense. This will fulfill the Azazel ritual perfectly, and thus complete perfectly the typical Day of Atonement (see NM15, under "Azazel," nm498).

Feast of Tabernacles Represents third and Final Harvest Great Multitude with white robes and palms (Rev 7:9) Third Order or Rank fulfills God all in all (1Cor 15:23-28)									Time after
Sabbath 15th of 7th mo.								Sabbath 22nd of 7th mo.	23rd day [1Ch 7:10]
1st day Living in temp. palm booths	2nd day Living in temp. palm booths	3rd day Living in temp. palm booths	4th day Living in temp. palm booths	5th day Living in temp. palm booths	6th day Living in temp. palm booths	7th day Living in temp. palm booths	8th day Great Last Day **Atonement** with God		9th "day" Forever Time >
< **Signifies** eight periods of the creation (seven millenniums & one shorter period) >									Forever >

Third Harvest: <u>Feast of Tabernacles</u>

Third and Final Resurrection to Immortal Life

nm627 » The typical Feast of Tabernacles, also called the Feast of Booths, occurred <u>after</u> the last harvest of the year:

- Leviticus 23:34 "Speak to the sons of Israel, saying, 'On the fifteenth of this seventh month is the Feast of Booths for seven days to the LORD. 35 'On the first day is a holy convocation; you shall do no laborious work of any kind. 36 'For seven days you shall present an offering by fire to the LORD. On the eighth day you shall have a holy convocation and present an offering by fire to the LORD; it is an assembly. You shall do no laborious work. 37 'These are the appointed times of the LORD which you shall proclaim as holy convocations, to present offerings by fire to the LORD-- burnt offerings and grain offerings, sacrifices and drink offerings, *each* day's matter on its own day-- 38 besides *those of* the sabbaths of the LORD, and besides your gifts and besides all your votive and freewill offerings, which you give to the LORD. 39 'On exactly the fifteenth day of the seventh month, <u>when you have gathered in the crops of the land</u>, you shall celebrate the feast of the LORD for seven days, with a rest on the first day and a rest on the eighth day. 40 'Now on the first day you shall take for yourselves the foliage of beautiful trees, palm branches and boughs of leafy trees and willows of the brook, and you shall rejoice before the LORD your God for seven days. 41 'You shall thus celebrate it *as* a feast to the LORD for seven days in the year. It *shall be* a perpetual statute throughout your generations;

you shall celebrate it in the seventh month. 42 'You shall live in booths for seven days; all the native-born in Israel shall live in booths, 43 so that your generations may know that I had the sons of Israel live in booths when I brought them out from the land of Egypt. I am the LORD your God.'" 44 So Moses declared to the sons of Israel the appointed times of the LORD. (Lev 23:34-44)

Temporary Booths

nm628 » This festival lasted eight days. The first day was an assembly, and the eighth day was an assembly. But during the first seven days, physical Israel was to live in temporary booths made up of branches of trees (Lev 23:40; Neh 8:15).

- Now on the first day you shall take for yourselves the foliage of beautiful trees, palm branches and boughs of leafy trees and willows of the brook, and you shall rejoice before the LORD your God for seven days. (Leviticus 23:40)

- So they proclaimed and circulated a proclamation in all their cities and in Jerusalem, saying, "Go out to the hills, and bring olive branches and wild olive branches, myrtle branches, palm branches and branches of other leafy trees, to make booths, as it is written." (Nehemiah 8:15)

- You shall live in booths for seven days; all those born in Israel shall live in booths (Lev 23:42)

All born in Israel were to live in booths for seven days. This Spiritually signifies the seven Spiritual days, the seven millenniums, and all of mankind born in the seven millenniums. Why? Because everyone will eventually become a Spiritual Israelite (see, "Seed Paper" [PR1]), this thus signifies that all ever to be born of God will be born of man during the first seven millenniums. During the seven millenniums all of mankind will have lived in their temporary bodies. After the seven millenniums, mankind as a whole will be typically complete, or joined with their Spirit. Two (the Spirit and the physical body) will be typically complete in the eighth Spiritual day. Therefore they will <u>not</u> be dwelling in their temporary tabernacles (old temporary bodies) after the seven millenniums, but they will be dwelling in their new bodies as a New Soul. But in the short eighth Spiritual day, they will be in their New Soul as a *begotten* son of God. Later they will be *born* of God (see "Begotten, Born Paper" [NM 5]).

100 Years

nm629 » The eighth day of the Feast of Tabernacles indicated the eighth Spiritual period of the Spiritual creation. This eighth Spiritual day will be shorter than the first seven Spiritual days as indicated by the eighth period of time in the Jubilee and Pentecost festival:

> (1) The Jubilee is the eighth year <u>after</u> seven periods of seven years;
>
> (2) The Pentecost is the eighth day <u>after</u> seven periods of seven days.

This shorter eighth period of time will last 100 years. The proof for the length is indicated by Isaiah 65:20 and other scripture (Ex 26:3, 7; 36:10, 14; 26:12; see Sacrifice Table below). Here it shows all people living to the age of 100 years. People will live that long once the typical new heaven and earth is created (Isa 65:17-19). This begins at Christ's return and lasts until the creation of the true new heaven and earth. Since there is to be an eighth Spiritual day of creation, as indicated by the pattern of various festivals and rituals of the Bible (see NM15), and since the life spans will be 100 years in the typical new heaven and earth, and since those resurrected after the seven millenniums will still be human beings, but with the Spirit joined to them, and since in the Biblical patterns the eighth unit in the cycle is pictured as being smaller than the previous seven units; then the eighth day of creation will be smaller than the previous ones, and it will last for 100 years, which is the length of human life in the typical new heaven and earth. Note: To understand the patterns and cycles of eight, see the paper, "Thousand Years and Beyond" [NM 15].

nm630 » When Jesus kept the feast of tabernacles, he on the eighth day ("the last day") stood up and said, "if any man thirst, let him come unto me, and drink" (John 7:37). By this he meant the water of the Spirit (John 7:39). Therefore this indicates the giving of the Spirit in the eighth Spiritual day of creation.

Feast of Tabernacles Fulfillment

nm631 » Notice that the Festival of Tabernacles was celebrated *after* the gathering of the harvest (Lev 23:39; Deut 16:13; Ex 23:16). This indicates in the higher sense the time after the gathering of all the harvest of mankind. Mankind is totally gathered together in the eighth Spiritual day, wherein all will be resurrected to life with the new Spirit given them. The festival is thus fulfilled perfectly after all have been harvested. Harvest, in the first two harvests, always meant being harvested to immortality and with that being placed back into the God.

After The Eighth Day

nm632 » Now notice: "and in the eighth day they made a solemn assembly: for they kept the dedication of the altar seven days, and the feast [of tabernacles] seven days [that is, the 7 days before this 8th day]. And **on the three and twentieth day** [23rd] of the seventh month he sent the people away into their tents, glad and merry in heart for the goodness that the LORD had showed unto David, and to Solomon, and to Israel his people. **Thus Solomon FINISHED the house of the LORD**, and the kings's house" (2Chron 7:9-11). Now the eighth day of the feast happens on the 22nd day of the seventh month. Solomon sent them back to their tents on the 23rd day (the 9th day after the feast began) when the house of God was finished.

House of God Finished

Since Christians are the antitypical temple and house of God (1Cor 3:16; 1Pet 2:5), and since all will eventually go into God (NM13), then when the house of God is finished (all in God), so too will be the creation, in its truest sense. Also we show in PR1 that the antitypical Solomon is Jesus Christ. Jesus Christ will build and finish the house of God for all things are built through Christ (1Pet 2:5; Col 1:16-19).

nm633 » Therefore the higher meaning of 2Chron 7:9-11 indicates that after the eighth Spiritual days of creation, then the family (house) of God will be finished. The house of God is the Spiritual house that Jesus Christ is to build. That is, the creation of God will be finished – all will be born of God – right after the eighth Spiritual day of creation. The next instant after the eighth Spiritual day would be the ninth Spiritual day. But since right after the eighth Spiritual day is when the true creation of the new heaven and earth will happen (Rev 21:5), then it is at that point that Jesus Christ (the Spiritual Solomon) will have finished the creation. At the end of the 100 year Great Last Spiritual Day of Creation (at the end of the eighth day of the feast of tabernacles) those who are still human will be born of God as in the description of 1Corinthians 15:52-55. At that time the new universe will be created and the immortal state of happiness will begin for everyone.

So when does this happen? This happens at the very end of the eighth day of the Feast of Tabernacles, on the very last moments of the last day of the Great Last Day (8th Spiritual day of creation), then the last group and all the universe will be created new, so that God will be all in all.

Jubilee Harvest Sowing the Seed			
6th Yr	7th Yr **Land Rest**	8th Yr = **Jubilee**	9th Yr
Seed sown	No seed sown people eat from 6th year's harvest	Seed sown, but not harvested this year; people eat from 6th year's harvest	People eat harvest sown in the 8th year
< Seed sown in 6th year produces enough for these three years (Lev 25:21) >			
Lev 25:3, 20-22	Lev 25:4-5, 20-21	Lev 25:11, 22	Lev 25:22

Jubilee Harvest: Sowing of the Seed

nm634 » The year of Jubilee is the year *after* seven Sabbatical years. A Sabbatical year occurs every seventh year wherein there is not to be sown any crop (Lev 25:3-5). Now the Jubilee year is the eighth year, the year after the seventh Sabbatical year. In the seventh year is the year people were not to sow at all, while in the Jubilee year the people were supposed to sow, but not to reap the fruit thereof *in* the Jubilee year (Lev 25:11,22). And "the Jubilee; it shall be holy unto you: you shall eat the increase thereof out of the field" (Lev 25:12). But you were not to eat the sown fruit, that you sowed in the eighth year, the Jubilee year (Lev 25:22, 11). One was to eat of the fruit sown in the sixth year until the ninth year, until the fruit sown in the eighth year came in (Lev 25:20-22). Therefore in the sixth year, God would give a blessing that would last for three years (Lev 25:21). Since one was not to sow in the eighth Jubilee year to reap or harvest that same year (Lev 25:11), yet one was to sow in that year (Lev 25:22, 1st part), then they did sow the land in the eighth Jubilee year. But thereafter *when* this crop came forth in the *ninth* year they were eating the increase *from* the eighth Jubilee year, but not *in* the eighth Jubilee year as the laws required (cf Lev 25:12, 11, 22).

Higher Meaning of the Sowing and Harvest

nm635 » The higher meaning of this indicates that during the sixth millennium enough of the good Spiritual crop of the earth (Christians, Mat 13:38-41) will be harvested to provide for the sixth, seventh, and eighth Spiritual days of creation. And on the seventh Spiritual day, the seventh millennium, there will be no souls sown with the good Spiritual seed (see "Thousand Years Paper" [NM 15]). But in the eighth Spiritual day (the 100 year Great Last Day of Creation) souls will be sown with the good Spiritual seed, but they will not be reaped in the eighth Spiritual day. They will be reaped on the ninth or thus at the very end of the eighth Spiritual day, when the antitypical new heaven and earth will be created. At that time the others will be Spiritually harvested. They will then be born of God like those who were born after the first harvest of the world (Rev 14:4, 14-20; Mat 13:38-41).

nm636 » The Feast of Tabernacles thus represents the full creation; it pictures the eight Spiritual days of Creation, and the GREAT Spiritual harvest of them.

Review: Three Orders of Creation

nm637 » God directed that all males must stand before the LORD during three of the annual festivals: "Three times in a year shall your males appear before the LORD your God in the place which he shall choose; in the feast of unleavened bread, and in the feast of weeks [Pentecost], and in the feast of tabernacles" (Deut 16:16). These three appearances indicate the three times when those of the creation will be born of God, or go back to God, or be resurrected immortal, or be infused to God and be one with God:

■ **(1)** During the Feast of Unleavened Bread the sheaf being waved indicated Christ ascending to his Spiritual Father, and thus being born of God. He was the first-born of God – the first fruit of creation (John 1:18; 1Cor 15:23).

■ **(2)** The Feast of Pentecost (50 weeks) indicates those of the first-fruits or *first-products* of God. That is, the first-ones after Christ returning to God (Rev 14:4; 1Cor 15:23).

■ **(3)** The Feast of Tabernacles indicates the gathering of the rest of mankind to God (1Cor 15:24-28; Rev 7:9, note the "palms in their hands").

Do Christians keep the above mentioned feasts or appointed times?

nm638 » Christians keep these feasts in a Spiritual manner and in the place where God has placed his NAME (note Deut 12:5, 18; etc.). A place where God has placed his NAME is inside true Christians. The NAME of God is in anyone who has the Spirit of God or the New Mind, for anyone with the Spirit of God is a child of God (Rom 8:16). Christ was the first to keep a festival in a Spiritual manner. Christ kept the Passover in a Spiritual manner. The New Testament Christians on the antitypical Day of Pentecost will be the first to keep the Pentecost in a Spiritual manner. All the rest will keep the Feast of Tabernacles in a Spiritual manner.

Sacrifices and Jesus Christ

nm639 » Sacrifices were an almost universal way for mankind to worship God up until the time of Christ. Then things began to change until today when most religions no longer practice sacrifices. The reason for this former universal practice of ritualistic sacrifice is hidden in history, but it has to do with the other-mind (NM20).

Quick Review of Sacrifices in the Bible

Cain and Abel

nm640 » The book of Genesis first mentions sacrifices in the story about Cain and Abel:

■ Now the man had relations with his wife Eve, and she conceived and gave birth to Cain, and she said, "I have gotten a manchild with *the help of* the LORD." 2 Again, she gave birth to his brother Abel. And Abel was a keeper of flocks, but Cain was a tiller of the ground. 3 So it came about in the course of time that Cain brought an offering to the LORD of the fruit of the ground. 4 Abel, on his part also brought of the firstlings of his flock and of their fat portions. And the LORD had regard for Abel and for his offering; 5 but for Cain and for his offering He had no regard. So Cain became very angry and his countenance fell. (Gen 4:1-5)

The account of the offerings of Cain and Abel shows that ritualistic sacrificing dated from almost the beginning. The custom of offering the firstlings and first-fruits had already begun. Cain's offering was grain and is called *minchah*, "a gift" or "presentation." The same term is applied to Abel's. There is no hint that the bloody sacrifice was in itself better than the unbloody one, but it is shown that sacrifice without a right attitude is not acceptable to God.

Noah

nm641 » The sacrifices of Noah followed and celebrated leaving the ark after the flood. He offered burnt offerings of the clean animals (Gen 8:20 ff). Remember he brought seven sets (male & female) of the clean animals onto the ark (Gen 7:2).

Abraham

nm642 » Abraham on his arrival at Shechem erected an altar (Gen 12:6-7). At Beth-el he also built an altar (12:8), and on his return from Egypt he worshiped there (Gen 13:4). At Hebron he built an altar (Gen 13:18). In Gen 15:1-18 he offers a "covenant" sacrifice, when the animals were slain, divided, the parts set opposite each other, and prepared for the appearance of the other party to the covenant. In Genesis 22 Abraham attempts to offer up Isaac his son as a burnt offering, but instead offered up a lamb for a burnt offering because of what the angel of God said to him (Gen 22:8, 11-13). What God really wanted was an obedient heart (Gen 22:12). Abraham continued his worship at Beer-sheba (Gen 21:33).

Isaac

nm643 » Isaac built an altar at Beer-sheba apparently to have regularly offered sacrifices (Gen 26:25).

Jacob

nm644 » Jacob poured oil upon the stone at Beth-el (Gen 28:18-22). After his covenant with Laban he offered sacrifices (*zebhachim*) and ate bread with his brethren (Gen 31:54). At Shechem, Jacob erected an altar (Gen 33:20). At Beth-el (Gen 35:7) and at Beer-sheba he offered sacrifices to Isaac's God (Gen 46:1).

Israelites in Egypt

nm645 » Sacrifices were not something new to the fathers of Israel. Therefore because Egyptians had a sacrificial system this meant that the Israelites were also accustomed to spring sacrifices, spring feasts, and fall feasts. Such sacrifices also had been found among the Arabs, Syrians and others. Such festivals were handed down even to the Romans. For example, according to Pliny Romans never ate their new corn or wine, till the priests had offered the first-fruits to the gods (*Clarke's Commentary*, vol 1, p. 417) which is similar to the fact that Israel could not eat of the first fruits until the sheaf of first fruits was waved by the priest (Lev 23:14, 10-14). According to Plutarch, the false god Bacchus had a festival like the feast of tabernacles, "they celebrated [the festival] in the time of vintage, bringing tables out into the open air furnished with all kinds of fruit, and sitting under tents made of vine branches and ivy" (*Clarke's Commentary*, vol 1, p. 587[1]; cf. Lev 23:35-41). As in most nations, in Egypt at these festivals sacrifices and food were offered to their gods. Apparently it was to some such feast Moses said Israel as a people wished to go in the wilderness (Ex 3:18; 5:3 ff; 7:16) to sacrifice to Yehowah. Pharaoh understood and asked who was to go (Ex 10:8). Moses demanded flocks and herds for the feast (Ex 10:9). Pharaoh wanted to keep the flocks (Ex

---

[1] "In imitation of this feast among the people of God, the Gentiles had their feasts of tents. Plutarch speaks particularly of feasts of this kind in honour of Bacchus, and thinks from the custom of the Jews in celebrating the feast of tabernacles, that they worshipped the god Bacchus, "because he had a feast exactly of the same kind called the feast of tabernacles, *skhnh*, which they celebrated in the time of vintage, bringing tables out into the open air furnished with all kinds of fruit, and sitting under tents made of vine branches and ivy."-PLUT. Symp., lib. iv., Q. 6. According to Ovid the feast of Anna Perenna was celebrated much in the same way. Some remained in the open air, others formed to themselves tents and booths made of branches of trees, over which they spread garments, and kept the festival with great rejoicings."

10:24), but Moses said they must offer sacrifices and burnt offerings (Ex 10:25 ff). The sacrifice of the Passover soon occurred thereafter, but according to the pattern that God manifested to Moses, not according to Egyptian traditions (Ex 12:1-17). [Some of the above material taken from *ISBE* (1915)]

Moses

Not Sacrifice, but Obedience

nm646 » One main difference between the sacrificial system initiated by Moses and the other systems was that the foundational principle was obedience (Ex 19:4-8). The main aspect in Israel's religion was obedience and loyalty to Jehovah, not sacrifices. God spoke about obedience, not sacrifices:

- Thus says the LORD of hosts, the God of Israel, "Add your burnt offerings to your sacrifices and eat flesh. 22 "For **I did not speak to your fathers, or command them in the day that I brought them out of the land of Egypt, concerning burnt offerings and sacrifices.** 23 "But this is what I commanded them, saying, **'Obey My voice, and I will be your God, and you will be My people**; and you will walk in all the way which I command you, that it may be well with you.' (Jer 7:21-23)

- Sacrifice and meal offering You [Yehowah] have not desired; My ears You have opened; Burnt offering and sin offering You have not required. (Psa 40:6)

- For You [Yehowah] do not delight in sacrifice, otherwise I would give it; You are not pleased with burnt offering. (Psa 51:16)

- To do righteousness and justice Is desired by the LORD more than sacrifice. (Prov 21:3)

- "What are your multiplied sacrifices to Me?" Says the LORD. "I have had enough of burnt offerings of rams, And the fat of fed cattle; And I take no pleasure in the blood of bulls, lambs or goats. (Isa 1:11)

nm647 » At Mount Sinai on the day that God asked Israel to obey his law, he didn't ask for sacrifices, he asked for obedience

- 'If you walk in My statutes and **keep My commandments** so as to carry them out, 4 then I shall give you rains in their season, so that the land will yield its produce and the trees of the field will bear their fruit. (Lev 26:3-4)

■ **These are the statutes and ordinances and laws** which the LORD established between Himself and the sons of Israel through Moses at Mount Sinai. (Lev 26:46)

■ "Then the LORD spoke to you from the midst of the fire; you heard the sound of words, but you saw no form-- only a voice. 13 "So He declared to you **His covenant which He commanded you to perform**, *that is*, the Ten Commandments; and He wrote them on two tablets of stone. 14 "The LORD commanded me at that time to teach you statutes and judgments, that you might perform them in the land where you are going over to possess it. (Deut 4:12-14)

■ Then Moses summoned all Israel and said to them: "Hear, O Israel, **the statutes and the ordinances** which I am speaking today in your hearing, that you may learn them and **observe them carefully**. 2 "The LORD our God made a covenant with us at Horeb. 3 "The LORD did not make this covenant with our fathers, but with us, *with* all those of us alive here today. 4 "The LORD spoke to you face to face at the mountain from the midst of the fire, 5 *while* I was standing between the LORD and you at that time, to declare to you the word of the LORD; for you were afraid because of the fire and did not go up the mountain. He said. (Deut 5:1-5)

■ Then you shall say to your son, 'We were slaves to Pharaoh in Egypt, and the LORD brought us from Egypt with a mighty hand. 22 'Moreover, the LORD showed great and distressing signs and wonders before our eyes against Egypt, Pharaoh and all his household; 23 He brought us out from there in order to bring us in, to give us the land which He had sworn to our fathers.' 24 "So **the LORD commanded us to observe all these statutes**, to fear the LORD our God for our good always and for our survival, as *it is* today. 25 "It will be righteousness for us if we are careful to observe all this commandment before the LORD our God, just as He commanded us. (Deut 6:21-25)

If Israel Brings an Offering

Paradoxes of Sacrifices

nm648 » Although God did not wish or require sacrifices, if Israel brought them He wanted them offered in the way or pattern he commanded: "If any man of you bring an offering unto the LORD, you shall bring your offering ...." (Lev 1:2; 2:1; 22:18; Ex 25:2) But on the other hand, God apparently used mankind's inclination towards sacrifices as a form of

worship in order to reveal Jesus Christ. When one studies the Old Testament scriptures on sacrifices, one finds the paradoxical situation of God not wanting sacrifices, yet commanding certain ways of performing these unwanted sacrifices. This may have something to do with the left and right side of God as mentioned in the *God Papers*. We cannot speak about this subject here.

Even Physical Circumcision...

nm649 » Also God was really looking for a circumcision of Israel's heart or mind, not their foreskins:

■ "So circumcise your heart, and stiffen your neck no longer. (Deut 10;16)

■ "Moreover the LORD your God will circumcise your heart and the heart of your descendants, to love the LORD your God with all your heart and with all your soul, so that you may live. (Deut 30:6)

■ "Circumcise yourselves to the LORD And remove the foreskins of your heart, Men of Judah and inhabitants of Jerusalem, Or else My wrath will go forth like fire And burn with none to quench it, Because of the evil of your deeds." (Jer 4:4)

Sinai's Covenant

nm650 » The covenant was made and the terms and conditions are then laid down by Moses and accepted by the people (Ex 24:3). After the ten commandments and covenant code were given, an altar is built, burnt offerings and peace offerings of oxen are slain by young men servants of Moses, not by priests, and blood is sprinkled on the altar (Ex 19:25-24:8). The Law was read, the pledge given, and Moses sprinkled the representatives of the people, consecrating them.

First Forty Days, Moses Shown the Patterns of Tabernacle

nm651 » After Moses, Aaron, Nadab, Abihu, and seventy of the elders of Israel saw a vision of God, Moses went up onto Mt Sinai for forty days (Ex 24:9-18). At that time God gave Moses the "pattern of the tabernacle, and the pattern of all the instruments thereof" (Ex 25:9,40; Heb 8:5).This pattern given to Moses was a pattern of the heavenly or spiritual things to come:

- "And almost all things by the law are cleansed with blood; and without shedding of blood there is no remission [sending away of sin]. Therefore it was necessary for the copy [Moses' Tabernacle] of the things in the heavens to be cleansed with these [physical] sacrifices, but the heavenly things themselves [are cleansed] with better sacrifices than these [physical sacrifices]. For Christ [the better sacrifice] did not enter a holy place made with hands, a *mere* copy of the true one, but into heaven itself, now to appear in the presence of God for us." (Heb 9:22-24)

- "For the Law, since it has *only* **a shadow of the good things to come** *and* not the very form of things, can never, by the same sacrifices which they offer continually year by year, make perfect those who draw near." (Heb 10:1)

A better or Spiritual sacrifice was needed to fulfill the physical things in the Tabernacle that foreshadowed things to come. In table form let us look at some of the patterns of Moses' Tabernacle, its furniture, and its sacrifices and manifest to you the things that have come and are coming:

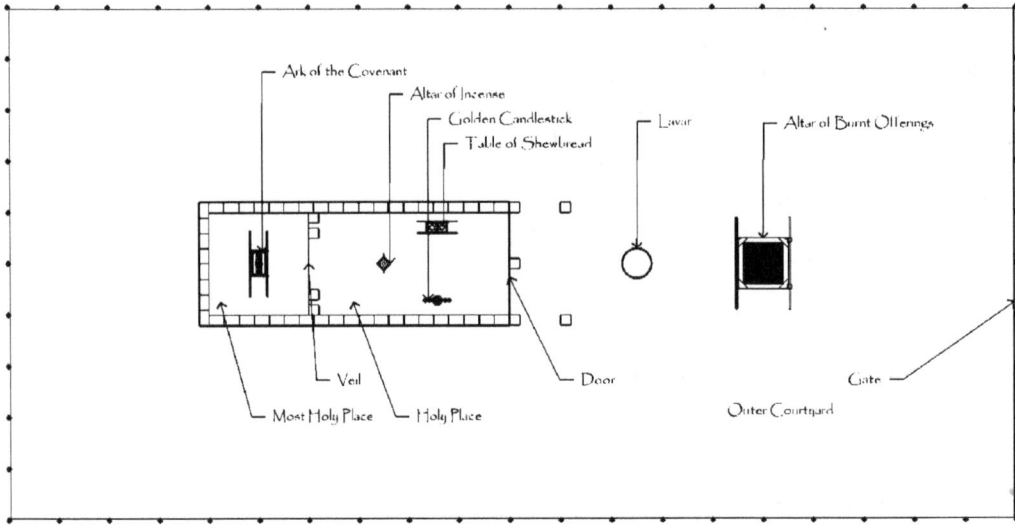

Drawing through Wikimedia Commons by Gabriel Fink

Tabernacle

nm652 »

Tabernacle, Sacrifices and Jesus Christ	
Physical Aspect	**Spiritual Aspect**
Tabernacle or Temple	**Tabernacle or Temple = Spiritual Body of Christ**
Holy of Holies 1. **Golden Ark of the Testimony**: Cherubs and Mercy Seat [west side of tabernacle] (Ex 25:10-22; 26:34)	**Ark = Christ Fulfilled (Now is Right Side)** Jesus Christ went into the Holy of Holies as the Right side of God (Heb 9:24; Acts 2:33-35; Heb 10:10-12; *God Papers*).
2. **Veil Between Most Holy and Holy Place** of Blue, Purple, and Scarlet (Ex 26:33)	**First Veil = Christ's Flesh** Luke 23:45: because the sun was obscured; and the **veil** of the temple was torn in two. Hebrews 10:20: by a new and living way which He inaugurated for us through the **veil**, that is, His flesh,

Tabernacle, Sacrifices and Jesus Christ	
Physical Aspect	**Spiritual Aspect**
nm653 » **Holy Place** 3. **Golden Altar of Incense** placed before the Holy of holies atoned with the blood of the sin offering once a year (Ex 30:1-10) [located west side of Holy Place]	**Incense = Prayer** Revelation 8:3: Another angel came and stood at the altar, holding a golden censer; and much **incense** was given to him, so that he might add it to the **prayers** of all the saints on the golden altar which was before the throne. 4 And the smoke of the **incense**, with the **prayers of the saints**, went up before God out of the angel's hand.
4. **Golden Table of Showbread**, (Ex 25:23-30); 26:35) [located north side of tabernacle]	**Bread = Christ's Spiritual Bread** The showbread was unleavened bread continually present in the Most Holy place of the tabernacle. Also see NM16 under "Spiritual Bread"
5. **Golden Candlestick** [south side of tabernacle], Lit Always by Aaron & Sons with Beaten Olive Oil (Ex 25:31-39; 26:35; 27:20-21) Oil was from beaten olives (Ex 27:2; Lev 24:2)	**Candlestick = Christ's Body (Church)** Revelation 1:20: The mystery of the seven stars which you saw in my right hand, and the seven golden **candlesticks**. The seven stars are the angels of the seven churches: and the seven **candlesticks** which you saw are the seven churches. For antitypical beaten oil see Isa 53:4-5; 1Pet 2:24.
6. **Tabernacle Covering** of Blue, Purple, and Scarlet *Ten Curtains* with Needlework of Cherubs (Ex 26:1)	**Ten Curtains = Ten Life-Span Era of the Kingdom** Ten curtains over the tabernacle indicate the ten 100 year life-spans or ten generations (1000 years) of the seventh period of time. See Deut 23:2-3 (non-Spiritual Israelites will not enter the kingdom for ten generations or ten 100 year life spans); Isaiah 65:20 (life spans in the kingdom will be for 100 years);
7. **Additional Tabernacle Covering** of Eleven Curtains of Goat [skin or hair] (Ex 26:7); More Tabernacle Coverings (above the Goat's hair) of Ram Skins Dyed Red and Badgers' Skins covered and hid the eleven curtains (Ex 26:14).	**Eleventh Curtain = 11ᵗʰ Life-Span Era** As these additional coverings hid and protected the tabernacle, the Spiritual tabernacle and truth have been hidden from the world. See NM15 for scripture that points to the hidden atonement period, the eighth period of time.
8. **Veil outside of the Holy Place** at the entrance of the Tabernacle of Blue, Purple, and Scarlet (Ex 26:36)	**Second Veil = Flesh of Christians** As the veil to the holy of holies was torn to manifest Christ's death (see # 2 above) in order for him to enter into it, so too those who enter into the holy place die (through Spiritual baptism) in order to go into it.

Tabernacle, Sacrifices and Jesus Christ

Physical Aspect	Spiritual Aspect
nm654 » Tabernacle Outer Court 9. Court Surrounds Tabernacle (Ex 27:9-1538:9-20) [There can be reasonable disagreement with what the outer court in Ezekiel and Revelation represented. Did it represent Moses' Tabernacle or not. We think it did, because the scripture on the "outer court" outside of scripture in Exodus (Ex 27:9-18; 38:9-20) is extremely vague as to the nature size, and location of it. When we understand that the first Tabernacle is a pattern for our time (see below), we understand this matter better.]	The outer court represented the non-Christians, who are not counted because God's holiness (Spirit) was not transmitted to them. "When they go out into the outer **court**, into the outer **court** to the people, they shall put off their garments in which they have been ministering and lay them in the holy chambers; then they shall put on other garments so that **they will not transmit holiness to the people** with their garments. (Ezek 44:19) "Then there was given me a measuring rod like a staff; and someone said, "Get up and measure the temple of God and the altar, and those who worship in it. 2"Leave out the court which is outside the temple and do not measure it, for **it has been given to the nations**; and they will tread under foot the holy city for forty-two months."(Rev 11:1-2)
10. **Brass Water Basin** (laver) for Aaron and his sons to wash hands and feet (Ex 30:17-21)	**Water = Spiritual Water or Spirit** Christ, was anointed and washed by the Spirit of God; his sons are washed and anointed by God's Spirit also. (John 7:37-39; 1 Cor 6:11; Heb 10:22)
11. **Brass Four Horned Altar** outside the entrance of the Tabernacle (Ex 27:1-8)	**Antitypical Horned Altar** The four horns symbolize the four beasts of Daniel and Revelation, which in turn represent Satan's kingdoms (PR2) in the four corners of the world. The sacrifices and death on this altar represent the death and sacrifices of all mankind under Satan's chaos: Revelation 12:9: … **Satan, who deceives the whole world**…. Romans 8:22: …**the whole creation groans and suffers the pains of childbirth together until now**. Christ releases the creation from Satan at his second coming by sending sin away at his coming (Rev 20:1-3; & see Azazel).
12. **Veil** of Blue, Purple, and Scarlet at Court Entrance	**Third Veil = Flesh of the Third Order** Entrance into the court outside the holy tabernacle also must be through the blue, purple, and scarlet veil (death). See # 2 and # 8 above.

Tabernacle, Sacrifices and Jesus Christ	
Physical Aspect	**Spiritual Aspect**
nm655 » **Priests' Garments**	
13. **Garments**, Robe, Coats, and Breeches for the Priests (Aaron & Sons) made of White Linen of Gold, Blue, Purple, and Scarlet trim and designs (Ex 28:1-8; 28:31-35)	**Garments = Spiritual Garments** Since Christ is the chief priest (Heb 7:21-22), The white represents holiness and the gold represents the Spiritual aspect of Christ and his body (See Rev 7:13-14) while the blue, purple, and scarlet represent the death of Christ.
14. **Two Onyx Stones** engraved with the Names of the son of Israel (Ex 28:9-10)	**Engraved Stones = Reveals that Christ Represents Spiritual Israel** Christ and his Church are the antitypical Priest and priests (Heb 7:21-22; Rev 1:6; 5:9-10). Christ (his Spiritual Body) is Spiritual Israel. (PR1; Rev 7:4-8)
15. **Breastplate** of Twelve Precious Stones representing the tribes of Israel (Ex 28:15-21)	**Breastplate = Reveals that Christ Represents Spiritual Israel** Christ and his Church are the antitypical Priest and priest (Heb 7:21-22; Rev 1:6; 5:9-10). Christ (his Spiritual Body) is Spiritual Israel. (PR1; Rev 7:4-8)
16. **Golden Plate** on Forehead of Priest engraved with "Holiness to the YHWH" (Ex 28:36-38)	**Engraved Insignia = Indicates God's Name is on Christ and His Body** Christ came in his Father's Name, and all in Christ's Spiritual Body are baptized into God's Name (John 17:11; Mat 28:19; John 20:31)
nm656 » **Priests**	
17. **Aaron and Sons** are the priests. See Exodus 28:1.	**Jesus and Sons = Body of Christ** Christ and his Church are the antitypical Priest and priests (Heb 7:21-22; Rev 1:6; 5:9-10).
nm657 » **Sacrifices and Offerings for Priests' Consecration**	

Tabernacle, Sacrifices and Jesus Christ	
Physical Aspect	**Spiritual Aspect**
18. Sacrifice of Bull and Two Rams without Blemish (Ex 29:1)	**Unblemished Sacrifice = Christ** Christ is the real unblemished sacrifice: "but with precious blood, as of a lamb **unblemished** and spotless, *the blood* of Christ." (1Peter 1:19)
19. **Basket of Unleavened Bread**, Cakes, and Wafers with Oil and made of Wheat (Ex 29:2)	**Christ's Spiritual bread** See # 4 above.
20. **Priests washed with Water** (Ex 29:4)	**Spiritual Baptism washes** (NM4)
21. **Anointed with Oil** (Ex 29:7) Oil was from beaten olives (Ex 27:2; Lev 24:2)	**Spiritual Oil of Christ** (Heb 1:9; 1John 2:20; 2:27). Christ was beaten for our sins
22. **Blood** of Bull put on Horn of Altar and **poured out** at bottom of altar (Ex 29:12)	**Christ's Blood Poured Out** Not blood of bulls and goats, but Christ's blood being poured out is forgiveness of sins for all including the four beasts and to all the creation (Heb 9:12-14, 28; NM13)
23. **Fat** of Bull, liver, two kidneys burned on the Altar (Ex 29:13)	**Fat = Evil Burned up on the Altar** Leviticus 7:25: For whoever eats the fat of the animal from which an offering by fire is offered to the LORD, even the person who eats shall be cut off from his people. In a sense, God's food is the fat and blood (Ezek 44:7; see *God Papers*). To destroy sin, God eats up the blood or life of the sinners and the "fat," which is the evil in mankind: Ezekiel 34:16 "I will seek the lost, bring back the scattered, bind up the broken and strengthen the sick; but **the fat and the strong I will destroy. I will feed them with judgment**. 17 "As for you, My flock, thus says the Lord GOD, 'Behold, I will judge between one sheep and another, between the rams and the male goats.... 20 Therefore, thus says the Lord GOD to them, "Behold, **I, even I, will judge between the fat sheep and the lean sheep.** Their heart is covered with **fat**, *But* I delight in Your law.(Ps 119:70) And their [wicked] body is **fat**. (Ps 73:4) But Jeshurun grew **fat** and kicked-- You are grown **fat**, thick, and sleek-- Then he forsook God who made him, And scorned the Rock of his salvation. (Deut 32:15)

Tabernacle, Sacrifices and Jesus Christ

Physical Aspect	Spiritual Aspect
24. Flesh, skin, dung of the sacrifices for sin were **burned outside of camp**: a sin offering (Lev 16:17; Ex 29:14)	**Christ Suffered Outside the Gate** Christ was the antitypical sacrifice and therefore Christ had to die outside of the gate of Jerusalem and outside (before) the kingdom on earth: Hebrews 13:11 For the bodies of those animals whose blood is brought into the holy place by the high priest *as an offering* for sin, are burned outside the **camp**. 12 Therefore Jesus also, that He might sanctify the people through His own blood, suffered outside the gate.
Laying on of Hands 25. First Ram killed after Priest placed his hands on head of Ram (Ex 29:18)	**Laying on of Hands** In the Old Testament, the placing of the hands on the animal represented placing the sins on the animal (Lev 16:21); the animal's death made atonement for sin by destroying sin with the animal. But in the New Testament Jesus laid hands on the little children (Mt 19:13,15; Mk 10:16) and on the sick (Mt 9:18; Mk 6:5, etc.), and the apostles laid hands on those whom they baptized that they might receive the Holy Spirit (Acts 8:17,19; 19:6), and in healing (Acts 12:17). Christians received the Spirit and sin was thus sent away when hands were laid on them (1Tim 4:14; 2Tim 1:6).
26. First Ram's blood poured out and is therefore burnt on altar: a burnt offering (Ex 29:18)	See # 22 above
27. Second Ram killed, blood placed on right side of Priest (Ex 29:19-20)	By blood being placed on the **right** side it foreshadowed Christ death and his going to the right side of God.
28. Second Ram's Blood and anointing oil sprinkled on Aaron and his sons (Ex 29:21)	It is the antitypical Aaron (Christ) and his sons (Body of Christ) that have forgiveness because of Christ's blood and sacrifice.
29. Second Ram's fat, oiled and unleavened bread waved before LORD (Ex 29:22-24)	These are types of the real sacrifice of Christ, who is the real unleavened bread and oil who was an offering offered (waved) to God (Heb 10:10-12).
30. Waved items are then burned on the Altar as a burnt offering (Ex 29:25)	As the waved items were burnt or destroyed so was Christ. Christ is the antitype of the burnt offering of Abraham (Gen 22:2-8; Heb 11:17-19).

Tabernacle, Sacrifices and Jesus Christ

Physical Aspect	Spiritual Aspect
Eating the Sacrifice 31. Aaron and sons shall eat the flesh (Ex 29:27) of the ram and the bread in the basket (which atonement was made for) by the door of the tabernacle to consecrate them and make them holy (Ex 29:32-33; Lev 10:14-15,17)	**Communion: Christians eat the Sacrifice** Christians who are the sons of Christ Spiritually eat the sacrificed body of Christ in this age before they go into the golden Holy Place. See NM16 under "Spiritual Bread" through "Communion" section.
Day of Atonement Sacrifices 32. A special "twin" goat ritual was performed on the day of Atonement (Lev 16)	**One died for sin; the other was sent away** never to appear as sin again. See NM15, under "Azazel" for more detail.
nm658 » Continual Daily Offering 33. **Daily offering** on the altar before the door (veil) of two lambs, one in the evening and one in the morning, continual burnt offering, throughout Israel's generations (Ex 29:38-42) The daily offering was offered on the altar before the door of the tabernacle, where God and his glory would meet the children of Israel and sanctify them along with the priests (Ex 29:42-44).	**Daily Sacrifices** These daily offerings of two lambs represent Christ's sacrifice for all the sins of mankind for all time, and also represent the sacrifices and suffering of mankind because of sin. It is a continuous suffering for mankind, until God sanctifies mankind. The sacrifices for mankind end at the coming of Christ when the door of the tabernacle (temple) is opened and the Saints are brought into the temple of God (Rev 15:5; 11:19; PR6; PR2 under "Daily Sacrifice").
nm659 » Sin offering only for sin through ignorance 34. See Lev 4:13; 5:2-4, 13	**All sin through ignorance** But even the killing of Christ was through ignorance (Luke 23:34; Acts 3:17; John 16:3; 1Cor 2:7-10; see NM13)
nm660 » God Meets Israel in His Glory 35. **God will meet Israel by the door** of the tabernacle at the altar and will sanctify Aaron, the priest, and Israel there with his glory (Ex 29:43-44)	**The door of the antitypical tabernacle opens** and God's great glory is revealed for all when he returns (Rev 15:5-8; 11:19) At the time the tabernacle or temple is opened is at the time of the wrath (Rev 11:18; see PR4-6) The tabernacle and temple are metonymical names for the same thing: The coming glorified Body of Christ.

Look and Dimensions of Tabernacle

Does it have any meaning?

nm661 » We figure the size of Moses' Tabernacle in the following way:

■ We know some things like the size of the tent coverings and the length of the boards (ten cubits) used to construct it (Ex 26:1-12). So the measurement may have been ten cubits wide, ten cubits high, and thirty cubits long or **10 x 10 x 30 cubits** in size with the Holy of Holies being 10 x 10 x 10 cubits and the Holy Place being 10 x 10 x 20 cubits. This can be checked by examining David and Solomon's temple which kept the same relative shape of Moses' Tabernacle, but the measurement was increased to twice the width, length, and height of Moses' Tabernacle (1Kings 6:16-20). The Holy of Holies (Oracle) was 20 x 20 x 20 cubits in size, while the Holy Place (house in front of the Oracle) was 20 x 20 x 40 cubits; thus, the all over size was **20 x 20 x 60 cubits** which was twice the size of Moses' Tabernacle.

Without the tent covering, looking down from the top it looked something like this:

Picture through Wikimedia Commons of "Degem Mischan made by Michaael Osnis"

With the tent coverings, Moses' Tabernacle and outer court area looked something like this:

Drawing through Wikimedia Commons by Aleksig6

Tabernacle's Measurement: Higher Meaning

Holy Of Holies' Cubic Measurement and Time

nm662 » One interpretation of the measurement of Moses' Tabernacle is that the cubic size of the Holy of Holies represented the length of Jesus Christ's kingdom on earth, since the Holy of Holies represents Christ since Christ went Spiritually into it after his death (Heb 9:8-11, 23-24). The cubic volume is 10 x 10 x 10 cubits or 1000 cubits, or if cubits here indicate years (one cubit for each year), then Christ's kingdom will last for 1000 years.

Holy Place' Cubic Measurement and Time

nm663 » One interpretation of the measurement of Moses' Tabernacle is that the cubic size of the Holy Place represented the length of time of the Church's age, since the Holy Place represents the Church with its candlesticks and incense of the prayers of the Church or saints (Rev 1:20; 8:3). The cubic volume is 10 x 10 x 20 cubits or 2000 cubits, or if cubits here indicate years (one cubit for each year), then the Church's age will last for 2000 years. The Holy Place was located just before the Holy of Holies, as the age of the Church is located in time just before the presence of God or the Holy of Holies.

Distance or Time between Israel and the Ark of the Covenant

nm664 » The ark of the Covenant held God's presence and glory (1Sam 4:21-22). As physical Israel was going into the promised land, they were commanded to follow at a distance of 2000 cubits behind the Ark (and the priests carrying it):

■ And they commanded the people, saying, "When you see the ark of the covenant of the LORD your God with the Levitical priests carrying it, then you shall set out from your place and go after it. "However, there shall be between you and it a **distance of about 2,000 cubits** by measure. Do not come near it, that you may know the way by which you shall go, for you have not passed this way before." (Joshua 3:3-4)

Using the same logic as the two previous items, if cubits here indicate years, we see here that in the higher meaning, the people of *Spiritual* Israel are 2000 years (cubits) behind Christ who first went in and sat down on the Ark on the right side of God's presence about 2000 years ago (Heb 9:8-11, 23-24). Spiritual Israel or the Church is 2000 years behind Christ (the first-one) on the way to a place they have not passed before (Joshua 3:4b). The place they were going was into the presence of God (1Thes 2:19; Judah 1:24) unlike the others who are punished away from the presence of God (2Thes 1:9; Rev 6:16). At the end of the age, after the wrath and great tribulation, the temple or tabernacle of God will be opened (Rev 11:19; 15:8) and the Spiritual Israel (Church) will be in the presence and glory of God, "Behold, the tabernacle of God is with men, and he will dwell with them ... and God himself shall be with them" (Rev 15:8; 21:3). "For we are the temple of the living God. As God has said: I will live with them and walk among them, and I will be their God, and they will be my people" (2Cor 6:16).

Do the Tabernacle Dimensions Mean Anything?

nm665 » Do the dimensions really mean anything? Wasn't the dimension of Solomon's temple different than Moses' Tabernacle? Wasn't the second temple at Christ's time also different in dimensions? Why should we think there is something important about Moses' Tabernacle?

■ First, Moses Tabernacle was made just as God commanded on Mt Sinai: "Make the tabernacle and all its furnishings exactly like the pattern I will show you" (Ex 25:9).

■ Second, look at what Paul said, "while the first tabernacle was yet standing, which was a type for the present time" (Heb 9:8-9). What this means is that Moses' Tabernacle was an example or type for the present era, the era of God's Spiritual Israel, the Church, his Body.

nm666 » Even though Solomon's temple was built by the pattern given by David with the help of "the Spirit" (1Chron 28:11-12), the direction for Moses' tabernacle was given by God with great emphasis so that Moses would be very careful how he constructed the Tabernacle and its furniture (Ex 25:9, 40; Num 8:4; Heb 8:5). Because Paul said that Moses' Tabernacle was a type for our time, we must give much more weight to the dimensions of Moses' Tabernacle versus Solomon's temple. In Solomon's temple the dimensions of the Holy of Holies and Holy Place were twice that of Moses' Tabernacle (see above).

Other Tabernacle or Temple Measurement Scripture

nm667 » Other scripture also projects that the dimensions of certain key objects also are to give us hints. For example, the wall in Revelation 21:17 was 144,000 cubits according "to the measurement of man, that is, angel." The number 144,000 happens to equal the number of the saints who are harvested at the end of the age (Rev 14:3-4; 7:4). And we know by other scripture, each man has an angel, his own angel (see, "Our Own Angel" above), so this verse points to the 144,000 with their own angel.

nm668 » Another scripture to look at is Revelation 11:1-2, where John was told to measure the temple of God "and them that worship therein," but to leave the outer court out of his measurement. When we study the Bible we see we can count or measure the number of the harvest, that is the 144,000 (Rev 14:3-4), but not the others outside of the symbolic "outer court" (see "Outer Court" above).

Spiritually Keeping The Passover

More Details

nm669 » How then is Spiritual Israel to keep the feast or appointed time of the Passover? "Therefore let us keep the feast, not with old leaven [the doctrines of man], neither with the leaven of malice and wickedness; but with the unleavened bread of sincerity and truth" (1Cor 5:8).

nm670 » When do we keep this feast or this appointed time? Spiritually we keep it always, since we are out of spiritual Egypt and are in one of the seven Spiritual days.

Sacrifice?

nm671 » What about sacrifice? Christ's death was for the new covenant with His Spirit that puts the law of God (love) in the minds of Christians, and their sins are no more remembered, they are forgiven. Therefore there is no more offering for sins by animals, for Christ's sacrifice takes away animal offerings (Heb 10:1-18), which were not pleasant to God anyway (Heb 10:5-6). Christians are living sacrifices (Rom 12:1) who sacrifice "the sacrifice of praise to God continually, that is, the fruit of lips confessing to his name" (Heb 13:15).

Spiritual Bread

nm672 » Do Christians eat unleavened bread for some religious reason? No! Christians do not do this because they Spiritually worship God as God asks us to worship (John 4:23-24). Since the higher or Spiritual meaning of unleavened bread is Truth (1Cor 5:8), and since God's Word is Truth (John 17:17; James 1:18); then Christ's Word is the unleavened bread which we eat. Therefore Christians eat Christ's Spiritual Word for the rest of the Spiritual seven days, they eat Spiritual food (1Cor 10:3) with the Spirit of Truth that comes from the Father through Christ (John 16:13; 15:26; 17:17, 22 etc.).

nm673 » Let's amplify on the eating of Spiritual unleavened bread. "I am the living bread which came down from heaven: if any man eat of this bread, he shall live into the age" (John 6:51). Let's see what context this statement was made in.

nm674 » The day after Jesus performed the miracle of feeding five thousand with a few loaves of bread, when it was near the time of the Passover festival (John 6:4-13), some people came to him (John 6:25-26). At that time Christ said to those who came to him:

- "You seek me, not because you saw the miracles, but because you did eat of the loaves, and were filled. Labor not for the food which perishes, but for the food which endures unto aeonian life, which the Son of man shall give unto you: for him has God the Father sealed" (John 6:26-27).

This was said around the Passover (v. 4), and Christ said to seek the bread that endures into aeonian life – the millennium and beyond.

Bread from Heaven

nm675 » Then he went on and said, "Moses gave you not that bread from heaven; but my Father gives you the true bread from heaven . . . . I am the bread of life: he that comes to me shall absolutely not hunger; and he that believes on me shall absolutely not thirst" (John 6:32, 35). Here Jesus speaks of Spiritual bread, and Spiritual drink.

Water from Heaven

nm676 » There is Spiritual food and drink as the following verses indicate: "And did all eat of the same Spiritual food, and did all drink the same Spiritual drink" (1Cor 10:3, 4). "Jesus answered and said unto her, whosoever drinks of this water that I shall give him shall not into the ages thirst; but the water [Spiritual] that I shall give him shall be in him a well of water springing up into aeonian life" (John 4:13-14). "Jesus stood and cried, saying, If any man thirst let him come unto me, and drink. He that believes on me, as the scripture has said, out of his belly shall flow rivers of living water (But this spoke he of the Spirit, which they that believe on him should receive.)" (John 7:37-39).

Living Bread

nm677 » Christ continued: "I am the living bread which came down from heaven: if any man eat of this bread, he shall live into the age: and the bread that I will give is my flesh, which I will give for the life of the world ... Whoso eats my flesh, and drinks my blood, has aeonian life; and I will raise him up at the last day ... He that eats my flesh, and drinks my blood, dwells in me, and I in him ... It is the Spirit that makes alive; the flesh profits nothing: the words that I have spoken unto you, they are Spirit, and they are life" (John 6:51, 54, 56, 63).

Eating His Flesh

nm678 » Notice that after Christ finished speaking about eating his flesh and drinking his blood, he said the words he had just spoken were Spiritual, and that they were life. He said we must eat his flesh, or his bread, since he said they were the same (v. 51), in order to live into the age of peace. And when we do eat his bread, or flesh, and drink his blood we would be in him

and he in us. Since these words were Spiritual, Christ meant that if we "ate" his Spiritual bread, or if we were nourished by his Spiritual body; then we would dwell in him, and he in us (cf Eph 5:30, 29; 1Cor 12:27, 12-13; Rom 8:9-17; John 1:5; see "Proof of Being a Christian"). When we are in Christ's Spiritual body (the Church) we are nourished by his Spiritual flesh, which is his Spiritual bread, and by his Spiritual blood.

nm679 » "Whoso eats my flesh [bread, v. 51], and drinks my [Spiritual] blood, has aeonian life . . . . I live through the Father [his Father is Spirit, John 4:24]: so he that eats me [his Spiritual bread, or Spiritual body, or Spiritual blood], even he shall live through me" (John 6:54, 57). The only way one can get aeonian life, and to live through Christ, is to be in the Body of Christ, which is the Church. And the only way one can be in the Church is to have the Spirit of God the Father (1Cor 12:12-13). It was by this Spirit that Christ lived; it is by this Spirit that we live. We can only eat his bread, flesh, and blood when we have the Spirit. When we have his Spirit we are "eating" his flesh.

Eat, This is my Body

nm680 » "Then it came [aorist – a verb of action, not time] towards the day of unleavened bread, when the Passover must be killed" (Luke 22:7). Or, "And towards the first day of unleavened bread, when they killed the Passover..." (Mark 14:12). It was the day just before the Passover – the 13th – that he sent his apostles to make ready for the Passover (Luke 22:8-13, see Notes). "And when the hour was come, he sat down, and the twelve apostles with him. And he said unto them, With desire I have desired to eat this Passover with you before I suffer: for I say unto you, I will not eat it, until it be fulfilled in the kingdom of God" (Luke 22:14-16; see the Greek). It was the evening of the 13th when he was eating, and he said he would not eat it, until it was fulfilled in the kingdom of God. Of course he did not eat the Passover that evening because:

- **(1)** it was one day *before* it was to be eaten according to scripture (John 18:28);

- **(2)** he said he would *not* eat it until it was fulfilled in the kingdom of God;

- **(3)** it was fulfilled perfectly only after he died as the antitypical Passover;

- **(4)** the meal of the 13th was not eaten in haste as the Passover was commanded to be eaten (Ex 12:11), for in this evening during this meal Christ stopped and washed all the apostles feet (John 13:1-12; see notes).

nm681 » But right after the evening of the 13th, in the night of the 14th right near or at sunset, Christ "took bread: and when he had given thanks, he broke it, and said, 'take, eat: this is my body, which is broken for you: this do in remembrance of me.' After the same manner also he took the cup, when he had the supper, saying, 'this cup is the new covenant in my blood: this do you, as often as you drink it in remembrance of me.' For as often as you eat this bread, and drink this cup, you do proclaim the Lord's death till he come ... For he that eats and drinks unworthily, eats and drinks judgement to himself, not recognizing the Lord's body" (1Cor 11:23-29).

What Is The Lord's Body?

nm682 » "For as the body is one, and has many members, and all members of that one body, being many, are one body: so also is Christ. For by one Spirit are we all baptized into one body . . .. Now you are the body of Christ, and members in particular . . .. For we are members of his body, of his flesh, and of his bones . . .. for his [Christ's] body's sake, which is the Church" (1Cor 12:12-13, 27, Eph 5:30; Col 1:24). Christ's body is the Church. Christians are members of Christ's body, we are of his flesh, blood, and bones. We are nourished by his body (Eph 5:29). His body, or Church, is Spiritual. We eat and drink his flesh, bread, or blood when we are nourished by it, if we are of his body. And we are only of his body when we are in the Church and have the Spirit of God in us. And if we have the Spirit, then we are of Christ's body – his flesh and blood – and are nourished by it Spiritually.

Communion and Breaking of Bread

nm683 » Therefore "the cup of blessing which we bless, is it not the common sharing of the blood of Christ? The bread which we break, is it not the common sharing of the body of Christ? For we being many are one bread, and one body: for we are all partakers of that one bread" (1Cor 10:16-17). That one bread is Christ – the bread of life (John 6:35, 48, 51). And that bread is Spiritual. We are in communion, or in sharing with Christ's body when we share the Holy Spirit (2Cor 13:14; 1Cor 12:12-13). Hence, we drink his blood and eat his flesh when we are Spiritually in the Church; and we are in

remembrance of Christ, and we do proclaim the death of Christ until he comes because we are Spiritually in his Church; and his Spirit and those in his Spirit project Christ's Spirit and proclaim Christ's death and gospel *always* until he comes (Mat 24:14). Christ was the real sacrificial lamb of God, the real sacrifice for sin (John 1:29; 1Pet 1:18-19). By Spiritually eating his Spiritual Body or Bread, we fulfill the Old Testament scripture where the priests ate parts of the sacrificed animals for a peace offering (Ex 29:32-33; Lev 10:14-15,17).

Unleavened v. Leavened Bread

nm684 » The unleavened bread Christians eat are the doctrines of Christ since the leaven are the doctrines of the Pharisees and of the Sadducees, which are merely the doctrines of mankind, which are hypocrisy (Mat 16:12; Mark 7:5-9; Luke 12:1). The unleavened bread that Christians eat is that truth (1 Cor 5:8). The bread we break is the bread of Christ – the doctrines of Christ with the sharing of his Spirit and the sharing of the Spirit's understanding of Christ's doctrines. The apostles continued after Christ's death "in breaking of bread" (Acts 2:42; 20:11). The Christians "come together to eat" (1Cor 11:33) the doctrines of Christ and share in the Spirit.

When Eating Wait for Each Other

nm685 » Paul in 1Corinthians 11:17-34, was telling them when they came together to eat Christ's bread to tarry "one for another," "for in eating every one takes before the others his own supper: one is hungry, and another is drunken." In other words, when Christians come together they should not go over some of the congregation's heads in the teaching of doctrine. If some are hungry for deeper doctrines on Christ, let them do it at home, and let the Church assembly be for the not so deep doctrines, or not so complex doctrines of Christ. Some doctrines may make some "weak" Christians "drunk." Yet to others the same teaching of these doctrines will not satisfy them because the teachers may not go into even more difficult aspects of the doctrines. Read 1Corinthians 11:17-34 in a Spiritual way, remembering that the bread eaten is the doctrines of Christ.

Break Bread

nm686 » Notice how Christ broke the bread: "And he commanded the multitude to sit down on the grass, and took the five loaves, and two fishes, and looking up to heaven, he blessed, and broke, and gave the loaves to his disciples, and the disciples to the multitude" (Mat 14:19;

15:36; Mark 8:6; etc). This antitypically signifies the Spiritual words originating from Christ who gave, and gives, it to his disciples then, and now, through his physical word and Spiritual word. His disciples gave, and now give, the bread of Christ's doctrines to the people.

Showbread and Candlesticks

nm687 » The bread is typically represented as the "showbread" or as translated – "loaves of the setting before" (see Luke 6:4-5). These loaves of bread were set before the candlesticks of the physical tabernacle (Lev 24:5-6; Ex 40:22-24). The candlesticks represent the Church (Rev 1:20). The olive oil used with these lamps represents the Spirit in the Church (1Sam 16:13; 1John 3:27). Thus, the "showbread" typified the bread of Christ and his doctrines put before the Church.

Olive Oil

nm688 » The oil in the candlesticks was "pure olive oil beaten for light" (Ex 27:20). This typified the oil (Spirit) that was given to the Church because Christ, the True Olive Tree (Rom 11:17-24) was beaten for us. The loaves of bread before the candlesticks were unleavened bread made with olive oil (Ex 29:2-3; Lev 2:4). This oil signifies the only way Christ's unleavened bread (doctrines) can be eaten; they must be eaten with oil of the Spirit, that is, with God's Spirit.

Sacrifices

nm689 » Christians are set apart by Christ's sacrifice (Heb 10:10). The "ram of consecration" is only one of the sacrificed animals that were representative of Christ's sacrifice (Lev 8:22-24; Ex 29:19-22). The eating of this ram of consecration, and the unleavened bread with it, typified the eating of Christ's body and bread (Ex 29:31-33, 2), for we are the Spiritual sons of Aaron because Christ is the antitypical high priest who has Spiritually taken over Aaron's office (Heb chaps. 8, 9, 10).

Law

nm690 » Christ's sacrificed blood is the blood of the new covenant (Luke 22:20). His blood makes it possible for us to receive God's laws into our hearts and minds (Heb 10:12-17). This is the antitypical event corresponding to Moses sprinkling blood on physical Israel when they received the laws of God (Ex 24:5-8). But physical Israel didn't receive the laws of God inside their minds and hearts at that time (Deut 29:4; 30:6 Jer 9:26). It is the Spiritual Israel of God which receives the law of God into their minds through the medium of God's Spirit. The Israel of God is

the Church (Gal 6:16). This new law is what Paul called the "law of God" (Rom 7:22; see "Other Mind Paper" [NM 21]). The old law that was done away with because of Christ's death, was the law of Moses that was only written on stone and not in the minds of man (Eph 2:15). Paul called this old law the "law of sin" (Rom 7:23, 25). The law given by Moses was a law of sin because Israel was without God's Spirit, and thus unable to keep the commandments in Truth (Rom 8:7). Physical Israel only had the other-mind or enemy spirit, which misleads people to the way of sin. This adverse spirit uses the written ten commandments to mislead people into the way or law of sin (Rom 7:8). The other-mind enjoys doing the opposite to the laws of love, and since it is inside the minds of mankind, this law of sin, this other-mind, transmits its desire to the world's people. But since to break the laws of love, or the system of love (see the "Freedom & Law Paper" [NM 17]), brings unhappiness, the person who breaks the laws is harmed thereby, yet feels inside his confused mind a twisted delight in breaking the laws of harmony. This delight comes from the other-mind or other spirit inside of him.

Sacrifice for All

nm691 » As the sacrifices of atonement were made for the Israelites and the stranger (Num 15:28-29), so too was Christ's sacrifice for ALL (1Tim 2:5-6).

Eat Leavened Bread: Be Cut Off

nm692 » As those who ate leavened bread in the typical festival of the Passover were cut off, so too are those in this age who eat leavened bread, or the doctrines of man with only the other-mind to guide them, they are cut off from Christ's sacrifice in this age (Ex 12:19).

Notes

Passover Scriptures

nm693 » The chief priests of the Jews had a meeting wherein they took counsel to kill Christ (John 11:47-53; Luke 22:2).

Because of those out to kill him, Christ no more could walk openly (John 11:54).

The Passover wherein Christ was killed was near, and many had come to Jerusalem for the festival and were wondering among themselves whether Christ would show up at the festival or not. The chief priests had given directions that if any knew where Christ was that they should point him out to them so the priests could take Jesus (John 11:55-57).

9th of NISAN, a Friday

nm694 » Then six days before the Passover (the 9th of Nisan), Christ came to Bethany which is only a few miles from Jerusalem (John 12:1; 11:17).

At that time they had a supper wherein Mary anointed the feet of Jesus, and wiped his feet with her hair (John 12:2-3; Mark 14:3; Mat 26:6-7).

Some of the disciples had indignation inside their minds at this act, and one named Judas Iscariot, said: "Why was not this ointment sold for three hundred pence, and given to the poor?" (Mark 14:4-5; Mat 26:8-9; John 12:4-6)

Then when Jesus understood what was being said, he spoke saying, "let her alone: that of the day of my burial may she keep it" (John 12:7). The Mary that anointed Christ was Mary Magdalene, who later brought this ointment to Christ's tomb on the day of his resurrection along with some spices she and others had bought and prepared the day before Christ's resurrection (Luke 23:56-24:1; Mark 16:1; Mat 28:1) (John 12:7-8; Mark 14:6-9; Mat 26:10-13).

10th of NISAN, a Saturday

nm695 » Now right after this supper, right after sunset, thus on the 10th of Nisan, Judas the betrayer of Christ went to the chief priests and said he would help them take Jesus in the absence of a great crowd. Because of this the priests were glad that Judas would betray Christ, and gave him 30 pieces of silver. Further, they consulted if they shouldn't also put Lazarus to death since many of the Jews believed in Jesus because Christ had previously resurrected Lazarus from the dead (Mark 14:10-11; Mat 26:14-16; Luke 22:3-6; John 12:9-11).

On the next day after Christ came to Bethany, which was the daylight hours of the same 24 hour day that Judas had betrayed Christ to the priests, Christ came into Jerusalem – 10th day of Nisan (John 12:12).

Christ rode on an ass into Jerusalem, and the people cried, "Hosanna; Blessed is he that comes in the name of the Lord [YHWH]." This was the 10th day of the Jews 1st month – Nisan (Mark 11:1-10; Mat 21:1-11; Luke 19:28-40).

In the 10th day of Nisan after he entered into Jerusalem, Jesus went into the temple (Mark 11:11).

Then in the evening just before sunset, Christ went into Bethany, a town a few miles from Jerusalem (Mark 11:11).

First week with two Sabbaths

11th of NISAN, a Sunday

nm696 » Now in the morrow (the next day, that is the 11th of Nisan), Christ came from Bethany back into Jerusalem (Mark 11:12-15).

It was the 11th, and again Christ goes into the temple in Jerusalem (Mark 11:15; Luke 19:45; Mat 21:12).

At this time Christ put the money changers out of the temple (Mat 21:12-16; Mark 11:15-18; Luke 19:45-46).

During this time period just before the Passover, Christ was teaching daily in the temple (Luke 19:47).

Then in the evening of the 11th he went out of Jerusalem again and went into Bethany (Mark 11:19; Mat 21:17).

12th of NISAN, a Monday

nm697 » After he stayed in Bethany, he came back into Jerusalem on the next day, the 12th (Mark 11:20; Mat 21:18).

When they returned into Jerusalem, the disciples with Christ noticed the tree Christ cursed the previous day (Mark 11:13-14), and how it already had dried up (Mark 11:21; Mat 21:19-20).

At this time on the 12th Christ entered again into the temple (Mat 21:23; Mark 11:27; Luke 20:1).

At this time on the 12th, Christ taught various parables (Mat 21:23-23:39; Mark 11:27-12:44; Luke 20:1-21:4).

Then Christ went out of the temple, and taught his disciples on the mount of Olives about the time of the end of the age (Mark 13:1-33; Luke 21:5-36: Mat 24:1-25:46).

At that time on the 12th of Nisan, Christ noted that after two days would be the feast of unleavened bread, which some call the Passover festival (Luke 22:1) (Mat 26:1-2; Mark 14:1).

At this time Christ mentioned that he is betrayed (for on the 10th remember Judas went to the chief priests to betray Jesus) (Mat 26:2).

The chief priests had decided at that time that Christ shouldn't be taken on the feast day (the 15th, Num 28:17), because there might be an uproar among the people (Mat 26:3-5; Mark 14:1-2).

But remember on the 10th Judas had come to the chief priests, and said he would betray Jesus (John 12:1-11; Luke 22:3-6; Mark 14:3-11; Mat 26:6-16).

After Christ had returned from the temple on the 12th, after he taught many parables (see above), and after he on Mount Olives had spoken of the end of the age that late evening of the 12th (or early on the 13th after sunset), he then stayed on Mount Olives in Bethany (note Luke 24:50 with Acts 1:12) (Luke 21:37).

13th of NISAN, a Tuesday

nm698 » The next day on the 13th, Jesus taught in the temple again after he abode in Bethany the night of the 13th, for at this time Jesus was teaching daily in the temple (Luke 19:47; Luke 21:38).

Before the feast of the Passover, on the evening of the 13th, after Christ had taught during the daylight of the 13th in the temple, he was again in Bethany, and was eating his supper, as he had been doing each evening since he had began teaching in the temple on the 10th (John 13:1).

The home he was staying in was that of Mary Magdalene, Martha, and Lazarus (John 12:1).

Jesus on the 13th had instructed the apostles to get a room and make it ready for the Passover meal, which was to occur on the 14th (Mat 26:17-19; Mark 14:12-16; Luke 22:7-13).

[[Let's correct a few verses that were mistranslated in many English translations of the Bible. These corrections were made using a Greek text. Mat 26:17 should read: "now *towards* the first [day] of unleavened [bread] approaches the disciples to Jesus ... " And Mark 14:12 should read: "and *towards* the first day of the unleavened [bread], when they kill the Passover, his disciples say to him . . .." And Luke 22:7 should read: "now it came *towards* the day of unleavened [bread], in which was needful to be killed the Passover." Therefore what these verses are saying is that *towards* or near the 14th day when the Passover was to be killed, the disciples had asked Christ where they would eat the Passover the next day. Now in Matthew 26:19, Mark 14:16, and Luke 22:13 the Greek verbal word translated "they made ready" is an aorist word that indicates a verbal action

without indicating the time of the action. Thus, it can mean action in the past, present, or future. According to the context of this verse this Greek word should have been rendered in the following manner: "were to make ready" the Passover in the certain house where Christ said to prepare it. Christ had ordered them to prepare for the Passover, but the events surrounding Christ's betrayal made it impossible to go and eat the Passover.]]

The evening of the 13th came and Christ was in Bethany at supper with his disciples. "And supper taking place..." (John 13:2; Mat 26: 20-21; Mark 14:17-18; Luke 22:14).

"Washing Feet" and Principle of Serving

nm699 » Now at this supper taking place on the evening of the 13th just before sunset, "the devil already having put into the heart of Judas" to deliver Christ up, Jesus "rises from supper, and laid aside his garments; and took a towel, and wrapped himself" (John 13:2-4). It was on the 10th of Nisan that Judas had gone to the chief priests, and now Judas was at this meal on the 13th. Christ thus rose from the meal and began to wash the disciples feet, and did wash their feet (John 13:4-11).

nm700 » Then Jesus sat down again to supper (John 13:12). After he sat down Christ told the disciples what he had just done, "you call me teacher and the Lord: and you say well; for so I am. If then, your Lord and teacher have washed your feet; you also ought to wash one another's feet. For I have given you an example, that you should do as I have done to you. Truly, truly, I say unto you, the servant is not greater than his Lord; neither he that is sent greater than he that sent him. If you know these things, happy are you if you do them" (John 13:12-17).

nm701 » Some who only look to the physical meaning of Christ's words have gotten the meaning from this that we should perform some kind of ritual on the Passover that has to do with washing others' feet. Somehow they think that evening that Christ instructed them in a ritual to be performed by Christians. But we are to worship God Spiritually (John 4:23-24), not with rituals.

nm702 » Christ was speaking of a principle of serving one another. He was saying if the Lord serves, then his disciples should also serve others, for the disciple is not above the Lord. While they were eating that evening some had a strife among themselves as to who was the greatest (Luke 22:24). Christ's answer to this strife was, "and he said unto them, The kings of the Gentiles exercise lordship over them; and they that exercise authority upon them are called benefactors. But you shall not be so: but he that is greatest among you, let him be as the younger; and he that is chief, as he that does serve. For whether is greater, he that sits at meat, or he that serves? Is not he that sits at meat? *But* I am among you as he that serves" (Luke 22:25-27).

nm703 » Jesus was saying that the kind of lordship exercised by man is not the kind of lordship he himself was performing. Christ who is to be King of kings *serves*, and so are the apostles to do so. The leaders of the world should serve the peoples' needs, but they do not, they oppress. Christ showed the way to lead and lord over people – He served them.

- "And whosoever will be chief among you, let him be your servant: Even as the Son of man came not to be ministered unto, but to minister, and to give his life a ransom for many" (Mat 20:27-28).

- "But he that is greatest among you shall be your servant. And whosoever shall exalt himself shall be abased; and he that shall humble himself shall be exalted" (Mat 23:11-12).

- "If any man desire to be first, the same shall be last of all, and servant of all" (Mark 9:35).

nm704 » Christ time after time taught the disciples that the greatest thing was to serve others (Luke 9:46-48; Mark 10:42-44). And again on the evening of the 13th he again by using the example of washing their feet, said that the greatest thing was to serve, not to lord over others. "It is more blessed to give then to receive" (Acts 20:35). The washing of the feet was a reiteration of the principle of giving and serving. We must try our best to follow this principle always, not just on one day of the year.

nm705 » During this meal Jesus broke the bread, and passed it around, and said he could "not eat it," the Passover, with them until it was fulfilled in the kingdom of God (Luke 22:15-20; Mark 14:22-24; Mat 26:26-29).

nm706 » And during this meal, Christ revealed who would deliver him up that night (14th, after sunset; after the supper they were eating on the 13th before sunset) (Mat 26:21-25; Mark 14:18-21; Luke 22:21-23; John 13:21-29).

14th of NISAN, a Wednesday

nm707 » **(1)** It was Judas Iscariot, and right after he took the piece of bread, which pointed him out as the betrayer (yet the apostles didn't understand), Judas immediately went out, "and it was night." That is, right after sunset Judas went out to bring the chief priests to take Jesus (John 13:30). They were in Bethany, which is on the side of Mount Olives, when they were eating this meal (see above).

(2) After they sang a hymn, they went onto Mount Olives, and brought two swords with them so scripture could be fulfilled (Mark 14:26; Mat 26:30; Luke 22:35-39).

(3) On Mount Olives Christ speaks of various matters to the apostles. (The scripture is vague as to whether these things were spoke still in the house in Bethany, or near the house, or somewhere on the mount of Olives.) (Mark 14:26-31; Mat 26:31-35; John 13:31-17:26)

(4) At Gethsemane (probably on the mount of Olives) he enters into a garden to pray (Mat 26:36-46; Mark 14:32-42; Luke 22:40-46; John 18:1).

(5) Judas knew where this garden was, for Jesus came often to it to pray (John 18:2).

(6) It was in this garden that Judas came with the chief priests and Pharisees, who came with lanterns and torches because it was at night on the 14th after sunset in the night (John 18:3; Luke 22:47; Mark 14:43; Mat 26:47).

(7) At this time Judas revealed Christ by greeting him with a kiss. Peter cut off an ear of a guard, but Christ healed the ear. Then *all* the disciples "forsook him, and fled" (Mat 26:47-56; Mark 14:43-52; Luke 22:47-53; John 18:4-11).

(8) Then the band of men with the chief priests bound Christ and brought him, and led him away to Annas *first*, for he was the father-in-law to Caiaphas, who was the high priest that same year (John 18:12-13).

(9) Then they took Christ to Caiaphas the high priest (John 18:24).

(10) During that time Peter denied Christ three times as Jesus foretold (John 18:15-27; Luke 22:54-65; Mark 14:66-72; Mat 26:69-75).

(11) At dawn of the new day light (Mat 27:1; Mark 15:1; Luke 22:66; John 18:28), the chief priests and the elders came to take Christ, and right after they had him before the high priest, they brought him to the judgement hall to Pilate (John 18:28-29; Luke 22:66-71; 23:1; Mark 15:1; Mat 27:1-2). Pilate sent Jesus to Herod Antipas because "he knew that he was of Herod's jurisdiction" (Luke

23:6-7). But Herod after mocking Jesus sent him back to Pilate (Luke 23:8-12).

(12) Then Jesus was tried and sent to be crucified (John 18:29-19:16; Luke 23:2-25; Mark 15:2-20; Mat 27:2-31).

(13) During that 24 hour day Judas the Betrayer killed himself (Mat 27:3-10; Acts 1:16-20).

(14) Then they crucified Christ (Mat 27:32-56; Mark 15:21-41; Luke 23:26-49; John 19:16-37).

(15) The Passover was prepared and killed on the 14th of Nisan, and also the 14th was a day to prepare for the 15th, which was an annual Sabbath wherein no work was to be done. Thus, because of a law that a body could not hang or remain on a tree (stake or wood cross) during the night, but must be taken down the very same day (Lev 21:23; cf Josh 8:29; 10:26-27). "The Jews therefore ... besought Pilate ... that he might be taken away" (John 19:31).

(16) Therefore Christ was quickly buried just before sunset, just before the annual Sabbath of the 15th (John 19:40-42; Luke 23:53-55; Mark 15:42; Mat 27:57-61).

(17) Right after the burial "they returned" to their houses (Luke 23:56, 1st part of verse).

15th of NISAN, a Thursday, 1ˢᵗ _Sabbath_ of a 7 day week

nm708 » This was day of the annual Sabbath, a "high day" or festival Sabbath for Jews (John 19:31; Mark 15:42; Luke 23:54; see text under "Passover").

16th of NISAN, a Friday

nm709 » After this one annual Sabbath (the 15th) some women bought spices, and they prepared these spices and the ointments that Mary Magdalene had saved (John 12:7). (Mark 16:1; Luke 23:56, middle part of verse).

17th of NISAN, a Saturday, Second _Sabbath_

nm710 » Jesus was then resurrected on the last of the evening, just before or at sunset, on the 7th day Sabbath. He laid in the tomb from late Wednesday evening to late Saturday evening, three days and three nights as he said he would in Matthew (Matt 28:2-4; 12:38-40; Matthew 27:63; and Mark 9:31)

Note: The resurrection of the dead saints occurred with the resurrection of Christ at the end of the 7th day Sabbath when the earth-quake happened. (Matt 27:51-54)

NOTE: See the _Chronology Papers_ for more information on Christ's death and resurrection.

Second week with two Sabbaths

18th of Nisan, a Sunday

Mat 28:1 "Now late in the week, as it began to dawn toward the first [day] of the [next] week, Jesus ascends to His Father sometime after sunset after greeting the women. In four days Jesus fulfills Genesis 4th day of creation: Gen 1:14-19.

19th of Nisan, a Monday, 2nd day of week

20th of Nisan, a Tuesday, 3rd day of week

21st of Nisan, a Wednesday, 4th day of week, **an annual Sabbath**, 1st Sabbath of the week

22nd of Nisan, a Thursday, 5th day of week,

23rd of Nisan, a Friday, 5th day of week

24th of Nisan, a Saturday, 7th day Sabbath, 2nd **Sabbath** of the week

NM 17: Freedom And Law

NM17 Abstract

In this paper we are going to learn what the Bible says about law and freedom. Are Christians under the law or are they under freedom? What happened at the council of Jerusalem concerning this matter? What is the new commandment? And what kind of love did Christ teach?

Freedom and Law: Definition of

The definition of Freedom is

nm711 » Freedom is the lack of restriction. Freedom to do anything, including evil, is bad because the harm connected with such freedom can destroy life. Forms of freedom that do not destroy or diminish life are good. Therefore freedom can either be good or bad. The system of love that we will explain later in this paper is a form of freedom that does no ill towards others (Rom 13:10).

The definition of Law is

nm712 » Law is restriction of activity and movement; law is regulated order. Law restricts random or chaotic movement and dictates movement that is orderly. Law can either be good or bad. Law made by men must be in harmony with the true nature of mankind or it is bad law. Good law is not self-contradictory. Too much law makes life less spontaneous and less individualistic. If the abundance of law is against the nature of mankind, it makes life less enjoyable. There is a law of love or system of love that brings harmony which we will explain later in this paper.

No Total Freedom; Life has Structure

nm713 » Total freedom in the cosmos would mean there would be no structure in the cosmos: therefore there is no total freedom in the cosmos because the cosmos has order. All life has structure; life is the order of elements. Our bodies have structure. Our bodies are forms of life. Our bodies have law. Certain parts of our body must work in certain ways. A heart must beat (pump blood); it cannot breathe or fly. Our heart functions in a certain way. It has order. It has structure. It has law. A heart by itself would be meaningless. But a heart with a body has meaning. It has purpose. It's purpose is to keep the blood moving through the body and thus nurturing the body and keeping it alive.

nm714 » Not only is there a law of our body, there is also a law for the interaction within our society. Human beings can die. We can be harmed by each other either on purpose or by accident. We can harm each other in degrees, from hurting someone's feelings, to breaking someone's arm, to killing someone. Because some human actions can harm ourselves and others, there is law against them. Governments make law to order the behavior of people. Good law is law that keeps people functioning together towards harmony and life. Bad law is law that causes disharmony and death. Since God created the universe and everything in it, then God knows the law of the universe and consequently also knows freedom. This is why we look at Biblical freedom and Biblical law.

Law in the Bible

nm715 » As indicated in the "Other Mind Paper"[NM 21], the first law uttered to man was, "but of the tree of the knowledge of good and evil, you shall not eat of it" (Gen 2:17). God didn't give man a bookcase of law books. He gave man one law. But mankind broke that law when Eve took and ate and gave it to Adam who also ate from the tree. From this one act of law-breaking, sin entered the world. Sin is lawlessness (1John 3:4). Sin is action or behavior *against* law. The law in the garden of Eden was that man shall not eat from the tree of knowledge of good and evil; mankind broke that law; sin entered the world.

nm716 » Now, just because God orally only gave this one law, does not mean that God was saying that killing others was okay, or that lying was okay. God articulated that one law to them. It was a very simple law. Mankind could eat from all the trees in the garden except the tree of knowledge of good and evil. But it was from this one simple law that the evil in man's mind was manifested. As shown in the "Other Mind Paper" [NM 21], this evil is the enemy mind or spirit that dwells and works inside the mind of mankind in this, the old age. This Other Mind is an evil mind that actually enjoys evil and law-breaking. From the first sin, the "law of sin" was shown to be in the world. The "law of sin" is an order or system of law-breaking. It is a system of going against the way or order that God created. God created everything to work in a certain way. When you do things the right way, you are doing it the way God intended it to be done. Doing things God's way is doing it the right way. When one buys a car he follows the manufacture's instructions on how to operate it and keep it running in good shape. If the car was made to operate through the use of the fuel we call gasoline, we use gasoline, because if we do not we would harm our car engine. But the "law of sin" is illogical and confused. The "law of sin" acts like a person who puts a different fuel in the car's tank, just for the thrill of destroying the car or of breaking the law of the car.

nm717 » An example of the law of sin at work is homosexuality. It is confusion for males to perform sexual-like acts with males; it is confusion for females to perform sexual-like acts with females. Males were created to fit and complement females; Females were created to fit and complement males. But the law of sin in some people actually makes them desire their own sex. This is not only confusion, it is against the natural order that was created by God and a form of spiritual foolishness and rebellion. "He that created them, in the beginning made them male and female" (Mat 19:4).

nm718 » Man by breaking the one simple law shown in Genesis 2:17 manifested the law of sin in its thinking. But the only law that is True law is the law that is in harmony with God's laws. Today at least some, if not much more than some, of the law in the law books of governments is corrupt and unlawful, mostly because the law makers do not understand or know God the creator.

Use of "Law" in the Bible

nm719 » The word "law" is used in different ways in the Bible. "Law" in the Bible may refer to all Old Testament scripture as a whole (John 10:34) , or to the five books of Moses (Gal 3:17-21), or sometimes only to the tables of stone written by the finger of God (Deut 24:12). The New Testament also talks about the "law of sin" (Rom 7:23), the law of the Gentile nations, etc. Therefore when studying "law" in the Bible one must be sure to know what law the scripture is speaking about. Careful attention must be given to context.

Law of Mount Sinai and the Promises

nm720 » Now, the main body of law that was given in the Bible was the body of law given to the nation of Israel from mount Sinai (Exodus chap. 19 ff; Deut 5 ff). This law not only had the ten commandments (Ex chap. 20; Deut 5:7-21) in them but also other laws and judgments (Ex chap. 21, 22, 23; Deut 12ff), and religious ordinances concerning offerings, the ark of testimony, the golden candlestick, the tabernacle, garments for the priests, sacrifices, the altars, atonement, the Sabbath as a sign, the holy feasts, etc (Ex 25-31 ff; Book of Leviticus; and the book of Deuteronomy). Before and after these laws were given, the people of Israel promised to follow them all: "*All* the words that the LORD has said we will do" (Ex 19:8; Deut 5:27). Israel gave its promise that they would do all these commandments of God (Exo 24:3, 7-8). God had promised Israel certain things (Ex 19:5-6; Deut chap 28) if they obeyed his voice. If Israel didn't follow all these laws they were cursed (Deut 11:26ff; 12:1; 26:16-18; 27:1, 10, 26; 27:26; 28:1, 15; 28:15ff; 30:15-20). But if they followed *all* these laws they would be blessed (Deut chap. 28).

Only Jesus Obeyed All the Laws and Received the Promises

nm721 » There has only been one Israelite who has followed all these laws of God. It is through this one, Jesus Christ, that all of mankind will be blessed by the promises of God (Rom 5:18-21). Jesus Christ is the one "to whom the promise was made" (Gal 3:19). This "promise" not only was the promise to Abraham, but the promises promised if Israel obeyed all of the law (Deut 28:1ff). The only way the people of Israel could be saved or could receive the True promises of God was for them to follow *all* the laws of God perfectly. This proved impossible for the people of Israel, and the curses for disobedience came on the head of the Israelites. It was impossible for them because they had the other-mind misleading them to destruction (see "Other Mind Paper" [NM 21]; Rom 8:7). It is only through Christ and his Spirit or mind of God that people can be saved or receive the promises of God (see Acts 4:12; John 14:6 & read all of this book). Before Christ, the only way people could be saved and receive the True promise of God was for them to belong to physical Israel and to follow *all* the laws of God (Gal 3:10; James 2:10). Those outside of Israel, the Gentiles, could not receive the promises of God (Eph 2:11-12). But now this condition has been abolished. Now even non-Israelites (Eph 2:11-12; Gal 3:13-14; Acts 4:12; Heb 11:39, 13; Rom 2:12-16; 3:9; 5:12, 18; 9:4-5) can receive all the promises first given to physical Israel (Eph 2:11-13).

We are free from the <u>letter</u> of the law to follow the <u>Spirit</u> of the law.

nm722 » What some believe about keeping the Old Testament law and about Grace depends on how they have been taught by their physical teachers. These teachings have become a part of their brain cells (See "Mindset Paper" in the Preface). Once you have these teachings imprinted on your brains it is difficult and sometimes painful to correct them. Although there are plenty of scriptures that say we are no longer under the letter of the Old Testament law, in order to be saved, nevertheless, some of us have been taught, at least in part, that the Old Testament law has *not* been abolished since Christ said he did not come to abolish it. With this in mind, lets review some of these scriptures to see what Paul meant when he said we are not under the law. Remember Paul was taught Spiritually and by the authority of Christ (Eph 3:1-11; 1Tim 1:11; Titus 1:3; 1Thes 4:15; 1Cor 15:3; Gal 1:11, 12). Christ called and changed Paul, a legalistic Jew, to start to teach to the Spiritual Church about the new Spiritual law of grace and love (Acts chap 9). The night of Christ's death, when he was speaking about the future (John 13:1-3; 15:26; etc.), he gave his Jewish followers his new commandment; he also called the law of Moses, "their law" (John 15:25), not the apostles' law. Here he was making the distinction between the physical Jew and the Spiritual Jew – the Jew led by the letter of the law versus the Jew led by the Spirit of the law

- **(1)** "Do not think I have come to abolish the Law or the Prophets; I have not come to abolish them but to **fulfill** [Strong's # 4137] them" (Mat 5:17). Christ did fulfill the law, he kept all of it, for he was sinless (John 8:46; 1Pet 1:19); since Christ was also a Spiritual Jew, he **fulfilled** [# 4138 from 4137] the law Spiritually through the Spiritual law of love (Rom 13:10); the law was for the physical-Israelite nation (Ex 19:1, 8; 20:1 ff), not the Gentiles (Rom 2:14). "What then, was the purpose of the law? It was added because of transgressions until the Seed to whom the promises referred had come" (Gal 3:19). "For we maintain that a man is justified by faith apart from observing the law. Is God the God of Jews only? Is he not the God of Gentiles too? Yes, of Gentiles too, since there is only one God, who will justify the circumcised by faith and the uncircumcised through the same faith. Do we, then, void the law, no we establish the law" (Rom 3:28-31).

- **(2)** "Before the Faith came, we were held prisoners by the law, locked up until Faith should be revealed. So the law was put in charge to lead us to Christ that we might be justified by Faith. Now that Faith has come, we are no longer under the supervision of the law" (Gal 3:23-25). This "Faith" is faith of the Spirit of God: the "Spirit of faith" (2Cor 4:13); "Faith by the same Spirit"(1Cor 12:9) "But if you are led by the Spirit, you are not under law" (Gal 5:18).

- **(3)** "What I am saying is that as long as the heir is a child, he is no different from a slave [he must follow the father's rules], although he owns the whole estate [through inheritance]. He is subject to guardians and trustees [of the inheritance law] until the time set by his father. So also, when we were children [of the evil world], we were in slavery under the basic principles of the world. But when the time had fully come, God sent his Son, born of a woman, born under law, to redeem those under law, that we might receive the full rights of sons. Because you are sons, God sent the Spirit of his Son into our hearts, the Spirit who calls out, 'Abba, Father.' So you are no longer a slave, but a son; and since you are a son, God has made you also an heir" (Gal 4:1-7).

- **(4)** "For through the law I died to the law so that I might live for God" (Gal 2:19).

- **(5)** "By abolishing in His flesh the law with its commandments and regulations" (Eph 2:15).

■ **(6)** "Having canceled the written code, with its regulations, that was against us and that stood opposed to us; He took it away, nailing it to the cross" (Col 2:14).

■ **(7)** "Now if the ministry that brought death, which was engraved in letters on stone, came with glory, so that the Israelites could not look steadily at the face of Moses because of its glory, fading though it was, will not the ministry of the Spirit be even more glorious? If the ministry that condemns men is glorious, how much more glorious is the ministry that brings righteousness! For what was glorious has no glory now in comparison with the surpassing glory. And if what was fading away came with glory, how much greater is the glory of that which lasts!

Therefore, since we have such a hope, we are very bold. We are not like Moses, who would put a veil over his face to keep the Israelites from gazing at it while the radiance was fading away. But their minds were made dull, for to this day the same veil remains when the old covenant is read. It has not been removed, because only in Christ is it taken away. Even to this day when Moses is read, a veil covers their hearts. But whenever anyone turns to the Lord, the veil is taken away. Now the Lord is the Spirit, and **where the Spirit of the Lord is, there is freedom**. And we, who with unveiled faces all reflect the Lord's glory, are being transformed into his likeness with ever-increasing glory, which comes from the Lord, who is the Spirit" (2Cor 3:7-18).

■ **(8)** "But thanks be to God that, though you used to be slaves to sin, you wholeheartedly obeyed the form of teaching to which you were entrusted. You have been set free from sin and have become slaves to righteousness" (Rom 6:17-18).

■ **(9)** "For example, by law a married woman is bound to her husband as long as he is alive, but if her husband dies, she is released from the law of marriage. So then, if she marries another man while her husband is still alive, she is called an adulteress. But if her husband dies, she is released from that law and is not an adulteress, even though she marries another man.

So, my brothers, **you also died to the law through the body of Christ**, that you might belong to another, to him who was raised from the dead, in order that we might bear fruit to God. For when we were controlled by the sinful nature, the sinful passions aroused by the law were at work in our bodies, so that we bore fruit for death. **But now, by dying to what once bound us, we have been released from the law so that we serve in the new way of the spirit, and not in the old way of the written code**" (Rom 7:2-6).

■ **(10)** "Christ is the end of the law so that there may be righteousness for everyone who believes" (Rom 10:4).

■ **(11)** "For I through the law am dead to the law, that I might live until God" (Gal 2:19).

■ **(12)** "For you are not under the law, but under grace" (Rom 6:14).

■ **(13)** "The law and the Prophets were proclaimed *until* John. Since that time, the good news of the kingdom of God is being preached, and everyone is forcing his way into it" (Luke 16:16; note Gal 3:19; 4:4-5).

■ **(14)** Now you, if you call yourself a Jew; if you rely on the law and brag about your relationship to God... A man is not a Jew if he is only one outwardly, nor is circumcision merely outward and physical. No, a man is a Jew if he is one inwardly; and circumcision is circumcision of the heart, by the Spirit, not by the written code" (Rom 2:17, 28-29; Note: there is a physical Jew and there is a Spiritual Jew [Rom 2:29; Rev 2:9; 3:9; Deut 30:6]; Christ was a physical and Spiritual Jew [Gal 4:4; Luke 2:40; Isa 42:1; 11:2]).

■ **(15)** "So I say, live by the Spirit, and you will not gratify the desires of the sinful nature.... But if you are led by the Spirit, you are not under law (Gal 5:16, 18).

■ **(17)** "He has made us competent as ministers of a new covenant – not of the letter but of the Spirit; for the letter kills, but the Spirit gives life" (2Cor 3:6).

■ **(18)** "In order that the righteous requirements of the law might be fully met in us, who do not live according to the sinful nature but according to the Spirit" (Rom 8:4).

■ **(19)** "The time is coming, declares the LORD, when I will make a new covenant with the house of Israel and with the house of Judah.... This covenant I will make with the house of Israel after that time, declares the LORD. I will put my law in their minds and write it on their hearts. I will be their God, and they will be my people" (Jer 31:31).

■ **(20)** "But now, we are delivered from the law, that being dead in which we were held, so that we may serve in newness of spirit, and not in oldness of letter" (Rom 7:6).

Council at Jerusalem

nm723 » In the fifteenth chapter of Acts we see more proof that the old testament law was done away with:

Some men came down from Judea to Antioch and were teaching the brothers: "Unless you are circumcised according to the custom taught by Moses, you cannot be saved." This brought Paul and Barnabas into sharp dispute and debate with them. So Paul and Barnabas were appointed, along with some other believers, to go up to Jerusalem to see the apostles and elders about this question. The church sent them on their way, and as they traveled through Phoenicia and Samaria, they told how the Gentiles had been converted. This news made all the brothers very glad. When they came to Jerusalem, they were welcomed by the church and the apostles and elders, to whom they reported everything God had done through them.

Then some of the believers who belonged to the party of the Pharisees stood up and said, "***The Gentiles must be circumcised and required to obey the law of Moses***."

The apostles and elders met to consider this question. After much discussion, Peter got up and addressed them:

"Brothers, you know that some time ago God made a choice among you that the Gentiles might hear from my lips the message of the gospel and believe. God, who knows the heart, showed that he accepted them by giving the Holy Spirit to them, just as he did to us. He made no distinction between us and them, for he purified their hearts by faith. Now then, why do

you try to test God by putting on the necks of the disciples a yoke that neither we nor our fathers have been able to bear? No! **We believe it is through the grace of our Lord Jesus that we are saved, just as they are.**"

The whole assembly became silent as they listened to Barnabas and Paul telling about the miraculous signs and wonders God had done among the Gentiles through them. When they finished, James spoke up:

"Brothers, listen to me. Simon [Peter] has described to us how God at first showed his concern by taking from the Gentiles a people for himself. The words of the prophets are in agreement with this, as it is written:

After this I will return and rebuild David's fallen tent. Its ruins I will rebuild, and I will restore it, that the remnant of men may seek the Lord, and all the Gentiles who bear my name, says the Lord, who does these things that have been known for ages.

It is my judgment, therefore, that we should not make it difficult for the Gentiles who are turning to God. Instead we should write to them, telling them to abstain from the pollution of idols, from fornication, from things strangled, and from blood. For Moses has been preached in every city from the earliest times and is read in the synagogues on every Sabbath."
[That is, Moses' law has been taught for generations, but no one truly followed it – except Christ who died because of it.]

Then the apostles and elders, with the whole church, decided to choose some of their own men and send them to Antioch with Paul and Barnabas. They chose Judas (called Barnabas) and Silas, two men who were leaders among the brothers. With them they sent the following letter:

The apostles and elders, your brothers, To the Gentile believers in Antioch, Syria and Cilicia: Greetings. We have heard that some went out from us without our authorization and disturbed you, troubling your minds by what they said. So we all agreed to choose some men and send them to you with our dear friends Barnabas and Paul – men who have risked their lives for the name of our Lord Jesus Christ. Therefore we are sending Judas and Silas to confirm by word of mouth what we are writing. It seemed good to the Holy Spirit and to us not to burden you with anything beyond the following requirements: You are to abstain from things sacrificed to idols, from blood, from things strangled, and from fornication. You will do well to avoid these things. Farewell.

The men were sent off and went down to Antioch, where they gathered the church together and delivered the letter. The people read it and were glad for its encouraging message. Judas and Silas, who themselves were prophets, said much to encourage and strengthen the brothers. After spending some time there, they were sent off by the brothers with the blessing of peace to return to those who had sent them. But Paul and Barnabas remained in Antioch, where they and many others taught and preached the word of the Lord (Act 15:1-35).

nm724 » Notice that not only was the contention here concerning physical circumcision, but the contention of some that the Gentiles are **"required to obey the law of Moses"** (Acts 15:5). The answer to the argument was that the Gentiles did not have to follow circumcision and other aspects of the law (circumcision was part of the Old Testament law [Lev 12:3]), but "you are to abstain from things sacrificed to idols, from blood, and from things strangled, and from fornication."

What Does It Mean That Christians Are Not Under The Law?

nm725 » "What then, shall we sin, because we are not under the law, but under grace?" (Rom 6:15) Paul asked this question, but he also answered it – no ("God forbid" – KJV). Christians were not taken out from under the

requirements of the law so they could sin. Christians were given liberty from a law that could not by itself bring salvation or justification: "Through Him [Christ] everyone who believes is justified from everything you could not be justified from by the law of Moses" (Acts 13:39). With the law of Moses, one had to follow perfectly *all* the commandments and orders (James 2:10; Gal 3:10). This law of Moses was prescribed through angels (Acts 7:38, 53; Gal 3:19; Heb 2:2). But we are no longer in subjection to angels, but to Christ (Heb 2:5ff). For with Christ and His Faith – the *Spirit* of Faith – you are free from the law of Moses. Be careful here. We are not free to sin. Do read on to see what is meant.

Liberty

nm726 » There is liberty in Christ: "our liberty which we have in Christ Jesus" (Gal 2:4). We are in liberty because "Christ has made us free" (Gal 5:1). "Where the Spirit of the Lord is, there is liberty" (2Cor 3:17). This *liberty* from the Spirit of Christ is actually a "law of liberty" (James 1:25; 2:12) or a system of liberty.

nm727 » This liberty is not the liberty to sin. "For brethren, you have been called unto liberty; only use not the liberty for an occasion to the flesh, but by love serve one another" (Gal 5:13). This law of liberty has nothing to do with the freedom to sin. Sure when you have the Spirit of God you do not have to follow the letter of all the Old Testament laws in order to be saved. (You are given the Spirit and you now follow the Spirit of the law.) But this does not make it right for you to kill, or for you to lie. Just because a Christian was released from the requirements of the Old Testament laws, does not mean the Old Testament laws were bad. As mentioned above the Old Testament laws had a curse for those who did not follow them *all*. In the truest sense, if the Jews didn't follow *all* the law, they were under a curse: "All who rely on observing the law are under a curse, for it is written: 'Cursed is everyone who does not continue to do everything written in the book of the law.'" (Gal 3:10)

nm728 » And one of the curses was death: "But it shall come to pass, if you will not listen unto the voice of the LORD thy God, to observe to do *all* his commandments and his statutes which I command you this day; that all these curses shall come upon you, and overtake you ... The LORD shall send upon you cursing, vexation, and rebuke, in all that you set your hand to do, until you be destroyed, and until you perish quickly" (Deut 28:15, 20).

nm729 » So the Old Testament commandments which were to be for life and happiness if Israel followed them all (Deut 28:1ff), were instead to death because Israel did not keep its word – they transgressed the written laws. Thus Paul said of this factor, "And the commandment, which was for life, I found for death [the curse of not following the law]. For sin, taking occasion by the commandment, deceived me, and by it killed me. So that the law is holy, and the commandment holy, and just, and good" (Rom 7:10-12).

nm730 » The commandments of the Old Testament are good, but the evil or the sin in man (the other-mind) took the good laws and used them to deceive mankind. The other-mind is an evil power that gets thrills from breaking laws (see the "Other Mind Paper" [NM 21]). The evil power took something that was good, the law, and used it to kill mankind by deceiving mankind into thinking that breaking these laws was fun. So the liberty of Christ is not the liberty to sin, "you have been called unto liberty; only use not liberty for an occasion to the flesh, *but by love serve one another*. For all the law is fulfilled in one word, even in this: You shall love your neighbor as yourself" (Gal 5:13-14). The freedom of Christians is the freedom from the written requirements of the old Testament. But these written requirements have been replaced with a Spirit – the very Spirit of love with the New Commandment of love.

New Law and Spirit Prophesied

nm731 » "And I scattered them among the heathen, and they were dispersed through the countries: according to their way and according to their doings I judged them" (Ezek 36:19). BUT, "I will take you from among the heathen, and gather you out of all countries, and will bring you into your own land ... . *A new heart also will I give you, and a new spirit will I put within you: and I will take away the stony heart out of your flesh, and I will give you a heart of flesh. And I will put my spirit within you, and cause you to walk in my statutes, and you shall keep my judgments, and do them*" (Ezek 36:24, 26-27; see Ezek 11:19-20).

nm732 » These scriptures show the promise of God's Spirit. It was through Jesus Christ that this promise came (see Acts 1:4; 2:33; Eph 1:13). And it is this Spirit of promise that frees us from the old law written on stone, and replaces the old law with a New Law written on the heart. Therefore Paul could write, "You [Christians] are our letter, written in our hearts, known and read by all men; being manifested that you are a letter of Christ, cared for by us, written not with ink, but with the Spirit of the living God, not on tablets of stone, but on tablets of fleshly hearts" (2Cor 3:2-3).

New Commandment

nm733 » God in the Old Testament said he would send a Spirit, and through that Spirit man would follow his laws, not by letter, but by Spirit: "I will put my Spirit within you and cause you to walk in my statutes" (Ezek 36:27). This Spirit is God's Spirit. This Spirit is Christ's Spirit. All of this book manifests that we are saved only if we have this Spirit or New Mind, which is the Spirit or Mind of God. And with this New Mind comes a New Law, a New Commandment. That New Commandment was, and is, Love. Not any kind of love, but Love with the power of the Spirit of God. "For, brethren, you have been called unto liberty; only use not liberty for an occasion to the flesh, but by **love** serve one another" (Gal 5:13). True liberty, true law keeping has something to do with love. As we will see, "Love" is the New Commandment.

nm734 » In the 24 hours or so before Christ was killed he gave his commandment: "A new commandment I give unto you, That you love one another; as I have loved you, that you also love one another" (John 13:34). This new commandment was love. Paul shows us what love is: "Owe no man any thing, but to love one another: for he that loves another has fulfilled the law" (Rom 13:8). As Paul shows, the true fulfilling of the law is love. And this love can be stated simply, "You shall love your neighbor as yourself. Love works no ill to his neighbor: therefore love is the fulfilling of the law" (Rom 13:9-10). Christ said basically the same thing, "Jesus said unto him, you shall love the Lord your God with all your heart, and with all your soul, and with all your mind. This is the first and great commandment. And the second is like it, you shall love your neighbor as yourself. On these two commandments hang all the law and the prophets" (Mat 22:37-40). All the law of the Old Testament hung on love – first the love of God, second the love of our fellow man.

nm735 » Christ gave a new law. But this law was not really new for all the old commandments hung on love (Gal 5:14). Christ emphasized this great law of love. And as we have seen it was because of Christ and the Spirit that He gives, that real Christians are freed from the old law (which was good, but had no power because the people had no Spiritual power to truly follow that law). The old law was replaced by the New Law, the New Commandment of Christ, the New Law of Christ, which is love.

nm736 » John has something to say about the new commandment: "Again, a new commandment I write unto you, which thing is true in him and in you: because the darkness is past, and the true light now shines ... He that loves his brother abides in the light" (1John 2:8, 10). John speaks of love: "Beloved, let us love one another: for love is of God; and every one that loves is born of God, and knows God. He that loves not, knows not God; *for God is love*" (1John 4:7-8). God is love. That is, God is Spirit (John 4:24). Thus, God's Spirit is love. Christians are Christians only if they have the Spirit of God. With this Spirit comes the love.

nm737 » What is this love? Is it sexual desire? Can you hate something and still love it? What kind of love was Christ speaking about? What is Biblical love?

System of Love

nm738 » The love Christ spoke of is a system of behavior, or a law of behavior. Paul helped to define it:

■ Love is patient, love is kind, and is not jealous; love does not brag and is not arrogant, does not act unbecomingly; it does not seek its own, is not provoked, does not take into account a wrong suffered, does not rejoice in unrighteousness, but rejoices with the truth; bears all things, believes all things, hopes all things, endures all things. Love never fails (1Cor 13:4-8, NASB).

nm739 » From the 1978 New International Version:

■ Love is patient, love is kind. It does not envy, it does not boast, it is not proud. It is not rude, it is not self-seeking, it is not easily angered, it keeps no record of wrongs. Love does not delight in evil but rejoices with the truth. It always protects [covers], always trusts, always hopes, always perseveres. Love never fails (1Cor 13:4-8, NIV).

nm740 » From the BeComingOne Bible:

■ <1Cor 13:4> Love is enduring, is kind; the Love is not envious {jealous, spiteful}; love is not boastful, is not conceited; <1Cor 13:5> is not rude {unbecoming}, is not self-centered, is not easily provoked {inflamed; angered}, is not numbering wrongs {of others}; <1Cor 13:6> is not rejoicing in iniquity, but is rejoicing in the truth; <1Cor 13:7> covers all (sin), believes all (good), hopes all (good), perseveres all (good). <1Cor 13:8> Love never fails.

nm741 » The fruit of God's Spirit of love is "love, joy, peace, patience, kindness, goodness, faithfulness, gentleness and self-control. Against such things there is no law" (Gal 5:22-23). There is wisdom from this Spirit of love: "But the wisdom from above is first pure, then peaceable, gentle, reasonable, full of mercy and good fruits, unwavering, without hypocrisy" (James 3:17).

nm742 » What did Christ say when he gave His New Commandment? "A New Commandment I give unto you, *that you love one another, as I have loved you*" (John 13:34). Christ who had the Spirit of God inside of him behaved and acted one way – with Love. Everything Christ did was the living of the system of love, or the law of love. Every behavioral quality mentioned as belonging to the Spirit of God in the New Testament is also a description of love, the law or system of love. Let's look at scripture for further descriptions of love.

nm743 » The so-called sermon on the mount in Matthew, chapter five, was one of Christ's amplifications on the system of love. In chapter five Christ showed how deep the law of love goes; it goes to the very heart of man. In love you should not only not kill, but not be angry with your brothers (Mat 5:21-22; 1John 3:15).

nm744 » Love does not give evil for evil (Rom 12:17). Love does not bless and curse at the same time (James 3:10). But love is "peaceable, and easy to be entreated, full of mercy and good fruits, without partiality, and without hypocrisy" (James 3:17).

nm745 » Love does not worry or fear (Mat 6:25-29). Love, loves its enemies yet not their ways (Mat 5:44; Prov 8:13). Love does not hate his brother (the person himself), yet if his brother's ways are of evil he hates his brother's ways, yet not the person himself (Prov 8:13 & Luke 14:26).

nm746 » Love does in deeds what it utters in tongue (1John 3:18 & James 1:22). Love does not overtly tempt others with objects that might lead them away from the truth (Rom 14:20-21), yet Love knows no thing in itself is bad (Rom 14:14 & Titus 1:15).

nm747 » Those with love will clothe themselves "with compassion, kindness, humility, gentleness and patience. Bear with each other and forgive whatever grievances you may have against one another. Forgive as the Lord forgave you. And over all these virtues put on love, which binds them all together in perfect unity" (Col 3:12-14).

nm748 » "Finally, all of you, live in harmony with one another; be sympathetic, love as brothers, be compassionate and humble. Do not repay evil with evil or insult with insult, but with blessing" (1Peter 3:8-9).

nm749 » "And to put on the new self, created to be like God in true righteousness and holiness. Therefore each of you must put off falsehood and speak truthfully to his neighbor, for we are all members of one body. 'In your anger do not sin': Do not let the sun go down while you are still angry, and do not give the devil a foothold. He who has been stealing must steal no longer, but must work, doing something useful with his own hands, that he may have something to share with those in need ..."

nm750 » Get rid of all bitterness, rage and anger, brawling and slander, along with every form of malice. Be kind and compassionate to one another, forgiving each other, just as in Christ God forgave you" (Eph 4:24-28, 31-32).

nm751 » "Brother, do not slander one another. Anyone who speaks against his brother or judges him, speaks against the law and judges it. When you judge the law, you are not keeping it, but sitting in judgment on it. There is only one Lawgiver and Judge, the one who is able to save and destroy. But you – who are you to judge your neighbor?" (James 4:11-12) Be slow to judge because only God has all the factors that enter into people's apparent misbehavior. This says nothing about obvious evil, where you have sound evidence of the evil. But we must be careful. Some people on one part of the earth believe one thing is wrong, while others on another part of the earth believe it is okay. Be careful. In areas of gray, it is better not to judge, than to make a mistake.

nm752 » "Do nothing out of selfish ambition or vain conceit, but in humility consider others better than yourselves. Each of you should look not to your own interests, but also to the interests of others" (Phil 2:3-4).

nm753 » "Therefore, as we have opportunity, let us do good to all people, especially to those who belong to the family of believers" (Gal 6:10).

nm754 » "If anyone has material possessions and sees his brother in need but has no pity on him, how can the love of God be in him? Dear children, let us not love with words or tongue but with actions and in truth" (1John 3:17-18).

Law of Love Is More Flexible

nm755 » In the kingdom of God when all people have the New Mind and the new immortal body, some kinds of behavior that are now evil, or considered evil by some, may not be evil in the New Cosmos. For example, you can jump off a high cliff without it being a sin (suicide) because you cannot get hurt. Your angel will glide you down safely.

nm756 » People with the New Mind and new body will have fewer social restrictions. Compared to today, their behavior will be more flexible because they do not see evil in as many places as those with the other-mind. Some of those of the other-mind see evil in the ankles of women, or in the wearing of red, or in having money, or in moderate drinking, or in other things that are not evil in, and of, themselves. Remember nothing God has made is evil in itself (Rom 14:14; Titus 1:15).

nm757 » People in the New Cosmos (new heavens and earth) with their new body of immortality will have fewer restrictions concerning their body because the body will be incapable of becoming diseased, will not age, cannot get pregnant, and will be truly beautiful to all. There will be no marriage in the New Cosmos between the new age people (Luke 20:35) because there will be no more child bearing. But if two agree to stay together as a couple for a million years or more, then this is, of course, alright. Relationships will last longer than any marriage ever lasted in the old age.

nm758 » Sexual intercourse in the old age foreshadowed two, the physical and spiritual, becoming one in the New Age. Sexual intercourse in the New Cosmos will be a memento of the creation, a reminder of the reason for the old age (see *God Papers*, Parts 5-7).

nm759 » Remember, "love works no harm to his neighbor." When people receive the New Mind and the new body, and begin to live in the new creation, the Spiritual law of love is flexible enough to take the new situation into account. But written law on stone is not as flexible.

What Law Did Christ's Spirit Free Us From?

nm760 » As shown above, those with the Spirit of God are freed from the Old Testament law and the curses in it. A Christian with the Spirit does not have to obey physically all those laws and rituals in order to be saved. But this does not mean a Christian can lie, or kill, commit adultery or do any other form of evil, for now they are under the law of love which goes deeper than the written laws in the Old Testament. But even though we are not under the law of the Old Testament, the law still projects to us what forms of behavior will harm us as individuals and as groups, and the law is an example to us and projects the antitype to us (see *God's Appointed Times [NM16]*).

Do Christians keep the Ten Commandments?

nm761 » Christians are not under the letter of the law, but they are under the Spirit of the law. In the New Testament scriptures we see the ten commandments reiterated in a deeper more Spiritual way in the Spiritual Church:

> "We ... preach unto you that you should turn from these vanities unto the living God" (Acts 14:15).

> "Little children, keep yourselves from idols" (1John 5:21).

> "But above all things brethren, swear not, neither by heaven, neither by the earth, neither by any other oath" (James 5:12).

Paul taught on the Sabbath (Acts 13:14; 16:13) even though there were no references in the Bible after Christ was resurrected that said that Christians had to keep the physical Sabbath. Christians did come together on the Pentecost (Acts 2:1), which is an annual Sabbath. We who are Spiritual Christians keep the Sabbath Spiritually (See next section).

> "Children obey your parents in the Lord: for this is right" (Eph 6:1).

> "Whosoever hates his brother is a murderer: and you know that no murderer has eonian life abiding in him" (1John 3:15).

> "Neither fornicators, nor idolaters, nor adulterers ... shall inherit the kingdom of God" (1Cor 6:9-10).

> "Steal no more" (Eph 4:238).

> "Lie not" (Col 3:9).

> "Covetousness, let it not be named among you" (Eph 5:3).

According to Lewis Sperry Chafer, in his book called *Grace*, "Under the teachings of grace, the appeal of the first commandment is repeated no less than fifty times, the second twelve times, the third four times, the fourth (about the Sabbath day) not at all, the fifth six times, the eighth six times, the ninth four times, and the tenth nine times" (p. 156; I have not confirmed his count of the scriptures, but I know that the ten commandments are reiterated many times in the scriptures of Grace.).

nm762 » Christ taught a more Spiritual law (Mat 5:20-48; 6:1-7:29) because unlike physical Israel, he had the very Spirit of God in him so that his words and actions were Spiritual (John 4:24; 6:63). For example, Christ's words about eating his body can only be understood correctly in a Spiritual manner (John 6:55-56, 63). Those churches that try to make *physical* sense out of these scriptures become foolish in their doctrine. Christ just before his death, when he was still under the law (remember he was born under the law [Gal 4:4], for his mother was a Jew), began to teach his followers about the coming new Spiritual commandment, a system of love (John 13:34; 15:12). See "Keep the Commandments?" in the "Proof Paper" [NM 10] for more information.

Do Christians Keep the Sabbath?

nm763 » Yes. The only question is *how* do you keep it.

- If you have the Spirit you keep it Spiritually.

- *If you are under the law*, then to be saved you must keep *all* of it physically – you cannot keep it Spiritually since you don't have the Spirit (Rom 8:7-10; Gal 5:18).

Old Law Led to New Law

nm764 » Israel put themselves under the law when they promised to keep **all** the laws of Moses (Ex 24:3, 7; 19:8; James 2:10). One (Jesus Christ) did keep all of the physical laws. And we are saved through Him by our faith, our Spiritual faith (Gal 3:23-24; "Proof Paper" NM 10). But physical Israel did not keep the law even though they promised to do so (Acts 7:53). The law was set up to lead us to our true leader (Gal 3:19). We were given a new commandment by this new leader, Christ. That new commandment is the commandment of love on which all the laws of

Moses hang (Mat 22:37-40). This new commandment only works with the new law's New Spirit. Those without the New Mind, the New Spirit, will say such things as, "It is okay to take these street drugs, I don't see anything in the Bible saying it is a sin to take them – so it is okay" (a person commenting to another about his street drug usage). Those of the *New Mind* will say, "I don't see any law in the Bible that tells me the street drug is wrong, but I know this drug will harm me, therefore I will not use it because not only do I not wish to hurt others, but also not to hurt myself." Therefore those with the New Mind do not need a library full of detailed laws to know how to behave; they know it intuitively, depending on their measure of the Spirit (Eph 4:7; see, nm336). This "love" is not sexual and it is not friendship love, but this New Love is a system of love, a system of behavior that Paul wrote about (Gal 5:22; 1Cor 13:4-8; etc). In this system of love there is no adultery, fornication, witchcraft, hatred, wrath, murders, drunkenness, and so forth (Gal 5:19-21).

New Way of Keeping the Sabbath

nm765 » We in the Spiritual Church do not have to keep all of the letter of the Old Testament law to be saved; we keep it Spiritually, which is a greater standard of morality. In reference to the Sabbath, we keep it always, we set each day aside to keep it holy, we seek to be holy always, not just on the seventh day. We fellowship, we congregate, and we, so to speak, are in church every day because we have the Spirit of Christ inside of us teaching us. But since we are also physical and need to fellowship with each other and need a day of rest, we gather when possible with others who have the New Spirit, on the seventh day and more often if possible, and at other days besides the seventh day if we are forced to by circumstance and state laws. Of course, if we controlled the laws in our country we would set aside the seventh day for rest and holiness. During the 1000 years the seventh day of the week will be the day of rest because this new age will be controlled by those of the New Spirit. See "Thousand Years and Beyond" paper [NM 15] and the "God's Appointed Times" paper [NM 16] to understand more about the Sabbath.

Do Christians Have To Follow the Laws of the Civil Government?

nm766 » Christians should attempt to follow all the laws and regulations of all nations on earth as long as the laws are not against the law of love. But if a government should try to force Christians by law to do things such as kill, or lie, or hate the truth, Christians need not and should not obey these laws. If laws are being applied wrongfully against us, we should try to correct the situation through the courts or in other ways.

nm767 » Real Christians can live under almost any kind of government including communism and dictatorships. All these governments are temporary. If you are not forced to kill, lie, hate, rape, bruise, and so forth, then Christians can live in these societies even if they forbid your Christianity. In such countries your example will teach. If such countries forbid you to teach, you may teach, howbeit in a more restricted matter, but be careful. If you teach in a discreet manner the authorities may let you be (Acts 4:18-20; 5:28-29). *No* country has the right to put you in jail for teaching Christ. But you must teach Christ with respect towards the rights of others. You can't teach Christ and be rude at the same time. If they do put you in jail for teaching Christ, you may escape if you can. You have done no wrong. (But maybe God through predestination put you in jail to teach those in jail; maybe this is *your* mission.) But be careful. Why affront the authorities when with a little thinking you can teach, and not provoke the authorities. In countries in which you can vote, vote for the best and less corrupt representative. If your country allows you to be honest, allows family, allows liveable wages for your work, allows you to in some way to teach or at least speak about your faith, then you can live in such a country.

Hate In The System of Love?

nm768 » Yes, for real Christians who have the law or system of love in them, do *hate* their life in the present system of this world (Rev 12:11; John 12:25; Luke 14:26; James 4:4). "If any one come to me, and *hate* not his father, and mother, and wife, and children, and brethren, and sisters, yea, and his own life also, he cannot be my disciple" (Luke 12:26). This does not mean you hate your father, mother, wife, children and so on because they are your father, mother, wife, children and so on, but you hate their wrong behavior: "The fear of the LORD is to *hate* evil: pride, and arrogance, and the evil way, and the froward mouth, do I *hate*" (Prov 8:13; Psa 97:10). You cannot have the Spiritual law of love and at the same time love evil behavior. When you have true love, you hate evil. If your brother behaves evilly, you hate his evil behavior, not because of anything else. If you hate your life in this world, you hate it because of evil behavior in this world, not for anything else. Each and everything God has created is good in itself (Rom 14:14; Titus 1:15). It is merely the misuse of these things that we hate, not the thing itself. "You adulterous people, don't you know that love ['friendship'- KJV] of the world [system] is hatred toward God? Anyone who chooses to be a friend [or lover] of the world [system] becomes an enemy of God" (James 4:4). In this present age the world is evil, how can you love a system of evil?If you have the Spiritual law of God, you hate your life in this evil world. You do not hate life itself, you do not hate the few moments of joy in this world, but generally you hate the world because of its evil.

How Can You Love Your Enemy?

nm769 » When you have the Spiritual law of love in your heart, you can and do love your enemy, *but* you do not like him (his wrong behavior). The love Christ spoke of was the system or law of love. "Love is patient, love is kind. It does not envy, it does not boast" (1Cor 13:4). When you love your enemy you are patient with him, you are kind to him, you do not envy him, you do not boast to him, etc. This is how you love your enemy. You do not like your enemy, you love him, or that is, you practice the system of love on him.

Difference Between Biblical Love and Other Love?

nm770 » Yes! Greek is the language of the New Testament of the Bible. In Greek there are at least three words that can be translated into "love" in the English language: (1) *eros* (which means, love between the sexes); (2) *phileo* (cherishing or friendship or high regard or even sexual love depending on the context); (3) *agapao* (system of love). The last two Greek words were used in the New Testament. But it was mainly the Greek word, *agapao*, that was used in the New Testament to describe Christ's love. *Agapao* was the Greek word Christ used when he asked us to love our enemies (Mat 5:44). It was also the Greek word Paul used to describe love (Rom 13:9-10). Christ's love is the **system of love**.

NM 18: Other Papers On Christianity

NM18 Abstract

In this paper we are going to discuss various subjects including parables, the meaning of "flesh" in the New Testament, Christian's Warfare, Miracles, justification, and so forth.

Parable of the Sower

First Group (hears, understands not)

nm771 » Turn to the parable of the sower in Matthew 13:1-8. This parable is also explained in Mark 4:3-25 and Luke 8:5-18. Here are pictured four groups of people who are being called or invited to the kingdom of God. Jesus explains this parable in Matthew starting in verse 19: "when any one hears the word of the kingdom, and understands it not, then comes the wicked one, and catches away that which was sown in his heart. This is he which received seed by the way side."

nm772 » This first group hears about the kingdom of God but really does not perceive or understand the significance of the words being spoken about the kingdom. Why don't they understand? "Every man therefore that has heard, and has learned of the Father, comes unto me [Christ]" (John 6:45). Christ said that those who have heard and have learned do come to him. What is meant by having heard and learned? As Matthew 13:14-17 and other verses throughout the Bible show, there are people who hear physically, but not Spiritually. Those who hear physically the word about the kingdom, but do not understand Spiritually are those not chosen by God and are not being called by God's Spirit at the time they hear about the kingdom. (But they will understand later.) "We are of God: he that knows God hears us; he that is not of God [does not have the Spirit] hears not us. Hereby know we the Spirit of truth [if they hear], and the spirit of error [if they do not hear, Spiritually]" (1John 4:6).

Second Group (hears, but does not endure – no root)

nm773 » "But he that received the seed [the word, Mark 4:14, 15] into stony places [stones are dead things], the same is he that hears the word, and immediately with joy receives it; Yet has he no root [Christ is the true root, Rev 22:16] in himself, but endures for a while: for when tribulation or persecution arises because of the word, by and by he is offended" (Mat 13:20-21).

nm774 » This second group hears physically the word (seed) about the kingdom of God, the New Age, but the word comes into a stony heart. Even though these see physically the importance of the kingdom of God and take the word of this kingdom with great joy at first, they do not endure because they have not the Spirit or root of Christ which is of course God's Spirit since all in the Body (Church) have one kind of Spirit (1Cor 12:13). These can't endure the tribulation because they haven't God's Spirit that allows one to overcome (1John 5:4). "They [those without the Spirit] went out from us, but they were not of us; for if they had been of us, they would have continued with us; but that they might be made manifest [by their act of going out] that they were not all of us" (1John 2:19).

Third Group (hears, not fruitful)

nm775 » "He also that received seed among the thorns is he that hears the word [physically]; and the care of this world, and the deceitfulness of riches, choke the word, and he becomes unfruitful" (Mat 13:22).

nm776 » The third group is being called or invited to the kingdom of God, and they hear physically, but the cares of the world choke the word of the New Age and they do not produce the fruit. As shown in the paper, "Prove Paper" [NM 10], those who are indeed real Christians do produce much fruit. Thus, this is one proof that this third group who are being called are not really Christian. The third group is the group of people who are being called, but they are not of the chosen (Mat 22:14). Another proof that this third group isn't made up of Christians is that they let the cares of the world interfere with the word of the kingdom of God. One of the tests of being a true Christian is whether one hates the world's wrong ways (1John 2:15). In fact how could a Christian let the cares of the world get in their way when they are supposed to hate their very life in the world? (John 12:25)

Fourth Group (hears, understands, bears fruit)

nm777 » The fourth group are the real Christians, the called, chosen, and predestinated. "But he that received seed into the good ground is he that hears [Spiritually] the word, and understands it; which also bears fruit [John 15:5, 8, 16], and brings forth, some a hundredfold, some sixty, some thirty" (Mat 13:23).

Parable of the Ten Virgins

nm778 » The parable of the ten virgins is in Matthew 25:1-12: "Then the kingdom of heaven shall be compared to ten virgins." The word "virgin" means, one put aside. This Spiritually speaks of the set-apart ("holy") people, the Christians (see Rev. 14:4).

nm779 » "Then the kingdom of heaven shall be compared to ten virgins [Christians], who took their lamps." The word "lamp" is Spiritually speaking about God's word, "your word a lamp unto my feet" (Psa. 119:105).

nm780 » "Then the kingdom of heaven shall be compared to ten virgins [Christians], who took their lamps [God's word] and went out to meet the bridegroom." The word "bridegroom" is Spiritually speaking about Jesus Christ (see Rev 19:7).

nm781 » "Then the kingdom of heaven shall be compared to ten virgins [Christians], who took their lamps [God's word] and went out to meet the bridegroom [Jesus Christ]. And five of them were wise [with the Spirit], and five were foolish [without the Spirit]. When they took their lamps [God's word], the foolish ones [non-Spiritual] did not take oil [the Spirit, see 1John 2:20,27; 2Cor. 1:21] with them. But the wise took oil [the Spirit] in their vessels [bodies, see Rom. 9:21-23; 1Thes 4:4; Acts 9:15] with their lamps [God's word]. And as the bridegroom [J.C.] delayed, they all nodded and went to sleep. And in the middle of the night there was a cry, Look, the bridegroom [J.C.] is coming! Go out to meet him. Then all those virgins [Christians, with and without the Spirit] rose up and trimmed their lamps. And the foolish ones said to the wise, share your oil [Spirit] with us, for our lamps are going out. But the wise [with Spirit] answered and said, lest there should not be enough for us and for you. But rather go to those who sell, and buy oil for yourselves. But as they went away to buy, the bridegroom [J.C.] came. And those who were ready went in with him to the wedding feast. And the door was shut. And afterwards the other virgins [without the Spirit] also came, saying, Lord, Lord, open to us! But he [J.C.] answered and said, truly I say to you, I do not know you."

nm782 » In context of the verses around Matthew 25:1-12, where it is speaking of the end of the old age, we can see here that the virgins of this parable are Christians, waiting for the kingdom of God, those with the Spirit (oil) and those without the Spirit. Since those with the Spirit are the only true Christians, then the five virgins (Christians) with the oil (Spirit) are the ones that truly belong to Christ and will be in his kingdom. But the others who call themselves Christians, yet do not have the oil (Spirit), are not known by Christ and will not be in his kingdom during the 1000 year age.

nm783 » This paper like others again tells us that the only true Christian is the one with the Spirit of God, the New Mind. This parable also prophesies that just before Jesus Christ's return, one-half of those in the group with the Real Christians will not have the oil – the Spirit.

Flesh

nm784 » How does Paul use the word "flesh" in his scriptures? Notice that "the flesh lusts against the Spirit, and the Spirit against the flesh. And these are contrary to one another" (Gal. 5:17). The Spirit and flesh are contrary to each other. What does this mean?

nm785 » Paul is speaking to Christians and he says: "But you are not in the flesh but in the Spirit if the Spirit of God dwells in you" (Rom 8:9). Real Christians are "not in the flesh" when they have the Spirit of God in them, or that is, the New Mind in them.

nm786 » Those who have the Spirit are real Christians or sons of God: "For as many as are led by the Spirit of God, these are the sons of God ... you received a spirit of adoption in which we cry, Abba, Father. The very Spirit bears witness with our spirit that we are children of God" (Rom 8:14-16). Those who have the Spirit are the children of God, or sons of God, or the real Christians. Remember the Spirit is contrary or against the flesh (Gal 5:17), and Christians are not "in the flesh" (Rom 8:9).

nm787 » Notice, "that is, the children of the flesh are not the ones who are the children of God" (Rom 9:8). Those called the children of the flesh, those of the flesh, those in the flesh, are not Christians. They are not the New Age people because they do not have the Spirit of the New Age, the New Mind (Rom 8:14-16).

nm788 » Those of the Spirit are contrary to those of the flesh because one group has the Spirit of God that leads them toward the ways of God (Rom 8:14,6; Gal. 5:22-23). But the other group of the flesh does not have the Spirit of God, thus they follow in the ways of confusion (note Gal 5:19-21). The group of the flesh only has the other-mind, that Other Mind with the twisted spirit, in their minds misleading them. But the children of God have the Spirit of God, the New Mind, which is stronger than the other spirit in them (1John 4:4), and the New Spirit leads them to overcoming wrong (1John 5:4). Yet the children of God, because they still have the spirit of error in them in this the old age will sometimes still make mistakes. Therefore Christians or the New Age people are not completely out of the flesh, or the ways of the flesh. They are fighting the good warfare in their minds (see "Warfare Paper" [NM 18]). They will only be completely outside the ways of the flesh when they are in the New Age.

nm789 » Thus Galatians 5:17 is dual: First, it speaks of the "flesh" that is still in Christians, and it says that the Spirit is contrary and against the ways of the flesh. And second, it speaks of the people of the flesh, those without the Spirit of God, and says they are contrary to those of the Spirit.

Chapter Eight of Romans

nm790 » *"There is therefore, now no condemnation to those in Christ Jesus, who walk not according to the flesh, but according to the Spirit"* (Rom 8:1).

Those in Christ, those with the Spirit or New Mind, have no condemnation. That is, they do not die for the thousand year judgement, but live in the kingdom of God during the

thousand years (see "Reward for Christians Paper" [NM 11]).

nm791 » *"For the law of the spirit of life in Christ Jesus set me free from the law of sin and of death"* (Rom 8:2).

Through Jesus Christ we are dead to the law of sin and Satan (Rom 6:10), and we are set free from this law of bondage (Rom 8:15) or the law of sin and death.

nm792 » *"For what the law was not able to do, in that it was weak through the flesh, God, in sending his own son in the likeness of sinful flesh, and about sin condemned sin in the flesh"* (Rom 8:3).

Jesus came and overcame the present world which is the spiritual enemy's world (see John 16:33). Jesus judged the prince of the world (the other spirit) (John 16:11). The law of Moses was weak because those of the flesh, those without the Spirit of God, could not keep the law correctly. Thus, the law became for them (the flesh) the starting point for Satan (the other-mind) to mislead them. Thus, the law of Moses to the flesh, those without the Spirit, was a law of sin.

nm793 » *"In order that the righteous demand of the law should be fulfilled in us, who walk not according to the flesh but according to the Spirit"* (Rom. 8:4).

Those of the Spiritual law of God, of God's Spirit, have performed the requirement of the law through Jesus Christ (see "Freedom & Law Paper" [NM 17]).

nm794 » *"For they that are according to the flesh set their mind on the things of the flesh, and they who are according to the Spirit on the things of the Spirit"* (Rom 8:5).

The Spiritual walk according to the ways of the Spirit as opposed to the others who walk according to the ways of the flesh.

nm795 » *"For the mind of the flesh is death, but the mind of the Spirit is life and peace"* (Rom 8:6).

Those with the New Mind, those with the Spirit of God, think on life and peace as opposed to the others with their thoughts dwelling on death and destruction.

nm796 » *"Because the mind of the flesh is enmity towards God, for it is not subject to the law of God, for neither can it be"* (Rom 8:7).

The mind of the flesh is the mind that does not have the Spirit of God, the New Mind. The mind of the flesh has continuous flash-thoughts on death, fear, destruction, confusion, etc. Because of this, the mind of the flesh is an enemy against the ways of God. The mind of the flesh isn't subject to the flash-thoughts of the New Mind. The New Mind gives flash-thoughts of life, trust, peace, truth, etc. The Other Mind, the mind of the flesh, is not subject to the good thoughts of the New Mind.

nm797 » *"And they that are in the flesh are not able to please God"* (Rom 8:8).

This does not mean that Christians (New Age people) who live in a fleshly body can't please God. The way to make sense out of this verse is to take it in context with chapter eight as a whole and with Romans 7:5. That is, "they that are in the flesh" are not real Christians.

nm798 » *"But you are not in the flesh but in the Spirit if the Spirit of God dwells in you. But if anyone has not the Spirit of Christ, he is not His"* (Rom 8:9).

Real Christians are not "in the flesh." That is, they are not "in the flesh" in the way that Paul uses the expression. Real Christians are in the Spirit.

nm799 » *"But if Christ is in you, the body indeed is dead because of sin, but the Spirit is life because of righteousness. But if the Spirit of him who raised up Jesus from among the dead dwells in you, he who raised up the Christ from among the dead will also make your death-doomed bodies live because of His Spirit that dwells in you. So, then, brothers we are not debtors to the flesh to live according to the flesh. For if you live according to the flesh, you are going to die. But if you by the Spirit put to death the deeds of the body, you will live. For as many as are led by the Spirit of God, these are the sons of God"* (Rom 8:10-14).

Notice the "if" or the hypothetical statement in verse 13. Paul is saying that *if* you do the work of the flesh you are proving you do not have the Spirit

because the Spirit does produce good works as John 15 and 1John manifest over and over again. When one does the works of the flesh he is proving he does not have the Spirit. One does good works: (1) to prove he has the Spirit and thus to prove that he will be saved or freed from the confusion; and (2) because when you have the Spirit you like to do good.

Christian's Warfare

nm800 » What kind of war should real Christian's fight? "For though we walk in the flesh, we do not war according to the flesh: For the weapons of our warfare are not carnal, but mighty through God" (2Cor 10:3, 4).

nm801 » Christian's warfare is not a war of flesh and blood; their weapons are not those of the flesh. The Bible calls the Christian's warfare: "a good warfare" (1Tim 1:18).

nm802 » Christians are called and asked to "fight the good fight of faith, lay hold on aeonian life, whereunto you art also called, and has professed a good profession before many witnesses" (1Tim 6:12).

nm803 » Not only were Christians called (invited), but they were chosen to fight the good warfare: "You therefore endure hardness, as a good soldier of Jesus Christ. No man that wars entangles himself with the affairs of this life, that he may please him who has chosen him to be a soldier" (2Tim 2: 3-4).

nm804 » In verse 5 it adds importantly: "if a man also strive for masteries, yet is he not crowned, except he strive lawfully." No Christian will be crowned (with life, James 1:12) except if he fights lawfully. This means by God's laws not man-made laws, for it is speaking about God's soldiers, not soldiers of the world's nations. One of the laws of God tells man not to kill. God's soldiers do not kill, they fight lawfully. How can Christians win without killing? We need to know what kind of war Christians are fighting to answer this question.

nm805 » *What kind of war are Christians fighting?* Who are the Christians fighting?: "because your adversary the devil, as a roaring lion, walks about, seeking whom he may devour" (1Pet 5:8). "For we wrestle not against flesh and blood, but against principalities, against powers, against the rulers of the darkness of this world, against spiritual wickedness in heavenlies" (Eph 6:12). It is Satan and his spiritual angels that real Christians are fighting against. In fact the word Satan means: enemy. Further, Christians are warring against Satan's law in man's mind, the law of sin (see Rom 7:22-25).

nm806 » Christian's have two spiritual qualities in their minds – God's Spirit (the New Mind) and a satanic one (the other-mind). Christians are fighting against the satanic one, yet it is not the human-Christian who is really fighting, but God's Spirit in the Christian's mind which is doing the spiritual fighting against the satanic spirit or mind (see "Other Mind Paper" [NM 21]).

Overcome, How?

nm807 » How does a Christian overcome or subdue this enemy inside his mind? One overcomes by resisting through their Spiritual power (the New Mind) the other-mind in their head (note James 4:7). One must resist the suggestive power of this other-mind or other spirit. One suggestive power of this spiritual quality is envy: "the spirit that dwells in us lusts to envy" (James 4:5). But how does a Christian resist this other-mind or other spiritual power?

nm808 » "Finally, my brethren, be strong in the Lord, and in the power of his might. Put on the whole armor of God, that you may stand against the wiles of the devil" (Eph 6:10-11).

nm809 » What is the armor of God? "Stand therefore, having your loins girt about with truth, and having on the breastplate of righteousness; and your feet shod with the preparation of the gospel of peace; above all, taking the shield of faith ... and take the helmet of salvation, and the sword of the Spirit, which is the word of God: Praying always" (Eph 6:14-18, see 1Thes 5:8). Thus, the armor of God consists of such qualities as faith, righteousness, the good news of peace, salvation, the word of God, and prayers. All these qualities come from, or work best with, God's Spirit. Only real Christians have God's Spirit.

nm810 » *Winner, God's Spirit*. We should know that God's Spirit in real Christians' minds will win: "You are of God, little children; and have overcome them: because greater is he that is in you [God's Spirit], than he that is in the world [the other spirit]" (1John 4:4). God's Spirit is stronger in Christians, than their other spirit in their minds; thus, God is able to overcome.

nm811 » "For all that is begotten of God overcomes the world: and this is the victory that overcomes, our Faith" (1John 5:4). Thus, those begotten with God's Spirit will overcome by their faith which is a manifestation of God's Spirit (see Gal 5:22 & Eph 2:8). In other words, those who have God's Spirit, the New Mind, will have the Faith that will give victory. Christians will overcome as their example Jesus Christ did: "I have overcome the world" (John 16:33). Christians were called and chosen to conform to the image of Christ, thus, to overcome (see Rom 8:28-30 & "Predestination Paper" [NM 8]).

nm812 » Christians are fighting the good warfare and have the armor of God and God's Spirit to fight the spiritual war. Since God's Spirit is greater in power than the adversary's spirit in man's mind, and since with "the holy Spirit of God, whereby you are sealed unto the day of redemption [Eph 4:30]," then we know that Christians as individuals will win the good warfare, if they truly have God's Spirit in them, and thus are real Christians (Rom 8:11).

New Jerusalem

nm813 » Many look upon the book of Revelation's account of the New Jerusalem that comes out of heaven as being a physical entity only. But God tells us to look to the higher meaning (Phil 3:18-19; Col 3:1-2). The truth of the matter is that the Bible is dual and this prophecy in Revelation speaks of a Spiritual Jerusalem.

nm814 » In Revelation 21:2 & 10 we read, "and 1 John saw the holy city out of heaven, prepared as a bride adorned for her husband .... And he carried me in spirit to a great and high mountain, and showed me that great city, the holy Jerusalem, descending out of heaven from God. Having the glory of God: and her light like unto a stone most precious, even like a jasper stone, clear as crystal." The chapter goes on to describe it in fuller detail.

nm815 » But now let's turn to Revelation 3:12. Here Jesus Christ is speaking to the Philadelphia Church. "Behold I come quickly: hold that fast which you have, that no man take your crown. Him that overcomes will I make a pillar in the temple of my God, and he shall go no more out: and I will write upon him the name of my God, and the name of the city of my God, New Jerusalem, which comes down out of heaven from my God: and I will write upon him my new name."

Temple

nm816 » We see here that Christ will make the overcomers pillars in the temple of God. Now of course this does not mean being made a stone – what kind of a reward would that be to be made a dead, stone pillar? No this pillar is a spiritual pillar in the spiritual temple of God. But then it says he will never go out of the temple. Again if this verse was a physical temple what kind of reward would that be, to be confined to a temple forever? No, again it speaks of a spiritual temple.

nm817 » Paul speaking to Christians said, "you are the temple of the living God" (2Cor 6:16). And again, "for through him [Christ] we have access by one Spirit unto the Father. Now therefore you are no more strangers and foreigners, but fellow citizens with the saints, and of the household of God; and are built upon the foundation of the apostles and prophets, Jesus Christ himself being the chief corner stone; in whom all the building fitly framed together grows into a Holy *temple* in the Lord. In whom you also are built together for a habitation of God through the Spirit" (Eph 2:18-22). And again in 1Peter 2:5, "you also, as lively stones, are built up a spiritual house." So the temple spoken about in Revelation 3:12 is a spiritual one – the Christians who overcome. And as there is a spiritual temple, so too is there a spiritual Jerusalem.

New Jerusalem and the Church

nm818 » Now we see in 1Peter 2:5 that it speaks of living stones that build-up to a spiritual house or temple. Peter calls the Christians, lively stones. Now turn to Malachi 3:16-17: "then they that feared the LORD spoke often one to another: and the LORD listened, and heard it, and a book of remembrance was written before him for them that feared the LORD, and that thought upon his name. And they shall be mine, says the LORD of hosts, in that day when I make up my JEWELS; and I will spare them, as a man spares his own son that serves him."

nm819 » With these two scriptures; and Malachi 3:3; Isaiah 54:11-13; Revelation 4:3; Ezek 28:13; and Isaiah 28:16 (Zion is the Church, Heb 12:22) taken in their context, we can conclude that the precious stones of Revelation 21:11, 19-21 are simply telling the reader that the new Jerusalem is Spiritual, for precious stones are symbolic of spiritual things. This Spiritual Jerusalem is the Church, for one proof that New Jerusalem is God's Church is revealed in Revelation 21:2, 9. New Jerusalem here is spoken of as a bride of the Lamb. Revelation 19:7-9 proves that the bride of the Lamb is the Church.

nm820 » For another proof note Hebrews 12:22 where it calls the Church "the city of the living God, THE HEAVENLY JERUSALEM." And in Galatians 4:26 it calls Jerusalem the "Jerusalem which is above is free, which is the mother of us all." Notice also the woman of Revelation 12:1 is clothed with the sun, and upon her head are twelve stars. Now we know that women are symbolic of the Church (Eph 5:22-25). And we know that the symbolic meaning of being clothed with the sun is to be clothed with the Spirit because (1) God is light (1John 1:5); (2) Christ the God is clothed with light (Mat 17:2; Psa 104:2) that makes him shine as the sun (Mark 9:3; Rev 1:16); and God is spirit (John 4:24). Thus, when one is clothed with light he is clothed with the Spirit (see Isa 30:1). The woman of Revelation 12:1 is God's Church.

nm821 » What are the twelve stars on her head? Stars are symbolic of angels (Rev 1:20). Notice the description of New Jerusalem – "and at the gates twelve angels, and names written thereon, which are the names of the twelve tribes of the children of Israel" (Rev 21:12). Thus, the stars or angels on the woman's (Church's) head are a tie-in with New Jerusalem (the Church) and its twelve angels of the tribes of Israel.

nm822 » But for a final proof that the New Jerusalem is the Church of God let's return to Revelation 3:12. Jesus speaking to the Church says he will write on them the NAME of the God, and the name of God's city – New Jerusalem. They will have the NAME of the God and New Jerusalem. Now we see God's Church in Revelation 14:1 having Christ's NAME written on their foreheads. New Jerusalem is the Church of God, either resurrected (Born of God), or, taken typically, those begotten of God (see "Begotten, Born Paper" [NM 5]).

Outward Show

nm823 »

- some do things for appearance so they may escape suffering. [Gal 6:12]

- But we do what is right even if it appears evil. [2Cor 13:7-8; 6:8; Mat 27:63]

- Christ who did everything right, (1) was called the deceiver [Mat 27:63], (2) was said by others that "he is beside himself" [Mark 3:21], (3) was called possessed by the prince of demons [Mark 3:22]

- Paul's speech was unpolished, but he was rich in knowledge. [2 Cor 10:10; 11:6]

- Paul did not speak in flattery. [1Thes 2:5]

- Christ did what was right even if it appeared wrong to others: (1) He did things against the traditions of the Jews [Mat 15:2]; (2) Christ rebuked the Jews even though they were offended by the reproof. [Mat 15:3-14]

- When Peter put on a show to satisfy some of the Jews, Paul rebuked Peter because he was being partial against the Gentile Christians. [Gal 2:11-14]

These verses and others indicate that we do what is right even if it may appear to others as evil. The verse, "abstain from all appearance of evil" (1Thes 5:22), is a bad translation from the Greek, it should read: "abstain from every form of evil."

Wisdom

nm824 »

- Real wisdom will rebuke all. [Luke 21:14-15]

- Man's wisdom is to be destroyed. [Isaiah 29:14; 1Cor 1:19]

- Those who reject God's word, reject the basis of wisdom. [Jer 8:9]

- Real wisdom is obtained through God's Spirit. [Eccl 8:1, 16-17; with 1Cor 2:9-14]

Reproof

nm825 »

- He that regards reproof shall be honored. [Prov 13:18]

- A fool despises his father's [God the Father] instruction, but he that regards reproof is prudent. [Prov 15:5]

- God asks us to turn at his reproof. [Prov 1:23-27]

- Instruction is grievous unto him that forsakes the way: and he that hates reproof shall die. [Prov 15:10]

- He that refuses instruction despises his own soul: but he that hears reproof gets understanding. [Prov 15:32]

- The ear that hears the reproof of life abides among the wise. He is one of the Spiritual wise. [Prov 15:31]

- Reprove one that has understanding [the Spirit of understanding], and he will understand knowledge. [Prov 19:25]

- A scorner loves not one that reproves him: neither will he go unto the wise. [Prov 15:12]

- Just because God is silent about the things one does, does not mean he approves. [Psalms 50:16-21]

- The reprover is hated, in the city a trap is laid for him. [Amos 5:10; Isa 29:21]

- One should not add to God's words, lest God reprove that one, and he be found a liar. [Prov 30:6]

- Open rebuke is better than secret love; he that rebukes a man, afterward shall find more favor than he that flatters with the tongue. [Prov 27:5; 28:23]

- Those that do wrong rebuke before all. [1Tim 5:20; note context]

Miracles; Healing; Wonders; Signs

nm826 »

- Miracles are performed through the power of the Spirit of God. [John 3:2; Acts 14:3; Acts 15:12; 19:11 Mat 12:28; 1Cor 12:9]

- Miracles are given as a gift of the Spirit. [1Cor 12:9-10]

- Miracles are given as God wishes them to be given. [1Cor 12:11]

- Miracles are a sign of the Church. [Mark 16:17, 20]

- Miracles or signs are indications of God's prophet. [Ex 4:1-9]

- Signs, wonders, and mighty works are done by apostles. [2Cor 12:12]

- The Spirit has power over *all* kinds of sickness and disease, and over the other spirit. [Mat 10:1]

- Stephen did great wonders and miracles because he was full of grace and power. [Acts 6:8]

- The Spirit of God can and shall cast out the other spirit from some people possessed by it, and eventually the whole world. [Mat 8:28-32; 12:28; Zech 13:2]

- No man can do miracles, "except God be with him." [John 3:2; 9:16, 33; Mark 9:39-40; 3:22-23]

- False miracles deceive the non-Spiritual. [Rev 19:20; 2Thes 2:9-12]

- True miracles are undeniable. [Acts 4:16; John 11:47]

- True miracles produce astonishment in those who see them performed. [Acts 3:1-10]

- Such impossible things as making the blind see and raising the dead are true miracles. [John 9:1-8; 11:39-44]

- Jesus Christ did his works in his Father's NAME [John 10:25, 38]

- Jesus was in his Father's NAME because he had his Father's Spirit inside him. [John 14:9-10; see *God Papers*]

- Healing, signs, and wonders are done through the NAME of Christ. [Acts 4:30]

Thus, one must be in the NAME of Christ in order for one to perform miracles, for miracles are performed through his name or through the power of being in his NAME.

- Healing sickness is like forgiving the sins of those of the Faith. [Mark 2:9-12; Luke 5:20; Mat 9:1-2]

- To have Faith is to have the Spirit of Faith, that is God's Spirit. [2Cor 4:13; Gal 5:22; Eph 2:8; 1Cor 12:9]

- Faith (true faith with the Spirit) is required in order to perform miracles, healings, signs, and wonders, etc. [Mat 17:20; 21:21; John 14:12; Acts 6:8]

nm827 » Faith (true faith with the Spirit) is required in those for whom the miracles are performed.

- "according to your faith" – Mat 9:29; "he had faith to be healed" – Acts 14:9; "your faith has saved you" – Luke 18:42, 43; "your faith has made you whole" – Mat 9:22; "I believe, help my unbelief" – Mark 9:22-24

Note: Those before the Pentecost did not have the Spirit, thus their "faith" was physical only, as were the miracles only physical: no one was spiritually healed before the Pentecostal events.

nm828 » Some signs that follow them that believe, thus those who have Faith of the Spirit, thus those who are in Christ's NAME are as follows:

- they cast out the other spirits or demons; they speak in new languages or tongues; they are sometimes not hurt by poison; they heal the sick by the laying on of the hands [Mark 16:17-18]

nm829 » In most cases the healer of a person touches the person being healed (for example Mat 8:15). Sometimes the healer indirectly heals the other person by sending pieces of cloth that he has touched to them, or the person to be healed touches the healer in Faith and then is healed (Acts 19:11-12; Mat 9:20-22; 14:36).

nm830 » In James 5:14 it speaks about anointing the sick with oil in the name of the Lord. Does this mean olive oil? Do we in the Spiritual Church, who worship God Spiritually (John 4:23-24), pour oil over those we are healing. No, we look to the higher sense of olive oil, the Spirit of God is the higher oil (cf 1John 2:20, 27; 1Sam 16:13). The elders of the Church anoint with the Spiritual oil – God's Spirit. It is God's Spirit that heals, not olive oil.

nm831 » It should be noted here that we are healed according to our Faith to be healed. Further we are given the measure of Faith to be healed. Sometimes this measure of Faith will not be given to us to be healed, for God does things that are, in the final analysis, best for us (Rom 8:28). If someone is sick, and dies because he didn't have the measure of Faith to be healed, then if he had the Spirit, it was best that he died. He is taken from the evil (note Isa 57:1); and thus we should be glad for he has run his race, and he is precious in God's eyes (Psa 116:15). But even though we should be glad that one has died from the wrongness of this world, we will be sad, for we will miss him. Yet we know he is away from the world's madness, thus we are glad for him.

Justification

nm832 » What does justification or being justified mean? In the Greek, which the New Testament was written in, the word translated justify means: to become right; to make right or just. In the inspired language justify means that those persons justified have been made right or just. Even the original meaning of the English word "justify" meant to make just. The English word has evolved to mean acquittal of past blame, but the word justify in its inspired language means to make right or just. The word itself says nothing about *how* one is made right or just. Just because a word through misuse has evolved to mean something does not mean we are to use its evolved meaning. "Justify" means to make right or just. We must ascertain how one is made just by the context the word is used in.

How Is One Justified?

nm833 » "Therefore having been justified by faith" (Rom 5:1). People are justified by faith. Again, "we conclude that a man is justified by faith" (Rom 3:28). People are justified by faith as the following verses also conclude: Gal. 2:16; 3:24; Rom 3:30; 4:5.

nm834 » People are justified by faith. But what kind of faith is the Bible speaking about? The true faith is the faith of the Spirit (1Cor. 12:9; 2Cor. 4:13; Gal. 5:22). "People are justified in the NAME of our Lord Jesus Christ, and in the Spirit of our God" (1Cor. 6:11). You are justified when you are in the NAME of Christ and when you are in the Spirit of God. When one has the Spirit they are in the NAME of Christ (see the "Baptism Paper" [NM 4]). When you are in the Spirit you have the true Spiritual faith. When you have this faith you are justified. What all this means is that when you have the Spirit of God you are justified, you are made right. You are made right because you have been given the good Spirit, the right Spirit, and with it you have the New Mind of love and harmony. With the New Mind you have been made right. You have put on the New Mind.

Richness

nm835 »

■ "There is the one that makes himself rich, yet has nothing: there is the one that makes himself poor, yet has great riches." [Prov 13:7]

■ "Give me neither poverty nor riches ... lest I be full, and deny you, and say, Who is the LORD [Jehovah]? or lest I be poor, and steal, and take the name of my God in vain." [Prov 30:8, 9]

■ What good is riches to the owner, except to look at? [Eccl 5:11]

■ The abundance of the riches will not permit the rich to sleep in peace. [Eccl 5:12]

■ If you can't eat or use your riches what good are they? [Eccl 5:19; 6:1-2]

■ "Take heed, and beware of covetousness [over-desire]: for a man's life consists not in the abundance of the things which he possesses." [Luke 12:15]

■ "Now when Jesus heard these things, he said unto him, Yet you lack one thing: sell all that you have, and distribute unto the poor, and you shall have treasure in heaven: and come, follow me." [Luke 18:22]

■ One should not glory in physical riches, but in the richness of knowing God. [Jer 9:23-24]

■ You can't trust in riches and also get into the kingdom of God; it is very difficult for the rich (who trust in their physical wealth) to get into the Kingdom. [Mark 10:24-25]

■ The rich are easily tempted to their destruction. [1Timothy 6:9]

■ In order for the rich to make it into the Kingdom of God they must be willing in certain cases to sell their goods and give it to the poor of the Church. [Luke 18:22; Rom 12:13; 1Tim 6:17-18; Luke 14:33]

nm836 » In the Bible Spiritually speaking, when the rich are mentioned derogatorily, it signifies the children of the enemy; when the poor are mentioned positively, it signifies children of the Spirit.

Joy

nm837 »

■ Christian sorrows are daily as those of the others in the world. [1Pet 5:9; Job 15:20; 31:3; Psa 13:2; 31;10-11; 34:19; 43:2; 44:9-19, 22; 80:4-6; Rom 8:17-18; 1Thes 2:14; 2Tim 3:12; 1Pet 4:1]

■ But the sorrows for Christians are only in the now; total Joy comes when Christ returns. [John 16:20-22; note Mat 5:4; & Isa 66:6-10 with Rev 12:1-2]

■ The world rejoices now. [John 16:20]

■ For folly is joy to those without wisdom. [Prov 15:21]

■ The wicked *seem* happy, they *seem* to prosper. [Jer 12:1; Mal 3:15]

■ But those with apparent joy *now* will mourn and weep at Christ's coming. [Luke 6:25 with Mat 25:30; 22:13; 24:51]

■ Christians will be joyful in God's presence. [Psalms 16:11]

■ We will have joy after we are freed from our captivity. [Psalms 53:6]

■ We will have joy after God's return, then the sorrow will turn to joy. [Jeremiah 31:13]

■ The sorrow shall flee away after God returns, then comes the songs of aeonian joy. [Isaiah 35:10]

■ At Christ's physical return is the aeonian joy, at that time sorrow and mourning shall flee away. [Isaiah 51:11]

■ Christians are recompensed at the resurrection of the just. [Luke 14:14]

■ When Christians awake or are resurrected they will then be satisfied. [Psalms 17:15]

■ Christian's joy is according to the joy of the harvest. [Isaiah 9:3; Rev 14:15; Mat 13:39]

■ Christians will be glad in God's salvation after they have waited for him. [Isaiah 25:9]

■ God will create Jerusalem or the Church (Rev 21:2, 9; Gal 4:26) a rejoicing, and her people a joy in the new heaven and earth. [Isaiah 65:17-19; Rev 21:1-4]

■ Physical and Spiritual Israel will be happy when God returns to his people at his coming. [Psalms 53:6]

■ At God's coming, then the mourning ends. [Isaiah 60:20]

■ In God's salvation will our hearts rejoice. [Psalms 13:5]

■ Now Christians sow in tears, but they shall reap in joy. [Psalms 126:5-6]

■ The joy Christ left with Christians is the joy of the Holy Spirit. [Acts 13:52; 1Thes 1:6; one of the fruits of the Spirit is Joy, Gal 5:22]

■ But this "joy" is not *full*, for the Spirit is given in measure (Eph 4:7, 16). The tribulation of the world (John 16:33) makes us suffer now, but full joy comes when Christ comes as the above scriptures indicate.

■ The "joy of the Lord" is in the kingdom of God. [Mat 25:21]

■ Rejoice because of the future hope. [Luke 6:22-23]

■ The Hope of the righteous shall be gladness. [Prov 10:28]

■ Christians rejoice for the Hope of the future great glory of God and his Kingdom. [Romans 5:2]

■ Christians rejoice in Hope. [Romans 12:12]

■ Christians rejoice now in their sufferings because of the hope of Christ's coming and the exceeding joy thereof. [1Peter 4:13]

Prayer

nm838 » The word "pray" from the Hebrew and Greek text of the Bible means: to ask; to wish for; to meditate; to pour out. Praying to God is pouring out of ones's self to God, and asking for something, or meditating about things, ideas, or ideals to God: it is communicating with God.

nm839 » All over the world today you see people "praying" to God. They get on their knees to pray; they lay flat on their faces to pray; they flog themselves to pray; they do penitence when praying; they squint their eyes and twist their faces when praying, the more they squint their eyes and twist their faces the more they think God will respond. What are they saying when they pray in this manner? Does the all powerful God need this kind of praying? The all knowing and all powerful God can read minds; He knows things you need *before* you ask for them (Mat 6:8). Is God like some kings of the old age who had a need for people to kneel before them, and to speak glowing words to them? Does God need our glowing words to help uphold His confidence in Himself? Does God need us to beg for help, when He knows we need His help? We are His children. Will He allow us to destroy ourselves permanently if we do not pray while squinting our eyes, twisting our faces, and flogging our bodies? Christ prayed in a physical way to teach us the Spiritual way of praying (John 11:41-42; etc.). There is a Spiritual way to pray.

Prayer Is Communication

nm840 » *In the physical or typical meaning of the Bible we learn the following.* We should admit our wrongs to him (Lev 5:5; Num 5:6-7; 1Sam 12:10; Psa 32:3, 5; Psa 51:3, 4; etc.). We should thank him for the present good and the coming Good (Phil 4:6; Acts 27:35; 1Cor 10:30-31; etc.). When we pray we should meditate on his Word (Joshua 1:8; Psa 1:2; 19:14; 104:34; 119:15-16; etc.). Prayer brings one closer to God, for it establishes a bond, or a relationship with him. Prayer is talking to God, is communicating with God. We can speak to God as if he is our friend (John 15:14-15).

What Do We Pray For?

nm841 » We ask only what God wills or wishes (Mat 26:39 & 1John 5:14). We can't receive what God has *not* prepared for us (Mat 20:20-23). We can find out about God's will through his Word, the Bible. So, in order to know what to ask for, we must know the Bible. To know the Bible we must read and study it. We should pray for the good of all, for others, and for us, not selfishly (Job 42:10; 1Kings 3:9-14; Mat 6:9-13). If we pray for ourselves it should be for strength to endure the world (Psa 31:3; 4:1), for wisdom to help others (1Kings 3:9-12), to overcome our weakness (Rom 12:21), and so forth. We do not pray for decaying physical wealth, for we know life is not the abundance of the things which we possess (Luke 12:15), for in this age wealth only brings worry and trials (Eccl 5:10-13; Mark 10:30). When we pray for physical things it is for what we need, such as our *daily* bread or food (Mat 6:11; Acts 2:44-46; 4:34-37).

Does God Hear Our Prayer?

nm842 » God hears the prayer of those who do his will, who are thus in the Spirit; He hears them always (1John 5:14; 3:22; John 9:31; 11:42).

When Does God Answer Prayer?

nm843 » The True God fulfills our prayer only if it is in harmony with the final Good. The good God does not fulfill prayer of harm to others, for we are to give good for evil, not evil for evil (Rom 12:14, 17). Many prayers in harmony with the word of God will ultimately be fulfilled

beginning at Christ's return to the physical world (Psa 102:2, 13; Luke 14:14; Psa 17:15; Isa 25:9). The whole book of Psalms in its higher or Spiritual meaning, projects to us that most Christian's hope and prayer will be fulfilled beginning at God's return to earth. God answers our prayer if we have the Faith to believe he will fulfill our asking (James 1:5-7; Mark 11:24). This Faith comes from the Spirit of God (Gal 5:22). This Faith is given by measure (Rom 12:3, 6). Therefore we believe enough to receive an answer only when we have the given Faith to believe.

nm844 » *In Summary.* God answers prayer only when it is for the good, when it is within the rules of creation, when it is for the best of the person according to the predestinated plans for that person (Rom 8:28).

Pray Always

nm845 » Daniel prayed three times a day (Dan. 6:10), David prayed three times a day (Psa 55:17), some prayed seven times a day (Psa 119:164), but Christ said to pray always (Luke 18:1), and Paul also said to pray without ceasing (1Thes 5:17). We should pray always (Eph 6:18). The Spirit in Christians in a sense enables them to be a continual prayer (Rom 8:26;Heb 13:15-16; Rom 12:1). Because praying to God is communicating with God and because we love God, Christians consciously will pray often. Since prayer is talking to God, one can "talk" to God almost anywhere, and in any position (Luke 23:46; Luke 22:41; Psa 4:4; Neh 8:6; etc.). We should not pray just to be seen (Mat 6:5), for we can pray secretly (Mat 6:6) within our minds anywhere and at any time. This is not to say that praying in public is wrong. Jesus prayed openly in front of some people in order to teach them (John 11:41-42). We should not use worthless repetitions (Mat 6:7), but pray with all our hearts (Luke 22:44; Ho 7:14; James 5:17).

nm846 » Praying is not magical or repetitious (Mat 6:7). Praying is merely talking to God, rightly. Thus, talking within his will or wish. Therefore talking with God about the good, for the good, and thanking him for the good now, and for the Good to come. God is not ritualistic. There is no set position in which to pray. There are no set words to say. We pray with almost any words as long as they convey the meaning of what we want to talk with God about. There are no set number of minutes that we must pray to him. We think on the good, and talk about the good always with him everywhere within our minds. At times it may be easier to concentrate when we go to a spot in private and pray or communicate with Him (note

Mark 1:35). Sometimes we might need to communicate all night as Christ did at times (Luke 6:12).

Who Do We Pray To?

nm847 » In the Old Testament they prayed to the Father, to the YHWH, the BeComingOne, also called Jehovah or Yehowah by some. Jesus taught the manner in which you should pray (Mat 6:9-13; 26:39). But near the end of his ministry Jesus told us that we could ask (pray) *in his name*, "that I will do, that the Father may be glorified in the Son" (John 14:13-14; 15:16; 16:23-26). Why did Jesus say that? Because Jesus was about to be given *all* God's power (John 16:15; Mat 28:18), He would be going into the glory of the God (Jn 13:31-32), He would be given God's Name (John 17:11), which is Christ's New Name (Rev 3:12). God's name was shown to Moses in the Old Testament. God's Name is YHWH or the BeComingOne, or the "One (who) Will-Be" (See GP 1). "And whatsoever you shall ask in my name, that will I do, that the Father may be glorified in the Son" (John 14:13). Jesus will not relay these prayers to the Father (John 16:26). Why? Because Jesus is now in a sense the Father (Isa 9:6; see GP 5: ¶ gp487), sitting as the very right side of God (Acts 2:34). The Spirit for Christians, predestinated before the creation (NM 8), is now given through Jesus (John 20:22; Act 2:33). Thus, He is now in a sense our Spiritual Father (Isa 9:6). He has the Name of God; He represents the God, who will fill all in all (Note 1Cor 15:28; GP 6). , for He will fill all in all (Eph 1:10, 23; GP 6), but now, not all that will be in Christ, is in Christ (Heb 2:8; see *God Papers*, GP 5). Who do we pray to? We pray to Jesus, for the Father has glorified Himself in Jesus. Our God, who is the "Will-Be-One" (YHWH) is now represented by Jesus Christ, the Right Hand of the God, who will fill all in all (see the *God Papers*).

Prayer's Higher Meaning. With the New Mind: we pray always; we are always in contact with God because we are ONE in the Spirit of God, which is the Spirit of Christ, which is the Spirit of Christians. God does not need anyone to get down on their knees to Him, squinting their eyes, and doing penitence to get his attention. God is not a monster, or a pretentious king. God is/will-be a friend to all. In our minds we pray to God (Jesus was called: "my lord and my God") at any time, any where, because we are in the Spirit, and in contact or communication with God through His Spirit, which is the Spirit of the Father, Son, and Holy Spirit – the one same Spirit who is in all who will-be in Him (GP 6:gp483).

Living in Common

nm848 »

■ Those who believed at first sold their possessions and goods and lived in common; each had only what he or she needed. [Acts 2:44-46; 4:32, 34-37]

■ Christ said in order to be complete ("perfect") one must sell all and give to the poor. [Mat 19:20-21; Luke 18:22; Mark 10:21]

■ The Spiritual poor are Christians. [Isaiah 66:2; 14:32; Mat 5:3; Luke 6:20; Rev 2:9]

■ To be a disciple of Christ one must forsake all. The higher meaning here is that we must forsake all of the old mind and its way in order to be a disciple of Christ in the truest sense. [Luke 14:33]

■ Christ came that we might have abundant *life*. [John 10:10]

■ But, "take heed, and beware of covetousness [desire to have more than one's share]: for a man's life consists not in abundance of the things which he possesses." [Luke 12:15]

■ Sell your possessions and give, thus providing a treasure in heaven or in the Spiritual world. [Luke 12:33]

■ For he that lays up treasures for himself is not rich towards God. [Luke 12:21]

■ Notice the parable of the treasure in heaven. [Mat 6:19-21]

■ Notice the parable of the treasure in the field and of selling all to buy it. [Mat 13:44]

■ Notice the parable of the pearl and of selling all to buy it. [Mat 13:45-46]

■ Notice a principle of sharing: "He that has two coats, let him impart to him that has none; and he that has meat, let him do likewise." [Luke 3:11]

■ Notice another principle of sharing: "But by an equality, that now at this time your abundance may be a supply for their want, that their abundance also may be a supply for your want: that there may be equality: As it is written, He that had gathered much had nothing over; and he that had gathered little had no lack." [2Cor 8:14-15; note Ex 16:18]

■ Christians labor in work that is good so they may have something to give to others in need in the Church. [Ephesians 4:28]

- Paul asked through Timothy that the physically rich of the Church do good so that they would be rich in good works; they must be ready to distribute, and must be willing to share. In this way they would be laying up in store for themselves a good foundation against the time to come, the time of the New Age, so that they may lay hold onto aeonian life. [1Timothy 6:18-19]

- Paul asked the Christians to do good and to share. [Hebrews 13:16]

- There is a difference between physical wealth and Spiritual wealth. [Revelation 3:17; 2:9]

- Those who make themselves poor for God are rich. [Prov 13:7]

- Wisdom and understanding are better than wealth. [Prov 16:16; 3:13-15]

- Better the poor that walk right, than the rich who do wrong. [Prov 28:6]

- The ungodly prosper in this age. [Psalms 73:12]

Book of Life

nm849 » What is the book of life? Whose names are in the book? The key to this truth is in Revelation 13:8. In the King James Version this verse reads: "and all that dwell upon the earth shall worship him [the beast], whose names are not written in the book of life of the lamb slain from the foundation of the world." So we see that all will worship the beast except those names written in the book of life of the lamb, Jesus Christ.

nm850 » This verse says that those written in the book of life were written into the book of the lamb (Jesus) who was as good as slain [Greek verb, perfect] before the foundation of the earth. This is confirmed in Revelation 17:8, "and they that dwell on the earth shall wonder, whose names were not written in the book of life from the foundation of the world, when they behold the beast." Thus, the book of life has names written in it since the foundation of the earth. But those who worship the beast haven't got their names in this book. Who are those with their names in the book of life?

Christians in The Book

nm851 » Those with their names in the book of life are Christians according to Paul: "and I entreat you also, true yoke-fellow, help those women which labored with me in the gospel, with Clement also, and with other my fellow laborers, whose names are in the book of life" (Phil 4:3). Thus, we see that the Christian's names are written in the book of life. And in Hebrews 12:23 we see that "the general assembly and church of the first-born" are written in heaven. Also in Luke 10:20 the seventy that were sent out two at a time were told to "rejoice because your names are written in heaven." Thus, we see that the followers of Christ have their names written in heaven. Being written in heaven is being written in the book of life.

nm852 » In 2Timothy 1:9 (in its Greek text) we see that those called were called "before the times of eons." And in Ephesians 1:4 we read "according as he has chosen us in Him before the foundation of the world, that we should be holy." Further, in Revelation 3:5 we read in Jesus Christ's message to the Sardis church: "he that overcomes, the same shall be clothed in white raiment: and I will not blot out his name out of the book of life." Surely, we can conclude from this that it is the true Christians, the called, chosen, and predestinated, who are the ones written in the book of life from and even before the world's foundation.

Good News

nm853 » Now here comes good news for real Christians. Notice in Revelation 3:5 Jesus says: "I will not blot out his name out of the book of life ... he that overcomes." True Christians are written in the book of life before the world began, and here Jesus said he would not blot out their names if they overcame. Please turn to Hebrews 10:39. There we see Paul speaking to true Christians and saying: "But we are not of them who draw back unto perdition; but of them that believe to the saving of the soul." And in 1John 5:4 we read, "for whatsoever is begotten of God overcomes the world." The reason for this is "because greater is he that is in you [the Spirit of God], than he that is in the world [the spirit of confusion]" (1John 4:4). God has predestinated real Christians "to be conformed to the image of his Son, that he might be first born among many brethren" (Rom 8:29).

nm854 » In Revelation 20:12 we read about the day of judgment: "and I saw the dead, small and great, stand before God; and the books [of the Bible] were opened: and another book was opened, which is the book of life: and the dead were judged out of those things which were written in the books [of the Bible], according to their works." Comparing this with verse 15 and other verses, we see that those of the "dead" during the day of judgement have not their names yet in the book of life. But they are to be put into the book of life when they are resurrected from their 1000 year judgment into the day of atonement (see, "Thousand Years and Beyond"). For all will be saved eventually as God clearly tells those who take God's word as truth.

nm855 » Go back to Revelation 13:8 and 17:8 and see that those who worshiped the beast were not in the book, for as we have shown above only those in the church of the first-born (Christians) are in the book of life now. Thus, those who worship the beast will, after the day of judgment, be put into the book.

nm856 » In Revelation 21:27, after describing the New Jerusalem it reads: "and there shall in no way enter into it any thing that defiles, neither works abomination, or a lie: but they which are written in the Lamb's book of life." So we see that no one can enter New Jerusalem except those written in the book of life (see "New Jerusalem Paper" [NM 18]). One must remember here, that all are under sin (Rom 3:9) and all have sinned (1John 1:8) except Christ (Heb 4:15). Thus, those who are written into the book of life have or will die to sin (Rom 6:10). The Christians are to die to sin through Spiritual baptism. The others will have died to sin by the time the millennium is through, then they too will be added to the book of life.

nm857 » One last proof that real children of God are those now written in the book of life is in Daniel 12:1. Here it says that "at that time your people shall be delivered, everyone that shall be found written in the book [of life] ... " The time setting of this verse is at Jesus Christ the Messiah's return.

nm858 » This book of life is the very "book of remembrance" (Mal 3:16-17).

NM 19: Reason Why

NM19 Abstract

Why is there evil? If God is all powerful, if God is good and if he created all things, then why is there evil? Could there be a reason for evil? Yes, there is a reason for evil and it has something to do with the law of knowledge. In fact, there cannot be good without evil. Good and evil are comparative qualities that need each other in order for us to know either quality.

Why is there Evil?

nm859 » Why is there evil in this life or age? Why is the world the way it is? Why is there disease? Why do children get sick? Why are there natural catastrophes? Why is there war? Why is there death? Why is there hunger? Why this world? Why the confusion and tears? Why has God "allowed" evil? Or, to be more blunt, since the most powerful being created the all, why has the Power, why has the God (YHWH) created evil?: "forming light and creating darkness, making peace and creating evil; I, the LORD [YHWH], do all these things" (Isa 45:7, see Hebrew text). In the book many call God's book, the Bible, it says that God created evil. The original text (Hebrew) says this, not some translation of the text.

nm860 » "And the LORD said, Behold, the man is become as one *from* us, to know good and evil" (Gen 3:22, see Hebrew; see Greek also). This comment was made right after mankind had broken God's first commandment by the influence of the serpent (see The "Other Mind" paper [NM 20-22] for more details). Thus scripture says that man was getting to know good and evil from the plurality ("us") of God (LORD or YHWH). From the "us" of God man is learning good and evil. There was/is a plurality to God (See the God Papers).

nm861 » In the middle of the garden of Eden was "the tree of KNOWLEDGE of good and evil" (Gen 2:17). It was a tree of good and evil, not just a tree of good or not just a tree of evil. It was not just an ordinary tree, but a tree of *knowledge*. After mankind took from the forbidden fruit from the tree of knowledge of good and evil, God said man was getting to KNOW good and evil (Gen 3:22). God then took away the tree of life and placed the cherubs to guard the way to the tree of life (Gen 3:23-24). The Hebrew word translated "*from* us" in Genesis 3:22 can also be translated "*out of* us" or even "*of* us" as it is translated in most English Bibles. Because of Adam and Eve's behavior mankind did at this time go "out of" the God, but also, since the God knows all, including good and evil, then mankind was becoming like ONE *of* the God (of the "us" [His hidden plurality]) by learning good and evil. "One" here can be translated "whole" since in history the word one was more likely to mean "whole" or "unity" rather than just the number one (See the God Papers under "One Yehowah"). Consequently, as events manifested, man was mostly left under the influence of the evil spirit of Satan, who was symbolized by the serpent of Genesis (see "Other Mind" paper [NM 20-22]). In the New Testament Paul said we were and are under the influence of the devil/Satan/evil powers and so forth (Eph 6:12).

nm862 » In my studies it has become obvious to me that the one basic reason that mankind was left under the influence of Satan was to learn good *and* evil. To learn, not just good, not just evil, but good and evil. But why is God allowing this evil age to go on until the appointed time? (Mat 24:3 & Acts 1:6-7) Maybe God is evil or partly evil and wants us to suffer under evil? Or is God too weak and can't stop the evil or confusion of our existence? Why doesn't the God of love, the all powerful ONE, stop the evil and the general confusion of the present age? Maybe, just maybe there is a logical and reasonable reason?

nm863 » If you read the *God Papers* you will see the scriptures that indicate that the true God is ALL MIGHTY. Thus, He has the power to stop the evil, if that is what He wishes. But God has allowed this kind of world because He knows man *must* suffer or live in an age of confusion and unhappiness in order to be happy. Does this statement shock you? What are we saying? Man must suffer. As we will show, a purpose of creation is for man to develop the cognition of good *and* evil. But the reason we are learning good *and* evil is that we cannot truly learn good without also learning evil. For us to even understand what is good, we must know evil. Further, we must all learn evil by living it, for experience is the best teacher. Scripture and the idea and definition of God tells us that God is all powerful and thus could have created a non-changeable paradise-like-environment at first with each human being physically perfect and unable to die. But God did not do this because he understood that man must first suffer in order to be happy. God cut off the tree of life from mankind in order to allow billions of people to learn good *and* evil. This is very important so do read on.

Why Know Evil

nm864 » Why is it important to know good *and* evil? Why know evil? Why live evil to know it? The main difference between a man and any other animal is his higher power to reason and know. So far, it is true he has misused this power, but, nevertheless, greater knowledge is what makes man greater than most other creatures of God.

nm865 » But why know evil at all? Why not just know good? Why do we need to know evil? Before we answer this we must know how one knows evil.

nm866 » *Experience Teaches*. In order to know something, to truly know something, you must live it. It takes experience with evil to know evil. Our very life today teaches us that. How can you know pain if you had never felt it? How can anyone explain pain to you if you have never felt pain? Just stop and think for a moment. Try to imagine that you have never felt pain. If someone showed you someone else in pain, would you know what it was to be in pain, if you had never *felt* it? As you looked, you would see this person with an expression on his face like he was in pain. But how can you know pain through the face of a person in pain? Remember you have never *felt* pain. Any outward sign of a person in pain is just that, a sign or symbol of pain. Just because you see someone in pain, it does not mean you *know* pain for remember you have never *felt* pain, or *experienced* pain. You must *feel* pain to know it.

nm867 » The same applies with evil. To truly know evil, one must live it. How would you explain misery to one who never felt or lived misery? How would you explain the pain of losing a loved one to someone who has never felt such a feeling? Now on this latter example, you could compare it with some other form of misery or pain. But, what if the person who you

were trying to explain this grief to, had never felt any grief, misery, or pain? You could never compare your grief of losing a loved one with anything that would allow that person to know of your misery. To obtain the knowledge of knowing evil, then, you must *live* it and *feel* it. To obtain the characteristic of knowing evil we must live in such a world as we now live in.

Know Evil To Know Good?

nm868 » But this is only a part of the overall picture. We must know evil to know good! Evil and good are inseparable! We must suffer evil to know good. Again, does that shock you? But why should it? Every day we live, we prove the principle that one cannot know good without real knowledge of evil. Every day that we obtain knowledge, we live this principle, and prove this principle. One cannot know good unless one knows evil. You cannot separate the knowledge of good and evil. The very Law of Knowledge tells us that. What is that law?

Law of Knowledge

nm869 » **Basic Definition of the Law of Knowledge can be stated as:**
*Knowledge of **A** is equal to and dependent on the knowledge of **non-A**.*
　　Where **A** can be any particular object, technique or belief;
　　n**on-A** is anything but that particular object, technique or belief.
It follows —
　　The depth of one's knowledge of **A** (and it truthfulness) is contingent upon the depth of one's knowledge of **non-A**; particularly, in the case of opposite qualities (light and darkness), you must know both qualities to know either; you must compare each with the other to know either.
In other words —
● To know **A** you must also know something to everything about **non-A**;
● The knowledge of **A** presupposes at least some knowledge of **non-A**;
● In order to know **A** you must compare **A** with **non-A**;
● the knowledge of **A** (and its truthfulness) is proportional to the knowledge of n**on-A**.

nm870 » **True Knowledge through the law of knowledge:**

The continuum from incorrect knowledge —> to absolute true knowledge

- The less one knows about **non-A**, the less one knows about the truthfulness of **A** and the more likely one's knowledge is incorrect.
- The more one knows about **non-A**, the more certain one knows the truthfulness of **A**.
- If one knows all that is **non-A**, one knows absolutely the truthfulness of **A**.

> (An omniscient being would know the full truth; less than omniscient beings would not know the full truth.)

nm871 » In explaining the Law of Knowledge, we will first deal with how one obtains knowledge of opposite qualities. Next we will explain in a more general manner, in its most broadest sense, how one obtains knowledge of anything.

Knowledge of Opposite Qualities

Blind: Light & Darkness

nm872 » To amplify on this law we will use the example of a blind person. Try to empathize with a person that was totally blind from birth. Try to put yourself in such a person's mind. Close your eyes and imagine yourself as being blind. Now such a person has never seen light. Light is the quality that allows one's eyes to see objects. Without light no one would see even if they had perfect eyes. Light is the quality that the totally blind person cannot perceive or comprehend.

nm873 » If you had never seen light, how would someone explain light to you? What choice adjectives would describe light to someone who has never seen light? To explain anything to someone who has never seen it, you have to use comparison, and say it is like this or like that. But there is no comparative quality in the universe that compares with light. It would be impossible for someone to explain light to you, let alone sight, if you had never seen light.

Knowledge of Each Presupposes Knowledge of Both

nm874 » Yet at the same time one truly does not know what *darkness* is until one has seen light. The very definition of dark is: "without light." Darkness means without light as light means "without darkness." Each definition is dependent on its opposite quality. A definition of something is a statement of the knowledge of that thing. To know light or darkness by their very definition presupposes

knowledge of each other. A blind person in order to know what darkness is, would have to see light. He knows darkness only if he sees light, for it is only then that he will understand what people were talking about when they spoke of darkness. The only reason that you can close your eyes, and call the result darkness, is because you have *seen* light. One cannot know darkness or light unless one has seen both and compared both qualities with each other.

nm875 » Thus, specifically in the case of opposite qualities, your knowledge of darkness ("A") is dependent upon your knowledge of light (opposite-"A"), and vice versa. Because they are opposite qualities, you must know both to know either quality, but in order to know either quality, you must compare each with the other.

nm876 » **Furthermore**, remembering that a blind person is blind because he cannot see light, it also follows that if there was only white light we would also be blind because we would not see or recognize any object, since in order to see anything, we need different shades of light and darkness, or more correctly since most of us see in color, in order to see anything, we need different shades of light and darkness and different hues of color.

Sound And Silence

nm877 » The same applies for sound and silence. If you had never heard sound, how would you know what silence was like? Sound and silence are opposite qualities as light and darkness are opposite qualities. You must know both to know either, and you must compare each with the other to know either. Since these two qualities are interrelated, one has to know both to know one. The very basic definition of sound ("without silence") and silence ("without sound") need the opposite quality to define it. To know sound or silence by their very basic definition presupposes knowledge of each other.

Hot and Cold

nm878 » The same can be said about hot and cold. "Hot" and "cold" are relative opposite qualities. One knows something is cold only so far as he has something hot to compare it with. You can place your hand into a container of water that is 90 degrees and it will feel warm to you. But if you place your hand into a container that is 110 degrees and keep it there for a while, and then place it again into the container of water of 90 degrees, the 90 degree water will then feel cool while before it felt warm. Your knowledge of hot or cold is obtained through contrast and comparison of both qualities. Knowledge of hot or cold presupposes knowledge of the other quality.

Life and Death

nm879 » Further, one does not know what life is until he has seen death. To have knowledge of life you must have knowledge of death. One is very aware of life only if one has seen or become aware of death.

nm880 » Adam and Eve didn't know death and that is one reason why they chose the tree of good and evil in the garden of Eden. Adam had never seen or felt the pain of losing a loved one. All he saw around him was life. This is very difficult for us to perceive today, for all around us are the living, the sick and dying as well as our remembrance of dead friends and relatives. Because of this we know a lot more about life and death than Adam and Eve. It is difficult for us to put ourselves into Adam's position.

Right and Left & More Examples

nm881 » The right side has no meaning unless there be a left side. You don't know what the meaning of right is until you know about left; you don't know what left is until you know what about right. You need knowledge about both to know either. You don't know something is "high" unless you know there is something "lower." You don't know something is "low" unless you know something is "higher." You don't know a "plus" quality until you know its "minus" quality. You don't know a "minus" quality unless you know its "plus" quality. You don't know light if you don't know darkness. But you can know light if you know darkness. You don't know or realize harmony, if you have never known confusion. Think on what is being said. If you had always lived in an environment where there was no confusion, where there was harmony, would you realize the goodness of that harmonic environment? Would harmony mean anything to you in such a harmonic environment? Can you really *appreciate* harmony if you have never lived in confusion?

nm882 » If you had good vision for forty years, and then lost your sight, you would truly know the value of sight, as does a blind person who miraculously gains his sight. But how does someone, after he loses his sight, come to appreciate the sight he once had?

Appreciation

nm883 » What does it mean to appreciate something? Webster's Dictionary says that to appreciate something one must: "recognize it gratefully; estimate its worth; estimate it rightly; be fully aware of it; and notice it with discrimination." Thus, when one comes to appreciate something (especially if it is good), one in fact comes to know that thing. To appreciate something is to really know it; to know something is to appreciate it.

nm884 » When one loses a loved one, one by the loss of the loved one knows the worth of the loved one. The same with good. One comes to know the worth of good only after he has lived in evil.

nm885 » How can we know joy, until we have lived sorrow? How can you really become happy unless you have been sad. How can we know good until we know evil? Opposite qualities need to be compared to each other to know either.

nm886 » The Law of Knowledge not only explains knowledge of opposite qualities, but also knowledge of everything capable of being known. The following is a short explanation of how we learn, not only about opposite qualities, but about everything.

The General Law of Knowledge

How Children Learn

nm887 » One way to understand the Law of Knowledge is to understand how a child learns. Children's simple generalizations reflect lack of differentiation. That is, a child's wrong generalization about *A* (cow) reflects lack of knowledge of the difference between a cow and all that is not a cow (*non-A*) such as other four legged animals.

nm888 » A child when he is first learning about four legged animals sometimes may mix up a cow and a horse, or a cow and a deer, or even a cow and a dog. This is because the child does not know what a cow is not. When parents first begin telling their child what a cow is, they point to a cow and say, "that is a cow." The child with the aid of other knowledge in his memory and his senses "sees" this living animal with four legs. Depending on how many other four legged animals are pointed out to him, he may mix the cow up with any or all other four legged animals.

nm889 » After a cow is pointed out to him he may call a horse a cow, after all, to the child a horse is a four legged living animal (not a two legged animal or a toy animal or stuffed animal) just like the one pointed out earlier by his parents. But the child is wrong. This four legged animal is a horse, not a cow. The child fails to differentiate between a cow and a horse. How does the parent correct the child? The parent says, "no, it is not a cow, it is a horse." The parent is telling the child what a cow is not. The parent by telling the child what is not a cow is helping the child to learn what is a cow. Normally, after the child learns that a horse is not a cow, he doesn't call a horse a cow again. But the child may call a deer or other four legged animals a cow. When the child does this he is again corrected, "no, it is not a cow, it is a deer." The child has learned something else is not a cow (*A*); he has learned one more of the *non-A's* (all else besides cows). The more the child learns about other four legged animals not being cows, the better he is able to understand what a cow is. A cow is a four legged animal of a certain size (a cow is not a dog because for one thing a cow is bigger than a dog, etc.), but it is not any other four legged animal: it is not a dog, it is not a horse, it is not a deer, it is not an elephant, it is not a bear, etc.

nm890 » But further the child from other knowledge knows a living cow is not a mountain, it is not dead (not a dead toy, not a dead stuffed animal, etc.), it is not a rock, it is not the sky, it is not a two legged animal, it is not an ant, it is not a fish, it is not fog, it is not a color, it is not a quality like "good," it is not a plant, it is not water, etc. The child knows more what a cow is, by the more he knows what a cow is not. **Thus, the knowledge of a cow (*A*) is dependent on the knowledge of what a cow is not (*non-A*); or the child knows more about what is a cow (*A*), by the more he knows what is a cow is not (*non-A*).**

The Color Green

nm891 » Let's take another example, the color **green**. The more we know what the color green is *not* the more we know the uniqueness of the color green. The only way to point green out is to show what green is *not*. Since most of us know what the color green is (because we know what green is not), we will again try to understand how a child learns about the color green.

GREEN a color is "A"

nm892 » **The knowledge of GREEN (A) is dependent upon the knowledge of all that is not green (non-A).**

- First "green" is a subdivision of color. Before a child can learn what the color "green" is, he must know what is color. In order for a child to understand "color" his parents tell him, "that thing is the *color* red, that thing is the *color* blue, that thing is the *color* orange, that thing is the *color* green, that thing is the *color* ...." Along the line of learning "color" the child comes to understand (through comparison) what "color" is *not*: the color blue on a wall is not the wall, it is not the *material* that makes up the wall such as wall board, or wood studs, or nails, etc., but the quality on the wall that we call "color" is the *color* of the wall. A child learns what color is by understanding what color is not. So before a parent can make a child understand what the "color" green is, the child has to understand what "color" is, by understanding what "color" is not.

Now assuming that the child knows what "color" is we will continue:

- We know GREEN by knowing what is *not* green (non-A). Thus the child comes to know GREEN by knowing what is not green.

What the color green is not (non-A)

nm893 » **Green Is Not:**

- *More generally green is not*: a tree, a bush, a rock, an animal, a fish, a man, the universe, the sun, the moon, our parents, a car, a road, atoms, space, form or shape, relative position in space, time, a dimension, or any other thing or quality except for a quality we call "color."

- *More specifically green is not*: red, blue, orange, purple, or any other color, but the color we call green.

To summarize, *GREEN* is A; *GREEN* is not non-A. We know *GREEN* (A) because we know what *GREEN* is not (it is not non-A).

God Has Created Evil? ...

nm894 » Considering the above it is not difficult to see why God (YHWH) has *created* evil (Isa 45:7): it is so we can know good, to know good's worth, to appreciate good, and to enjoy good. The reason we must suffer the effects of evil is so we can know, to truly know good. To know what is good we must have something to compare good with. God through his wisdom has given mankind a time for good and a time for evil (Eccl 3:1-8), so as to know each. Thus in this way mankind comes to realize the value of good and harmony. God has given us joy to balance against adversity, so as to know joy (Eccl 7:14). To be able to know goodness, one must know evil: "For in much wisdom is much grief: and he that increases knowledge increases sorrow" (Eccl 1:18); "Sorrow is better than laughter: for by the sadness of the face the heart is made to be good" (Eccl 7:3).

Should We Then Seek Evil?

nm895 » Considering the above, then does this mean we should seek evil? No! Once we come to realize how bad evil is, then evil has served its purpose as the comparative quality to good. In good is where the happiness lives, not in evil. We in this age are mainly learning more about evil than good. There are moments of joy and happiness in this world which allow us to partially perceive just how bad evil is, and at the same time allow us to perceive how precious good is.

New Mind And True Knowledge

nm896 » As we are trying to communicate in this book, the best way to truly perceive good is only with God's Spirit – the New Mind. Through God's Spirit man begins to renew his knowledge and mind to the ways of good (Col 3:10; Rom 12:2). Before man receives God's Spirit, man is like a blind man: he lives in darkness, yet comprehends it not, for the blind do not know light. "And the light shines in the darkness; and the darkness comprehended it not" (John 1:5). Why? Because this world is Spiritually blind, this world or this age cannot perceive their sad state of affairs. This age and most people in it, do not and cannot know how bad this age really is until they receive God's Spirit – the New Mind, which is the Spirit of truth (John 14:17). This age only partially perceives how bad this age is, and this only because there is some joy in this age to compare with the average state of affairs. But those who have received God's Spirit know ever so much more just how bad this age is (Rev 12:11).

Two Forces

nm897 » Scripture project to us that there are two spiritual forces or mental forces in the world today: God's and Satan's. God's Spirit is "A" and Satan's is contrary or opposite to "A." Your knowledge of God ("A") is dependent upon your knowledge of Satan (opposite-"A"); To know God ("A") you must compare Him with Satan (opposite-"A"). Mankind will only have the knowledge of good and evil after they live under the bondage of Satan's rule and under the harmony of God's rule. That is why all who are eventually born of God will and **must** live under Satan's spiritual law of confusion *and* under God's Spiritual law of harmony.

nm898 » All must suffer evil. So that "they that sow in tears shall reap in joy" (Psa 126:5). The tears come first for man, the joy is the dessert of the creation. We learn unhappiness or the knowledge of sin through Satan's way. And it is through this knowledge of sin that we are able to truly know good, for then we have something to compare with God's way and his law of harmony. It is through God's Spirit and His law in our minds that we see the good (the light). And it is because of our former blindness (Spiritually speaking) concerning the good (light) that we are able to comprehend the worthiness of the good. Mankind is like a blind person who has lived in darkness (Satan's way) yet really didn't know how bad it was until he gained (or will gain) his sight (through God's Spirit) and was made able to comprehend the light (good), then all became understandable to him.

Two Forces Help Us to Distinguish and Know

nm899 » "Except they give a distinction in the sounds, how shall it be known what is piped or harped?" (1 Cor 14:7) Except that there be a period of time to distinguish between good and evil, how else would mankind learn or understand what is good?

nm900 » Since the knowledge of God depends upon the knowledge of Satan, then man must have a period under the way of Satan and a period under the way of God in order to understand the Goodness and worth of God and His way. "A time to love and a time to hate; a time of war and a time of peace" (Eccl 3:8). "Better is the end of a thing than the beginning thereof" (Eccl 7:8).

nm901 » *Mankind in School*. One could say that mankind is going through a learning process. Mankind is in school. Man is going through a process of discriminating between plus and minus qualities. Mankind is learning to discriminate between good and evil, by living each. Man is living each for it is impossible to teach it through words. How can you know pain through words? How can you teach a blind person what light is by words? No, man must *feel* pain to know pain, and the blind must *see* light to know light. But further, the blind must see light to know darkness, for our very definition of light ("without darkness") and our basic definition of darkness ("without light") projects to us that opposite qualities need each other to *know* either one of the qualities. A totally blind person even though he lives in darkness, does not know darkness until he sees light. We only know darkness because we have seen light. To know what is darkness one must have something to compare it with.

nm902 » We know something is "up" only because we see something below it in position. If everything were of the same height, there would be no "up" or "down."

nm903 » The same principle holds true for pain and non-pain, or sound and silence, or for that matter clean air and smog. But, what is important to us in this paper is that this principle holds true for good and for evil. If you only had lived in an environment of harmony, how would you know it was a good environment? You would have nothing to compare it with. You would be like a person who lived all his life at the top of a hundred story building in a room without any window or way to go downstairs. Even though you have 99 levels below you, you do not know you are at the top, for you do not know there is a down.

Law of Knowledge

(Pertaining to Opposite Qualities)

Both sides complement the other and give meaning to each other;
you must know both qualities to know either: you must compare each with the other to know either

One Side	Opposite Side
love	hate
light	darkness
right	left
front	back
up	down
affection	contempt
good	evil
peace	war
kind	unkind
forgiving	unforgiving
thankful	unthankful
reconciliatory	revengeful
lawful	lawless
hope	hopeless
truthful	liar
fairness (impartial)	unfairness (partiality)
temperance	overindulgence
honorable	dishonorable
unpretentious	pretentious
elegant	crude
patient	impatient
harmony	disharmony
sympathetic	unsympathetic

NM 20: Other-Mind

NM20 Abstract

We all have positive thoughts, but are there negative flash thoughts that pop into people's minds? Why do people sometimes have inappropriate or strange thoughts that seem to just pop into their minds? Why does the mind have these unwanted/intrusive thoughts? Are people in control of all their thoughts? If not, why not?

White Bear and Unwanted Thoughts

nm904 » To start, let me introduce Richard Restak. Restak maintains a private medical practice in neurology and neuropsychiatry in Washington, D.C. where he is also a Clinical Professor of Neurology at George Washington Hospital University School of Medicine and Health. He has to date [2011] written 18 books on various aspects of the human brain; two were on The New York Times Best Sellers List. His first bestseller, *The Brain* (1984), was also the first companion book he wrote for a PBS series. *The Mind* (1988) was his second bestseller.

nm905 » This same Richard Restak wrote a review of Daniel M. Wegner's book about unwanted thoughts. Wegner is a professor of Psychology at Harvard University.[1] In this article Restak wrote:

> As a child, the Russian novelist Fyodor Dostoyevski "once challenged his younger brother to remain standing in a corner until he could stop thinking of a white bear. In this homespun experiment, the child learned something important about the human mind: We do not so much control our thoughts as we are controlled by thoughts that we don't want to think.
>
> Indeed, just about all mental illnesses – obsessions, compulsions, depressions,

phobias, anxiety reactions, post-traumatic stress disorders, self-control problems such as addiction and eating disorders, schizophrenia and other psychoses – along with just plain everyday emotional distress – are marked by problems in the area of mental control.

> There is a paradox here, too; namely, the more effort that one expends not to think something, oftentimes the more difficult it is to expel it from our consciousness."

("Honey of an approach to problems of the mind," review by Richard Restak of the book: *White Bears and Other Unwanted Thoughts: Suppression, Obsession, and the Psychology of Mental Control*, by Daniel M. Wegner, found in *Washington Times*, July 17, 1989, p. E9)

As Restak said, "we do not so much control our thoughts as we are controlled by thoughts that we don't want to think." And "just about all mental illnesses" are marked with problems of mental control. And in context of Wegner's book, this lack of control is the lack of control of *unwanted* thoughts. There are many thoughts each day that flash into people's minds: only some are unwanted.

Lately there is evidence that some mental problems have something to do with biochemical imbalances, as in the case of depression. But this factor has nothing to do with specific individualized unwanted thoughts. Depression is a *feeling* that comes from a certain part of the brain, and is unlike the thought or idea of a white bear. The white bear in the mind of Dostoyevski's brother was a concept, not a feeling. Although depression can be initiated by unwanted thoughts (death of a loved one), depression can just as well be initiated by a biochemical imbalance. There is a difference. The thought of white bear came before Dostoyevski's brother's unwanted thoughts of the white bear; the depression of a biochemical nature comes from an imbalance in the brain first, then the feeling of depression occurs.

Yes, good and positive thoughts enter our minds, from time to time, all day; other times, not so positive thoughts enter our minds. Sometimes these thoughts are like an intermittent breeze: lingering for awhile, disappearing for awhile. But sometimes the thoughts are like a fire that burns in the brain and cannot be extinguished. Like a simmering fire the thoughts may burn every so slightly in the back of people's minds, or the thoughts may

---

1

http://www.wjh.harvard.edu/%7Ewegner/backbio.htm

rage like a wild fire that incapacitates the individual.

Wegner writes in his book about unwanted thoughts:

> Most people report having at least one thought that won't go away. In a study conducted early in this century [1922], one psychologist found that many of the students in his classes admitted to having "fixed ideas" that could not be eliminated.[3] In a San Antonio study, when 180 people were asked to write down an unwanted thought, almost every person had one or more to mention.[4] They reported that their thought was "distressing," and occurred from once a day to every few minutes. Similarly, researchers in England report that people have "normal" obsessions that parallel in several ways the "abnormal" obsessions individuals seek psychotherapy to eliminate.[5] The fact that most people report such thoughts may provide a bit of solace to those of us who think we're odd for worrying. But this fact also indicates that there is indeed a general human problem in the area of mental control.
>
> Unwanted thoughts turn up in a variety of psychological disorders. Of course, they are in center stage when people suffer from obsessions (recurrent unwanted thoughts) or compulsions (recurrent unwanted actions). But having trouble with thoughts that won't go away is characteristic also in many cases of depression, phobic or anxiety reactions, posttraumatic stress disorders, self-control problems such as addictions and eating disorders, and even in psychotic reactions such as schizophrenia. It is not surprising that mental control is rare when people have very severe problems, because the extremes of mental disorder are almost defined in terms of control lapses. However, unwanted thoughts themselves do not define a particular form of psychological disorder. Rather, they occur at all points in the spectrum from normal to abnormal, cutting across different kinds of disorders rather than distinguishing them from one another. (pp. 6-7)

...

> What thoughts do people express the desire to avoid? The contents of such a list will vary, of course, with the time and customs of the people, with their sex and age and habitat. In 1903 in France, for example, the renowned psychiatrist Pierre Janet reported the obsessions of his patients in five major groups: sacrilegious thoughts, urges to commit crimes, shame about one's behavior, shame about one's body, and hypochondria.[21] These obsessions are thoughts that the people were thinking *too much*, and so qualify as very unwanted. Many of these topics are still favorites today....

(*White Bears and Other Unwanted Thoughts: Suppression, Obsession, and the Psychology of Mental Control*, pp 6-7, 20)

Robert L. Leahy, Ph.D., who is a Clinical Professor of Psychology at Weill-Cornell Medical School and Director of the American Institute for Cognitive Therapy, writes about unwanted thoughts and says everyone has these crazy thoughts:

> Thinking about your thoughts
>
> Three rules are important.
>
> 1. Everyone has crazy and disgusting thoughts
>
> 2. Thoughts are not the same thing as reality
>
> 3. Thought-suppression doesn't work.
>
> Research on people without anxiety disorders shows that almost 90% of them have "bizarre" thoughts---thoughts about contamination, harm, religious impropriety, losing control, sexual "perversion"---you name it, we all have thought about it before."

("Those Damn Unwanted Thoughts," *Psychology Today*, June 1, 2009)

Unwanted Thoughts Are Intrusive Thoughts

From a well documented article in the free web encyclopedia called *Wikipedia* we see that "unwanted thoughts" are called "intrusive thoughts":

> Intrusive thoughts are unwelcome involuntary thoughts, images, or unpleasant ideas that may become obsessions, are upsetting or distressing, and can be difficult to manage or eliminate.[1]

> ...

> According to Lee Baer (a specialist at the OCD clinic of Massachusetts General Hospital), intrusive thoughts, urges, and images are of inappropriate things at inappropriate times, usually falling into three categories: "inappropriate aggressive thoughts, inappropriate sexual thoughts, or blasphemous religious thoughts."

> ...

> Many people experience the type of bad or unwanted thoughts that people with more troubling intrusive thoughts have, but most people are able to dismiss these thoughts.[1] For most people, intrusive thoughts are a "fleeting annoyance."[5] London psychologist Stanley Rachman presented a questionnaire to healthy college students and found that virtually all said they had these thoughts from time to time, including thoughts of sexual violence, sexual punishment, "unnatural" sex acts, painful sexual practices, blasphemous or obscene images, thoughts of harming elderly people or someone close to them, violence against animals or towards children, and impulsive or abusive outbursts or utterances.[6] Such bad thoughts are universal among humans, and have "almost certainly always been a part of the human condition."[7]

> ...

> When intrusive thoughts occur with obsessive-compulsive disorder (OCD),

patients are less able to ignore the unpleasant thoughts and may pay undue attention to them, causing the thoughts to become more frequent and distressing.[1] The thoughts may become obsessions which are paralyzing, severe, and constantly present, and can range from thoughts of violence or sex to religious blasphemy.[5]

> ...

> The possibility that most patients suffering from intrusive thoughts will ever act on those thoughts is low. Patients who are experiencing intense guilt, anxiety, shame, and upset over these thoughts are different from those who actually act on them. The history of violent crime is dominated by those who feel no guilt or remorse; the very fact that someone is tormented by intrusive thoughts and has never acted on them before is an excellent predictor that they will not act upon the thoughts.

> ...

Inappropriate aggressive thoughts

> Intrusive thoughts may involve violent obsessions about hurting others or themselves.[16] They can include such thoughts as harming an innocent child, jumping from a bridge, mountain or the top of a tall building, urges to jump in front of a train or automobile, and urges to push another in front of a train or automobile.[4] Rachman's survey of healthy college students found that virtually all of them had intrusive thoughts from time to time, including:[6]

> 1. Causing harm to elderly people

> 2. Imagining or wishing harm upon someone close to one's self

> 3. Impulses to violently attack, hit, harm or kill a person, small child, or animal

> 4. Impulses to shout at or abuse someone, or attack and violently punish someone, or say something rude, inappropriate, nasty or violent to someone.

These thoughts are part of being human, and need not ruin the quality of life.[17] Treatment is available when the thoughts are associated with OCD and become persistent, severe, or distressing.

...

Inappropriate sexual thoughts

Sexual obsessions involve intrusive thoughts or images of "kissing, touching, fondling, oral sex, anal sex, intercourse, and rape" with "strangers, acquaintances, parents, children, family members, friends, coworkers, animals and religious figures," involving "heterosexual or homosexual content" with persons of any age.[18]

Like other unwanted intrusive thoughts or images, everyone has some inappropriate sexual thoughts at times, but people with OCD may attach significance to the unwanted sexual thoughts, generating anxiety and distress. The doubt that accompanies OCD leads to uncertainty regarding whether one might act on the intrusive thoughts, resulting in self-criticism or loathing.[18]

One of the more common sexual intrusive thoughts occurs when an obsessive person doubts his or her sexual identity. As in the case of most sexual obsessions, sufferers may feel shame and live in isolation, finding it hard to discuss their fears, doubts, and concerns about their sexual identity.[12]

...

Blasphemous religious thoughts

Blasphemous thoughts are a common component of OCD, documented throughout history; notable religious figures such as Martin Luther and St. Ignatius were known to be tormented by intrusive, blasphemous or religious thoughts and urges.[20] Martin Luther had urges to curse God and Jesus, and was obsessed with images of "the Devil's behind".[20][21] St. Ignatius had numerous obsessions, including the fear of stepping on pieces of straw forming a cross, fearing that it showed disrespect to Christ.[20][22] A study of 50 patients with a primary diagnosis of obsessive-compulsive disorder found that 40% had religious and blasphemous thoughts and doubts—a higher number than the 38% who had the obsessional thoughts related to dirt and contamination more commonly associated with OCD.[23] One study suggests that content of intrusive thoughts may vary depending on culture, and that blasphemous thoughts may be more common in men than in women.[24]

According to Fred Penzel, a New York psychologist, some common religious obsessions and intrusive thoughts are:[13]

1. sexual thoughts about God, saints, and religious figures such as Mary

2. bad thoughts or images during prayer or meditation

3. thoughts of being possessed

4. fears of sinning or breaking a religious law or performing a ritual incorrectly

5. fears of omitting prayers or reciting them incorrectly

6. repetitive and intrusive blasphemous thoughts

7. urges or impulses to say blasphemous words or commit blasphemous acts during religious services.

["Intrusive Thoughts," *Wikipedia*, Febrary 9, 2011]

Thoughts, Sometimes Negative

nm906 » Why do people sometimes get recurring thoughts which they can't seem to control? Why are people unreasonably afraid of some things? Why do we sometimes, in bitter arguments, say offensive things that we do not really mean? Why do people sometimes unreasonably criticize other people? If we do say something good about a person we may, in some situations, add something negative. Our thoughts are sometimes negative. And sadly from these negative thoughts come the confusion and misbehavior that makes for unhappiness. These thoughts, flash into people's minds in a split second, and some of them stay in our minds as unwanted thoughts because of the negative or persistent aspect of them. They are unwanted because of the way they affect us.

Flash Thoughts, Positive or Negative

nm907 » These unwanted thoughts pop into people's mind in a flash, a split second, and therefore can be called "flash-thoughts." A *positive* flash-thought is like when a great idea or thought pops into one's mind. A *negative* unwanted flash-thought is a thought that not only irritates, but *may* in certain cases lead to or result in behavior that hurts or harms our self or someone else.

But why do people get these flash-thoughts or ones like them? Do our parents teach them to us? Does society teach them to us? And the question must be asked, if people are actually in control of their minds, why do they get these thoughts, or why can't they rid themselves of these thoughts? Why can't people just order their brains to stop these thoughts as they order their brains to move their arms or fingers? Are we in control of our negative thoughts? Not too many positive thoughts are unwanted, if any. Why the unwanted thoughts?

More Examples: Sexual Fantasies Thrusts into the Open

nm908 » Many of these flash-thoughts may be embarrassing if revealed and so people generally don't talk about them. For example, erotic sexual fantasies were seldom talked about openly in most societies. Yet these fantasies existed and manifested themselves in the underground, in houses of prostitution and in banned pornographic writings, pictures, drawing or paintings. Due to the liberalization of the Western cultures and even in some of the religions, we now see books and magazines published with accounts of these sexual fantasies (*Forum* magazine; Nancy Friday's *My Secret Garden: Women's Sexual Fantasies*, or her *Forbidden Flowers*, and her *Beyond My Control: Forbidden Fantasies*. etc). The table of contents in Nancy Friday's *Beyond My Control: Forbidden Fantasies* list fantasies that pertain to: domination, masturbation, incest, exhibitionism/ voyeurism, S&M, threesomes and so forth (found on Amazon's "Look Inside" feature). The readers of Nancy Friday's books did not even know other women had sexual fantasies. Her female reader's thought their sexual fantasies were a manifestation of something wrong with them because they had

erotic sexual fantasies. In Friday's own words from her book:

> I loved original work [research] and always had sexual fantasies. As I've noted before, when I approached several eminent therapists and psychoanalysts and asked their opinion of my research, I was repeatedly told: 'Women do not have sexual fantasies. Men do.'

According to Friday, after reading her books women felt liberated from guilt knowing other women also had them in one form or another:

> But "we don't have to act on the fantasies to feel this way [liberated]. Some, fully realized, would become nightmares. Nor share them with our partners" ("Author to Reader" in *Beyond My Control*).

Why would they be nightmares? The nature of these sexual fantasies can sometimes be unusual and even perverse in a religious sense, or in the sense of decorum or propriety. So where do these sexual fantasies come from? Do we teach our children sexual fantasies? Or do we learn of these fantasies through books? Some of these women according to Friday never saw a pornographic magazine. They came up with these thoughts even though some came from sheltered or conservative religious cultures. So where did these thoughts come from? Where do other thoughts come from that could be called negative or twisted? Some if acted out would be destructive and could land the person to jail. Where do these thoughts come from?

Children and "No"

nm909 » Children are innocent little creatures, right? But the caretakers of them see that they can be at times cunning, selfish, lying, aggressive, violent, and so forth. Some say this is because they learn from their parents' behavior and their environment. This is true to a certain extent, but it is not the true answer.

nm910 » When children are very young parents notice that sometimes they are very interested in doing things they were told not to do. The parents say "no" to the child, and the "no" makes the child want to do it even more. Now think about this. Do we teach our children to do what we want them not to do? Do we set them aside each day and tell them, "now Johnny, every time I tell you not to do something, you should do it anyway and take great pleasure in doing what I have told you not to do"? Of course not, we never teach our children to disobey. We teach them to obey us and we back it up with various forms of discipline. Yet, they continue at times to disobey us, even after we discipline them, and they, to judge by their facial expressions, take pleasure in disobeying us. Where do they learn this misbehavior? Who teaches them? We are not talking about children with neurological disorders; we are talking about healthy normal children.

In Control?

nm911 » Are we in FULL control of our minds and consequently our behavior? The psychologists, psychiatrists, priests, ministers, rabbis and so forth know people are *not* in full control of their minds or behavior, and that is the reason for their occupations: their patrons come to them for guidance and help. If anyone wishes to move any part of their body, let's say a hand, they merely command their mind to move their hand when and where they want it to move. Only those who are physically impaired or disabled can't control their physical actions. If we are in control of our bodies, why are we not in control of our thoughts and our behavior? Some say we are in full control of our mind. If everyone is in full control of their mind then mankind could easily rid themselves of all falsehood, all guilt, all negativity in all forms. It would be easy to mold our children and our society. But it is not easy, is it? There *is* something more to all this.

The Dark Side of Man, "Devil within us"

nm912 » In *The Washington Times*, June 22, 1989, p. F4, Dr. Richard M. Restak, a neurologist and neuropsychiatrist, an author of "The Brain" and "The Mind," reviewed a book by Ronald Markman and Dominick Bosco, *Alone with the Devil: Famous Cases of a Courtroom Psychiatrist*, and in the review he writes:

> "Most crimes – even grisly murders – are not committed by mentally ill people, but by people just like you and me." He quotes with approval a statement by Linda Kasabian, a member of the Charles Manson "family," found guilty of the Sharon Tate-LaBianca murders: "I believe that we all have a part of the Devil within us – it's just a matter of bringing it out." Dr. Markman admits, "We all do have a willingness – even an appetite – to kill within us. All it takes is the right

combination of factors to raise it to the surface."

This may seem like an extreme statement. But Dr. Restak is not an extremist, but an informed neuropsychiatrist. From others we hear, "The devil made me do it" excuses. Others speak of the "dark" side of man. The religious speak of the spiritual dark side. Mark Twain wanted very much to write the whole truth about his life, but even Mark Twain couldn't bring himself to write about *his* dark side as the introduction of Mark Twain's new autobiography manifests. He tried to put off his autobiography for 100 years so he could write about his dark side. He never could write the truth about this side of him. (see Introduction, *Autobiography of Mark Twain*, Vol 1, Pub.: 2010)

The Imp of the Mind and "Bad Thoughts"

From a review in Publishers Weekly of a review of Lee Baer book, *The Imp of the Mind: Exploring the Silent Epidemic of Obsessive Bad Thoughts.*

> Specializing in the diagnosis and treatment of obsessive-compulsive disorder, psychologist Baer (an associate professor at Harvard) turns the spotlight on a little-known [by the general public] but common form of obsession, "bad thoughts." According to Baer, these "intrusive" thoughts fall into a few basic types: violent, sexual and blasphemous words, and images of a religious nature. Borrowing from Edgar Allan Poe, Baer blames such mental torment on "the imp of the perverse," that little devil inhabiting all human minds, cross-culturally and across time, "who makes you think the most inappropriate thoughts at the most inappropriate times." For most people, the imp proves no more than a "fleeting annoyance" most of the time, but for Baer's patients, these impish thoughts create extreme fear, guilt and worry. Attempting to suppress them only makes them stronger, leading the afflicted to avoid places, people and situations that provoke them. A new mother who obsessively thinks about harming her infant, for example, may increasingly avoid daily caretaking activities. Tending to be perfectionist and "overly conscientious," these people are highly unlikely ever to act

on their bad thoughts, Baer explains. [From Publishers Weekly Jan. 15, 2000]

Professor Lee Baer calls these intrusive thoughts, "the imp of the perverse." But since the 1970s I've been calling these thoughts, the thoughts of the "other-mind."

Other-Mind

nm913 » Considering the above information, other studies, my observation and others' observations, since the early 1970s:

> I have come to the conclusion that *there is something like another mind in our brain feeding thoughts to our brain, many times unwanted thoughts. I call this the "other-mind."*

The other-mind is what I call the phenomena of those unwanted and many times negative thoughts that seem to annoy peoples' minds. Scientists do not know *why* people have them or *how* people get these unwanted thoughts, just like they don't know *why* there is gravity or *how* gravity works: science can only *describe* and list the unwanted thoughts and describe gravity in words or through mathematics.

Granted that in certain cases neurological disabilities or chemical imbalances can cause or lead to problems pertaining to mind and thoughts, but still the question remains: why don't these disabilities and chemical imbalances cause *positive* mental thoughts?

Lost of Control

Even though I believe that everyone has the "other-mind" feeding everyone unwanted thoughts, most people do not carry out these intrusive thoughts because most have some control over their behavior. But some do lose control. We see it every day in our newspapers. One example appears on CNN's web site today:

> (CNN Feb. 12, 2011) -- An unemployed New Yorker fatally stabbed three people, slashed at least four others, hit and killed one man with a car and hijacked two vehicles before being wrestled to the ground early Saturday while trying to break into the cab of a subway car, polic said.

Maksim Gelman, 23, was arrested aboard a north-bound train in Manhattan around 9 a.m. Saturday, 28 hours after he allegedly began a spree across three New York City boroughs, Police Commissioner Ray Kelly said.

"It's so horrendous and bizarre," Kelly told reporters Saturday afternoon. "We have no reason that we can give you as to why he did this."

When we come to understand that the unwanted thoughts of the "other-mind" cannot harm us, unless we allow it, and that by knowing that these thoughts do not radiate from our very selves, the shame and guilt can be mitigated. Also others with these unwanted thoughts can be understood and empathized with in a much more appropriate, if not tolerate manner.

NM 21: "Other-Mind" – Its Beginning

What is and Why is there the Phenomenon of the Other-mind?

nm914 » There is evidence through scripture and through human experience that there are supernatural phenomena. If the answer to the puzzlement of the other-mind comes from the supernatural, then science will not and cannot find the answer. Science has to do with nature, not the supernatural. If there are supernatural phenomena, if there is a God, then science, by its very definition, cannot find the answer.

nm915 » Through study I believe that there are supernatural phenomena and that there is a God being, who created the universe and who put some knowledge about himself into a text called the Holy Bible. Some of our other papers describe in detail more about this God Being and about the great debate of the genesis of the universe.

nm916 » From scripture, we believe that the *other-mind* is not physical; it is spiritual. That is, the *other-mind* is a mental phenomenon that does not belong to our physical body, but is an invisible (thus spiritual) power *in* our minds. Something that is spiritual is something that is not easily detectable by sight, touch, smell, or by our other senses. Spirit is analogous to air: you cannot see air but we see the effect of its wind; spirit is invisible but we see the effects on mankind's behavior. The spiritual is detectable mentally by many. Some of you believe in the spiritual. If you think of yourself as Christian, you believe in evil spirits and good spirits or angels. Some of you interested in the occult

believe in the spiritual dimension also, although your perception of the spiritual is not like the Christian perception. The New Testament of the Bible speaks of demons doing harm to people. The Old Testament speaks of lying spirits. Some of you from other religions or backgrounds also believe in the spiritual or invisible world. Almost all peoples in history have had beliefs about the spiritual world. Even those who do not believe in the spiritual believe nevertheless in invisible forces that they call by such names as gravity, black holes, protons, etc. In the psyche of man there is knowledge of a spiritual dimension which is evil and affects mankind in various ways. So what does the Bible have to say about the subject of unwanted thoughts?

nm917 » Genesis is the first book of the Bible. In this first book is the Biblical story of the beginning of the universe ("heavens and earth") and the beginning of mankind. There were seven days of creation. On the sixth day of creation "God created man in his image, in the image of God created he him; male and female created he them" (Gen 1:27; see *God Papers* for information on the "Image of God").

Spirit Inside Mankind

nm918 » An amplification on the creation of mankind is mentioned in chapter 2 of Genesis. Here it tells us God created man out of the dust of the earth on the sixth day. The word translated as Adam or man is a Hebrew word meaning "to be red or reddish."

> "And Jehovah God [LORD God] formed the man, dust from the ground, and *breathed* into his nostrils (the) *breath* of life." [Gen 2:7]

Now notice that Jehovah "breathed" into man the "breath" of life. In the English version of the Old Testament the word "breath" is translated from either the Hebrew word *neshamah* or *ruah*. These words differ slightly in meaning, both signifying sometimes "wind" then sometimes "breath." The word translated into "breath" in Genesis 2:7 is the Hebrew word *neshamah*.

nm919 » Both Hebrew words, *neshamah* and *ruah*, are translated as "spirit" in various places in the English translations of the Bible. And in the book of John in the New Testament, Jesus Christ tries to explain spirit to some: "the wind blows where it wishes, and you hear the sound of it, but you do not know from where it comes and where it goes" (John 3:8). In other translations it has for John 3:8, the "spirit breathes" instead of the "wind blows." This is so because in the Greek New Testament the Greek word translated in English Bibles as "wind" or "spirit" is the same word. Thus in the Old Testament and the New Testament "wind" and "spirit" are interchangeable words. And Jesus Christ tries to explain spirit by comparing it to the wind that blows. And in John 20:22 it reads:

> "and having said this he [Jesus] breathed on them and said to them, receive the Holy Spirit."

Notice the similarity between Jesus Christ breathing into his followers, which signifies them receiving the Holy Spirit, and Jehovah in Genesis breathing into man, which signifies them receiving the breath or spirit of life. We thus take Genesis 2:7 as indicating a dual sense. In one sense, the physical sense, Jehovah breathed into man the literal breath or air of life and man became a breathing and living life. In the second sense, the spiritual sense, Jehovah was putting a spirit into man after he created man. What spirit was put into man at that time? And what is meant by spirit? Also see information in NM 22.

Angels Are Spiritual Messengers

nm920 » Spirits and angels are connected because angels are spirits (Heb 1:7). And angels are messengers because the word "angel" is translated from a Hebrew word in the Old Testament and a Greek word in the New Testament which means messenger, or one who is sent. Not only do we know that angels are messengers by the meaning of the word, but also because of their activities in the Bible. Angels are spiritual beings sent to do the will of the God. They affect people and nations by putting thoughts into people's minds. The thoughts of confusion are administered by the angels of evil, or the other-minds, or shall we say the demons. (Do not jump to conclusions yet, do read on.) I do not care to use the word evil or demon because of the connotations these words create in some people's minds. But what I wish to communicate here is that the other-mind that works in each of us is a spiritual being or angel (messenger) who puts negative flash-thoughts into our minds, sometimes very subtle thoughts. An evil angel is a messenger of evil; he brings base thoughts to our brains. There are now some good angels or messengers who do bring good thoughts into some people's minds in this age, but for the most part most people in this age only have the other-mind inside their mind. This is the reason the present age is an evil age. Not only do we have to fight a hostile environment, we also must fight the negative thoughts from the other-mind.

nm921 » Now to answer the question above which was, "what spirit was put into man?" At the beginning of creation a spirit was put into the mind of man by Jehovah. (The word, Jehovah, is translated "LORD" in some English Bibles.) By looking around us today and by knowing something about the other-mind as I explained it previously, we know that the spirit or angel (messenger) that was placed in mankind's mind at the beginning was the other-mind, that confused and evil mind. We can begin to see the Biblical proof of this by reading chapters two and three of the book of Genesis. Let's go over the scripture in those chapters.

Genesis: Chapters Two and Three

nm922 » In Genesis 2:16-17 it reads,

> "And Jehovah God commanded the man saying, You may freely eat of every tree in the garden; but of the tree of knowledge of good and evil you may not eat, for in the day that you eat of it, dying you shall die."

In the day that man ate from the tree of knowledge of good and evil he would die according to God. Now next God created out of man a woman (in the beginning God created the man first before He created the woman) for God said that it wasn't good for man to be alone (Gen. 2:18-22). After God brought woman to man, the man said,

> "This now at last is bone from my bone, and flesh from my flesh. For this it shall be called Woman, because this has been taken out of man."

Then the story continues,

> "Therefore a man shall leave his father and his mother and shall cleave to his wife, and they shall become one flesh. And they were both naked, the man and his wife, and they were not ashamed."

Thus we see that they weren't ashamed, they had no shame even though they were naked. But next in chapter three of Genesis something happens.

First Lie

nm923 » In chapter three of Genesis it speaks of the "serpent" who deceived mankind into taking fruit from the "tree of knowledge of good and evil." Now in the book of Revelation (12:9) we see that this "serpent" is none other than the one called the Devil and Satan. He was the father of lies (John 8:44). We also see that in the Biblical rendition on this serpent in Genesis that this serpent spoke to Eve, the wife of Adam. Today we do not see any serpents or snakes talking to women. Because of such stories in the Bible some have come to think of the Bible as a book of tales with little or no truth. By reading our papers and by reading some of the books we recommend which document the soundness of the Bible as a true document, you will find out for yourself that the Bible is not a book of tales but a book of facts and true history with Spiritual insight and prophecy of future events.

Figures of Speech

How Satan Spoke to Eve

nm924 » What does the Bible mean when it says that a serpent spoke to the woman? Are we to take this literally that this serpent in the beginning spoke to mankind? I do not believe so just as we are not to take it literally that trees clap their hands or that mountains and hills shall break forth into singing (Isa 55:12). The Bible uses all kinds of figures of speech. The Bible uses similes, "his eyes were as a flame of fire" (Rev 1:14). The Bible uses metaphors, "tell that fox" (Luke 13:32). The Bible uses metonyms, "if the house be worthy" (Mat 10:13). The Bible uses synecdoches, "all the world should be taxed" (Luke 2:1). The Bible uses personifications, "the earth mourns and fades away" (Isa 24:4). The Bible uses apostrophes, "O death, where is thy sting?" (1Cor. 15:55) The Bible uses hyperboles, "the light of the sun shall be sevenfold" (Isa 30:26). The Bible uses allegories, "this Hagar is Mount Sinai in Arabia."(Gal 4:24) The Bible uses parables, "behold, a sower went forth to sow" (Mat 13:3). The Bible also uses irony, riddles, and fables (1Kings 18:27; Rev 13:18; and Judges 9:8 ff & 2Kings 14:9 ff). So we can see that the Bible is rich in its use of language (see *Figures of Speech Used in the Bible*, by Bullinger). The serpent did not literally speak to Eve, only in a figurative way did the serpent speak to Eve. I will now show you this.

Serpent Tests Eve

nm925 » In chapter three we see a serpent that

> "was crafty above every animal of the field which Jehovah God had made. And he said to the woman, Is it true that God has said, You shall not eat from any tree of the garden? And the woman said to the serpent, We may eat of the fruit of the trees of the garden, but of the fruit of the tree which is in the middle of the garden, God has said, You shall not eat of it, nor shall you touch it, lest you die. And the serpent said to the woman, Dying you shall not die, for God knows that in the day you eat of it, your eyes shall be opened and you shall be as God, knowing good and evil. And the woman saw that the tree was good for food, and that it was pleasant to the eyes, and that the tree was desirable to make wise, and she took of its fruit and ate; and she also gave to her husband with her, and he ate. And the eyes of both of them were opened, and they knew that they were naked, and they sewed leaves of the fig-tree and made girdles for themselves. And they heard the sound of Jehovah God walking up and down in the garden at the breeze of the day. And the man and his wife hid themselves from the face of Jehovah God in the middle of the trees of the garden. And Jehovah God called to the man and said to him, Where are you? And he said, I have heard your sound in the garden, and I was afraid, for I am naked, and I hid myself. And He [Jehovah] said, Who told you[1] that you were naked? Have you eaten of the tree of which I have commanded you not to eat? And the man said, The woman whom you gave to be with me, she has given to me of the tree, and I ate. And Jehovah God said to the woman, What is this you have done? And the woman said, The serpent deceived me and I ate." [Gen. 3:1-13]

---

[1] Doesn't this seem to indicate that God knew an outsider ("who told you") influenced Adam and Eve?

Figuratively Speaking

nm926 » Notice what is revealed in these verses. First, *before* mankind took from the tree of knowledge of good and evil, they had no shame (Gen 2:25). Next a snake or serpent came along and spoke to the woman. Now we do not have to take this literally. As I explained above the Bible is very lively in its use of the language. The Bible uses many different kinds of figures of speech. We know trees do not clap their hands because for one reason trees do not have hands. Yet in Isaiah 55:12 it speaks of trees that will clap their hands. We also know that serpents or snakes do not speak. At least they do not speak like mankind. (Of course the serpent in the garden may have actually spoken, for God can do that if he had wished it to be so. But because the Old Testament is so filled with figures of speech, and because we have no proof that serpents literally speak in the way mankind speaks, we therefore won't take this part of the Bible literally just as we won't take Isaiah 55:12 literally.)

nm927 » What happened was this: The woman was near the tree of knowledge of good and evil when a physical serpent was winding itself around the tree and through the serpent's various motions and movements the serpent brought the woman's attention to the fruit on the tree. But it was the spiritual serpent inside the head of Eve that spoke to her. In other words, what tempted Eve that day was the spirit that Jehovah put inside mankind (Gen 2:7). That spirit is the other-mind that even today is inside mankind's mind. It was the other-mind inside Eve that spoke to her that day. Notice in Revelation, chapter 12:9, it reads that "the old serpent, called the Devil, and Satan, which misleads the whole world." As explained in the "Duality Paper" we are to take the higher or spiritual meaning of the Bible in order to ascertain the Spiritual truths of the Bible. The higher meaning of the serpent in Genesis is the spiritual serpent which is the Devil or Satan. One can say that the spirit that Jehovah put into man in the beginning was the spirit of Satan or the Devil, or that is, the spirit of evil and confusion, or as we are calling it in this paper, the other-mind that now lives in mankind. The actual spirit ("breath") that God put into mankind was the spirit that started evil by testing Eve (See also NM 22). This is difficult to understand, but do read on to see more proof.

Law and Sin

nm928 » When mankind took from the tree of knowledge they were breaking the law that God had given to them. The law was simple, do not take any fruit from the tree of knowledge of good and evil. Mankind could take from all the other trees in the garden, but somehow they desired the forbidden tree very much. Even though God promised death as the reward for taking from the tree of knowledge of good and evil, mankind still took from that tree. The serpent lied to Eve about the death they would receive, "You shall not surely die" (Gen 3:4) The serpent mixed one lie with the truth to deceive Eve, a very subtle deception that played on Eve's desire for knowledge. And so as to not displease his wife, Adam listened to his wife and also sinned.

Romans Chapter Seven

nm929 » Notice the following statement that Paul made in Romans, chapter seven of the New Testament. What Paul said in chapter seven fits very well what happened in the Genesis' account:

> **But sin, seizing the opportunity afforded by the commandment** [of Gen 2:17]**, produced in me** [Paul represents mankind] **every kind of covetous desire** [for the tree]**. For apart from law, sin is dead. Once I was alive apart from law** [before Gen 2:17]**; but when the commandment** [Gen 2:17] **came, sin sprang to life** [the spiritual serpent sprang to life tempting Eve] **and I died.** [that is mankind was given death because they broke the law in Genesis 2:17] **I found that the very commandment that was intended to bring life actually brought death.** [God's warning to man was intended to keep man from death and bring him to life] **Did that which was good, then, become death to me? By no means! But in order that sin might be recognized as sin, it produced death in me through what was good, so that through the commandment sin might become utterly sinful. We know that the law is spiritual; but I am** [mankind] **unspiritual, sold as a slave to sin. I do not know what I am doing, For what I want to do I do not do, but what I hate I do. And if I do what I do not want to do, I agree that the law is good. As it is, it is no longer I myself who do it, but it is sin living**

in me. I know that nothing good lives in me, that is, in my sinful nature. For I have the desire to do what is good, but I cannot carry it out. For what I do is not the good I want to do; no, the evil I do not want to do – this I keep on doing. Now if I do what I do not want to do, it is no longer I who do it, but it is sin living in me that does it. So I find this law at work: When I want to do good, evil is right there with me. For in my inner being I delight in God's law; but I see <u>another law</u> at work in the members of my body, waging war against the law of my mind and making me a prisoner of the <u>law of sin</u> at work within my members." [Rom 7:8-23]

Law of Sin

nm930 » Paul in these verses can be substituted for mankind as a whole with one exception. Paul had the good Spirit of God in him, but in this age most people only have the evil spirit in them, for most people will not receive their own good spirit until a later time. Yet we see Paul fighting the *law of sin* inside him. Paul wants to do good, he has the Spirit of God inside him leading him to good, but Paul finds another law inside him, and that law is the law of sin. The other-mind in mankind's head in this age works like a law of evil. The other-mind is an evil angel or spiritual messenger inside man's mind misleading man, and this other-mind operates like a law, that is, it operates like a law of sin or evil.

Sin and Satan

nm931 » Sin is defined in the Bible as lawlessness (1John 3:4). As Revelation 12:9 says the serpent in Genesis was Satan who has deceived the whole world. Actually "sin" is just another name for Satan. The words "sin" and "Satan" are interchangeable. As light, life, and love are used in the Bible for God (John 1 & 1John 4:8) so too can darkness, death, and sin be used for Satan. God is love, but Satan is sin. Now when you transpose in Romans, chapter seven, the word Satan for sin, you can see more clearly how this chapter adds to the rendition in Genesis. It was the spirit of Satan that Jehovah put into mankind in the beginning(Gen 2:7). But this spirit didn't show himself evil until the commandment came (Gen 2:17). Then this spirit in mankind took as an occasion the

commandment (Roman 7:8,11) and deceived man (Gen 3:1-5) and thereby killed man. As Jesus Christ said: "You belong to your father, the devil, and you want to carry out your father's desire. He was a murderer from the beginning [the spirit of Satan in a sense killed man by deceiving man in Genesis], not holding to the truth, for there is no truth in him" (John 8:44). It is the spirit of Satan inside mankind that is causing most of the problems in this age. It is the other-mind that is contrary to our eventual happiness and is the main cause of our present problems.

Other-Mind Is a Spirit and Power of Evil

Our struggle is not against flesh and blood

nm932 » It is Satan or the spiritual other-mind that misleads the whole world (Rev 12:9). It is the power of Satan or the power of the other-mind which we must turn from (Acts 26:18). It is the devil's schemes, deceiving spirits, and things taught by demons of which we must be careful (Eph 6:11; 1Tim 4:1). "Our struggle is not against flesh and blood, but against the rulers, against the authorities, against the powers of this dark world and against the spiritual forces of evil in the heavenly realms" (Eph 6:12). "He who does what is sinful is of the devil, because the devil has been sinning from the beginning" (1John 3:8). As we have shown you it was the spirit of Satan that put the thought in Eve's mind of breaking God's first law. This spirit of Satan is the spiritual other-mind that works in us.

Other-Mind also Called...

nm933 » The no-good thing living in man is that sin or Satan living and working in man (Romans 7:18,20). It is the spirit in us that lusts after evil (James 4:5). This spirit is also called the spirit of slumber (Rom 11:8), it is called the spirit of bondage (Rom 8:15), it is called the spirit of sleep (Isaiah 29:10), it is called the lying spirit (1Kings 22:21-23), it is called the unclean spirit (Zech 13:2), it is called the spirit of whoredoms and it causes people to err (Hosea 4:12; 5:4), it is called the spirit of this world (1Cor 2:12), it is called the haughty spirit (Prov 16:18), it is called the perverse spirit (Isa 19:14), it is called the troubled spirit (Gen 41:8), it is called the spirit of jealousy (Num 5:14), it is called the spirit of man (1Cor 2:11), and so on.

Spirit of Man

Satan in Peter?

nm934 » It is the spirit of man that is within mankind that knows the thoughts of man (1Cor 2:11). Since the Bible looks upon man in this age as evil, then this spirit of man must be evil. What is this spirit in man that thinks like man? "Out of my sight, Satan! You are a stumbling block to me; you do not have in mind the things of God, but the things of men" (Mat 16:23). Now in this last verse Jesus was speaking to Peter, for "Jesus turned and said to Peter, Out of my sight, Satan!" Jesus was talking to the physical Peter, but he called him Satan. Why? Jesus called Peter "Satan" because he was talking to the spirit of man which lived inside Peter and which was misleading Peter at that time.

Fight against Satan's Power

nm935 » In Luke 22:3 and John 13:27 we read where Satan even caused Judas to err. "The evening meal was being served, and the devil had already put it into the heart of Judas Iscariot, son of Simon, that he should betray Him." It wasn't Judas himself that conceived of the idea of betraying Jesus, it was the spirit of Satan inside him that did it. It is Satan's spiritual power that takes captive at his will, not mankind's nature (2Tim 2:26; Acts 26:18). It is the snare and wiles of the spirit of the devil that people must guard against, not human nature (Eph 6:11; 1Tim 4:1). It is the other-mind or Satan that we are fighting against, "not flesh and blood" (Eph 6:12). It is Satan who fills the heart to lie (Acts 5:3). There is a spirit in man that yearns to envy and this spirit is of the devil (James 4:5,7).

Worthless Mind

nm936 » Mankind was given the spirit of Satan in the beginning although it didn't show itself as evil until after God gave man a law. Because Satan's spirit is a law-breaking-spirit, it took the law to show the nature of Satan's spirit. Satan's spirit is the other-mind that works in us. It is the worthless mind described in Romans 1:28-31:

> "God gave them up to a mind that was not fit for any good, to do those things which were not right, being filled with all kinds of unrighteousness: fornication, wickedness, covetousness, malice, full of envy, murder, quarrels, deceit, evil habits, whisperers, slanderers, God-haters, insolent, proud, braggarts, deviser of evil things, disobedient to parents, without understanding, impossible to trust, without natural love, unforgiving and without mercy."

Man's Heart and the Other-Mind

Evil from Within

nm937 » It is because of the other-mind, that "man's heart is evil from his youth" (Gen 8:21). "For from within, out of the heart of men, proceed evil thoughts, adulteries, fornications, murders, thefts, covetousness, wickedness, deceit, lasciviousness, an evil eye, blasphemy, pride, foolishness" (Mark 7:21,22). "And an evil man out of the evil treasure of his heart brings out that which is evil. For out of the abundance of the heart his mouth speaks" (Luke 6:45). It is out of the abundance of the heart of mankind that man acts. And in man's mind is the other-spirit which continuously feeds man with unsound thoughts.

nm938 » This age is worse than many people think. False pride, false affections, false accusations, false documents, false theories, false love, false words, and so forth are so much a part of this age that most of us are somewhat unaware of the asininity and falsehood of this age. The great religions and theories of this age are built on the confusion of the other-mind and the falsehood of the other-mind.

Mind of this Age of Evil

nm939 » According to the Bible the people of this age are the children of the devil (1John 3:10), children of disobedience (Eph 2:2), children of the wicked one (Mat 13:38), children of the world (Luke 16:8; 20:34), and so forth. The father of these children is the devil according to Jesus Christ (John 8:42-44). The people of this age are the children of confusion because their spiritual father is the great other-mind that lives and feeds negative flash-thoughts into the mind of mankind. So this is why we must be careful:

> Beloved, **believe not every spirit**, but try the spirits whether they are of God. (1John 4:1a)

New Mind for New Age of God

nm940 » In the next age, the New Age, the spirit of the other-mind will be put out of man. For when the demons ("devils") or when the other-mind is cast out and locked up, "then the kingdom of God [the New Age] has come upon you" (Mat 12:28). "In that day" Jehovah will cause the unclean spirit or the other-mind to pass out of the land (Zech 13:2). "That day" is the New Age that is coming upon us. In the New Age people will be truly free from the other-mind because it will be put out of people's minds. The New Age begins at the physical return of the Messiah.

See the "Reason Why" paper in the *God Papers* [GP 7] for the reason God put this other-mind into man in the beginning. See the *God Papers* to understand who or what is God and what will happen to the spiritual other-mind.

Review

nm941 » There is a power of confusion and evil in our physical minds. We call this power the, "other-mind." It is a spiritual power that flashes negative thoughts into our minds. This spiritual power was placed in our minds at the beginning of the creation by the one called Jehovah. The "other-mind" didn't manifest itself as evil until a law was given by Jehovah, then the other-mind showed its evil nature. Ever since the beginning the other-mind has fed negative flash-thoughts into mankind's mind. Because of these invading thoughts, man has behaved often in a negative manner since the creation. But a New Age is coming when the other-mind will be put out of mankind's mind, and at that time an utopian system will begin.

nm942 » With the knowledge of the other-mind, we can better understand our misbehavior in this age, and through understanding it we can better control it. The best way to control our misbehavior is to have the New Mind. But better still is to rid ourselves of the other-mind – the old mind – and obtain the New Mind. We wait for this to happen to us all.

NM 22: Spirit of Man Given by God

Paradoxes of the Spirit of Man

Spirit put in Mankind

Who put the spirit in man?

NM22 Abstract

There is a "spirit in man." What kind of spirit is it? Where did it come from?

Paradoxes of the Spirit of Man

nm943 » It is hard to understand how God, who is good, can give mankind the **spirit of man** which has led to so much evil on the earth. This is yet another paradox of God. From the *God Papers* we read:

> How can God be love (1 John 4:8), and also a killer? In scripture the LORD says, "I kill and I make alive; I wound, and I heal" (Deut 32:39; 1Sam 2:6). Yet the Bible says that the God is good to all (Psa 145:9). How can God be good to all and also a killer? How can God predestinate some to wrath and destruction (Rom 9:21-23; Jude 4; Prov 16:4; 1Peter 2:8), and some to mercy and glory? (Rom 9:21-23; Eph 1:4-5; etc.) Not only is God love, but He is *all* powerful (Gen 17:1; Rev 1:8). In his all powerfulness He even *created* evil: "I make peace, and create evil: I the LORD do all these things" (Isa 45:7). These are some of the Biblical paradoxes of God. Just how can God be love and also a killer, or how or why has He created evil? According to the Biblical definition of love (1Cor 13:4-8), killing or evil isn't one of the qualities of love. Yet, according to the Bible, God is love and in someway has killed and in someway has created evil (From GP 1, ¶ gp10).

nm944 » Because of the paradoxes pertaining to God, some simply do not believe in God, or deny the Biblical God, or some are forced to not admit or not see the plentiful scriptures concerning God's "evil" side. As we find out in the *God Papers*, the secret to unlocking the paradoxes of God is the meaning of the NAME of God, the phenomenon of time, and the phenomenon of predestation. His NAME manifests to us *how* God can be good without actually doing evil in this old age. **Hint**: there is a left and right side of God. Satan, the left side of God (that is, the left cherub in the Holy of Holies [Ezek 28:14]; both cherubs were made from one piece; see *God Papers*, especially parts 8 and 9) is now doing all the evil in this age. The other hint is that God predestinated everything <u>before</u> the cosmos, thus before good, before evil, before law, and consequently before sin, so God's predestination is without sin. Soon the right side of God will take control of the earth away from Satan and bring good to the earth.

Spirit Put in Mankind

nm945 » Did God really put the **spirit of man** inside mankind? In the *Other Mind Papers* (PR21) we mention the spirit of man as being put into mankind at the beginning by God, that is, by Yehowah Elohim. In this paper we will show you the Biblical verses. We will also use the KJV with Mr. Strong's numbers behind each word or phrase. Strong's numbers are inclosed in "< >."

nm946 » Notice where the Bible first indicates **when** God gave man the spirit of man:

> KJV Genesis 2:7: And the LORD God formed man *of* the dust of the ground, and **breathed** into his nostrils the **breath** of life; and man became a living soul.

> KJV Genesis 2:7: And the LORD <03068> God <0430> formed <03335> (08799) man <0120> *of* the dust <06083> of <04480> the ground <0127>, and **breathed** <05301> (08799) into his nostrils <0639> the **breath** <05397> of life <02416>; and man <0120> became a living <02416> soul <05315>.

nm947 » Notice also that the "LORD God" **breathed** into man the **breath** of life. The Hebrew word translated "breathed" is Strong's number, 05301. The Hebrew word translated "breath" in "breath of life" is Strong's number, 05397.

nm948 » From Strong's concordance we see that the number 05301 means:

> **5301** נָפַח naphach {naw-fakh'} ... 1) to breathe, blow, sniff at, seethe, give up or lose (life) 1a) (Qal) to breathe, blow

1b) (Pual) to be blown 1c) (Hiphil) to cause to breathe out

nm949 » From Whittaker's Revised BDB lexicon, Strong's number 05301 means:

ו conjunction נפח verb: qal imperf waw consec 3rd pers masc sing **B6438** [נָפַח] **vb. breath, blow -- Qal** *breath, blow*; sq. b also *blow* into it (to scatter it); sq. לִפַחַת כָּלָיו אֵשׁ : עַל *to blow fire upon it* (ore, for melting), so fig. *and I will blow upon you with* (בְּ) *the fire of my wrath*; נ נָפְשָׁה *she hath breathed out her life* (of a mother, cf. **Hiph.**); abs. סִיר נָפוּחַ *a blown* (i.e. well-heated, boiling) *pot.* **Pu.** לֹא נֻפַּח אֵשׁ *a fire not blown* (by any human breath). **Hiph.** נֶפֶשׁ בְּעָלֶיהָ הִפַּחְתִּי *the life of its* (the land's) *owners I have caused them to breathe out*; אוֹתוֹ וְהִפַּחְתֶּם *and ye have sniffed at it* (in contempt). **(pg 655)**

nm950 » Strong's number 05397 means:

5397 נְשָׁמָה נ n°shamah {nesh-aw-maw'} ... 1) breath, **spirit** 1a) breath (of God) 1b) breath (of man) 1c) every breathing thing 1d) **spirit** (of man)

nm951 » From Whittaker's Revised BDB lexicon Strong's number 05397 means:

נְשָׁמָה common noun fem sing const **B6589** נְשָׁמָה **n.f. breath -- 1.** *breath* of God as hot wind kindling a flame; as destroying wind; as cold wind producing ice; as creative, giving breath to man. **2.** *breath of* man; breath of life נשמת חיים; as breathed in by God it is God's breath in man; and it is characteristic of man. **3.** syn. of נֶפֶשׁ in נפש כָּל-נִשְׁמָה *every breathing thing.* **4.** *spirit* of man. **(pg 675)**

Notice that one meaning of this word is "**spirit**." The breath of life could have been translated the spirit of life.

nm952 » There was actually a dual sense to Genesis 2:7: one physical; one spiritual. Notice Genesis 2:7's close resemblance to John 20:22:

KJV John 20:22: And when he had said this, he **breathed** on *them*, and saith unto them, Receive ye the Holy **Spirit**:

KJV John 20:22: And <2532> when he had said <2036> (5631) this <5124>, he **breathed** on <1720> (5656) *them*, and <2532> saith <3004> (5719) unto them <846>, Receive ye <2983> (5628) the Holy <40> **Spirit** <4151>:

nm953 » Christ blew on them saying, "receive the Holy Spirit." The word "blew" or "breathed" is Strong's number 1720 which means:

1720 ἐμφυσάω

emphusao {em-foo-sah'-o} ... 1) to blow or breathe upon

nm954 » The word translated "**Spirit**" (# 4151) in John 20:22 means:

4151 πνεῦμα pneuma {pnyoo'-mah} ... 1) a movement of air (a gentle blast 1a) of the wind, hence the wind itself 1b) breath of nostrils or mouth 2) the **spirit**, i.e. the vital principal by which the body is animated 2a) the rational spirit, the power by which the human being feels, thinks, decides 2b) the soul 3) a **spirit**, i.e. a simple essence, devoid of all or at least all grosser matter, and possessed of the power of knowing, desiring, deciding, and acting 3a) a life giving **spirit** 3b) a human soul that has left the body 3c) a **spirit** higher than man but lower than God, i.e. an angel 3c1) used of demons, or evil **spirits**, who were conceived as inhabiting the bodies of men 3c2) the **spiritual** nature of Christ, higher than the highest angels and equal to God, the divine nature of Christ 4) of God 4a) God's power and agency distinguishable in thought from his essence in itself considered 4a1) manifest in the course of affairs 4a2) by its influence upon the souls productive in the theocratic body (the church) of all the higher spiritual gifts and blessings 4a3) the third person of the trinity, the God the **Holy Spirit** 5) the disposition or influence which fills and governs the soul of any one 5a) the efficient source of any power, affection, emotion, desire, etc.

nm955 » In Genesis 2:7 Yehowah Elohim (Jesus Christ's Father, see GP 2) breathed or blew into Adam's nostrils and thereby gave him the breath of life. The same word "breath" (Strong's number 05397) in "breath of life" is also translated **spirit** in Proverbs 20:27 ("spirit" of man) in the KJV. And as we saw above it can and does mean spirit.

nm956 » In John 3:8 **spirit** (Strong's # 4151) and **wind** (#4151) are used interchangeably depending on which English translation you use and in what context it is used:

> KJV John 3:8: The **wind** bloweth where it listeth, and thou hearest the sound thereof, but canst not tell whence it cometh, and whither it goeth: so is every one that is born of the **Spirit**.

> KJV John 3:8: The **wind** <4151> bloweth <4154> (5719) where <3699> it listeth <2309> (5719), and <2532> thou hearest <191> (5719) the sound <5456> thereof <846>, but <235> canst <1492> <0> not <3756> tell <1492> (5758) whence <4159> it cometh <2064> (5736), and <2532> whither <4226> it goeth <5217> (5719): so <3779> is <2076> (5748) every one <3956> that is born <1080> (5772) of <1537> the **Spirit** <4151>.

> YLT John 3:8: the **Spirit** where he willeth doth blow, and his voice thou dost hear, but thou hast not known whence he cometh, and whither he goeth; thus is every one who hath been born of the **Spirit**.'

> NAB John 3:8: "The **wind** blows where it wishes and you hear the sound of it, but do not know where it comes from and where it is going; so is everyone who is born of the **Spirit**."

nm957 » John 3:8 and 20:22 were put in the Bible by God for a reason. In context with Genesis 2:7, it is telling us that as Yehowah Elohim in the beginning put the spirit of man in mankind, but in the end times Jesus Christ (through the power of God given to him) will give mankind the new and better spirit, the Holy Spirit (John 20:28). The Holy Spirit was given to mankind on the first Pentecost after Jesus Christ was resurrected (Acts 1:5; 2:1-4). This was the Holy Spirit prophesied to be given to mankind (Isa

44:3; Ezek 11:19; 36:26; Joel 2:28). But before this Holy Spirit was given there was and is even today (1997) in mankind the spirit of man that God (Yehowah Elohim) gave to mankind in the beginning.

nm958 » There is a spirit of man that is in man (Job 32:8; Proverbs 20:27). It is man's spirit (Job 34:14; Eccl 3:21). This is an evil spirit (1Sam 16:16; see, "Other Mind Paper" [NM 21]).

Yehowah Elohim Gave the Evil Spirit to Mankind

nm959 » Yehowah Elohim, who created the heavens and the earth, he gives **breath** and **spirit** to them that walk on the earth (Isa 42:5).

> KJV Isaiah 42:5: Thus saith God the LORD, he that created the heavens, and stretched them out; he that spread forth the earth, and that which cometh out of it; he that giveth **breath** unto the people upon it, and **spirit** to them that walk therein:

> KJV Isaiah 42:5: Thus saith <0559> (08804) God <0410> the LORD <03068>, he that created <01254> (08802) the heavens <08064>, and stretched them out <05186> (08802); he that spread forth <07554> (08802) the earth <0776>, and that which cometh out <06631> of it; he that giveth <05414> (08802) **breath** <05397> unto the people <05971> upon it, and **spirit** <07307> to them that walk <01980> (08802) therein:

nm960 » The **spirit** and souls Yehowah Elohim made (Isa 57:16). God made the spirit. Yehowah (YHWH or the BeComingOne) formed or created the "**spirit** of man within him":

> KJV Zechariah 12:1: The burden of the word of the LORD for Israel, saith the LORD, which stretcheth forth the heavens, and layeth the foundation of the earth, and formeth the spirit of man within him.

Not only did God make the spirit he made the "spirit of man *within* him."

nm961 » It is the **spirit of man** that is within mankind that knows the thoughts of mankind, while the things pertaining to God no one can understand except they that have the Spirit of God (1Cor 2:11).

Battle Against the Spiritual, Not the Physical

nm962 » Since angels are spirits (Heb 1:7), since the meaning of "angel" is messenger, then the way evil entered the world is through the spirit of man within mankind (Also see "Other Mind" paper [NM 21]). That is why we fight against the spiritual world and not the physical world. Our battle with sin and evil on earth has to do with fighting a spiritual battle against the other-mind and its power, not against flesh and blood (Eph 6:12). That is, the flesh is neutral pertaining to good and evil, but the evil spirit or the spirit in man in this the old age is the evil we fight against. The good Spirit or the Holy Spirit or the Spirit of Christ is what is given to mankind to fight against evil. It is this good spirit that causes mankind to repent of its old ways.

Repentance

nm963 » The change of mind in **repentance** is the change from the ruling spirit of man to the new ruling Spirit of God. When we receive the Spirit of God we get a spirit that is stronger than the spirit of man inside of us; strong enough to overcome the old mind and its ways of sin (1John 4:4).

Go to the **"Other Mind Paper"** [NM 21] to read more about the spirit of man (¶ nm928ff).

NM 23: Judging

Judge Not

Christ Judges

Brothers not Fools

Who are Fools

Paul Judged Also

Christians to Judge the World

NM23 Abstract

What did the Bible mean when it said, "do not judge"? Does this mean we are not to judge anyone? If this is true, why did Jesus judge some people's behavior, and call them on their wrong behavior? Are we to have tolerance for the evil around us?

Judge Not

nm964 » There are scriptures that appear to some to say that Christians are not supposed to judge others at all. What do the scriptures say?

- Matthew 7:1: "Do not judge, so that you may not be judged. 2 For with the judgment you make you will be judged, and the measure you give will be the measure you get. 3 Why do you see the speck in your neighbor's eye, but do not notice the log in your own eye? 4 Or how can you say to your neighbor, 'Let me take the speck out of your eye,' while the log is in your own eye? 5 You hypocrite, first take the log out of your own eye, and then you will see clearly to take the speck out of your neighbor's eye."

nm965 » What this says is that you should be very careful how you judge. Do not be a hypocrite if you judge. But does this really say and mean, judge not at all, ever? These scriptures cannot possibly mean not to discriminate (judge) between good and evil because the whole Bible tells us to do so. So what is meant here?

nm966 » Let's look at Christ and Paul to see what Christ meant in Matthew 7.

John 8:15: You judge by human standards; I judge no one. 16 Yet even if I do judge, my judgment is valid; for it is not I alone who judge, but I and the Father who sent me.

Christ here says that others judge by human standards, but he does not judge by human standards.

Christ Judges

nm967 » Christ does not say he never judges, for as we will see he did/does judge.

John 8:26: I have much to say about you and **much to judge**;

John 5:30: I can do nothing on my own. **As I hear, I judge**; and my judgment is just, because I seek to do not my own will but the will of him who sent me.

John 7:24: Do not judge by appearances, but **judge with righteous judgment**.

2Timothy 4:1: In the presence of God and of **Christ Jesus, who is to judge** the living and the dead

Here it says that Christ does judge. But his judgment is righteous for one thing he does not seek his own gain (See "God's Wrath" papers to understand righteous judgment.).

Brothers not Fools

nm968 » Do not be angry with your brother or call him a fool:

Matthew 5:22: **But I say to you that if you are angry with a brother or sister, you will be liable to judgment**; and if you insult a brother or sister, you will be liable to the council; **and if you say, 'You fool,' you will be liable to the hell of fire.**

nm969 » But Christ called people fools many times:

Matthew 23:17: "You blind fools!" (See v. 19 & Luke 11:40)

nm970 » Christ judged people and called them hypocrites:

Luke 12:56: "You hypocrites!" (Mat 15:7; 16:3; 22:18; 23:13)

nm971 » Christ even called a few of his apostles, "O fools" (Luke 24:25).

nm972 » What did the scripture say about calling people fools?

> Matthew 5:22: **But I say to you that if you are angry with a brother or sister, you will be liable to judgment**; and if you insult a brother or sister, you will be liable to the council; **and if you say, 'You fool,' you will be liable to the hell of fire.**

Do not be angry with your brother, do not call him a fool.

nm973 » Now we know from scripture that Christ's "brothers" and "sisters" are those in the Spirit.

> Matthew 12:50: For whosoever shall do the will of my Father which is in heaven, the same is my **brother,** and sister, and mother.

> Matthew 7:21: Not every one that says unto me, Lord, Lord, shall enter into the kingdom of heaven; but he that does the will of my Father which is in heaven.

As we see in this book you can do the will of the Father only when you have the Spirit. When Christ called a few of his apostles fools (Luke 24:25) they had not yet received the Spirit. (They received it later at the first Pentecost after Christ's resurrection.) At the time Christ called them fools, they were not yet Christ's Spiritual brothers. Thus, Christ was not wrong when he called them fools, since anyone without the Spirit, in their behavior is a fool.

Who are Fools?

nm974 » Who are fools, who are the foolish?

> Matthew 7:26: And everyone who hears these words of mine and does not act on them will be like a foolish man who built his house on sand.

Those who do not Spiritually hear his words and thus cannot act on them are fools, are foolish, and anything they do is built on sand instead of the Rock.(¶ nm28)

Ten Virgins and who was Foolish

nm975 » In the parable of the ten virgins, which were foolish?

> Matthew 25:1: Then the kingdom of heaven will be like this. Ten bridesmaids took their lamps and went to meet the bridegroom. 2 **Five of them were foolish**, and **five were wise**. 3 When the foolish took their lamps, they took no oil with them; 4 but **the wise took flasks of oil with their lamps**. 5 As the bridegroom was delayed, all of them [all ten] became drowsy and slept. 6 But at midnight there was a shout, 'Look! Here is the bridegroom! Come out to meet him.' 7 Then all those bridesmaids got up and trimmed their lamps. 8 **The foolish said to the wise, 'Give us some of your oil, for our lamps are going out.'** 9 But the wise replied, 'No! there will not be enough for you and for us; you had better go to the dealers and buy some for yourselves.' 10 And while they went to buy it, the bridegroom came, and those who were ready went with him into the wedding banquet; and the door was shut. 11 Later the other bridesmaids came also, saying, 'Lord, lord, open to us.' 12 But he replied, 'Truly I tell you, I do not know you.'

nm976 » The foolish had no oil in their lamps, that is, they did not have the Spirit, and thus were "fools." The olive oil they used in the Old Testament to anoint their kings typified the new oil of the Spirit (1John 2:20, 27). The lamp is God's word, "Your word is a lamp to my feet, and a light unto my path" (Psa 119:105). The foolish do not have the Spirit to go with the word of God, and thus at Christ's coming the Lord will say to the foolish without the Spiritual oil, "I know you not" (Mat 25:12).

Paul Judged Also

nm977 » Paul judged obvious wrongs, and asked his physical churches to also judge against them:

> 1Corinthians 5:9: I wrote to you in my letter not to associate with sexually immoral persons – 10 not at all meaning the immoral of this world, or the greedy and robbers, or idolaters, since you would then need to go out of the world. 11 But now I am writing to you not to associate with anyone who bears the name of brother or sister who is sexually immoral or greedy, or

is an idolater, reviler, drunkard, or robber. Do not even eat with such a one. 12 For what have I to do with judging those outside? Is it not those who are inside that you are to judge? 13 God will judge those outside. "Drive out the wicked person from among you."

Christians to Judge the World

nm978 » Paul also said that since Christians are to judge the world, why cannot they judge wrong among themselves instead of going to the outside:

1Corinthians 6:1: When any of you has a grievance against another, do you dare to take it to court before the unrighteous, instead of taking it before the saints? 2 Do you not know that the saints will judge the world? And if the world is to be judged by you, are you incompetent to try trivial cases? 3 Do you not know that we are to judge angels-- to say nothing of ordinary matters? 4 If you have ordinary cases, then, do you appoint as judges those who have no standing in the church? 5 I say this to your shame. Can it be that there is no one among you wise enough to decide between one believer and another, 6 but a believer goes to court against a believer-- and before unbelievers at that?

nm979 » Paul also cautioned others to be careful how they judged each other:

Romans 14:10: Why do you pass judgment on your brother or sister? Or you, why do you despise your brother or sister? For we will all stand before the judgment seat of God.

Romans 14:13: Let us not therefore judge one another any more: but judge this rather, that no man put a stumbling block or an occasion to fall in *his* brother's way.

In context Paul was speaking about so-called clean and unclean food. And so he said, "I know and am persuaded in the Lord Jesus that nothing is unclean in itself; but it is unclean for anyone who thinks it unclean" (Rom 14:14). There are some things that appear evil to some, but according to the law of love are not evil.

nm980 » Paul said we are to judge for ourselves what is right. We must be very careful how we do this:

1Corinthians 10:15: I speak as to sensible people; judge for yourselves what I say.

1Corinthians 11:13: Judge for yourselves: is it proper for a woman to pray to God with her head unveiled?

Colossians 2:16: Therefore do not let anyone condemn you in matters of food and drink or of observing festivals, new moons, or sabbaths.

James 5:9: Beloved, do not grumble against one another, so that you may not be judged. See, the Judge is standing at the doors!

nm981 » Some social customs that have no bearing on the law of love may be considered evil in some parts of the world, avoid appearing evil if you can (1Cor 10:23, 27-33). Always walk in love (Eph 5:2; Rom 13:8, 10, 13-14).

nm982 » Accusation against Spiritual elders should not be accepted except with two or three witnesses:

1Timothy 5:19: Never accept any accusation against an elder except on the evidence of two or three witnesses. 20 As for those who persist in sin, rebuke them in the presence of all, so that the rest also may stand in fear.

nm983 » When we think we find fault with a brother, we must bring our complaint first to the brother for clarification. We may have seen something that was not there.

nm984 » We must be careful how we judge, especially against a Spiritual brother. Why? Because **if** we call a real Spiritual brother a fool, it may be proof that we do not have the Spirit, and thus, we may be in danger of being destroyed in the fire to come (Mat 5:22; 1John 3:15; see "Last Judgment" paper [NM 24]). Those with the Spirit will recognize their brothers; those without the Spirit will not.

NM 24: Last Judgment

NM24 Abstract

In many religions there is a day of judgment, scales of Justice, a lake of fire, and hell. In this paper we are going to learn what the Bible says about judgment and hell. Is there a hell, and will people go to hell? Because of much confusion on these ideas, we will take a close look at these subjects.

nm985 » Before we see what the Bible says about the Judgment we will briefly examine some other ideas about it. We ourselves have heard various legends pertaining to hell and the judgment of the dead so we will recognize some of the following ideas. But the only ideas that count are the ones actually taught in the Bible, not the ideas that are alleged to be in the Bible.

Day of Judgment: An Old Idea

nm986 » The idea of the day of judgment has existed in most religions at least as far back as ancient Mesopotamia and Egypt. One book by S. Brandon, called *The Judgment of the Dead*, finds the idea of the judgment in ancient Egypt, Mesopotamia, the Greco-Roman culture, among Hebrews, among Christians, among Islamites, among Iranians, among Hindus and Buddhists, and among those in China and in Japan. Brandon believes that, "it is in Egypt that the earliest evidence is found of the idea that judgment awaited a man after death" (p. 6).

nm987 » Actually, this idea can be traced from the Biblical statement about not eating from the tree of knowledge of good and evil, "for in the day that you eat thereof, in dying, you will die" (see Hebrew text). Thus, If you sin, then when you die, you will really be dead, for you will be part of those called the "dead" in the Bible. The idea of the judgment of the dead is indeed very old, and not new.

Scales of Justice

nm988 » Brandon after expounding for pages about various ancient Egyptian writings about the weighing of good deeds and bad deeds of people at death writes about the tale of Senosiris, "Those whose misdeeds outweigh their good deeds are delivered to Amait, the bitch belonging to the lord of Amentit, so that their bodies and souls are utterly destroyed. Those who pass the awful test are conducted to heaven. The man, whose good and bad deeds equally balance, is placed among the dead furnished with amulets who serve Sokarosiris." Brandon goes on, "Belief in a judgment after death, symbolized by the balance or scales, can be traced on into the Roman period of Egyptian religion, and, as the curious *History of Joseph the Carpenter* shows, it passed in turn unto Coptic Christianity. The idea of weighing the deeds of men had already been adopted into Jewish apocryphal literature, and the variant concept of the weighing of souls had entered into Greek thought, as we shall see. Ultimately the idea found expression in mediaeval Christian art, with the archangel Michael assuming the role of 'Master of the Balance' which Thoth had held in ancient Egypt." (p . 45, *The Judgment of the Dead*)

Pit of Fire; Lake of Fire

nm989 » "In the *Amduat*, which purports to describe the underworld, these 'enemies,' represented either in human form or by hieroglyphs denoting 'shadows' or 'souls,' are shown in pits of fire" (p. 46). In the *Book of the Dead* it speaks of the " double Lake of Fire" (p. 186, 342 in Budge's Dover ed.).

nm990 » In *The Legends of the Jews*, Louis Ginzberg writes about what Enoch saw in hell, "He saw there all sorts of tortures, and impenetrable gloom, and there is no light there, but a gloomy fire is always burning. And all that place has fire on all sides, and on all sides cold and ice, thus it burns and freezes. And the angels, terrible and without pity, carry savage weapons, and their torture is unmerciful" (Vol. I, p. 132). In Ginzberg's books there are about 60 or more sections that speak about the Jewish legends concerning hell.

nm991 » The *Apocalypse of Peter*, a once popular book, which dates from approximately the early second century, shows the horrors of hell: "hanging by their tongues, and those were they that blasphemed the way of righteousness, and under them was laid fire flaming and tormenting them ... And in another place were gravel-stones sharper than swords or any spit, heated with fire, and men and women clad in filthy rags rolled upon them in torment" (Brandon, p. 116-117). These and other torments were to last forever, for the sinners, "shall be tormented eternally, for God willeth it so" (Brandon, p. 117).

Second Death

nm992 » "With these indications of the perpetual torments of the damned must also be set the idea of 'second death,' about which concern is shown in the *Book of the Dead*. From this conflict of eschatological imagery we can, accordingly, only safely deduce that the Egyptians believed that some awful fate awaited those whose hearts were found to be not right with *Maat* in the judgment after death" (p. 46-47). In the *Book of the Dead*, there are sections that deal with what one must do to escape the second death. These sections start out with the phrase, "chapter of not dying a second time [second death] in the netherworld." (p. 105, 184, 341 in Budge's Dover ed.)

Judgment After Death

nm993 » Greek Thought. In Brandon's opinion, "there was a significant body of opinion which affirmed that after death men would be judged on their conduct here. This belief was authorized by the poetry of Homer, which was truly the 'Bible of the Greeks.'" (p. 87) From Plato's *Republic* we see Socrates fearing what may come after his death. At one time Socrates "laughed at the tales about those in Hades, of punishment to be suffered there by him who here has done injustice. But now his soul is tormented by the thought that these may be true" (P. 87).

Are the above descriptions of a forever torture in a hell fire, and other tales even worse, really what hell is all about?

Trial and the Thousand Years

nm994 » Among the Greeks, if Plato's writings are any indication, was the idea of the judgment or trial of the souls after death: The souls are "on the termination of their first life, brought to trial; and, according to their sentence, some go to the prison-houses beneath the earth, to suffer for their sins, while others, by virtue, of their trial, are borne lightly upwards to some celestial spot, where they pass their days in a manner worthy of the life they have lived in their mortal form. But in the **thousandth year** both divisions come back again to share and choose their second life, and they select that which they severally please" (Plato, *Phaedrus*).

Notice that the idea of the millennium was also in Greek thought in Plato's time. And notice that *after* 1000 years both divisions would come back again (are resurrected) and do as they please.

Koran: Those on the Right and Left

nm995 » In the Islamic holy book, the Koran, the tales of hell and judgment are told. "The sinners will be in punishment of hell, to dwell therein" (Sura 43:74). At the end of the world, "when the earth will be shaken to its depths, and the mountains shall crumble to atoms" there will be those on the right and left of God, those on the right have "fruits in abundance" they are given "shade," "thrones," and virgin companions, but those on the left go off into "the midst of a fierce blast of fire and in boiling water" (Sura 56:4, 8, 27ff, 41ff).

Universal Salvation

nm996 » Not all shared in the idea of the forever judgement to hell. The great Alexandrian scholar Origen (about 185-254 AD) believed in universal salvation. "According to his doctrine of *apokatastasis*, in the end all souls, and even the demons, would be purified and reunited with God. Although Origen's views greatly influenced many Eastern Christians, Christians generally found it easier, and, it would seem, more congenial, to believe in both a Purgatory and a Hell where sinners would suffer physically the most horrible and revolting tortures that a morbid imagination could devise" (Brandon, p. 118).

nm997 » In Ginzberg's books on the legends of the Jews, he relates the legends of how God revealed to Enoch "that the duration of the world will be seven thousand years, and the eighth millennium will be a time when there is no computation, no end, neither years, nor months, nor weeks, nor days, nor hours" (Vol. I, p. 135).

nm998 » In some forms of Zoroastrianism all men are saved after the Good ultimately prevails over the Evil, "and all men become immortal for ever and everlasting" (Brandon, p. 163).

nm999 » Hosea Ballou in his book, *The Ancient History of Universalism*[1] attempts to show the belief in universal salvation from Titus Flavius Clemens (190 AD) to Origen (230 AD) and those who agreed with him, and onto the other traces of it in history to about 1498 AD. So the belief in universal salvation is not new. Some today who believe in it are called universalist. Ballou was pastor of the Universalist Church in Roxbury in the early 1800's.

Great White Throne Judgment

nm1000 » The popular Christian belief concerning the judgment of the dead is not new and can be stated as follows:

> At the return of Christ there will be a resurrection of the righteous and the wicked. (The Pre-millenarians teach a double resurrection: one of the just at the return of Christ, and another of the unjust a thousand years later.) Christ will judge those resurrected. This is called the great white throne judgment. All people will be judged at that time. This final judgment will be at the end of the world. Those judged the "wicked" will be consigned to hell or the lake of fire, and will suffer pains in body and soul for ever and ever. Those judged righteous will be given heaven as a reward. This heavenly state will be eternal. The resurrected will enjoy the fullness of life in communion with God for ever and ever.

---

[1] Published in Boston, 1828

Questions

nm1001 » Is there a judgment of the dead? What is the judgment of the dead? Does this judgment last forever? Is there a hell fire? Who goes into the hell fire? Are people tortured forever? Can people's flesh burn forever? Will people be tortured? When is the Judgment? What happens to Satan and his angels? Is there a heaven?

Biblical Teaching on Judgment

Know Three Things

nm1002 » In order to understand the real last judgment mentioned in the Bible you must know three things.

1. **That God judges righteously**. See "God's Wrath" [PR4] to understand God's righteous judgment. In short, "the LORD is known by the judgment which he executes: the wicked is snared in the work of his own hands" (Psalms 9:16). In short, God's wrath and judgment is righteous: he lets those who do evil destroy themselves.

2. **That the word translated "forever" or "everlasting" or "eternal"** in the English Bible literally means age or aeon or aeonian. See our "Age Paper" [NM 7] for proof on this matter. In short, there are ages in God's plan.

3. **There is a difference between immortal life and aeonian life.** Knowledge is distinguishing between facts that superficially may seem to be the same (see below & NM 11).

Resurrection of the Dead and the Judgment

nm1003 » There is Biblical doctrine pertaining to the resurrection and judgment (Heb 6:2).

> "Indeed, just as **the Father raises the dead and gives them life, so also the Son gives life to whomever he wishes.** 22 The Father judges no one but has given all judgment to the Son, 23 so that all may honor the Son just as they honor the Father. Anyone who does not honor the Son does not honor the Father who sent him. 24 Very truly, I tell you, anyone who hears my word

and believes him who sent me has aeonian life, and does not come under judgment, but has passed from death to life. 25 "Very truly, I tell you, the hour is coming, and is now here, when the dead will hear the voice of the Son of God, and those who hear will live. 26 For just as the Father has life in himself, so he has granted the Son also to have life in himself; 27 and he has given him authority to execute **judgment**, because he is the Son of Man. 28 Do not be astonished at this; for the hour is coming when all who are in their graves will hear his voice 29 and will come out-- those who have done good, to the **resurrection of life**, and those who have done evil, to the **resurrection of judgment**" (John 5:21-29).

Christ given the power of the Resurrection and Judgment

nm1004 » So in John 5 we have scripture that says the Father gave the authority to Christ ("Son of Man") to execute judgment. Remember the Father gave Christ *all* the power (Mat 28:18). One of these powers was to execute judgment and to give life to "whomever he wishes." In other words, Christ has the power to resurrect people. In verse 29 it speaks about the resurrection of life and the resurrection of judgment. Those who have done good will be in the resurrection of life, but those who have done evil will be in the resurrection of judgment. To be resurrected, you must have first died. So in the book of Acts 24:15 we see Paul saying "I have a hope in God – a hope that they themselves also accept – that there will be a resurrection of both the righteous and the unrighteous." Paul called this the resurrection of the dead (Acts 24:21; 1Cor 15:21). Jesus Christ himself was resurrected from the dead by the Father (1Pet 1:3; see GP 5). Daniel spoke about the resurrection of the dead, "And many of them that sleep in the dust of the earth shall awake, some to life of olam, and some to shame and contempt of olam." (Dan 12:2).

Two Groups; Two Destinies

One on the Right, One on the Left

nm1005 » Notice that there are two groups or divisions: one of the just and good, and one of the unjust and evil (John 5:29; Acts 24:15; Dan 12:2).

"When the Son of Man comes in his glory, and all the angels with him, then he will sit on the throne of his glory. 32 All the nations will be gathered before him, and he will separate people one from another as a shepherd separates the sheep from the goats, 33 and he will put the sheep at his right hand and the goats at the left. 34 Then the king will say to **those at his right hand**, 'Come, you that are blessed by my Father, **inherit the kingdom** prepared for you from the foundation of the world; 35 for I was hungry and you gave me food, I was thirsty and you gave me something to drink, I was a stranger and you welcomed me, 36 I was naked and you gave me clothing, I was sick and you took care of me, I was in prison and you visited me.' 37 Then the righteous will answer him, 'Lord, when was it that we saw you hungry and gave you food, or thirsty and gave you something to drink? 38 And when was it that we saw you a stranger and welcomed you, or naked and gave you clothing? 39 And when was it that we saw you sick or in prison and visited you?' 40 And the king will answer them, 'Truly I tell you, just as you did it to one of the least of these who are members of my family, you did it to me.' 41 Then he will say to those **at his left hand**, 'You that are accursed, depart from me into the **aeonian fire** prepared for the devil and his angels.' 46 And these will go away into aeonian punishment, but the righteous into aeonian life" (Mat 25:31-41, 46).

One group, the just, at Christ's right hand, is given the kingdom, the other group on the left hand side is sent to the aeonian fire.

Lake of Fire

Second Death is the Lake of Fire

nm1006 » Notice in the book of Revelation, 20:11-15:

> "Then I saw a **great white throne** and the one who sat on it; the earth and the heaven fled from his presence, and no place was found for them. 12 And I saw the dead, great and small, standing before the throne, and books were opened. Also another book was opened, the book of life. And the dead were judged according to their works, as recorded in the books. 13 And the sea gave up the dead that were in it, Death and Hades gave up the dead that were in them, and all were judged according to their works. 14 Then Death and Hades were thrown into the **lake of fire**. This is the **second death**, the lake of fire; 15 and anyone whose name was not found written in the book of life was thrown into the lake of fire."

nm1007 » Those not written in the book of life are thrown into the lake of fire, and this is the second death – the lake of fire (Rev 20:14). Notice that all were judged according to their works (Rev 20:13; see NM 12). Those who overcome evil "shall inherit all things, and I will be his God, and he shall be my son. But the **fearful, and unbelieving**, and abominable, and murderers, fornicators, and sorcerers, and idolaters, and all liars, shall **have their part in the lake which burns with fire** and brimstone [lake of fire], which is the second death" (Rev 21:7-8). Thus the lake of fire is the second death.

Lake of Fire: More Information

nm1008 » As we have just seen in the book of Revelation, chapters 20 and 21, there is going to be a "lake of fire," and the lake of fire is the second death. The lake of fire is not a myth. This is not a myth. But at the beginning of this paper we wrote about the mythical *Book of the Dead* wherein a lake of fire is also mentioned. Also in many of the "myths" of ancient civilizations, pits of fire, hell of fire, lakes of fire, and other such descriptions are mentioned in connection with the fate of those judged evil in the day of judgement, or at the last judgment. There seems to be a collective-subconscious idea about some fiery fate for those who are evil, or for those who do not do enough good works while they are on earth. These myths are not the truth; they only have some aspects of the truth. The spirits of this age know something about a future judgment for them (Mat 8:28-29). Do they somehow project this subconsciously to mankind? (See, "Other Mind" paper [NM 21])

Scriptures that Point to the Last Judgment

nm1009 » The following verses, through type and antitype (NM15; Nm16; PR1; etc.), in their higher meaning all point to the real last judgment:

> 2Peter 3:10 But the **day of the Lord** will come like a thief, and then the heavens will pass away with a loud noise, and **the elements will be dissolved with fire**, and the earth and everything that is done on it will be disclosed.

> Zephaniah 1:18 Neither their silver nor their gold will be able to save them on the **day of the LORD's wrath**; in the **fire of his passion** the whole earth shall be consumed; for a full, a terrible end he will make of all the inhabitants of the earth.

> Malachi 4:1 See, the day is coming, burning like an oven, when all the arrogant and all evildoers will be stubble; the **day that comes shall burn them up**, says the LORD of hosts, so that it will leave them neither root nor branch.

> Isaiah 29:6 you will be visited by the LORD of hosts with thunder and earthquake and great noise, with whirlwind and tempest, and the **flame of a devouring fire.**

> Ezekiel 38:22 And I will judge against him with pestilence and with blood; and I will rain upon him, and upon his bands, and upon the many people that *are* with him, an overflowing rain, and **great hailstones, fire, and brimstone**.

> Isaiah 30:30 And the LORD will cause his majestic voice to be heard and the descending blow of his arm to be seen, in furious anger and a **flame of devouring fire**, with a cloudburst and tempest and **hailstones**.

> Ezekiel 13:13 Therefore thus says the Lord GOD: In my wrath I will make a stormy wind break out, and in my

anger there shall be a deluge of rain, and **hailstones** in wrath to destroy it.

Jeremiah 10:10 But the LORD is the true God; he is the living God and the everlasting King. At **his wrath the earth quakes**, and the nations cannot endure his indignation.

Ezekiel 31:16 I made the nations quake at the sound of its fall, when I **cast it down to Sheol [hell] with those who go down to the Pit**; and all the trees of Eden, the choice and best of Lebanon, all that were well watered, were consoled in the world below.

Revelation 16:18 And there came flashes of lightning, rumblings, peals of thunder, and a **violent earthquake**, such as had not occurred since people were upon the earth, so violent was that earthquake.

Revelation 11:19 Then God's temple in heaven was opened, and the ark of his covenant was seen within his temple; and there were flashes of lightning, rumblings, peals of thunder, an **earthquake**, and **heavy hail**.

Job 38:22 have you entered into the treasures of the snow? or have you seen the treasures of **the hail**, Job 38:23 Which **I have reserved against the time of trouble**, against the day of battle and war?

Revelation 16:21 and huge **hailstones**, each **weighing about a hundred pounds**, dropped from heaven on people, until they cursed God for the plague of the hail, so fearful was that plague.

Revelation 11:13 At that moment there was a **great earthquake**, and a tenth of the city fell; seven thousand people were killed in the earthquake, and the rest were terrified and gave glory to the God of heaven.

Revelation 8:5 Then the angel took the **censer and filled it with fire** from the altar and threw it on the earth; and there were peals of thunder, rumblings, flashes of lightning, and an **earthquake**.

Revelation 18:8 therefore her plagues will come in a single day-- pestilence and mourning and famine-- and she

[spiritual Babylon] will be **burned with fire**; for mighty is the Lord God who judges her."

Revelation 19:20 And the beast was captured, and with it the false prophet who had performed in its presence the signs by which he deceived those who had received the mark of the beast and those who worshiped its image. These two were thrown alive into the **lake of fire** that burns with sulfur.

Revelation 20:10 And the devil who had deceived them was thrown into the **lake of fire** and sulfur, where the beast and the false prophet were, and they will be tormented day and night into the ages of ages [not, forever and ever].

Revelation 21:8 But as for the cowardly, the faithless, the polluted, the murderers, the fornicators, the sorcerers, the idolaters, and all liars, their place will be in the **lake that burns with fire** and sulfur, which is the second death.

nm1010 » The scriptures above and others point to the great fire, great earthquake, great hailstones, great wind, great thunder, great lightning, thus, a **great tribulation**. This is the Last War. I write about in the "Last War and God's Wrath" paper [PR5]. See an outline of scripture concerning this Last War in "God's Wrath, An Outline" [PR6]. This Last War has to do with a great atomic war with a massive fiery wind storm caused by the abundance of weapons going off in close proximity to each other.

Hell Fire: Who is it For?

nm1011 » There is a lake of fire, but who is it for? Is the fire for humans? Myths say that humans in bodily form will be tortured in a hell fire forever and ever for the evil they committed on earth. Does the Bible teach this?

"Then he will say to those at his left hand, You that are accursed, depart from me into the aeonian **fire prepared for the devil and his angels**" (Mat 25:41).

"And **the devil** who had deceived them **was thrown into the lake of fire and sulfur**, where the beast and the false prophet are, and they will be tormented day and night into eons and eons" (Rev 20:10).

"He seized the dragon, **that ancient serpent, who is the Devil and Satan**, and bound him **for a thousand years**, and **threw him into the pit**, and locked and sealed it over him, so that he would deceive the nations no more, until the thousand years were ended. After that he must be let out for a little while" (Rev 20:2-3).

Jesus Christ said that the fire of hell was prepared or made for the devil and his angels. In the book of Revelation it prophesies about the Devil (Satan that old serpent) being sent to a pit of fire for one thousand years.

nm1012 » Peter and Jude also wrote about Satan and his angels being reserved unto the day of judgment.

> 2PET 2:4: For if GOD spares not [the] angels who sin, but casts them chained into the hell [Tartarus] of darkness to be kept [there] for judgment;

> JUDE 1:6: And angels who had not kept their own original state, but had abandoned their beginning, for [the] judgment of [the] great day; chained perpetually[1] under gloomy darkness, he keeps [them].

And In Proverbs, "The LORD has made all things for a purpose, Yes, even the wicked for the day of evil" (Prov 16:4).

nm1013 » This is much like the heavens and earth being kept in store for the day of judgment:

> 2PET 3:7: But the present heavens and the earth by his word are laid up in store, kept for fire unto a **day of judgment** and destruction of ungodly men.

nm1014 » And this day of judgment will be with fire so hot the elements will melt:

> 2PET 3:12: waiting for and hastening the **coming of the day of the GOD**, by reason of which [the] heavens, being on fire, shall be dissolved, and [the] elements, burning with heat, shall melt?

---

[1] Strong's # 126 (from #104) = always, continual, perpetual, not necessarily "forever." See NM 24.

ISA 64:1: Oh that you [God] would rip the heavens, that you would come down, that the mountains might flow down **at your presence**. ISA 64:2 As **when the melting fire burns**, the fire causes the waters to boil, to make your name known to your adversaries, that the nations may tremble at your presence! ISA 64:3 When you did terrible things which we looked not for, you came down, the mountains flowed down at your presence. ISA 64:4 For since olam past men have not heard, nor perceived by the ear, neither has the eye seen, O God, beside you, what he has prepared for him that waits for him.

PSA 97:1 The BeComingOne [YHWH] reigns; let the earth rejoice; let the multitude of isles be glad thereof. PSA 97:2 Clouds and darkness are round about him: righteousness and judgment are the habitation of his throne. PSA 97:3 A **fire goes before him**, and burns up his enemies round about. PSA 97:4 His lightning enlightened the world: the earth saw, and trembled. PSA 97:5 The **hills melted like wax** at the presence of the BeComingOne, at the presence of the Lord of the whole earth. PSA 97:6 The heavens declare his righteousness, and all the people see his glory.

Names for the Judgment

nm1015 » A day will come when God will Judge the world. Various phrases are used in the Bible to describe it.

- Day of the LORD's anger (Zeph 2:2)

- Wrath of the LORD of hosts (Isa 13:13)

- Day of His fierce anger (Isa 13:13)

- Day of the LORD's wrath (Zeph 1:18)

- Great day of wrath (Rev 6:17; Psalms 110:5)

- Day of the Lord (2Pet 3:10; 1Thes 5:2)

- Day of judgment (Mat 10:15; 2Pet 2:9; 3:7)

- Day of the Lord Jesus (1Cor 1:8)

- Great and dreadful day of the LORD (Mal 4:5)

- Great day of the LORD (Zeph 1:14)

- Day of the Lord Yehowah [YHWH], a day of vengeance (Jer 46:10)

- Day of wrath and revelation of the righteous judgment of God (Rom 2:5)

There are many places in the Bible where it speaks about the day of judgment for the enemies of physical Israel. But in the Spiritual view or higher meaning they pertain to the Last War, and the Last Judgment against the enemies of Spiritual Israel.

When is the Judgment?

nm1016 » The next question is, *when* will this Judgment begin? "Jesus Christ, who shall judge the living and the dead at his appearing and his kingdom" (2Tim 4:1). At the appearance of his kingdom Christ will judge the living and the dead.

Right and Left Side

nm1017 »

MAT 25:31 But **when the Son of man comes in his glory**, and all the angels with him, then shall he sit down upon his throne of glory, 32 and all the nations shall be gathered before him; and **he shall separate them** from one another, as the shepherd separates the sheep from the goats; 33 and he will set the sheep **on his right hand**, and the goats on his left. 34 Then shall the King say to those on his right hand, Come, blessed of my Father, inherit the kingdom prepared for you from the world's foundation: MAT 25:41 Then shall he say also to **those on the left**, Go from me, cursed, into aeonian fire, prepared for the devil and his angels: MAT 25:46 And these shall go away into aeonian punishment, and the righteous into life aeonian.

Last Trumpet

nm1018 » This judgement of the right and left side occurs when Christ comes with his angels at the last trumpet,

"And he shall send his angels with a great sound of a **trumpet**, and they shall gather together his elect..." (Mat 24:31).

"For the Lord himself, with a cry of command, with the archangel's call and with the sound of God's **trumpet**, will descend from heaven, and the dead in Christ will rise first" (1Thess 4:16).

"In a moment, in the twinkling of an eye, at the **last trump**: for the trumpet shall sound, and the dead shall be raised incorruptible, and we shall all be changed, for this corruptible must put on incorruption, and this mortal put on immortality" (1Cor 15:52-52).

"Then the seventh angel blew his **trumpet** [last trumpet], and there were loud voices in heaven, saying, 'The kingdom of the world has become the kingdom of our Lord and of his Messiah, and he will reign into the ages of the ages.'" (Rev 11:15)

When are those on the right and left separated? When is the last trumpet, when the angels help gather the elect, when immortality is given, when the kingdoms of this world become the kingdom of God?

Last Day

nm1019 » The Bible mentions the *day* of the Lord, the *day* of wrath, the *day* of anger, the *day* of judgment. And the Bible also mentions the **last day**:

John 6:39 And this is the will of him who sent me, that I should lose nothing of all that he has given me, but raise it up on the **last day**.

John 6:40 This is indeed the will of my Father, that all who see the Son and believe in him may have aeonian life; and I will raise them up on the **last day**."

John 6:44 No one can come to me unless drawn by the Father who sent me; and I will raise that person up on the **last day**.

John 11:24 Martha said to him, "I know that he will rise again in the resurrection on the **last day**."

John 12:48 The one who rejects me and does not receive my word has a judge; on the **last day** the word that I have spoken will serve as judge,

nm1020 » What happens on this last day? There is a resurrection for those who are drawn Spiritually, who see Spiritually, and believe Spiritually. But those who reject Christ are judged by the words Christ spoke (John 12:48; Ho 6:5).

nm1021 » When is this resurrection? From the "last trumpet" section above we see that the resurrection is at the last trumpet, just as the kingdom of God begins its reign on earth. "Jesus Christ, who shall judge the living and the dead at his appearing and his kingdom" (2Tim 4:1).

How will Christ Judge?

nm1022 » At Christ coming in his kingdom he will judge (2Tim 4:1). He will let his words be the judge (John 12:48; Ho 6:5). One is judged according to their works (See NM 12). He judges righteously (PR4). Note the books (Bible) being opened in Revelation 20:12, "and the dead were judged out of those things which were written in the books, according to their works." In other words, those who are "just" according to their works and are on the right are given the kingdom and aeonian life (Mat 25:33-34, 46); those who are "unjust" according to their works and are on the left are sent away from the glory of the kingdom and given aeonian punishment (Mat 25:41, 46). This is the righteous judgment, you are judged according to your works. The evil other-minds will be destroyed in the lake of fire of their own creation – the results of the Last War (see "Last War and God's Wrath" paper [PR5]).

> "The Lord Jesus is revealed from heaven with his mighty angels in flaming fire [caused by Satan's war], inflicting vengeance on those who do not know God and on those who do not obey the gospel of our Lord Jesus. These will suffer the punishment of aeonian destruction, separated from the presence of the Lord and from the glory of his might" (2Thess 1:7-9).

nm1023 » The judgment and the resurrection occur at the very time Christ comes with the power of his kingdom.

> DAN 7:26 But the **judgment shall sit**, and they [Saints] shall take away his [Satan & Beast's] dominion, to consume and to destroy it unto the end. DAN 7:27 And the kingdom and dominion, and the greatness of the kingdom under the whole heaven, shall be given to the people of the saints of the most High, whose kingdom is a kingdom of olam, and all dominions shall serve and obey him.

nm1024 » This in context with Daniel 7:12-14, 21-22, pictures the time when Christ and his saints take the kingdom away from Satan, and when the judgment is set to happen, which is at the time the Beast and its kingdom are destroyed (See *Beast Papers*; PR2, PR3).

> "And they went up on the breadth of the earth, and surrounded the camp of the saints and the beloved city: and fire came down and devoured them. And the Devil who deceived them was cast into the lake of fire and brimstone, and where the Beast and the false prophet [are]; and they shall be tormented day and night into the ages [aeons] of ages [aeons]. And I saw a great white throne, and him that sat on it, from whose face the earth and the heaven fled, and place was not found for them" (Rev 20:9-11; see Rev 20:15 & 19:19-21).

Last Day = Thousand Years

nm1025 » There are patterns in the Bible. One of them is the cycle of seven: Six similar units; the seventh different (See NM 15). In the beginning was the pattern of the week: six days of work; one day of rest (Gen 1:1-2:3; Exo 20:8-11). There is type and antitype in the Bible ("Duality Paper"). As Paul made much of the scriptures about Adam and Eve and Christ being the second Adam and the Church being the second Eve[1] (1Cor 15:45; Eph 5:23-32; etc), I will also make a great deal about the antitypical week of creation, in which each "day" equals 1000 years. When Peter was talking about the judgment, he wrote about a day for the Lord is as a thousand years (2Pet 3:7-8). Revelation 20:2-5 speaks about a rest from Satan lasting for 1000 years. Putting this together and knowing about type and antitype, we conclude that the "last day" Christ was speaking about in the book of John pertaining to the resurrection and judgment is the "day" of one-thousand years as mentioned in the book of Revelation.

When the Beast system is destroyed is when the judgment occurs, and when the kingdom of God takes over. This is also the time of the gathering of the nations to fight against Christ and his saints.

---

[1] Christ the second Adam; the Church the second Eve returns and becomes one with her Husband: the marriage of the Bride and Lamb (Rev 21:2, 9). Etc.

Gathering of the Nations

nm1026 » The following pertain to the time of the gathering of nations, Gog and Magog:

EZE 38:2 Son of man, set your face against **Gog, the land of Magog**, the chief prince of Meshech and Tubal, and prophesy against him, EZE 38:14 Therefore, son of man, prophesy and say unto **Gog**, Thus says my Lords the BeComingOne; In that day when my people of Israel dwells safely, shall you not know it? EZE 38:15 And you shall come from your place out of the north parts, you, **and many people with you**, all of them riding upon horses, a great company, **and a mighty army**: EZE 38:16 And you **shall come up against my people of Israel**, as a cloud to cover the land; it shall be in the end of days, and I will bring you against my land, that the nations may know me, when I shall be sanctified in you, O Gog, before their eyes. EZE 38:17 Thus says my Lords the BeComingOne; are you he of whom I have spoken in old days by my servants the prophets of Israel, which prophesied in those days many years that I would bring you against them? EZE 38:18 And **it shall come to pass at the same day when Gog shall come against the land of Israel**, says my Lords the BeComingOne, **that my fury shall come up in my face.** EZE 38:19 For in my jealousy and in the **fire of my wrath** have I spoken, Surely in that day there shall be a great **shaking** in the land of Israel; EZE 38:20 So that the fishes of the sea, and the fowls of the heaven, and the living creatures of the field, and all creeping things that creep upon the earth, and all the men that are upon the face of the earth, **shall shake at my presence**, and the mountains shall be thrown down, and the steep places shall fall, and every wall shall fall to the ground. EZE 38:21 And I will call for a sword against him throughout all my mountains, says my Lords the BeComingOne: every man's sword shall be against his brother. EZE 38:22 And **I will judge against him** with pestilence and with blood; and I will rain upon him, and upon his bands, and upon the many people that are with him, an overflowing rain, and **great hailstones, fire, and brimstone.** EZE 38:23 Thus will I magnify myself, and sanctify myself; and I will be known in the eyes of many nations, and they shall know that I am the BeComingOne.

REV 20:8 and [Satan] shall go out to deceive the nations which [are] in the four corners of the earth, **Gog and Magog**, to **gather** them together **to the war**, whose number [is] as the sand of the sea. REV 20:9 And they went up on the breadth of the earth, and **surrounded the camp of the saints** and the beloved city: and fire came down and devoured them. REV 20:10 And the **devil** who deceived them was **cast into the lake of fire and brimstone**, and where the beast and the false prophet; and they shall be tormented day and night into the ages [aeons] of ages [aeons] REV 20:11 And I saw a **great white throne**, and him that sat on it, from whose face the earth and the heaven fled, and place was not found for them.

REV 19:19 And I saw the **Beast** and the **kings of the earth** and their **armies gathered** together **to make war** against him that sat upon the horse, and against his army. REV 19:20 And the beast was taken, and the false prophet that [was] with him, who worked the signs before him by which he deceived them that received the mark of the beast, and those that worship his image. Alive were both **cast into the lake of fire** which burns with brimstone; REV 19:21 and the rest were slain with the sword of him that sat upon the horse, which goes out of his mouth; and all the birds were filled with their flesh. REV 20:2 And he [Christ] laid hold of the dragon, the ancient serpent who is [the] Devil and **Satan, and bound him a thousand years**, REV 20:3 and cast him into the abyss, and shut [it] and sealed [it] over him, that he should not any more deceive the nations until the thousand years were completed; (after these things he must be loosed for a little time.) REV 20:4 And I saw thrones; and they sat upon them, and **judgment was given to them** [Christ & Saints, Dan 7:22]; and the souls of those beheaded on account of the testimony of Jesus, and on account of the word of the GOD; and those who had not done homage to the Beast nor to his image, and had not received the mark on their forehead and hand; and they lived and reigned with the Christ a thousand years: Rᴇᴠ 20:5 (the rest of the dead did not live until the thousand years had been completed.) This [is] the first resurrection. Rᴇᴠ 20:6 Blessed and holy he who has part in the first resurrection: over these the second death has no power; but they shall be priests of the GOD and of the Christ, and shall reign with him a thousand years. Rᴇᴠ 20:7 And when the thousand years have been completed, Satan shall be loosed from his prison.

nm1027 » **Gathered Against Christ**. Here we have Gog and Magog, with the gathered nations and their armies going against Christ and his "army." Christ's army is, "And the armies which [are] in the heaven followed him upon white horses, clad in white, pure, fine linen" (Rev 19:14).

Who are those Dressed in White

nm1028 » And who are those dressed in white?

> Rᴇᴠ 7:13 And one of the elders answered, saying to me, These who are clothed with white robes, who are they, and whence came they? Rᴇᴠ 7:14 And I said to him, My lord, you know. And he said to me, These are **they who come out of the great tribulation**, and have washed their robes, and have made them white in the blood of the Lamb. Rᴇᴠ 7:15 Therefore are they before the throne of the GOD, and serve him day and night in his temple, and he that sits upon the throne shall spread his tabernacle over them. Rᴇᴠ 7:16 They shall not hunger any more, neither shall they thirst any more, nor shall the sun at all fall on them, nor any burning heat.

Thus, those clothed in white are the saints (Rev 3:5, 18; 6:11). They came out of the great tribulation.

Great Tribulation

nm1029 » There are two senses to this tribulation: (1) The greatest and last tribulation at the very end, and (2) the whole tribulation since the time of Adam (cf. Rom 8:22; Rev 1:9; 1Thess 3:4; 2Cor 7:4;). By reading our newspapers and by reading history, we know that the whole world has been in tribulation: and we know this is the result of sin.

To review. We see that the time of the last judgment is right at the time Christ comes to take over the kingdom or rulership from Satan. Christ does not destroy the world, the world destroys itself, for Christ comes to save not destroy (see God's Wrath papers [PR4, PR5, PR6]).

Hell-Fire; Mankind; Real Judgment

nm1030 » We know by various scriptures that the hell-fire will be for Satan and his angels (Mat 25:41; Rev 20:2-3, 10). But according to myth some of mankind will suffer **in** a fire forever for their sins. How can this be since physical bodies will burn-up in a fire? This whole idea is nonsense. They make God out to be sadistic. But our God is Love. (The *Parable of Lazarus* is logically explained in the "All Saved" paper [NM 13]) Yet Augustine in his book called, the *City of God*, quotes from John 5:29, Matthew 13:41-43, and Matthew 25:46 as proof of "the perpetual punishment of those condemned with the devil." In Augustine's answer to the question "whether bodies can survive in a burning fire," he says, "what proof then can I offer to convince unbelievers that it is possible for human bodies, endowed with soul and life, not merely never to be decomposed by death, but also to outlast the torments of eternal flames? They refuse to accept from us an appeal to the power of the Almighty, but press us to cite a precedent by way of argument. We can reply that there are animals, which are certainly liable to destruction, since they are mortal, but still survive in the midst of flames" (Book 21.1). This animal Augustine is speaking about is the so-called "fire-salamander" which never existed, except in myth. Later Augustine goes on and says, "we say that there will be living human bodies which will always burn and suffer, yet will never die" (21.5, Loeb Classical Lib, p. 31). He goes on about this, "In many things it is unclear to us what His will is. This [human bodies in hell, never dying], however, is very certain, that none of the things which He has willed is impossible, and we believe His predictions since we cannot question either his strength or his truth" (Loeb Lib, p. 31).

nm1031 » **God's Will**. Augustine tries to argue with things in nature, but fails, so he falls back on *his* corrupt idea of God's will. But what is God's will? "God our Savior, who wishes <u>all</u> to be saved, and to come to the knowledge of the truth" (1Tim 2:3-4). And the Lord is "not willing that any should perish, but that all should come to repentance" (2Pet 3:9). Because Christ does his Father's will (Heb 10:7; John 6:38), then what Christ wills is also what God wills. As shown in the paper called "All Saved" [NM 13], God willed that all will eventually be saved, and He will make this happen, irrespective if Augustine or you or anyone else believes otherwise. See "All Saved" paper [NM13] for details.

nm1032 » By his own arguments, we can see that Augustine knew that the word translated into "forever" in many Bibles can mean an "age" or "long period of time" and not an eternal period of time. In order to disprove that the Greek word *aionios* or the Latin word *aeternus* means forever and not aeonian, Augustine agues thus,

> Then what sort of reasoning is it, to take the eternal [*aionios*] punishment of the wicked as a fire of long duration and believe that eternal life is without end? For Christ said in the very same place, including both in one and the same sentence: "So these will go into eternal punishment, but the righteous into eternal life." If both are eternal, then surely both must be understood as "long," but having an end, or else as "everlasting," without an end. For they are matched with each other: in one clause eternal punishment, in the other eternal life. But to say in one and the same sentence: "Eternal life shall be without end, eternal punishment will have an end," is utterly absurd. Hence, since the eternal life of the saints will be without end, eternal punishment also will surely have not end, for those whose lot it is.

(See "Context Argument Two in the "Age Paper" [NM 7] for a fuller explanation of this.)

nm1033 » **What Augustine does here is mix-up two things**: (1) Christians are given immortal life when they are resurrected at Christ's coming; and (2), they are given aeonian life in the same age that the "dead" and Satan are given their aeonian punishment.

nm1034 » Notice that Paul in the so-called resurrection chapter in the first book of Corinthians, mentions Christian's immortality:

> 1Corinthians 15:42: So it is with the resurrection of the dead. What is sown is perishable, what is raised is imperishable [immortal].

1Corinthians 15:51: Listen, I will tell you a mystery! We will not all die, but we will all be changed, 52 in a moment, in the twinkling of an eye, at the last trumpet. For the trumpet will sound, and the dead will be raised imperishable, and we will be changed. 53 For this perishable body must put on imperishability, and this mortal body must put on immortality. 54 When this perishable body puts on imperishability, and this **mortal body puts on immortality**, then the saying that is written will be fulfilled: "Death has been swallowed up in victory."

Thus Christians will receive immortality as a gift. There is a difference between immortal life and aeonian life (NM11).

Three Orders or Divisions

nm1035 » **Three Orders:** This resurrection chapter is also talking about the sequential *order* in which mankind will be given immortal life.

> 1Corinthians 15:20 But in fact Christ has been raised from the dead, the first fruit of those who have died. 21 For since death came through a human being, the resurrection of the dead has also come through a human being; 22 for as all die in Adam, so **all will be made alive in Christ**. 23 But **each in his own order: Christ the first fruit, then at his coming those who belong to Christ. 24 Then comes the end**, when he hands over the kingdom to God the Father, after he has destroyed every ruler and every authority and power.

This thus pictures the three times that mankind will be resurrected from the dead and given immortal life. This fulfills the typical three times physical Israel was ordered by God to stand before him (See NM 16).

The first was Christ

nm1036 »

> 25 "Very truly, I tell you, the hour is coming, and is now here, when the dead will hear the voice of the Son of God, and those who hear will live. 26 **For just as the Father has life in himself, so he has granted the Son**

also to have life in himself; 27 and he has given him authority to execute judgment, because he is the Son of Man. 28 Do not be astonished at this; for the hour is coming when all who are in their graves will hear his voice 29 and will come out-- those who have done good, to **the resurrection of life**, and those who have done evil, to the resurrection of judgment (John 5:25-29).

Christ was resurrected from the dead by the power of his Father (Rom 1:4; 1Pet 1:3; GP 5). He was first, "If then you see the Son of man ascending up where he was the first?" (John 6:62; see Greek text) Christ was the "beginning of the creation of God" or the *new* creation of God (Rev 3:14; Gal 6:15), "the first born of all creation" (Col 1:15), thus "first born of the dead" "who is the beginning" (Col 1:18) "He is the first-born among many brethren" (Rom 8:29).

Second Order

nm1037 » The second time mankind will be resurrected to immortal life is at Christ's coming (1Cor 15:23). This is the resurrection of life (John 5:29). This is the group on Christ's right at his coming who are given the kingdom and rule during the 1000 years (Rev 20:1-6). Satan and his angels, on the left side, are counted as unjust, and are put into the bottomless pit for 1000 years (Rev 20:2-3, 11-15; 21:8; 2Thess 1:9). Those counted as good are given immortal life and also the 1000 year aeonian life (Rom 2:7; NM 11). We explain the difference between the immortal life and the aeonian life in our paper called "Reward for Christians." [NM 11] They will rule with Christ during the 1000 years (Rev 2:26; 3:21; 5:10; Dan 7:27). This is the first of the last two resurrections to immortal life (Rev 20:5b).

Third Order

nm1038 » But the others "will suffer the punishment of aeonian destruction, separated from the presence of the Lord and from the glory of his power" (2Thess 1:9). While the "just" are enjoying the wonders of the kingdom during the 1000 years, the "unjust" will be taken from the kingdom for 1000 years. It is at the "end," after the 1000 years, when the "unjust" are resurrected (Rev 20:5a; 1Cor 15:24). This is the resurrection of the dead (John 5:29).

Conclusion:

nm1039 » The "just" are given immortal life and given the 1000 year aeonian life, but the "unjust" go for the 1000 year aeonian punishment (Mat 25:46 w/ Rev 20:1-5). This is the aeonian judgment (Heb 6:2). The judgment for sin is death (Ezek 18:11-13; Gen 2:17; Rom 6:23). "The wages of sin is death, but the gift of God is aeonian life, through Jesus Christ" (Rom 6:23). What God said in Genesis 2:17 was that if they ate from the forbidden tree they would in dying, be dead: "in dying, you will die" or "in dying, you will be dead" (see Hebrew text or BCB). What God was doing was projecting to us the group called the "dead" in the Bible (Luke 9:60; 2Tim 4:1; 1Pet 4:6).

In Short: Judgment is

nm1040 » Those who overcome will inherit all things (Rev 20:7; 1Cor 3:21-22; 2Pet 1:3). Those who do not overcome or are unjust or are evil are sent to the lake of fire to be destroyed from the presence of the Lord and his glory (Rev 20:10, 15; 21:8; Mat 13:41-42; 25:41; 1Pet 3:7, 10-11; 2Thess 1:9), for the 1000 years (Rev 20:4-5). Those who overcome eat from the tree of life, are not hurt by any second death, given a new Name, rule over nations in the 1000 years, will not be taken out of the book of life, will have God's Name, and will sit in/on Christ's throne (Rev 2:7, 11, 17; 26; 3:5, 12, 21 cf. 21:8). But after the 1000 year judgment there will be the coming together of all, so that God will be all in all (1Cor 15:28; NM15; NM16; GP6).

nm1041 » Christians will also judge, so to speak, the angels and the world at Christ's coming (1Cor 6:2-3; Dan 7:26-27). Their good works judge-down the evil works of the world, Satan, and his angels.

Remaining questions:

What happens to Satan and his angels? (See, NM 13,14, 15; GP 6, 7, 8) Is there a Heaven? (See, NM 25)

NM 25: Kingdom of God

What Others Say

nm1042 » *Is there a heaven?* In many churches members look upon going to heaven as going above the clouds and beholding God always while being in a state of heavenly bliss. Not too much detail is given, except it is a much better place to be than hell. According to many, if you go to hell, you are in a fire suffering forever (Last Judgment). But if you go to heaven you are blessed forever and you will live in one of the heavenly mansions in cities paved with gold. If you get to heaven there will be no work, no suffering, no day, no night. Everyone eats from the tree of life and worships God forever enjoying the beatific vision.

nm1043 » In the Koran, one description of heaven for the males is having a continuous supply of virgin "companions," with alcoholic drink and delicate food to eat – a continuous orgy and feast (Sura 56:4-41ff). This belief is not particular to the Koran, for many others also believe that the pleasures forbidden them on earth will be their permissible pleasure in heaven. But are the sinful "pleasures" denied on earth, the rewards of heaven? Or does this have more to do with the other-mind's pleasures being projected into mankind's physical mind in this age? (See "Other Mind" paper [NM 21])

What the Bible Says

But what does the Bible say about the "reward" for Christians? Do they go to heaven?

nm1044 » *Go to Heaven?* Now since the word "heaven" or "heavenly" is used interchangeably with "spirit" or "spiritual" (1Cor 15:44-49; see Duality Paper), then at Christ's coming, when we are infused into the spiritual dimension, we will, so to speak, go to heaven or go to the spiritual dimension. But at Christ's coming is also the end of the Beast of Revelation's rule ("Beast-System Paper" [PR2]), and the beginning of God's rule through his Christ and his saints (Dan 7:11-14, 24-27; Rev 19:20-20:4; 2Thes 1:7, 10). At Christ's coming there will be clouds in the sky (Dan 7:13; Mat 24:30; Rev 1:7; Ezek 32:7; Isa 13:10; Joel 2:31; 3:15; Amos 8:9; Rev 6:12-13; Zech 14:7) from the effects of the Last War [PR5], the two witnesses will be resurrected and ascend into heaven in a cloud (Rev 11:12), the other Christians also will ascend into heaven at that same time to meet Christ in the air (1Thes 4:13-17; 2Thes 2:1; 1Cor 15:23, 51-55) to be with Christ always, and because at that time Christ will come down to earth to rule, the Christians will bring Christ down (Rom 10:6) to rule on earth for the 1000 years (Rev 20:4; 5:10; 11:15), for Christ's kingdom will rule on the earth (Rev 5:10 Greek text; Psalm 37:29, 9; Mat 5:5). Thus, "heaven" as commonly thought of, is not the "reward" Christians receive, per se. Remember here that any "reward" Christians receive is not earned; we receive all from God as a gift (See NM 8-11).

nm1045 » In this book we have given you, here and there, some aspects of the heavenly "rewards." Let's review and put the facts together, so in one place we can see what these "rewards" will be:

1. **Aeonian Life**: This is life in the 1000 year period. This is explained in the "Reward for Christians" paper. [NM 11] This is different from the immortal life.

2. **Immortality**: This is explained also in the "Reward for Christians" paper and in the "Does All Mean All?" paper [NM11; NM 14]. Immortal life is life in an immortal physical body as a New Soul. It is different from the aeonian life, which is life with Christ in the 1000 year period during which others of the old age are in judgment (Last Judgment).

3. **Life in the Kingdom of God**: There are three orders of going into the Kingdom (1Cor 15:20-24). The first was Christ the man. The second "order" occurs at the return of Christ. The third order occurs *after* the 1000 year period. Life in the Kingdom of God is life under the rulership of God in peace and harmony. What the Kingdom of God will be like in the present universe is outlined in NM26.

4. **Life Forever in the New Heaven and Earth or New Universe**: Most fully this New Universe will start *after* the small age of 100 years which comes after the 1000 years of the Kingdom of God under Christ (NM 26 #18). It may be as impossible to understand this as it would for a child in the womb to understand how it will be after he/she is born. Our lives up to the time of the creation of the New Universe are in preparation for life in the New Universe.

5. **Pleasure in the New Life**: We will have great pleasure in the New Life only because we have had great suffering in the old life. Pleasure and pain are comparative qualities: both need each other for either to exist (See "Reason Why" [NM 19], and GP 7 of the *God Papers*). Our very bodies, our New Souls, will be in and of themselves pleasurable (See "New Body" in GP6). We will live both in the physical dimension and the spiritual dimension (GP 6). We will not need to travel by foot, or by car, or by plane, or by space ship, but by and through the spiritual dimension if we so choose ("New Body" in GP6).

6. **Happiness Forever in the New Life**: Because of the very law of knowledge, we will be happy in the New Life ("Reason Why" NM19). Happiness is a comparative quality. You cannot be happy, if you have never known unhappiness. Happiness and unhappiness are comparative qualities: both need each other for either to exist (See "Reason Why" [NM 19]).

7. **Love and Freedom in the New Life**: God will be all in all (1Cor 15:28). "God is love. Whoever lives in love, lives in God, and God in him" (1John 4:16). "God lives in us and his love is made complete in us" (1John 4:12). We will all live in the Freedom and Harmony of the System of Love (See "Freedom and Law" [NM 17]).

Jesus Christ: "You are my friends if you do what I command... My command is this: Love each other as I have loved you" (John 15:12, 14). "Yes, I am coming soon" (Rev 22:20).

NM 26: Plan of Creation, An Outline

NM26 Abstract

In this paper we will give a short outline of our main beliefs.

A Short Outline of Our Beliefs

Here follows an outline of our belief concerning the plan of creation, which we ascertained through scripture, patterns in scripture, and logic. Read Introduction to this book, to understand our premises and to better understand why we use the Bible and believe that the Bible reveals the truth. Better yet, read all of the many papers we offer for more details. It is almost impossible to understand this outline in full without studying all the *BeComingOne Papers*. Please do not jump to conclusions.

1. Before the Beginning the All Powerful God Predestinated Events.

nm1046 » Scripture indicated that God predestinated events *before* the creation. Christians were chosen before the foundation of the world (Eph 1:4). Jesus Christ, what he would do and what would happen to him, was foreordained before the foundation of the world or cosmos (1Pet 1:19-20; Act 2:23; 3:18; 4:27-28). God who is all powerful predestinated everything (NM 8; NM 9; etc.). Even evil was predestinated (Rom 9:21-23; Jude 1:4; Prov 16:4; 1Peter 2:8; etc.). God in his all powerfulness even commands Satan, the leader of evil (Job 1:12; 2:6). In his predestination God predestinated everything <u>before</u> the cosmos began, <u>before</u> good (as we know it), <u>before</u> evil (as we know it), <u>before</u> law (as we know it), and consequently, <u>before</u> sin (as we know it).

2. In The Beginning God Created The Cosmos; In The Six Literal Days God Created The First Cosmos From The Laws And Matter Created On The Very First Day.

nm1047 » In the beginning God created the heavens and the earth, that is, the cosmos (Gen 1:1; Isa 48:3, 13; Heb 4:3). After the creation of the heavens and the earth, or the cosmos (Gen 1:1), the earth was not yet finished: "the earth was without form and void" (Gen 1:2). God continued to create during the first six days of creation and rested on the seventh day (Gen 1: 3-2:3). It was only when the real God creates and finishes everything that it is pronounced good: "everything that he had made ... was very good" (Gen 1:31; 2:1). What the true God creates is good, not in confusion, for God is not the author of confusion (1Cor 14:33; see Isaiah 45:18; 1Tim 4:4; Deut 32:4). This creation week typified the longer spiritual creation of approximately 7,000 years. Peter called a day of God being equal to 1000 years (2Peter 3:8). The book of Revelation speaks of a 1000 year period when Satan is put away. This 1000 year period is like a rest period for mankind, a rest without evil (NM15; NM16).

3. Man Went Out of God To learn Good And Evil; The Kingdom of The Adversary Begins.

nm1048 » After man was created, a spiritual adversary misled man to break a law of God (Gen 2:16-17; 3:1-24). This spiritual adversary was "the old serpent, called the Devil, and Satan, which deceives the whole world" (Rev 12:9). See the "Other Mind Paper" [NM 21] for more information on this apparent myth concerning the "serpent." Mankind by breaking this first law of God was sent out of the Garden of Eden and, "Behold, the man is become as one out of [or from] us, to *know* good and evil" (Gen 3:22). After mankind sinned in the Garden they were sent out to learn good and evil (see the "Reason Why" paper [NM 19] and see GP 7 of the *God Papers* for the reason for evil). There is a great reason why the all powerful God allowed evil.

4. The Breaking of God's First Law Manifests the Other-Mind Inside Mankind.

nm1049 » In the "Other Mind Paper" [NM 21] we learn why man broke the first law of God and why this present age is an age of confusion and hate. It was the other-mind inside of man's mind that misled man into breaking God's first law. Because of this the world has been held in "the bondage of corruption" and "the whole creation groans and travails in pain" because of the other-mind's power (Rom 8:21, 22; Rev 12:9; 13:2,3; see the "Other Mind Paper" [NM 21]).

5. God Promised a Savior To Save Mankind

nm1050 » When God spoke to the "serpent" after he had misled man into taking the forbidden fruit, God said he would send a child of the woman ("her seed") and that he "shall bruise your [serpent's] head and you [serpent] shall bruise his [the seed's] heel" (Gen 3:14, 15). See the "Other Mind Paper" [NM 21] to understand the serpent of Genesis.

There was a child that was born of a woman (Gal 4:4) who will "bruise Satan," and he is Jesus Christ (Rom 16:20). "Jesus" is a word that means, Jehovah's *Savior*, and "Christ" is a word that means, anointed. Christ Jesus is the, "Anointed Savior of Jehovah." He is "the Savior of the cosmos" (1John 4:14).

Jesus Christ is the savior of the world, who is/will save the world from the spiritual adversary's influence that now is misleading mankind towards destruction. See the *God Papers* to understand who is Jesus Christ.

6. The Messiah Savior, Jesus Christ Came in the Flesh, and Died.

nm1051 » But Jesus Christ was a man ("the *man* Christ Jesus," 1Tim 2:5); he came in the flesh as a man (1John 4:1-3). Jesus Christ was a man born to be savior of the world, yet he did not save the world when he was alive. He died on a stake and was buried (John 19:16-17, 33, 40-41). While he was on earth, before his death, he had the Spirit of God in him, leading him and giving him power over evil (GP 4). In fact, this "Spirit of God" was the actual angel of God who helped Israel in the Old Testament (GP 3; GP 4; GP 5).

7. But Jesus Christ Was Resurrected and Became God.

nm1052 » After Jesus was resurrected, Thomas called him "My Lord and my God," for he was in God (John 20:28; see *God Papers*). Jesus who was born a man actually went into God and sits at God's right hand, as the right side of God (GP 5). The Bible talks about human beings actually becoming like God – that is, being *born* of God (1John 3:9; 5:18). If one is born of mankind, we know he is a man. If one is born of God, we know he is a God-like person – a son or daughter of God. The Bible actually calls man, Gods (John 10:33-35). In some way mankind will become God-like. See the *God Papers* to understand who or what is God, and in what way and how Jesus became God. It is by Jesus

Christ's new life as God that He will save us (Rom 5:10).

8. Jesus Christ Promised to Return to Earth.

nm1053 » Time after time Christ promised to return to the earth (Mat 24:27; 30-31, 42; Mark 13:26; Luke 12:42-43; 13:28-29; 17:24; 18:8; 19:12; 21:27; John 14:3).

9. When Will Christ Return? – At The End of The Age.

nm1054 » Notice what Christ's students asked him: "Tell us, when shall these things be? and what shall be the sign of your coming, and the *end* of the age?" (Mat 24:3) What age is being spoken about here? It is the present age of confusion that is being spoken about. See *End of the Age* [NM7] for more details.

10. Christ The God Will Come to Save Mankind at the End of the Age of the Adversary's Spiritual Rulership.

nm1055 » Christ will come as the savior of mankind; he will save man from destroying themselves: "for the son of man [Christ] is *not* come to destroy men's lives, but *to save them*" (Luke 9:56). There is going to be a great Last War at the end of the age, and if Christ does not come to save, then *no* flesh will be saved out of this war (Mark 13:20; see "Last War and God's Wrath" paper [PR4; PR5; PR6]).

11. At The Same Time Christ Will Judge Mankind.

nm1056 » The judgment of God is probably the most misunderstood doctrine of the Bible. What is God's judgement?: "The LORD is known by the judgment which he executes: the wicked is snared in the work of his *own* hands" (Psa 9:16). In other words, Christ the God will come to save the world after an atomic *Last War* has started before he returns to the physical dimension. And God's judgment is to let the works of man's OWN hands (their own weapons) begin to destroy mankind, then He saves mankind from their own madness. Man thus actually judges his own self, but God comes to save. See "Last War and God's Wrath" paper [PR5] for information on the Last War.

There will be a 1000 year judgment (see "Thousand Years and Beyond Paper"[NM 15]). This is the judgment of the dead. Some people have mixed this judgment up and turned it into a "hell" theory, or "lake of fire" theory.

12. Christ The God Will Come to Set Up The Kingdom of God on Earth.

nm1057 » Christ will return to set up the kingdom or rulership of God that will rule all the earth. Notice what prophecy shows: "the kingdoms of this world are become the kingdoms of our Lord, and of his Christ; and he shall reign into the ages of ages" (Rev 11:15). At His return the kingdoms of the world will be His. Since Christ is God, then these kingdoms will become the kingdoms of God.

nm1058 » Notice in Daniel 7:14 where it prophesies of Christ's return, "and there was given him dominion, and glory, and a *kingdom*, that *all* people, nations and languages, should serve him." The kingdom of God will rule people, *all* people. This is the time God will live with mankind and rule mankind (Rev 20:4). This kingdom of God will rule *on* earth with the saints, "and they shall reign ON the earth" (Rev 5:10). Notice that the symbolic stone, which is a symbol of Christ, smote the image, which represents the world's kingdoms, "and filled the whole earth" (Dan. 2:35).

nm1059 » The Messiah is coming to set up a kingdom *on* the earth to bring mankind into a utopia. God will then be King of kings (Rev 19:16).

13. But What Is The Kingdom of God?

nm1060 » A kingdom is a kingdom. It is a system of rulership. The expression, kingdom of God or the kingdom of heaven, is used about 150 times in the New Testament of the Bible. And tens of times in the Old Testament it speaks about God setting up a kingdom on earth. In the book of Isaiah it continually speaks of the LORD (BeComingOne, that is "Jehovah" or YHWH) coming to earth to save his people and setting up a utopian society.

nm1061 » Some think the Church is the kingdom of heaven or of God. But God's Bible says that when the kingdom is set up there will be no misery, no wrong, no tears (Isa 2:4; 11:9; Zeph 3:15; Rev 21:4). In the truest sense, the kingdom has not come to the earth. The Church now is merely a typical kingdom of God with only some of the world's people in it. Not until the Messiah comes with His peace, will there be true peace.

nm1062 » We know the kingdom of God is here only when we see Christ coming in clouds (the smoke of the Last War), for "every eye shall see him" (Rev 1:7). All will eventually *see* him.

Since God is coming to save the world from cosmocide, and since he comes to set up a system of peace, and since everyone will literally see him; then we know he has not come yet and the kingdom is not set up on earth yet, for we still have war and we do not see anyone on earth like Christ. If you study the resurrected Christ you see pictured a person that could appear from nowhere and disappear out of sight. But when he appeared he was flesh and blood just like us. But he could change his body at will into an invisible body that could go through walls and any other matter. There is no one on earth who can walk through walls or appear and disappear at will; thus, Christ has not returned yet. If any one says to you that he is Christ, ask him to kindly walk through the wall (of course walk through the wall without harming it like Christ did). See the *God Papers* to understand who or what is God. Christ's main message throughout his ministry on earth was about the coming of the kingdom of God: "Now after that John was put in prison, Jesus came into Galilee, preaching the gospel of the kingdom of God" (Mark 1:14). He preached the gospel of the kingdom of God. The word "gospel" is translated from a Greek word that means *good news*. Christ was teaching people about the good news of the kingdom of God. It is good news because when the kingdom of God is set up there will be complete peace (Isa 2:4).

nm1063 » Paul the apostle also taught about the kingdom of God (Acts 19:8; 20:25; 28:23, 31).

nm1064 » Christ sent 70 men preaching the kingdom of God (Luke 10:9). He sent the apostles to preach the kingdom of God (Luke 9:1a-2). In fact one of the most important things that the Church has to do in this age is teach about the kingdom (Mat 24:14; Mark 16:15; Col 1:23).

14. What Will Happen During The Kingdom of God?

nm1065 » As Isaiah 2:4 says the first thing that will happen is "they shall beat their swords [weapons] into plowshares, and their spears into pruning hooks: nation shall not lift up sword against nation, neither shall they learn war any more" (Isa 2:4). This will happen because man's mind will change the instant Christ the God returns physically to earth. God will put away the spiritual adversary out of man's mind (Zech 13:2; Rev 20:1-3; see the "Other Mind Paper" [NM 21]). Man's attitude will change. He will no longer desire violence.

What Are Some of the Specific Changes God Will Make When He Returns?

nm1066 » **A. *Cities Rebuilt*.** Just before Christ the God's return there will be a Last War that will be destroying the earth (see "Last War and God's Wrath" paper [PR5]). The cities destroyed in this war will be rebuilt (Amos 9:14).

nm1067 » **B. *The Earth Repopulated*.** After this Last War many of the people alive now will be dead. The earth will be repopulated (Ezek 36:10-11).

nm1068 » **C. *The Earth's Surface Renewed*.** After the Last War the world will have atomic waste. But Christ the God will renew the surface of the world (Psa 104:30; Isa 61:4). This is the *typical* creation of the *new* heaven and earth.

nm1069 » **D. *Trees And Forests Will Be Everywhere*** even in the places where there are now deserts (compare Isa 41:19; 29:17; 60:13; 51:3).

nm1070 » **E. *Great Highways Will Be Built*** (Isa 19:23; 11:16; 40:3).

nm1071 » **F. *Children Will Play With Animals That Are Now Wild*.** Animals will all be tame in the kingdom of God (Isa 11:6-7).

nm1072 » **G. *Everyone Will Own His Own Land*** with the products of the land for their personal use (Amos 9:14; Isa 65:21-22; Micah 4:4).

nm1073 » **H. *Rain Will Fall At The Right Time In The Right Amounts*** so as to produce fantastic crops (Ezek 34:26; Isa 30:23; Jer 31:5, 12; Amos 9:13; Ezek 36:29-30).

nm1074 » **I. *A New Clear Language*** will be given to all mankind; there will be *one* language (Zeph 3:9).

nm1075 » **J. *There Will Be No More Sickness*;** all diseases will be cured (Isa 58:8; 33:24; 35:5-6; Jer 31:13-14). There will be no more sickness because Christ the God will heal them (Jer 30:17).

nm1076 » **K. *There Will Be True Freedom*.** When people speak of freedom they mean not total freedom, but freedom from things, activities, and restrictions that make them unhappy. What makes people unhappy is sin. Sin should be looked upon as a way of destruction. War is sin because in war people kill. Killing is against the laws of harmony and love. There will never be peace until people stop sinning, that is, stop making war. "Glory to God in the highest, and on earth peace among men of good will" (Luke 2:14). Peace will only come when men are of peace, that is, when men are of "good will." He that kills with the sword or weapon will be killed by the sword (Rev 13:10). Man kills, therefore man will die by the sword until he stops killing. Man will have true freedom in God's kingdom because God will automatically change man's attitude from conflict to peace.

nm1077 » **L. *All Men Will Be Prosperous During The Kingdom*.** Man will be prosperous because man during God's kingdom will be in harmony with the ways of harmony. Man will be putting his total output of work towards useful and constructive projects and thus will be in a much better economic situation than today when much energy is directed towards destruction (war) and towards the repairing of the destruction. There won't be any sick, mentally or physically, and this will allow more energy to be directed towards a prosperous economy.

15. 1000 Years of the Kingdom of God.

nm1078 » The Bible speaks about an "everlasting" kingdom of God (2Pet 1:11; Dan. 7:27). But the translators of the Bible have *mistranslated* such words as "forever," "everlasting," and "eternal." The Bible was written mostly in two languages – Hebrew and Greek. The original words that were translated into "forever," "everlasting," and "eternal" actually mean aeonian, not forever (see the "Age Paper" [NM 7]). In other words, almost everywhere in the Bible where you see "forever," "everlasting," and "eternal" should read aeonian, age, or ages. Aeonian is a far cry from forever, for in order for it to be aeonian it must have a beginning and it implies the possibility of an end. But the word everlasting means it has no possibility of an end. Thus, when scripture is correctly translated, God's kingdom is an aeonian kingdom, and there are ageS within it not everlastingS or eternitieS within it. But unlike most ages, God's great age does not end, yet within this great age there are other ages (see "Age Paper" [NM 7]). One age within the ages of the Kingdom of God is the 1000 years (Rev 20:3, 5). Also see NM11 to learn more about *Ages in God's Plan* .

16. God's Kingdom under Christ, Who Is King of Kings (Rev 19:16), Will Last Through an Age of 1000 Years (Rev 20:4).

nm1079 » During this age some men "born" of God will rule as servants with Christ (Rev 20:4). These who are to be born of God, who were human beings during the time *before* Christ's physical return, can be labeled as true Christians. These real Christians will rule as *servants* of mankind, not as warlords (read Mat 20:25-28), during this 1000 year age.

17. But After the 1000 Year Age, The Kingdom of God Will Continue Into The Next Age – The Great Spiritual Last Day.

nm1080 » The next age in the kingdom of God is an age of 100 years, which we call the Great Last Spiritual Day of Creation. It is a spiritual day of unity with God. *ALL* people who ever lived will be resurrected and live during this 100 year period. This age is a period of typical equal rulership and authority between all peoples – "the God, all things in all" (1Cor 15:28).

This period after the millennium has been completely overlooked by most people. See "God's Appointed Times Paper" [NM 16] and the "Thousand Years and Beyond Paper" [NM 15] for more information on the 100 year period.

18. Then After The 1000 Year Age, And After The 100 Year Age, Comes The End of The Spiritual Creation When The Antitypical New Heaven And Earth Will Be Created (1Cor 15:24-28; Rev 21:5).

nm1081 » The antitypical creation is unlike the typical creation of the new heaven and earth. The typical creation of the new heaven and earth begins at Christ's physical return at the beginning of the 1000 year period (see 14 C above). The antitypical creation of the new heaven and earth happens at the time the totally *new* universe is created (note Rev 21:5). In that time everyone and everything will have been created new and will live in harmony and freedom from then and onward for God's age is an age without end (See "Age Paper" [NM 7]).

Prophecy Papers

Biblical Prophecy

by

Walter R. Dolen

The *Prophecy Papers* is part of the *BeComingOne Papers*

1993 Cumulative Edition

and subsequent editions

Trade Hardback:

ISBN 1-877981-36-2

First Printing July 2000

2nd: 2011 Printing

3rd: October 2012 Printing

(spelling correction)

4th 2023 Printing

Newest Version:

8781619180499

word for word exactly like earlier versions

except for typo corrections and the added material

BeComingOne Publications

https://becoming-one.org/books.htm

"I am the BeComingOne [יהוה]; that is my name! See the former things have taken place, and new things I declare; before they spring into being I announce them to you." [Isaiah 42:9]

PR1: Seed Paper

Physical Promises to Israel

Spiritual Promises to Israel

Ur of the Chaldeans

Promises and Prophecies

Concerning Mankind and Israel

pr1» In the first book of the Bible God gave the patriarchs of Israel (Abraham, Isaac, & Jacob) certain promises that He would perform through their children. We will examine these promises in two ways — physically and Spiritually, for these promises are dual. Each promise given to the patriarchs has a typical and an antitypical fulfillment for the Bible is dual in meaning. The Bible has its physical and Spiritual fulfillment. It is the physical that prefigures the Spiritual.

pr2» Among other things, in examining the scriptures on the promises given to the patriarchs we will be able to ascertain where the Real Israel is today. We will deal with the physical promises first, then in part two we will cover the Spiritual promises. Both the physical and Spiritual promises are important and must be understood correctly.

Physical Promises to Israel

Abraham

pr3» God called Abram, whose name was later changed to Abraham, in the land of Ur of the Chaldees (Acts 7:2-4). God asked Abram to go out of his land and from his nativity "unto a land that I will show you."

[Gen 12:1; Abram or Abraham was born in the "Ur of the Chaldeans" in northwestern Mesopotamia in today's northern Syria and southeastern Turkey. See Notes, "Ur of the Chaldeans"]

pr4» And God said: "I will make you a great nation, and I will bless you, and make your name great; and you shall be a blessing: and I will bless them that bless you, and curse them that curse you: and in you shall *all* families of the earth be blessed" (Gen 12:2-3). Abram departed Ur and went to Haran. When Abram's father died he departed Haran: "So Abram departed ... And Abram passed through the land unto the place of Sichem, unto the plain of Moreh. And the Canaanite was then in the land. And the LORD appeared unto Abram, and said, Unto your SEED will I give this land."(Gen 12:4, 6-7)

pr5» Therefore, the first promises to Abraham (Abram) were:

- of him God would make a *great nation* (Gen 12:2)

- in him would *all* the families of the earth be blessed (Gen 12:3)

- his seed (offspring) would inherit the land of Canaan (Gen 12:6-7)

pr6» Next Abraham (Abram) was promised that his seed would have this land in the Middle East (Canaan) for a distance as far as he could see to the north, south, east, and west (Gen 13:14-15). He said his seed would obtain the land for an agelasting period, not "for ever" as it is mistranslated in most Bibles (See "Age Paper" [NM 7]). Therefore for an age of unknown length Abraham's seed (offspring) would have this land of the Middle East.

pr7» At this same time God promised, "I will make your SEED as the dust of the earth: so that if a man can number the dust of the earth, then shall your seed also be numbered"(Gen 13:16). Therefore the offspring of Abraham would be a great number of people. This promise of a great number of offspring is reiterated in Genesis 15:5; 22:17; 26:4; etc. (as the stars of heaven), and in Genesis 22:17; 32:12 (as the sand of the sea). And in Genesis 24:60 it says the number of offspring would be thousands of myriads (KJV, "millions"). Hence we see Abraham's children would become a *great* nation that would have a great population, and that all families (nations) of the earth would be blessed therein: "in thee shall all families of the earth be blessed" (Gen 12:3).

pr8» When God changed Abram's name ("exalted father") to Abraham ("father of a multitude") He said, "my covenant is with you, and you shall be a father of many nations ... and I will make you exceeding fruitful, and I will make nations of you, and kings shall come out of you" (Gen 17:4,6). Therefore Abraham's offspring would be rich (fruitful) and kings would come out of them.

pr9» Also note in Genesis 17:8 that the possession of the Middle East is for an *AGE* (KJV, "everlasting"), *not* everlasting. As it turned out they only did possess this land for a certain age although this promise is a dual one and later we will explain this better.

pr10» Next as a *token* of this covenant between God and Abraham, "Every man child among you shall be circumcised" (Gen 17:10-11). This was an *AGE*lasting covenant, not "everlasting" as it is mistranslated (Gen 17:13). As it turned out this *token* of the covenant did only last for an age, for physical circumcision was cut off as a requirement when Spiritual circumcision was installed through Christ (Col 2:11-12; Acts 15:5-29).

Isaac

pr11» Next we see this covenant between Abraham and God was passed on to the son of Sarah — Isaac (Gen 17:19,21). It is through Isaac that the SEED of Abraham would be called (Gen 21:12).

pr12» After Abraham obeyed God to the point of attempting to sacrifice his son Isaac (Gen 22:1-14), God again reiterated the promises to Abraham (Gen 22:16-18):

- to multiply his SEED as the stars of heaven;

- that in his SEED the nations would be blessed;

- and that his SEED would possess the cities (KJV, "gates" as in "city gates"; see Septuagint and cf. Deut 12:21; Mic 1:9 of his enemies.)

pr13» Later God again promised the same thing, but this time directly to Isaac, "I will make your seed [offspring] to multiply as the stars of heaven, and will give unto your seed all these countries [or lands or cities of his enemies — the Canaanites]; and in your seed shall *all* the nations of the earth be blessed; Because that Abraham obeyed my voice, and kept my charge, my commandments, my statutes, and my laws" (Gen 26:5). Notice the reason Abraham's seed was given these promises is because Abraham followed the directions of God. Again God appeared to Isaac and said, "I am with you, and will bless you, and multiply your seed for my servant Abraham's sake" (Gen 26:24). Abraham's faith was proven through his deeds (Rom 4:13-16; James 2:17-24, see "Proof Paper" [NM 10]). Therefore what God promised to Abraham was passed on to Isaac.

Jacob (Israel)

pr14» Isaac's blessing in turn was passed on to Jacob. Isaac's wife Rebekah had twins — Esau and Jacob (Gen 25:21-26). Now Esau was the first born of the twins, thus the rights of Isaac was passed on to Esau. But Jacob got Esau to sell his birthright to him (Gen 25:29-33). So the rights that went with being first born was passed on to Jacob. Thus, Jacob was the seed of Isaac in whom God would bestow his promises.

pr15» Further, Jacob took the blessing Isaac wanted to give Esau — his first born — whose right it was to have these blessings. Yet Esau sold his birthright and God told Rebekah that her younger child (Jacob) would become the greater nation than the elder (Esau), and that Esau would serve Jacob (Gen 25:23). Genesis 27:1-36 shows how Jacob took the blessing away from Esau. Genesis 27:27-29 pictures the blessing given to Jacob in the form of a prophecy, "Therefore God give you of the dew of heaven [good weather], and fatness of the earth [good land], and plenty of corn and wine [good crops]: let people serve you and nations bow down to you [thus other nations would serve Jacob's prophesied nation in one way or another]...."

pr16» Again God promises Jacob what he promised to Abraham and Isaac, "your seed shall be as the dust of the earth, and you shall spread abroad to the west, and to the east, and to the north, and to the south; and in you and in your seed shall *ALL* the families of the earth be blessed" (Gen 28:14). Hence, the *whole* world would benefit from Jacob's SEED (cf. Acts 3:25). To the four corners of the earth Jacob's SEED would physically bless the world.

Jacob's Name Changed

pr17» "And God appeared unto Jacob again ... and God said unto him, your name is Jacob: Your name shall not be called any more Jacob ['supplanter'] but Israel ['ruling with God'] shall be your name: and he called his name Israel. And God said unto him, I am God Almighty: be fruitful and multiply; a nation and a company of nations shall be of you, and kings shall come out of your loins; and the land which I gave Abraham and Isaac, to you I will give it, and to your seed after you I give the land" (Gen 35:9-12). This confirms again that the promises of Abraham was passed on to Jacob whose name was changed to Israel. Thus every time one sees Jacob or Israel in the prophecies he knows it speaks of the same nation or nations who grew up out of Jacob. A list of Jacob's children to whom the promises were passed on is in Genesis 35:23-26. Some of the blessings that were passed on to these children are noted in Genesis 49:1-28.

Joseph, Ephraim and Manasseh

pr18» Joseph, one of the sons of Israel (Jacob), was sold by his brothers and was brought into Egypt. Joseph became a great leader under the Pharaoh. Jacob moved over into Egypt because there was a famine in the earth, and Joseph in Egypt promised Jacob his father and his sons food and land in Egypt (Gen 45:17-20). In Egypt "Jacob [Israel] said unto Joseph, God Almighty appeared unto me at Luz in the land of Canaan, and blessed me, and said unto me, 'Behold, I will make you fruitful, and multiply you, and I will make of you a multitude of people; and will give this land to your seed after you, for an agelasting [KJV. 'ever-lasting'] possession. And now your [Joseph's] sons, *Ephraim* and *Manasseh*, which were born to you in the land of Egypt before I came unto you into Egypt, *ARE MINE*; as Reuben and Simeon, they [Ephraim and Manasseh] shall be MINE. And your issue [children], which you begettest after them [Ephraim and Manasseh], shall be yours..." (Gen 48:3-6).

pr19» Notice that Jacob called Joseph's two children, *Ephraim* and *Manasseh*, his — he adopted them as his own children (v.5). BUT the rest of Joseph's children are Joseph's children (v.6).

pr20» Next Jacob blessed Ephraim and Manasseh, "and let my name [Israel or Jacob] be named on them and the name of my fathers Abraham and Isaac; and let them grow up into a multitude [a great population] in the midst of the earth" (Gen 48:16). Notice that ABRAHAM and ISAAC names would be on Ephraim and Manasseh. Further Jacob qualified this blessing by saying that *Manasseh* would become "a people, and he also shall be great . . " (Gen 48:19). But he added "truly his younger brother [EPHRAIM] shall be greater than he, and his seed shall become a multitude of nations" (Gen 48:19). This is a very important qualification of God's promises.

pr21» God had promised to make of Abraham's offspring a great nation (Gen 12:2; 18:18; 35:11) *AND* a company or multitude or congregation of nations (Gen 17:4; 35:11; 48:4). Now this promise is qualified: *Manasseh* is to become a great nation ('people'); and Ephraim is to become a company or multitude of nationS.

pr22» Further notice something about Jacob's blessing, "And he blessed them [Joseph's sons — Manasseh and Ephraim] that day, saying In you [Manasseh and Ephraim] shall Israel be blessed..." (Gen 48:20; see *Septuagint*). So Israel will be blessed through Joseph's sons.

pr23» *Judah*. Let's notice a few more promises made to the children of Israel. Judah, one of Jacob's sons was promised that "the scepter [the right of rulership] shall not depart from Judah ... until Shiloh [CHRIST, see *Young's Analytical Concordance*] come" (Gen 49:10). And, "Now the sons of Reuben the first-born of Israel [Jacob], (for he was the first-born; but, forasmuch as he defiled his father's bed, his birthright was given unto the sons of Joseph the son of Israel ... for Judah prevailed above his brethren, and of him the chief ruler; but the birthright was Joseph's)" (1 Chron 5:1-2). So we see the chief ruler would come from Judah, and that the birthright was given to Joseph's sons — Ephraim and Manasseh. And we see the scepter, or the right of rulership in Israel, would not depart from Judah until Christ comes.

Now let's show the *physical* fulfillments of these promises of God to the children of Abraham, Isaac, and Jacob — the children of Israel.

Fulfillment of Physical Promises

pr24» After Jacob had brought his family into Egypt because of the lack of food elsewhere, Israel (his children) grew into a nation.

Moses & Joshua

pr25» Moses brought Israel out of Egypt. God gave his commandments of the covenant to Moses who passed it on to the children of Israel. Moses was leading Israel into the promised land of Canaan. It took 40 years for Israel to enter the land of Canaan. Moses did not bring Israel into this land but Joshua did. God showed the land to Moses before he died of old age, and God "buried him in a valley in the land of Moab ... But no man knows of his sepulcher unto this day" (Deut 34:6, l-7). The movement of Israel into the promised land under Moses and Joshua is shown in the books of Exodus, Leviticus, Numbers, Deuteronomy, and Joshua. Moses appointed judges to judge Israel (Exo 18:13-27; Deut 1:9-18; 16:18-20).

Judges

pr26» After the death of Joshua, judges were set up to rule the people under the chief judge — God (Judges 2:16; 3:9-10). The whole theme of the book of Judges is that, "every man did that which was right in his *own* eyes" (Jud 17:6; 21:25), not in the eyes of God. In other words the people did as *they* pleased, not what God pleased. "In those days there was no king in Israel every man did that which was right in his own eyes" (Judges 21:25). But, "there is a way that seems right unto a man; but the end thereof are the ways of death" (Prov.16:25). The people of Israel were following the ways of death during the time the judges ruled Israel.

Samuel, "Make us a King"

pr27» Samuel was a judge in Israel "and it came to pass, when Samuel was old, that he made his sons judges over Israel.... And his sons walked not in the ways [of God], but turned aside after money, and took bribes, and perverted judgement. Then all the elders of Israel gathered themselves together, and came to Samuel unto Ramah, and said unto him, Behold, you are old, and your sons walk not in the ways [of God]: now *make us a king to judge us like all the nations*" (1 Sam 8:1, 3-5). Thus the elders of Israel were asking Samuel, the head judge, to make them a king to rule Israel.

pr28» Samuel was greatly displeased by the elders request, "and Samuel prayed unto the LORD. And the LORD said unto Samuel, Listen unto the voice of the people in all that they say unto you: for they have not rejected you, but they have rejected me, that I should not reign over them" (1 Sam 8:6-7).

pr29» God told Samuel to listen to the people and make them a human king. It was God they rejected not Samuel. God was the king of Israel up to this time. Notice that when Solomon sat on the throne of David he actually was sitting on God's throne (cf 1 Chron 29:23; 2 Chron 9:8). God was considered king up to this time when the elders of Israel asked Samuel to make them a human king.

pr30» After Samuel was told to let the people have their way, God asked Samuel to also warn the people of the effects of having a human king instead of God as king. For such a king would take their sons for soldiers and civil servants, and would levy great taxes and generally oppress the people (1 Sam 8:9-18).

pr31» "Nevertheless the people refused to obey the voice of Samuel; and they said, Nay; but we will have a king over us; that we also may be like the nations; and that our king may judge us, and go out before us, and fight our battles ... And the LORD said to Samuel, Listen unto their voice, and make them a king..." (1 Sam 8:19-22).

Saul

pr32» Saul was anointed Israel's first king (1 Sam.10:1). Saul was a Benjamite and physically impressive (1 Sam 9:1-2). But Saul did not follow in the ways of God, thus Saul was rejected by God (1 Sam 13:1-14; 15:1-26; 16:1). Samuel was sent to Saul, "And Samuel said to Saul, You have done foolishly; you have not kept the commandment of the LORD thy God, which He commanded you; for now would the LORD have established the kingdom upon Israel, for agelasting. But now your kingdom shall not continue: the LORD has sought him a man after his own heart, and the LORD has commanded him to be captain over His people, because you have not kept that which the LORD commanded you" (1 Sam 13:13-14).

David

pr33» The man God appointed to take Saul's place as king was David. (1 Sam 16:1, 13 & 2 Sam 2:4, 5:3) Although Saul was physically impressive he was a bad leader, for "he feared the people, and obeyed their voice: instead of God's" (1 Sam 15:24). "The LORD said unto Samuel, Look not on his [Saul's] countenance, or on the height of his stature; because I have refused him: for the LORD sees not as a man sees; for man looks on the outward appearance, but the LORD looks on the heart" (1 Sam 16:7). Although David made mistakes when he was made king after Saul died, he had the right attitude in his mind (2 Sam chap. 11 & 12). When David's mistakes were pointed out to him he would acknowledge them and turn away from them (2 Sam 12:9-13). Read of David's attitude in Psalm's 51. The Bible projects to us, by using the physical David as an example (1 Cor 10:11), the heart or attitude that God wants in everyone (note: 1 Kings 11:4,6). And that attitude is one of admitting mistakes and correcting these mistakes when ascertained. Read all the scripture on David to see this attitude in action (also note: Job 33:27-28; Prov 28:13; Luke 15:21-24; 1 John 1:9).

pr34» David was anointed king over Judah after Saul's death (2 Sam 2:4), then later he was anointed king over Israel (2 Sam 5:3). In all he reigned over Judah 7 and 1/2 years, and 33 years over Israel and Judah together (2 Sam 5:5). Notice that here the Bible deals with Judah and Israel *separately*.

Solomon

pr35» The next king over Israel was David's son — Solomon, "then Solomon sat *on the throne of the* LORD as king instead of David his father, and prospered, and all Israel obeyed him.... And the LORD magnified Solomon exceedingly in the sight of all Israel, and bestowed upon him such royal majesty as had not been on any king before him in Israel" (1 Chron. 29:23,25). So Solomon actually sat on the throne of God in the nation of Israel and had great wealth.

pr36» To see what happened next in Israel's history let's quote directly from the Bible:

■ "King Solomon was a lover of women, and besides Pharaoh's daughter he married many foreign women, Moabite, Ammonite, Edomite, Sidonian, and Hittite, from the nations with whom the LORD had forbidden the Israelites to intermarry, 'because,' he said, 'they will entice you to serve their gods.' But Solomon was devoted to them and loved them dearly. He had seven hundred wives, who were princesses, and three hundred concubines, and they turned his heart from the truth. When he grew old, his wives turned his heart to follow other gods, and he did not remain wholly loyal to the LORD his God as his father David had been. He followed Ashtoreth, goddess of the Sidonians, and Milcom, the loathsome god of the Ammonites. Thus Solomon did what was wrong in the eyes of the LORD and was not loyal to the LORD like his father David. He built a hill-shrine for Chemosh, the loathsome god of Moab, on the height to the east of Jerusalem, and for Molech, the loathsome god of the Ammonites. Thus he did for the gods to which all his foreign wives burnt offerings and made sacrifices.

■ The LORD was angry with Solomon because his heart had turned away from the LORD the God of Israel, who had appeared to him twice and had strictly commanded him not to follow other gods; but he disobeyed the LORD's command. The LORD therefore said to Solomon, '**Because you have done this and have not kept my covenant and my statutes as I commanded you, I will tear the kingdom from you** and give it to your servant. Nevertheless, for the sake of your father David I will not do this in your day; I will tear it out of your son's hand. Even so not the whole kingdom; I will leave him one tribe for the sake of my servant David and for the sake of Jerusalem, my chosen city.'" (1 Kings 11:1-13, *NEB*)

pr37» So because of the wrongs of Solomon, his kingdom was to depart from him, but because of a promise to David his father the kingdom would depart from only Solomon's son, not Solomon himself. Yet not *all* the kingdom would be taken from his son.

pr38» Now God said he would give the kingdom to one of his servants (1 Kings 11:11), and *Jeroboam* was that servant:

Jeroboam

pr39» "Jeroboam son of Nebat, one of Solomon's courtiers, an Ephrathite from Zereda, whose widowed mother was named Zeruah, rebelled against the king. And this is the story of his rebellion. Solomon had built the Millo and closed the breach in the wall of the city of his father David. Now this Jeroboam was a man of great energy; and Solomon, seeing how the young man worked, had put him in charge of all the labour-gangs in the tribal district of Joseph. On one occasion Jeroboam had left Jerusalem, and the prophet Ahijah from Shiloh met him on the road. The prophet was wrapped in a new cloak, and the two of them were alone in the open country. Then Ahijah took hold of the new cloak he was wearing, tore it into twelve pieces and said to Jeroboam, Take ten pieces, for this is the work of the LORD and God of Israel: **I am going to tear the kingdom from the hand of Solomon and give you ten tribes**. But one tribe will remain his, for the sake of my servant David and for the sake of Jerusalem, the city I have chosen out of all the tribes of Israel. **I have done this because Solomon has forsaken me**; he has prostrated himself before Ashtoreth goddess of the Sidonians, Kemosh god of Moab, and Milcom god of the Ammonites, and has not conformed to my ways. He has not done what is right in my eyes or observed my statutes and judgments as David his father did. Nevertheless I will not take the whole kingdom from him, but will maintain his rule as long as he lives, for the sake of my chosen servant David, who did observe my commandments and statutes. But I will take the kingdom, that is the ten tribes, from his son and give it to you. One tribe I will give to his son, that my servant David may always have a flame burning before me in Jerusalem, the city which I chose to receive my name" (1 Kings 11:26-36, NEB).

pr40» "After this Solomon sought to kill Jeroboam, but he fled to King Shishak in Egypt and remained there till Solomon's death" (1 Kings 11:40, *NEB*).

Israel and Judah Split

pr41» Notice carefully that *ten* tribes were given to Jeroboam (1 Kings 11:31,35). Only one tribe was given to Solomon sons (V.32, 36, 13). Jeroboam was given the kingdom of Israel (1 Kings 11:37) — ten tribes of it.

pr42» So after Solomon died his son Rehoboam ruled. Then Jeroboam came back out of Egypt where he had fled from Solomon and joined in a revolt against Rehoboam — Solomon's son (1 Kings 12:1-18). "So Israel rebelled against the house of David unto this day" (v.19). Then Israel made Jeroboam "king over all Israel: there was none that followed the house of David, but the tribe of Judah only" (v.20). Only the tribe of Judah followed

Solomon's son "with the tribe of Benjamin" who lived near Judah (v.21). It was the tribe of Judah (with Benjamin's tribe) that was not torn from Solomon's son and that did not revolt against the kingship of Solomon's son, Rehoboam.

pr43» Right after this revolt by the ten tribes of Israel, Solomon's son, Rehoboam, "assembled all the house of Judah, *with* the tribe of Benjamin ... to fight against the house of Israel..." (1 Kings 12:21). Notice the Bible calls the ten tribes under Jeroboam — the house of Israel.

pr44» Judah with the tribe of Benjamin under Rehoboam is called collectively, Judah: "Speak unto Rehoboam, the son of Solomon, KING OF JUDAH . . " (v.23). From this point on the tribes of Judah (Jews) and Benjamin were a separate nation from Israel.

pr45» The kingdom of Israel and Judah fought wars against each other from this point of separation and onward (cf. 2 Kings 16:1-6; 2 Chron. 16:1; etc.). In 2 Chronicles chapters 11 to 36, it shows the separate history of Judah as a distinctive nation apart from the nation of Israel. 1 Kings chapter 12 to 2 Kings chapter 25, shows each nation's history. These books of the Bible treat Judah and Israel as separate nations from the time of Solomon's death onward. (Note Ezek 37:19-22)

Israel Scattered

pr46» Now right after Jeroboam was made king of Israel, he began to change the laws of God to satisfy his own purpose. He made two calves of gold and said, "behold your gods, O Israel" (1 Kings 12:28). He changed the feast of tabernacles from the *seventh* month to the *eighth* month.

pr47» God then sent a prophet to Jeroboam (through his wife) telling him that because he did not keep His laws (for he changed a festival from the seventh month to the eighth month; etc.), and because he made molten images, and so on; that God would cut off Jeroboam's offspring (1 Kings 14:5-10) and "the LORD shall smite Israel, as a reed is shaken in the water, and he shall root up Israel out of his good land ... and shall scatter them beyond the river, because they have made their groves, provoking the LORD to anger. And he shall give Israel up because of the sins of Jeroboam, who did sin, and who made Israel to sin" (1 Kings 14:15-16).

pr48» Finally, after many years, "the king of Assyria took Samaria and carried Israel into Assyria" (2 Kings 17:5-6). Because of Israel's sins,

Israel was taken captive into Assyria (2 Kings 17:7-8). In verses 9-17 it lists some of the wrong things that Israel did. "Therefore the LORD was very angry with Israel, and removed them out of his sight: *there was none left but the tribe of Judah only*" (v. 18). Only Judah (the Jews) was left with the tribe of Benjamin who dwelled with them in the land of Israel.

pr49» Because Israel followed in the ways of Jeroboam, they were removed to Assyria (v. 21-23). Then the king of Assyria brought in other peoples to fill up Samaria and they learned the ways of Jeroboam from one of the priests who was carried away but who returned to Samaria (2 Kings 17:24-34).

Judah Scattered

pr50» Next Judah was warned that it too would be going into captivity if it didn't turn away from its sin (see 2 Kings 21:1-14). Actually Judah sinned even more than Israel (Jer 3:6-11). "And the LORD said, I will remove Judah also out of my sight, as I have removed Israel, and will cast off this city Jerusalem which I have chosen, and the house of which I said, My name shall be there" (2 Kings 23:27).

pr51» And thus Judah went into captivity by the kingdom of Babylon (2 Kings 24 & 25). Now this fulfills Moses prophecy, "I call heaven and earth to witness against you this day, that you shall soon utterly perish from off the land whereunto you go over Jordan to possess it; you shall *not* prolong your days upon it, but shall utterly be destroyed. And the LORD shall scatter you among the nations, and you shall be left few in number among the heathen, whither the LORD shall lead you" (Deut 4:26-27). All twelve tribes thus went into captivity. The ten tribes of the kingdom of Israel were taken captive by Assyria, and the kingdom of Judah (with the tribe of Benjamin) went into captivity by Babylon.

Promises That Did and Did Not Come True

pr52» Now let's look at what promises of God came true up to the Babylonian captivity:

- Israel did possess the land of Canaan (the Middle East) for an agelasting time as promised in Genesis 13:15 and 17:8. For over 900 years they possessed this land before being driven out (note Joshua 21:43).

- Israel did have kings as the seed of Abraham was prophesied to have in Genesis 17:16 and 35:11.

- Israel did grow to a great population: "Your fathers went down into Egypt with threescore and ten persons; and now the LORD your *God has made you as the stars of heaven for multitude*" (Deut 10:22; 26:5; 1:10, see Neh chapter 9).

- Abraham's SEED was to possess the gates or cities of its enemies (Gen 22:17; 24:60) and that nations would bow down to it (Gen 27:29), and that it would spread out to the west, east, north, and south (Gen 28:14). This being typically fulfilled by David when he took the Philistines (west), and the Moabites (east), and Hadadezer - king of Zobah (north), and the land of Edom (2 Sam 8:1-3, 13-14). Also Solomon was paid tribute (1 Kings 10:25).

- That Israel did become a great nation with great glory could have been said to come true (relative to that time) in the reigns of David and Solomon, but only in a typical sense.

- That in Israel *all* families of the earth were blessed was only *very* typically true through the era of David and Solomon (note 1 Kings, chapter 10; Matt 6:29).

pr53» All these "fulfillments" were typical and only imperfectly represent the true fulfillment. We as Christians are to look to the higher meaning of scripture (see "Duality Paper" [BP4]).

Spiritual Promises of God

pr54» We have just seen in the first part of this paper the physical promises of God to Israel. These promises were typically fulfilled. These promises prefigured the real or intended promises.

pr55» But an antitypical fulfillment of prophecy will happen, "the kingdom of heaven is like unto a man that is a householder, which brings forth out of his treasure things new [Spiritual] and old [the physical]" (Matt 13:52). The kingdom of God will be set up on earth at Christ the God's physical return. This kingdom will bring in the old (the old physical blessings) and the new (the Spiritual blessings of God's Spirit).

Seed

pr56» In most of the promises given to Abraham, Isaac, and Jacob, the promises were pertaining to their "seed." This can and does mean their children. But the Bible uses the word "seed" in another and special way: "Now to Abraham and his seed were the promises made. He does *not* say, and to seeds, as of many; but as of one, and to your seed, which is Christ ... And if you be Christ's, then you are Abraham's seed, and heirs according to the promise" (Gal 3:16, 29). Hence Spiritually speaking Christ is *the* seed, and those of Christ are heirs according to the promise, they are counted as the seed of Abraham and they are the Spiritual Israel (cf Gal 6:16). And other scripture indicates one does not have to be a physical Israelite to become a Spiritual Israelite, for through being Spiritually baptized into Christ one becomes the real seed of Abraham (Gal 3:27-29, 16):

- "and if you are Christ's, then you are Abraham's seed, and heirs according to the promise."

Thus, all the promises made in the Bible to the physical seed of Abraham, Isaac, and Jacob (Israel) will come true to anyone Spiritually baptized into Christ's body. That is, will come true in a Spiritual sense.

pr57» Let me in outline form list many of the promises made to the "seed" of Abraham, Isaac, and Israel (Jacob). **We now know from the just mentioned scriptures that all these promises pertain to all people *in* Christ through Spiritual baptism.** These promises will be fulfilled in the truest sense beginning at Christ the God's physical return.

Promises To The *Seed*

pr58»

- The **land** of the Middle East is the Seed's. [Gen 12:7; 15:18; 17:8]

- The seed have the land for an **agelasting** time. [Gen 13:15; 17:8]

(But the new age will never end unlike the ages before it. See "Age Paper" [NM 7].)

- The seed is to be the heir of Abraham's **promises** (note Gal 3:29). [Gen 15:3-4]

- The seed is to be **numbered as the stars.** [Gen 15:5; 22:17; 26:4; Ex 32:13]

(Stars are symbolic of angels, Rev 1:20. The seed will equal the number of angels. See the *God Papers*, GP6, to understand this.)

- The seed is to be Spiritually **circumcised.** [Gen 17:9-14]

(That is Spiritually fulfilled in Spiritual baptism, Col 2:11-12; Phil 3:3.)

- In the seed all the nations will be **blessed** (cf Acts 3:25). You are in the seed when you are in Christ, for Christ is the seed. [Gen 22:18; 26:4; 28:14]

- The promises were given to Abraham because he **kept God's ways**; the same with the seed (Christians). One must follow Abraham's ways, which are God's ways to be the seed of Abraham (cf John 8:37-40). [Gen 26:5; 22:18]

- The **seed will spread around the world**, that is to the east, west, north, and south. [Gen 28:14; 13:14]

(The seed, the Christians, will be heirs of the kingdom of God that will spread around the world after Christ returns (cf Isa 2:2; Dan 2:44; 7:14, 27; Rev 11:15).

- The seed are to be **kings.** [Gen 17:6; 35:11]

(And the resurrected Christians, the seed, will be kings after Christ comes [cf Rev 1:6; 5:10; 20:4].)

- Israel, the seed, are to become a **holy nation** and a kingdom of **priests.** [Ex 19:5-6]

(Christians are now typically this (1 Per 2:9). Later they will antitypically be a holy kingdom of priests [Rev 5:10; 1:6].)

- The seed are and were **chosen**. [Deut 4:37; 10:15; Psa 105:6; 1 Chron 16:13; Isa 44:1]

(Christians are thus chosen [Eph 1:4].)

- The seed are to be **circumcised in heart** (mind, attitude). [Deut 30:6]

(Christians are Spiritually circumcised. They have the New Mind through Spiritual baptism [Phil 3:3; Col 2:11-12].)

- The seed will not pass through the **fire** to Molech. [Lev 18:21]

(This pictures that Christians will not pass through the lake of fire as Satan and his children will. Christians will not be dead during the 1000 year lake of fire [see "Thousand Years" paper (NM 15)].)

- The seed are shown **mercy**. [2 Sam 22:51]

(Hence Christians are the vessels of mercy. [Rom 9:23, see "Predestination Paper" (NM 8)]])

- The seed is to **inherit the earth**.

(Thus the Christians (the meek), who are the seed, will inherit the earth [Matt 5:5].) [Psalm 25:13]

- The seed will have a **throne** for an agelasting period. [Psalm 89:4, 29, 36]

(The Christians have their throne for an agelasting time. The Christian throne is the Spiritual throne — the Spirit or New Mind. See *Mew Mind Papers*. But God's age does not end like other ages [See the "Reward" Paper" (NM 11), and "Age Paper" (NM 7)].)

- The seed will "**rain**" on the earth at Christ's coming. [Isaiah 30:23]

(The coming of Christ the God with the resurrected saints is symbolically pictured as rain [cf Isa 45:8; Ho 10:12; Isa 32:15]. This is the "early rain" [James 5:7]. The "latter rain" is at the true end of creation.)

- The seed, the Christians, are God's **friends** (John 15:14). [Isaiah 41:8]

- The seed will be **gathered** at Christ's return. [Isaiah 43:5-6]

(Thus Christians will be gathered from the graves of the earth at Christ's return [cf Matt 24:31].)

- The seed will inherit the Gentiles, **all nations**. [Isaiah 54:3]

(This is pictured in Revelation 11:15; Dan 7:27; etc.)

- **Christ, the Seed, His Seed, and Seed's Seed**. God's word will not depart out of Christ, who is *the* seed, or Christ's seed (Christians), or the seed's seed. [Isaiah 59:21]

(The Christians are to be a Spiritual wife and mother [Isa 54:1-17, 13; Gal 4:27; Rev 19:7; 21:2, 9] while Christ will be a Spiritual husband and father [Isa 54:5; Isa 9:6; 22:21] in the kingdom of God. Together they will Spiritually have children [Isa 54:13; 44:3 and 65:23; 61:9; 59:21; Psa 102:28; etc.]. These children are those who are Spiritually begotten and/or born of God's Spirit [New Mind] in the new age [after the 1000 years].)

Other Promises

pr59» **We have just seen through the outlined scripture that the promises to Abraham's children (his seed), are Spiritually the same promises given to Christians.** Christians are of the seed of Abraham through Christ, Who is *the* seed of Abraham. But how are Christians the seed of Abraham through Christ? And what about the promises that David would "forever" have someone sitting on his throne (2 Sam 7:10-16), and what about the promise that the chief ruler would come from Judah (1 Chron 5:2), and what about the promise that the scepter (rulership) would not depart from Judah until Christ comes? (Gen 49:10) Let's examine these items in some detail.

Judah and the Scepter

pr60» As 1 Chronicles 5:2 said, the chief ruler would be of the tribe of Judah: He would be a Jew. And the chief ruler did come out of Judah, He was and is Christ. Christ will be King of kings when He returns to earth (Rev 19:16). And Christ was a Jew, for He was a seed of David who was a Jew of the tribe of Judah (Rom 1:3; Mat 1:1-17).

pr61» In Genesis 49:10 it says, "the scepter shall not depart from Judah ... until Shiloh [Christ] come." Notice the right of rulership for Israel would not depart from Judah. Christ has already come; thus for Genesis 49:10 to come true, the scepter need only be with Judah until the first coming of Christ. Today the scepter of physical Judah need not exist on earth for Genesis 49:10 to be fulfilled. So if we can trace this scepter to Christ's time, then we have established the fulfillment of another prophecy. **Thus we will trace the scepter up to Christ for this reason, but more importantly we**

will trace this to understand how one *in Christ is a seed of Abraham.*

pr62» Also the *Spiritual* scepter of Christ will still exist and will exist up to Christ's physical return: "I am with you always, even unto the end of the age" (Matt 28:20). This speaks of Christ's Spirit being with the Spiritual Church until the end of the old age.

Promises to David

pr63» Now let us examine the scripture concerning the promise to David that his seed would "forever" rule on the throne of Israel. First we will list some of these promises, then we will try to trace the physical fulfillment. We will ascertain herein that these physical or typical fulfillments didn't come true perfectly. As with all typical or physical Biblical fulfillments of God's word, they never come *perfectly* true. God is a Spirit and speaks Spiritually to us. The Spiritual or antitypical or second fulfillment of prophecy always comes true perfectly. Thus the promises to David will come true perfectly antitypically, but only imperfectly true typically.

pr64» "I will set up *your seed* after you, which shall proceed out of your bowels, and I will establish *his* kingdom. He shall build a house for my [God's] name, and I will establish the throne of *his* kingdom for agelasting" (2 Sam 7:12-13).

[Here we see where David's seed will establish the kingdom, and build God's house (the Church) and establish the throne for an agelasting period. Notice not "forever," but an agelasting throne. A throne is the symbol of ruling power. Solomon, David's son (his seed), established the kingdom for an age as a type of the antitypical agelasting kingdom of God under Christ, who is the true seed of David. But of course the New Age will not end. See "Age Paper" [NM 7].]

pr65» "I [God] will be *his* [the seed's] father, and he will be my son. If he commit iniquity, I will correct him with the rod of men, and with the stripes of the children of men" (2 Sam 7:14).

[Here Solomon fulfills this, yet Christ the true seed of David was God's son in a truer sense than Solomon. Solomon is the typical Christ, yet Christ as the antitypical seed of David fulfills this position much better than Solomon. Although Christ didn't commit iniquity,

God has "made him to be sin for us, who knew no sin ... and he was numbered with the transgressors" (2 Cor 5:21; Isa 53:12).]

pr66» "And your house and your kingdom shall be established for agelasting before you: your throne shall be established for agelasting" (2 Sam 7:16).

[Again we see this kingdom will last for an age. Solomon's kingdom did last for an *age*, but Christ the true seed's kingdom will last for an age of 1000» years under him as king of kings, and last beyond the 1000 years because his kingdom has no end, unlike previous ages (Luke 1:33, see "Age Paper" [NM 7]). Christ is the true SEED of David Who will rule over the seed of Abraham, Isaac, and Jacob (note Jer 33:15).]

Also notice these promises reiterated in Psalm 89:

- ▪ "Your seed will I establish agelasting" (v. 4).

[Christ, the seed, will be established for the great age.]

pr67» When one reads about Solomon, one can see that Solomon is a physical type of Christ. He prefigures Christ. Solomon did many things typically and physically that Christ did/is/will-do Spiritually and antitypically. For example, Solomon built God's physical temple and established a rich, but small, kingdom for Israel. Christ the God is now building the Church, the true temple (1 Cor 3:16), and will establish the true kingdom of happiness and wealth beginning at his physical return.

Conditions

pr68» Solomon was the imperfect, typical seed. He didn't establish the *great* kingdom, because he didn't fulfill the conditions that were needed to be the leader of this great kingdom: "Hear the word of the LORD, O king of Judah, that sits upon the throne of David, you, and your servants, and your people that enter in by these gates: Thus says the LORD; Execute judgment and righteousness, and deliver the spoiled out of the hand of the oppressor: and do *no* wrong, do *no* violence to the stranger, the fatherless, nor the widow, neither shed innocent blood in this place. For *if* you do this thing indeed, then shall there enter in by the gates of this house kings sitting upon the throne of David, riding .... But if you will not hear these words, I swear by myself, says the LORD, that this house [Judah, verse 1] shall become a desolation" (Jer 22:2-5).

pr69» Notice these conditions, "do *no* wrong, do *no* violence." There has only been *one* person who ever lived that never did any violence, and that person was Christ: "he had done no violence" (Isa 53:9 with 2 Cor 5:21). It was Christ who only fulfilled these conditions for sitting on the throne of David, which is also God's throne (1 Chron 29:23; 2 Chron 9:8).

pr70» Notice other reiterations of this condition: 1 Kings 2:4; 6:12-13; 8:25; 9:4-7; 2 Kings 21:8; 1 Chron 22:13; 2 Chron 6:16. And as Jeremiah 22:5 said, the house of Judah would be desolate unless these conditions were fulfilled.

pr71» Notice because these conditions were never fulfilled by Solomon, his kingdom was taken away, for he didn't perfectly fulfill these conditions (1 Kings 11:6-12). "Howbeit I will not rip away all the kingdom; but will give one tribe to your son for David my servant's sake, and for Jerusalem's sake which I have chosen" (1 Kings 11:13). This last verse refers to the promise made to David that his seed (Christ) would establish the kingdom. In verse 36 it says what verse 13 did say, but in a different way: "And unto his son will I give one tribe, that David my servant may have a light all the days before me in Jerusalem ...."

pr72» Notice this verse didn't say that this "light" would all of the days *rule* in Jerusalem, but would be there in Jerusalem. Christ was physically born as a seed of David (Rom 1:3). This "light" (physical) in Jerusalem was the ancestors of Christ who lived in and around Jerusalem up to Christ's time, thus enabling Christ, the true seed, to be born of the physical seed of David. Even though most of Israel was scattered throughout the earth, a remnant lived in and around Jerusalem thus enabling Christ the true seed of Abraham, Isaac, Jacob, and David to be born as a seed of them (Mat 1:1-17).

Branch

pr73» Notice the prophecy concerning a person who would fulfill the conditions needed to establish the kingdom: "Behold, the days come, says the LORD, that I will raise unto David a righteous Branch, and a king shall reign and prosper, and shall execute judgment and justice in the earth .... And there shall come forth a rod out of the stem of Jesse, and a Branch shall grow out of his roots; and the spirit of the LORD shall rest upon him, the spirit of wisdom and

understanding, the spirit of counsel and might, the spirit of knowledge and of the fear of the LORD; and he shall not judge after the sight of his eyes, neither reprove after the hearing of his ears" (Jer 23:5; Isa 11:1-3).

Christ: The Branch, The King, The Savior

pr74» Christ is this king; Christ is this Branch of Jesse; Christ is *the* seed of David. Therefore Christ fulfills or will fulfill the promises of God to David, Abraham, Isaac, and Jacob.

pr75» God promised David that Christ would sit on His throne (Acts 2:30). Typically, Solomon fulfilled this, but Christ is the true and intended king who was to sit on David's throne as Acts 2:30 and as Luke 1:31-33 prove: "and, behold, you shall conceive in your womb, and bring forth a son, and shall call his name Jesus [Savior]. He shall be great, and shall be called the Son of the Highest: and the Lord God shall give unto him the throne of his father David: and he shall reign over the house of Jacob into the ages; and of his kingdom there shall be no end."

pr76» Christ was born to be king (John 18:37). Speaking about Christ, "for unto us a child is born, unto us a son is given: and the government shall be upon his shoulder: and his name shall be called Wonderful, Counselor, the mighty God, the duration Father, the Prince of peace. Of the increase of his government and peace there shall be no end, *upon the throne of David, and upon his kingdom, to order it, and to establish it* with judgment and with justice from henceforth [from his birth, Isa 9:6; Luke 1:31-33] and for *olam* [the great age]" (Isa 9:6-7). It is Christ, not Solomon, who is the SEED to establish the kingdom; He was and is establishing it from His birth and onward. At His physical return He will take the kingdom of this world and make it His (Dan 7:9-14, 27).

Christ versus Satan; David versus Saul

pr77» Christ was even a king of Israel when he was only a human (John 1:49-50). But like David who was anointed king of Israel, Christ must wait until Satan's kingdom destroys itself much as David had to wait for Saul to destroy himself (1 Sam 16:1-3, 13 to 31:6). Christ was born to be king (John 18:33, 36-37).

pr78» After Christ's death and resurrection Paul tells us Christ is now "crowned with glory and honor" (Heb 2:9). Christ is now sitting on the throne of his father (Rev 3:21). This is dual: Christ is on his physical father's (David's) throne (Luke 1:32) which is also his Spiritual Father's throne or God's throne (note, 1 Chron 29:23; Rev 3:21). But as with David, who was appointed to be king of Israel, who had to wait for the anointed Saul to kill himself before he could take over the throne and rule, so too with Christ, He must wait until Satan, the anointed cherub (Ezek 28:14), destroys himself.

pr79» Satan who rules the world now is the antitypical of Saul. As Saul destroyed himself (1 Sam 31:1-6), so too will Satan and his kingdom destroy themselves in the Last War (see "Last War and God's Wrath" paper [PR5]). As David took over rulership of the kingdom after Saul destroyed himself, so too will Christ take over rulership when Satan destroys himself at the appointed time.

pr80» Although Christ is now the anointed king, He can't take over until Satan destroys himself. The typical example is Saul and David. Christ now has the throne, but is waiting for the end of Satan's kingdom. Christ received this right of rulership through being the seed of David's, and because he fulfilled the conditions of the right of rulership.

Spiritual Seed of Abraham

pr81» Today those in Christ are also the seed of Abraham because Christ is the seed of Abraham, and when we have Christ's Spirit (which is God's), then we are a part of Him. If Christ is a seed of Abraham, then so are we. As this paper clearly shows, Christ is the true SEED of Abraham, thus we are of that true SEED when we are in Jesus Christ.

Spiritual Jews

pr82» Today, those in Christ are also Jews because Jesus who came from the physical tribe of the Jews (through David) was a Jew. When we have Christ's Spirit, then we are part of him, we are in Him. We are Spiritual Jews. Thus, Paul writes: "For he is not a Jew that is one outwardly ... but he is a Jew that is one inwardly" (Rom 2:28, 29). Here Paul tells us that a physical Jew ("one outwardly") is not a real Jew, but the Spiritual Jew ("one inwardly") is a real Jew. A Spiritual Jew is a real Christian. Thus in the book of Revelation when it speaks of a "Jew," Spiritually it speaks of a real Christian (note Rev 2:9; 3:9; 7:5ff).

Spiritual Virgins

pr83» Jesus Christ was a sexual virgin: He never married, he kept all the laws of God, thus he never had sexual intercourse outside of marriage. He died a sexual virgin. Thus, following the logic above, those Spiritually in Christ are also considered Spiritual virgins. Therefore the "virgins" mentioned in the book of Revelation are real Christians (note Rev 14:4).

Christ The Mediator

pr84» Christ is the mediator between God and man. Christ is part man (son of man), and part God (son of God). He has both the physical and Spiritual essences in Himself (see the *God Papers*). If we have Christ's Spirit, since He is both Spirit and flesh, then we also have His flesh, we are *both* a son of Abraham, and a son of Christ the God when we have His Spirit.

Summarize: Jews and Israelites = Christians

By studying the above and scripture, after Christ's resurrection the only real Jews or Israelites were/are Christians. When you are in Christ, you are a Jew, you are an Israelite. Because the Jews did not bring forth the fruit required of them, their kingdom and their identity were in a sense taken from them and given to another people who will bring forth fruit required of them (Mat 3:10;7:17-19;12:33;13:8;21:19-21;21:34 [in context];John 15:4-8;Gal 5:22; etc.). But the Christians receive their fruit of good works through the Spirit of Christ: they are in Christ. Whatever Christ fulfilled, Christians will fulfill because they are in Christ. What this means is that the truest sense of the Old Testament prophecy will come true through Christians (the real Jews), not through physical Jews.

Notes for PR1

Three Orders/Divisions

pr85» The *three orders* or divisions of mankind can be seen typically in Joseph and his two sons, Manasseh and Ephraim. *Joseph*, the one set apart from his brethren (Deut 33:16), represents Jesus Christ, who was the first to be born of God (1 Cor 15:22-28). *Manasseh*, the one that was to become a great nation, represents the second group to be born of God, which are the Christians who lived in the old age. *Ephraim*, the one that was to become a multitude of nations, represents the third group to be born of God, which are the multitudes of peoples who will be born of God at the end of creation. They at that time will also become Christians — they will go into the Spiritual Body of Christ.

pr86» In these three groups will the great promises of God be Truly fulfilled. The promises to Israel are the promises that will be fulfilled to all of mankind through Jesus Christ our Lord, who has the NAME of God — the BeComingOne. (see in the *New Mind Papers* the "All Saved" paper [NM 13] and also "Three Orders of Creation" in "God's Appointed Times and Seasons" paper [NM 16])

Ur of the Chaldeans

pr87» Abraham (Abram) came from the "Ur of the Chaldeans" (Gen 11:31). The "Ur of the Chaldeans" mentioned in the Bible is not the city identified as "Ur" by many today. The contemporary "Ur" is hundreds of miles southeast of Haran and Ebla on *this* side of the Euphrates river. That is, on the side (this side) of the Euphrates river nearest Jerusalem. But the Biblical "Ur of the Chaldeans" was *across* the river Euphrates in northwestern Mesopotamia somewhere near Haran. Archeological finds and Biblical proof indicate this.

pr88» This "Ur of the Chaldeans" is most likely the same Ur that is mentioned to be "in the territory of Haran" in the Ebla Clay tablets discovered in 1975, not the Ur on *this side* of the river Euphrates southeast of the Mesopotamia region. The contemporary "Ur" is located in the southeastern territory of the Sumero-Akkadian Empire and had a different culture and language than Abraham's.

Ur of Abraham located near Haran in Mesopotamia

pr89» There is Biblical proof that the city of "Ur" that Abraham came from was near Haran:

- Abram, Lot, and Terah "went forth with them from *Ur of the Chaldees*, to go into the land of Canaan; and they came unto *Haran*, and dwelt there" (Gen 11:31).

- "Thus says the LORD God of Israel, your fathers dwelt *on the other side of the river*

[Euphrates] in old time ... and I took your father Abraham from the other side of the river [Euphrates]" (Joshua 24:2, 3).

pr90» Thus, Abraham came from the *other* side of the river (from Jerusalem's viewpoint). That river being the Euphrates. But the contemporary so-called "Ur of Chaldeans" is on *this* side of the river (from Jerusalem's viewpoint).

- "The God of Glory appeared unto Abraham, when he was in Mesopotamia, *before* he dwelt in Haran" (Acts 7:2).

pr91» Abraham came from Mesopotamia. The word Mesopotamia means "*the country between the rivers*" (*Unger's Bible Dict.*, "Mesopotamia"). These rivers being the Euphrates and Tigris. But the contemporary "Ur" is not located *between* the rivers. Originally the word "mesopotamia" stood only for the northwestern region between the rivers Euphrates and Tigris (*Unger's Bible Dict.*, 3rd Ed., "Mesopotamia"). Albert Clay mistakenly wrote in 1907:

- "In former years Urfa, not far from Harran, was identified as the ancestral city of the patriarch [Abraham], but it is now [1907] fifty years since Rawlinson identified the mounds known as Mugayyar, in the southern part of the valley, as the home of Abraham. Ur is a very ancient city" (*Light on the Old Testament from Babel*, by Albert T. Clay, pub. 1907).

The former identification of Urfa as the area where the old Ur was located is much closer than the new and wrong identification of Ur.

Ur of the Chaldeans by the river Chebar

pr92» Abraham came "out of the land of the Chaldeans" (Acts 7:4). This is what Genesis 11:31 and other verses say:

- "And they went forth with them *from Ur of the Chaldees*" (Gen 11:31).

- "You the LORD God, who did choose Abram, and brought him forth *out of Ur of the Chaldees*, and gave him the name of Abraham" (Neh 9:7).

- "I am the LORD that brought you *out of Ur of the Chaldees*" (Gen 15:7).

- "The *land of the Chaldeans by the river Chebar* .." (Ezek 1:3).

pr93» In Abraham's time the Biblical land of the Chaldeans where Ur was located was northwestern Mesopotamia. It was close to the Armenians (*Ramses II and His Time*, by I. Velikovsky, pp. 170, 168ff; note Gurney, *Hittites*, Chap VI; see endnote). The Biblical Chaldean language was the Aramean or Syriac language (Dan 2:4). The Biblical river Chebar was "in the land of the Chaldeans" (Ezek 1:3). This river C*hebar may* be the present day river K*habor* in northeast Syria near Haran and U*r*fa and south of Armenia in Turkey. Later the Chaldeans moved southward to Babylon and were known in Ezekiel's time as the "Babylonians of Chaldea, the land of their nativity" (Ezek 23:15). But in contemporary literature the "Ur of Chaldeans" is located hundreds of miles in a southeastern direction from the Chaldeans' northwestern Mesopotamian homeland.

Land of Abraham's Nativity was the real Ur of the Chaldeans

pr94» The real Ur of the Chaldeans was the land of Abraham's nativity or birth:

- God told Abraham in Mesopotamia to "get you out of *your country, and from your kindred*" (Acts 7:3).

- Haran was a brother of Abram [Gen 11:27], and Haran died "in the land of his nativity [Hebrew, "his (place of) birth"], in *Ur of the Chaldees*" (Gen 11:28).

- When Abram went out of Ur he was told by God, "get out of your country, and from your kindred [Hebrew, "your (place of) birth"]" (Gen 12:1). And "get you out of your country, and from your kindred [Greek, "relations"]" (Acts 7:3).

- "The LORD ... took me from my father's house [family], and from the land of my kindred [Hebrew, '(place of) birth']" (Gen 24:7).

pr95» Abraham's birthplace, his nativity, was in *northwestern* Mesopotamia in the land of Syria or Aram or Padan-Aram (rivers of Aram):

- "And Abraham said unto his eldest servant of his house ... you shall go unto *my* country, and to *my* kindred, and take a wife unto my son Isaac ... and the servant took ten camels ... and went to *Mesopotamia* unto the city of Nahor [the name of Abraham's brother] ... and Isaac was forty years old when he took Rebekah to wife, the daughter of Bethuel the *Syrian of Padan-Aram*, the sister to Laban the Syrian [Hebrew — "Aramite"]" (Gen 24:2, 4, 10; 25:20).

pr96» Syrians are Aramites or Arameans who lived between the rivers (Euphrates & Tigris), in the land of Aram, northwestern Mesopotamia. This area was Abraham's birthplace. Abraham was called a Hebrew (Gen 14:13). Abraham the Hebrew came from across the Euphrates river. One of the fathers or patriarchs of Moses was Jacob (Israel) who was perishing from famine in Palestine before he went down to Egypt (Gen chap 42ff). Thus, Moses said, "My father was a perishing *Aramean*, and he went down to Egypt" (Deut 26:5, see Hebrew text).

pr97» *Review.* Abraham went forth out of Ur of the Chaldeans "to go unto the land of Canaan, and they came unto Haran and dwelt there" (Gen 11:31). This "Ur" was his homeland, his birthplace. He was born there with his brother Haran (Gen 11:27-28). When God spoke to Abraham in Ur of the Chaldeans, in Mesopotamia, he told Abraham to move away from his birthplace, his relatives, his kindred, and from his fathers house (family) (Gen 12:1; Acts 7:2-4, see above). Because Abraham spoke in a Semitic tongue, because his own country was the Ur of Chaldeans, because one of the Chaldeans' languages was a Semitic tongue (Dan 2:4 — "Syriac" or "Aramaic"), this is one reason why we can say that the real "Ur of Chaldeans" was located in northwestern Mesopotamia near Ebla and Haran (the name of Abraham's brother). It was not the southeastern "Ur" with its different language and culture. This southeastern "Ur" is actually spelled, "Urim" not *Ur* (*The Sumerians*, S. N. Kramer, pp. 28 & 298). But the "Ur of the Chaldeans" was probably the "Ur" mentioned in the Ebla tablets that was located "in the territory of Haran" (Ebla Tablets, p. 42; *Ebla*, by Bermant and Weitzman, 1979, p. 190; *Riv. Bibl.* [1977], p. 236).

Ebla Tablets' Proof

pr98» A Professor Paolo Matthiae of the Rome University has been excavating the Tell Mardikh (Ebla) since 1964. In 1968 he discovered a statue bearing the name Ibbit-Lim, a king of Ebla. The kingdom of Ebla was known to a few because Ebla is mentioned in Sumerian, Akkadian, and Egyptian texts (*Ebla Tablets*, pp. 11-12). Professor Giovanni Pettinato, University of Rome, is the epigrapher working on the tablets. He has written the book, *The Archives of Ebla* (1981) and wrote in such journals as *Biblical Archaeologist* (May, 1976).

pr99» The reports on the Ebla tablets reveal that the culture of Ebla had a Semitic language, "a forerunner of all the Canaanite dialects, which include Ugaritic, Phoenician, and Hebrew" (*National Geographic*, Dec, 1978, p. 749; *The Archives of Ebla*, by Giovanni Pettinato, 1981, pp. 56, 65). Many of the personal names in the Ebla tablets closely resemble Hebrew names: *Abramu* (Abraham), *Esaum* (Esau), and *Saulum* (Saul) (Nat. Geo. Dec, 1978, p. 736). The old city state of Ebla with its Semitic language was only about 100 miles from Haran, while the other and more southern "Ur" (Urim) with its different language was about 600 miles away — a large distance in those days. Along with the Biblical proof, we conclude that Abram, who spoke in a Semitic tongue, came from the "Ur of Chaldeans" which was much closer to Haran and Ebla than the southeastern "Ur." It was in this northwestern area where a Semitic culture existed. It was from this area that Abraham came from.

Ur was in the Territory of Haran

pr100» Clifford Wilson in his paperback book called *Ebla Tablets* writes of his disappointment on finding "a city of Ur is referred to in the trade tablets. It is described as being 'in the territory of Haran.'" (p. 42) Not only does the Ebla tablets mention Ur, but they say it is in the territory or locality of Haran. This is further proof that Abraham's city of Ur was near Haran in northwestern Mesopotamia, not the contemporary "Ur" hundreds of miles southeast of Haran and Ebla.

pr101» But Mr. Wilson was "somewhat disappointed." Why was he disappointed? "I am the producer of a number of audio-visuals on Bible backgrounds, and one of them is based on Sir Leonard Wooley's findings at the city of Ur" (Wilson, p. 42; and see C.L. Woolley, *Ur of the Chaldees*, 1929). It was the southeastern "Ur," first identified by Henry Rawlinson in the middle 1800's, that Wooley helped to popularize as being the "Ur of Chaldeans." Instead of Mr. Wilson seeing that he made a mistake, instead of reviewing the Biblical data as we have, Wilson comes up with a weak excuse to retain the contemporary "Ur" as the Ur of Abraham (p. 44).

Chaldeans Language Confusion

pr102» At one time the so-called "Syriac" language (dan 2:4) or the "Aramaic" language was called Chaldee. Notice "Chaldee" in such books as *The New Englishman's Hebrew and **Chaldee** Concordance*, or the *Hebrew and **Chaldee** Lexicon* by Gesenius. Before the mistaken identification of the southeastern "Ur" for the Biblical "Ur of Chaldeans," the Aramaic tongue was identified with the Chaldeans. "It [Aramaic] was formerly inaccurately called Chaldee (Chaldaic) because spoken by the Chaldeans of the book of Daniel (2:4-7:28). But since the Chaldeans are known to have generally spoken Akkadian, the term Chaldee has been abandoned" (*Unger's Bible Dict.*, "Aramaic").

pr103» Unger calls the former identification inaccurate because of the contemporary identification of the southern "Ur" as being the Biblical "Ur of Chaldeans." But this contemporary identification was mistaken as this paper makes clear. Outside of the Ebla evidence, the internal evidence of the Bible should have made it plain to Sir Leonard Woolley and the others that the "Ur of Chaldeans" was not some foreign culture to the Semitic Abraham, but Abraham's own culture and homeland.

pr104» It is of interest to note that the Chaldeans used at least two languages: the language used in Babylon was Akkadian-Babylonian, and the Syriac or Aramean language (*Ramses II*, p. 171 & Dan 2:4ff). The city state of Ebla also used two or more languages in their writings: the Semitic Paleo-Canaanite language, and the "Sumerian script, with Sumerian logograms adapted to represent Akkadian words and syllables" (*Ebla Tablets*, p. 24). "The schematic presentation of the verbal, nominal, and pronominal systems warrants classifying Eblaite in the West Semitic group .... For this reason I prefer to classify Eblaite as a Canaanite Language, thanks to its close relationship with Ugaritic, Phoenician, and Biblical Hebrew .... Eblaite becomes a chronological companion of Old Akkadian of the East Semitic group" (Pettinato, *...Ebla*, p.65). But "the bilingualism of the tablets is only apparent. Though 80 percent of the words are Sumerian and only 20 percent are Eblaite, all of them were read as Eblaite. The Sumerian terms are in reality logograms which the scribes translated without difficulty into their own language when they read them" (*...Ebla*, by Pettinato, p. 57). It should be noted that there is no recorded evidence that the southeastern "Ur" had a Semitic culture or wrote with a Semitic script. Although one must be careful. The famous H.C. Rawlinson in about the 1850s designated the Sumerian language as the "Akkadian" or "Scythian or Turanian" language (*The Sumerians*, p. 20). So you must be careful when studying old writings concerning the Sumerian language.

Babylonians and the Hittites

pr105» Velikovsky tries to connect the Babylonians with the Hittites (*Ramses II*, chapters IV ff). Gurney in his book, *The Hittites*, may in someway connect them:

> ■ "Akkadian. This is the name now universally given to the well-known Semitic language of Babylonia and Assyria; to the Hittites, however, it was known as 'Babylonian'. It was widely used in the Near East for diplomatic correspondence and documents of an international character, and the Hittite kings followed this custom when dealing with their southern and eastern neighbours. Many Hittite treaties and letters are therefore wholly in Akkadian and were available in translation long before the great bulk of the archive of Boghazkoy had been deciphered. In addition, as mentioned above, Akkadian words are common in texts written in Hittite, but it is generally held that this is a form of allography .... Two languages only — Hittite and Akkadian — were used by the Hittite kings for their official documents" (Chap VI, pp. 125, 117).

pr106» *Summarize.* The above Biblical evidence clearly indicates that the real "Ur of Chaldeans" was the Semitic speaking one, located in northwestern Mesopotamia. This "Ur" is mentioned in the Ebla tablets as being near Haran. The Ebla culture used a Semitic language and had similar names as the ones used by the Hebrews. The culture of Ebla was located near the city of Haran and near northwestern Mesopotamia at approximately the same time as Abraham lived. It was Abraham, a Semitic speaking Hebrew, who left his own homeland, where his relatives lived so as to go into the land of Canaan. Abraham's homeland was the "Ur of Chaldees" which is also close to or the same as Padan-Aram, located in northern Mesopotamia. From his homeland, Abraham went to Canaan, by first going through and living in Haran for a while. But the contemporary "Ur" is located far from northwestern Mesopotamia; it had a different culture than Abraham's. Thus, this southern Ur (*Urim*) is not Abraham's own country.

PR2: Beast-System Paper

pr107» Who or what is the Beast? The "Beast" is described primarily in the 12th, 13th, and 17th chapters of Revelation, and in the book of Daniel. Notice carefully this Biblical description:

- And I stood upon the sand of the sea and saw a beast rise up out of the sea, having *seven heads* and *ten horns*, and upon his horns ten crowns, and upon his heads the names of blasphemy. And the beast which I saw was like unto a *leopard*, and his feet were as the feet of a *bear*, and his mouth as the mouth of a *lion*: and the *dragon* gave him his power, and his throne, and great authority (Rev 13:1-2).

pr108» This description of the "Beast" is symbolic. The very word "beast" is a symbol. We need to interpret the symbols in the Bible concerning the Beast, for they stand for something real. And when we know what all these symbols mean, we will know who or what the "Beast" is. The Bible interprets its own symbols and tells us what these symbols represent.

pr109» In the seventh chapter of Daniel, we find these same symbols described again. We see the beasts with the "seven heads," the "ten horns," and we see the "lion," the "bear," and the "leopard." And in the book of Daniel it tells us what these symbols represent.

pr110» God had given Daniel understanding in dreams and visions (Dan 1:17):

- Daniel said: In my vision at night I looked, and there before me were the four winds of heaven churning up the great sea (Dan 7:2, NIV).

As in Revelation, the "beasts" came up out of the sea:

- *Four great beasts*, each different from the others, came up out of the sea (Dan 7:3, NIV).

Four Beasts

pr111» And Daniel had a dream and vision in which he saw four great beasts:

- **(1)** The *first* was like a *lion*, and it had the wings of an eagle. I watched until its wings were torn off and it was lifted from the ground so that it stood on two feet like a man, and the heart of a man was given to it (Dan 7:4, NIV; note Dan 4:33-34, 'eagle feathers').

- **(2)** And there before me was a *second beast*, which looked like a *bear*. It was raised up on one of its sides, and it had three ribs in its mouth between its teeth. It was told, 'Get up and eat your fill of flesh!' (Dan 7:5, NIV)

- **(3)** After that, I looked, and there before me was *another beast*, one that looked like a *leopard*. And on its back it had four wings like those of a bird. This beast had *four heads*, and it was given authority to rule (Dan 7:6, NIV).

- **(4)** After that, in my vision at night I looked, and there before me was a *fourth beast* — terrifying and frightening and very powerful. It had large iron teeth; it crushed and devoured its victims and trampled underfoot whatever was left. *It was different from all the former beasts, and it had ten horns* (Dan 7:7, NIV).

pr112» So the first beast was like a "lion," the second was like a "bear," the third like a "leopard," and the fourth was so dreadful and terrible it could not be compared to any wild beast known to inhabit the earth, and it has ten horns.

Four Beasts = Four Kingdoms

pr113» Notice Daniel 7 where we see the *interpretation*:

■ These great beasts, which are four, *are four kings* which shall arise out of the earth (Dan 7:17).

pr114» The word "king" is synonymous with *kingdom*, and is used in the sense that the king represents the kingdom over which he rules, for in verse 23 we read:

■ The fourth Beast shall be the fourth *kingdom* upon the earth (Dan 7:23).

Notice also the word "kingdom" is used to explain the beasts in verses 18, 22, 24, and 27.

Seven Heads

pr115» There was only one head for the lion, one for the bear, but the third beast, the leopard, had *four heads*:

■ After that, I looked, and there before me was *another beast*, one that looked like a *leopard*. And on its back it had four wings like those of a bird. This beast had *four heads*, and it was given authority to rule (Dan 7:6).

The fourth beast also had just one head (Dan 7:19-20, 'in his head'). Thus, there are seven heads on the Beast of Revelation.

Seventh Head With Ten Horns

pr116» Notice again that the fourth beast or kingdom is the seventh head. But out of this seventh head or kingdom comes the ten horns:

■ "Then I wanted to know the true meaning of the *fourth beast* ... And concerning the ten horns in its head..." (Dan 7:19, 20).

■ "The *fourth beast* is a fourth kingdom that will appear on earth ... And as to the ten horns, *out of this kingdom* shall arise ten kings" (Dan 7:23, 24).

pr117» This fourth beast with ten horns was different from the previous three kingdoms or beasts:

■ "The fourth beast shall be the fourth kingdom upon earth, which *shall be diverse* from all kingdoms, and shall devour the whole earth, and shall tread it down, and break it in pieces" (Dan 7:23).

■ "After this I kept looking in the night visions, and behold, a fourth beast, dreadful and terrifying and extremely strong; and it had large iron teeth. It devoured and

crushed, and trampled down the remainder with its feet; and *it was different* from all the beasts that were before it, and it had ten horns" (Dan 7:7, NASB).

pr118» In the book of Revelation we note:

■ "And I saw a beast coming up out of the sea, having ten horns and seven heads, and on his horns were ten diadems [crowns], and on his heads were blasphemous names. And the beast which I saw was like a *leopard*, and his feet were like those of a *bear*, and his mouth like the mouth of a *lion*" (Rev 13:1, 2, NASB).

pr119» The fourth beast of Daniel is the Beast of Revelation. The Beast pictures the previous three beasts incorporated within it (Rev 13:2). It has the *leopard*, which is the third kingdom (Dan 7:6, see above). It has the *bear*, which is the second kingdom (Dan 7:5, see above). It has the *lion*, which is the first kingdom (Dan 7:4, see above). It is different from the kingdom before it because it includes all of them in some sense, and because it has *ten horns*. The fourth kingdom will be as, if not more, rich, powerful, magnificent, than all the kingdoms before it. It will be so powerful that it will "devour the whole earth" (Dan 7:23). ["The whole earth" here means all the known earth at the time of the Biblical writer; in antitype it literally means all the earth.]

Ten Horns = Ten Kingdoms or Nations

pr120» Out of the great and dreadful fourth beast grew ten horns:

■ It was different from all the former beasts, and it had ten horns (Dan 7:7).

■ Then I wanted to know the true meaning of the fourth beast ... I also wanted to know about the ten horns on its head ... (Dan 7:19, 20).

pr121» What do the "horns" represent? Notice Dan 7:24: "and the ten horns out of this kingdom *are ten kings that shall arise*." Notice the ten horns, come *out of a fourth kingdom*. Since *king* in these prophecies stands for the *kingdom* he represents, and since the words are used interchangeably (Dan 7:17, 23), it follows that these ten horns are kingdoms or nations within the fourth kingdom or beast. These ten nations are represented by the ten "toes of the feet" of the image in Daniel 2:31-35, 42.

pr122» *Ten crowns.* Notice the ten horns of Revelation 13:2 had "ten crowns." These same horns with crowns on them are the same horns or kings indicated in Revelation 17:12: "and the ten horns which you saw are ten kings, which have received no kingdom yet [note, v. 10]; but receive power as kings one hour [Greek, "short period of time"] with the beast." When these kings receive power with the Beast for "one hour," or a short period of time, is when the Beast has ten crowns on its ten heads.

Ten to Seven

pr123» The Beast has ten horns in its seventh head, which is represented by the *ten* crowns of Revelation 13:1. But the Beast loses three. The following scriptures show that the Beast loses three horns or nations by them being "plucked up," "subdued," "broken," or "wounded to death."

pr124» Something strange happens to the fourth Beast: he loses three horns, or three kings, or three nations, or breaks three toes:

- "While I was thinking about the horns, there before me was *another horn*, an ignoble one {'little' #2192, 6810, & 6819: small in size or dignity; ignoble}, which came up among them; and *three of the first horns were uprooted* before it. This horn had eyes like the eyes of a man and a mouth that spoke boastfully. " (Dan 7:8)

- "And of the ten horns that were in his head, and of the other [horn] which came up, *and before whom three fell*" (Dan 7:20).

- "And the ten horns out of this kingdom are ten kings that shall arise: and another [kingdom] shall rise after them; and he shall be different from the first, and *he shall subdue three kings*" (Dan 7:24).

- "And the fourth kingdom shall be strong as iron ... and whereas you saw the feet and toes, part of potters' clay, and part of iron, *the kingdom shall be divided* ... the kingdom shall be partly strong, and *partly broken*" (Dan 2:40-42).

- "Having seven heads and ten horns, and upon his horns *ten crowns* ... and I saw *one of his heads* [the seventh, with 10 horns] *as it were wounded to death*" (Rev 13:1, 3; see below, 'Eighth Head?').

pr125» Putting these scriptures together we see that the fourth kingdom or Beast, with its ten nations, will have three of its member nations: plucked up, subdued, broken, or wounded to death. Three nations will be

subdued. They will be subdued by the "horn," that "shall subdue three kings" (Dan 7:24, 8, 20).

pr126» *Eighth Head?* "And here is the mind which has wisdom. The seven *heads* are seven mountains [or kingdoms], on which the woman sits. And there are seven kings [or mountains]; five are fallen and one is [remains of the third kingdom], and the other [7th mountain/kingdom] is not yet come; and when he comes [the 7th], he must continue a short space [time]" (Rev 17:9-10). Notice that the seventh kingdom ("mountain") when it comes will only continue a short time. But, "the Beast that *was, and is not*, even he is the *eighth* [kingdom or mountain], and is *out* of the seven [kingdoms or mountains] and goes into perdition [the lake of fire, Rev 19:19-20]" (Rev 17:11).

pr127» Now since the seven mountains or kingdoms are equated to the seven heads (Rev 17:9), then these are the heads that the woman sits on. But these verses say an eighth mountain will arise, and it is this mountain or kingdom that will go into the lake of fire. And it says this eighth is the Beast that was, and is not. And since it is the eighth mountain, then it must be the eighth head. But the Beast who was, and is not, has only seven heads (Rev 17:7).

pr128» *Deadly Wound*. Notice that in Revelation 13:3 that a head was as wounded to death, but its deadly wound was healed. This deadly wound will happen to the seventh mountain or kingdom of Revelation 17 with its ten horns or nations. And this "deadly wound" will subdue three of the ten horns (nations) of this seventh kingdom ("mountain") which is the ten-nation Beast (Dan 7:24). Thus, the eighth mountain or kingdom is the *healed* Beast of Revelation 13:3. It had ten horns, but three nations ("horns") will be subdued by the mean horn (Dan 7:20, 24).

pr129» In other words, "the Beast that *was* [the ten-nation Beast], and *is not* [the ten-nation Beast], even he is the eighth, and is out of the seven, and goes to perdition" (Rev 17:11). This Beast *was* [the ten-nation Beast], and *is not* [the beast with ten horns or nations], and *yet is* [the beast, but with only seven nations]. The "deadly wound" destroys the kingdom that *was* by subduing three nations ("kings"); the healing of this deadly wound creates a kingdom with seven nations as opposed to ten nations as before. The seven-nation Beast "is not" like the ten-nation Beast, "yet is" the same beast, but with three nations subdued.

First End-of-the-age Beast

pr130» At first, the seventh head of the Beast will have ten horns or nations. The seventh head, which is the fourth beast, is the head *after* the four heads of the third beast or kingdom. The seventh head is the fourth beast. This fourth beast is the *first* end-of-the-age Beast of Revelation with ten ruling or crowned horns (Rev 13:2; 17:12).

Second End-of-the-age Beast

pr131» But this first end-of-the-age Beast has a deadly wound whereby *three* horns, or kings, or toes, or nations are subdued. After this deadly wound the second Beast of Revelation comes alive with three of its former horns put down (Rev 13:3ff).

pr132» *Image of the Beast.* The seven-nation Beast is the "image" or likeness of the first Beast of Revelation (the ten-nation Beast). Therefore the "image of the Beast" is the seven-nation Beast (Rev 13:11-15). This is the true end-of-the-age Beast that rules superior in the last 1260 days of the kingdom of Satan.

1260 Days: the Last Days Before Christ Rules

pr133» The casting down of the three nations will be the "deadly wound" (see Rev 13:3). Notice that it is *after* the "deadly wound" when the Beast of Revelation, pictured in chapter 13, begins to speak "a mouth speaking great things and blasphemies" for 42 months or 3 and ½ years (Rev 13:5). This is what happens after three kings are subdued:

- "and he shall speak great words against the most high [blasphemy] ... until a time and times and dividing of time [3 and ½ years]" (Dan 7:25, 24).

pr134» The seven-nation end-of-the-age Beast rules the earth for 42 months; or a time, times, and dividing of time; or 1260 days (Rev 13:5; Dan 7:25; Rev 12:6, 14). It is during these 1260 days that the seven-nation Beast will speak blasphemies against God, and will seek to destroy God's Church (Rev 13:5-6; Dan 7:24-25; 12:7; Rev 12:6, 14).

pr135» The book of Revelation speaks *often* of a time period of 1260 days just before Christ returns:

- The 42 months of Revelation 13:5;
- and the 1260 days of Revelation 12:6;
- the time [1 year] times [2 years] and half a time [½ year] of Revelation 12:14;
- the 1260 days of Revelation 11:3;
- and the 42 months of Revelation 11:2

are the 1260 day periods mentioned in Revelation.

pr136» It is during these 1260 days or 42 months just before Christ physically returns that the Beast, in the truest sense, will reign great in the world (Rev 13:5; Dan 7:25; Dan 12:7).

pr137» After the 1260 days of the 'healed' Beast, then Christ returns and the 1000 years of the Kingdom of God begins. Note in context Daniel 7:8-9, 24-27; 2:42-45. We see in these scriptures the fourth beast of Daniel rules until he is destroyed and the Most High and his saints begin their rule. Christ begins his rule after the 1260 days as the above scriptures prove. Therefore the seven-nation Beast exists in the 1260 day period just before Christ returns.

Four Beasts or Kingdoms

pr138» These four kingdoms or beasts are also described in the second chapter of Daniel. King Nebuchadnezzar of the Chaldean Empire, who had taken the Jews captive, had a dream, the meaning of which God revealed to Daniel.

pr139» The dream is described in Daniel 2:31-35. The king saw a great image. Its head was of gold, its breast and arms of silver, its belly and thighs of brass, its legs of iron and its feet and toes were part iron and part clay. Finally, a stone, not with man's hands (but supernaturally) was cut out of the mountain, and this stone smote the image upon his feet and toes, thereby smashing the entire image together (v. 35a). Then the stone that smashed the image became a great mountain and filled the whole earth:

- "This is the dream; and we will tell the *interpretation* thereof to the king ... You art this head of gold. And after you shall arise another kingdom inferior to you, and another third kingdom of brass, which shall bear rule over all the earth. And the fourth kingdom shall be strong as iron: forasmuch as iron breaks in pieces and subdues all things; and as iron that breaks all these, shall it break in pieces and bruise .... And in the days of these kings shall the God of heaven set up a kingdom which shall not for olams be destroyed ... it shall break in pieces and consume all these kingdoms, and it shall stand for olams" (Dan 2:36-40, 44).

Nebuchadnezzar Until The Stone

pr140» There are four world-ruling kingdoms that begin with Nebuchadnezzar's kingdom. His kingdom was the Chaldean Empire which took away the Jews to Babylon. From Nebuchadnezzar's kingdom, which was the symbolical first beast of Daniel 7:4, or the symbolical tree of Daniel 4:20-23, or the golden head of the image of Daniel 2:32, 38, grew up or raised up the remaining kingdoms. And out of the fourth kingdom was to be ten kings or nations, which are symbolized by the ten toes on the image of chapter two of Daniel, or the ten horns of the fourth beast of chapter 7 of Daniel, or the ten horns of Revelation's Beast (Rev 13:1; 17:3, 12). The fourth kingdom will exist before Christ the Stone, and His kingdom will take over the world's governments, and fill the whole earth with the kingdom of God (cf Dan 2:32-45; Dan 7:2-14, 25-27).

pr141» *Christ Is The Stone*. The interpretation of the stone smashing the image on its toes is found in the 44th verse: "And in the days of these kings shall the God of heaven set up a kingdom, which shall not for olams ['never'] be destroyed: and the kingdom shall not be left to other people, but it shall break in pieces and consume all these kingdoms, and it shall stand for olams." The Stone is Christ and his kingdom. The interpretation of the Stone is given many places in the Bible, "Jesus of Nazareth ... *is the stone* which was set at naught of you builders, which is become head of the corner" (Acts 4:10-11).

pr142» The time when the Stone (Christ) takes over the kingdoms of the world "in the days of these kings," that is, the kings being the 'toes' of the image of Daniel, chapter two (Dan 2:43-44, 34). At Christ's physical return is when He shall destroy the Gentile kingdom (the Beast of Revelation) in righteous warfare (Rev 19:11). The Stone will smite the image (the Gentile kingdoms) in righteousness, for "in righteousness he does judge and make war" (Rev 19:11). And righteous judgment or warfare is explained in Psalm 9:16, "the LORD [YHWH] is known by the judgment which he executes: the wicked is snared in the work of his *own* hands." In other words Christ the Stone will smite the image, by merely letting them or those of the Gentile kingdoms destroy themselves. This event is pictured in many places in the Bible. Here are two of them: "and was given to him that sat thereon to take peace from the earth," [How] "and that they should kill one another" (Rev 6:4). And, "for nation shall rise against nation, and kingdom against kingdom" (Matt 24:7). This is the Last War. See our paper on "Last War and God's Wrath" [PR5] for more on the Last War.

Fourth Beast

pr143» As we have seen the fourth beast is the seventh head of the seven-head and ten-horn Beast of Revelation 13:1-2. Out of this seventh head comes ten horns or nations who will rule as ten nations for "one hour," a short time period with the Beast (Rev 17:12). But the beast that causes the greatest trouble rules for 1260 or 42 months just before Christ's physical return. Just before these 1260 days three of the ten nations are subdued by the Beast (Dan 7:24-27, see above).

Identity of the First Three Beasts

pr144» Who were the first three beasts or kingdoms? Let's look at scripture and history to see.

pr145» *Image*. In the second chapter of Daniel, it identifies the head of the great image as king Nebuchadnezzar or the kingdom of Babylon (Dan 2:38). But in Daniel 2 it does not identify the second, third, and fourth kingdoms by name (Dan 2:39-40).

pr146» *Tree*. In the fourth chapter of Daniel, it identifies king Nebuchadnezzar with the tree, but it does not identify the bands of *iron* and *brass* left after the king is cut down. Remember, the third and fourth kingdom of the image in Daniel, chapter two, was made of brass and iron (Dan 2:39-40; see PR3 concerning this tree).

pr147» *Four Beasts*. In Daniel, chapter seven, it does not identify any of the beasts by name, except the clue of the eagle feathers (cf. Dan 7:4 w/ 4:33-34, 19-32).

pr148» *Seven-Head, Ten-Horn Beast*. In Revelation, chapters 12, 13, and 17 it does not identify the name of the kingdom involved except to tie this Beast of seven heads and ten horns with Daniel's beasts by its leopard, bear, and lion; Daniel's beasts were also like a lion, bear, and leopard (cf. Rev 13:2 w/ Dan 7:3-7) and they had/have seven heads and ten horns (see above).

pr149» *Ram, Goat*. But in Daniel, chapter 8, it identifies the second and third kingdom, or second and third beast, or the second through sixth head:

■ "There stood before the river a RAM which had horns: and the **two** horns were high; but one was higher than the other, the higher came up last. I saw the RAM *pushing westward, and northward, and southward*, so that no beasts might stand before him..." (Dan 8:3-4).

■ "The RAM which you saw having two horns are the kings of MEDIA and PERSIA" (Dan 8:20).

pr150» Thus the RAM with two horns (nations or kingdoms) is the Media-Persian empire with the two nations of Media and Persia. From its homeland it pushed westward (as far as Greece), northward (as far as the Caucasus Mts.), and southward (into Egypt), and while going westward took the land of Israel. It destroyed the Babylonian empire (Dan 5:24-31; see secular historical records). Through comparing various Biblical scripture, we can identify this RAM with the second beast of Daniel 7 and with the silver chest and arms of the great image of Daniel 2. *Media-Persia is thus identified, through comparison, as the second head of the seven headed beast.*

pr151» In Daniel, chapter 8, we read:

■ "and as I was considering, behold, a he-GOAT came *from the west* ... and the-GOAT had a notable horn between his eyes ... and he came to the RAM that had two horns ... and I saw him come close unto the RAM ... and smote the RAM, and broke his two horns ... Therefore the he-GOAT grew very great: and when he was strong, the great horn [see v. 5] was broken; and *in its place came up four notable ones toward the four winds of heaven*" (Dan 8:5-8).

■ "And the rough GOAT is the king of GRECIA: and the great horn that is between his eyes is the first king. Now that being broken, whereas four stood up in its place, *four kingdoms shall stand up out of the nation*, but not in his power" (Dan 8:21-22).

■ One of the last kings of Persia "shall stir up all against the realm of GRECIA. And a mighty king [Alexander the Great] shall stand up, that shall rule with great dominion, and do according to his will. And when he shall stand up, his kingdom shall be broken, *and shall be divided toward the four winds of heaven*; and not of his posterity, nor according to his dominion which he ruled; for his kingdom shall be plucked up, even for others beside these" (Dan 11:2-4).

pr152» Thus, the he-GOAT was the kingdom of Macedonia. History identifies this empire as the one under Alexander the Great who came from the west and destroyed the Media-Persian empire and took its territory. After Alexander "grew very great," after "he was strong," he "was broken" and four kingdoms came up in the place of Alexander's kingdom.

pr153» Alexander the Great died shortly after he conquered a great territory from the Indus river near India to Greece and south into Egypt. The four kingdoms that came from Alexander's kingdom came from four of his generals not from his posterity (his physical heirs). After more than 20 years of struggle among successors of the empire, by the year 301 BC (according to conventional chronology and history) four generals and their territories were:

■ **(1)** CASSANDER reigning over Macedon and Greece;

■ **(2)** LYSIMACHUS reigning over Thrace and Bithynia (Asia Minor);

■ **(3)** PTOLEMY reigning over Egypt and Palestine;

■ **(4)** and SELEUCUS reigning over Syria, Babylonia, and territory to the east as far as the Indus river (Seleucus took over Syria and Babylon from general Antigonus, but Antigonus was never called "king" in the territory he controlled militarily. Antigonus was a power behind some who through struggle were attempting to be successors to Alexander's rule; see *Babylonian Chronology* by Parker and Dubberstein, 1971, pp 19 & 20).

pr154» *Greek Culture*. These four kingdoms did not have the unified power of Alexander's empire: "but not in his power" (Dan 8:22). But because of Alexander's conquest, the culture of Greece was spread to the four winds or four directions of the compass. A unified currency was established, the Attican coin measure, and thus a world economy or vast economic area was created. Greek became the universal language (*Koine*) over a wide area, cities were established, and libraries built. The *Septuagint* or a Greek text of the Bible, was also a result of the third beast — the Grecian Empire. *It is these four kingdoms of the four generals that are counted as the four heads of the third beast* (Dan

7:6), *which are the third, fourth, fifth, and sixth head of the seven headed beast.*

Babylon — The First Beast

pr155» The four beasts of chapter 7 of Daniel are the same four ruling Gentile powers that are identified by the interpretation of the "image" of Nebuchadnezzar's dream (see above). The *first was Nebuchadnezzar's kingdom*, the Chaldean Empire, called "Babylon" after the name of its capitol city (Dan 2:32, 38; and see above). This kingdom symbolically had one head.

Media-Persia — The Second Beast

pr156» The *second kingdom* which followed we know from history. It was the Persian Empire, often called Media-Persia, composed of Medes and Persians. Notice the Biblical description of the Medes taking over the kingdom of Babylon. This kingdom symbolically had one head. See Daniel 5:28-31; 5:1-31; 8:3-4, 20, and see above.

Greece With Four heads — Third Beast

pr157» The *third world kingdom* was Greece, or Macedonia under Alexander the Great, who conquered the great Persian Empire. But Alexander died after his swift conquest, and his four Generals divided his vast Empire into four regions: Macedonia and Greece; Thrace and Western Asia; Syria and territory east to the Indus River; and Egypt. These Generals were the "four heads" of the third beast of Daniel 7:6. Notice the scripture on this in Daniel 7:6; 8:5-8, 21-22, and see above.

Six Heads = Babylon, Media-Persia, And Greece

pr158» The first three beasts or kingdoms have six heads. Thus the fourth beast or kingdom must have the seventh head.

Who Or What Is The Seventh Head?

pr159» *Roman Empire?* There are presently many theories on the identity of the seventh head or the fourth world kingdom of Daniel. Some say it is the revival of the Roman Empire, but this is not so. Some argue that Rome was too important to be left out of Daniel's vision of world-empires, but Daniel never mentions Rome. It is argued that the terrible character of the fourth kingdom is best fulfilled by Rome, but I say it is best fulfilled by a nuclear armed nation(s). Some argue that the Roman theory is favored by the statement in Dan 2:44, "in the days of these kings [fourth kingdom with ten "toes" or ten "horns"] shall the God of heaven set up a kingdom," for the Roman Empire was ruling Palestine when Christ first appeared. But remember, even after Christ's resurrection the kingdom of God had not yet been set up for at that time his apostles asked him, "will you at this time restore again the kingdom" (Acts 1:6). Christ's apostles were asking about restoring Israel's kingdom, which was the kingdom of God in a typical sense (see PR1). His apostles at that time were thinking physically about physical Israel, but Christ when he spoke of the Kingdom of God was referring to Spiritual Israel (PR1). Christ had not yet at that time set up the Kingdom of God on earth. Prophesy tells us the kingdom will only be set up when the time of the Gentile kingdoms is ended and Christ returns (Dan 7:23-27; 2:44; Luke 21:24; Rev 11:2; PR2 & PR3; see *Biblical Hermeneutics*, by Milton S. Terry, chapter 22). Let's look to scripture for the answer.

Mean and Base Horn

pr160» From Daniel 7:8, 20 we see that the "mean horn" {'little'= # 2192, 6810, 6819, mean, ignoble, base, diminished in size or behavior} with the mouth that spoke great things and ruled the 1260 days before Christ's return is the same horn spoken about in Daniel 8:9 and is "the king of fierce countenance" (Dan 8:23). This horn subdues three kings (Dan 7:8, 20, 24). This is the horn, which grows exceeding great, *toward the south, and toward the east and toward the pleasant land*" (Dan 8:9). The pleasant land is the land of Israel. It was Alexander the Great that came from the west to conquer Media-Persia (Dan 8:5). It is the debased horn that comes, and in some way grows great in reference toward the South, East, and pleasant land (Dan 8:9).

pr161» **It is the horn that comes "out of *the* one from them**" (Dan 8:9, Hebrew text). **The "them" being the four notable ones or four kingdoms that spread out toward the four winds of heaven** (Dan 8:8, 22). It is in the "latter time of *their* kingdom" that the evil king will appear (Dan 8:23). "Their kingdom" was the Grecian empire. But this kingdom was spread out to the four winds of heaven (Dan 8:8; 11:4). Yet out of the one of the dissipated kingdoms will come the base horn.

pr162» In an imperfect sense, Antiochus IV Epiphanes was 'the one of them' ("them" being the four kingdoms that came from Alexander's kingdom). And it was this kingdom that was great towards the south (Egypt), east (to the Indus R.), and the pleasant land or the land of Israel (see *Encyclopedia of Biblical Prophecy*, by Payne, pp 389-390; *The Interpretation of Prophecy*, by Tan, pp. 323ff; Clarke's commentary, Vol 4; etc.). Antiochus IV Epiphanes made great havoc in the land of Israel (Josephus, *Antiquities of the Jews*, Book 12, Chap 5 ff). But this is not the true meaning of these scripture, for one thing it never had ten horns (kingdoms) united to it.

Unity Power

pr163» The phrase, 'the one,' from 'out of *the one* from them' in Daniel 8:9 has two senses in Hebrew. Antiochus IV and his Seleucidan Empire was in a sense from the former four kingdoms of Greece under Alexandra the Great: he was one from them. But the Hebrew text also indicates a *unity* of the remains of the Alexandrian Empire:

- "out of *the* one from them," or "out of the unity from them" (see Hebrew text). In the Hebrew text 'the one' (#259, *ha 'echad*) means 'the **unity** from them.' (Hebrew text: compare Gen 32:8 (9); 34:16; 41:25; Exo 12:49; 24:3;Jud 20:1, 8; 1 Sam 11:7; 2 Sam 2:25; etc.)

pr164» The real sense of this scripture is that there will be a unification of the dispersed four kingdoms of the four generals of Alexander. Out of the *unity* of the old and dispersed Grecian empire comes the Beast. The unification character of the Beast is indicated by the previous empires being included in some way in the Beast of Revelation. Thus we see that the Beast in Revelation 13:1-2 included Babylonian (lion), Media-Persian (bear), and the Grecian empire (leopard). The second kingdom, the Persian one, took over the territory of the first kingdom, Babylon. The third kingdom (Grecian empire) took over the second kingdom's territory. The fourth kingdom, with its ten nations, will in turn in some way control the territory of the former kingdoms. But when we look at the maps of these kingdoms we see that the second kingdom took somewhat different territories than the first, and the third took somewhat different territories than the second. Thus, the fourth kingdom's territory will not have *exactly* the same territories as the beasts before it. **The fourth beast with ten horns will control or have great influence in the land through a treaty or as a league of nations united for some common purpose.**

pr165» The seventh head is then a **unity** head or unifying head. The seventh head has ten nations or horns. Its existence is just before the time of Christ's kingdom. The Bible does *not* give us the names of these ten kingdoms, but it does give the **number** of the nations who will join into an end-of-the-age united league. It is out of the unity of these kings that the Beast will come. It is out of the unity of the ten nations that the Beast will show itself, and this Beast will subdue three of the ten nations exactly 1260 days before Christ's return. This Beast becomes great in reference to the east, the south, and the pleasant land. But in a higher sense the whole world is the future land of Spiritual Israel. In the truest sense it is the spiritual Beast that becomes great in *all* the earth (see PR3 & "Seed Paper" [PR1]).

Kings of the North and South

pr166» In a sense the kings of the north and south are represented by the two legs on the great image of Daniel, chapter two. The king of the north being the Seleucidan kingdom. The king of the south being the Ptolemaic kingdom. Both these kings struggled over the land of Israel. In Daniel, chapter 11, is a description of the struggle between these two kingdoms (see *Encyclopedia of Biblical Prophecy*, by Payne, pp 389-390; *The Interpretation of Prophecy*, by Tan, pp. 323ff; Clarke's commentary, Vol 4; etc.).

pr167» But *antitypically* the king of the north and king of the south indicate the divided kingdom of the Beast. *All* the kingdoms of the world in the old age belong to the influence of the spiritual Beast. Thus all the kingdoms are divided against each other (see "Last War and God's Wrath" paper [PR5]). Typically, chapter 11 of Daniel occurred over decades, but antitypically the battles will be in the true time of the end before Christ's return, especially at the very end — the Last War. The Beast will be great in the "glorious land" (Dan 11:16, 24, 41, 45; 8:9) and the "place" of the sanctuary (Dan 8:9, 11). But in an antitypical or higher sense, the "place" of the sanctuary are the real Christians — those with God's Spirit inside them. The Beast will go into the physical place of the physical sanctuary, but in the higher sense of the scripture the Beast goes against the Spiritual sanctuary. Chapter 11 can be interpreted in a dual sense (see PR3 and "Last War" paper [PR5]).

Identity of the Fourth Beast

pr168» We have seen that the first end-of-the-age Beast is an accumulation of world powers into a ten-nation league of nations, a united league of ten nations united by treaty, written or oral. The Babylonian (lion), Media-Persian (bear), and Grecian (leopard) kingdom were old-world powers who came to control the land of Israel. The third beast (Grecian empire) was spread out to the four winds. But near the end-of-the-age it will be unified. That is, a unification of *world powers* will occur, since the first three beasts represent known world powers. This unification will be of ten nations.

Identified by the Number of Nations Unified

pr169» As of now, the identity of each of the ten nations cannot be ascertained. They *may* include the people or territory of Greece, or Babylon, or Persia, but not necessarily so. The scripture says the "Beast" becomes great in reference to the south, to the east, and to the glorious land (Dan 8:9). Thus, it becomes great to the land of Israel, and to the east and south of it. Until the ten nations join together no one will be able to identify the ten nations exactly. But before they join you can be sure of the NUMBER of the Beast or the number of the nations of this united force or empire. At first there will be ten; later, after 30 days there will be seven (see 1290 Days, below). **Thus the identity is ascertained, not by the *name* of the nations, but by the *number* of nations that are unified**.

pr170» The unification of *ten* nations is but the first of an end-of-the-age Beast. There is a second end-of-the-age Beast and it has *seven* nations. The seven-nation league comes into existence 1260 days before Christ's return.

When Is The Beast Set Up?

pr171» Now we know what is the Beast of Revelation. The Beast in the truest sense will be fulfilled during the 1260 day period just before Christ physically returns. But *when* is the ten-nation league going to be set up? We know by our study that during the 1260 day period only *seven* nations will be with the individual Beast, who will be the shadow of Satan himself (see above & PR3). Sometime *before* the 1260 days a ten-nation league must be set up that will give its power to the system of the Beast (Rev 17:12-13). At the 1260th day before Christ appears physically, the seven-member Beast will take over; but when is the ten-nation league to be set up?

pr172» Now the Bible is dual, type and antitype. The prophecy of the Bible is dual: typical and real fulfillments. It is the time of the end when the prophetic fulfillments will happen in the truest sense.

Daily Sacrifice

pr173» Now in Daniel it speaks about the mean horn and the DAILY SACRIFICE, and the "abomination that makes desolate" (Dan 8:9, 13, 23-25; 11:31; 12:11). The "abomination that makes desolate" is also spoken of by Christ when he was telling His disciples about His return (Matt 24:15). What is the "daily sacrifice," and the "abomination that makes desolate?"

pr174» Now typically we know what the daily sacrifice was. It was the Israelite's twice daily religious sacrifice of animals (Num 28:1-8). Antiochus IV Epiphanes in a sense stopped these sacrifices (Josephus, *Antiquities of the Jews*, Book 12, Chap 5 ff). Now what did these sacrifices represent?

pr175» As Hebrews 10:1 and 8:4-5 projects, these sacrifices were "the example and shadow of heavenly things," or spiritual things since the Bible uses the words "heaven" and "spiritual" interchangeably (1 Cor 15:44-49; see "Duality Paper" [BP4]). What is the heavenly, or Spiritual, or antitypical meaning of the daily sacrifices?

pr176» Now the Bible speaks about Christians being "*living* sacrifices," or "*spiritual* sacrifices" (Rom 12:1; 1 Pet 2:5). Further the Bible calls Christians "sheep for the slaughter" (Rom 8:36) in that they suffer for doing what is right, as Christ suffered, or was sacrificed for what was right (1 Pet 3:14, 17; 4:1; 3:17; Heb 10:14; 9:23, 26; Isa 53:10; 2 Cor 5:21).

pr177» Christians and Christ were/are "*living* sacrifices," or "*Spiritual* sacrifices." They were/are living sacrifices because they have the **living** Spirit inside them. The sacrifices or sufferings of Christians are somewhat like the others (Job 15:20; 31:3) except Christians have the living Spirit, and they suffer for their good (1 Pet 3:14, 17).

pr178» Yet there are also *dead* sacrifices offered by those labeled the "dead" by the Bible. The old sacrifices were for the remembrances of sin (Heb 10:3). These sacrifices were for remembrances of sin, for many of mankind's problems, sacrifices, and sufferings are because of his sins. It is wrong behavior that causes mankind's ills. And sin is merely wrong behavior, or behavior that man does to others that he does not really want done to himself. The "dead" sacrifices are the sacrifices of those with only the dead spirit, the other-mind.

pr179» Therefore the antitypical "daily sacrifice" is the daily suffering of all mankind. And when Daniel speaks of the daily sacrifice being taken away (Dan 8:11; 11:31; 12:11), he means, Spiritually or antitypically, the daily sacrifice of mankind by the hand of sin (or Satan) will be taken away. We are to understand God by taking the higher, or Spiritual, or antitypical meanings of Biblical scripture since God is Spirit and speaks to us Spiritually (John 4:24; 6:63; Phil 3:19; Col 3:1-2). And the time when the "daily sacrifice" is taken away is at Christ's physical return, for it is at that time and after that time that man's daily sacrifice will be taken away.

Now we know the truest meaning of what Daniel was speaking of when he spoke about the daily sacrifice being taken away. And we know when it will be taken away — at Christ's physical return.

pr180» *Abomination of Desolation.* In Daniel 12:11 it speaks about the "abomination that makes desolate" being "set up" ["to give" or "to bestow" in Hebrew]. The Hebrew word translated "abomination" means "a detestable thing." Now this "abomination" is an "abomination that makes desolate." It is a detestable thing that destroys.

pr181» In Daniel 8:9-13, 23-25, it speaks of the "mean horn" as destroying the daily sacrifice through transgression, and speaks of the "*transgression of desolation.*" And Daniel 9:26-27 speaks about an end-of-the-age war that makes desolation. And in Daniel 11, which antitypically speaks of the "latter days" (note Dan 10:14 & Dan 11:40), it speaks about the "abomination that makes desolate" and about the abomination taking away the daily sacrifice. Therefore by putting these verses together, Daniel is saying that the *abomination is* the despicable horn, or as we show in PR2 and 3 (*Beast Papers*) he is also the end-of-the-age Beast described in Revelation. He is a detestable thing, for he speaks lies, kills, and plays like he is God, yet he is merely the shadow of Satan. And the Abomination makes desolate by the means of the end-of-the-age war in which he is a principal participate (Dan 9:26-27; see God's Wrath). But because after this Last War, Christ with his peace comes, and thereby sin and evil is put away, then, in a sense, the Abomination of Desolation through the Last War also takes away man's daily sacrifice.

pr182» Or, it is through the LAST great sacrifice of mankind in this Last War (Isa 34:6; Zeph 1:8; Ezek 39:17; Rev 19:17-21) that man's sacrifice will be stopped by Christ the God, the Messiah (Matt 24:22; John 3:17) with His righteous wrath (see "God's Wrath" paper [PR4]).

1290 Days

pr183» The "abomination" is the *individual* Beast or the *system* of the Beast. And according to Daniel 12:11 it is set up 1290 days from (away from) when the daily sacrifice is taken away. The real daily sacrifice is taken away at Christ's return. Since the abomination can't be "set up" *after* Christ returns, then it is set up *before* Christ returns. It is set up *before* Christ's physical return. It is set up (or given or bestowed on the world) 1290 days *from* when the daily sacrifice is taken away, or from before the time of Christ's return. (It is the abomination or detestable thing that destroys that is set up on the 1290th day before Christ's return.)

pr184» But we have seen that for 1260 days the Beast of Revelation will rule before Christ returns, and that this Beast will be of seven nations. Therefore it is the ten nations, or a first end-of-the-age Beast that will be set up 1290 days before Christ physically returns. And it will exist for 30 days as ten nations with one mind (Rev 17:13). This is the "one hour" (a short period of time) that the ten nations will be together. But these nations' power goes to "the horn that looked more imposing than the others and that had eyes and a mouth that spoke boastfully" and the seven-member league (Dan 7:20, NIV). This leader takes control through his deceit and craftiness (Dan 7:23-25) 1260 days before Christ returns, as explained in PR2 & 3 (Beast Papers). Thus, the ten-nation league will rule for 30 days, *from* the 1290th day before Christ physically returns *to* the 1260th day before Christ returns. This **"backward count"** of the 1290 days is not an unusual method of counting. For example, the Greeks and Romans counted days in their month backward at times:

- "In this [Grecian] system, the count in the last decade [of the month] was backwards, i.e., counting from high to low towards the end of the month ... " (*Greek and Roman Chronology*, P. 60).

■ "The designation of the days within the month was made by a peculiarly Roman system. The first day of the month was called *Kalendae*, the 5th (or 7th in a 31 day month), was called the *Nonae*, and the 13th (or 15th in a 31 day month) was called the *Idus*. These are the named days, and other days in the month were designated by counting back from these named days, counting inclusively" *(Greek and Roman Chronology,* p. 154).

2300 Evenings and Mornings Sacrifices or 1150 Days

pr185» In Daniel, chapter 8, it speaks of "evening morning two thousand and three hundred" (Dan 8:14, 26; v. 14 "day" = "evening-morning" in Aramaic [Hebrew]). In context with verses 11, 12, and 13, verse 14 is speaking about the daily sacrificeS: one in the evening and one in the morning. "This is the offering made by fire which you shall offer to the LORD; TWO lambs ... The one lamb shall you offer in the morning, and the other lamb shall you offer at even" (Num 28:3-4; see Exo 29:38ff). There were *two* sacrifices each day. Verse 14 spoke of 2300 evening and morning sacrifices: One in the evening and one in the morning equals two per day. 2300 evening-morning sacrifices equals 1150 days.

pr186» The 2300 evening-mornings in verse 14 is the answer to the question between the two holy men in verse 13: "How long shall be the vision concerning the daily sacrifice, and the transgression of desolation, to give both the sanctuary and the host to be trodden under foot?" (Dan 8:13) But in verse 9 and 11 it speaks of the evil horn going towards the pleasant land and casting down the *place* of the sanctuary. Since we know that the daily sacrifices, in its highest meaning, is taken away at Christ's coming, then the scriptures concerning the 2300 evening-morning sacrifices speak of the 1150 days *before* Christ's return when the place of the sanctuary is "cast down" or "trodden under foot."

1335th Day

pr187» There is also something important to happen on the 1335th day before Christ returns when he who reaches this date is blessed (Dan 12:12).

[**NOTE**: The 1335th day mentioned in Daniel 12:12 is the 1335th day before the antitypical daily sacrifice is to be taken away. This sacrifice is taken away at Christ's return. See the Beast Papers [PR2, PR3]and the "End of the Age" paper [PR7] to understand this: to understand the 1335th day one must understand our rendering of the 1290th and 1260th day.]

The Four Beasts of Daniel 7:3-7

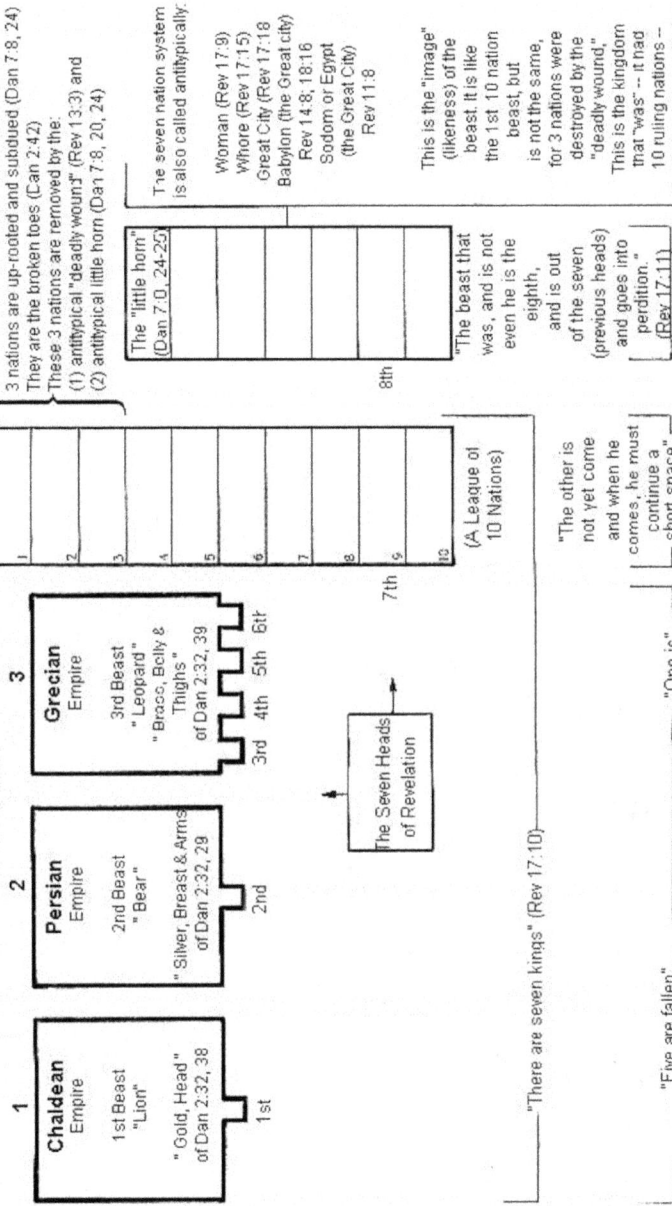

1

Chaldean Empire

1st Beast "Lion"

"Gold, Head" of Dan 2:32, 38

1st

2

Persian Empire

2nd Beast "Bear"

"Silver, Breast & Arms" of Dan 2:32, 29

2nd

3

Grecian Empire

3rd Beast "Leopard" "Brass, Belly & Thighs" of Dan 2:32, 39

3rd 4th 5th 6th

4

(A League of 10 Nations)

1 2 3 4 5 6 7 8 9 10

The "little horn" (Dan 7:0, 24-25)

8th

"The beast that was, and is not, and is he the eighth, and is out of the seven (previous heads) and goes into perdition." (Rev 17:11)

The Seven Heads of Revelation

"There are seven kings" (Rev 17:10)

"One is"

7th

"The other is not yet come and when he comes, he must continue a short space"

"Five are fallen"

3 nations are up-rooted and subdued (Dan 7:8, 24). They are the broken toes (Dan 2:42) These 3 nations are removed by the: (1) antitypical "deadly wound" (Rev 13:3) and (2) antitypical little horn (Dan 7:8, 20, 24)

The seven nation system is also called antitypically:

Woman (Rev 17:9) Whore (Rev 17:15) Great City (Rev 17:18 Babylon (the Great city) Rev 14:8; 18:16 Sodom or Egypt (the Great City) Rev 11:8

This is the "image" (likeness) of the beast. It is like the 1st 10 nation beast, but is not the same, for 3 nations were destroyed by the "deadly wound". This is the kingdom that "was" -- it had 10 ruling nations -- "and is not, and yet is" -- it does not have 10 nations, but seven nations, after 3 were subdued; Thus, it is not the same, exactly, yet it is the same kingdom.

From a 1989 book

Becoming-One Papers

Beast Chart

The Four Beasts
of Daniel 7:3-7

Chaldean	Persian	Grecian	Fourth Beast
1	2	3 4 5 6	7

Seven Heads of Rev 13:1

7th Head

8th Head (Rev 17:11)

3 rooted up, subdued, broken (Dan 7:8,24; 2:42) by deadly wound/little horn (Rev 13:3; Dan 7:8,20,24)

— First —
— United —
— League —

— Second —
— United —
— League —

The seven nation system is also called antitypically: Woman (Rev 17:9); Whore (Rev 17:15; Great City (Rev 17:18) Babylon ("Great City" Rev 14: 8; 18:16; Sodom or Egypt (Rev 11:8)

Ten Horns
Ten Toes
with
Ten Crowns
(Dan 7:7;
2:42; Rev
13:1)

"Little Horn"
7 Heads
with
7 Crowns
(Rev 12:3)

This is the "image" (likeness) of the Beast. It is like the 1st United League, but not the same because 3 nations were destroyed by the "deadly wound." This is the kingdom that "was" (it _had_ ten ruling nations) "and is not, and yet is" (it does not have 10 nations, but 7 nations after 3 were subdued: thus it is not the same exactly, yet it is the same kingdom).

"There are seven kings" (Rev 17:10)

"Five are fallen"

"One is"

"The other is not yet come and when he comes he must continue a short space" (Rev 17:10)

The Beast that was, and is not, even he is the eighth, and is out of seven [previous kingdoms] and goes to perdition" (Rev 17:11)

30 Days	1260 Days
1290 Days	

4024

PR3: Beast-Man Paper

False-Prophet

Antichrist

Mark of the Beast

Beast and Church in the 1260 Days

Tree of Daniel

Individual Beast

System of the Beast versus Beast-man

pr188» From PR2, we see that the first Beast of Revelation will have ten nations with ten leaders or "kings." These ten leaders of the ten-nation league will be in a council of ten, they "receive power as kings one hour [a short period of time] *with* the Beast" (Rev 17:12). This "Beast" represents:

- (1), the whole *system* of the Beast (the league of nations);

- and (2), the man who is called the 'little' (diminished in size or behavior) horn by the Bible.

These ten leaders will share their power with each other (with the system of the Beast), and with the Beast-man. Thus the ten leaders are, in some way, in power with the Beast-man. But they "shall give their power and strength unto the Beast" (Rev 17:13; 17:17). At first the power of these nations will be given to the *system* of the Beast. Yet, because of this, later, at some point the Beast-man will gain all power. At some point the individual Beast will have the authority of the system of the Beast.

Last False-Christ

Abomination of Desolation

pr189» Although there are many who have/do/will claim to be Christ (Mat 24:5,24), there will be only one last false-Christ,

- "And I beheld another beast coming up out of the earth; and he had two horns like a lamb [false Lamb of God], and he will speak as a dragon [Satan Rev 12:9]" (Rev 13:11).

This same person will eventually claim to be God (2 Thes 2:4, 1-12; Dan 11:37; 8:11); he is the false-Christ of Revelation 13:11 who will make fire come down out of heaven (Rev 13:13 – the fire of the Last War and other wars before it). He is also called the abomination of desolation by Daniel and Christ (Dan 11:31; Mat 24:15; Mark 13:14) and by comparing Daniel 11:31 with Daniel 8:9-11 we see this same person is also the "little horn." He will come up among a diminished group, with seven kings or leaders instead of ten (Dan 11:23), and he will give life to the image of the Beast (Rev 13:15). The image of the Beast is the eighth kingdom, the league of seven nations — that is of the likeness or "image" of the seventh kingdom (see PR2 of the *Beast Paper*; Rev 17:10-11).

pr190» Now this false-Christ, or this individual Beast-man, after he receives all the power of the system of the Beast will rule as the speaker of the system of the Beast for 1260 days (Rev 13:5, 15; Dan 7:25). Thus, 1260 days before Christ the God physically returns, the false-prophet, who will eventually claim to be God, will exercise the power of the first Beast. The first Beast is the ten-nation league that exists immediately before this individual and his dictatorial-like rulership (Rev 13:12, 5). Those who do not honor this Beast-man will be destroyed (Rev 13:15). Thus, this is one reason the Church will be destroyed at the very end by the Beast (see PR6).

Little Horn

pr191» This Beast-man is the mean "horn" that is greater, and he is the one that will do the speaking for the whole kingdom (the system of the Beast). He is the horn with the eyes and mouth (Dan 7:20; Rev 13:5). He speaks for the system of the Beast. He will do all the speaking because at some point the other leaders of the ten member council or league, passively or overtly, will give him the power of the system of the Beast. The "diminished horn" or 'little[1] horn' in another sense is also the system of the Beast — the seven-nation Beast, because it was made smaller or diminished by three nations from its former ten nation status.

Man with the Number Six Hundred Sixty Six – 666

pr192» The individual Beast-man will have a name that calculates to six hundred and sixty six (Rev 13:18). Not 6 - 6 - 6, not three 6's in a row, but as in the Greek text, six hundred sixty six. In the Greek language and in the Hebrew language, each letter had a numerical value, with the

---

[1] See Hebrew text

exception that Hebrew vowels were not written and thus they had no numerical value. You can find the value of these letters by looking them up in a reliable dictionary, or reliable Greek and Hebrew grammars, or in various manuals of style such as the University of Chicago's *A Manual of Style* (1969):

■ "Numbers, when not written out, are represented in ordinary Greek text by the letters of the alphabet, supplemented by three special characters .... The entire series of Greek numerals is shown in table 9.3." (pp. 232-233)

Beast-man = Last Antichrist = Last False-Prophet = Little Horn = Man of Sin = Abomination of Desolation = Man with the Number 666

Metonymical Names for the Same Person

pr193» The Bible uses many different names to depict the same evil person. When you use many different names to picture the same person, you are using metonymy. Metonymy is a figure of speech "by which one name is used instead of another, to which it stands in a certain relation" (*Figures of Speech Used in the Bible*, by E.W. Bullinger, p. 538). The last false-Christ (Rev 13:11; 19:20), is also the last false-prophet, is also the Beast-man, and is the man with the number six hundred and sixty six (Rev 13:18). By comparing scripture we see he is also the "little horn" who will do major destruction to people, nations, and eventually to the whole world (Dan 7:1-25; 8:24; 9:27; 11:31,38-44; Rev 13:4,13). It is because of the great destruction caused by his misuse of power that the world will say, "Who is like unto the Beast? Who is able to make war with him?" (Rev 13:4) It is his *lying* wonders or "miracles" of his great fire-destruction against some nations that will put fear into much of the world and cause it to follow him (2 Thes 2:9; Rev 13:13-17, 7). The evil behavior of the Beast-man reveals the power behind him. The verses about the Beast-man are dual: in the typical sense, they indicate the physical man of sin, but in the antitypical or spiritual sense, they reveal the power behind the man, and that power is Satan. The physical false-Christ or prophet is merely the shadow of Satan (2 Thes 2:9; Dan 8:23-24; 11:36-39), he is the "lamb" who speaks as Satan or for Satan (Rev 13:11; 12:9). In so many words, he will say he is the Christ (note the "lamb" of Rev 13:11), this "man of sin" will say he is God (remember Christ is God) while standing in the holy place

or temple (2Thes 2:4; Dan 11:36-38; Mat 24:15; Isa 14:12-14; Ezek 28:17).

pr194» This Beast-man will eventually say he is God after he is given the power of the Beast system for the forty two months or 1260 days (Rev 13:5; Dan 7:25; see "1260 Days..." in PR2). At the very end he will say he is God through his words and actions (2 Thes 2:4). During his 1260 day rulership he will hope to change the times and the law (Dan 7:25). Those are some of the blasphemies he will say and do during his 1260 day rulership (Rev 13:5-6; Dan 7:25; 11:36). Yet even though this Beast-man is alive today, even up close to the time he is to do these things, he will not realize he will do these exact things (Isa 10:5, 7, antitypical meaning). But he, through his great desire for power and through his deceit, will contrive to put himself into a position of great power with the help of the evil power. It is the power of Satan that will enable and empower the individual Beast-man and the system of the Beast (Rev 13:4; Hab 1:11, 5-10; 2 Thes 2:9). But the Beast-man willfully goes along with this evil power within him, and thus he is sent to the fire to be destroyed at the very end of Satan's kingdom (Rev 19:20; 20:10).

Antichrist

pr195» There are many antichrists:

■ "Children, it is the last hour; and just as you heard that the antichrist is coming, even now many antichrists have appeared; from this we know that it is the last hour" (1John 2:18)

We know it is the last hour or times after "many antichrists have appeared." But there is a specific Antichrist that will come: "you heard that [the] antichrist is coming." Since "anti" is translated from a Greek word that means either against or before, then this specific Antichrist will be against Christ and will exist just before Christ comes.

pr196» We see the Antichrist, with his many different names, warring, in words and deeds, against the Church and against Christ:

■ I kept looking, and that horn was waging war with the saints and overpowering them.... He will speak out against the Most High and wear down the saints of the Highest One, and he will intend to make alterations in times and in law; and they will be given into his hand for a time, times, and half a time. (Dan 7:21,25)

■ And through his shrewdness He will cause deceit to succeed by his influence; And he will magnify *himself* in his heart, And he will destroy many while *they are* at ease. He will even oppose the Prince of princes, But he will be broken without human agency. (Dan 8:25)

■ And he opened his mouth in blasphemies against God, to blaspheme His name and His tabernacle [Church], *that is*, those who dwell in heaven. 7 It was also given to him to make war with the saints [Church] and to overcome them, and authority over every tribe and people and tongue and nation was given to him....And I saw the beast and the kings of the earth and their armies assembled to make war against Him [Christ] who sat on the horse and against His army. (Rev 13:6-7; 19:19)

pr197» This last Antichrist, is also called by other metonymical names such as:

■ the "little horn" that speaks (Dan 7:8,11,20-25; 8:9-12, 23-25)

■ the "abomination of desolation" (Dan 11:31-45; 9:27; 12:11; Mat 24:15; Mark 13:14)

■ the "man of sin" (2Thes 2:4)

■ the "son of perdition" (2Thes 2:4)

■ the "false-prophet" (Rev 19:20 w/ 13:11-18)

■ the "Beast" (Rev 13:3-7,11-14; 19:20)

■ man with the number, six hundred sixty six (Rev 13:18)

70 Weeks

69 Weeks

pr198» In Daniel 9:24-27 it speaks about the 70 weeks. As we indicate in the "Last War and God's Wrath" paper [PR5], these 70 weeks can indicate either 70 weeks of days or 70 weeks of years. In the "Last War and God's Wrath" paper we show you the interpretation of the **70 weeks of _years_** [Dan 9:2, 24-27; Mat 18:22] In this paper we will show you the interpretation of the **70 weeks of _days_**. Let's look at the scripture:

■ ‹24› Seventy weeks are determined {divided} concerning your people and upon your holy city, to finish the transgression, and to seal up sins, and to make atonement for iniquity, and to bring in righteousness of olams, and to seal up the vision and prophecy, and to anoint (the) holy of holies [most Holy]. ‹25› Know therefore and understand, that from the going forth of the commandment {word} to restore and to build Jerusalem unto

(the) anointed prince {leader, ruler, 'one in front'} shall be seven weeks, and threescore and two weeks: the street shall be built again, and the moat {trench}, even in troublous times. ‹26› And after the weeks, threescore and two, shall (the) anointed one be cut off, but not {nothing} for himself {'and not for him'}: and the people of the prince {leader, ruler}, who is to come [see Isa 55:4], shall destroy the city and the sanctuary; and his end thereof shall be with a flood, and unto the end of the battle, desolations are decreed. ‹27› And he shall confirm a covenant with many for one week: and in the midst of the week he shall cause the sacrifice and the offering to cease, and upon the wing [extremity] of abominations (comes) the desolator that makes desolate [11:31], but at the full end that (which is) decreed shall be poured out upon the desolator (Dan 9:24-27, BCB).

pr199» This can be interpreted two ways and should be interpreted both ways. The anointed one is interpreted by many as Jesus Christ, but the 'he' in verse 27 is interpreted by some as the Antichrist. They believe that the Antichrist will make an agreement (covenant) for one week of years, that is for seven years, but in the middle of the seven years he will break this agreement (p. 223, Biederwolf). In such books as *The Millennium Bible*, by William E. Biederwolf (pp. 217ff) various interpretations are listed, both the Messianic and anti-Messianic interpretations. (See PR11 In *Prophecy Papers: Biblical Prophecy*, Expanded Edition, 2018, where we reproduce these pages.) In fact, this obscure and difficult scripture has to do with both the Messiah and the anti-Messiah. The real Messiah, Jesus Christ, was born approximately 2,000 years ago (see *Chronology Papers*). He died after three and one-half years of teaching, on a Wednesday, a Passover on the 14 of Nisan of the Jewish Calendar. At His death, in one sense sacrifice and the offering ceased (Dan 9:27) because His was the perfect and fulfilling sacrifice (study Hebrews chapter 9 & 10; note Heb 10:18 & Zech 11:10). Christ was cut off after three and one-half years of confirming a new covenant (Heb 8:6ff; see *Chronology Papers*; *Handbook of Biblical Chronology*, Finegan, pp. 280-285; see Dan 9:26; Isa 22:25; Zech 11:10). His two witnesses will fulfill the rest of the seven years in their 1260 days (Rev 11:3-4; Zech 4:11-14; see "Two Witness" paper [PR8] & "Last War" [PR5, pr284 ff]). Christ began his three and one-half years of confirming the new covenant starting at his 'Baptism' which was **69 weeks of years (483 *years*)** [Dan 9:2, 24-27; Mat 18:22] after a word had gone out to rebuild Jerusalem (see *New Mind Papers* & "Last War" paper [PR5]). Jesus Christ fits the scripture in Daniel 9:24-27. But also the Antichrist will fit these scriptures. The Antichrist may be a participant of an agreement or covenant. He will be the false-Christ who will manifest himself in the last 1260 days of the Old Age (Rev 13:11-18, 'like a lamb'; 1 John 2:18; 2 John 7; 2 Thes 2:3-4; Matt 24:15; Dan 11:36-39, 31-32; 8:9ff). There will be an order to restore and build Jerusalem **69 weeks of *days* (483 *days*)** before the false Messiah comes (Dan 9:25-26). This Antichrist will sit down in the 'temple' of the God pretending that he is the God (2 Thes 2:4; Dan 11:36; 8:23ff; Ezek 28:2; etc.). He is the false messiah projected in one meaning of Daniel 9:25-26. He will kill the two witnesses **3 ½ days** after he declares his false Messiahship and **3 ½ days** before the Real Messiah comes and saves the world from total destruction. As noted before, the real Messiah is also projected in Daniel 9:25-26. The vagueness of this scripture has something to do with the cherubs, the right and left side of the True God (see *God Papers*).

Mark of the Beast

pr200» "And he caused all, both small and great, rich and poor, free and bond, to receive a mark in their right hand, or in their foreheads: And that no man might buy or sell, save he that had the mark, or the name of the beast" (Rev 13:16-17). As indicated in our *God Papers*, and in our *New Mind Papers*, when one is in the NAME of God, he has the Spirit of God. Therefore, when one has the *name* of the Beast, he has the spirit of the Beast. As just shown this spirit is Satan's. When one has the *name of the Beast*, he has the spirit of the Beast. But furthermore, in the highest sense, when many have the *mark of the Beast*, they have the spirit of the Beast "in their forehead." The spirit of evil is the mark of the beast; it is the power behind the Beast. Those who worship the Beast, do so because they have the mind of Satan (see "Oher Mind" paper [NM 21] for information on Spirit and mind). It is those with God's Spirit who have "the victory over the Beast [Satan], and over his image, and over his mark [spirit]" (Rev 15:2-3; 20:4). By knowing the true meaning of the mark of the Beast, many sections in the book of Revelation make sense.

Monetary System

pr201» The system of the Beast may initiate a monetary system whereby people have to have some kind of physical mark on their foreheads or right hand in order to buy or sell (Rev 13:16-17). Although the true and higher meaning of the mark of the Beast is the *spirit of the Beast*, we must not willingly submit to the physical mark if there is something evil or nefarious about it.

Beast and Church in the 1260 Days

pr202» It is during the 1260 days that the seven-nation end-of-the-age Beast will speak "great things," or "great words against the Most High," or "great things and blasphemies," or "in blasphemy against God, to blaspheme his NAME, and his tabernacle, and them that dwell in heaven" (Dan 7:8, 20, 25; Rev 13:5-6). Sometime during this period the Beast:

- will *persecute* the woman"; will *blaspheme* God's tabernacle; "will make war with the Lamb ... and they that are with him"; will "wear out the saints of the most High"; will "make war with the saints"; will "destroy the mighty and the holy people"; will cause "the sanctuary and the host to be trodden under foot"; and so forth (Rev 12:13; 13:6; 17:14; Dan 7:25, 21; 8:24, 13).

pr203» As our "Last War God's Wrath" paper [PR5] indicates, the true woman, tabernacle, those with the Lamb, the saints, the holy people, and the sanctuary are the true Church of God. As the "God's Wrath, An Outline" paper [PR6] indicates the Beast makes final war against the Church at the end of the 1260 days, for during the 1260 days the Church is in the wilderness (Rev 12:6, 14), that is the wilderness of the people (Ezek 20:35-36), that is in the wilderness of spiritual Egypt (Ezek 20:36; Rev 11:8; see "God's Wrath, An Outline" paper [PR6]). And during these days the two witnesses will teach (Rev 11:3, see Two Witnesses [PR8]). Only near the end of the 1260 days, or 3 ½ days before Christ returns, are the two witnesses killed, and thereafter at Christ's return is the resurrection (Rev 11:11-12) at "the same hour" as the destruction of the city or Church (Rev 11:13) and the beginning of God's kingdom (Rev 11:15). But during this time period (and before it) some in the Church:

- "And those who have insight among the people will give understanding to the many; yet they will fall by sword and by flame, by captivity and by plunder, for days. Now when they fall they will be granted a little help, and many will join with them in hypocrisy. And some of those who have insight will fall, in order to refine, purge, and make them pure, until the end time; because it is still to come at the appointed time" (Dan 11:33-35, NASB).

Tree of Daniel

Kingdom of the Enemy

pr204» In the fourth chapter of the book of Daniel we read about a TREE that is "in the midst of the earth, and the height of it was great. The tree grew, and was strong, and the height of it reached to heaven, and the sight of it to the end of all the earth: The leaves of it were fair, and the fruit of it were much, and in it was food for all: the beast of the field had shadow under it, and the fowls of the heaven dwelt in the branches, and all flesh was fed from it" (Dan 4:10-12).

pr205» It reads in the next three verses about a "holy one" from heaven coming down to earth and breaking down this tree but leaving the "*stump of his roots in the earth*, even with a band of *IRON* and *BRASS*." This is to prove "that the most High rules in the kingdom of men, and gives it to whomsoever he will" (Dan 4:17). Then in verse 22, Daniel interprets the meaning of this tree: "It is you, O King, that has grown and become strong: for your greatness is grown, and reached to heaven, and your dominion to the end of the earth" (Dan 4:22). Thus, this tree is symbolic of the king's kingdom (Babylon), and it is cut down by God to show the world who controls the affairs of the world.

pr206» But notice that the most High leaves "*the stump of his roots in the earth*." The stump of the roots of the king or his kingdom of Babylon would remain in the earth. Now we know that after the Babylonian kingdom, the Persian Empire "raised up itself" (Daniel 7:5). Therefore we can reasonably conclude that the stump of the Babylonian kingdom, that was left in the earth after the tree or kingdom of Babylon was destroyed, did grow into another tree or kingdom which is the second Beast or world kingdom. Why can we reasonably conclude this?

Band of Iron and Brass

pr207» When the "holy one" cut down Nebuchadnezzar's tree or kingdom, he left "the stump of the roots, even *with a band of IRON and BRASS*" (Dan 4:15, 23). What is this band of IRON and BRASS?

pr208» In the book of Daniel the second chapter, Daniel explains the image that Nebuchadnezzar saw in a dream: "You are this head of gold. And after you shall arise another kingdom inferior to you, and another *third kingdom* of *BRASS*, which shall bear rule over all the earth. And the *fourth kingdom* shall be strong as *IRON*" (Dan 2:38-40). The "image" of Daniel chapter 2, was made up of a head of *gold*, its breast and arms of *silver*, its belly and thighs of *BRASS*, and its legs of *IRON* (Dan 2:31-33). This image is representative of all the world-ruling kingdoms as are the four Beasts of Daniel chapter seven. But notice that the third kingdom was of *brass*, and the fourth kingdom was of *iron*. Remember that when the tree of Daniel 4 was cut down, in its stump or roots was a band of *IRON* and *BRASS*.

pr209» In other words, in the stump or roots of the TREE was the life for: (1), the third world ruling kingdom, the BRASS of the "image" is the BRASS band of the roots of the tree; and (2), the fourth kingdom, the IRON of the "image" is the IRON band of the roots of the tree.

Who or What Is the Stump?

pr210» Remember that the last kingdom in the chain of Gentile kingdoms will be the Beast described in the book of Revelation with its seven heads and ten horns. And in Revelation 16:13 we read of a spirit coming out of the Beast. Who is this spirit?

pr211» "And there appeared another wonder in heaven; and behold a great red [fiery] DRAGON, having SEVEN HEADS AND TEN HORNS, and seven crowns upon his heads ... And the great DRAGON was cast out, *that old serpent, called the Devil, and Satan*, which misled the whole world" (Rev 12:3, 9). So we see that Satan has misled the whole world? Also we see that Satan is described as having seven heads and ten horns – the same as the Beast of Revelation, chapter 13, which is nothing but an accumulation of the Beasts in Daniel 7:4-7.

pr212» "And I stood upon the sand of the sea, and saw a Beast rise up out of the sea, having

SEVEN HEADS AND TEN HORNS ... And the Beast which I saw was like unto a *leopard* [3rd Beast of Dan 7:6], and his feet were as the feet of a *bear* [2nd Beast of Dan 7:5], and his mouth as the mouth of a *lion* [1st Beast of Dan 7:4, which is synonymous with the tree in Daniel, chapter 4]: and *the dragon gave him his power, and his throne, and great authority*" (Rev 13:1-2). The dragon or Satan gave his power to the Beast. We see here that both Satan and the Beast have seven heads and ten horns, and that part of the Beast of Revelation 13 is the first Beast of Daniel 7, which is synonymous with the tree of Daniel 4, and that Satan gives the Beast its power.

Kingdom of Satan

　　pr213» By putting together all the information above, how could anyone otherwise conclude than that the Beast is the kingdom of Satan and that it has misled the whole world. And furthermore, that the symbolic tree of Daniel 4 is part of Satan's kingdom, and that the stump of the roots left from the first tree had grown again and was cut down again each time one of the world kingdoms had come up after its predecessor had fallen. Since the roots of a tree are the foundation of a tree, then we see from Revelation 13:2 that Satan is the foundation of the Beast. So we can conclude that the stump of the roots of the tree in Daniel 4 is nothing other than Satan and his spiritual power. Satan is the root of the tree or kingdom of Satan. The kingdoms of this age are Satan's as we see in Matthew 4:8 where Satan shows Jesus his "kingdoms of the world, and the glory of them." But we see in Revelation that the Beast is finally destroyed and the kingdom of God rules from then on (Rev 19:20; note Dan 7:9, 14). See the "Thousand Years and Beyond" paper [NM 15] for details on the seven times of Daniel, chapter 4.

　　pr214» *Seven Crowns*. Notice that the great red dragon, Satan (Rev 12:9), had seven heads and ten horns and that his heads had *seven* crowns on them (Rev 12:3). The seven crowns indicate Satan's power over all seven heads of all the four beasts of Daniel.

　　pr215» **In the truest sense the Beast of Revelation represents Satan's kingdom and power throughout all the ages of his kingdom** (Rev 13:2; 12:3, 9; 17:15, 18; 18:2-3, 24; see "Last War and God's Wrath" paper [PR5]).

PR4: God's Wrath, What is it?

Righteous Judgment

Definition of God's Wrath

pr216» What is God's wrath? The Bible speaks about "the great day of his wrath," "the wrath of the Lamb," and "the wrath of God," and so forth (Rev 6:16-17; 15:1). People picture in their minds after reading the book of Revelation a God pouring out fire and damnation on the people of the earth. They read about the seven seals, the seven trumpets, and the seven plagues of Revelation and see a wrathful God getting even for man's wrongs. What is God's wrath? By only reading the book of Revelation can you know what the Bible means when it speaks of God's wrath?

> ■ "For as the heavens are higher than the earth, so are my ways higher than your ways, and my thoughts than your thoughts. For my thoughts are not your thoughts, neither are your ways my ways, says the LORD" (Isa 55:9, 8).

pr217» God's ways and thoughts are *not* man's ways and thoughts. Man's heart is "deceitful above all things" (Jer 17:9). What man thinks about God is many times quite different from what the truth is. What man does is to project his own ideas onto God. When God says through the Bible that there is a "wrath of God," man thinks of *man's wrath* and projects this onto God's behavior. But "the wrath of man works *not* the righteousness of God" (James 1:20). God's ways are not like man's ways (Isa 55:8-9). God's wrath isn't what the deceitful heart of man (Jer 17:9) thinks it is.

pr218» Man teaches as the true doctrines the commandments of men (Mark 7:7), not the commandments of God. Man teaches as true the precepts of men (Isa 29:13) as opposed to the precepts of God. Man teaches what comes "out of their own hearts" as "the word of the LORD" (Ezek 13:2).

pr219» For example, God says "their *fear* toward me is taught by the precept of men" (Isa 29:13). Remember God's thoughts and ways are *not* like man's (Isa 55:8-9). Then what is the fear of God? The Bible interprets itself:

> ■ "The fear of the LORD is to hate evil: pride and arrogance, and the evil way" (Prov 8:13; Psa 97:10).

pr220» The fear of God is to hate the wrong ways. Not to hate those who do evil, for nothing of itself is evil (Rom 14:14), but to hate the wrong ways of man. This is the fear of God. We hate evil behavior, for it is what harms us, not the God. Isn't this almost opposite to what many teach as the fear of the LORD? Man thinks it means fearing God as one fears to be killed.

pr221» What we are saying here is that God's ways and thoughts are not like man's. What some verses of the Bible at first glance seem to say is on more research quite different. The Bible uses many metaphorical expressions to describe God. More than some of the Old Testament is poetry. To find out the Spiritual essence you must have the New Mind or New Spirit.

Does this mean that God has purposely made the Bible difficult?

pr222» No, for in the places where God through his word has come right out and said something, man has changed the meaning of the words used in order to suit man's own motives. For example, there are words in over 400 verses in the Bible that mean "agelasting" which has evolved through misusage to be wrongly translated as "eternal." This twisting of the scriptures has caused much misunderstanding of the Bible. Thus, when God comes right out and says something, man changes the meaning of the words. *So God has used symbolism, not necessarily to hide, but more correctly to preserve the truth for the latter days.*

pr223» Most people have read God's word without the New Mind. When people speak of God's wrath, they think of it as if it were man's wrath. They make God out to be a hypocrite, for God has told us to give good for evil (Rom 12:21), and to love our enemies (Matt 5:44). What does it mean to love our enemies? Does this mean pouring fire over their heads as some picture God doing at his return to the physical dimension? If we can know what love is, we know how to love our enemies. What is love?

pr224» *Love Is.* "Love is patient; love is kind and envies no one. Love is never boastful, nor conceited, nor rude; never selfish, not quick to take offense. Love keeps no score of wrongs; does not gloat over other men's sins, but delights in the truth. There is nothing love cannot face; there is no limit to its faith, its hope, and its endurance. Love never comes to an end" (1 Cor 13:4-8, NEB).

pr225» Now if God is love as the Bible pictures him (1 John 4:8), then will God lose his *patience* with his enemies and kill them with fire at the end of the age? Remember, God is the almighty, the All Powerful. God created the whole heaven and earth by his words. All God has to do to correct his enemies is to *change* them. All He has to do is give them the New Mind, and instead of being bad, they will be good. Will God lose his patience? Will God be angry with his enemies? In other words is God's wrath like man's wrath? If it is, then God is like us, a hypocrite. He tells us to love our enemies, he tells us to give good for evil, then at the end he does what we shouldn't do, at least this is what many say God will do.

Will God come to destroy his enemies, or will he come to do good, to love his enemies? Let's let God, through his word, answer.

Christ Comes to *Save* Mankind

pr226» One of Jesus Christ's names means savior or Jehovah's savior (Jesus): "the Father sent the Son to be the savior of the world" (1 John 4:14). And, "you shall call his name Jesus [savior]: for he shall save his people from their sins" (Matt 1:21). And, "except those days ['the last days'] should be shortened, there would be no flesh saved: but for the elect's sake those days shall be shortened" (Matt 24:22). "And except that the Lord had shortened those days, no flesh would be saved: but for the elect's sake, whom he has chosen, He [God] has shortened the days" (Mark 13:20). Does this sound like God is coming to destroy?

pr227» When some didn't receive Christ the disciples asked, "do you wish that we command fire to come down from heaven and consume them even as Elijah did? But he [Christ] turned, and rebuked them, and said, You know not what manner of spirit you are of [at that time the disciples hadn't yet received the Spirit of God]. For the son of man is not come to destroy men's lives, but to save them" (Luke 9:54-56).

pr228» God through Jesus Christ will come to save mankind from destroying himself as these scriptures show. Then what is God's wrath? What does the Bible mean by God's wrath?

Righteous Judgment

pr229» The day of wrath is also called a day of trouble, of distress, of desolation, of darkness (Zeph 1:15), and further called a day or hour of judgment (Joel 3:12; Rev 14:7; 18:8, 10). God's wrath is the judgment on the society that the other-mind (Satan's influence) has created on earth. It is the appointed day of righteous judgment (Act 17:31). Paul calls the day of wrath, "the day of wrath and revelation of the righteous judgment of God" (Rom 2:5). And, "the LORD: for he comes, for he comes to judge the earth: he shall judge the world with righteousness" (Psa 96:13; 98:9; 1 Chron 16:33). "And in righteousness he does judge and make war" (Rev 19:11).

pr230» Notice God comes to judge the world righteously. God's wrath, God's judgment is righteous, not unrighteous like man's wrath (see James 1:20). What is righteous wrath or judgment? "For all your commandments are righteousness" (Psa 119:172). Righteous wrath, righteous judgment does not include direct mass killing of inherently weak human beings by the all-powerful God as some accuse God of planning for the end of this age. Evil will be punished. But not in the manner projected by many. What is God's wrath, or God's judgment at his return to the physical dimension?

- "The LORD is known by the judgment which he executes: the wicked is snared in the work of his own hands" (Psa 9:16).

pr231» That is a Biblical definition. In other words the wicked judge their own self, they are snared in the works of their *own* hands.

- Again, "the heathen are sunk in the pit that *they* made: in the net which *they* hid is their own taken" (Psa 9:15).

- "The wicked in his pride does persecute the poor: let them be taken in the devices that *they* have imagined" (Psa 10:2).

- The "transgressors shall be taken in their *own* naughtiness" (Prov 11:6).

- "The wicked is snared by the transgression of *his* lips" (Prov 12:13).

- "Woe unto the wicked it shall be ill with them: for the reward of *his* hands shall be given him" (Isa 3:11).

- "According to *their* deeds, accordingly he will repay" (Isa 59:18).

How Will God Repay Their Deeds?

pr232» "Will you render me a recompense [reward]? and if you recompense me, swiftly and speedily will I *return* your recompense upon your own head ... For the day of the LORD is near upon all the heathen: as you have done, it shall be done unto you: your reward shall return upon your own head" (Joel 3:4; Obad 1:15; check context).

pr233» When one comes to understand that at Christ's coming all the peoples of the world will try to destroy Christ and his people (Rev 17:14, "and they that are with him;" Rev 19:14, 8, "his army" clothed in white; see Matt 25:40), then one understands God's "wrath." As we will prove in the notes and in PR4 and PR5, the world will go mad through the other-mind's (Satan's) influence, which man agrees with, and through circumstances (Rev 16:13-16; 12:4, 13; Jer 51:7; 25:15, 16) the world will be burning up because an instant super-war will have begun. At that time God will *save* mankind from their madness, or thus from their sins (Matt 1:21). Man will be in the process of destroying the world, and their own deeds (use of war weapons) are blowing them off the earth when Christ comes to save (cf Isa 29:6; Matt 24:22; John 3:17). Their *own* deeds (war, nuclear bombs, etc.) are returning on their *own* heads, for all nations will go mad at that time, as the Bible emphatically manifests, and will be committing cosmocide.

pr234» Thus, "the LORD is known by the judgment which he executes: the wicked is snared in the work of his *own* hands" (Psa 9:16). God comes to save, for He is love. The transgression of the built-in law of the universe (cause and effect) will destroy the transgressors of that law, or will destroy them, if God doesn't come to save. How is God's "wrath" poured out on the earth? It is poured out by man himself against himself.

pr235» We will now show you many verses in the Bible where it describes nations and peoples destroying each other during the Last War. When read in the higher sense (see "Duality Paper" [BP4]) the following verses are in context at the day of the LORD:

- "Violence in the land, ruler against ruler" (Jer 51:46).

- "Evil shall go forth from nation to nation" (Jer 25:32).

- "The Egyptians against the Egyptians: and they shall fight every one against his brother, and every one against his neighbor; city against city, and kingdom against kingdom" (Isa 19:2).

- "For nation shall rise against nation, and kingdom against kingdom" (Matt 24:7).

- "Every man's sword shall be against his brother" (Ezek 38:21).

- "By the swords of the mighty will I cause your multitude to fall" (Ezek 32:12).

- "And the horses and their riders shall come down, every one by the sword of his brother" (Hag 2:22).

- "They shall lay hold every one on the hand of his neighbor, and his hand shall rise up against the hand of his neighbor" (Zech 14:13).

- "And the ten horns which you saw upon the beast, these shall hate the whore, and shall make her desolate and naked, and shall eat her flesh, and burn her with fire" (Rev 17:16).

pr236» The horns of the beast destroy the whore (the woman of Rev 17:1, 4). And who is the woman? The woman is the great city (Rev 17:18). And what is the great city? The great city is Babylon (Rev 14:8). Since the "Beast" represents spiritual Babylon, then in other words, the Beast of Revelation destroys itself.

pr237» Now why will mankind begin to destroy themselves? "And in the latter time of their kingdom [man's kingdom], when the transgressors are come to the full..." (Dan 8:23). At the End transgression will be full. The ultimate end of transgression of the built-in laws of the universe is death. Mankind has chosen death and will die, or more correctly will die if the Spiritual Jesus (the SAVIOR) does not come to save the world (Matt 24:22).

pr238» *What is God's Wrath?* It is man's own transgression of God's built-in laws. If some of mankind kill, some of mankind will be killed. If some lie, then some will be lied to. And so on and so forth. See the "Freedom & Law Paper" [NM 17].

pr239» God himself does not have the "freedom" to be wrathful towards man, as man thinks of wrath. God's "wrath" is not like man's wrath (James 1:20). **God's wrath is to let the built-in laws of the universe deal with man's misbehavior** (Psa 9:16). When man goes against God's built-in laws, they are in fact following the ways of destruction. And it is the ways of destruction that destroy man. The laws of God, in and of themselves, do not destroy. But when one goes against them, they are in fact following the way of death and destruction. There is actually a law of sin or a way of destruction (Rom 7:23, 25). And it is the following of this law of sin that destroys man.

pr240» Man does not sin against God when they sin, but against themselves (Prov 8:36). Each sin is a transgression of the built-in laws of the universe. And each transgression of these laws is "rewarded" by an effect. And the effects of these transgressions are all around us today. The misery of life was/is/will-be caused by the transgression of the universe's built-in laws. And as we have noted the transgression of these laws, is in reality, the positive doing of the law/way of destruction, the law of sin.

Spiritual Influence

pr241» As we noted so far that mankind will begin to destroy itself in the Last War. What is the reason man will perform this madness? As we've shown in the "Oher mind Paper" [NM 21], mankind is now being misled by enemy spirits in their minds. Not only is the present age the age of man, but it is also the age of Satan who is the enemy of mankind. The present age is the age of man/Satan. It is the wrath or madness of satanic spirits at the time of the Last War that will mislead man into beginning to destroy themselves. Notice the following Biblical verses for the proof that it is the satanic spiritual power (the other-mind) that is behind man's madness:

- Now, *why* will Egypt fight against itself, and its neighbors; city against city, and kingdom against kingdom? (Isa 19:2; Matt 24:7) [The antitypical Egypt is Satan's kingdom, see PR5]

- *Because* the SPIRIT of Egypt (Satan) is a perverse spirit and has "caused Egypt to err in every work thereof" (Isa 19:14, 13).

- Notice that the wrath against typical Egypt, when God led typical Israel out of Egypt, was caused by "evil angels" (Psa 78:49, 43-51). Thus, from the antitypical Egypt (Satan's kingdom) the antitypical Israel (the Church, Gal 6:16) will be saved from the wrath of the satanic spirits, or the wrath of the devil (Rev 12:12).

pr242» *Wine*. Notice further proof that the Last War's madness is caused by the satanic spiritual influence in man's mind that man now agrees with:

- "Babylon [Satan's kingdom, see the Beast Papers] is fallen, is fallen, that great city, *because* she made all nations drink of the wine of the wrath of her fornication [spiritual fornication]" (Rev 14:8).

- "Babylon has been a golden cup in the LORD's hand, that made all the earth drunken [of Babylon's cup]: the nations have drunken of her wine; therefore the nations are mad" (Jer 51:7).

pr243» What is this *wine* that the nations are drunk on that makes them mad and causes them to fall? (Rev 14:8-10; Jer 51:7-8)

- "Stay yourselves, and wonder; cry you out, and cry: they are drunken, but not with wine; they stagger, but not with strong drink. For the LORD has poured out upon you the *spirit* of deep sleep, and has closed your eyes" (Isa 29:9-10).

pr244» This spirit of deep sleep is the satanic spirit that God gave to man (see "Oher Mind Paper" [NM 21]). It is the mad Satanic spirit that man is drunk on. This is the opposite wine (spirit) from which God's children drink, the new wine (Zech 9:17; Luke 5:37). Notice once again: It is "the *spirit* of the kings of the Medes" that is raised up against Babylon for the "vengeance of his [God's] temple" (Jer 51:11).

pr245» On the day of the wrath besides great madness there will be great fear (Rev 6:15-16; Ezek 32:10, 7-9) because of the cruel angels that are sent through Egypt (Satan's kingdom) (see Ezek 30:2-3, 9; Prov 17:11, "angel" is translated "messenger" in the KJV). Also note and compare with the previously mentioned examples, Revelation 16:12-16.

pr246» Now we have shown you one more dimension of the cause of the wrath of Satan (God's "Wrath"): The satanic influence of Satan's spirits that are misleading mankind (Rev 12:9). Now let's show another facet concerning the question as to why man destroys himself. That is, man will destroy himself, if there is not a Jesus (Savior) to save mankind from his own madness.

Pride

pr247» At the End of the old age will come "the son of perdition," (2 Thes 2:3) or the "false prophet" (Rev 19:20; 13:11-18), or the "little horn" (Dan 7:8, 20, 24-25; 8:9-12, 23-25). If you will read the verses quoted you will see that this antichrist has one obnoxious quality — pride. This false prophet is not only one person, but he epitomizes mankind as a whole at the end of the age — filled with pride of *self*. Notice how God describes mankind at the end of the old age: "in the last days perilous times shall come. For men shall be lovers of their *own* selves ... proud ... heady, high-minded" (2 Tim 3:1, 2, 4). And what is pride a shadow of?

pr248» "How art you fallen from heaven, O Lucifer ['O wail'], son of the morning! ['son of darkness'] how are you cut down to the ground, which did weaken the nations! For you have said in your heart, I will ascend into heaven, I will exalt my throne above the stars of God: I will sit also upon the mount of the congregation, in the sides of the north: I will ascend above the heights of the clouds; I will be like the most High" (Isa 14:12-14). Lucifer is another name for Satan. Notice how he wants to become "the most High," and notice how much *pride* he manifests in these verses.

Why is false pride destroying mankind?

- when pride, then shame (Prov 11:2)

- the way of a fool is right in his own eyes (Prov 12:15)

- contention comes through pride (Prov 13:10)

- pride before destruction, a haughty spirit before a fall (Prov 16:18)

- before destruction a haughty man (Prov 18:11-12)

- high looks, and a proud heart is transgression (Prov 21:4)

pr249» Why are pride and high looks wrong?

- "For who makes you to differ from another? and what have you that you did not receive? now if you did receive it, why do you boast, as if you had not received it?" (1 Cor 4:7)

pr250» In other words, all of us have received what we have mentally, physically, and spiritually from the outside — from our parents (genes, training), from our teachers, or from our spiritual fathers (God, Rom 8:14, 16; or Satan, John 8:44).

pr251» Why is pride wrong? It is wrong because man receives *all* from outside of him. Even man's life has nothing to do with him, it comes from outside of him. The trouble with mankind's pride is that man is lying to himself, if he has pride in himself, because everything

that mankind is comes from outside of man. Mankind is lying to himself when he has pride in himself. And the problem with lying to oneself is that one loses reality. Reality is what is real, or true. Because mankind has false pride, he has lost contact with reality — the true or real. In other words, man has lost contact with God (the truth) because of his false pride in himself. Man has lost contact with the ways of happiness (God's law of love), thus, he is in confusion. Mankind is confused when it comes to finding happiness because he has lost the truth (God). And because mankind is in confusion, he is destroying himself. Another name for God's law is harmony. Another name for confusion is disharmony. Because man has false pride, he has lied to himself concerning the truth (God). Because mankind has lied concerning the truth, he has not found the truth (God). Because he has not found the truth, he is in disharmony. Because man is in disharmony, he is destroying himself.

Destruction Comes From God Himself?

pr252» Now some verses in the Bible seem to say that fire comes from God, instead of from man/Satan's kingdom fighting against itself. Let's examine some of these verses in question:

- "Our God shall come, and shall not keep silence, a fire shall devour *before* [immediately preceding; he comes to *save* mankind from that fire] him" (Psa 50:3).

Now notice one mistranslation:

- "And fire came down *from God* and devoured them" (Rev 20:9).

Some translators thought they were clarifying this verse by saying the fire came from God, for they knew not about the super weapons of the latter days. But that phrase ('from God') is not in most old Greek texts of the New Testament. But even if the phrase 'from God' does belong here it means that because God has predestinated all things that are now happening, in a sense the fire does come from God (see "Predestination Paper" [NM 8]).

Anger of the Lord

pr253» Now in this part we showed you how God is a person of love. And a person of love is a person who loves even his enemies. But the Bible speaks of the ANGER of God.:

- "The LORD will not spare him, but then the *anger* of the LORD and his jealousy shall smoke against that man, and all the curses that are written in this book shall lie upon him" (Deut 29:20).

pr254» What is the *anger* of God that destroys? Is the True God a destroyer? Notice 2 Samuel 24:1:

- "and again the *anger* of the LORD was kindled against Israel, and *he* moved David against them to say, Go, number Israel and Judah." But this verse is qualified by 1 Chronicle 21:1 which points out that the "he" of 2 Samuel 24:1 is Satan. Thus, by comparing we see that the "*anger* of the LORD" is Satan — not the True God. The anger of God is Satan as we proved by comparing the just mentioned verses. Hence, each time one sees the "anger of the LORD" in the Bible he knows the Bible speaks of Satan. (See the *God Papers* for a deeper understanding of the anger of the God.)

pr255» Notice how Sodom and Gomorrah were destroyed (read Gen 19:24-29). Now did God destroy these cities? Is God a destroyer? (In a sense the God is somehow or someway a destroyer and a creator of evil. [Deut 32:39; Isa 45:7] But in this age it is true to say that Satan is the destroyer and the true God the healer.) It was God's "anger" and "wrath" that overthrew Sodom and Gomorrah: "like the overthrow of Sodom, and Gomorrah, Admah, and Zeboim, which the LORD overthrew in his *anger*, and in his *wrath*" (Deut 29:23). As we have just shown you the "anger" of the LORD is Satan. And in *God's Wrath* we showed you that God's "wrath" is in reality Satan's wrath.

pr256» Since in some sense God's power is everywhere, everything that exists is *of* God. This is one reason the Bible speaks of the anger *of* God, the wrath *of* God, and so forth. But the Bible qualifies this, and tells us that the true God only uses his power for good. It is Satan who is the destroyer. Since Satan is a spirit being who came into being *from* God, then Satan is also *of* God (see the *God Papers*).

pr257» The writers of the Bible wrote as if they knew that everything was *of* God. Thus, the writers of the Bible told their readers that God destroyed Sodom, and so forth without qualifying that it was Satan that did the destroying. This cannot be understood in the fullest sense unless you have read and understood the *God Papers*.

PR5: Last War and God's Wrath

Faulty Translations

pr258» In PR5 we examine the book of Revelation. This book of the Bible is highly symbolic. It is a very difficult book to understand because of its symbolism. But the Bible does interpret its own symbols. In order to ascertain the meaning of the symbols of Revelation one must be knowledgeable of the rest of the Bible, for the symbolism can be found scattered throughout the Bible. For example, information on the Beast of Revelation can be found in the book of Daniel (see Beast Papers [PR2, PR3]). In order to understand the book of Revelation, one must understand the rest of the Bible. One must know something about type and antitype. One must know about looking for the higher *meaning* in scripture. The truest meaning in the book of Revelation is its antitypical or Spiritual meaning. The same can be said of the rest of the Bible, but it is especially important in understanding the book of Revelation.

Aorist Verbs & Other Timeless Verbs

pr259» One must also know that the book of Revelation is full of the *aorist* verbs. The aorist verb is a verb of action, not time. An aorist verb by itself tells us nothing about the *time* of the action. It speaks of action without denoting the duration of the action or time of the action:

- "The aorist stem presents action in its simplest form (*a-oristos* 'undefined'). This action is simply presented as a point by the tense. This action is timeless ... The aorist is

a sort of flashlight picture, the imperfect a time exposure" (pp. 824, 1380 in, *A Grammar Of The Greek New Testament*, by A.T. Robertson; see also such books as, *Do It Yourself Hebrew and Greek*, by E.W. Goodrick, pp. 4.4-4.5).

pr260» In fact, in Greek, the aorist, present, and perfect are timeless:

- "These ideas (punctiliar, durative, perfected state) lie behind the three tenses (aorist, present, perfect) that run through all the moods ... The present is also timeless in itself as in the perfect ... These three tenses (aorist, present, perfect) were first developed irrespective of time. Dionysius Thrax erred in explaining the Greek tenses from the notion of time, and he has been followed by a host of imitators. The study of Homer ought to have prevented this error" (p. 824, A.T. Robertson).

- "The terms aorist, imperfect, and perfect (past, present, future) are properly named from the point of view of the state of the action, but present and future are named from the standpoint of the time element. There is no time element in the present subjunctive, for instance. But the names cannot now be changed, though very unsatisfactory" (pp. 825-826, A.T. Robertson).

Because of errors some Greek verbs were misnamed; many today read the idea of *time* into Greek verbs, when they should read the state of the *action*.

Simultaneous Events

pr261» We will in this paper make it plain that many events in the Book of Revelation (the seals, trumpets, and vials, etc.) do not depict *sequential* events. In a way the Book of Revelation was written in the same manner as the movie *Mystery Train* (1989, Directed by Jim Jarmusch).

pr262» The movie *Mystery Train* portrays three separate stories of people in a hotel in Memphis Tennessee during one night through three sequential movie scenes, but it does not tell its viewers that these three separate adventures happened on the same night. At first the viewers think they are separate adventures on separate nights.

pr263» The first story was about a Japanese couple's trip to Memphis to visit Graceland, the home of Elvis. The movie shows the Japanese couple coming to Memphis on a train and their escapades before renting a hotel room (27) and after leaving the hotel room. The second story was actually of two separate women, one from Rome, who was stranded in Memphis, and the other a woman who had just broken up with her boy friend. Both women by chance end up in room 25 in the same hotel as the Japanese couple, but the movie does not indicate that they were spending the same night in the hotel as the Japanese couple. The third story was about three men, one of whom happens to be the boy friend of one of the two women in room 25. They eventually hid that same night in room 22 after one of them shot a man at a local liquor store. It isn't until the movie gets into the third story that you begin to see that all three groups are in the same hotel on the *same* night.

pr264» You begin to see that each group was in the hotel on the same night by subtle hints in the movie:

- the repetition of the same song by Elvis (*Blue Moon*);

- the reiteration of events and conversation between the manager and the bellhop

- the same train passing by in each story

- the Japanese couple being overheard by the two women

- and then the gun shot from room 22 that all the separate groups re-act to in different ways, all without any interaction between the three groups of people renting different rooms in the same hotel the same night.

pr265» The three stories were told sequentially, but they occurred in parallel time periods. This is what is happening in the Book of Revelation. Again and again with subtle hints Biblical evidence projects to us that the seals, trumpets, and vials happen all at once. This was one reason John used the aorist verb in describing events in the book of Revelation. The use of the aorist verb and other timeless verbs (verbs of action not time) helped John to describe many *simultaneous* events. Under the subtitle 'But Simultaneous Action is Common also,' A.T. Robertson in his Grammar states: "Indeed this simultaneous action is in exact harmony with the punctiliar meaning of the aorist tense." And he states, 'in many examples only exegesis [interpretation] can determine whether antecedent or coincident action is intended ....' Many events in the book of Revelation were *written* sequentially, but they actually transpired in parallel time periods. The Greek text of the Book of Revelation superficially appears sequential only as the movie *Mystery Train* superficially appears sequential. An understanding of the Greek aorist verbs (and other timeless verbs) and the discernment of the reiteration of simultaneous events in the Book of Revelations makes it plain that the Book of Revelation describes many simultaneous events not just sequential events. In the rest of this paper we will make this plain.

Day of the Lord

pr266» Furthermore, the whole vision of John was concerning the "day of the Lord." The "day of the Lord" has two senses: the regular seventh day — the weekly Sabbath; or the antitypical seventh day — the 1000 years. John was in Spirit "*in* the Lord's day," and was told to "write the things which you [John] saw and the things which are and the things which shall be after these things" (Rev 1:10, 19). In John's vision he saw things that occurred in the 1000 years — the antitypical Sabbath (Rev 20:2-3). He also saw things that happened before the 1000 years (Rev chap. 2 & 3). And he saw things that happened after the 1000 years (Rev 20:7; 21:1, 4, 6; 5:13 & note Psalm 148 150 & "All Saved" paper [NM 13]; etc).

pr267» In order to understand the book of Revelation we need to synthesize most of the Bible. Why? "Son of man, what is that proverb that you have in the land of Israel, saying, The days are prolonged, and every vision fails? Tell them therefore, Thus, says the Lord GOD; I will make this proverb to cease, and they shall no more use it as a proverb in Israel; but say unto them, *The days are at hand*, **and the effect of every vision**" (Ezek 12:22-23). *All* the visions of the Bible will be fulfilled in the last years of the old system of man. Visions in the past that appeared to fail will come true soon.

pr268» The vision of the Beast in Revelation and Daniel are for the end time (Dan 12:4, 9). But we know the time is truly *near* when we truly hear and understand all the book of Revelation (Rev 22:10; 1:3).

Here Some, There Some

pr269» Another principle a person needs to understand is related in Isaiah 28:9-10: "Whom shall he teach knowledge? and whom shall he make to understand doctrine? them that are weaned from the milk, and drawn from the breasts. **For precept must be upon precept, precept upon precept; line upon line, line**

upon line; here a little, and there a little." The truth of Revelation is scattered throughout the Bible. We must put all these scriptures together. This is much like the truth of Jesus Christ's first coming. Scripture on this event is scattered throughout the Bible. When all the prophetic scriptures on Jesus Christ are put together we see just how much of the Bible pointed to Jesus Christ (see, *Messianic Prophecies of the Bible*, by Lockyer; etc.). Yet the Jews in Jesus Christ's time could not see these prophecies as pointing to Jesus Christ. We must not be like those Jews. We must be the Jews with the New Mind that see all the scriptures that point to the Last War and the return of Christ.

Higher Meaning

pr270» Furthermore one must understand that the Bible is dual — there is a shadow/physical/type of the real/Spiritual/antitype. This is manifested by comparing the following verses (Rom 1:20; Heb 8:5, Heb 9:23-24; 10:1; and all of Hebrews 8:1 to 10:21). And this is proven by the repeated consistency of duality throughout the Bible (see "Duality Paper" [BP4]). God through the Bible asks us to look for and seek after the higher or spiritual or real or antitypical meaning in the Bible. Compare the usage of "heaven," "earth," and "spiritual" in Isaiah 55:9 with 1 Cor 15:44-49; Phil 3:18-19 with Col 3:1-2. Thus, "heaven" = spiritual, and we should look to the above/heavenly or spiritual meaning of scripture (see "Duality Paper" [BP4]).

pr271» What is an example of taking the higher meaning? Notice in the book of Hebrews 12:22-23 where it calls the Spiritual Church: "mount Zion," or the "city of the living God," or the "heavenly Jerusalem." In other words, these terms are metonymical for the Church. Most of the time when we read Zion, Jerusalem, or city of God we know it means literally a city, as well as Spiritually, the Spiritual city or Church. This is duality: type (the physical city) and antitype (the Spiritual city or Church). Also the Church is called the holy temple (Eph 2:21), thus, anytime the Bible uses the word "temple" (meaning holy temple) we know it can mean the physical and/or Spiritual temple. The Spiritual temple is the Spiritual Church of God. But God wants us to look to the heavenly, or Spiritual, or antitypical meaning (Col 3:1-2; John 4:24; 6:63). When Revelation 11:1 says to measure the temple of God, we then know it means antitypically to go measure the Church or count its members.

pr272» Now God's wrath and the day of the Lord are mentioned throughout the Bible. We need to tie-in the events of the Old Testament with the things in the book of Revelation because of the principle in Ezekiel 12:22-23, and because we would be taking Revelation out of context from the whole Bible if we did not use the rest of the books to amplify the book of Revelation.

Great City

pr273» In Revelation 11:8, speaking of a few days before Christ's return, it says the two witnesses "shall lie in the street of the *great city*, which spiritually [or antitypically] is called *Sodom* and *Egypt*, where also our Lord was crucified." Now we know the Lord was not crucified in the physical Sodom or Egypt; it was near physical Jerusalem. Yet Christ was killed in the *antitypical* Egypt, or Sodom, or great city which is Satan's spiritual kingdom. Therefore the physical Jerusalem was in the spiritual kingdom of Satan, for all the kingdoms and lands of the world are now, in this old age, Satan's kingdoms (Matt 4:8-9).

pr274» Notice that the spiritual or antitypical Sodom and Egypt are both called "the great city" (Rev 11:8). The spiritual Sodom, Egypt, and the great city are each just different names for each other. They are metonymical terms. Each term speaks of the same thing. What is the great city?

pr275» "And the *woman* which you saw is that *great city*, which reigns over the kings of the earth" (Rev 17:18). The antitypical woman who sits on the Beast *is* the great city (Rev 17:1, 9, 18). Who is the Beast? The Beast is Satan's kingdom (see the "Beast-Man Paper" [PR3]. Thus, the woman is metonymical to the spiritual great city, which is the antitypical Sodom and Egypt. Yet we know women are symbolical to churches (Eph 5:22-25). Thus, the woman on the Beast symbolizes the church essence of the church/kingdom of Satan, and she is metonymical to the spiritual great city, Sodom, and Egypt.

pr276» Notice also in Revelation 14:8, "Babylon is fallen, is fallen, that *great city*, because *she* made all the nations ..." Babylon is also the "great city."

pr277» Therefore according to the book of Revelation, the spiritual great city, Babylon, Sodom, Egypt, and the antitypical woman on the Beast are metonymical for each other.

pr278» Notice after it says the woman is the great city, that it describes the great city as Babylon (Rev 17:18; 18:1-24). In describing Babylon in chapter 18 it says, "and saying, Alas, alas, that *great city*, that was clothed in fine linen, and purple, and scarlet, and decked with gold, and precious stones, and pearls!" (Rev 18:16) Compare this with the description of the *woman* in Revelation 17:4, "and the woman was arrayed in purple and scarlet color, and decked with gold and precious stones and pearls." The city and the woman are clothed alike. This is another proof that the antitypical woman and Babylon are one and the same — each term merely adds more detail to the other term.

pr279» Just as the Bible uses many terms to describe Satan (the devil, dragon, serpent, etc.), it has used many terms to describe the same system of Satan. These terms are metonymical names for Satan's "Beast" system.

Metonymical Names For The Antitypical Beast, Or The Kingdom of Satan

pr280»

- Great City [Rev 17:18; 11:8]

- Woman or Whore (the great city) [Rev chap 17; 17:18]

- Babylon (the great city) [Rev 14:8; 18:16]

- Sodom (the great city) [Rev 11:8]

- Egypt (the great city) [Rev 11:8]

- All Kingdoms Are Satan's — The Beast

pr281» If you have read the papers on the "Beast" you know that the Babylonian kingdom was the first Beast of Daniel, chapter 7, which is pictured as part of the "Beast" of Revelation, chapter 13. Further, we know that the Beast of Revelation will encompass at once all the characteristics of all the "Beasts" of Daniel because the book of Revelation's Beast encompasses the same characteristics as Daniel's four Beasts (Rev 13:2; Dan 7:4-6, "lion," "bear," and "leopard"). Thus, since Babylon was the first Beast of Daniel, then we know the antitypical Babylon will belong to the antitypical Beast of Revelation, for the Beast of Revelation will encompass all the Beasts of Daniel at once. **The Beast of Revelation in its truest meaning encompasses all of Satan's kingdoms.** Therefore the Beast is the spiritual or antitypical great city, Sodom, Egypt, Babylon, and the Woman. All these terms describe the

same church/kingdom of Satan. Each term adds another detail to the overall description of the final Beast kingdom. The woman in Revelation 17 merely tells us that the system of Satan also includes a church. Revelation calls this church, the synagogue of Satan (Rev 2:9, 3:9).

pr282» *Church of Satan*. A church is an assembly of people who worship the same god: the church of Satan worship their god Satan even though they *think* they are worshiping the true God. To be in the church of Satan you need not ever go *to* church; an atheist with the spirit of Satan is in the church of Satan and worships Satan.

pr283» *A Principle of Interpretation*. Now since the Bible is dual, then what the Bible says concerning the day of the Lord or God's wrath towards the physical or typical Egypt, Sodom, and Babylon, can be used to synthesize the story about the Last War. We know now that when the Bible describes the "wrath" on the physical Egypt and Babylon, in the antitypical meaning, it is speaking about the "wrath" on the Beast of Revelation, for the antitypical Egypt and Babylon belong to the Beast of Revelation.

Seventy Years And Seventy Weeks

pr284» Let us further synthesize the Old Testament with the New Testament through the "seventy weeks" and "seventy years." In Jeremiah 25:9-14 it speaks about the land of Israel and the nations around it serving Babylon during a seventy year period, but "when **seventy years** are accomplished, I will punish the king of Babylon, and that nation." All that Jeremiah has prophesied against Babylon will happen after the seventy years.

Wine Cup

pr285» But the next verse speaks about the wine cup that all nations are to drink. And in Jeremiah 51:7-8 it speaks about this same wine cup, calling it a golden cup. Also in Revelation 18:3; 14:8-10 it speaks of the same cup poured out on Babylon. By reading Jeremiah 25:27-33 and comparing this with Revelation, one knows that this great slaughter of nations didn't happen to the *full* extent shown in Jeremiah 25:15-33. Only a typical event happened. The true slaughter of nations in this chapter of Jeremiah will happen in the Last War. There is an antitypical seventy years, and after those seventy antitypical years then the cup of wrath will be poured out to the *full* extent mentioned in Jeremiah, chapters 25 and 51.

Cyrus

pr286» "To fulfill the word of the LORD by the mouth of Jeremiah, until the land had enjoyed her sabbaths: for as long as she lay desolate she kept the sabbath, to fulfill **threescore and ten years**. Now in the first year of Cyrus king of Persia, that the word of the LORD spoken by the mouth of Jeremiah might be accomplished, the LORD stirred up the spirit of Cyrus king of Persia, that he made a proclamation throughout all his kingdom, and put it also in writing, saying, Thus says Cyrus king of Persia, All the kingdoms of the earth has the LORD God of heaven given me; and he has charged me to build him a house in Jerusalem, which is in Judah. Who is there among you of all his people? The LORD his God be with him, and let him go up" (2 Chron 36:21-23; see Ezra 1:1-3).

pr287» Thus, after the seventy typical years Cyrus came to build a house for the LORD God of Israel in Jerusalem (Ezra 1:3). Now the Bible is dual and God tells us to look to the higher meanings. Is there an antitypical meaning to this verse? Notice that the Hebrew word translated "Cyrus" means *sun* and that this is the symbol of the returning Christ (Mal 4:2; see *God Papers* on sun and moon symbolism). Further the antitypical house of God which Cyrus was to build is the Church (Eph 2:19). And who is building this house? — Christ (Heb 3:3-6). Is Cyrus a typical Christ? Does this mean that after the antitypical seventy years that the antitypical Cyrus (Christ) will return and claim "all the kingdoms of the earth" for himself? (2 Chron 36:23) Then will the antitypical Cyrus (Christ) be King of kings? (Rev 19:16) Is this a dual story? We answer yes, for all the Bible is dual. And there is an antitypical Cyrus and he will destroy the antitypical Babylon as did the typical Cyrus destroy the typical Babylon. But Christ, the antitypical Cyrus, will destroy the antitypical Babylon righteously as explained in the *God's Wrath*, PR4.

pr288» Notice: "that says of Cyrus, He is my shepherd, and shall perform all my pleasure: even saying to Jerusalem, You shall be built; and to the temple, Your foundation shall be laid" (Isa 44:28).

pr289» What is the antitypical Jerusalem and the antitypical temple? They are the Church (Heb 12:22; Eph 2:21; see "New Jerusalem" in [NM 18]). What are the foundations that are to be laid?

pr290»

- The righteous are an agelasting foundation (Prov 10:25)

- Zion (the Church, Heb 12:22) a foundation, a tried stone (Isa 28:16)

The wife (the Church, Rev 19:7) of Christ will be laid with stones of fair colors, and her foundations with sapphires (Isa 54:11).

pr291» Cyrus is a shepherd. Who is the chief shepherd? Christ is the chief shepherd (1 Pet 5:4). What does the Hebrew word translated Cyrus mean? It means the *sun*. Who is the antitypical sun? Or what is the sun symbolic of? It is symbolic of Christ (Mal 4:2). Then in the higher or antitypical meaning of Isaiah 44:28 it says Christ will come to do what Cyrus did typically after the seventy years of Israel's captivity? Isn't Cyrus a typical Christ?

pr292» "Thus says the LORD to his *anointed*, to Cyrus" (Isa 45:1). Who is the LORD's true anointed? It is Christ, the Messiah, who is the anointed. Compare Psalm 2:2-7 with Hebrews 1:5, 9 to prove Christ is the Anointed of the LORD. Thus, Cyrus is a typical Christ. And after the seventy antitypical years, Christ will return with God's "wrath" as explained previously in PR4.

pr293» What we are doing remember is tying-in the Old Testament's Babylon, Egypt, and Cyrus with what is about to come on this old age. We are tying-in the Old Testament with the book of Revelation in order to have a detailed description of the Last War as well as to explain the book of Revelation. Now we need to know what the antitypical seventy years are.

Antitypical Seventy Years

pr294» "In the first year of his reign [Darius], I Daniel considered by books the number of the years, whereof the word of the LORD came to Jeremiah the prophet, that he would accomplish *seventy years* in the desolation of Jerusalem" (Dan 9:2).

pr295» Daniel is speaking about the same seventy years we have been writing about. Then Daniel begins to pray for his people asking forgiveness (v. 3-19). And while he was praying Gabriel came to him (v. 20-21). And Gabriel said he came to give Daniel skill in understanding.

pr296» Now Daniel was seeking through prayer two things: (1) for mercy on Israel (v. 18, 13); and (2) for understanding of the truth (v. 13). He had considered the seventy years mentioned in Jeremiah. Now what was there to consider if these seventy years were a literal seventy years of desolation? It is true that *some* came back to Jerusalem after this seventy years (Ezra). But did the following happen to the full extent: "that after seventy years be accomplished at Babylon I will visit you, and perform my good word toward you, in causing you to return to this place" (Jer 29:10, 11-14). Sure it is true in one sense. But many Jews didn't return and God hasn't visited his people in the truest sense of Jeremiah 29:10-14. When will the LORD visit his people? He will visit at the day of the LORD when the enemies have destroyed themselves and He begins to gather all the peoples of Israel back to their land (Obad 1:15; Isa 13:6-14:3; Joel 2:31-32).

pr297» Daniel was considering and perceiving the seventy years of Jeremiah. Then Daniel began to seek through prayer (v. 3) the true understanding of these seventy years (v. 13). While he was praying Gabriel came to give him understanding (v. 21-22).

pr298» Gabriel: "at the beginning of your supplication [v. 3] the commandment came forth, and I am come to show you [the truth of the seventy years as explained just previously]; for you art greatly beloved: therefore understand the matter, and consider the vision. *Seventy sevens* [KJV, "weeks"] *are determined upon your people and upon your holy city* [Jerusalem], *to finish the transgression..."* (v. 23-24). Let's stop here.

pr299» In many translations it uses the word "weeks" for a Hebrew word [#7620 — *shabuwa*] that means, *sevens*. It is like the word, *heptad*, which means a group of seven. In Hebrew, the seventy *sevens* could mean seventy sevens of days (70 weeks) or seventy sevens of years.

pr300» Why was Jerusalem desolate: "Yea all Israel have transgressed your law, even departing, that they might not obey your voice; therefore the curse is poured upon us" (Dan 9:11). The seventy *years* of desolation was caused by transgression and Daniel was praying for mercy (v. 18, 13), and understanding as to the truth. Gabriel came to give understanding, and said seventy *sevens* are determined for the people of Israel to *finish* their transgression. Thus, if the seventy years of desolation was caused by transgression, then the desolation in the truest sense will not end until the seventy sevens are completed.

pr301» *483 Years or 69 weeks*. Now Gabriel goes on to say an anointed one would come after sixty-nine sevens (KJV, "weeks") from the "commandment to restore and to build Jerusalem" (v. 25). This is dual. The scripture in

Daniel 9:24-27 is ambiguous. It can be understood in two ways: (1) anti-Messianic; and (2) Messianic. That is, the anointed prince can be either the Messiah Prince or the coming false anointed prince. By reading, *The Millennium Bible*, by William E. Biederwolf, (pages 219-225, Baker, 1972; aka *The Second Coming Bible*) you can see some of the two different interpretations given these scriptures down through history. The Messianic interpretation has the Messiah coming after 69 weeks of years: 483 years "from the going forth of the commandment." By various interpretations this commandment was: (1) Cyrus' (Isa 44:28; Ezra 6:14); (2) Darius' degree in the second year of his reign (Ezra 6:12) (3) The degree of Artaxerxes in his seventh year (Ezra 7:1, 7, 11 ff); (4) or the degree of Artaxerxes in his twentieth year (Neh 2:1,7). According to the Messianic interpretation, since after sixty-nine sevens of days Christ did not come, then Daniel 9:25 means sixty-nine sevens of years. The sixty-nine sevens of years equals 69 times 7 years, or 483 years. After 483 years the Messiah appeared as the prophecy in Daniel indicated. This is when Christ was baptized. See our *Chronology Papers*.

pr302» *7 Years Cut in Half*. And the Messiah was to "confirm the covenant with many for one seven: and in the midst of the seven [seven days or seven years] he shall cause the sacrifice and the oblation [offering] to cease" (v. 27). Furthermore in the middle of the seven (7 years) of "confirming the covenant" he was cut off (Dan 9:26, 27; Isa 22:25; Zech 11:10).

(The "he" in verses 26 and 27 refers to the Messiah; the "covenant" being the new covenant. Matt 26:28; Mal 3:1; Jer 31:31-33; Heb 9:26; Heb 10:7-9; see *Chronology Papers*. In the second sense 'he' refers to the Antichrist. See "69 Weeks" in the "Beast-Man Paper" [PR3])

pr303» Christ's first ministry of confirming the covenant lasted 3 ½ years. Thus, 69 ½ sevens of years of the 70 have been fulfilled. A week of years is seven years. Therefore one-half of seven years is 3 ½ years. At Christ's death this prophecy was cut off since 3 ½ years after Christ died the transgression of Israel didn't cease as Daniel 9:24 indicated. There are still 3 ½ years left in Daniel's prophecy.

pr304» Notice that "the end [of the prophecy] thereof shall be with a flood, and unto the end of the war [the Last War] desolation are determined" (v. 26). Further, note that Christ the Messiah would confirm the covenant for one seven, or one seven year period (Dan 9:27). But so far he has taught only for 3 ½ years. **Where are the other 3 ½ years**? If we find it, we will know when the transgression of Israel will cease and when the Last War will take place. We will also know when the end of the antitypical seventy years of Jeremiah will end since in explaining the seventy years Gabriel used seventy sevens. (The 70 antitypical years are the 70 sevens, or 70 sevens of years.) Also if we can locate these 3 ½ years we will know *how* Christ will confirm the covenant during these 3 ½ years.

Two Witnesses In Their 1260 Days

pr305» Now who will confirm the covenant of Christ these last 3 ½ years? We know it isn't Christ himself, for he won't return to earth until *after* the 3 ½ years (see Beast Papers [PR4, PR5, PR6]). There will be an agent or agents for Christ these 3 ½ years. Who?

pr306» "And I will give power unto my two witnesses, and they shall prophesy a thousand two hundred and threescore days [3 ½ years], clothed in sackcloth. These are the two olive trees, and the two candlesticks standing before the God of the earth" (Rev 11:3-4).

pr307» It will be the two witnesses who will preach 1260 days (3 ½ years) "before the God of the earth." Or as Zechariah 4:14 says, "these are the two anointed ones, that stand by the Lord of the whole earth." These two persons who come in the spirit of Moses and Elijah will teach as agents for Christ (see the "Two Witnesses" paper [PR8]).

pr308» It is during these 3 ½ years of teaching by the two witnesses that the Church will be in the wilderness (Rev 12:6, 14), will be in trial (Dan 11:33-35; 12:10; Rev 3:14-19) for the 3 ½ years (Dan 12:7). Hence, the Church will be in the spiritual wilderness for 3 ½ years during the Beast's misrule (Rev 13:5; Dan 7:25; 12:7; Rev 12:6, 14). And during this 3 ½ years of misrule by the Beast the Church will be tried: some during the 3 ½ years will be physically tried, but all will be physically tried at the end of the 3 ½ years, and all will be spiritually tried throughout the 3 ½ years (as well as throughout their life) by the spirit of Satan, that other mind (see "Oher Mind" paper [NM 21]).

pr309» What is the antitypical Jerusalem? It is the Church (Heb 12:22). What is the antitypical Babylon which makes Jerusalem desolate? The antitypical Babylon is the Beast. Thus, the antitypical Jerusalem or the holy city (the Church) will be as good as desolate during these 3 ½ years under the antitypical Beast (Rev 11:2). They will physically be under the Beast (Satan's kingdom) and spiritually tried by it (see "Oher Mind Paper" [NM 21]). And at the end of the 3 ½ years they will be physically desolated (see below, see PR6).

pr310» *Hence*, by putting this together, the last part of the seventy sevens of years will be the 3 ½ years when the Beast of Revelation, the antitypical Babylon, will rule the earth. (The spirit of the Beast rules all the earth; the seven nation Beast rules great in the earth and great in the Middle East.) And it is after these 3 ½ years, that Cyrus (Christ) will return and become King of kings.

To Review

pr311» So far then, we have tied-in the seventy years of Jeremiah and the seventy weeks of Daniel to the story of Revelation. Also we have shown that Cyrus was a typical Christ. We've shown you that the two witnesses are agents of Christ to confirm the covenant in Christ's place the last 3 ½ years. We have shown you that physical Egypt and Babylon were types of the Beast of Revelation, which is the kingdom of Satan. We have shown you that the antitypical woman of Revelation 17 is a dimension of the Beast, both church and kingdom at once. We are thus showing you the Spiritual meaning of the Bible.

Old Wars as Pattern of the Last War

pr312» All the battles of the Old Testament happened as examples, but during the last 3 ½ years they will happen again in the antitype, or truest sense. Much of the Bible is for the next few years, that is, the years just before and during the 3 ½ years or 1260 days (1 Cor 10:11; Heb 10:1; Ezek 12:22-23).

pr313» Thus, after this 3 ½ year period, which fulfills the antitypical seventy years of Jeremiah or the seventy sevens or weeks of Daniel, then Christ (Cyrus) will come to build the temple (Church/ kingdom) of God (2 Chron 36:23; Isa 44:28), and to subdue the nations (Isa 45:1; Rev 17:14). He will come at the Last War (Dan 9:26) to *save* mankind from destroying themselves (Matt 24:22; John 3:17). He will *not* come to destroy nations, but to save them from their own transgression as shown previously in PR4. Transgression will be at its full as Daniel explained (Dan 8:23). The ultimate end of transgression is death (Deut 30:15). Thus, when the world has reached its height of transgression God will come to *save*.

pr314» But Christ comes only when the world is in the process of destroying itself. One-third of the world's population at that point will have died in the war (Rev 9:18), *and* one-fourth of the earth will have been destroyed by the Last War (Rev 6:8).

All Nations Destroyed

pr315» Now Jeremiah 25:12-13 said *all* the nations written about in his book would receive all that God pronounced against them in Jeremiah's book by the completion of the seventy years. As explained, the end of the antitypical 70 years is after the 3 ½ years just before Christ's rule. The nations included in the destruction includes the antitypical Babylon, Egypt, the Medes, all the kings or rulers of the North, and so forth (Jer 25:12-26). All these nations will drink of the wine cup of fury (Jer 25:15-17); there will be a great slaughter throughout the land (Jer 25:26, 32-33).

pr316» In Ezekiel it speaks about a group of nations from the north who will come down and gather for God's wrath (see Ezek 38:15, 18-20). These are the peoples of Gog, Meshech, and Tubal. Along with these will come Persia, Cush, Phut, Gomer, and Togarmah (Ezek 38:2-6; see Dan 11:44 & Rev 16:12). They all will be destroyed at Christ's coming according to the higher meaning of Jeremiah 25:12-38 and according to Ezekiel 38:18-22; 39:2, 11. These nations are also the same nations as shown in Revelation 16:12-16; 9:11-18; 19:19; 20:8-10, and which Daniel 11:44 speaks about.

Pattern of the Last War Throughout Bible

pr317» Yet not only are all the nations spoken about in Jeremiah to be destroyed, but all nations mentioned in all the prophecies. The nations' destruction is mentioned in most of the Bible, and these prophecies speak about all nations being destroyed because of God's "wrath." Thus, we can use most of the Bible to describe the details of the Last War (see Ezek 12:22-23).

Day of Trouble

pr318» For example, much of the book of Psalms speaks about the day of the LORD. In the book of Psalms time and time again it speaks of a "day of trouble" with the LORD coming and saving God's people. All these scriptures can add to the details of the Last War.

Wicked and Evil Ones

pr319» Then in the book of Job it speaks about the wicked ones and their fate. All these scriptures can also be used to describe the Last War. In fact anywhere in the Bible where it speaks about evil or the wicked's fate can be used, for many of the Biblical prophecies are all at once completed on the day of the LORD (Ezek 12:22-23). The day of the LORD is the climax of evil, the fullness of transgression, and also its *end*. It is also the beginning of God's rulership

All Evil and War Destroyed

pr320» The following verses in their higher meaning or antitypical meaning prove there will be no more evil from the time God takes over:

- they shall not learn war any more [Isa 2:4; Mic 4:3]

- violence no more [Isa 60:18]

- no evil seen any more [Zeph 3:15]

- affliction *not* to rise a second time [Nah 1:9]

pr321» Thus after this climatic Last War there will be no more confusion, no more war, no more evil. As Psalms 37:38 and Jeremiah 51:48-49 show the transgressors will be destroyed *together*.

How Long Will The War Last?

pr322» But how long will this Last War last. Some have said that God's "wrath" will be poured out over a period of time. As we've shown you in PR4, God's wrath is the wrath of Satan, his age, and his people. Some say the Last War may take a year or more.

pr323» But this can't be, for when antitypical Babylon (the Beast) is destroyed so too will the other nations be destroyed (Jer 25:12-26; Jer 51:48-49). The Beast will be destroyed after 3 ½ years of misrule (Rev 13:5; 19:20). It is during these 3 ½ years that the Church will be in the wilderness (Rev 12:6, 14). The Church of this old age is not the utopia or the New Age: the Church consists of those belonging to the New Age because they have the Spirit of the New Age. Since no one (except Christ) is allowed to enter the temple (kingdom of God, or the New Age, see notes) until the last plague is poured out (Rev 15:8), the last plagues will be poured out at the *end* of the 3 ½ year period.

Wrath: All At Once

pr324» How long is the Last War? How long is God's wrath? There are at least 12 verses in the Bible that say in their higher meaning that the end of this age and its way will come in *one day*, *one hour*, in an *instant*, *at once*, *swiftly* and *speedily* as a overwhelming flood (Nah 1:8; Zeph 1:18; Mal 3:5; Isa 29:5; Isa 42:14; Luke 18:8; Rev 18:8, 10, 19; Dan 9:26; Isa 47:11; Joel 3:4; Isa 10:17). The destruction will come just before or near the beginning of the day of the LORD. But it will happen in an instant, at once. Hence *the seals, trumpets, and plagues of Revelation will happen all at once.*

Greater Detail

pr325» Now let's go into some greater detail. After this we will construct what will happen in the 3 ½ year period and thus explain the book of Revelation. Let's now continue to simplify and synthesize the book of Revelation.

Angel and Angels of Revelation

pr326» Notice that the shout of the voice of Christ sounds like a *trumpet* (cf 1 Thes 4:16; Rev 4:1; 1:10). God's voice also is described as *thunder* (Job 37:4-5). And his voice is as the sound of *many waters* (Ezek 43:2). And the "seven thunders" of Revelation 10:3 is nothing other than the seven trumpets, for God's voice is like a trumpet, thunder, and many waters. These terms are merely metonymical of each other.

pr327» The angel of Revelation 10:1-4 that uttered "with a voice, as when a lion roars: and when he had cried [with *a* voice], seven thunders uttered their voices," is Christ's *own* angel.

pr328» Notice this angel had a rainbow on his head, and his face was as it were the sun, and his feet as pillars of fire (Rev 10:1). This rainbow is of God's throne (Rev 4:3; Ezek 1:28), not just any angel's throne. This angel's face looked like the sun which is the way Christ's face is described (Rev 1:16). He swore by him who created the heaven and earth (Rev 10:6).

pr329» Another proof yet that this angel is Christ's own is to compare the following verses:

- "A lamb [Christ] as it had been slain, having seven horns and seven eyes, which are the seven Spirits of God sent forth into all the earth" (Rev 5:6).

pr330» All these seven eyes (Spirits) are on the Lamb, and the Lamb is Christ (John 1:29). These seven Spirits are of Christ. Note also:

- "Seven lamps of fire burning before the throne, which are the seven Spirits of God" (Rev 4:5).

pr331» Now spirits are angels (Heb 1:7). Thus, these seven Spirits are seven angels. Also, the seven lamps (flames) must burn on a lamp stand. And sure enough there are seven golden candlesticks (lamp stands), and these hold the seven lamps or spirits or angels (Rev 1:20). And as Revelation 1:20 shows these seven candlesticks indicate the seven churches of Revelation, chapters 2 and 3. These seven lamps (flames) are the seven spirits or angels of the seven churches. Notice Hebrews 1:7 where flames of fire are equated to spirits or angels. These seven spirits or angels are shown throughout the book of Revelation blowing trumpets, talking to John, and so forth. Now we know these seven angels are the seven eyes of the Lamb (Christ); they are of Christ (cf. Rev 4:5). **That is, they are of Christ's Spiritual Body that will eventually fill all in all** (1 Cor 12:12, 27).

pr332» Notice Zechariah describes these seven same eyes upon *one stone* (Zech 3:9). In Zechariah 4:7 it calls this same stone the *headstone*: "and he shall bring forth the headstone thereof with shouting." The head or chief stone is Christ (Eph 2:20). Thus, this is another proof that these seven eyes or angels are of one, Christ. Revelation uses seven angels to describe Christ or the Spiritual Body of Christ. Hence, the angel of Revelation 10:1-4 is Christ and his voice sounds like seven thunders, and his voice is metonymical for the sound of a trumpet or many waters. Remember there is Christ the individual and the Body of Christ with many individuals.

Things We Should Understand

pr333» Revelation is a poetical, symbolical rendition of the end-of-the-age events. It seems mystical only if the reader doesn't understand Biblical symbolism. The Bible uses many metonymical terms to describe the same things or events. Revelation is understandable only *if* one knows that the seals, trumpets, and plagues happen all at once. And the only way one knows this is to use the whole Bible to unveil the book of Revelation. But the only way to properly do this is to study the whole Bible.

pr334» Next one must know all the terms used to describe the Church, for example the first-fruits, the bride, the woman of Revelation 12:1-2; the temple, new Jerusalem, etc. Also, one must know that the seven angels are merely the manifestations of Christ, and that his "voice" is metonymical for many waters, trumpets, thunders, and shouts. What he speaks with his voice is the Word of God, and the physical Word of God is in the Bible. Further one must know that the lake of fire is the beginning and aftermath of an instant atomic (and other super weapons) war that is beginning to burn up the earth at the return of Christ to the physical dimension, and it is this fire that will burn for an agelasting time (Matt 25:41), and this age is for one-thousand years (Rev 20:1-3).

pr335» One must know that many prophecies will be fulfilled to the fullest extent at the end of the age (Ezek 12:22-23). And we must know the Beast of Revelation is the kingdom of Satan/man which will at once encompass all the qualities of the four Beasts of Daniel 7. And we must know that the Revelations's woman, great city, Babylon, Egypt, and Sodom are all a descriptive part of the end-of-the-age's Beast. Thus, what the Bible says about the typical Babylon, Egypt, and so forth can be used to describe the end-of-the-age events since many prophecies are to be fulfilled near or at the end of the age.

God's Throne

pr336» Now let's identify one spiritual meaning of God's throne. Revelation 3:21 pictures Christ sitting on a throne of his Father. In Isaiah 66:1 God describes his throne as being heaven: "the heaven is my throne." But what is heaven representative of? It represents the Spiritual dimension. Compare the use of "heaven" and "earth" with "spirit" and "flesh" in, Isaiah 55:9; Phil 3:18-20; Col 3:1-2; 1 Cor 15:44-49; and Hebrews 9:23-24.

pr337» Now Christ says we can sit on his throne, which is his Father's (Rev 3:21), which is heaven, which is the Spiritual dimension. Thus, Christians can put on the Spiritual dimension as Christ did (see the *God Papers*). Also notice that Christ's throne is like a fiery flame (Dan 7:9), and fiery flames are equated to spiritual beings (Heb 1:7). What is God's throne? It is Spiritual life. This is one higher meaning of God's throne. In other words, God's throne is *not* like man's throne, just as the fear of God is different from what man thinks. Those ruling in the 1000 year age won't be sitting around on thrones, but they will act and behave as Christ did on earth. They will be the *servants* of the new world.

Seals, Trumpets Plagues, and 1260 Days

pr338» Now let's synthesize the seals, trumpets, and plagues (vials) of Revelation.

- The Church of God (Rev 3:7-13) will be in the spiritual wilderness of the Beast for 1260 days (Rev 12:6, 14).

- The Beast is to rule these 1260 days (Dan 7:25; Rev 13:5).

- The Beast is destroyed after these 1260 days (Rev 14:8-10; Rev 16:19; 19-20).

- These 1260 days (3 ½ years) are the last ½ week of Daniel's prophetic seventy weeks (Dan 9:24-27).

- It is also the last half of Christ's "week" of confirming the covenant with many, through his two witnesses (Dan 9:26-27, & see above).

- It is the 3 ½ years that the Church (Rev 3:14-22) is tried (Rev 3:18-19; Dan 11:33-35; 12:7, 10; Mal 3:2-3, see previous qualifications).

- The end of the seventy weeks of years of Daniel will come on as a flood, with war (Dan 9:26).

- "Then, in the end [of the 70 weeks; at the end of the 3 ½ years], what has been decreed concerning the desolation will be poured out" (Dan 9:27, *NEB*).

What is poured out?

- the vials of Revelation 16

- the "wine of wrath" or "cup of wine" (Rev 14:10; 16:19; 18:6, 8; Jer 51:7-8; 25:15-17)

- the angel's censer (Rev 8:5)

pr339» Thus, God's "wrath" is poured out at the end of the age. This word is translated "consummation" in Daniel 9:27 in the KJV and "end" in the NEB, but in Hebrew it means, *full* end. Therefore, at the *full* end of the 3 ½ years the wine of wrath is poured out. Then the Beast will be destroyed while the Church will be saved (Dan 12:1), and the rulership of God will commence (Rev 11:15).

pr340» This is also confirmed by Jeremiah's seventy years, and the seventy weeks of Daniel 9. At the end of these antitypical seventy years, Cyrus (Christ) will physically return and be King of all nations.

pr341» Thus, Christ can't return physically until after the 3 ½ years. The Church is not "born" of God until after the 3 ½ years. The Beast isn't destroyed until after the 3 ½ years. The cup of wrath (vials) is not poured out until after the 3 ½ years.

pr342» Yet Revelation 11:15 says after the seventh trumpet (which is, in the word flow of Revelation, before the vials of Revelation 16), Christ will come and take over the nations, and the Church will be "born" of God. But as we've shown above these vials are poured out after the 3 ½ years, and at that time the Church is to become the kingdom of God, and at that time Christ is to return physically. Is this another Biblical contradiction? No!

Wrath All at Once

pr343» Notice that in the higher or antitypical meaning of the following scriptures that the wrath is to happen all at once:

- as an overrunning flood [Nah 1:8; Dan 9:26]

- speedy riddance [Zeph 1:18]

- in one day [Isa 10:17; 47:9; Rev 18:8]

- at an instant, suddenly [Isa 29:5; 1 Thes 5:3; Isa 47:11; Eccl 9:12; Jer 51:8; Psa 73:19]

- destroy and devour at once [Isa 42:14]

- as in a moment [Psa 73:19]

- avenge them speedily [Luke 18:8]

- as a whirlwind [Psa 58:9-10]

- in one hour [Rev 18:10, 19]

- swiftly, and speedily [Joel 3:4]

Then all nations will be destroyed *together*:

- the transgressors are destroyed together [Psa 37:38]

pr344» Because of Babylon's sin against the Church (Israel), then in or "at Babylon [the antitypical Babylon] shall fall the slain of all the earth," thus, of all the nations (Jer 51:49, 48).

No War Or Evil Will Be Around After The Kingdom of God Takes Over:

pr345»

- they shall not make war anymore [Isa 2:4; Mic 4:3]

- violence no more [Isa 60:18]

- affliction not to rise a second time after the end of the wrath [Nah 1:9]

- No one shall see evil anymore [Zeph 3:15]

- hail (from the super weapons — Rev 16:21) to sweep away the lies [Isa 28:17]

Summarize

pr346» Thus, by putting these verses above together, the seals, trumpets, and vials must:

- happen all at once, in an instant;

- no war or evil will be thereafter;

- all destroyed together, that is, evil and the nations' power will all be destroyed together;

- all this will happen *after* the 3 ½ years;

- Christ will physically return after the 3 ½ years to *save* mankind from his own wrath against himself (Matt 24:22; Luke 9:56);

- the Church will be born after the 3 ½ years;

- and the kingdom of God takes over after the 3 ½ years.

pr347» The seals, trumpets, and vials picture events that happen all at once, in a very short period of time. They are not sequential events. They merely amplify with different words and events this greatest of great points in time. This is the reason the book of Revelation is filled with the *aorist* verb (and other timeless verbs). This verb is one of action, not of time. The aorist verb and other timeless verbs cannot be translated correctly into English, thus the confusion.

Similarities of the Seals, Trumpets, and Vials

pr348» Now let's show the similarities between the seals, trumpets, and vials as well as show you the meaning of the symbolism of these events. After this, in PR6, we will list many groups of verses so you can help further to confirm for yourself what we have been putting forward in this paper.

Seals

First Seal

pr349» The first seal was opened and John heard, "as it were the noise of thunder" (Rev 6:1). As we've shown before God's "voice" sounds like, or is metonymical for, thunder, many waters, and trumpets. Thus, another meaning of the noise of thunder is the sound of a trumpet. What John heard was the first trumpet when the seal was opened. Remember John was "in Spirit on the Lord's day." John was transfigured to the day of the Lord. This is spiritual Lord's day which exists for 1000 years (see "Thousand Years And Beyond" paper [NM 15]).

pr350» Then in verse two John sees a white horse with one on it with a bow, "and a crown was given unto him: and he went forth conquering, and to conquer."

Four Horses

pr351» These four horses mentioned in the sixth chapter of Revelation are the same horses as in Zechariah 1:8 and 6:1-8. Notice these four horses are commissioned to go "to and fro through the world" (Zech 1:10-11; 6:5, 7). And these four horses are called four spirits or winds of heaven, which go forth from standing before the Lord of all the earth (Zech 6:5).

pr352» Now in the book of Job it speaks about Satan coming at certain times and standing before the LORD (Job 1:6; 2:1). Notice that Satan is like the four horses or spirits, he goes to and fro through the earth (Job 1:7). Satan's job, one might say, is to go to and fro through the world trying mankind. Now what is the meaning of these four horses of Revelation and Zechariah?

pr353» Notice the four horses are associated with the "Beasts" in Revelation chapter 6. These four "Beasts" should be translated "living creatures." These are the same four living creatures as the ones described in Ezekiel 1:5-28; 10:1-22. These living creatures describe a part of the throne of God. This throne of God describes the total characteristics and powers of God as the Beast of Revelation describes the total powers (rulership) of Satan (see the *God Papers*, GP9). But if this is so, what part of God's power do the living creatures signify?

pr354» They signify the four Beasts of Daniel 7, for again in Daniel "Beast" should be translated "living creature." These living creatures of God's throne are the living creatures ("Beasts" — KJV) of the seventh chapter of Daniel. These living creatures are the world-ruling kingdoms of the earth. And these kingdoms are a part of God's power (throne) because all kingdoms were given their predestinated power by God (John 19:10-11; Dan 4:32-37; 2:20-21; etc.).

pr355» *God's Power Over the World's Kingdoms.* It is God who has the overall power over the world's kingdoms. Notice that God cut Nebuchadnezzar off as a leader of Babylon, "to the intent that the living may know that the most High rules in the kingdom of men, and gives it to whomsoever he will, and sets up over it the basest of men" (Dan 4:17). And again, "by me [God] kings reign, and princes decree justice. By me princes rule, and nobles, even all the judges of the earth" (Prov 8:15-16).

pr356» God is the overall head ruler of the world's nations in that all power comes from God (Dan 4:25; Jer 27:5). But the true God (the Becoming-One) himself does not do any wrong, for the true God *is* love (1 John 4:8).

[*Remember* here that all the works were done from the beginning (Heb 4:3). The creation is like a wound up clock ready to ring at the appointed time. God through knowing ahead of time the outcome has planned it in such a way so as to best form within man the true knowledge of right and wrong. So in a sense, since God planned it this way he has the authority over the parts therein. Yet he himself has not done any harm, but Satan — the opposite force of the True God — has been doing the destruction. What God did was create a situation whereby, through cause and effect, he was able to determine the outcome. Since he created man's mind with its limits, and created the degree of effect for each cause, and created the parts of the creation (man, spirit, earth, etc.); he knows the outcome as

does one who builds a clock and then sets it to ring at a certain time (see the *God Papers*).]

pr357» Now we showed you how the four horses go "to and fro through the earth," (Zech 6:7) just as Satan goes "to and fro" through the earth (Job 1:6-7). And we pointed out that the four horses are associated with the four living creatures ("Beasts" — KJV) in Revelation 6, which are the same living creatures as the ones described in Ezekiel, chapters 1 and 10. Yet we know that the four living creatures ("Beast") of Daniel 7 are the successive rulership of Satan's kingdom since Satan is at this time the ruler of the world (under God who has allowed Satan's rulership for a higher purpose). Therefore Satan goes "to and fro" through the earth trying it by the medium of the world's kingdoms which are the four living creatures ("Beasts") of Daniel 7. And since the four horses are associated with the four living creatures, and also go throughout the world, then the four horses are just another metonymical name for the four living creatures.

pr358» The four horses are like the four living creatures ("Beasts") of Daniel 7 which are the four living creatures of Ezekiel 1 and 10 and the four living creatures of Revelation. These four horses, which indicate the four Beasts of Daniel 7, will at the end of the age be the final Beast pictured in Revelation 13:1-2.

pr359» *Four Winds; Four Horses.* These same four horses, which are called the four spirits (winds) of heaven (Zech 6:2-5), are also indicated by the four horns of Zechariah 1:18-21 which are the "horns of the Gentiles" (v. 21).

pr360» *Four Horns of Brazen Altar.* Now "horns" in the Bible are symbolic of kingdoms (see Dan 7:24). These are the four world ruling kingdoms of Daniel 7. These four kingdoms are also indicated by the four horns on the Brazen altar of Moses' tabernacle. The fire and sacrifices of this altar indicate two things:

- the trial and sacrifices the whole world has been going through because of these four horns (kingdoms of Satan);

- the lake of fire, by which these four Beasts will be destroyed, thus fulfilling God's righteous judgment (Psa 9:15-16).

pr361» Notice the angel takes a censer full of fire off the altar and casts it to the earth (Rev 8:5). This pictures the lake of fire being poured on the earth through the atomic Last War. But let's get back to the first seal.

First Seal

pr362» Who is riding on this first horse, the horse being symbolic of Satan's kingdom? Notice he has *A* crown, and he went to conquer. It is Christ with his golden crown (Rev 14:14) on his white horse, and his making war in RIGHTEOUSNESS (Rev 19:11).

pr363» Christ is making war in righteousness, as explained before in PR4, by letting the wicked destroy themselves. Christ is pictured on top of this horse because all power and authority is his (Matt 28:18); he has the power over this horse (kingdom); he allows man to begin to destroy himself, yet he will save man from man's own madness. The first seal, thus, pictures Christ allowing the horse to begin to make war.

Second Seal

pr364» Now the second seal shows us how the war is fought "they should kill one another." As shown previously mankind will make war on themselves. They will go mad at the end and begin to commit cosmocide.

Third Seal

pr365» The third seal pictures judgment — the scales or balances. Also the third vial pictures judgment (Rev 16:4-7). This is judgment of nations (Joel 3:12). This is the righteous judgment of God on Babylon (see, Rev 14:7; 18:10). This seal also pictures the famine of Babylon which comes upon it in one day, in one hour, or one moment of time (Rev 18:8, 10).

Fourth Seal

pr366» The fourth seal pictures Death (satan) and the destruction of one-fourth of the earth. This indicates the scope of the destruction by the Last War. It is over one-fourth of the earth's surface. All the destruction was caused by the sword (war), hunger, and death.

Fifth Seal

pr367» The fifth seal is reflective and qualitative. It pictures all those saints killed under the power of the horses (Satan's kingdoms). Daniel 7:21, 24; 11:33-35; and Daniel 12:7, 10 picture the same thing. They are told to rest a while until the full number of the saints are killed.

[*Remember* this is the fifth seal. The whole instantaneous wrath is amplified into seven parts of three sets of descriptions — the seals, trumpets, and vials; yet the Bible also describes events leading up to this time, and events happening after it (Rev 1:19). The end of the wrath isn't until the action of the last (7th) seal, or trumpet, or vial is completed.]

Sixth Seal

pr368» Now the sixth seal shows the sun becoming black. This is described elsewhere in the Bible, but it speaks of it as the sun and moon being darkened (Isa 13:10), and Ezekiel 32:7 says why it will be darkened: "and when I put you [Pharaoh — a type of Satan] out, I will cover the heaven, and make the stars thereof dark; [*how*?] I will cover the sun with a cloud." The same clouds are pictured in the fifth trumpet as the smoke that came out of the bottomless pit (caused by the bombs) that made the air and sun dark (Rev 9:2).

pr369» Also in the sixth seal it speaks about the stars falling on the earth. Now if any of the stars ever fell on the earth, the Earth would blow up. This is symbolic. God tells us to look to the higher meaning (Col 3:1-2). Stars are representative of angels (Rev 1:20). This merely speaks about the one-third of the total angels who belong to Satan being cast into the bottomless pit at the Messiah's return (Rev 12:4; 20:1-3; Isa 34:4; 14:12, 15; Ezek 28:7-8, 17; 31:16).

pr370» Verse 14 speaks of *every* mountain and island being moved out of their places. Now this can't be because only one-fourth of the earth will be affected by the Last War (Rev 6:8). There is a higher meaning here. Now mountains are symbolic of nations or kingdoms (Rev 17:9-10). Although the earth will be greatly shaken by this Last War (Isa 24:19-21), the higher sense of this verse tells us every kingdom of this world will be put down; then the rulership of God will take over (Rev 11:15).

pr371» Verses 15 and 16 show people hiding themselves in caves and so forth. This pictures the great fear surrounding this day of wrath (Isa 2:10, 19, 21; Ezek 32:10). One reason for this fear is that about three days before this the two witnesses were killed. For 3 ½ years these two have been teaching what the world was about to do to itself — commit cosmocide. They have been telling them that the New Age or the kingdom of God is coming. Revelation 11 pictures the world as if it were relieved that these prophets died. The world will probably hope what they were saying was wrong, yet subconsciously (the other-mind) they will know that these prophets are right. The spirit in man (the other-mind) knows it will at some time be tormented (Matt 8:29; Mark 5:7). But also they think they will be *destroyed* in the lake of fire (Mark 1:24; Job 15:22).

pr372» Man, and the enemy spirit in man, as the hours tick off after the prophets are killed will remember their last words that the Messiah will come within 3 ½ days after they are killed (Rev 11:9-12). The spirit in man knows it will be tormented once the Messiah returns. This is the reason for the great fear as the hours tick away toward Christ's physical return. This is why people go into caves to hide from the Lord. Even though the two witnesses will tell the world that Christ will come to save or free mankind, they will not believe it, for they have the spirit of fear in them (2 Tim 1:7; Rom 8:15). Not only was Christ's first stay on this earth marred, but his return will be even more so, "his visage was so marred more than any man" (Isa 52:14). People now think God will come with fire and damnation, and this idea won't change much over the next few years.

Seventh Seal

pr373» Now when the seventh seal is opened, "there was silence in heaven about the space of half an hour" (Rev 8:1). This "half an hour" is translated from a Greek word. It merely indicates a very short period of time. This is also pictured in Revelation 7:1-3. As the Last War begins, in the middle of it God will stop it for a short moment and seal his elect, by writing his NAME on their foreheads (Rev 14:1). They at that moment are born of God in the middle of the lake of fire caused by the super weapons. Now let's explain the trumpets and vials.

Trumpets

pr374» Before the first trumpet is blown we see an angel with a golden censer filled with fire out of the altar. This altar has four horns indicating the four Beasts of Daniel 7 or the four horses of Revelation and Zechariah. On this altar a continuous fire was always going (Lev 6:13). This pictures the continuous fire or trial of mankind caused by Satan's spiritual kingdom — the other-mind's power. A censer is nothing but a cup. This golden censer can be looked upon as the golden cup that is poured upon Babylon (Rev 14:7-10). Notice the golden censer of fire is

poured out on the earth (Rev 8:5). Anything poured out of a cup or censer falls all at once. This pictures the fire of the lake of fire falling on the earth all at once. When the bombs go off there will be "thunderings, lightning, and an earthquake" as Revelation 8:5 shows and Revelation 11:19; 16:18, 20-21; Isa 29:6. The hail stones are merely caused by the unbelievable Last War.

pr375» Notice that Peter says on this day that, "the heaven being on fire shall be dissolved, and the elements shall melt with fervent heat" (2 Pet 3:12, 10). God has reserved a cause and effect law for this great Last War. If the elements of the sky are burning as Peter tells us on that day, then there must be great heat generated. There must be a certain temperature when reached that will cause the sky to be set on fire. In Revelation 16:21 it says the hail stones weigh as much as a talent, which is approximately 100 pounds. The great heat will cause matter to explode, "have you seen the treasures of the hail, which I have reserved against the time of trouble, against the day of battle and war?" (Job 38:22-23) No we have never seen it, but Revelation 16:21 describes it as does Ezekiel 38:22.

First Trumpet

pr376» Now notice what happened after the trumpet is blown, "and there followed hail and fire mingled with blood" (Rev 8:7). This hail is the same hail we just spoke about, and the fire causes the hail, and the fire will come from man's own weapons for God's wrath is man's wrath against himself as we've shown you in PR4.

pr377» Now what happens when one is in a lake of fire caused by man's wrath against himself? "Their flesh shall consume away while they stand upon their feet, and their eyes shall consume away in their holes, and their tongue shall consume away in their mouth" (Zech 14:12). "And their blood shall be poured out as dust, and their flesh as the dung" (Zeph 1:17). Some have accused the true God of doing such acts, but we see herein how these things will happen.

pr378» Further in Revelation 8:7 we see where one-third of the trees are to be burnt up and the grass also. Now trees are symbolic of kingdoms as well as of the people in these kingdoms (Isa 10:18-19). This is merely a symbolical way of describing the fact that one-third of those living before the day of wrath will die on that day (Rev 9:15, 18). Notice, "by these three was the third part of man killed, [1] by the fire, [2] and by the smoke, [3] and by the brimstone" (Rev 9:18).

pr379» Revelation 8:7 also says all the grass was burnt up. Thus this means that all the grass around where this war will be located will burn up (over one-fourth of the earth, Rev 6:8), and since grass is symbolic to people (Isa 40:6), then all the people in this area of the earth will burn up except those supernaturally saved (Matt 24:22).

Second Trumpet

pr380» Now the second trumpet sounds and we see a great mountain burning that is cast into the sea. Read Jeremiah 51:24-25 where it identifies this mountain as Babylon. Since we know it doesn't mean the old Babylon but the Babylon described in Revelation (Satan's kingdom), then we know Satan's kingdom will be destroyed by burning and then at the same time thrown into the sea. The "sea," is used interchangeably in the Old Testament with the pit — bottomless pit (see Ezek 27:32; 28:8). Actually the bottomless pit or lake of fire of Revelation 20:1-3, 10 is an antitypical event of what happened to the Pharaoh and his troops — the Red Sea buried them (Ex 14:27). The lake of fire will bury the antitypical Egypt, Babylon, and the Pharaoh.

pr381» Next we see a third part of the sea becomes blood (Rev 8:8-9). The higher meaning here again reiterates that one-third of all the angels are being destroyed (see next section).

Third Trumpet

pr382» Then when the third trumpet sounds we see a star falling to the earth. The higher meaning here pictures an angel falling to the earth, for stars are symbolic of angels (Rev 1:20). This angel is identified as Satan the Dragon who brings down one-third of the total angels (stars) to the earth (Rev 12:3-4, 9, 12). This is also pictured in Revelation 6:13 and in Isaiah 14:12ff.

pr383» Notice that this star called Wormwood fell on a third part of the rivers and that third part became wormwood. What does that mean? What is the higher or antitypical meaning? Moving or running water is symbolic to spirit (John 7:38-39). Thus, the star or angel called Wormwood (Satan) fell on a third of the rivers of water (spirit) and they became wormwood; they became Satan's angels (Rev

12:4). "And many men died of the waters [spirits of Satan] because they were made bitter" (v. 11).

pr384» Thus, now we know what the following verse means, "I will feed them, even this people, with wormwood [Satan], and give them water [spirit] of gall to drink" (Jer 9:15; 23:15). And now we know by putting Revelation 8:10-11 and Jeremiah 9:14 together what Peter meant when he said to Simon of Samaria, "for I perceive that you art in the gall of bitterness," that is, he was bitter with the Wormwood (Satan) that was inside him misleading him.

Fourth Trumpet

pr385» Then the fourth angel sounded the trumpet, and one-third of the sun, moon, and stars were smitten, and made dark. What is the higher meaning here? Now darkness is symbolic to Satan as light is symbolic to God (1 John 1:5) since Satan and God are opposite qualities as darkness is opposite to light (see Col 1:13). Thus, one-third of the sun and moon and stars were darkened by Satan. Now we're aware that one-third of the stars (angels) were darkened by Satan's way, but what does it mean when it says the sun and moon were darkened?

pr386» If you have read the paper on the symbolic meaning of the sun and moon you know they are symbolic of God and Jesus Christ the man. Now we know if a third of the sun (God) is darkened then we know a third of the moon will also be darkened, for the moon merely reflects the sun's light. Now at the beginning the stars (angels) were created by God. Since angels are made of spirit and God is spirit, the angels were thus made out of the material of God. One-third of the darkened angels were at one time a part of God's Spirit. And if you read the *God Papers* you will see that God doesn't consider himself complete until the end of creation when *all* things are gathered into Christ. This includes all the angels. Now since the sun is symbolic of Christ, what is meant by it being one-third darkened, is that, one-third of God (the completed God) is darkened. Not until these darkened angels are made light again will the Body of Christ be completed. In other words, the fourth trumpet merely reiterates the fact that one-third of the angels are of Satan, and through the knowledge about the sun's symbolic meaning, we know God considers one-third of his potential is also darkened.

Fifth Trumpet

pr387» Then the fifth trumpet sounds at this point (Rev 9:1), and we see a star fall to earth with the key of the bottomless pit. Stars are symbolic to angels, thus, the higher meaning here means that an angel came to earth with the key. Now this same event is shown in Revelation 20:1-3. This bottomless pit elsewhere in the Bible is called hell (cf Ezek 32:24, 27). The bottomless pit is hell, the grave. Who has the key to the bottomless pit, to hell? "I am he that lives, and was dead; and, behold, I am alive into the ages of ages, and have the keys of hell and of death" (Rev 1:18). God has the keys to the pit — to hell. We have already shown you that the seven angels are merely manifestations of Christ, so the angel with the key in Revelation 9:1 and 20:1 is Christ.

pr388» Revelation 9:2 shows the pit and smoke coming out. Now we know that God will not initiate the lake of fire, but those nations who will be fighting among themselves will initiate the fire. And since the weapons that could cause the heavens to burn and the elements to burn (2 Pet 3:10, 12) are atomic in nature, this Last War will begin and end with atomic weapons and other weapons flying everywhere. The deeds of this war will fall on man's own head (Obad 1:15). Yet this war is begun by the spiritual madness of Satan who knows his time is short: "Woe to the inhabiters of the earth and of the sea! for the devil is come down unto you, having great wrath, because he knows that he has but a short time" (Rev 12:12).

pr389» Satan and his angels through their influence in the minds of mankind will cause man to go mad at the end of this age. The smoke coming out of the pit (Rev 9:2) is the smoke from the aftermath of the atomic weapons. This is where the clouds of smoke come from that cover the stars (Isa 13:10; Ezek 32:7).

pr390» "The heathen are sunk down in the pit that *they* made ... Hell from beneath is moved for you [Satan, Babylon, Lucifer, see context] to meet you at your coming ... All they [nations] shall speak and say unto you [Satan], Art you also become weak as we? art you become like unto us? Your pomp is brought down to the grave, and the noise of your vials: the worm is spread under you, and the worms cover you. How art you fallen from heaven, O Lucifer, son of the morning ['darkness' — from Strong's # 7837 & 7835]! How art you cut down to the ground, which did weaken the nations!" (Psa 9:15; Isa 14:9-12) This

pictures Satan after he is cast down to the earth (Rev 12:12), after his rulership is taken away. Remember all these things happen at once: the Last War (Matt 24:7; Dan 9:26-27), Satan's power being taken away (Rev 20:1-3), and the kingdom of God taking over (Rev 11:15-18).

pr391» "And out of the smoke came locusts *into* the earth [pit]" (Rev 9:3). The correct translation shows locusts passing away into the earth. Now these locusts are identified in Nahum 3:15-18 as the troops of the king of Assyria. This same king is the one described in Isaiah 10:5-19. This king prefigured the leader of the Beast of Revelation 13, and is also the false-prophet. This king is also a shadow of Satan himself (2 Thes 2:9). Thus, not only do these locusts indicate the troops of this Assyrian king, but also the troops of what this king represents on earth — Satan's troops, his angels. These locusts at once picture the Assyrian troops and Satan's angels. They are destroyed by the effects of the Last War. One effect of the Last War being the smoke.

pr392» Now Revelation 9:5 should be translated, "and to them [the locusts] it was given that they should not kill them [man], but that they [man] will be tormented five months: and their torment as the torment of a scorpion, when he strikes a man." This pictures man being tormented by the symbolic locusts (Satan's angels) for five months *before* the Last War. This can't be after the Last War because there will be no evil after this instant war. Man will be mentally tormented for five months before God returns. But, of course, since the garden of Eden man has been tormented by Satan (see "Oher Mind" paper [NM 21]).

pr393» Verse six shows mankind seeking death, but afraid to kill himself. Why? The spirit of man that misleads mankind is a spirit of fear. Even though these spirits are making men miserable, men will be afraid to take their own lives, for they don't know that there is no hell-fire for themselves. There is a hell-fire for the spiritual evil angels, but man because of these confused spirits in their minds think that they themselves will be tormented in this fire (see the "Thousand Years and Beyond" paper [NM 15]).

pr394» Notice the crowns of these locusts' heads in Revelation 9:7. Compare this with Nahum 3:17.

pr395» Notice the proof that these locusts are symbolic of Satan's angels: "and they had a king over them, which is the angel of the bottomless pit, whose name in Hebrew is Abbaddon, but in the Greek tongue has his name

Appollyon." These are two of the many names the Bible uses to describe Satan. Appollyon means destroyer. Satan is the destroyer of the world. His way is of destruction (Isa 14:12, 20).

Sixth Trumpet

pr396» Then comes the sixth trumpet (Rev 9:13-21). Now in verse 14 it speaks of four angels. Who or what are these four angels? Notice in Zechariah 6:5 it identifies four horses as four spirits or winds since the Hebrew word means both spirit and wind. These four winds or spirits are shown in Revelation 7:1, "the four winds of the earth." Here, they are being held back until the saints are sealed. These four winds are the four horses, which are the four Beasts, which are the kingdom of Satan. Thus, in Revelation 7:1 it pictures the kingdom of Satan (the four winds or spirits or horses or Beasts) being held back from destroying in the Last War until the saints are sealed. Yet since angels are spirits (Heb 1:7), we know the four angels of Revelation 9:14 are symbolic of the four spirits (winds) of Zechariah 6 and Revelation 7, or the four horses of Revelation 6. Therefore these four angels picture, in a way, what is being said in Revelation 7:1. These four angels of Satan's rulership are being held back from the end-of-the-age's wrath until the appointed time of wrath or judgment (Acts 17:31).

pr397» Now the Beasts of Daniel and/or Revelation specifically describe the kingdom of Satan that has or had rulership in or around Jerusalem. **Yet the whole world is a part of Satan's kingdom.**

pr398» Notice what Christ prophesies about Satan's kingdom at the end: "every kingdom divided against itself is brought to desolation; and a house divided against a house falls. If Satan also be divided against himself, how shall his kingdom stand?" (Luke 11:17, 18)

pr399» Next we see the four angels (the horses of Rev 6) being let loose, and a great army is allowed to cross the Euphrates river. This army was prepared for this battle for over a year (v. 15). It is by and with this army that a third of mankind and angelkind in the Last War will be destroyed or as good as destroyed. Who are these nations?

pr400» Ezekiel 38 and 39 picture these same nations as does Jeremiah 50 and 51, and Isaiah 13:6-22. These are the nations of the north, the Medes, Ma-gog, Meshech, etc. Somewhere today these nations or peoples exist. Along with these and other nations and peoples, are the ones who

come from the east and north towards Jerusalem (Dan 11:44). These nations are a part of Satan's rulership. (In Biblical symbolism, the "east" represents the future — the direction of the Sun's coming day light; and the "north" represents the spiritual *left*, the spiritual evil side. See Hebrew words for "east" and "north" in the Lexicon.)

pr401» Notice what Ezekiel 38:10-11 says about these nations, "thus says the Lord GOD; It shall also come to pass, that at the same time [the 'latter years,' v. 8] shall things come into your mind [the mind of these national leaders], and you shall think an evil thought: and you shall say, I will go up to the *land* of unwalled villages [the Middle East, see v. 8]" (see also Jer 51:28) They having been prepared to attack in a certain year, month, day, and hour (Rev 9:15). **In a Spiritual sense, the spiritual Beast is in the minds of Spiritual Jerusalem as the other-mind. See *Other Mind Paper.* [NM 21]**

pr402» This great army of the north (Ezek 38:15) will not only attack the *land* of the Middle East, but also the people of God's Church who will be within the strike of the army (Ezek 38:16; Mic 4:11; Jer 6:23; Isa 54:15). **Since in the Spiritual sense the land of Spiritual Israel is the whole earth [*Seed Paper*], then God's Church need not be located in the Middle East at the very End.** Actually this is a main reason this war will happen. Remember Satan's spiritual influence rules this world (Rev 13:2). But Satan will be committing cosmocide when he tries to stop the Church from being born. Revelation 12:2-4 pictures Satan trying to destroy the child that is about to be born. Isaiah 66:6-9 helps to identify this child as the born Church.

pr403» Thus, there are two reasons these nations will come against this land at the end:

- to spoil the rich Babylon (Jer 50:10)

- to try and destroy the Church at the same time (Rev 12:2-4; Isa 54:15).

pr404» Revelation 9:18 shows the destruction caused by the sides of Satan's kingdom coming against each other. Verse 20 says those not killed in these plagues did not repent or change their minds. But remember this is only the sixth trumpet, there is still one more. All these trumpets, seals, and vials happen at once. But after the completion of these, the spirit of man, the other-mind, will be locked up in the pit so that they will not mislead man again (Rev 20:1-4; Zech 13:2). These spirits in man's mind are what causes man to hate God's way of harmony even though God's way is the way of peace. Thus, after these spirits are taken out of man's mind, one knows "the kingdom of God is come upon you" (Luke 11:20).

pr405» Follow these verses to see exactly what will happen *after* the Last War: Revelation 20:1-3; Micah 7:16-17; Isaiah 52:14-15; Zephaniah 3:11, 15; Isaiah 60:2-4, 18; Isaiah 59:19; Isaiah 24:13-15; 17:7.

pr406» Chapter 10 to 11:13 of Revelation are inset or parenthetical chapters (see "Two Witnesses Paper" [PR8]).

Seventh Trumpet

pr407» The seventh trumpet (rev 11:15) shows the Church being born of God, and the kingdom of God taking over rulership (note, 1 Cor 15:52-55; 1 Thes 4:16-17; Rom 10:6-7). Notice at the sounding of the seventh trumpet the door of the temple is opened which, as we explain in the notes to this paper, is the door into the finished Spiritual temple (Rev 11:19). At that same time the earthquakes, thunder, and hail go off due to the atomic Last War.

pr408» Again we repeat, all the seals, trumpets, and vials happen all at once. They merely amplify this moment of time and tell a few events that lead up to them, like the northern army preparing to make war for a year (Rev 9:15).

Vials

pr409» Now let's quickly cover the vials of chapter 16. But instead of going over again what we have already put forth, let us just say the vials picture the damage done by the Last War — the blood, fire, etc. These verses picture evil spirits gathering the nations to fight this last battle. They are gathered at the "Armageddon" which was prefigured in the land of the Middle East many years ago. But the real Armageddon battle will be the final sacrifice:

- "And it shall come to pass in the day of the LORD's sacrifice, that I will punish the princes, and the king's [Satan's] children, and all such as are clothed with strange apparel. In the same day also will I punish all those that leap on the threshold, which fill their masters houses with violence and deceit" (Zeph 1:8-9).

pr410» This pictures the last great sacrifice on the brazen altar. The lake of fire is the antitypical meaning of the fire that always burned on this altar (Ex 38:1-5; Lev 6:13). The gathering of Satan's kingdoms (horns) towards Jerusalem is the antitypical meaning of the four horns around this brazen altar. This final sacrifice of the LORD is how the antitypical daily sacrifice is taken away. But as shown in PR4, it is actually Satan's sacrifice, for *his* ways cause trials and destruction. This is the *last* sacrifice for man at the hands of Satan's kingdom. The four horns of the altar represent the four kingdoms of Satan. The four kingdoms are described in Daniel 7.

What Is Important

pr411» It isn't really that important to know exactly where each army will be, or if such and such a nation mentioned in the Bible is modern day such and such. Why? It is because all nations will fight or be a casualty in the last war, so there is no need to identify them. All we need to know is that the seals, trumpets, and vials are the same event, an end-of-the-age war — the Last War. Much of Revelation describes and qualifies this Last War, and the events leading up to it, and the events right after it.

pr412» The important thing to know is that at the time of that war one-third of the people are killed (afterward one-third of the angels are locked up), one-fourth of the earth is destroyed, and Christ comes at that point to *save* the world (Matt 24:22). The Last War happens 1260 days *after* the Beast-man takes control of the Beast system (see "Beast-Man Paper" [PR2] and the "End of the Age" paper [PR7]). God's wrath is man's wrath on their *own* system.

Notes for PR5

Church In Symbolism

pr413» Now notice Revelation 15:8, where it says, "no man was able to enter into the temple, till the seven plagues of the seven angels were fulfilled." But we quoted before that the Church was the temple (Eph 2:19-22). What is Revelation 15:8 saying then? If there is a Church now on the earth, then men are already of, or in, the temple of God, yet Revelation 15:8 says no man can enter the Temple until the last plagues which are described in the 16th chapter of Revelation.

pr414» To clear this up we need to know that the Church is made up of Spiritually *begotten* people. At the last trumpet then they will be *born* of God (1 Cor 15:52-55). The Church can be looked upon as a people in the womb of its mother (the Church) who are growing into a born child of God. The Church is now in the process of being built, only when it is born will it be complete. Notice Isaiah 66:6-9 where it pictures a pregnant woman just before birth, and then giving birth. It indicates in verse eight that a whole nation was to be born at once. This nation is called the "holy nation" by Peter (1 Pet 2:9) after he called this same nation a "spiritual house" (v. 5). And he indicates in context of these verses that this nation, this spiritual house, is the Church. Thus, the Church is the nation to be born at once, and is now a begotten people in its mother's womb.

pr415» Notice that Paul calls "Jerusalem which is above ... the mother of us all" (Gal 4:26). What is this Jerusalem which is above? It is the "heavenly Jerusalem" which is the "church of the first-born" (Heb 12:22-23). And this "heavenly Jerusalem," or as Galatians calls it, the Jerusalem which is above (in heaven, thus the heavenly Jerusalem), is also described in Revelation 21:2, "New Jerusalem, coming down from God *out of heaven*, prepared as a bride adorned for her husband."

pr416» This pictures the heavenly Jerusalem coming down from heaven; it also calls this city the *bride*, or in verse nine the "Lamb's wife." Ephesians 5:22-33 says that wives or women are symbolic to the Church. Thus, here is another proof that the Church is a mother — the wife of the Lamb (Christ). But further, we see that it is the New Jerusalem, the temple, the house of God, the Lamb's wife, and the mother of us all. In other words, there are many names in the Bible that describe the same Church.

pr417» This same mother Church or wife Church is described in Revelation 12:1-2. Here it pictures it about to deliver its child. As we noted in Isaiah 66:6-9 this "child" of Revelation is the Church. Isaiah 66:7 calls this child a "man child." This same man child is shown in Revelation 12:5 (see also Isa 26:17-20). This "man child," pictures Christ's Spiritual Body (the Church) in its antitypical meaning. Thus again we see the begotten Church as a child in a woman's womb growing into a born child.

pr418» Yet we have shown God's temple as being symbolic to the Church. But note in Ephesians 2:21 that the Church is in the process of growing into a "holy temple." From the English translation in the *Interlinear Greek — English New Testament* (Zondervan Pub) verse 22 reads, "in whom also you are being built together for a habitation of God in Spirit." Also in this same translation it says the Church members "are being built up a house spiritual, a priesthood holy" (1 Pet 2:5). In other words, the temple (Church) is in the process of being built. It is not yet built, it is like a baby in a womb, it is growing into the finished product, the born Church.

pr419» Hence, the temple of Revelation 15:8 is the true temple (the *born* Church). This is the reward that Christ is to bring back to the earth, the Church (or more correctly the first-products of the Church). No person can enter this Spiritual temple, in the truest sense, until the plagues are poured out.

pr420» Thus, what Revelation 15:8 is saying is that no one will be born of God until after the last plagues. This is proof that God's kingdom does not take over until after the last plagues. Yet Revelation 11:15 says that after the seventh trumpet, the kingdom of God will take over rulership. And 1 Corinthians 15:52 shows after the seventh trumpet ("last trumpet"), the begotten Church will become the born Church. Of course, this is because the seven seals, trumpets, and plagues of Revelation are the same events. These are merely metonymical terms for the same events. And the description given is an amplification of one instant of time, for the verses we've shown to you previously prove the Last War happens all at once. Revelation merely amplifies that one point in time as well as gives information of the events up to that time, and after that Last War.

PR6: God's Wrath: An Outline

Outline Review

pr421» **(1) There is one-half of a week (of years) remaining of Daniel's seventy weeks, which equals three and one-half years as explained previously in PR5.**

During these 3 ½ years:

■ The physical Beast will rule a great part of the world; The spiritual Beast rules all the earth. (Rev 13:5; Dan 7:25; Rev 12:9)

■ The Church will be in the spiritual wilderness. (Rev 12:6, 14)

■ The Church will be in a spiritual trial or tribulation. (Dan 12:7, 10: 11:33-35; 7:25; Rev 12:17; 3:18; 11:2; see also the "Oher Mind" paper [NM 21])

■ The two witnesses will teach during their 1260 days. (Rev 11:3, 4; Zech 4:11-14)

■ It will be a period of great tribulation for the whole world. (Isa 13:4-11; Matt 24:31; etc.)

■ During this time the great false-prophet will say he is God (Rev 13:11-18; 2 Thes 2:3-9; Dan 7:25; Dan 8:11, 23-25; Dan 11:37).

pr422» **(2) Then at the *full* end of the 3 ½ years will come the Last War:**

■ "and the end thereof with a flood, and unto the end of the war that makes desolation is determined" (Dan 9:26).

■ "and until the end war to cut off the arrangement of destruction" (Dan 9:26; trans. from the Septuagint).

> (A proof that this war comes at the *full* end of the last 3 ½ years comes from Daniel 9:27 in the word translated "consummation" in the KJV, which was translated from a Hebrew word meaning — **Full** or **Complete** end. Read last part of Daniel 9:26 & 27. These verses are speaking of the same time.)

pr423» **(3) The nations will be gathered at the end:**

■ "All the nations of the earth be gathered together against it." (They will be gathered against Jerusalem, the physical and Spiritual one.) [Zech 12:3; Joel 3:2, 11, 9; Isaiah 13:4; 66:18;

> Zeph 2:1; 3:8; Mic 4:11-12; Matt 25:31-32; Rev 19:19]

■ The modern day Gog, Tubal, Persia, Ethiopia, Phut, Gomer, Togarmah, and so forth will gather against Jerusalem. [Ezek 38:7-8, 15-16; Rev 20:8; 16:12; Rev 16:14-16; Jer 1:14; 50:9, 41; Jer 51:28; Isa 13:17; Jer 50:29; Jer 51:11; Dan 11:44]

■ This is the gathering of the nations for the great last sacrifice — Jehovah's sacrifice. [Zeph 1:7-9; Ezek 39:17; Rev 19:17-18]

■ Gathering together the tares that are to be burnt. [Matt 13:30, 40, 42; Isa 10:17; Isa 27:4]

■ Gathering together of God's wheat; but the chaff of the wheat is burnt up. [Matt 3:12; Luke 3:17]

■ The gathering of fish to cast into the furnace (fish=people, Matt 4:19; Hab 1:14). [Matt 13:47, 50]

■ The gathering together of the guests for the wedding — both the good and the bad guests. [Matt 22:10]

■ The nations will gather like a flood. [Isaiah 59:19; 17:12-13]

pr424» **(4) Then after the gathering of the nations, the Last War will begin and the nations will fight against each other; people against people, city against city:**

■ "nation shall rise against nation." (Matt 24:7)

■ "violence in the land, ruler against ruler." (Jer 51:46)

■ "evil shall go forth from nation to nation" (Jer 25:32).

■ Concerning spiritual Egypt (see Rev 11:8): "the Egyptians against the Egyptians: and they shall fight every one against his brother, and every one against his neighbor; city against city, and kingdom against kingdom" (Isa 19:2).

■ "Every man's sword shall be against his brother" (Ezek 38:21).

■ "I will overthrow the throne of kingdoms ... and their riders shall come down, every one by the sword of his brother" (Haggai 2:22).

■ "and they shall lay hold every one on the hand of his neighbor, and his hand shall rise up against the hand of his neighbor" (Zech 14:13).

■ "By the swords of the mighty will I cause your multitude to fall" (Ezek 32:12).

■ The beast will destroy itself for the antitypical "whore" is the "woman" of Revelation 17, which is the great city (v. 18), which is the system of Babylon (Rev 14:8; 18:16), which is the antitypical Babylon (See Beast Papers [PR2 PR3], Rev 17:16-17).

pr425» (5) Why will these nations fight against each other? Isn't this madness?

Wouldn't it be the end of mankind if all nations came against each other, especially with the modern weapons of mankind? Then the world must go mad at the end of the present age if these things are to happen.

■ "In that day, says the LORD, I will smite every horse with astonishment, and his rider with MADNESS" (Zech 12:4).

■ "Babylon has been a golden cup in the LORD's hand, that made all the earth drunken: the nations have drunken of her *wine*; therefore the nations are MAD" (Jer 51:7; see also Jer 25:15-16).

■ "Babylon is fallen ... because she made all nations drink of the wine of the wrath of her fornication" (Rev 14:8; see also, Rev 16:19; 18:3). The higher meaning here is that the spiritual Babylon, that system of spiritual evil that lives in all mankind, has all the people drunk on its spiritual wine of confusion and evil.

pr426» What is the "Wine" of Wrath?

■ "Stay yourselves ... and cry: they are drunken, but not on wine ... For the LORD has poured out upon you the *spirit of deep sleep*, and has closed your eyes" (Isaiah 29:9-10).

■ It is "the *spirit* of the kings of the Medes" that will be raised up against modern day Babylon for the "vengeance of his temple." (Jer 51:11)

■ This is the opposite wine or spirit that God's people are or will drink — the new wine. (Zech 9:17; Luke 5:37)

■ The spirit of the Adversary will make the world go mad at the Last War. "For the devil is come down upon you, having great wrath, because he knows that he has but a short time" (Rev 12:12; see also, Isa 14:17, 12).

pr427» (6) How long will the Last War last?

Take the antitypical meaning of the following items for the answer:

■ One day [Isa 10:17; Zech 3:9 Rev 18:8]

■ one hour [Rev 18:10, 19]

■ as an overrunning flood [Nah 1:8 & Dan 9:26]

■ speedily or as a speedy riddance [Luke 18:8; Zeph 1:18]

■ Babylon suddenly falls [Jer 51:8]

■ suddenly [Isa 47:11; Eccl 9:12]

■ at an instant, suddenly [Isa 29:5-6; 1 Thes 5:3; Psa 73:19]

■ destroyed & devoured at once [Isa 42:14]

■ swiftly, speedily [Joel 3:4; Mal 3:5]

■ as in a moment [Psalms 73:19]

pr428» (7) Who will be destroyed by the Last War?

■ The transgressors are destroyed *together*; The real transgressors are the spiritual evil minds. (Psa 37:38; Isa 1:28; see "Oher Mind" paper [NM 21])

■ In spiritual Babylon, "shall fall the slain of *all* the earth" (Jer 51:48-49). Remember all the nations will be gathered against each other at the Last War.

■ At the same time Gog comes against "the *land* of Israel, says the Lord GOD, my fury shall come up in my face. For in my jealously and the fire of my wrath have I spoken, Surely in that day there shall be a great shaking [through the weapons of mankind] in the land of Israel" (Ezek 38: 18-19; cf Zeph 3:8). But all the earth belongs to Spiritual Israel.. ("Seed Paper" [PR1]) Gog thus comes against *all* the land of Spiritual Jerusalem — the whole earth.

- "And the slain of the LORD shall be at that *day* from one end of the earth even unto the other end of the earth" (Jer 25:33).

pr429» **(8) Therefore all nations will have gathered, and one-third of all the transgressors will be destroyed at once** (see PR5).

Here follows is a list of some of the various nations to be destroyed as powers, as well as in population. For example only one-sixth of the modern Ma Gog, Meshech and Tubal will be saved after the Last War (Ezek 39:1-2). The Bible uses the original name of the families that grew into nations, but today these families or nations described in the Bible have different names, yet they are the same peoples the Bible is prophesying against.

- List of nations to be destroyed as powers and in population: [Zeph 2:9; Jer 25:17-29; Ezek 30:4-5; 32:22, 24, 26, 30, 31-32; Ezek 31:18, 14; 39:11; Joel 3:2, 12, 14; Isa 14:12, 21-22, 25; Isa 30:30-33]

Thus, all nations will be gathered and all destroyed as powers along with the invisible power behind these nations — the spiritual adversary, Satan. It will be an instantaneous Last War destruction.

pr430» **(9) Now if all nations fight against each other, and the war happens in an instant, then how else can it happen besides it being an Atomic Last War?** Atomic weapons will be used in this war along with other such destructive weapons.

pr431» **(10) When the nations come at once against the Beast he will panic** (See, Jer 50:41-43; 51:28-29). **AND:**

- "But tidings out of the East and out of the north shall trouble him: therefore he shall go forth with great fury to destroy, and utterly to sweep away many" (Dan 11:44).

But in this great madness the Beast will destroy itself (Rev 17:16-17). All these nations will destroy themselves. This is God's righteous end-of-the-age judgment on the nations of this age (see, Psa 9:15-16):

- "For the day of the LORD is near upon all heathen: as you have done, it shall be done unto you: Your reward shall return upon your own head." (Obadiah 1:15, see Joel 3:4)

pr432» **(11) But there will be another reason besides the nations mad plan (Ezek 38:10-12; Isa 13:17) initiated by the spiritual mind of wrath in their minds. There is the spiritual reason:**

- It is Satan's wrath against God's people (Rev 12:12, 2, 4).

pr433» Not only are the nations gathered against the antitypical Babylon (the Beast), but the Bible emphatically says the nations are also gathered against the Church or the people of God. This is the spiritual reason the mad spirits of Satan gather the nations to fight (Rev 16:12-16; Jer 51:11; Isa 19:2, 3, 14; and Rev 20:8-9):

- Read Jeremiah 6:22-29. Note that the "daughter of Zion" is the Church (Heb 12:22).

- Read Revelation 19:19; 17:14; 20:9. Note respectively in each verse: "his army" with verse 14 and 8; "and they that are with him" and "the camp of the saints."

- God in Isaiah 29:8 describes what will become of the dream of the nations (the subconscious dream of the satanic spirits in their minds) "that fight against mount Zion." Again, the higher meaning for Zion is God's Church (Heb 12:22).

- The nations will gather against Jerusalem, and will be destroyed "in that day." (Zech 12:2, 2, 4, 8, 9) Those of the city of Jerusalem will be "as God" for then they will be God's sons and daughters. Remember the higher meaning for Jerusalem is God's Church or people (Heb 12:22 and the section on "New Jerusalem" in [NM 18]).

- This is the "day of trouble" for God's people (Hab 3:16; Psa 102:2, 13-17; Dan 12:1; Isa 33:2; Isa 26:16, 17-20; Jer 30:6-8; remember the higher meaning of Israel is God's Church, Rev 7:4; Gal 6:16).

- But, the Lord "shall defend them" and Zion (God's Church) will be cheerful with its new wine. (Zech 9:9 Zech 9:14-17; see also Isa 33:3; 34:8; Isa 31:4-5)

pr434» Here are other verses that show the nations gathering against the Church:

- Read Isaiah 54:15, notice that the barren woman is the barren woman that Paul speaks about in Gal 4:26-27.

- Read Ezekiel 38:16 and compare it with Revelation 20:9. Remember the higher meaning of "my people Israel" is the saints or the Church, Rev 20:9; Rev 7:4; Gal 6:16.

- Read Micah 4:11, 13; Isaiah 10:32. Remember "Zion" is the Church. See Hebrews 12:22.

- Read Isaiah 37:3, 11-12 and notice the talk by the Assyrian king against the fact that God's Church will be Born of God. This Assyrian king is representative of the end-of-the-age's false-prophet.

pr435» Notice God through his word asks the rhetorical question:

- "We [the Church, Rev 12:2] have been with child, we have been in pain, we have as it were brought forth wind; we have not worked any deliverance in the earth" (Isa 26:18).

- "Shall I bring to the birth, and not cause to bring forth: says the LORD" (Isa 66:9).

pr436» Notice the Church's reaction to the blasphemy of the false-prophet (Isa 37:3, 11-12) against the fact that the Church will bring forth. Read Isaiah 37:22-23 and you will see Faith in action:

- KJV Isaiah 37:22 This *is* the word which the LORD hath spoken concerning him; The virgin, the daughter of Zion, hath despised thee, *and* laughed thee to scorn; the daughter of Jerusalem hath shaken her head at thee. 23 Whom hast thou reproached and blasphemed? and against whom hast thou exalted *thy* voice, and lifted up thine eyes on high? *even* against the Holy One of Israel.

pr437» God through his word answers the negative talk by the false-prophet and His own rhetorical question of Isaiah 66:9:

- God will bring forth his people, they will be born of God all at once. That is, the "first fruits" will be born all at once; the rest will be born of God later (Isa 45:8; 66:8; Rev 11:15 with 1 Cor 15:52-55 & 1 Thes 4:16-17; etc.).

pr438» Notice just how close God's Church is to being destroyed:

- The "children of Israel" (the Church) "were *ready* to perish in the land of Assyria." Note the context — "in that day"

and "trumpet shall be blown" (Isa 27:13). The land of Assyria was the land of physical Israel's enemy at that time. The antitypical "land of Assyria" for the antitypical Israel is the land of their enemy (Satan) — the whole earth.

- "The sinner of Zion [one-half of those at the end of this age who think they are of the true physical organized Church] are afraid; fearfulness has surprised the hypocrites. WHO AMONG US SHALL DWELL WITH THE DEVOURING FIRE?" (Isa 33:14; note Psa 1:5)

- "Fear, and the pit, and the snare, are upon you, O inhabitant of the earth. And it shall come to pass, that *he who flees from the noise of fear shall fall into the pit.*" (Isa 24:17-18; note Luke 17:31-33; & 2 Chron 20:13-17)

pr439» In other words, just as the Church is ready to be physically destroyed by the flames of the Atomic Last War, God will defend it and will save and free it. This is salvation. They will be born of God. They will thus not be hurt by the fire of the Last War. But those among them who *say* they were in the Church will fear and thus begin to run. Yet the war begins and ends speedily and they will die in the lake of fire. This is what Paul was physically speaking about when he said:

- "Every man's work shall be made manifest: for the day shall declare it, because it shall try every man's work of what sort it is. If any man's work abide which he has built thereupon, he shall receive a reward. If any man's work shall be burned, he shall suffer loss: but he himself shall be saved, yet so as by fire" (1 Cor 4:13-15). This last part speaks of the agelasting fire for the satanic angels which is their fire baptism, and the agelasting death baptism for the humans who die in this fire. (see, "Thousand Years" paper [NM 15])

pr440» Because the Last War happens at once and the atomic fire at once, those in the Church killed by this fire will not know it, for:

- "In a *moment*, in the twinkling of an eye, at the last trump: for the trumpet shall sound, and the dead shall be raised incorruptible [immortal], and shall be changed ... Death is swallowed up in victory" (1 Cor 15:52, 54).

pr441» The Church will die by the fire for only as long as an instant (see Isa 54:8; 10:24-25), then they will become infused to the Spiritual and will be lifted into the clouds (1 Thes 4:16-17) to meet Christ to bring him down to earth (Rom 10:6) to rule on the earth (Rev 5:10).

Notice how in Matthew it describes this:

- "For nation shall rise against nation ... All these are the beginning of travail" (Matt 24:6-7).

pr442» This pictures the beginning of the instant Atomic Last War. It is the beginning of the Church's travail. Isaiah finishes the picture:

- "for as soon as Zion travailed she brought forth her children" (Isa 66:8). YET:

- "*Before* she travailed, she brought forth; before her *pain* came, she was delivered of a man child" (Isa 66:7).

pr443» Thus, at the very beginning of the travail, which is the Last War (Matt 24:6-7), just before the pain or hurt, but *as* the fire is destroying her, she is born of God. Remember that Atomic weapons go off in a flash; all this happens in an *instant* (note Psa 58:9-10).

pr444» This brings memories back as to what happened when the physical Israel was escaping Egypt. Just as the Pharaoh's troops were about to reach the people of Israel, the water of the Red Sea destroyed the Pharaoh's troops and saved Israel (Ex 14:10, 13-14, 21-23, 27-30). Notice Amos 9:5; 8:8; Isaiah 17:13-14; Daniel 9:26.

pr445» **(12) And from this Last War the earth shall be moved with a GREAT earthquake:**

- Jeremiah 10:10; 49:21: 50:46; Isaiah 13:13; 29:6; Ezekiel 31:16; 38:19, 20; Joel 2:10; 3:16; Haggai 2:21 Zech 14:5 Revelation 16:18; 11:19, 13; 8:5; 6:14; especially see Isaiah 24:18, 20

pr446» And there are many scriptures that show the destruction of the Last War, called the lake of fire. We'll only give a few more:

- 2 Peter 3:10, 12; Zeph 1:18; 3:8; Mal 4:1-2; Isaiah 34:3; 29:6; Zech 14:12; Ezek 38:22

pr447» **(13) At the very moment of the Last War, God's kingdom or Spiritual rulership will take over (Rev 11:15-19).** And after this instant war, and the taking over of the world by the rulership of God, then evil will *not* rise a second time:

- I will smite my hands together and I will cause my fury to rest. [Ezek 21:17]

- they will not learn war again [Isa 2:4; Mic 4:3]

- violence no more [Isaiah 60:18]

- affliction not to rise a second time [Nah 1:9]

- we shall see no more evil [Zeph 3:15]

- the hail will sweep away the lies [Isaiah 28:17]

- they shall *not* hurt nor destroy in *all* God's holy mountain or kingdom [Isaiah 11:9]

- no oppressor shall pass through them any more [Zech 9:8]

- God's kingdom and peace will not depart or be removed [Isa 54:10]

PR7: End of the Age

If we won't know the Day, Why Watch?

As in Noah's Day

End on a Holy Feast Day

Feasts of Israel Pre-Figured

Which Year?

End of the Old Age; Beginning of the New Age

pr448» The disciples came to Christ in private and asked, "tell us, when shall these things be [the destruction of Temple]? and what shall be the sign of your coming, and of the end of the world [*age*]?" (Matt 24:3) The disciples asked Christ when the age would end, and when his return would be. Christ then gave them some signs of his coming (Matt 24:4-44).

Father Only Knows the Date?

pr449» But notice with these signs, Christ said, "but of that day and hour [of his return] knows no man, no, not the angels of heaven, but my Father only" (Matt 24:36). And again he said, "therefore be you also ready: for in such an hour as you think not the Son of man comes" (Matt 24:44).

We Will Know the Date

pr450» From the two just quoted verses and other verses people conclude that no one will know the date of Christ's return. But this overlooks many other verses that tell us we will know the date of His return. Those who say we do not know, or will never know the date of His return are overlooking scripture that proves we (meaning the Church) will know the date. **But the Church will only come to know the date for sure some 3 ½ years before the date.**

pr451» We'll show several reasons why Matthew 24:36 and 24:44 are taken out of context with other scriptures that say we will know the date of the end of the age.

Principles of Biblical Study

pr452» First we must know a very important principle on ascertaining the doctrines of the Bible: "Whom shall he teach knowledge? And whom shall he make to understand doctrine? Them that are weaned from the milk, and drawn from the breasts. For precept must be upon precept, precept upon precept; line upon line, line upon line; here a little, and there a little" (Isa 28:9-10).

pr453» To understand doctrine one must study all the details of the doctrine found throughout the Bible. You must take a line here, and a line there from the whole Bible and put it together in order to understand it. But shall one just take any line here, and any line there, to ascertain or figure out doctrine? No, they should take the lines pertaining to each doctrine. But further they should examine *all* the scriptures on any one subject before putting the verses together. And the only way to do this is to study the whole Bible. If we want to know what the Bible says about the nature of God, we must study every scripture pertaining to God. If we want to know what the Bible says about the return of Jesus Christ, we must study every scripture pertaining to His return.

Christ Given All The Power

pr454» In order to understand the mistaken notion that we will not know the date, we want you to note the following fact:

- When Christ spoke about only his Father knowing the date (Matt 24:36), he was still a man not yet resurrected to God. If you understand who Christ is now, you will understand how significant this fact is in understanding Christ's words (see *God Papers*). But even if one does not understand who Christ is now, or does not want to believe it; he will have a hard time discounting a statement made by Christ after he was resurrected to God: "All power is given unto me [Christ] in heaven and in earth" (Matt 28:18; see Luke 10:21-22).

pr455» Christ has been given *all* the power of his Father. If Christ has all the Spiritual power of his Father, then he must also know the date.

Christians to Receive Power

pr456» Notice in Acts 1:7-8, Christ was speaking to the apostles after his resurrection to the Father, "And he said unto them, It is not for you to know the times or the seasons, which the Father has put in his own power." But Christ was given all his Father's power (Matt 28:18), therefore Christ has the power to know the times and the seasons — to know when the end of the age is. To continue in Acts 1:7-8: "But you shall receive power after that the Holy Spirit is come upon you." On the Pentecost the power was given to the disciples to know (Acts 2), for at that time they received the Holy Spirit. Therefore the apostles could know the date after they received the power of the Spirit.

Christians to Receive Knowledge

pr457» Notice, "I have yet many things to say unto you, but you cannot bear them now. Howbeit when this, the Spirit of truth, is come, it will guide you into *all* truth ... it will show you things to come. This [the Spirit of truth] shall glorify me: for it shall receive of mine, and show it unto you. All things that the Father has are mine: Therefore said I, that it [the Spirit of truth] shall take of mine, and show it unto you" (John 16:12-15).

pr458» These verses clearly say that the Spirit will show the apostles all that is Christ's. The Father had the power to know the times (Acts 1:7), but Christ was given all the power of the Father (Matt 28:18), for "all things that the Father has are mine" said Christ (John 16:15). Through the Spirit and scripture Christ will show us about the things to come (John 16:13).

Church Knows Hidden Secrets

pr459» *Notice the scriptures about the Church knowing the hidden secrets.* The Church will know because the Spirit shall reveal it through the scriptures:

- "For nothing is secret, that shall not be made manifest; neither any thing hid, that shall not be known and come abroad" (Luke 8:17).

- "In that hour Jesus rejoiced in Spirit, and said, I thank you, O Father, Lord of heaven and earth, that you have hid these things from the wise and prudent, and have revealed them unto babes: even so, Father; for so it seemed good in your sight. ALL THINGS are delivered to me of my Father: and no man knows who the Son is, but the Father; and who the Father is, but the Son, and he to whom the Son will reveal Him" (Luke 10:21-22).

- "Fear them not therefore: for there is nothing covered, that shall not be revealed; and hid, that shall not be known" (Matt 10:26).

- "For there is nothing hid, which shall not be manifested; neither was any thing kept secret, but that it should come abroad. If any man have ears to hear, let him hear" (Mark 4:22-23). This indicates that the secrets will be known.

- "Behold, the former things are come to pass, and new things do I declare: *before* they spring forth I tell you of them" (Isa 42:9).

- "I have much to say to you, more than you can now bear. But when he, the Spirit of truth, comes, he will guide you into all truth. He will not speak of his own; he will speak only what he hears, and he will tell you what is yet to come. He will bring glory to me by taking from what is mine and making it known to you. All that belongs to the Father is mine. That is why I said the Spirit will take from what is mine and make it known to you" (John 16:12-15, NIV).

- "But the Counselor, the Holy Spirit, whom the Father will send in my NAME, *will teach you all things* and remind you of everything I have said to you" (John 14:26, NIV; see 1 John 2:27). The verb translated in the English as, "I have said," is an aorist verb, which is a verb of action, not of time. Thus, this Spirit will remind ("will put into the mind" — Greek future verb) Christians of what Christ says (or said, or will say). As John 16:12-15 indicates, Christ did not at that time give all the truth to them, but said that in the future the Spirit of truth would lead them into all the truth. In the book of Acts it shows Christ revealing new truth to them (Acts 15:7 cf 10:28 & 10:1-33 & 11:7ff; Acts chap 15 — the teaching that physical circumcision was not needed; etc.). And the one who comes in the Spirit of Elijah will eventually lead the Church into all the truth (see "Two Witnesses" paper [PR8]).

Spirit Reveals Hidden Wisdom and Knowledge

Spirit Reveals

pr460» The secrets will be revealed through the Spirit (John 16:12-15). And, "but we [of the Church] speak the wisdom of God in a mystery, even hidden wisdom, which God ordained before the world unto our glory ... But God has revealed them [the hidden wisdom] unto us by his Spirit: for the Spirit searches all things, yea, the deep things of God ... Which things also we speak, not in words which man's wisdom teaches, but which the Holy Spirit teaches; comparing spiritual things with spiritual" (1 Cor 2:7, 10, 13).

Times known

pr461» One of the things that the Spirit will teach is the time of the end of the age. In Acts 1:7 it speaks of the power to know "the times and the seasons" and this power the Father has. But then in the very next verse (Acts 1:8) it says, "you shall receive power." They were to receive the power of the Father — His Spirit. It is through this power of the Spirit that ALL THINGS would be revealed because what the Father had was given to the Son (Matt 28:18; John 16:15; etc.). And the Son was to show these things to his servants (John 16:15, 13; Matt 11:27; Luke 10:22; 8:17).

pr462» The New Testament Church did know the typical end of the age: "but of the times and the seasons, brethren, you have no need that I write unto you" (1 Thes 5:1). Why? "And that, knowing the time" (Rom 13:11). They knew the times. This doesn't mean they knew the date of Christ's return, although they could have known it even then. More than likely they knew that this was the evil age and that the new age was fast approaching. But further they knew that there was an antitypical rest or Sabbath (Heb 4:1-10). And they knew God's days are counted as 1,000 years (2 Pet 3:8). Also they knew the typical sabbath rest was on the seventh day of the week. Thus they could have easily have figured that the antitypical rest or sabbath was the seventh 1,000 year "day." And through the chronology of the Bible they could see that Christ appeared near the 4,000th year of man. Thus, they could have figured up that the date of Christ's return was still about 2,000 years off (see *Chronology Papers*).

Daniel's Vision

pr463» Notice what Daniel was told about the prophecies given him, "but you, O Daniel, shut up the words, and seal the book, even to the time of the end" (Dan 12:4).

pr464» Daniel's prophecy is locked up until the time of the end of the age of evil. Thus at the time of the end of the age is when the appointed time of the end of the age will be revealed. All secrets are to be made known before they happen per the verses we have already quoted. And at the time of the end is when Daniel's book will be opened completely, for at the time of the end is when *all things* shall be restored (Mal 4:5; Amos 3:7; Matt 17:11; see "Two Witnesses" paper [PR8]).

Mistakes of Others' Interpretation

pr465» We will return to Daniel in a moment after we go over a couple of parables. These parables are put forth by some when they attempt to prove that we will not know the date of Christ's physical return. We will show that these parables are taken out of context.

pr466» First before we go over these parables note the following facts:

- What Christ said in Matthew 24 is also written about in Mark and Luke.

- The parable he spoke in Matt 24:44-51 was uttered by Christ *before* he repeated it to the disciples on the occasion mentioned in Matthew 24. (Luke 12:40-48) Thus at the time it was repeated to the disciples, the disciples had already heard it. We will explain why this is important in a moment.

- There is evidence in studying Matthew, Mark, Luke, and John that Christ repeated the same parables or teaching almost everywhere he went. When an account of any parable is given by one writer, it may be retold only in part. **We must compare each parable or teaching with all accounts of the same parable in order to understand the full extent of the words Christ used to explain these parables.** For example, Matthew in writing about a parable may have left out a certain sentence that Christ spoke pertaining to the parable, while Luke in his account may mention this missing sentence. Therefore we must put both accounts of Matthew and Luke together in order to understand what Christ said. If we don't do this we, in effect, are taking Christ's words out of context.

pr467» Now let's look at what is written in Matthew 24:37-51. It is very important that we comprehend what we are about to study. These verses, if taken alone, are out of context with what Christ said. Christ spoke these parables on many occasions, like any minister today repeats and repeats things, so Christ in his ministry repeated his parables.

pr468» Further in any one book on the sayings of Christ, each author gives the story from a different view point, and at times leaves out important sayings of Christ. But each thing each author (Matt, Mark, Luke, and John) says is true, yet to find out all the things Christ spoke concerning a parable, one must compare all the books that mention the parable to get the full information. We need the full information because we don't want to take Christ's words out of context.

If we won't know what day the Lord comes back, why watch?

pr469» This is the Basic question of which we will now answer. First lets look at Matthew 24:37-51:

- Mat 24:37 As it was in the days of Noah, so it will be at the coming of the Son of Man. [38] For in the days before the flood, people were eating and drinking, marrying and giving in marriage, up to the day Noah entered the ark; [39] and they knew nothing about what would happen until the flood came and took them all away. That is how it will be at the coming of the Son of Man. [40] Two men will be in the field; one will be taken and the other left. [41] Two women will be grinding with a hand mill; one will be taken and the other left. [42] "Therefore keep watch, because you do not know on what day your Lord will come. [43] But understand this: If the owner of the house had known at what time of night the thief was coming, he would have kept watch and would not have let his house be broken into. [44] So you also must be ready, because the Son of Man will come at an hour when you do not expect him. [45] "Who then is the faithful and wise servant, whom the master has put in charge of the servants in his household to give them their food at the proper time? [46] It will be good for that servant whose master finds him doing so when he returns. [47] I tell you the truth, he will put him in charge of all his possessions. [48] But suppose that servant is wicked and says to himself, 'My master is staying away a long time,' [49] and he then begins to beat his fellow servants and to eat and drink with drunkards. [50] The master of that servant will come on a day when he does not expect him and at an hour he is not aware of. [51] He will cut him to pieces and assign him a place with the hypocrites, where there will be weeping and gnashing of teeth.

Correct the Translation

pr470» Matthew 24:37-51 basically shows us that as in the days of Noah, so will be the days before the return of Christ — people will be engrossed in the every day things of life and will not be looking and noticing the times (notice Mark 13:33-34; Luke 21:34-35). Then in verse 42 Christ said, "*watch therefore because you do not know on what day your Lord will come.*" Let's translate this verse directly from the Greek text, using the Greek Lexicon published by Zondervan, and Thayer's Lexicon:

- The word "watch" is in the second person plural. Therefore should have been translated, "You be watchful," or "you be awake," or "you be attentive."

- The word translated "because" in the NIV or "for" in the KJV can just as well be translated, "that." In fact in the *New International Version* (NIV) of the Bible, out of 1298 times this Greek word (*hoti*) was translated, it was translated "that" more often than any other English word: 492 times translated "that"; 205 times translated "because"; and 145 times translated "for" (*The Niv Exhaustive Concordance*, Greek word #4022, p. 1766).

- The word translated "hour" in the KJV should have been translated "day" and is translated "day" in the NIV. See footnote *f* for Matt 24:42 in the *Interlinear Greek New Testament* by G.R. Berry, Published by Zondervan in 1969.

Therefore a correct translation of Matthew 24:42 from the Greek is:

- "you be watchful therefore: that you do not know what day your Lord will come."

As in Noah's Day so at the Coming of the Lord

pr471» Christ was warning that in the time of Noah before the flood people behaved as always until the flood took them away, and so will it be at the end of the age: "and they knew nothing about what would happen until the flood came and took them all away. That is how it will be at the coming of the Son of Man" (24:39). He said at His return it would be the same with people not recognizing the times. He warns them not to be like those in Noah's day: "you be watchful therefore: that *you* do not know what day your Lord will come."

pr472» It was because Noah *knew* what time the flood would come that he prepared the ark. Christ was merely telling them to be careful so that they won't be like those of the flood who did not prepare themselves for the flood like Noah did. *He wasn't saying that all would not know the time of his coming, but he said that they should not be in the position of __not__ knowing the time of his coming because then they would be like those dying from the flood.*

pr473» After this Christ went on and told about how happy will be the servant who is doing His work when He comes, and the future weeping of the unfaithful one or ones who are not doing God's work, but eating and drinking. The parallel here is that Noah was the faithful servant who knew the time of the flood and prepared the ark, but those who did not know, but kept on as always, were destroyed by the flood.

Wicked Servant Will Not Know the Time

pr474» Christ wasn't saying His true servants would not know the date, but that the unfaithful ones, or one, would "say *in his heart*, My lord delays his coming" (Matt 24:48). Inside his mind, the unfaithful servant will not believe the date or time of Christ's coming like those who did not believe Noah's warnings of the flood. "The master of that servant [wicked or unfaithful one] will come on a day when he [the wicked one] does not expect him [Lord] and at an hour he is not aware of" (Mat 24:50).

pr475» Matthew 24:44 should be correctly translated as, "and because of this [V. 43] you be ready: that in such an hour as you suppose not, the Son of man will come." In other words, be careful that *you* don't suppose wrongly about the time of His coming. Why? You will be like those who were destroyed by the flood, if you do not know the time.

Christians Not Asleep

pr476» Now we will show you an important verse that qualifies these sayings of Christ. If you look in Robertson's, *Harmony of the Gospels*, you see that Christ said some more things after he spoke what was recorded in Matthew 24:42. After Christ warned them about being watchful for his return, Christ adds, "lest coming suddenly he [Christ] find you *sleeping*" (Mark 13:36).

Christians not Asleep

pr477» But notice Paul's words about true Christians:

- "But of the times and seasons, brethren, you have no need that I write unto you. For yourselves know perfectly that the day of the Lord so comes as a thief in the night. For when they [non-Christians] shall say, Peace and safety, then sudden destruction comes upon them [like the flood came on those who did not believe Noah], as travail upon a woman with child; and they shall not escape [like those of the flood did not escape]. **But you, brethren, are not in darkness, that the day should overtake you as a thief**. You are all the children of light, and the children of the day: we are not of the night, nor of darkness. Therefore let us not sleep, as do others; but let us watch and be sober" (1 Thes 5:1-6).

pr478» If we are of the day, if we are of the children of light, if we are faithful servants; *then* we will not sleep and be drunk with the world's way. We will know the times so that we will not be destroyed like those of the flood.

All Must Watch

pr479» After Christ said be watchful "lest coming suddenly he find *you* sleeping," he said, "and what I say unto you I say unto ALL, watch" (Mark 13:37). Christ was not only speaking to his apostles at that time, but he was speaking to *all*, to all mankind. All mankind must be watchful lest they be asleep when Christ physically returns and the flood of it all takes them away to destruction.

pr480» Notice the confirmation on our last statement. Christ in Luke 12:37-40 is again speaking about watching. He was speaking to the disciples. "Then Peter said unto him, Lord, speakest you this parable unto us, or even to *all*?" As Mark 13:37 said, "what I say unto you I say unto *all*, watch."

Thus a close study of what we have put forward proves the true Christians will know the time of the end of the age, but others will not perceive the time.

Times of the Gentiles

pr481» In Luke's account of what Christ told his disciples about the time of the end of the age, Christ said Jerusalem would be "trodden down of the Gentiles, *until the times of the Gentiles be fulfilled*" (Luke 21:24). And Paul said that physical Israel would be blinded to the truth, "*until the fullness of the Gentiles be come in*" (Rom 11:25). This physical blindness of physical Israel also applies to all the other peoples, because they do not have the New Mind.

pr482» What are the times of the Gentiles? In context of Luke's writings, Christ will come right after the fullness of the times of the Gentiles (Luke 21:24-27). In the truest sense of Paul's writing, Israel's blindness will not cease until Christ physically returns. Therefore the times of the Gentiles must be fulfilled before Christ returns, then Christ shall immediately return.

pr483» What are the times of the Gentiles, and when will they be fulfilled? When we answer this question we will know when Christ shall return.

End of the Times on a Feast Day

pr484» In Colossians 2:16-17 we read, "Let no man therefore judge you in food, or in drink, or in respect of a *holyday, or of the new moon, or of the sabbath days: which are a foreshadow of things to come.*" The holydays Paul was speaking of are enumerated in the Old Testament. These holydays are foreshadows of things to come. Notice the pattern of some holydays or ceremonies:

Festivals of Israel Pre-Figured the Real Events

1. Sabbath

Type:	Six days of work; one day of rest (Lev 23:3; Gen 2:1-3)
Anti-type:	Six 1000 year-days, then a Sabbath of rest (2 Pet 3:8; Heb 4:1-10; Rev 20:4; NM 16)

2. Passover

Type:	Passover lamb (Lev 23:5; Exo 12:5-6)
Anti-type:	Christ the real Passover Lamb of God (John 1:29; 1 Cor 5:7)

3. Sheaf of First Fruits

Type:	Sheaf of barley was waved before the BeComingOne to be accepted for the people *after* the weekly Sabbath (Lev 23:10-11; "on the day after the Sabbath" - Lev 23:11)
Anti-type:	Christ ascended to his Father on Sunday morning after the Sabbath (Col 1:15, 18; 1 Cor 15:20, 23; CP 4; NM 16)

4. Pentecost

Type:	Feast of Weeks after Wheat Harvest (Lev 23:15-22)
Anti-type:	Not fulfilled perfectly yet, only in an imperfect way when the New Testament Church was given the Spirit on the first Pentecost after Christ went to his Father at the time of the sheaf of first fruits (Acts 2; NM 16).

pr485» The first, second, third, and fourth ceremonies or holydays we cover in *New Mind Papers*, part 15 and 16 which are mentioned in Leviticus, chapter 23, and other places in the Bible. The second and third ceremonies have all come true antitypically (see "God's Appointed Times" paper [NM 16]). The first holyday was the Sabbath. It will be fulfilled in the seventh 1000 year period (see "God's Appointed Times" paper [NM 16]). The 4th festival is the Feast of Pentecost. This occurred typically after the harvest of first fruits. The next resurrection will be the resurrection of fruits as explained in NM16. The 2nd and 3rd feasts perfectly foreshadows their antitypical fulfillment. *In fact on the very day on which the typical events were celebrated, is when the antitypical event occurred.* For example the foreshadowed Passover lamb sacrifices were performed on the 14th of the first month of the Hebrew's calendar. Christ the real Passover died also on the 14th of the first month. See *Chronology Papers* CP4 for more information on this Passover.

pr486» The next festival that hasn't yet been fulfilled perfectly in an antitypical way is the Feast of Pentecost. The next big prophesied event is the return of Christ and the resurrection of the saints. The foreshadowed festival of the Pentecost happened after the harvest of wheat. Thus, Christ's return, the antitypical event of the Feast of Pentecost, will happen on the day of the Pentecost in the Hebrew's calendar.

pr487» It should also be noted that the Sabbath holyday described in Lev 23:3 has not yet come true in its Spiritual or antitypical sense. As shown in the "God's Appointed Times" paper [NM 16] and the Thousand Years and Beyond" paper [NM 15], the physical Sabbath represents the 1000 year Sabbath.

pr488» Thus, so far we have learned that the DATE of the return of Christ, which is the beginning of the rule of the kingdom of God on earth, which is the end of the old age, which is the beginning of the NEW age, which is the beginning of the 1000 year Sabbath, will occur on the Pentecost.

Which Year?

pr489» We know that the fulfillment of the next feast day, Feast of Pentecost, will happen on the very day the Bible scripture tells us it will happen on. See NM16 for details. But we do not know which *year*.

pr490» There is one way to ascertain the DATE:

- "Concerning the coming of our Lord Jesus Christ and our being gathered to him … that day will not come until the rebellion [apostasy] occurs and the man of lawlessness is revealed, the man doomed to destruction" (2 Thes 2:1-4, NIV).

pr491» The coming of Christ will *not* occur until a rebellion occurs and the manifestation of the man of sin. There is a dual sense here. The antitype of the rebellion is the final war of Satan's angels against God's angels (see "God's Wrath" paper). The type is the apostasy of those who think they are in the Church (see "Great Falling Away" paper [PR10]). In context this man of sin is the mean horn of the Beast (see "Beast-System Paper" [PR2]). The two senses of the man of sin are: (1) his first manifestation at the 1290/1260 day periods; (2) the truest sense is when he declares his godhood (see "Beast-Man Paper" [PR3]). Thus, the coming of Christ will only occur after the apostasy and the manifestation of the Beast person and system. From the "Beast-System Paper" [PR2] we see this Beast system will last for 1260 days after 3 of its ten "kings" are subdued. *Therefore we know the DATE of the coming of the NEW age exactly 1260 days before that date.* We will recognize the 1260th day *before* Christ's return by the events of that day: That is, on that day the ten nation Beast will become a seven nation Beast (see "Beast-System Paper" [PR2]).

pr492» What this means at this time is that we do not yet for sure know the *year*, but we do know the day of the month. We also know through chronology that the seventh 1000 year period is near (see *Chronology Papers*). By using astronomical computer programs we can ascertain with reasonable accuracy future dates when the Pentecost of the Hebrew Calendar will occur. We use the observational calendar, not the modern calculated calendar, because in Christ's time and before His time dates were ascertained primarily by observation and only secondarily by calculation (see *Chronology Papers*). Today, following these premises various dates are possible. One thing that makes this difficult is that the new moon or new month starts when we *see* the first crescent of moon, not how we calculate it now. No matter how far fetched it may sound, the earth and moon through some astro-catastrophe could change their path and thus make our present calculations wrong. We will wait to see the events that will happen on the 1290th and 1260th day before we attempt to speak about the future.

pr493» **We are not setting any date here.** Through scripture we are getting close to ascertaining a date. But only when the 1260th day manifests itself will we know for certain. Any date must be tentative until the 1260th day manifests itself.

pr494» **Note**: Like others who have tried to determine the date of Jesus Christ's return, we have speculated on certain dates, hoping for the early return of our Lord. But even in this speculation we emphasized that it was only tentative, that we must wait for the 1260th day to manifest itself (5-9-97, WRD).

PR8: Two Witnesses

Elijah

Two Olive Trees

Anointed Ones

Who is the Other Witness?

Notes: 30 Years and John

pr495» At the end of the age before Christ's physical return someone shall come and restore all things (Matt 17:10-11). John the Baptist was merely the typical Elijah, but there shall be an antitypical Elijah who shall restore all things. He comes *"before* the great and dreadful day of the LORD" (Mal 4:5). Therefore before he comes the Truth shall not be full in the Church. But when he comes he shall restore all things, yet they will do "unto him whatsoever they listed, as it is written of him" (Mark 9:13). Elijah was rejected in the scripture by physical Israel, the true Elijah will also be rejected before Christ returns by those who will eventually become a part of Spiritual Israel. He will endure and restore the Truth, the Spiritual sacrifice of the Spiritual altar, to Spiritual Israel, which is the Church (1 Kings 18:30 with Heb 13:10; Matt 17:10-11). Spiritual Israel, is the Israel of God, which is the Church (see "Seed Paper" [PR1]).

Elijah will come

pr496» But who are the two witnesses who "shall prophesy a thousand two hundred and threescore days" (Rev 11:3). Although many have identified one of the two witnesses, as he who is to come in the spirit of Elijah, as promised by Malachi 4:5-6, most don't know who the other witness is or, that is, what the second witness will represent Spiritually from the Old Testament. Some believe they will be Elijah and Elisha. Now is the time of the end of the old age, and somewhere on this earth the two witnesses are alive. Herein we will identify *not* the names of the two witnesses, but what Spirit they will or are coming in, that is, what two Old Testament people the two modern day witnesses will fulfill. One of these two witnesses is Elijah, not the Elijah of the Old Testament, but a person who will do the major things that the Elijah of the Old Testament did. This person will be the antitypical Elijah.

pr497» Let's prove that one of the two witnesses is Elijah, and let's identify the other witness, and at the same time we'll break the symbols of Revelation chapter 11.

Two Olive Trees / Two Witnesses

pr498» Notice, "these are the two olive trees, and the two candlesticks standing before the God of the earth" (Rev 11:4). The two witnesses are called the two olive trees.

pr499» "And two olive trees by it, one upon the right side of the bowl, and the other upon the left side thereof ... What are these two olive trees upon the right side of the candlestick and upon the left side thereof ... What be these two olive branches which through the two golden pipes empty the golden oil out of themselves? ... Then said he, These are the two anointed ones, that stand till the Lord of the whole earth" (Zech 4:3, 11, 12, 14).

pr500» These two stand on the right and left of the candlestick. But the Church is the candlestick (Rev 1:20). And the Church is the body of Christ (1 Cor 12:12). Notice that Christ is in the midst of this same symbolic candlestick (Rev 1:13; 2:1). Thus, these two stand on the right and left of Christ, they "stand till the Lord of the whole earth" (Zech 4:14).

pr501» Further they stand "immediately preceding the God of the earth" (Rev 11:4). These two teach for 3 ½ years immediately preceding the return of Christ (Rev 11:3,7,11,12,14,15).

Purpose of the Two

pr502» What exactly is the purpose of having these two witnesses? There are at least two reasons:

- "At the mouth of two witnesses, or three witnesses, shall he that is worthy of death be put to death; but at the mouth of one witness he shall not be put to death" (Deut 17:6). In other words, the way of man/Satan will be destroyed only if there are at least two witnesses to speak out against man's way.

- "And he shall confirm the covenant with many for one week" (Dan 9:27). Here, it speaks about Christ confirming the new covenant for the last prophetic week (seven years) of Daniel's 70 prophetic weeks (See "Last War and God's Wrath" paper [PR5]). But Christ has only taught for 3 ½ years to confirm the covenant, not the full seven years (see *Chronology Papers*). But it is through the two witnesses that Christ will do this. This is the second main reason for the two witnesses — to fulfill the last 3 ½ years of the 7 years that Christ was to confirm the covenant. These two "stand till the Lord of the whole earth," as agents of Christ (Zech 4:14).

Anointed Ones

pr503» Next notice these two are the "anointed ones." But from the Hebrew, it should read — the two sons of the oil. And from the Septuagint it reads — the two sons of the fatness. Now the anointed one is Christ. "Christ" is a translation from a Greek word that means, *anointed*. The two witnesses are the sons of the anointed one through the Spirit in them (Rom 8:16). Yet not only are they sons in this sense, but they are commissioned like sons who take the place of their father while he is away. Christ will be the Father when He returns (Isa 9:6), and these two are His sons who act as His agents while He is away, until He comes.

Power of the Two

pr504» Now "if any man will hurt them, fire proceeds out of their mouth, and devours their enemies" (Rev 11:5). Now this is symbolical, as is much of the book of Revelation. What is the symbolic meaning of fire?

- "the tongue is a fire" (James 3:6)
- God's word is like fire (Jer 23:29)
- God's tongue is a devouring fire (Isa 30:27)
- there is a fire of God's mouth (2 Sam 22)

pr505» Thus, the fire that comes out of these two agents of Christ is God's word. Any one who tries to hurt them will be devoured by the fire of the truth that comes from their mouth.

pr506» Further Revelation 11:5 says that any who harms them will be likewise harmed. These two are eventually killed near the end of the 3 ½ years (3 ½ days before Christ returns) by the Beast of Revelation (Rev 11:7), but the Beast in return is killed (Rev 19:19-20).

pr507» Next they have power to smite the earth with plagues, etc (Rev 11:6). But as Christ did the will of his Father (John 14:31; 12:49; 6:38), so too will these two do *only* the will of God. And the will of the God is manifested in the Bible. The plagues shown in the Bible are what they

have power over. Or more correctly, they will tell the world what is to happen before it happens and they will ask for the will of God to be done (1 John 5:14).

Rain

pr508» Notice that they have "power to shut heaven, that it rain not in the days of their prophecy." Does this mean it will not rain for 3 ½ years? No! God tells us to look to the higher meaning (Phil 3:18-19; Col 3:1-2). Revelation is filled with symbolism. The "rain" in this verse is symbolic of a higher meaning. This event happened typically in the days of Elijah: "Elijah was a man subject to like passions as we are, and he prayed earnestly that it not rain: and it rained not on the earth [land of Israel] by the space of three years six months" (James 5:17). This will happen in its antitypical or Spiritual fulfillment during the last 3 ½ years of man/Satan's misrule. Notice that "rain" is symbolic of God's word (see Isa 55:10-11; Deut 32:2). Christ said that "in the days of Elijah, when the heaven was shut up three years and six months ... great *famine* was throughout all the land" (Luke 4:25). It so happens that the higher meaning of "famine" is the famine "of hearing the words of the LORD" (Amos 8:11). Thus, there will be a famine of God's true word during this 3 ½ years, except for the two witnesses' teachings (Rev 11:6,3). "There shall not be dew nor rain these years, but according to my [Elijah's] word" (1 Kings 17:1).

Elijah is One of the Witnesses

pr509» Since what is to happen when the two witnesses teach (famine of God's word, except for the witnesses' teachings), has happened before typically with Elijah, and since Elijah is to come before the day of the LORD (Mal 4:5); then one of the two witnesses must be Elijah.

Who Is the Other Witness?

pr510» We want you to note that these two stand before the Lord (Zech 4:14), they are on the right and left side of the candlestick, which is the Church, which is the body of Christ. Therefore, they stand on the right and left side of Christ (Zech 4:11; Rev 1:20; 1 Cor 12:12). And also they are the two "anointed ones."

Two at Christ's Ascension

pr511» Now notice Acts 1:9-11, from the New International Version: "After he said this, he was taken up before their very eyes, and a cloud hid him from their sight. They were looking intently up into the sky as he was going, when suddenly two men dressed in white stood beside them. 'Men of Galilee,' they said, 'why do you stand here looking into the sky? This same Jesus, who has been taken from you into heaven, will come back in the same way you have seen him go into heaven.'"

pr512» Who were these two men? These men were in bright-white clothes, and they said Christ would come in the same manner as the people saw him go up into heaven. When Christ went up these two men were there. When he comes again, he will come down when these two men are there. Who were these two men in bright-white clothes?

pr513» "Truly I say unto you, There be some standing here, which shall not taste of death, till they see the Son of man coming in his kingdom" (Matt 16:28). Now the kingdom of God in its truest sense has not yet come, so what did Christ mean by this statement? [Note: The kingdom of God has come in one sense. That is, the Spirit of the kingdom has come into man through Christ.]

Two at Transfiguration

pr514» "And after six days Jesus takes Peter, James, and John his brother, and brings them up into a high mountain apart, and was transfigured before them: and his face did shine as the sun, and his raiment was white [or as bright] as the light. And, behold! there appeared unto them *Moses* and *Elijah* talking with him" (Matt 17:1-3).

pr515» Thus, none of the disciples of Christ saw the kingdom of God coming literally, but they saw it by the means of the transfiguration. But what were they to see? — "the Son of man coming in his kingdom." And what did they see? They saw Christ with Moses and Elijah.

Scripture Together

pr516» What did the two men say when Jesus went into heaven? (Acts 1:9-11) They said he would come in a like manner. That is, he will come with the two men being present. And at the transfiguration of his coming to his kingdom Moses and Elijah were beside him.

pr517» Now notice Revelation 11:12. After the two witnesses were killed (v.7) their bodies laid around for 3 ½ days (v. 9). Then *after* these 3 ½ days they then were resurrected (v. 11), "and they ascended up to heaven in a cloud, and their enemies beheld them" (v. 12).

pr518» This isn't any special resurrection. "For the Lord himself shall descend from heaven with a shout ... and with the trump of God: and the dead in Christ shall rise first" (1 Thes 4:16).

pr519» Revelation 11:11-12 pictures the resurrection of the dead. The witnesses went up into a cloud "and their enemies beheld them." That is, they beheld Christ *and* the two witnesses.

pr520» And how was Christ's coming shown in the transfiguration? He was with *Moses* and *Elijah*.

pr521» Who stands by the Lord? The two olive trees, or the two witnesses stand by the Lord (Zech 4:14, 11).

pr522» Who was mentioned with Elijah in the prophecy of Malachi 4:5? "Remember you the law of *Moses* my servant" (Mal 4:4).

Moses and Elijah

pr523» Through comparing what has been shown you, how could the two witnesses be anyone else other than the two who will come in the antitype of Moses and Elijah? There are no other persons even implied besides Moses and Elijah. The Bible seems to point out the two witnesses as coming in the Spirit or antitype of Moses and Elijah.

pr524» This makes sense, for the typical Moses led the typical Israel out of the typical Egypt. Today, the antitypical Moses will lead the antitypical Israel (the Church, Gal 6:16; Rev 7:4) out of the antitypical Egypt or Satan's kingdom (see Rev 11:8; and "Last War and God's Wrath" paper [PR5]).

pr525» And the antitypical Elijah will come to challenge antitypical Israel (those who *say* they are in the Church) who will have "ministers" who will be between two opinions (Baal's & God's, 1 Kings 18:21-22), and this antitypical Elijah will "slay" these ministers with the sword of God, which is God's word (1 Kings 18:40). Thus, the antitypical Elijah will repair the altar (1 Kings 18:30), and restore all truth (Matt 17:11). This will help fulfill Daniel's word that at the end of the age "knowledge shall be increased" (Dan 12:4). **Note**: We know that the truest sense of the antitypical Israel is that all people ever born will eventually be in the antitypical Israel, the full Church or Body of Christ (see "All Saved" paper [NM 13], etc.). And we know that there are groups of people who *say* they are the Church who may only have a minority with the Spirit. Thus, there are two senses to the antitypical Elijah's challenge: (1), He will challenge all the people of the world at the end of the old age, who are between opinions on who or what is the true God; and (2), in this challenge of all the people he will challenge groups (churches) who say they are in the Church. But the real Church is the remnant of Israel.

pr526» What is another proof that the antitypical Elijah will challenge the *antitypical* Israel at the end of the age and not just the typical Israel?

pr527» First, we are to look to the higher meanings of the Bible (see "Duality Paper" [BP4]). Since we know the antitypical Elijah will come and do similar things that the typical Elijah did, then if the typical Elijah challenged the typical Israel's religion, the antitypical Elijah must challenge the antitypical Israel's religious belief: all religious belief in the world.

Further Proof

pr528» Notice further proof in Revelation, chapters two and three. In these chapters Christ the God speaks not only to each of the seven churches, but also to *all* the churches — that includes the Philadelphia church (Rev 3:7-13). Revelation speaks of the Church and "them that hold the doctrines of Balaam, who taught Balak to cast a stumbling block before the children of Israel, to eat things sacrificed unto idols, and to commit fornication" (Rev 2:14).

pr529» The higher meaning here is spiritual fornication with modern idols and doctrines. This event happened while the typical Moses was leading Israel to the promised land (Num ch 22-23). This again will happen, but this time to the antitypical Israel as Revelation 2:14 says.

pr530» Further proof is in Revelation 2:20. Note "Jezebel" in this verse. Her kind of fornication is what the typical Elijah went up against (1 Kings 18:17; 19:2; etc.). The antitypical Elijah will go against the spiritual fornication of Jezebel. But notice not all of the people who say they are in the Church will have fallen into spiritual fornication (Rev 2:14, 24). God "will fight against them with the sword of" his "mouth" (Rev 2:16). God will fight through his prophet, the antitypical Elijah (cf Rev 2:16 with 1 Kings 18:40; "sword" = God's Word, Eph 6:17).

The Two Witnesses Are Coming In The Spirit Or Antitypical Sense of Moses And Elijah.

pr531» Elijah will come at the end of the age to rebuke the world and those who *say* they are in the Church with God's word! Moses and Elijah will become the two witnesses in the 3 ½ years right before the Messiah returns with His peace and harmony.

pr532» *Elijah* means: "God [Eli] (is) Yehowah [Jah]" or "God (is) He (Who) Will-Be," or "God (is) (the) BeComingOne" since Yehowah or Jehowah means, "He (Who) Will-Be," or the "BeComing-One."

pr533» *Moses* means: "drawn from the water."

Notes for PR8

30 Years

pr534» Now the sons of Kohath had to be 30 years old to do the work of the tabernacle (Num 4:3-4), for they took care of the holy things (Num 4:4; 3:30-31). Antitypically, Jesus Christ who is in charge of the holy things (the Church) was not set aside to *begin* his Spiritual work until he was 30 years old (Luke 3:21-23).

pr535» Now John was born six months before Christ (Luke 1:36, 26-66; see *Unger's Bible Dict.* under "John the Baptist"). John was in the wilderness teaching and preparing the ways before Christ's coming (Luke 1:80, 76-77). John began openly preaching when he was 30 years old (Luke 3:1-2), just like Christ began to preach at about 30 years old. Since John was 6 months older than Christ, then John's work lasted for 6 months before Christ came to him to be baptized. During these 6 months, John preached about the kingdom of God (Matt 3:2). Then Christ took over preaching the kingdom (Mark 1:14-15). Since John's commission was to prepare the people of Israel *before* Christ came, then when Christ came his official commission was over. Therefore since the commission of the typica1 John was *before* Christ came to teach Israel about the kingdom of God, then the antitypica1 John (Elijah) will teach before Jesus Christ's return to set up the kingdom of God.

PR9: Seven Churches of Revelation

Speaks to the End Time

Days at Hand When Book Fully Opened

pr536» As all the Bible is dual so too are chapters 2 and 3 of Revelation. The book of Revelation is "the Revelation of Jesus Christ, which God gave unto him, to show unto his servants things which must shortly come to pass" (Rev 1:1). These things mentioned in the book have not come true in the truest sense. Christ has not returned, Satan has not been locked up, etc. It further says happy is "he that reads, and they that hear the words of this prophecy" (v. 3). This doesn't mean merely hearing it physically, but hearing it Spiritually, for there are people who only hear physically (Mark 4:11-12). Yet there are some who have a Spiritual ear to hear (Rev 2:7; Matt 13:16). Furthermore Revelation 1:3 speaks of those who hear, and at the time that they hear: "the time is at hand." And in Ezekiel 12:22-23 it speaks of the days or time being at hand, "and the effect of every vision." When Revelation was written the true day of the Lord was not at hand. Therefore at the time it was written they didn't understand the book of Revelation in its truest sense. It is only when the days are at hand that the whole book of Revelation shall be opened up fully.

Book to All the Churches in Christ

pr537» Typically John was told to "write a book, and send it unto the seven churches which are in Asia; unto Ephesus, and unto Smyrna, and unto Pergamos, and unto Thyatira, and unto Sardis, and unto Philadelphia, and unto Laodicea" (Rev 1:11). Notice this *one* book that was to go to these seven churches in Asia in the order written. Hence this one book was carried first to Ephesus, then to Smyrna, and so forth until it came to the Laodicean church. But moreover notice that in each message to each church it says the book spoke to the churches (Rev 2:7, 11, 17, 29; 3:6, 13, 22). The book didn't just speak to one church, but to all the churches. Therefore not only does this book speak to each church, but speaks to all the churches that are in Christ.

Book Speaks to the End Time and Thereafter

pr538» The things written about in this book John wrote were the things that were in his time, "and the things which shall be hereafter" (Rev 1:19). Concerning the seven churches, this indicates the churches as they were at John's time, and the things that were to come concerning the Church. The book spoke of the state of the churches of Christ at that time and what was to come.

pr539» Antitypically, we are in the time when the book is opened fully, and the effect of every vision is to come true soon. **Therefore antitypically the book of Revelation now is speaking of the present state of the churches**, and "the things which shall be hereafter." The book will be perfectly fulfilled shortly. All the things mentioned about the seven churches concern the present Church of God at the time the book was opened up, and shortly thereafter.

Chapters Two and Three Speak of the Condition of Today's Church

[section written on April 4th, 1971; revised January23, 1998 {some singular words changed to plural}]

pr540» At this writing there are false apostles in with those of the true Church, or those who say they have a commission by God to perform ("two witnesses"). But some in the Church "found them liars" (Rev 2:2).

1. Some have fallen from their love of the first, and are asked to turn, yet at the same time they hate the false doctrines of the Church. These will endure, and labor through the NAME of Christ, for they have the Spirit of Christ inside him (Rev 2:4, 5, 6, 3).

2. These in the Church are relatively poor physically, but rich Spiritually (Rev 2:9).

3. The Church will go through a trial; some will be put in prison (Rev 2:10).

4. The Church is dwelling or will dwell where Satan's throne is, or power is. ~~The modern day Antipas will be killed where Satan dwells~~ (Rev 2:13). This Greek verse is very awkward and with the help of a Hebrew text of Rev 2:13, this verse is

probably referring to a *future* faithful witness(es) like the faithful witness of God (Jesus) in the days of Herod Antipas (20 BC - c. 39AD), Verse Spiritually is probably referring to the last 3 ½ year period as are the other verses pertaining to the Church in chapter 2 and 3.

5.   The present Church holds false doctrines (Rev 2:14-15).

6.   But they will be reproved soon after they have been given a period to change their minds (Rev 2:16, 21, 3:3).

7.   Those who continue in the false doctrines will die the death, the agelasting death (Rev 2:23, 18-23).

8.   Some in the Church do not hold in their minds these false doctrines (Rev 2:24).

9.   At this time the present Church has a name that lives, but is dead Spiritually. They have "dead" leaders, biased behavior, etc (Rev 3:1).

10.   Those of the Church are asked to "strengthen the things which remain, that were about to die: for I have not found your works complete before God. Remember therefore how you have received and heard, and hold fast." The Church will be reproved, it is asked to remember how they received this truth, to hold it, and to strengthen it (Rev 3:2, 3).

11.   Again it says some in the present Church are Spiritually in Christ (Rev 3:4).

12.   Now here it shows the door being opened to the one who shall reprove the Church. This one has not denied Christ's NAME (Rev 3:7-8).

13.   The Church will have a little while strength. Thus they will receive a little while power (Rev 3:8).

14.   The scripture about the Laodicean Church pictures those now in the Church, physically, who are comparatively rich physically but not Spiritually. Those with false assurance in goods will be reproved, for as many as God loves, he rebukes (Rev 3:14-19).

15.   The door to the Church is now being knocked on. (Rev 3:20)

16.   Those who overcome will be saved (Rev 3:21).

PR10: Great Falling Away

Forty Days

Tithe of God's People

pr541» In order to find truth in the Bible we look for patterns; we look in a Spiritual way for Spiritual patterns (See *Duality Paper*). Paul found patterns in the Bible and manifested them to his physical churches through his letters (Epistles). Throughout our papers we do our best to manifest the patterns found in the Bible. Before the day of the Lord there is to be a great falling away (2 Thes 2:3, 1 Tim 4:1). What does this mean? If we are alive at the end, will we recognize this falling away or departure from the faith?

- "For the time will come when they will not permit [bear] sound doctrine, but they will heap up to themselves teachers who tickle the ear according to their own lusts. And they will turn away their ear from the truth and will be turned aside to myths" (2 Tim 4:3-4).

- "Now the Spirit speaks expressly, that in the latter times some shall depart from the faith, giving heed to seducing spirits, and doctrines of devils; Speaking lies in hypocrisy..." (1 Tim 4:1-2).

- "Concerning the coming of our Lord Jesus Christ and our being gathered to him... Don't let anyone deceive you in any way, for that day will not come until the rebellion occurs [falling away] and the man of lawlessness is revealed, the man doomed to destruction" (2 Thes 2:1, 3).

This lawless man is the Beast man, the man who will rule the Beast system, and is in a shadow of Satan. Satan will be inside this lawless man misleading him during the last 1260 days (See *Beast Papers*, PR2, PR3). But what about the "falling away" or the "rebellion?"

pr542» Those who say they are in the Church at the end (but are not) will not permit or will have cast aside sound doctrine, and will be or will have fallen away to myths. A myth is a belief that is not true. Pertaining to Biblical doctrine, a myth is a belief in something not Biblical, something not found in the Bible. The so-called doctrine about the "Trinity" is an example of something not found in the Bible and something that is impossible (See *God Papers*,

Part 5). Now, what is going to happen at or near the end? At the end of the age before Christ's physical return someone will come and restore all things (Matt 17:10-11) through the power of the Spirit (John 14:16-17, 26). In a way the scriptures in John, Chapter 14, concerning the "Spirit of Truth" applied to Paul, and in a certain degree applies to all those with the Spirit, according to the power given them, but also this scripture applies to the Spiritual Elijah. There will be an antitypical and Spiritual Elijah who will restore all things. He comes "before the great and dreadful day of the lord" (Mal 4:5). Therefore before he comes the Truth shall not be full in the Church or anywhere on earth. But when he comes he shall restore all things, yet they will do "unto him whatsoever they listed, as it is written of him" (Mark 9:13). At first the physical Elijah was rejected in the scripture by physical Israel; the Spiritual Elijah will also be rejected. Yet he will endure and restore the Truth, the Spiritual sacrifice of the Spiritual altar, to Spiritual Israel, which is the Church (1 Kings 18:30 with Heb 13:10; Matt 17:10-11; see New Mind Papers). Thus when Elijah comes "they will not permit" or bear or admit or hold up or conceive his sound doctrine (2 Tim 4:3; Strong's #430 from #2192). Yet the Bible says he will overcome them and restore all the truth to the Spiritual Church and teach the truth to all the earth as one of the two witnesses (See *Two Witnesses*).

Antitypical Ezekiel

pr543» Remember we look for patterns in the Bible (see *Duality Paper*). Notice in Ezekiel, chapters two to four, there is a prophecy that has happened typically, but not antitypically. In chapter 2 God speaks to the Son of man and sends him to Israel. God tells him not to be afraid of their words (v. 6), tells him to speak God's words even if they forbear (v. 7), tells him to eat the book of words he has been given (Ezek 2:8-10; 3:1-3; see Rev 10:8-11). He was to go to the house of Israel and give them the words of the book whether they hear or not (Ezek 3:1-11). He was restricted not to go out among them, and he would not reprove them at first (Ezek 3:25-27). Spiritually speaking, Israel is the Church. This speaks of someone who will restore the truth, but at first they would not Spiritually hear him. Today, we only see some of the true doctrines of the Spiritual Church through the writings of Paul, James, Peter, John, and the others of the New Testament. Most of these were letters. Paul in person, surely, had more detailed information regarding doctrine. If we had doctrinal papers written by Paul, there would be much less disagreement among those who call themselves Christians. (Of course, this could have allowed more of those who say they are Christians, but don't have the Spirit, to fool others who do have the Spirit.) Yet since Paul was not the Spiritual Elijah, he may not have held all the truth that God will allow us to learn before Christ's return.

Great Falling Away

pr544» Elijah was also to go to Israel, but Israel would not Spiritually hear him, and for a while would not answer him (1 Kings 18:21). The Son of man (Ezekiel) was to speak to Israel "whether they will hear, or whether they will forbear." Also the Son of man was to shut himself within his house (3:24), while he sieged Jerusalem (the Church) and prophesied against it (Ezek 4:3, 7). But he would not reprove it at first (Ezek 3:26). In prophecy, Ezekiel and Elijah's stories point to the soon coming events at the end-of-the-age, much as different physical persons and events of the Old Testament pointed to Jesus Christ. If you look at the scripture that pointed to the first coming of Christ, you can see why physical Israel did not recognize Christ's first coming. These scriptures are spread throughout the Old Testament and sometimes only one part of a scripture pointed to the first coming, while the other part pointed to his second coming. This was confusing to those without the Spirit. It is easy for us now to understand these scriptures because they were fulfilled by Christ, and because we have the Spirit.

390 & 40 Days

pr545» The antitypical Elijah-Ezekiel will 'siege' the antitypical Jerusalem (Ezek 4:2,7), the Church, for two periods of 390 and 40 days (Ezek 4:4-9), whether they listen to the words of the siege or not (Ezek 3:27). During the 390 day period he was to eat a grain diet of about ten ounces and a liquid of about four cups, from time to time. But he was to have another period of 40 days:

- "and when you have completed these [390 days] you shall lie down on your right side and bear *a second time* the iniquity of the house of Judah, **forty days**, a day for each year, I assign you" (Ezek 4:6, literal translation).

pr546» He already bore the iniquity for 390 days (Ezek 4:4-5). But now he must for the *second time* ("again") bear or put-up with the iniquity.

The people of Israel would not listen to Ezekiel for the 390 days and Ezekiel would have to endure them for another 40 days. These typical or physical events may have happened one right after another, but this doesn't mean the antitypical events will happen likewise. This is much like ambiguous scriptures that pointed to Christ: one part of the scripture may have pertained to Christ's first coming; the other to His second coming (Dan 9:26-27). At this time these scriptures are not clear.

Forty Days /Forty Years

pr547» Before the day of the Lord there is to be a great falling away (2 Thes 2:3, 1 Tim 4:1). This "falling away" begins after the siege, after the "40 days." The antitypical Elijah's mouth and power to *reprove* is closed during the period (Ezek 3:25-26; 4:8), but his mouth will be opened at the end (Ezek 9:6; 24:21, 24-27; 33:21-22; 29:21). After the "siege" of forty days he will overcome and begin to restore the truth to the Church/Israel. Remember also that Moses after **forty days** came down from mount Sinai, where he received the ten commandments (Ex 34:28), and then he began teaching Israel the law. And remember that after Christ's water baptized he had a 40 day period of trial. After Christ's 40 day trial he began to teach Israel (Luke 3:22; 4:1-2, 13-15). Remember that Elijah had a period of forty days before he went out and before he had his 7,000 in Israel (1 Kings 19:8-19). These forty day periods are also connected in someway to a forty *year* period: "a day for each year" (Ezek 4:6). Moses received the physical Ten Commandments for Israel from the angel of God twice during two *forty day* periods (Ex 24:18; 32:15-19; 34:27-28; Deut 9:18-10:10) at the beginning of the *forty year* period of Israel and Moses' wandering in the wilderness (Ex 20:1ff); Israel died by multitudes during the forty years because of their rebellion:

> "Your sons shall be shepherds for forty years in the wilderness, and they will suffer for your unfaithfulness, until your corpses lie in the wilderness. According to the number of days which you spied out the land, forty days, for every day you shall bear your guilt a year, forty years, and you will know my opposition." (Num 14:33-34)

Moses reiterated the Law at the very end of the forty years (Deut 5:1ff), and Israel went into the promised land after their **forty years** in the wilderness. Notice how forty days are connected to forty years. So it will be with the antitypical

Moses, he will receive the commandments (Spiritual law and truth) during a forty day period. This forty day period will be somehow connected to a forty year period because a day is counted as a year and the forty years are connected to the wondering of Israel in the wilderness. Today the whole world is a wilderness and those who will belong to the Spiritual Israel (Church) will have gone through their own forty years of wondering near the end-of-the-age in order to fulfill the type and antitype of scripture. After a "forty day- year" period the antitype Moses and Elijah will begin to teach Spiritual Israel the truth. But many will not want to admit the sound doctrine (2 Tim 4:3) and will fall away (1 Tim 4:1), and begin to persecute the true Christians (Luke 21:12, 16-17; Matt 24:10) after Elijah is revealed (Matt 11:12-14). Therefore after the siege by the antitypical Elijah-Ezekiel, then will occur the "falling away."

One-Third

pr548» At the end of the prophesying siege by the typical Ezekiel notice what happened: "when the days of the siege are fulfilled," then a third part will fall by the sword [the word of God, Eph 6:17], and another third by pestilence and famine [the effects of the sword and trials thereof] will be consumed (Ezek 5:2, 12). These are the two-thirds who are to fall away: "and it shall come to pass, that in all the land, says the LORD, two parts therein shall be cut off and die, but the third shall be left therein" (Zech 13:8). Two-thirds "leave" or are cut off from the Church/Jerusalem. They are cut off because, "they went out from us, but they were not of us; for if they had been of us, they would have continued with us" (1 John 2:19).

pr549» Notice what Zechariah says about the one-third that is left: "And I will bring the third part through the fire, and will refine them as silver is refined, and will try them as gold is tried" (Zech 13:9; see Ezek 5:3-4). Malachi shows this same refining (Mal 3:3). This refining is a "trial of your faith" (1 Pet 1:7; 1 Cor 3:12-14). Also Revelation 3:18 pictures trial or refining by fire of the Church. (**There is a higher meaning here** concerning Satan's angels and God's angels: the one-third and two-thirds of the angels versus two-thirds and the other one-third of mankind. See *God Papers* [GP7 & GP8]).

Tithe of God's People

pr550» Immediately at Christ's return one-tenth of Jerusalem (the Church) will fall (Rev 11:13). That is, one-tenth who will be cut off are those who *say* they are in the Church before the "falling away." Seven thousand were killed (Rev 11:13). But Elijah is to have 7,000 when he comes (Rom 11:2-4): he is one of the two witnesses. Furthermore, one-half of the Church (Jerusalem) was not cut off from the city while the other half are sent away from the city (Zech 14:2). Thus, those cut from the city and sent away from the city are the 50 percent of those who say they are in the Church (just before Christ returns physically), but actually do not have the Spirit (Oil) and thus are sent into the fire (Matt 25:1-12; 7:23; 13:41-42).

pr551» Putting this together, therefore there will be 14,000 (7,000 & 7,000) or so who will be physically in the Church at the very end, but only one-half of them will Spiritually be in the Church. These are the 7,000 for Elijah, "the residue of the people shall not be cut off from the city" (Zech 14:2).

pr552» Revelation 11:13 says that those who fell were one-tenth of the city. This one-tenth are the tithe of those in the physical Church near the end of the old age when the end-time events begin to come to pass.

- All tithes of herd and flock, every tenth one that passes under the shepherd's staff, shall be holy to the LORD (Lev 27:32).

Antitypical John

pr553» John the Baptist is a type of the Elijah (Matt 11:12-14; Luke 1:16-17). As John cried in the wilderness to make ready, for the time is at hand, so will the antitypical Elijah. As John did this until his time for publicly taking office, or his manifestation to Israel, so too shall Elijah until he publicly takes office or his manifestation to the Spiritual Israel (Matt 3:1-13; Luke 3:4; 1:80, "showing" is translated from a Greek word meaning: to openly take an office; or manifestation; or showing forth).

pr554» While John was in the wilderness his food was wild honey, which antitypically means God's word (Psa 19:9-10; 81:16; 119:103). This is the angel's food that Elijah ate (1 Kings 19:5-7; Psa 78:24-25; John 6:32, 63).

pr555» But after Elijah's time in the wilderness or as we have seen after the antitypical "390 day" period and the "40 day" period, then he will take office. As John was to prepare a people to make straight the way of God, so too the antitypical Elijah (Luke 1:16-17, 76-77, 3:4-5). He teaches that *all* will see the salvation of God (Luke 3:6; 16:16). And as John taught Israel six months before Christ and His 3 ½ years so too will the antitypica1 John (Elijah) teach before the 3 ½ years mentioned in the book of Revelation.

Two-thirds of Mankind Cut Off

pr556» What will happen to the Church first will also happen to the whole world, for the end-time destruction will begin with God's sanctuary (Ezek 9:6; 1-11; 24:21). A still higher meaning is that the two-thirds cut off are the two-thirds of mankind who were cut off from the Body of Christ during the old age of Satan. (See *God Papers* [GP7])

pr557» [**Comment**: Before I ran across this pattern in the Bible in 1971, I thought of true Christians being numbered in the millions near the end of the age. But if this paper's rendition of the "falling away" is close to the truth, then there are only thousands in the Church near the end. I have searched far and wide for a physical Church of God that held numerous Spiritual Christians, and I have yet to find one. This paper speaks about a physical church(es) counted as Spiritual Jerusalem (the Church), but only about one-tenth of it are real Christians. The Bible speaks about three groups or three orders being resurrected from the dead (1 Cor 15:20-28):

- the first was Christ;

- the second the 144,000 (Rev 7:4), the sealed servants of God;

- the third the "great multitude, which no man could number, of all nations..." (Rev 7:9).

The 144,000 is a very small number to come out of the billions of people who have ever lived on the earth for the first six thousand years of man's history. So maybe 7,000 plus would be about the right number to exist near the end. Too small for me, but what is the purpose of the 144,000 anyway? God made them kings and priests: and they shall reign on the earth (Rev 5:9-10). The kings, queens, and priests of this old age are the type of the antitype. These old kings and queens ruled ruthlessly. Those born of God from this evil age will rule as kings and queens during the 1000 years, and will rule with love. The 144,000 is about the right number of

people to help rule billions on the earth during the 1000 years. So maybe the 7,000 as a group is not so few after all. This 7,000 is one-tenth of another group(s) who say they are Christians. But who is this group(s) of 70,000? Time will tell. Hopefully, the time is short. Remember, we don't fear the end, because this is when our Messiah comes to save the world, and then and only then will peace come to the earth.]

[**Note**: This paper was first written in 1971]

The following taken from my 1977 book: The last sentence starting with
                                                   "Approximately 70,000...." is probably incorrect.

The chances are if one makes it through these trials shortly to come on the Church he is blessed, for approximately 50 percent who are physically in the Church at that time will have the Spirit of God leading them.
    [NOTE: The 1335th day mentioned in Daniel 12:12 is the 1335th day before the antitypical daily sacrifice is to be taken away. This sacrifice is taken away at Christ's return. See the Beast Papers and the Date Paper to understand this: to understand the 1335th day one must understand our rendering of the 1290th and 1260th day.]

Immediately as Christ returns one-tenth of Jerusalem (the Church) will be cut off. (Rev 11:13) These are the 50 percent of the Chruch who will not have the Spirit ("oil") when Christ returns. (Matt 25:1-12) They number 7,000 or so. (Rev 11:13) But Elijah is to have 7,000 when he comes (Rom 11:2-4): he is one of the two witnesses. Furthermore, one-half of the Church ("Jerusalem") is to fall at Christ's coming. (Zech 14:2) Therefore there will be 14,000 (7,000 & 7,000) or so who will be physically in the Church at the end in the Middle East near the old city of Jerusalem, but only one-half of them will Spiritually be in the Church. These are the 7,000 for Elijah, "the residue of the people shall not be cut off from the city." (Zech 14:2)

Now Revelation 11:13 says that the 7,000 cut off are one-tenth of the city. That is it is one-tenth of the physical population of the Church after the time of the antitypical Elijah's 40 day period, for after Elijah completed his 40 days God said he had 7,000. (1 Kings 19:8, 18) When the 40 days are completed is when the great falling away will occur. Thus, since it was one-tenth of the city who are to be cut off at Christ's return (Rev 11:13), and this one-tenth is only one-half of those alive at the time of his return (Zech 14:2), and since Elijah has 7,000 where he is witnessing (1 Kings 19:18; Rev 11), and since Elijah had also 7,000 after his 40 day period (1 Kings 19:8, 18); then there is to be 14,000 physically in the Church at Christ's return, with 7,000 (or one-half) being Elijah's and 7,000 (or one-half) being without the Spiritual oil (Matt 25:1-10), and the 7,000 cut off is one-tenth of those who existed in the Church at the time of Elijah's reproof of the Church, which occurs after the 40 days. Approximately 70,000 will be physically in the Church after the 40 day period.

WHEN IS THE 390 DAY AND 40 DAY PERIODS?

Now when is the 390 day period, and when is the 40 day period? The 390 day period has alread been perfectly fulfilled. One for exactly a 390 day period has fulfilled this prophecy. With no idea about this 390 day prophecy this person: (1) quit his job at the beginning of this period; (2) made a complete study of the Bible during this time period; (3) ate exactly ten onces of grain per day (90 to 95% of the days, "time to time") during this period (note Ezek 4:9-10); (4) prophesied against

PR11: Information on the Beast and his Name

Christ came in his Father's Name

Antichrist comes in his Father's name

Beast as Superman?

Deadly Wound

How do you calculate the number of the Beast?

Things to know when Calculating the Number of the Beast

Tables on Calculating the Number

How important is a name?

pr558» **Importance of God's Name.** God's Name is very important. Even one of the ten commandments warns against taking his Name in vain (Ex 20:7; Deut 5:11). In the first part of the God Papers (GP1: "Great Significance of the NAME") we point out from the Bible how important God considers his Name. If you haven't studied this section, you should study it, so you can better understand how important God considers his Name to be (See GP1c).

Christ came in his Father's Name

pr559» **Christ's Name and Person Denied.** "I have come in my Father's **Name,** and you do not accept me; if another comes in his own **Name,** you will accept him" (John 5:43).

Not only has the world not accepted the true Messiah, it has also rejected his Name. As shown in the first part of the God Papers, God's Name as revealed by God in Exodus 3:14 is the "BeComingOne" or "YHWH" or "He (who) Will-Be," or in his own words, "I will be." His Name is not the "Lord," or "I am" as commonly misrepresented in today's Bibles. Just two verses above Exodus 3:14 in Exodus 3:12 the *same* Hebrew word that many use to "prove" that God's Name means "I am" is translated as "I will be." This same Hebrew word appears in about 42 verses in the Bible, where it is mostly translated as "I will be" ("I will be" in GP1). I will be or He will be is God's Name, not I am.

Last Antichrist comes in his Father's Name

pr560» We know that there will have been many antichrists before the real and last Antichrist comes (1John 2:18). Christ came in his Father's Name and was rejected, but Satan will come inside a man and be identified by the number of this man's name (Rev 13:18, 11) and yet he will be accepted and followed by the world (Rev 13:8, 3). Even though the Beast's identity will be obvious (he has the number of the Beast; he is connected to the ten/seven nation league), he will be accepted and followed, and his identity as the Beast-man will be denied by his followers who will claim he is a leader doing "good" for the world. As Satan pretends to be a minister of light, "even Satan disguises himself as an angel of light" (2 Cor 11:14), so will the Beast-man appear to those without the New Mind to be "good." He looks like a "lamb," but in reality Satan the dragon speaks and acts through him (Rev 13:11). Although he is the real Antichrist and real false-prophet those of the world willingly believe his lies and are deceived (Rev 19:20; 2 Thes 2:9-12).

pr561» **The number of the Beast helps to identify him.** By carefully comparing scriptures of Daniel and Revelations we know that there is an *individual* Beast with Satan inside of him influencing his activities, and there is a *system* of the Beast that starts out with ten nations in some kind of league (See PR2 & PR3). It is the individual Beast that has a name that can be figured or calculated to be six hundred and sixty-six (Rev 13:18). But in order for this man to be the Beast-man he must also be the voice or leader of a group of nations that starts out with ten nations, then later the group will only have seven nations because three of them will have been destroyed (PR2 & PR3). We know from prophesy that this Beast-man will be connected with the system of the Beast for a 1260 to 1290 day or 3 and ½ year period (Rev 13:5; Dan 7:25; PR2 & PR3). This period has not yet started so we cannot call any man the Beast-man yet (first written 11-28-1998; edited again in June 2019). We will continue to watch this.

Beast as Superhuman?

pr562» **Miracles.** Now some may think that a man with the number cannot be the last Antichrist because he is not superhuman. But the Beast-man will **not** be superhuman or actually perform real miracles:

"lying wonders... God shall send to them [who worship the Beast] strong delusion, **that they should believe a lie...** and **deceived them** that dwell on the earth by these [lying] miracles" (2 Thes 2:9, 11; Rev 13:14)

pr563» Satan cannot preform real miracles or resurrect himself from the dead, for only God has the power of resurrection from death (John 5:26; 11:25), and He will **not** give that power to Satan for any reason or for any period. That is why Satan deceives the world with **lying** wonders or miracles. Satan cannot preform real miracles.

pr564» The man we had identifed[1] with the number is not the Beast-man (11-28-98; or June 2019). He will only be the Beast-man when and if he enters the 1290-1260 day period. But we will know this ahead of time, for God promised to show his servants ahead of time (Isa 42:9 Luke 8:17; 10:21-22; Mark 4:22-23).

Deadly Wound

pr565» Some say that the Antichrist (that is, Beast-man; that is, false-prophet; that is, Anti-Messiah) must be superhuman since he recovers from a deadly wound (Rev 13:3, 12). But the scripture speaks of the Beast "as if" or "like" it was wounded to death[2] (Rev 13:3). The scripture is not speaking of a literal death of a person and then a resurrection thereafter. You may say that there are two "deadly wounds": one for the Beast-man; one for the Beast-system.

There is a "deadly wound" to the system of the Beast:

pr566» The following section is from Beast-system Paper [PR2]:

Deadly Wound. Notice that in Revelation 13:3 that a head was as wounded to death, but its deadly wound was healed. This deadly wound will happen to the seventh mountain or kingdom of Revelation 17 with its ten horns or nations. And this "deadly wound" will subdue three of the ten horns (nations) of this seventh kingdom

("mountain") which is the ten-nation Beast (Dan 7:24). Thus, the eighth mountain or kingdom is the *healed* Beast of Revelation 13:3. It had ten horns, but three nations ("horns") will be subdued by the mean horn (Dan 7:20, 24).

In other words, "the Beast that *was* [the ten-nation Beast], and *is not* [the ten-nation Beast], even he is the eighth, and is out of the seven, and goes to perdition" (Rev 17:11). This Beast *was* [the ten-nation Beast], and *is not* [the beast with ten horns or nations], and *yet is* [the beast, but with only seven nations]. The "deadly wound" destroys the kingdom that *was* by subduing three nations ("kings"); the healing of this deadly wound creates a kingdom with seven nations as opposed to ten nations as before. The seven-nation Beast "is not" like the ten-nation Beast, "yet is" the same beast, but with three nations subdued.

(The above section is from Beast-system Paper [PR2])

There must also be a "deadly wound" for the Beast-man.

pr567» The fact that the "deadly wound" happens to *one* of the Beast's heads, and the fact that Beast-man only has one head, means the "deadly wound" refers to the system of the Beast, and that it is only a "deadly wound" to the Beast-man in the sense that part of the Beast-man's power base was destroyed or wounded as just explained above. It is after the "deadly wound" that the mouth or voice of the system of the Beast speaks for the system for 42 months (Rev 13:3-5).

Superhuman: To Summarize

pr568» Satan and his followers cannot perform real miracles. Some of the things he will do, like make fire come down from heaven (atomic bombs), are wonders and may have seemed miraculous to John in his vision, but they are the wonders that any nuclear-armed power can perform.

---

[1] William J. Clinton

[2] USA President Trump's bullet to his head (ear) was within an inch of killing him (July 13, 2024) is an example of such a wound.

How to Calculate the Number of the Beast

Calculate from Hebrew or Greek or Both?

pr569» Now some say that only the Greek language of the Bible can be used to calculate the number of the Beast, since the book of Revelation was written in Greek, or, that is, appears to have been written in Greek by John. But was it? There is some evidence that at least part of the New Testament of the Bible may have been first written in Hebrew and then translated into Greek after the Hebrew nation was destroyed by the Romans. After all, were not the first leaders of the Church Jewish or Hebrew? Even Paul was a Hebrew.

pr570» In the fourth century, Jerome in his *Concerning Illustrious Men* wrote about the book of Matthew being first written in the Hebrew language and only later translated to Greek. (See part 1 of the *God Papers* for more information on this.) Although Jerome spoke specifically about the book of Matthew, the book of Revelation and other books, such as Mark, may have also been written first in Hebrew. The book of Revelation has been called "the most Hebraistic in thought and language of all the New Testament books" (*Hermeneutics of the New Testament*, by A. Immer, 1873, pp. 132-144; taken from p. 126 of *Biblical Hermeneutics*, Milton S. Terry). Because the Catholic Church has done its best to rid Christianity of anything Jewish, it is almost impossible now to know how many books in the New Testament were first written in Hebrew, or first written in Greek and then translated into Hebrew for the Jewish Christians to read. There are those falsely saying[1] that there are many leaders with the number 666, and thus helping to blind the world because of their lies (2Thes 2:11-12). Also the modern "rapture" myth helps to blind people. Most modern rapture believers will not recognize the Beast or acknowledge him because they mistakenly think that they will be raptured before the Beast comes. Since they are not

---

[1] There are Internet sites where people claim that numerous individuals and leaders not only have the number 666 in Hebrew and Greek, but also in English, yet English has no numerical value for its letters. When you look at their material, the names are manipulated to fit the number of the Beast, it is undocumented, inaccurate, and often simply made-up to fit their own wishes – in other words worthless.

raptured they, of course, discount anyone with the number.

Things to know when Calculating the Number of the Beast

666. "This calls for wisdom: let anyone with understanding calculate the number of the beast, for it is the **number of a <u>man</u>. Its number is six hundred sixty-six**." (Revelation 13:18)

Not a number of a title, nor of a computer, nor an address. Scripture says the number of the Beast is the number of a man, not a system, not money, not a computer system, not a title, not a law, not an address, not anything but the number of a male human being. The book of Revelation was speaking about calculating or adding up the numbers of a man's name.

Irenaeus. Almost 1900 years ago Irenaeus in his *Against Heresies* told fellow believers that the number of the Antichrist was six hundred and sixty-six, and that it was figured by knowing the value of each letter of the name, "since [Greek] numbers also are expressed by letters."

Calculate number of the name by adding the numerical values of each letter of the name. Unlike modern languages which use the Arabic numerical system (0, 1, 2, 3, 4, 5, 6, 7, 8, 9), languages around the time of the book of Revelation and before, either spelled out the numbers or used letters that had numerical values. Each letter in Hebrew (except vowels) or Greek has a numerical value. Latin has its Roman numeral system. So by calculating or adding up the numerical values of the letters of a person's name, you will get the number of his name.

Not three sixes in a row. The number of the Beast-man is six hundred sixty six and it is not just three 6s in a row (6+6+6=18), but six hundred sixty-six. Three sixes in a row is not the number of the Beast-man.

Not the gematria system. We are not speaking about gematria. Gematria was a part of the Jewish mysticism which dealt with mystical aspects of words' numerical value. The Truth of the Bible is not found by mysticism. People are using the word "gematria" when they should be using the phrase, "numerical value."

Use your own mind and study the following table:

Hebrew Alphabet with numerical values			English Transliteration values		Greek Alphabet with numerical values		
letter sign	letter name	num.value	Hebrew	Greek	letter sign	letter name	num. value
א	aleph	1	guttural	a	α	alpha	1
ב	beth	2	b (bh)	b	β	beta	2
ג	gimel	3	g (gh)	g	γ	gamma	3
ד	daleth	4	d (dh)	d	δ	delta	4
ה	he	5	h	e	ϵ	epsilon	5
ו	waw	6	w	w [old usage]	ϛ	digamma or stigma	6
ז	zayin	7	z	z	ζ	zeta	7
ח	cheth	8	h	e	η	eta	8
ט	teth	9	t	th	θ	theta	9
י	yod	10	y (j)	i (j or y)	ι	iota	10
כ	kaph	20	k (c)	k (c)	κ	kappa	20
ל	lamed	30	L	L	λ	lambda	30
מ	mem	40	m	m	μ	mu	40
נ	nun	50	n	n	ν	nu	50
ס	samekh	60	s	x	ξ	xi	60
ע	ayin	70	guttural	o (short)	ο	omicron	70
פ	pe	80	p	p	π	pi	80
צ	sadi	90	s		-	koppa	90
ק	koph	100	q	r	ρ	rho	100
ר	resh	200	r	s	σ	sigma	200
ש	shin	300	s (sh)	t	τ	tau	300
ת	taw	400	t (th)	y, u	υ	upsilon	400
				ph	φ	phi	500
				kh (ch)	χ	chi	600
				ps	ψ	psi	700
				o (long)	ω	omega	800
					-	sampi	900

For more details about Biblical prophecy go to our web site where you will find more information on prophecy, God, Christianity, etc.

Https://beone.ws
https://becoming-one.org
https://becomingonechurch.org
https://b1-church.org

https://walterdolen.ws
https://walterdolen.com

etc.

Books available by the Author:

My God is the Becoming-One
New Mind and Christianity
Prophecy Papers
6000 Years of Mankind: Chronology Papers
Harmony of the Good News
Becoming-One Papers
Becoming-One Bible: Old and New Testament

Male & Female: Complementary Partnership
Einstein: Light, Time and Relativity

Available at https://beone.ws/books.htm

Available on the World Wide Web:
Amazon, Barnes and Noble, Walmart, etc.

Also available in digital formats:
https://b1-church.org/products

Index

Note: due to the limitations of our software some of the following may give incorrect page numbers